W9-AUI-421

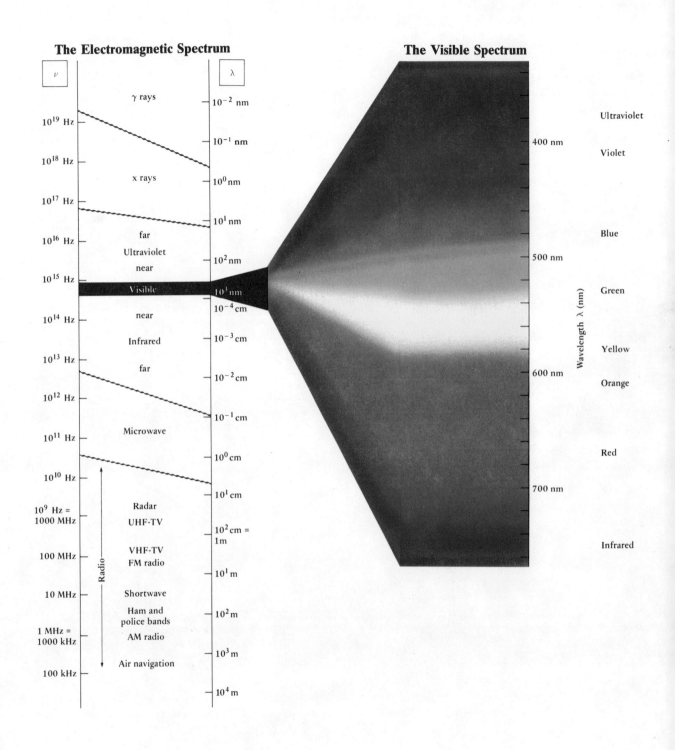

The Electromagnetic Spectrum

ν

λ

γ rays

10^{19} Hz

10^{-2} nm

10^{18} Hz

10^{-1} nm

x rays

10^{17} Hz

10^0 nm

10^{16} Hz

far

10^1 nm

Ultraviolet

near

10^2 nm

10^{15} Hz

Visible

10^3 nm

10^{-4} cm

near

10^{14} Hz

10^{-3} cm

Infrared

10^{13} Hz

far

10^{-2} cm

10^{12} Hz

10^{-1} cm

Microwave

10^{11} Hz

10^0 cm

10^{10} Hz

10^1 cm

10^9 Hz = 1000 MHz

Radar
UHF-TV

10^2 cm = 1m

100 MHz

VHF-TV
FM radio

10^1 m

Radio

10 MHz

Shortwave

10^2 m

Ham and police bands

1 MHz = 1000 kHz

AM radio

10^3 m

100 kHz

Air navigation

10^4 m

The Visible Spectrum

Ultraviolet

400 nm

Violet

Blue

500 nm

Green

Wavelength λ (nm)

Yellow

600 nm

Orange

Red

700 nm

Infrared

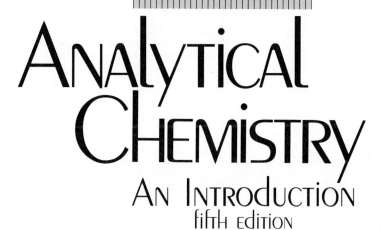

Analytical Chemistry
An Introduction
fifth edition

DOUGLAS A. SKOOG
Stanford University

DONALD M. WEST
San Jose State University

F. JAMES HOLLER
University of Kentucky

SAUNDERS GOLDEN SUNBURST SERIES

Saunders College Publishing
Philadelphia Ft. Worth Chicago San Francisco
Montreal Toronto London Sydney Tokyo

Copyright © 1990 by Saunders College Publishing, a division of Holt, Rinehart and Winston, Inc.

All rights reserved. No part of this publication may be reproduced or transmitted in any form or by any means, electronic or mechanical, including photocopy, recording, or any information storage and retrieval system, without permission in writing from the publisher.

Requests for permission to make copies of any part of the work should be mailed to Copyrights and Permissions Department, Holt, Rinehart and Winston, Inc., Orlando, Florida 32887.

Text Typeface: Times Roman
Compositor: Bi-Comp, Inc.
Acquisitions Editor: John Vondeling
Managing Editor: Carol Field
Project Editor: Merry Post
Copy Editor: Irene Nunes
Manager of Art and Design: Carol Bleistine
Art Director: Christine Schueler
Art Assistant: Doris Bruey
Text Designer: LMD Services for Publishers
Cover Designer: Lawrence R. Didona
Text Artwork: Larry Ward
Director of EDP: Tim Frelick
Production Manager: Charlene Squibb

Cover Credit: The molecule on the cover is a complex ion formed between copper(II) and ethylenediaminetetraacetic acid (EDTA). Titrations involving the formation of metal complexes of this type are discussed in Chapter 14. The computed-generated photo on the cover was provided by Dr. Jeffrey W. Hare, a Staff Scientist with Biodesign, Inc., Princeton, New Jersey.

Printed in the United States of America

Analytical Chemistry: An Introduction

ISBN 0-03-029924-1

Library of Congress Catalog Card Number: 89-043651

0123   0   39   987654321

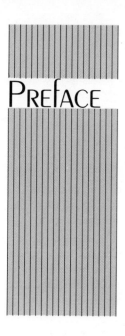

Preface

The fifth edition of *Analytical Chemistry: An Introduction*, like earlier editions, is an abbreviated version of the authors' other text, *Fundamentals of Analytical Chemistry*.[1] A primary audience for earlier editions of the current text has been students in fields such as medicine, geology, engineering, and the physical sciences whose only exposure to analytical chemistry is a one-semester or one-quarter introductory course. The text has also found use in sophomore-level courses for chemistry majors where fewer laboratory experiments and less detailed treatment of descriptive topics is desired. Finally the text has been used in freshman-level laboratory courses. This edition is designed to serve these same audiences.

Analytical Chemistry: An Introduction provides a rigorous background in chemical principles that are particularly important to analytical chemistry. Students will gain an appreciation for the difficult task of judging the accuracy and precision of experimental data and will see how these judgments can be sharpened by the application of statistical methods. The text introduces a wide range of techniques of modern analytical chemistry and teaches laboratory skills that will give students confidence in their ability to obtain high-quality analytical data.

Users of earlier editions will find numerous changes not only in content but in format and style as well. One significant content change is the exclusive use of molar concentrations in volumetric calculations. Our decision to abandon normal and formal concentrations brings the text in line with current practice in most chemical journals and monographs devoted to analytical chemistry. We also have put increased emphasis on including units in chemical calculations and using dimensional analysis to check their correctness.

---

[1]D. A. Skoog, D. M. West, and F. J. Holler, *Fundamentals of Analytical Chemistry*, 5th ed. Philadelphia: Saunders College Publishing, 1988.

Other technical additions and changes include the introduction of the operational definition of pH, the use of the IUPAC convention of always treating reference electrodes as anodes, the introduction of a brief chapter on molecular fluorescence, inclusion of a description of diode-array spectrometers, the deletion of a section on voltammetry, and expansion and clarification of the material on rounding data.

Other additions include a series of boxed and highlighted Features that contain derivation of equations, explanations of more difficult theoretical points, and historical background notes. Instructors may make these materials optional or required reading depending upon the rigor of their courses.

We have also introduced boxed and highlighted Computer Applications, which are short computer programs for solving various types of examples in the text. Some of these programs are written in BASIC; the remainder use "Eureka: The Solver," a software package available to educational institutions at modest cost.

We have made numerous style and format changes in order to make the text more readable and student friendly. Throughout, we have tried to use shorter sentences, simpler words, and the active voice. We have also shortened chapters and increased their number so that students are not so overwhelmed by the amount of material to be mastered in one sitting. In addition we have significantly increased the number of line drawings and graphs and have added photographs of modern equipment.

Another innovation is the use of marginal notes as study aids throughout the text. These notes include highlighted definitions, points of particular importance, line drawings and photographs of equipment, historical notes, and thought-provoking challenge questions.

The problem sets at the end of chapters have been entirely rewritten. Answers appear at the end of the text for the asterisked ones (approximately half). A solutions manual is available with step-by-step solutions to all the problems. About 70 transparencies of text illustrations are available.

We wish to acknowledge with thanks the comments and suggestions of the following who have reviewed the manuscript at various stages in its production: Professor Royce C. Engstrom of the University of South Dakota, Vermillion; Professor Dennis H. Evans of the University of Delaware, Newark; Professor Neil Danielson of Miami University, Oxford; Professor Gary Bender of the University of Wisconsin, La Crosse; Professor Tom Bydelak of the University of Minnesota, Duluth; Professor Wolfgang Bertsch of the University of Alabama, Tuscaloosa; Professor James Long of the University of Oregon, Eugene; Professor John Allison of Michigan State University, East Lansing; Professor William Kurtin of Trinity University, San Antonio; Professor John P. Walters of St. Olaf College, Northfield; Professor David Keeling of California Polytechnic State University, San Luis Obispo; Professor Howard Strobel of Duke University, Durham; and Professor Frank A. Guthrie of Rose-Hulman Institute of Technology, Terre Haute. We also wish to thank Professor C. Marvin Lang and Gary Shulfer of the University of Wisconsin at Stevens Point for providing the postage stamp illustrations found in the marginal notes and color plates of the book.

In addition, we thank Professor Elizabeth W. Kleppinger of Berea College, who was kind enough to read page proofs of the entire text.

We also want to thank the various persons at Saunders College Publishing for their friendly assistance in completing this project in record time. Among these are Project Editor Merry Post, Copy Editor Irene Nunes, Art Director Christine Schueler, and Production Director Charlene Squibb. Our thanks also to our Acquisitions Editor John Vondeling and Associate Editor Kate Pachuta.

Douglas A. Skoog
Donald M. West
F. James Holler

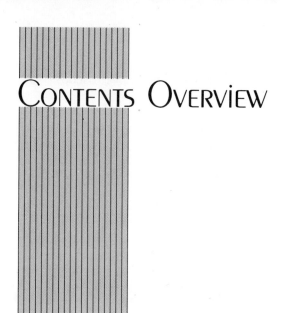

CONTENTS OVERVIEW

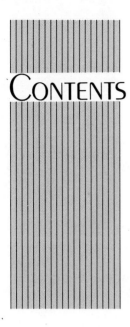

CONTENTS

CHAPTER 1

INTRODUCTION

Analytical chemistry involves separating, identifying, and determining the relative amounts of the components making up a sample of matter. Qualitative analysis reveals the chemical identity of the analytes. Quantitative analysis tells us the relative amount of one or more of these analytes in numerical terms. Qualitative information is required before a quantitative analysis can be undertaken. A separation step is usually a necessary part of both qualitative and quantitative analyses.

The principal topics covered in this text are quantitative methods of analysis and methods of analytical separations. We will refer occasionally to qualitative methods, however.

Those components of a sample that are to be determined are often referred to as analytes.

1A THE ROLE OF ANALYTICAL CHEMISTRY IN THE SCIENCES

Analytical chemistry has played a vital role in the development of science. For example, in 1894 Wilhelm Ostwald wrote,

Analytical chemistry, or the art of recognizing different substances and determining their constituents, takes a prominent position among the applications of science, since the questions which it enables us to answer arise wherever chemical processes are employed for scientific or technical purposes. Its supreme importance has caused it to be assiduously cultivated from a very early period in the history of chemistry, and its records comprise a large part of the quantitative work which is spread over the whole domain of science.

Since Ostwald's time, analytical chemistry has evolved from an art into a science with applications throughout industry, medicine, and all the sciences. To illustrate, let us list a few examples. The amount of hydrocarbons, nitrogen oxides, and carbon monoxide present in automobile exhaust gases must be measured to determine the effectiveness of smog-control devices. Quantitative measurements for ionized calcium in blood serum help diagnose parathyroid disease in human patients. Quantitative determination of the nitrogen content of foods establishes their protein content and thus their nutritional value. Analysis of steel during its production permits adjustment in the concentrations of such elements as carbon, nickel, and chromium to achieve a desired strength, hardness, corrosion resistance, and ductility. The mercaptan content of household gas supplies is monitored continually to assure that the gas has a sufficiently obnoxious odor to warn of dangerous leaks. Modern farmers tailor fertilization and irrigation schedules to meet changing plant needs during the growing season. They gauge these needs from quantitative analyses of both the plants and the soil in which they grow.

Quantitative analytical measurements also play a vital role in many research areas in chemistry, biochemistry, biology, geology, and the other sciences. For example, chemists unravel the mechanisms of chemical reactions through reaction-rate studies. The rate at which reactants are consumed or products formed in a chemical reaction can be calculated from quantitative measurements made at equal time intervals. Quantitative analyses for potassium, calcium, and sodium ions in the body fluids of animals permit physiologists to study the role these ions play in nerve-signal conduction and muscle contraction and relaxation. Materials scientists rely heavily upon quantitative analyses of crystalline germanium and silicon in their studies of the behavior of semiconductor devices. Impurities in these devices are in the concentration range of 1×10^{-6} to 1×10^{-10} percent. Archaeologists identify the source of volcanic glasses (obsidian) by measuring the concentrations of minor elements in samples taken from various locations. This knowledge in turn makes it possible to trace prehistoric trade routes for tools and weapons fashioned from obsidian.

Many chemists and biochemists devote a significant part of their time in the laboratory to gathering quantitative information about systems of interest to them. For such investigators, analytical chemistry serves as a tool for their scholarly efforts.

1B   A CLASSIFICATION OF QUANTITATIVE METHODS OF ANALYSIS

We compute the results of a typical quantitative analysis from two measurements. One is the weight or volume of sample to be analyzed. The second is the measurement of some quantity that is proportional to the amount of analyte in that sample. This second step normally completes the analysis.

Chemists classify analytical methods according to the nature of this final measurement. In a *gravimetric* method, the mass of the analyte or of some compound chemically related to it is determined. In a *volumetric* method, the volume of a solution containing sufficient reagent to react completely with the analyte is measured. *Electroanalytical* methods involve the measurement of such electrical properties as potential, current, resistance, and quantity of electricity. *Spectroscopic* methods are based upon measurements of the interaction between electromagnetic radiation and analyte atoms or molecules or upon the production of such radiation by analytes. Finally, a group of miscellaneous methods should be mentioned. These include the measurement of such properties as mass-to-charge ratio (*mass spectrometry*), rate of radioactive decay, heat of reaction, rate of reaction, thermal conductivity, optical activity, and refractive index.

Electromagnetic radiation includes X-ray, ultraviolet, visible, infrared, micro-wave, and radio-frequency radiation.

1C STEPS IN A TYPICAL QUANTITATIVE ANALYSIS

A typical quantitative analysis involves the sequence of steps shown in Figure 1-1. In some instances, we can leave out one or more of these steps. Ordinarily, though, all of them play an important role in the success of an analysis.

The first 23 chapters of this text focus on the last three steps in Figure 1-1. In the measurement step, we determine one of the physical properties mentioned in the previous section. In the calculation step, we compute the relative amount of the analyte in the samples. In the final step, we evaluate the quality of the results and estimate their reliability.

We now describe each of these steps to give you an overview of a quantitative analysis.

1C-1 Selecting a Method

Selecting which method to use to solve an analytical problem is a vital first step in any quantitative analysis. The choice is sometimes difficult, requiring experience as well as intuition on the part of the chemist. An important consideration in selection is the accuracy required. Unfortunately, high reliability nearly always requires a large investment of time. The chosen method is most often a compromise between accuracy and economics.

A second consideration related to economic factors is the number of samples to be analyzed. If there are many samples, we can afford to spend a good deal of time in preliminary operations such as assembling and calibrating instruments and equipment and preparing standard solutions. If we have only a single sample or a few samples, however, we may find it more expedient to select a procedure that avoids or minimizes such preliminary steps.

Finally, the choice of which method to use is always governed by the complexity of the sample being analyzed and by the number of components in the sample.

1C-2 Sampling

To produce meaningful information, an analysis must be performed on a sample whose composition faithfully represents that of the bulk of material from which the sample was taken. Where the bulk is large and inho-

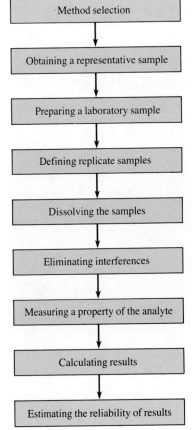

Figure 1-1

Steps in a quantitative analysis.

mogeneous, great effort is required to get a representative sample. Consider, for example, a railroad car containing 25 tons of silver ore. Buyer and seller must agree on the value of the shipment based primarily upon its silver content. The ore is inherently heterogeneous, consisting of many lumps that vary in size as well as in silver content. The assay of this shipment will be performed upon a sample that weighs about one gram. For the analysis to have significance, this small sample must have a composition that is representative of the 25 tons (approximately 22,700,000 g) of ore in the shipment. The task is to isolate 1 g of material that accurately reflects the average composition of the nearly 23,000,000 g of bulk sample. Obtaining such a representative sample is a difficult undertaking that requires a careful, systematic manipulation of the entire shipment.

Many sampling problems are easier to solve than the one just described. Regardless of whether the sampling problem is simple or complex, however, the chemist must have some assurance that the laboratory sample is representative of the whole before proceeding with an analysis.

> An assay is the process of determining how much of a given sample is the material indicated by the analyte's name. For example, a zinc alloy is assayed for its zinc content.

1C-3 Preparing a Laboratory Sample

After sampling, solid materials frequently are ground to decrease particle size, mixed to ensure homogeneity, and stored for various lengths of time before the analysis begins. During each of these steps, absorption or desorption of water may occur, depending upon the humidity of the environment. Because any loss or gain of water changes the chemical composition of a solid, it is a good idea to carefully dry samples at the start of an analysis. Alternatively, the moisture content of the sample can be determined at the time of the analysis in a separate analytical procedure.

1C-4 Defining Replicate Samples

Most chemical analyses are performed on replicate samples whose weights or volumes have been determined by careful measurements with an analytical balance or with a precise volumetric device. Obtaining replicate data on samples improves the quality of the results and provides a measure of their reliability.

Detailed instructions on techniques for measuring the weight or volume of samples appear in Sections 27D and 27G.

> Replicate samples are portions of a material of approximately the same size that are carried through an analytical procedure at the same time and in the same way.

1C-5 Preparing Solutions of the Sample

Most analyses are performed on solutions of the sample. Ideally, the solvent should dissolve the entire sample (not just the analyte) rapidly. The conditions of dissolution should be sufficiently mild so that loss of the analyte cannot occur. Unfortunately, many materials that must be analyzed are insoluble in common solvents. Examples include silicate minerals, high-molecular-weight polymers, and specimens of animal tissue. Conversion of the analyte in such materials to a soluble form can be a difficult and time-consuming task. Methods for decomposing and dissolving samples are described in various parts of Chapter 28.

1C-6 Eliminating Interferences

Few chemical or physical properties important in chemical analysis are unique to a single chemical species. Instead, the reactions used and the properties measured are characteristic of a group of elements or compounds. This lack of truly specific reactions and properties adds greatly to the difficulties faced by the analyst. These difficulties arise because a scheme must be devised to isolate the species of interest from all others in the sample that can influence the final measurement. Substances that prevent the direct measurement of the analyte concentration are called *interferences*. It is important to separate interferences prior to the final measurement step in most analyses. No hard and fast rules can be given for eliminating interferences. The resolution of this problem can be the most demanding aspect of an analysis. Chapters 24 through 26 describe separation methods.

Techniques or reactions that work for only one species of analyte are said to be specific. Techniques or reactions that work for only a few analytes are selective.

1C-7 Calibration and Measurement

All analytical results depend on a final measurement X of a physical property of the analyte. This property must vary in a known and reproducible way with the concentration C_A of the analyte. Ideally, the measurement of the physical property is directly proportional to the concentration. That is,

$$C_A = kX$$

where k is a proportionality constant. With two exceptions, analytical methods require the empirical determination of k with chemical standards for which C_A is known. The exceptions are gravimetric and coulometric methods, discussed in Chapters 5 and 19, respectively. The process of determining k is an important step in most analyses and is termed a *calibration*.

1C-8 Calculating Results

Computing analyte concentrations from experimental data is ordinarily a simple and straightforward task, particularly with modern calculators or computers. Such computations are based on the raw experimental data collected in the measurement step, the stoichiometry of the chemical reaction of interest, and instrumental factors. These calculations are illustrated throughout this book.

1C-9 Evaluating Results and Estimating Their Reliability

Analytical results are incomplete without an estimate of their reliability. The experimenter must provide some measure of the uncertainties associated with computed results if the data are to have any value. Chapters 3 and 4 present detailed methods for carrying out this important final step in the analytical process.

CHAPTER 2

A REVIEW
of SOME BASIC CONCEPTS

"SI" is the acronym for the French "Systeme International d'Unites."

This French postage stamp commemorates the Meter Convention of 1875. The stamp shows some signatures from the Treaty of the Meter, the seven SI units, and the definition of the meter as 1,650,763.73 wavelengths of the orange light given off by Kr-86. Since the stamp was issued in 1975, the meter has been redefined as the distance that light travels in a vacuum during 1/299,792,458 of a second.

This chapter reviews several topics that you may have studied before and that are of particular importance in analytical chemistry.

2A   SOME IMPORTANT UNITS OF MEASUREMENT

2A-1 SI Units

Scientists throughout the world are adopting a standardized system of units known as the *International System of Units* (SI). This system is based upon the seven fundamental base units shown in Table 2-1. Numerous other useful units are derived from these base units.

Table 2-1
SI BASE UNITS

Physical Quantity	Name of Unit	Abbreviation
Mass	kilogram	kg
Length	meter	m
Time	second	s
Temperature	kelvin	K
Amount of substance	mole	mol
Electric current	ampere	A
Luminous intensity	candela	cd

In order to express small or large measured quantities in terms of a few digits, a variety of prefixes are used with the SI base units and other derived units. As shown in Table 2-2, these prefixes multiply the unit by various powers of ten. For example, the wavelength of yellow radiation used for determining sodium by flame photometry is about 5.9×10^{-7} m, a value that can be expressed more compactly as 590 nm (nanometers); the volume of a liquid injected onto a chromatographic column is often roughly 50×10^{-6} L, or 50 μL (microliters); and the amount of memory on some computer hard disks is about 20,000,000 bytes, or 20 Mbytes (megabytes).

In this course, we will often determine the amount of chemical species by weighing. For such measurements, metric units of kilograms (kg), grams (g), milligrams (mg), or micrograms (μg) are used. Volumes of liquids are measured in units of liters (L), milliliters (mL), and sometimes microliters (μL). The liter is an SI unit that is *defined* as exactly 10^{-3} m³. The milliliter is defined as 10^{-6} m³, or 1 cm³.

2A-2 The Mole

The *mole* (mol) is the SI unit for the amount of a chemical species. It is always associated with a chemical formula and represents Avogadro's number (6.022×10^{23}) of the species represented by that formula. The *formula weight* (fw) of a substance is the weight in grams of 1 mol of that substance. Formula weights are calculated by summing the atomic weights of all the atoms appearing in a chemical formula. For example, the formula weight of formaldehyde, CH_2O, is

$$\text{fw } CH_2O = \frac{1 \text{ mol C}}{\text{mol } CH_2O} \times \frac{12.0 \text{ g}}{\text{mol C}} + \frac{2 \text{ mol H}}{\text{mol } CH_2O} \times \frac{1.0 \text{ g}}{\text{mol H}}$$

$$+ \frac{1 \text{ mol O}}{\text{mol } CH_2O} \times \frac{16.0 \text{ g}}{\text{mol O}}$$

$$= 30.0 \text{ g/mol } CH_2O$$

and that of glucose is

$$\text{fw } C_6H_{12}O_6 = \frac{6 \text{ mol C}}{\text{mol } C_6H_{12}O_6} \times \frac{12.0 \text{ g}}{\text{mol C}} + \frac{12 \text{ mol H}}{\text{mol } C_6H_{12}O_6} \times \frac{1.0 \text{ g}}{\text{mol H}}$$

$$+ \frac{6 \text{ mol O}}{\text{mol } C_6H_{12}O_6} \times \frac{16.0 \text{ g}}{\text{mol O}}$$

$$= 180.0 \text{ g/mol } C_6H_{12}O_6$$

Thus, 1 mol of formaldehyde weighs 30.0 g, and 1 mol of glucose weighs 180.0 g.

2A-3 The Millimole

Sometimes it is convenient to calculate with millimoles (mmol) rather than moles, where the millimole is 1/1000 of a mole. The weight in grams of one millimole, an amount called the milliformula weight (mfw), is likewise 1/1000 of the formula weight.

Table 2-2

PREFIXES FOR UNITS

Prefix	Abbreviation	Multiplier
giga-	G	10^9
mega-	M	10^6
kilo-	k	10^3
deci-	d	10^{-1}
centi-	c	10^{-2}
milli-	m	10^{-3}
micro-	μ	10^{-6}
nano-	n	10^{-9}
pico-	p	10^{-12}

The angstrom (Å) is a non-SI unit of length that is widely used to express the wavelength of very short radiation, such as X-rays (1 Å = 0.1 nm = 10^{-10} m). Thus, typical X-radiation lies in the range of 0.1 to 10 Å.

One mole of a chemical species is 6.022×10^{23} atoms, molecules, ions, electrons, ion pairs, or subatomic particles.

$6.022 \times 10^{23} =$

1 mmol = 10^{-3} mol

1 mfw = 10^{-3} fw

2A-4 Calculating the Amount of a Substance in Moles or Millimoles

The two examples that follow illustrate how the number of moles or millimoles of a species can be derived from its weight in grams or from the weight of a chemically related species.

Example 2-1

The number of moles of a species A is given by

$$\text{no. mol A} = \frac{\text{wt A}}{\text{fw A}}$$

The number of millimoles is given by

$$\text{no. mmol A} = \frac{\text{wt A}}{(\text{fw A})/1000} = \frac{\text{wt A}}{\text{mfw A}}$$

Remember that the units for the amount of a species are moles and millimoles.

How many moles and millimoles of benzoic acid (fw = 122.1 g) are contained in 2.00 g of the pure acid?

If we use HBz to represent benzoic acid, we can write

$$1 \text{ mol HBz} = 122.1 \text{ g HBz}$$

$$\text{amount HBz} = 2.00 \text{ g HBz} \times \frac{1 \text{ mol HBz}}{122.1 \text{ g HBz}} = 0.0164 \text{ mol HBz}$$

To obtain the number of millimoles, we divide by the milliformula weight:

$$\text{amount HBz} = 2.00 \text{ g HBz} \times \frac{1 \text{ mmol HBz}}{0.1221 \text{ g HBz}} = 16.4 \text{ mmol HBz}$$

Example 2-2

In doing calculations of this kind, you should always include all units, as we do throughout this chapter. This practice often reveals errors in setting up equations.

What weight of Na^+ (fw = 23.00 g) is contained in 25.00 g of Na_2SO_4 (fw = 142.0 g)?

We first calculate the number of moles of Na_2SO_4:

$$\text{amount Na}_2\text{SO}_4 = 25.00 \text{ g Na}_2\text{SO}_4 \times \frac{1 \text{ mol Na}_2\text{SO}_4}{142.0 \text{ g Na}_2\text{SO}_4}$$

$$= 0.17606 \text{ mol Na}_2\text{SO}_4$$

Since 1 mol of Na_2SO_4 contains 2 mol of Na^+,

$$\text{amount Na}^+ = \frac{2 \text{ mol Na}^+}{\text{mol Na}_2\text{SO}_4} \times 0.17606 \text{ mol Na}_2\text{SO}_4$$

$$= 0.35211 \text{ mol Na}^+$$

$$\text{wt Na}^+ = 0.35211 \text{ mol Na}^+ \times \frac{23.00 \text{ g Na}^+}{\text{mol Na}^+} = 8.099 \text{ g Na}^+$$

2B   SOLUTIONS AND THEIR CONCENTRATIONS

2B-1 Concentration of Solutions

Chemists express the concentration of solutes in solution in several ways. The most important of these are described in this section.

Molar Concentration

The molar concentration (C_A) of a solution of a chemical species A is the number of moles of that species contained in one liter of the solution (*not in one liter of the solvent*). The unit of molar concentration is *molarity*, M, which has the dimensions of mol L^{-1}. Molarity also expresses the number of millimoles of a solute per milliliter of solution:

$$C_A = \frac{\text{no. mol solute}}{\text{no. L solution}} = \frac{\text{no. mmol solute}}{\text{no. mL solution}} \tag{2-1}$$

Example 2-3

Calculate the molar concentration of solute in an aqueous solution that contains 2.30 g of ethanol (C_2H_5OH, fw = 46.07 g) in 3.50 L.

We first calculate the number of moles contained in 2.30 g of ethanol:

$$\text{no. mol} = 2.30 \text{ g } C_2H_5OH \times \frac{1 \text{ mol } C_2H_5OH}{46.07 \text{ g } C_2H_5OH} = 0.04992 \text{ mol } C_2H_5OH$$

To obtain the molar concentration, $C_{C_2H_5OH}$, we divide the number of moles of ethanol by the volume:

$$C_{C_2H_5OH} = \frac{0.04992 \text{ mol } C_2H_5OH}{3.50 \text{ L}} = 0.01426 \text{ mol } C_2H_5OH/L = 0.0143 \text{ M}$$

Analytical Molarity. The *analytical molarity* of a solution gives the *total* number of moles of a solute in one liter of the solution (or the total number of millimoles in one milliliter). That is, the analytical molarity specifies a recipe by which the solution can be prepared. For example, a sulfuric acid solution that has an analytical concentration of 1.0 M can be prepared by dissolving 1.0 mol, or 98 g, of H_2SO_4 in water and diluting to exactly 1.0 L.

Equilibrium, or Species, Molarity. The *equilibrium*, or *species, molarity* expresses the molar concentration of a particular species in a solution at equilibrium. In order to state the species molarity, it is necessary to know what happens to the solute when it is dissolved in a solvent. For example, the species molarity of H_2SO_4 in a solution with an analytical concentration of 1.00 M is 0.00 M because the sulfuric acid is entirely dissociated into a mixture of H_3O^+, HSO_4^-, and SO_4^{2-} ions; essentially no H_2SO_4 molecules are present in this solution. The equilibrium concentrations and thus the species molarity of these three ions are 1.01, 0.99, and 0.01 M, respectively.

Equilibrium molar concentrations are often symbolized by placing square brackets around the chemical formula for the species. Thus, for our solution of H_2SO_4 with an analytical concentration of 1.00 M, we can write

$$[H_2SO_4] = 0.00 \text{ M} \qquad [H_3O^+] = 1.01 \text{ M}$$
$$[HSO_4^-] = 0.99 \text{ M} \qquad [SO_4^{2-}] = 0.01 \text{ M}$$

> Analytical molarity describes how a solution has been prepared. It gives the total moles of a solute in one liter of solution.

> Equilibrium, or species, molarity gives the number of moles of a particular species in one liter of solution.

> Some chemists prefer to distinguish between species and analytical concentrations in a different way. They use molar concentration for species concentration and formal concentration (F) for analytical concentration. Applying this convention to our example, we can say that the formal concentration of H_2SO_4 is 1.00 F, whereas its molar concentration is 0.00 M.

> In this example the analytical molarity of H_2SO_4 is given by $C_{H_2SO_4} = [SO_4^{2-}] + [HSO_4^-]$ because these are the only two sulfate-containing species in the solution.

Example 2-4

Calculate the analytical and equilibrium molar concentrations of the solute species in an aqueous solution that contains 285 mg of trichloroacetic acid (Cl_3CCOOH, fw = 163.4 g) in 10.0 mL (the acid is 73% ionized in water).

Employing HA as the symbol for Cl_3CCOOH, we write

$$\text{no. mol HA} = 285 \text{ mg HA} \times \frac{1 \text{ g HA}}{1000 \text{ mg HA}} \times \frac{1 \text{ mol HA}}{163.4 \text{ g HA}}$$

$$= 1.744 \times 10^{-3} \text{ mol HA}$$

From the definition of molar concentration,

$$C_{HA} = \frac{1.744 \times 10^{-3} \text{ mol HA}}{10.0 \text{ mL}} \times \frac{1000 \text{ mL}}{1 \text{ L}} = 0.174 \frac{\text{mol HA}}{\text{L}} = 0.174 \text{ M}$$

Because all but 27% of the acid is dissociated into H_3O^+ and A^-, the species concentration of HA is

$$[HA] = 0.174 \frac{\text{mol}}{\text{L}} \times 0.27 = 0.047 \frac{\text{mol}}{\text{L}} = 0.047 \text{ M}$$

The molarity of H^+ as well as that of A^- is equal to the analytical concentration of the acid minus the species concentration of undissociated acid:

$$[H_3O^+] = [A^-] = C_{HA} - [HA] = 0.174 - 0.047$$
$$= 0.127 \text{ mol/L} = 0.127 \text{ M}$$

Note that the analytical concentration of HA is the sum of the species concentrations of HA and A^-:

$$C_{HA} = [HA] + [A^-]$$

Example 2-5

Describe the preparation of 2.00 L of 0.108 M $BaCl_2$ from $BaCl_2 \cdot 2\ H_2O$ (fw = 244 g).

We see that 1 mol of the dihydrate yields 1 mol of $BaCl_2$. Therefore, to produce this solution, we need

$$2.00 \text{ L} \times \frac{0.108 \text{ mol } BaCl_2 \cdot 2\ H_2O}{L} = 0.216 \text{ mol } BaCl_2 \cdot 2\ H_2O$$

The weight of $BaCl_2 \cdot 2\ H_2O$ is then

$$0.216 \text{ mol } BaCl_2 \cdot 2\ H_2O \times \frac{244 \text{ g}}{\text{mol } BaCl_2 \cdot 2\ H_2O} = 52.7 \text{ g } BaCl_2 \cdot 2\ H_2O$$

Therefore we would dissolve 52.7 g of $BaCl_2 \cdot 2\ H_2O$ in water and dilute to 2.00 L to prepare the 0.108 M solution.

Example 2-6

Describe the preparation of 500 mL of 0.0740 M Cl^- solution from solid $BaCl_2 \cdot 2\ H_2O$ (fw = 244 g).

The number of moles of Cl^- in the solution is given by

$$\text{amount } Cl^- = \frac{0.0740 \text{ mol } Cl^-}{L} \times \frac{1\ L}{1000\ mL} \times 500\ mL = 0.0370 \text{ mol}$$

Because 1 mol of the salt contains 2 mol Cl^-, we need only half the number of moles, or

$$\text{amount } BaCl_2 \cdot 2\ H_2O = 0.0370 \text{ mol } Cl^- \times \frac{1 \text{ mol } BaCl_2 \cdot 2\ H_2O}{2 \text{ mol } Cl^-}$$

$$= 0.0185 \text{ mol } BaCl_2 \cdot 2\ H_2O$$

$$\text{wt } BaCl_2 \cdot 2\ H_2O = 0.0185 \text{ mol } BaCl_2 \cdot 2\ H_2O \times \frac{244 \text{ g } BaCl_2 \cdot 2\ H_2O}{\text{mol } BaCl_2 \cdot 2\ H_2O}$$

$$= 4.51 \text{ g } BaCl_2 \cdot 2\ H_2O$$

Therefore we would dissolve 4.51 g of $BaCl_2 \cdot 2\ H_2O$ in water and dilute to 500 mL to prepare the 0.0740 M solution.

The number of moles of the species A in a solution of A is given by

$$\text{no. mol A} = C_A \times V_A$$

where V_A is the volume of the solution in liters.

Percent Concentration

Chemists frequently express concentrations in terms of percent (parts per hundred). Unfortunately, this practice can be a source of ambiguity because a solution's percent composition can be expressed in several ways. Three common methods are

$$\text{weight percent (w/w)} = \frac{\text{weight solute}}{\text{weight soln}} \times 100\%$$

$$\text{volume percent (v/v)} = \frac{\text{volume solute}}{\text{volume soln}} \times 100\%$$

$$\text{weight/volume percent (w/v)} = \frac{\text{weight solute, g}}{\text{volume soln, mL}} \times 100\%$$

Note that the denominator in each of these expressions refers to the *solution* rather than to the solvent. Note also that the first two expressions do not depend on the units employed (provided, of course, that there is consistency between numerator and denominator). In the third expression, units must be defined because the numerator and denominator have units that do not cancel. Of the three expressions, only weight percent has the virtue of being temperature-independent.

Weight percent is frequently employed to express the concentration of commercial aqueous reagents. For example, hydrochloric acid is sold as a 37% solution, which means that the reagent contains 37 g of HCl per 100 g of solution (see Example 2-11).

Volume percent is commonly used to specify the concentration of a solution that was prepared by diluting a pure liquid compound with another liquid. For example, a 5% aqueous solution of methanol *usually* means a solution prepared by diluting 5.0 mL of pure methanol with enough water to give 100 mL.

Weight/volume percent is often employed to indicate the composition of dilute aqueous solutions of solid reagents. For example, 5% aqueous silver nitrate *often* refers to a solution prepared by dissolving 5 g of silver nitrate in sufficient water to give 100 mL of solution.

To avoid uncertainty, always specify explicitly the type of percent composition being discussed. If this information is missing, the user must decide intuitively which of the several types is involved. The potential error resulting from a wrong choice is considerable. For example, commercial 50% (w/w) sodium hydroxide contains 763 g of the reagent per liter, which corresponds to 76.3% (w/v) sodium hydroxide.

Parts Per Million and Parts Per Billion

For very dilute solutions, parts per million (ppm) is a convenient way to express concentration:

$$C_{ppm} = \frac{\text{weight solute}}{\text{weight solution}} \times 10^6 \text{ ppm}$$

where C_{ppm} is the concentration in parts per million. Note that the units of weight in the numerator and denominator must agree.

For even more dilute solutions, 10^9 ppb rather than 10^6 ppm is used in the equation above to give the results in parts per billion (ppb).

The term "parts per thousand" (ppt) is also encountered, especially in oceanography.

You should always specify the type of percent when reporting concentrations in this way.

A handy rule in calculating parts per million is to remember that, for dilute aqueous solutions whose densities are approximately 1.00 g/mL, 1 ppm = 1.00 mg/L. That is,

$$C_{ppm} = \frac{\text{weight solute (mg)}}{\text{volume soln (L)}} \quad (2\text{-}2)$$

$$C_{ppb} = \frac{\text{weight solute}}{\text{weight solution}} \times 10^9 \text{ ppb}$$

Example 2-7

What is the molarity of K^+ in a solution that contains 63.3 ppm of $K_3Fe(CN)_6$ (fw = 329.3 g)?

Because the solution is so dilute, it is reasonable to assume that its density is 1.00 g/mL. Therefore, according to Equation 2-2,

$$63.3 \text{ ppm } K_3Fe(CN)_6 = 63.3 \text{ mg } K_3Fe(CN)_6/L$$

$$\frac{\text{no. mol } K_3Fe(CN)_6}{L} = \frac{63.3 \text{ mg } K_3Fe(CN)_6}{L} \times \frac{1 \text{ g}}{1000 \text{ mg}}$$

$$\times \frac{1 \text{ mol } K_3Fe(CN)_6}{329.3 \text{ g } K_3Fe(CN)_6}$$

$$= 1.922 \times 10^{-4} \frac{\text{mol}}{L} = 1.922 \times 10^{-4} \text{ M}$$

$$[K^+] = \frac{1.922 \times 10^{-4} \text{ mol } K_3Fe(CN)_6}{L} \times \frac{3 \text{ mol } K^+}{1 \text{ mol } K_3Fe(CN)_6}$$

$$= 5.77 \times 10^{-4} \frac{\text{mol } K^+}{L} = 5.77 \times 10^{-4} \text{ M}$$

Solution–Diluent Volume Ratios

The composition of a dilute solution is sometimes specified in terms of the volume of a more concentrated solution and the volume of the solvent used to do the diluting. The volume of the former is separated from that of the latter by a colon. Thus, a 1 : 4 HCl solution contains four volumes of water for each volume of concentrated hydrochloric acid.

This method of notation is frequently ambiguous in that the concentration of the original solution is not always obvious to the reader. Unfortunately, under some circumstances 1 : 4 is interpreted to mean dilute one volume with three volumes. Because of such uncertainties, you should avoid using solution–diluent ratios.

p-Functions

Scientists frequently express the concentration of a species in terms of its *p-function,* or *p-value*. The p-value is the negative logarithm (to the base 10) of the molar concentration of a species. So, for the species X,

$$pX = -\log [X]$$

For the chemical species X, $pX = -\log [X]$.

The best known p-function is pH, where $pH = -\log [H^+]$.

As shown by the following examples, p-values offer the advantage of allowing concentrations that vary over ten or more orders of magnitude to be expressed in terms of small positive numbers.

Example 2-8

Calculate the p-value for each ion in a solution that is 2.00×10^{-3} M in NaCl and 5.4×10^{-4} M in HCl.

$$pH = -\log [H^+] = -\log (5.4 \times 10^{-4})$$
$$= -\log 5.4 - \log 10^{-4} = -0.73 - (-4) = 3.27$$

To obtain pNa, we write

$$pNa = -\log (2.00 \times 10^{-3}) = -\log 2.00 - \log 10^{-3}$$
$$= -0.301 - (-3.000) = 2.699$$

The total Cl^- concentration is given by the sum of the concentrations of the two solutes:

$$[Cl^-] = 2.00 \times 10^{-3} M + 5.4 \times 10^{-4} M$$
$$= 2.00 \times 10^{-3} M + 0.54 \times 10^{-3} M = 2.54 \times 10^{-3} M$$
$$pCl = -\log 2.54 \times 10^{-3} = 2.595$$

Note that in Example 2-8 and in the one that follows, the results are rounded according to the rules listed in Section 3F-2.

Example 2-9

Calculate the molar concentration of Ag^+ in a solution that has a pAg of 6.372.

$$pAg = -\log [Ag^+] = 6.372$$
$$\log [Ag^+] = -6.372 = 0.628 - 7.000$$
$$[Ag^+] = \text{antilog } (0.628) \times \text{antilog } (-7.000)$$
$$= 4.246 \times 10^{-7} = 4.25 \times 10^{-7}$$

In SI units, density is expressed in units of kg/m^3.

Specific gravity is the ratio of the mass of a substance to the mass of an equal volume of water.

SPECIFIC GRAVITIES OF
CONCENTRATED ACIDS AND BASES

Reagent	Concen-tration, % (w/w)	Specific Gravity
Acetic acid	99.5	1.05
Ammonia	27	0.90
Hydrochloric acid	37	1.18
Hydrobromic acid	48	1.51
Nitric acid	70	1.42
Perchloric acid	70	1.66
Phosphoric acid	85	1.69
Sulfuric acid	95.5	1.83

2B-2 Density and Specific Gravity of Solutions

"Density" and "specific gravity" are terms often encountered in the analytical literature. The *density* of a substance is its mass per unit volume, whereas its *specific gravity* is the ratio of its mass to the mass of an equal volume of water at 4°C. Density has units of kilograms per liter or grams per milliliter in the metric system. Specific gravity is dimensionless and so is not tied to any particular system of units. For this reason, specific gravity is widely used in describing items of commerce. Since the density of water is approximately 1.00 g/mL and since we use the metric system throughout this text, density and specific gravity are used interchangeably.

Example 2-10

Calculate the molar concentration of HNO_3 (fw = 63.0 g) in a solution that has a specific gravity of 1.42 and is 70% HNO_3 (w/w).

Let us first calculate the grams of acid per liter of concentrated solution:

$$\frac{1.42 \text{ kg reagent}}{\text{L reagent}} \times \frac{10^3 \text{ g reagent}}{\text{kg reagent}} \times \frac{70 \text{ g } HNO_3}{100 \text{ g reagent}} = \frac{994 \text{ g } HNO_3}{\text{L reagent}}$$

Then

$$C_{HNO_3} = \frac{994 \text{ g } HNO_3}{\text{L reagent}} \times \frac{1 \text{ mol } HNO_3}{63.0 \text{ g } HNO_3} = \frac{15.8 \text{ mol } HNO_3}{\text{L reagent}} = 16 \text{ M}$$

Example 2-11

Describe the preparation of 100 mL of 6.0 M HCl from a concentrated reagent that has a specific gravity of 1.18 and is 37% (w/w) HCl (fw = 36.5 g).

Proceeding as in Example 2-10, we calculate the molarity of the concentrated reagent:

$$C_{HCl} = \frac{1.18 \times 10^3 \text{ g reagent}}{\text{L reagent}} \times \frac{37 \text{ g HCl}}{100 \text{ g reagent}} \times \frac{1 \text{ mol HCl}}{36.5 \text{ g HCl}} = 12.0 \text{ M}$$

The number of moles of HCl required is

$$\text{amount HCl} = 100 \text{ mL} \times \frac{1 \text{ L}}{1000 \text{ mL}} \times \frac{6.0 \text{ mol HCl}}{\text{L}} = 0.600 \text{ mol}$$

$$\text{vol concd reagent} = 0.600 \text{ mol HCl} \times \frac{1 \text{ L reagent}}{12.0 \text{ mol HCl}}$$

$$= 0.0500 \text{ L or } 50.0 \text{ mL}$$

Thus we would dilute 50 mL of the concentrated reagent to 600 mL.

The solution to Example 2-11 is based upon the following useful relationship, which we will be using countless times:

$$V_{\text{concd}} \times M_{\text{concd}} = V_{\text{dil}} \times M_{\text{dil}} \qquad (2\text{-}3)$$

where the two terms on the left are the volume and molar concentration of a concentrated solution that is being used to prepare a diluted solution having the volume and concentration given by the corresponding terms on the right. This equation is based upon the fact that the number of moles in the diluted solution must equal the number of moles in the concentrated reagent. Note that the volumes can be in milliliters or liters as long as the same units are used for both solutions.

Equation 2-3 can be used with liters and moles per liter or with milliliters and millimoles per milliliter:

$$L_{\text{concd}} \times \frac{\text{mol}_{\text{concd}}}{L_{\text{concd}}} = L_{\text{dil}} \times \frac{\text{mol}_{\text{dil}}}{L_{\text{dil}}}$$

$$\text{mL}_{\text{concd}} \times \frac{\text{mmol}_{\text{concd}}}{\text{mL}_{\text{concd}}} = \text{mL}_{\text{dil}} \times \frac{\text{mmol}_{\text{dil}}}{\text{mL}_{\text{dil}}}$$

2C STOICHIOMETRIC CALCULATIONS

Stoichiometry is defined as the weight relationships among reacting chemical species. This section provides a brief review of stoichiometry and its applications to chemical calculations.

A balanced chemical equation is a statement of the combining ratios, or stoichiometry, (in units of moles) that exist between reacting substances and their products. Thus, the equation

$$2 \text{ NaI(aq)} + \text{Pb(NO}_3)_2(\text{aq}) = \text{PbI}_2(\text{s}) + 2 \text{ NaNO}_3(\text{aq})$$

indicates that 2 mol of aqueous sodium iodide combine with 1 mol of aqueous lead nitrate to produce 1 mol of solid lead iodide and 2 mol of aqueous sodium nitrate.[1]

Example 2-12 demonstrates how the weight of reactants in a chemical reaction is related to the weight of the products. A calculation of this type is a three-step process involving (1) transformation of the known weight of a substance in grams to the corresponding number of moles, (2) multiplication by a factor that accounts for the stoichiometry, and (3) reconversion of the data in moles back to the metric units called for in the answer (see Figure 2-1).

The stoichiometry of a reaction is the relationship among the number of moles of reactants and products as shown by a balanced equation.

The physical state of substances in chemical equations is often indicated by the letters (g), (l), (s), and (aq), which refer to gaseous, liquid, solid, and aqueous solution states, respectively.

[1]Here it is advantageous to depict the reaction in terms of chemical compounds. If we wish to focus on reacting species, the net ionic equation is preferable:

$$2 \text{ I}^-(\text{aq}) + \text{Pb}^{2+}(\text{aq}) = \text{PbI}_2(\text{s})$$

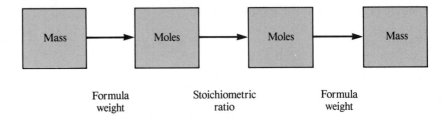

Figure 2-1
Flow diagram for making stoichiometric calculations.

Example 2-12

(a) What weight of $AgNO_3$ (fw = 169.9 g) is needed to convert 2.33 g of Na_2CO_3 (fw = 106.0 g) to Ag_2CO_3? (b) What weight of Ag_2CO_3 (fw = 275.7 g) is formed?
(a) $Na_2CO_3(aq) + 2\ AgNO_3(aq) \rightarrow Ag_2CO_3(s) + 2\ NaNO_3(aq)$

$$\text{Step 1: amount } Na_2CO_3 = 2.33\ \text{g } Na_2CO_3 \times \frac{1\ \text{mol } Na_2CO_3}{106.0\ \text{g } Na_2CO_3}$$

$$= 0.02198\ \text{mol } Na_2CO_3$$

Step 2: The balanced equation reveals that

$$\text{amount } AgNO_3 = 0.02198\ \text{mol } Na_2CO_3 \times \frac{2\ \text{mol } AgNO_3}{1\ \text{mol } Na_2CO_3}$$

$$= 0.04396\ \text{mol } AgNO_3$$

$$\text{Step 3: wt } AgNO_3 = 0.04396\ \text{mol } AgNO_3 \times \frac{169.9\ \text{g } AgNO_3}{\text{mol } AgNO_3}$$

$$= 7.47\ \text{g } AgNO_3$$

(b) no. mol Ag_2CO_3 = no. mol Na_2CO_3 = 0.02198 mol

$$\text{wt } Ag_2CO_3 = 0.02198\ \text{mol } Ag_2CO_3 \times \frac{275.7\ \text{g } Ag_2CO_3}{\text{mol } Ag_2CO_3}$$

$$= 6.06\ \text{g } Ag_2CO_3$$

Example 2-13

What weight of Ag_2CO_3 (fw = 275.7 g) is formed when a 25.0-mL portion of 0.200 M $AgNO_3$ is mixed with 50.0 mL of 0.0800 M Na_2CO_3?
   Mixing these two solutions will result in one (and only one) of three possible outcomes:

(a) an excess of $AgNO_3$ will remain after reaction is complete.
(b) an excess of Na_2CO_3 will remain after reaction is complete.
(c) an excess of neither reagent will exist (that is, the number of moles of $AgNO_3$ present just before the reaction begins is exactly equal to twice the number of moles of Na_2CO_3).

As a first step, we must establish which of these situations applies by calculating the amounts of reactants (in chemical units) available at the outset. Initial amounts are

$$\text{amount AgNO}_3 = 25.0 \text{ mL AgNO}_3 \times \frac{1 \text{ L AgNO}_3}{1000 \text{ mL AgNO}_3}$$

$$\times \frac{0.200 \text{ mol AgNO}_3}{\text{L AgNO}_3}$$

$$= 5.00 \times 10^{-3} \text{ mol AgNO}_3$$

$$\text{amount Na}_2\text{CO}_3 = 50.0 \text{ mL Na}_2\text{CO}_3 \times \frac{1 \text{ L Na}_2\text{CO}_3}{1000 \text{ mL Na}_2\text{CO}_3}$$

$$\times \frac{0.0800 \text{ mol Na}_2\text{CO}_3}{\text{L Na}_2\text{CO}_3}$$

$$= 4.00 \times 10^{-3} \text{ mol Na}_2\text{CO}_3$$

Because each CO_3^{2-} ion reacts with two Ag^+ ions, $2 \times 4.00 \times 10^{-3} = 8.00 \times 10^{-3}$ mol $AgNO_3$ is required to react with the Na_2CO_3. Since too little $AgNO_3$ is present, situation (b) prevails, and the amount of Ag_2CO_3 produced is limited by the amount of $AgNO_3$ available. Thus,

$$\text{wt Ag}_2\text{CO}_3 = 5.00 \times 10^{-3} \text{ mol AgNO}_3$$

$$\times \frac{1 \text{ mol Ag}_2\text{CO}_3}{2 \text{ mol AgNO}_3} \times \frac{275.7 \text{ g Ag}_2\text{CO}_3}{\text{mol Ag}_2\text{CO}_3}$$

$$= 0.689 \text{ g Ag}_2\text{CO}_3$$

Example 2-14

What will be the analytical molar Na_2CO_3 concentration in the solution produced when a 25.0-mL portion of 0.200 M $AgNO_3$ is mixed with 50.0 mL of 0.0800 M Na_2CO_3?

We saw in Example 2-13 that formation of 5.00×10^{-3} mol of $AgNO_3$ requires 2.50×10^{-3} mol of Na_2CO_3. The number of moles of unreacted Na_2CO_3 is then given by

$$\text{amount Na}_2\text{CO}_3 = 4.00 \times 10^{-3} \text{ mol Na}_2\text{CO}_3$$

$$- \left(5.00 \times 10^{-3} \text{ mol AgNO}_3 \times \frac{1 \text{ mol Na}_2\text{CO}_3}{2 \text{ mol AgNO}_3}\right)$$

$$= 1.50 \times 10^{-3} \text{ mol Na}_2\text{CO}_3$$

By definition, the molarity is the number of moles of Na_2CO_3 per liter. Thus,

$$C_{\text{Na}_2\text{CO}_3} = \frac{1.50 \times 10^{-3} \text{ mol Na}_2\text{CO}_3}{(50.0 + 25.0) \text{ mL}} \times \frac{1000 \text{ mL}}{1 \text{ L}} = 0.0200 \text{ M Na}_2\text{CO}_3$$

2D   QUESTIONS AND PROBLEMS

2-1. Define
  *(a) millimole.
  *(b) milliformula weight.
  (c) parts per million.

2-2. What is the difference between species molarity and analytical molarity?

*2-3. Give two examples of units derived from the fundamental SI base units.

2-4. Simplify the following quantities using a unit with an appropriate prefix:
  *(a) 1.5×10^6 Hz.
  (b) 2.62×10^{-9} g.
  *(c) 6.23×10^4 μmol.
  (d) 4.0×10^9 s.
  *(e) 96,494 C.
  (f) 47,000 kg.

*2-5. How many Na^+ ions are contained in 4.62 g of Na_3PO_4?

2-6. How many K^+ ions are contained in 5.96 mol of K_3PO_4?

*2-7. How many moles are contained in
  (a) 6.84 g of B_2O_3?
  (b) 296 mg of $Na_2B_4O_7 \cdot 10\ H_2O$?
  (c) 8.75 g of Mn_3O_4?
  (d) 67.4 mg of CaC_2O_4?

2-8. How many millimoles are contained in
  (a) 64 mg of P_2O_5?
  (b) 12.92 g of CO_2?
  (c) 30.0 g of $NaHCO_3$?
  (d) 764 mg of $MgNH_4PO_4$?

*2-9. How many millimoles of solute are contained in
  (a) 2.00 L of 2.76×10^{-3} M $KMnO_4$?
  (b) 750 mL of 0.0416 M KSCN?
  (c) 250 mL of a solution that contains 4.20 ppm of $CuSO_4$?
  (d) 3.50 L of 0.276 M KCl?

2-10. How many millimoles of solute are contained in
  (a) 175 mL of 0.320 M $HClO_4$?
  (b) 15.0 L of 8.05×10^{-3} M K_2CrO_4?
  (c) 5.00 L of an aqueous solution that contains 6.75 ppm of $AgNO_3$?
  (d) 851 mL of 0.0200 M KOH?

*2-11. How many milligrams are contained in
  (a) 0.666 mol of HNO_3?
  (b) 300 mmol of MgO?
  (c) 19.0 mol of NH_4NO_3?
  (d) 5.32 mol of $(NH_4)_2Ce(NO_3)_6$ (fw = 548.23 g)?

2-12. How many grams are contained in
  (a) 6.21 mol of KBr?
  (b) 10.2 mmol of PbO?
  (c) 4.92 mol of $MgSO_4$?
  (d) 12.8 mmol of $Fe(NH_4)_2(SO_4)_2 \cdot 6\ H_2O$?

2-13. How many milligrams of solute are contained in
  *(a) 26.0 mL of 0.150 M sucrose (fw = 342 g)?
  *(b) 2.92 L of 5.23×10^{-3} M H_2O_2?

  (c) 737 mL of a solution that contains 6.38 ppm of $Pb(NO_3)_2$?
  (d) 6.75 mL of 0.0619 M KNO_3?

2-14. How many grams of solute are contained in
  *(a) 450 mL of 0.164 M H_2O_2?
  *(b) 27.0 mL of 8.75×10^{-4} M benzoic acid (fw = 122 g)?
  (c) 3.50 L of a solution that contains 21.7 ppm of $SnCl_2$?
  (d) 21.7 mL of 0.0125 M $KBrO_3$?

2-15. Calculate the p-value for each ion:
  *(a) Na^+, Cl^-, and OH^- in a solution that is 0.116 M in NaCl and 0.125 M in NaOH.
  (b) Ba^{2+}, Mn^{2+}, and Cl^- in a solution that is 3.80×10^{-3} M in $BaCl_2$ and 2.22 M in $MnCl_2$.
  *(c) H^+, Cl^-, and Zn^{2+} in a solution that is 1.50 M in HCl and 0.120 M in $ZnCl_2$.
  (d) Cu^{2+}, Zn^{2+}, and NO_3^- in a solution that is 4.32×10^{-2} M in $Cu(NO_3)_2$ and 0.101 M in $Zn(NO_3)_2$.
  *(e) K^+, OH^-, and $Fe(CN)_6^{4-}$ in a solution that is 3.79×10^{-6} M in $K_4Fe(CN)_6$ and 4.12×10^{-5} M in KOH.
  (f) H^+, Ba^{2+}, and ClO_4^- in a solution that is 2.75×10^{-4} M in $Ba(ClO_4)_2$ and 4.44×10^{-4} M in $HClO_4$.

2-16. Calculate the molar H_3O^+ concentration of a solution that has a pH of
  *(a) 9.21   *(c) 0.45   *(e) 7.32   *(g) −0.21
  (b) 4.58   (d) 14.12   (f) 6.76   (h) −0.52

2-17. Calculate the p-functions for each ion in a solution that is
  *(a) 0.0100 M in NaBr.
  (b) 0.0100 M in $BaBr_2$.
  *(c) 3.5×10^{-3} M in $Ba(OH)_2$.
  (d) 0.040 M in HCl and 0.020 M in NaCl.
  *(e) 5.2×10^{-3} M in $CaCl_2$ and 3.6×10^{-3} M in $BaCl_2$.
  (f) 4.8×10^{-8} M in $Zn(NO_3)_2$ and 5.6×10^{-7} M $Cd(NO_3)_2$.

2-18. Convert the following p-functions to molar concentrations:
  *(a) pH = 8.67   *(e) pLi = −0.321
  (b) pOH = 0.125   (f) pNO_3 = 7.77
  *(c) pBr = 0.034   *(g) pMn = 0.0025
  (d) pCa = 12.35   (h) pCl = 1.020

*2-19. Sea water contains an average of 1.08×10^3 ppm of Na^+ and 270 ppm of SO_4^{2-}. Calculate
  (a) the molar concentrations of Na^+ and SO_4^{2-} given that the average density of sea water is 1.02 g/mL.
  (b) the pNa and pSO_4 for sea water.

2-20. Average human blood serum contains 18 mg of K^+ and 365 mg of Cl^- per 100 mL. Calculate
  (a) the molar concentration for each of these species; use 1.00 g/mL for the density of serum.

*Answers to the asterisked problems are given in the answers section at the back of the book.

(b) pK and pCl for human serum.

*2-21. A solution was prepared by dissolving 10.12 g of $KCl \cdot MgCl_2 \cdot 6 H_2O$ (fw = 277.87 g) in sufficient water to give 2.000 L. Calculate

  (a) the molar analytical concentration of $KCl \cdot MgCl_2$.

  (b) the molar concentration of Mg^{2+}.

  (c) the molar concentration of Cl^-.

  (d) the percent (w/v) of $KCl \cdot MgCl_2 \cdot 6 H_2O$.

  (e) the millimoles of Cl^- in 25.0 mL of this solution.

  (f) ppm K^+.

  (g) pMg for the solution.

  (h) pCl for the solution.

2-22. A solution was prepared by dissolving 367 mg of $K_3Fe(CN)_6$ (fw = 329 g) in sufficient water to give 750 mL. Calculate

  (a) the molar analytical concentration of $K_3Fe(CN)_6$.

  (b) the molar concentration of K^+.

  (c) the molar concentration of $Fe(CN)_6^{3-}$.

  (d) the weight/volume percentage of $K_3Fe(CN)_6$.

  (e) the millimoles of K^+ in 50.0 mL of this solution.

  (f) ppm $Fe(CN)_6^{3-}$.

  (g) pK for the solution.

  (h) pFe(CN)₆ for the solution.

*2-23. A 7.88% (w/w) $Fe(NO_3)_3$ (fw = 241.86 g) solution has a density of 1.062 g/mL. Calculate

  (a) the molar analytical concentration of $Fe(NO_3)_3$.

  (b) the molar NO_3^- concentration.

  (c) the grams of $Fe(NO_3)_3$ contained in each liter.

2-24. A 15.0% (w/w) $NiCl_2$ (fw = 129.6 g) solution has a density of 1.149 g/mL. Calculate

  (a) the molar concentration of $NiCl_2$.

  (b) the molar Cl^- concentration.

  (c) the grams of $NiCl_2$ contained in each liter.

*2-25. Describe the preparation of

  (a) 500 mL of 6.50% (w/v) aqueous ethanol (C_2H_5OH, fw = 46.1 g).

  (b) 500 g of 6.50% (w/w) aqueous ethanol.

  (c) 500 mL of 6.50% (v/v) aqueous ethanol.

2-26. Describe the preparation of

  (a) 2.50 L of 18.0% (w/v) aqueous glycerol ($C_3H_8O_3$, fw = 92.1 g).

  (b) 2.50 kg of 18.0% (w/w) aqueous glycerol.

  (c) 2.50 L of 18.0% (v/v) aqueous glycerol.

*2-27. Describe the preparation of 750 mL of 6.00 M H_3PO_4 from the commercial reagent that is 85% H_3PO_4 (w/w) and has a specific gravity of 1.69.

2-28. Describe the preparation of 900 mL of 3.00 M HNO_3 from the commercial reagent that is 69% HNO_3 (w/w) and has a specific gravity of 1.42.

*2-29. Describe the preparation of

  (a) 500 mL of 0.0750 M $AgNO_3$ from the solid reagent.

  (b) 1.00 L of 0.315 M HCl from a 6.00 M solution of the reagent.

  (c) 600 mL of a solution that is 0.0825 M in K^+, starting with solid $K_4Fe(CN)_6$.

  (d) 400 mL of 3.00% (w/v) aqueous $BaCl_2$ from a 0.400 M $BaCl_2$ solution.

  (e) 2.00 L of 0.120 M $HClO_4$ from the commercial reagent [60% $HClO_4$ (w/w), sp gr 1.60].

  (f) 9.00 L of a solution that is 60.0 ppm in Na^+, starting with solid Na_2SO_4.

2-30. Describe the preparation of

  (a) 5.00 L of 0.150 M $KMnO_4$ from the solid reagent.

  (b) 4.00 L of 0.175 M $HClO_4$ from an 8.00 M solution of the reagent.

  (c) 400 mL of a solution that is 0.0500 M in I^-, starting with MgI_2.

  (d) 200 mL of 1.00% (w/v) aqueous $CuSO_4$ from a 0.218 M $CuSO_4$ solution.

  (e) 1.50 L of 0.215 M NaOH from the concentrated commercial reagent [50% NaOH (w/w), sp gr 1.525].

  (f) 1.50 L of a solution that is 12.0 ppm in K^+, starting with solid $K_4Fe(CN)_6$.

*2-31. What weight of solid $La(IO_3)_3$ (fw = 663.6 g) is formed when 50.0 mL of 0.150 M La^{3+} is mixed with 75.0 mL of 0.202 M IO_3^-?

2-32. What weight of solid $PbCl_2$ (fw = 278.10 g) is formed when 100 mL of 0.125 M Pb^{2+} is mixed with 200 mL of 0.175 M Cl^-?

*2-33. Exactly 0.1120 g of pure Na_2CO_3 was dissolved in 100.0 mL of 0.0497 M $HClO_4$.

  (a) How many grams of CO_2 were evolved?

  (b) What was the molarity of the excess reactant ($HClO_4$ or Na_2CO_3)?

2-34. Exactly 50.00 mL of a 0.4230 M solution of Na_3PO_4 was mixed with 100.00 mL of 0.5151 M $AgNO_3$.

  (a) What weight of solid Ag_3PO_4 was formed?

  (b) What was the molarity of the unreacted species (Na_3PO_4 or $AgNO_3$) after the reaction was complete?

*2-35. Exactly 75.00 mL of a 0.3333 M solution of Na_2SO_3 was treated with 150.0 mL of 0.3912 M $HClO_4$ and boiled to remove the SO_2 formed.

  (a) How many grams of SO_2 were evolved?

  (b) What was the concentration of the unreacted reagent (Na_2SO_3 or $HClO_4$) after the reaction was complete?

2-36. What weight of $MgNH_4PO_4$ precipitated when 200.0 mL of a 1.000% (w/v) solution of $MgCl_2$ was treated with 40.0 mL of 0.1753 M Na_3PO_4 and an excess of NH_4^+? What was the molarity of the excess reagent (Na_3PO_4 or $MgCl_2$) after the precipitation was complete?

*2-37. What volume of 0.01000 M $AgNO_3$ would be required to precipitate all the I^- in 200.0 mL of a solution that contained 2.643 ppt KI?

2-38. Exactly 750.0 mL of a solution that contained 650.1 ppm of $Ba(NO_3)_2$ was mixed with 200.0 mL of a solution that was 0.04100 M in $Al_2(SO_4)_3$.

  (a) What weight of solid $BaSO_4$ was formed?

  (b) What was the molarity of the unreacted reagent [$Al_2(SO_4)_3$ or $Ba(NO_3)_2$]?

Chapter 3

Errors
in Chemical Analysis

It is impossible to perform a chemical analysis in such a way that the results are totally free of errors or uncertainties. Our goal is to keep these errors at a tolerable level and to estimate their size with acceptable accuracy. In this chapter we explore the nature of experimental errors and their effects on analytical results.

The effect of errors in analytical data is illustrated in Figure 3-1, which shows results for the quantitative determination of iron(III). Six equal portions of an aqueous solution known to contain exactly 20.00 ppm of iron(III) were analyzed in exactly the same way. Note that the results range from a low of 19.4 ppm to a high of 20.3 ppm of iron(III). The average $\bar{x}$ of the data is 19.8 ppm.

Every measurement is influenced by many uncertainties, which combine to produce a scatter of results like that shown in Figure 3-1. Measurement uncertainties can never be completely eliminated, and so the true value for any quantity is unknown. The probable magnitude of the error in a measurement can be evaluated, however. It is then possible to define limits within which the true value of a measured quantity lies at a given level of probability.

It is seldom easy to estimate the reliability of experimental data. Nevertheless, we must make such estimates *because data of unknown reliability are worthless*. Moreover, results that are not especially accurate may be of considerable value if the limits of uncertainty are known.

Unfortunately, there is no simple and widely applicable method for determining data reliability with absolute certainty. It often requires as

Parts per million (ppm), that is, 20.00 parts of iron(III) per million parts of solution.

The true value of a measurement is never known exactly.

20

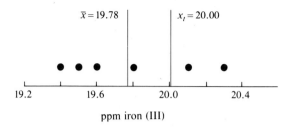

Figure 3-1

Figure 3-1
Results from six replicate determinations for iron in aqueous samples of a standard solution containing 20.00 ppm of iron(III).

much effort to assess the quality of experimental results as it requires to collect them. Reliability can be evaluated in a number of different ways. Experiments designed to reveal the presence of errors can be performed. Standards of known composition can be analyzed and the results compared with the known composition. A few minutes in the library consulting the literature of analytical chemistry can be profitable. Calibrating equipment enhances the quality of data. Finally, statistical tests can be applied to the data. None of these options is perfect, and so in the end we have to make *judgments* as to the probable accuracy of our results. These judgments tend to become harsher and less optimistic with experience.

One of the first questions to answer before beginning an analysis is, "What is the maximum error I can tolerate in the result?" The answer to this question determines how much time you will spend on the analysis. For example, a tenfold increase in reliability may take hours, days, or even weeks of added labor. *No one can afford to waste time generating data that are more reliable than is needed.*

In this chapter we consider the types of errors encountered in chemical analysis, methods for recognizing errors, and techniques for estimating and reporting their size.[1]

3A DEFINITION OF TERMS

Chemists usually carry two to five portions (*replicates*) of a sample through an analytical procedure. Individual results from a set of measurements are seldom the same (Figure 3-1), and so a central, or "best," value is used for the set. We justify the extra effort required to analyze several samples in two ways. First, the central value of a set should be more reliable than any of the individual results. Second, variation in the data should provide a measure of the uncertainty associated with the central result. Either the *mean* or the *median* may serve as the central value for a set of replicate measurements.

3A-1 The Mean and Median

Mean, arithmetic mean, and *average* ($\bar{x}$) are synonyms for the quantity obtained by dividing the sum of replicate measurements by the number of

---

[1]For detailed discussions of error analysis, see A. Currie, in *Treatise on Analytical Chemistry*, 2nd ed., I. M. Kolthoff and P. J. Elving, Eds., Part I, Vol. 1, Chapter 4. New York: Wiley, 1978; and J. Mandel, *ibid.*, Chapter 5.

The symbol Σx_i means to add all of the values x_i for the replicates.

measurements in the set:

$$\bar{x} = \frac{\sum\limits_{i=1}^{N} x_i}{N} \tag{3-1}$$

where x_i represents the individual values of x making up a set of N replicate measurements.

The *median* of a set of replicate data is the middle result when the data are arranged by size. There are equal numbers of data that are larger and smaller than the median. For an odd number of results, the median can be evaluated directly. For an even number, the mean of the middle pair is used.

The *median* is the middle value in a set of data that has been arranged by numerical value.

Example 3-1

Calculate the mean and median for the data shown in Figure 3-1.

$$\text{mean} = \bar{x} = \frac{19.4 + 19.5 + 19.6 + 19.8 + 20.1 + 20.3}{6}$$

$$= 19.78 \approx 19.8 \text{ ppm Fe}$$

Because the set contains an even number of measurements, the median is the average of the central pair:

$$\text{median} = \frac{19.6 + 19.8}{2} = 19.7 \text{ ppm Fe}$$

Ideally the mean and median are identical. Frequently they are not, however, particularly when the number of measurements in the set is small.

Computer Application 3-1
CALCULATING A MEAN

The most common computer language in the world is BASIC. Almost all computers have a BASIC interpreter. There are several dialects of BASIC, but by far the most common is Microsoft® BASIC. Because of its wide availability, we will use this dialect for a number of our computer applications. We do not intend to teach you how to program, but we will give you some examples of how computers can be used to solve problems and perform calculations in analytical chemistry. As our first example, we present a short program that calculates the mean of a set of numbers.

All lines in programs written in BASIC begin with a number. Although their purpose is not obvious in this program, we will use the numbers for reference in our discussion.

```
10 SUMX = 0
20 INPUT N
30 FOR I = 1 TO N
40 INPUT X
50 SUMX = SUMX + X
60 NEXT I
70 XBAR = SUMX/N
80 PRINT "The mean is ";XBAR
90 END
```

Statement 10 initializes the variable SUMX to zero. Statement 20 accepts the number of data from the keyboard. Statements 30–60 constitute a FOR–NEXT loop that accepts the N data one at a time in statement 40. The N data are summed into the variable SUMX in statement 50. When all N data have been typed into the computer and accumulated in SUMX, the program proceeds to statement 70, which calculates the mean. Finally, the mean value XBAR is displayed on the computer screen by the PRINT command in statement 80, and the program terminates. Notice that the variable names in the program correspond to the variables in Equation 3-1.

If you have a computer available, enter the program and try running it. Use the data in Example 3-1 to test the program. As an exercise, try modifying statements 20 and 40 so that they prompt you to enter the data.

3A-2 Precision

Precision describes the agreement between two or more measurements that have been made *in exactly the same way*. There are several ways to express precision.

> Precision is the closeness of a measurement to other measurements made in the same way.

Standard Deviation (s)

The *standard deviation* is a statistical term scientists and engineers use as a measure of precision. For small sets of data, we calculate the sample standard deviation s using the following equation:

$$s = \sqrt{\frac{\sum_{i=1}^{N}(x_i - \bar{x})^2}{N - 1}} \qquad (3\text{-}2)$$

where $x_i - \bar{x}$ is the *deviation from the mean* of the ith measurement.

> Equation 3-2 applies to small sets of data. It says find the deviations of the x_i's from the mean $\bar{x}$, square them, sum them, divide the sum by $N - 1$, and take the square root. The quantity $N - 1$ is called the number of degrees of freedom. Many scientific calculators have the standard-deviation function built in. Computer Application 3-2 shows how a short program written in BASIC can perform the same function.

Variance (s^2)

The *variance* is simply the square of the standard deviation:

$$s^2 = \frac{\sum_{i=1}^{N}(x_i - \bar{x})^2}{N - 1} \qquad (3\text{-}3)$$

> The deviation from the mean is obtained by subtracting the mean of a set of measurements from a measurement.

Table 3-1

METHODS OF EXPRESSING PRECISION AND ACCURACY*

Fe Concentration, ppm x_i		Deviation from Mean $|x_i - \bar{x}|$	$(x_i - \bar{x})^2$
x_1	19.4	0.38	0.1444
x_2	19.5	0.28	0.0784
x_3	19.6	0.18	0.0324
x_4	19.8	0.02	0.0004
x_5	20.1	0.32	0.1024
x_6	20.3	0.52	0.2704
$\Sigma x_i = 118.7$			$\Sigma(x_i - \bar{x})^2 = 0.6284$

Mean = $\bar{x}$ = 118.7/6 = 19.78 ≈ 19.8 ppm Fe

Standard deviation = $x = \sqrt{\dfrac{\Sigma(x_i - \bar{x})^2}{N - 1}} = \sqrt{\dfrac{0.6284}{6 - 1}}$

$$= \sqrt{0.1257} = 0.354 \approx 0.35 \text{ ppm Fe}$$

Variance = $s^2 = \dfrac{\Sigma(x_i - \bar{x})^2}{N - 1} = \dfrac{0.6284}{6 - 1} = 0.1257 \approx 0.13 \text{ (ppm Fe)}^2$

Spread, or range = w = 20.3 − 19.4 = 0.9 ppm Fe

Relative standard deviation = $\dfrac{s}{\bar{x}} \times 1000$ ppt = $\dfrac{0.354}{19.78} \times 1000 = 17.9 \approx 18$ ppt

Coefficient of variation = CV = $\dfrac{s}{\bar{x}} \times 100\% = \dfrac{0.354}{19.78} \times 100\% \approx 1.8\%$

Absolute error† = 19.78 − 20.00 = −0.22 = −0.2 ppm Fe

Relative error = $\dfrac{19.78 - 20.00}{20.00} \times 100\% = -1.1\%$

*For source of data, see Figure 3-1.

†Sample known to contain 20.00 ppm Fe.

Table 3-1 illustrates how to obtain the standard deviation and the variance for the data in Figure 3-1. The standard deviation is 0.35 ppm Fe, the variance is 0.13 (ppm Fe)², and the number of degrees of freedom is five. The significance of these three terms in statistics is discussed in Section 4A.

Note that the standard deviation has the same units as the data and the variance has the units of the data squared. Scientists and engineers tend to use standard deviation rather than variance as a measure of precision because it is easier to relate the precision of a measurement to the measurement itself if they both have the same units. The advantage of using variance is that variances are additive, as we will see later in this chapter.

An Alternative Expression for Standard Deviation

To compute s with a calculator that does not have a standard-deviation key, the following rearrangement of Equation 3-2 is easier to use:

$$s = \sqrt{\dfrac{\displaystyle\sum_{i=1}^{N} x_i^2 - \dfrac{\left(\displaystyle\sum_{i=1}^{N} x_i\right)^2}{N}}{N - 1}} \tag{3-4}$$

Example 3-2

The following data were collected by finding the absorbance (x_i) of replicate solutions of a colored complex of iron (Chapter 20). Determine the mean and standard deviation for the five data.

To apply Equation 3-4, we calculate Σx_i^2 and $(\Sigma x_i)^2/N$:

Solution	x_i	x_i^2
1	0.752	0.565504
2	0.756	0.571536
3	0.752	0.565504
4	0.751	0.564001
5	0.760	0.577600
	$\Sigma x_i = 3.771$	$\Sigma x_i^2 = 2.844145$

$$\bar{x} = 3.771/5 = 0.7542 \approx 0.754$$

$$\frac{(\Sigma x_i)^2}{N} = \frac{(3.771)^2}{5} = 2.8440882$$

$$s = \sqrt{\frac{2.844145 - 2.8440882}{5 - 1}} = \sqrt{\frac{0.0000568}{4}} = 0.00377 \approx 0.0038$$

Note that the difference between Σx_i^2 and $(\Sigma x_i)^2/N$ in Example 3-2 is very small. If we had rounded these numbers off before subtracting them, a serious error would have appeared in the computed value of s. To avoid this source of error, never use Equation 3-4 to calculate the standard deviation of numbers containing five or more digits. Use Equation 3-2 instead.[2] Note also that handheld calculators and small computers with a standard-deviation function usually employ a version of Equation 3-4. Therefore expect large errors in s when these devices are used to calculate the standard deviation of data that have five or more significant figures.

This is a general problem. Any time we compute the difference between two large numbers that are approximately equal to each other, the result will have a relatively large uncertainty.

Relative Standard Deviation (RSD)

Chemists frequently quote standard deviations in relative rather than absolute terms. We calculate the relative standard deviation by dividing the absolute standard deviation by the mean of the set of data.

Relative standard deviation is often expressed in parts per thousand (ppt) or in percent. For example,

$$RSD = (s/\bar{x}) \times 1000 \text{ ppt}$$

---

[2]In most cases, the first two or three digits in a set of data are identical to each other. As an alternative to using Equation 3-2, then, these identical digits can be dropped and the remaining digits used with Equation 3-4. For example, the standard deviation for the data in Example 3-2 could be based on 0.052, 0.056, 0.052, and so forth, or even on 52, 56, 52, and so forth.

The coefficient of variation is the percent relative standard deviation.

When the relative standard deviation is multiplied by 100%, it is called the *coefficient of variation* (CV).

$$CV = (s/\bar{x}) \times 100\% \qquad (3\text{-}5)$$

The relative standard deviation for the data in Table 3-1 is 18 ppt. The coefficient of variation for the set is 1.8%.

Computer Application 3-2
CALCULATING A STANDARD DEVIATION

This program has several things in common with our first program in Computer Application 3-1. First, statements 20 and 30 initialize the variables SUMX# and SUMX2# to zero. Notice the # after each variable in the program; this symbol at the end of a variable name indicates that the computer is to treat the variable in *double precision*. Double precision means that the computer will carry out arithmetic operations accurately to 15 or 16 decimal digits. Variables that do not end with a # are *single-precision* variables that are accurate to only 6 or 7 decimal digits. We use double precision to ensure that the difference between $\sum x_i^2$ and $(\sum x_i)^2/N$ in statement 110 retains enough significant figures.

```
10 INPUT "Type the number of points.";N
20 SUMX# = 0
30 SUMX2# = 0
40 FOR I = 1 TO N
50 PRINT "x(";I;") = ";
60 INPUT X#
70 SUMX# = SUMX# + X#
80 SUMX2# = SUMX2# + X#^2
90 NEXT I
100 XBAR# = SUMX#/N
110 S# = SQR((SUMX2#-(SUMX#^2/N))/(N-1))
120 PRINT "Mean = "XBAR#"   s = "S#
130 END
```

Statement 30 prompts the user for the number of data in the set. The FOR–NEXT loop of statements 40 to 90 accepts the data one at a time from the keyboard. The program sums the data x_i and the squares of the data x_i^2 in statements 70 and 80. Statement 100 calculates the mean $\bar{x}$, and statement 110 calculates the standard deviation s. Finally, statement 120 prints the results on the computer screen.

Run the program and test it using the data of Example 3-2. Add a statement 115 to calculate the relative standard deviation, and modify statement 120 to print it.

Relative standard deviations often give a clearer picture of data quality than do absolute standard deviations. As an example, suppose that a sample contains about 50 mg of copper and that the standard deviation of a copper determination is 2 mg. The CV for this sample is 4%. For a sample containing only 10 mg, the CV is 20%.

Spread or Range (w)

The *spread,* or *range,* of a set of data is the difference between the largest value in the set and the smallest. Thus, the spread of the data in Table 3-1 is 0.9 ppm Fe.

3A-3 Accuracy

Figure 3-2 illustrates the basic difference between accuracy and precision. *Accuracy* indicates the closeness of a measurement to the true or accepted value and is expressed by the *error.* Accuracy measures agreement between a result and its true value. *Precision* describes the agreement among several results measured in the same way. Precision is determined simply by replicating a measurement several times. In contrast, accuracy can never be determined exactly because the true value of a quantity can never be known exactly. An accepted value is used instead.

Absolute Error

The *absolute error E* in the measurement of a quantity x_i is given by the equation

$$E = x_i - x_t \tag{3-6}$$

where x_t is the true, or accepted, value of the quantity. Returning to our example in Table 3-1, we see that the absolute error is -0.2 ppm Fe. Note that with absolute error we retain the sign (in contrast to measures of precision). Thus, the negative sign in this example shows that the experimental result is smaller than the accepted value.

Relative Error

Often, the *relative error E_r* is a more useful quantity than the absolute error. The percent relative error is given by the expression

$$E_r = \frac{x_i - x_t}{x_t} \times 100\% \tag{3-7}$$

The relative error is also expressed in parts per thousand (ppt). Thus the relative error for the mean of the data in Table 3-1 is reported either as -1.1% or as -11 ppt.

Figure 3-2
Accuracy and precision.

Low accuracy, low precision

High accuracy, low precision

Low accuracy, high precision

High accuracy, high precision

The term "absolute" has a different meaning here than it does in mathematics. The absolute value in mathematics . means the magnitude of a number *ignoring its sign.* As we shall use it, the absolute error is the difference between an accepted value *and an experimental result, including the sign of the difference.*

3A-4 Types of Errors in Experimental Data

The precision of a measurement is readily determined by comparing data from carefully replicated experiments. Unfortunately, an estimate of the accuracy is not so easy to come by. To determine accuracy, we have to know the true value, and this is exactly what we are looking for.

It is tempting to assume that if we know an answer precisely, then we also know it accurately. The danger of this assumption is illustrated in Figure 3-3, which summarizes results for the determination of nitrogen in two pure compounds. The dots show the absolute errors of replicate results obtained by four analysts. Note that analyst 1 obtained relatively high precision and high accuracy. Analyst 2 had poor precision but good accuracy. The results of analyst 3 are surprisingly common: the precision is excellent, but the numerical average for the data is quite inaccurate. Both precision and accuracy are poor for the results of analyst 4. Figure 3-3 shows all four sets of results graphically.

Figures 3-1 and 3-3 show that chemical analyses are affected by at least two types of errors. One type, called *indeterminate* (or *random*) *error*, causes data to be scattered more or less symmetrically around a mean value. Refer again to Figure 3-3, and notice that the scatter in the data for analysts 1 and 3 is significantly less than that for analysts 2 and 4.

The precision of the data reflects the indeterminate errors in an analysis. For example, the relative standard deviation of 18 ppt Fe computed in Table 3-1 is a valid measure of the indeterminate errors associated with this method for determining iron.

A second type of error, called *determinate* (or *systematic*) *error*, causes the mean of a set of data to differ from the accepted value. For example, the data in Figure 3-1 have a determinate error of about -0.22 ppm Fe. The results of analysts 1 and 2 in Figure 3-3 have little determinate error,

> Indeterminate errors are errors that affect the precision of measurements.

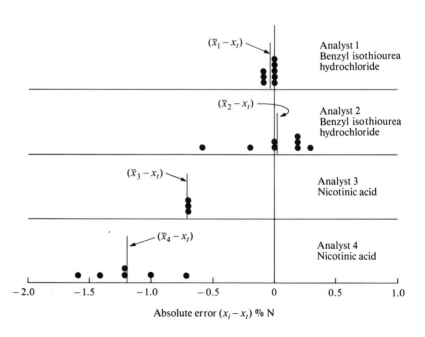

Figure 3-3

Absolute error in the micro-Kjeldahl determination of nitrogen. Each dot represents the error associated with a single determination. Each vertical line labeled $(\bar{x}_i - x_t)$ is the absolute average deviation of the set from the true value. (Data from C. O. Willits and C. L. Ogg, *J. Assoc. Offic. Anal. Chem.*, **1949**, *32*, 561. With permission.)

but the data of analysts 3 and 4 show determinate errors of about -0.7 and -1.2% nitrogen, respectively.

A third type of error is *gross error*. Gross errors differ from indeterminate and determinate errors. They usually occur only occasionally, are often large, and may cause a result to be either high or low. Gross errors lead to *outliers*—results that differ markedly from all other data in a set of replicate measurements. There is no evidence of a gross error in Figures 3-1 and 3-3.

> Gross errors usually affect only a single result in a set of replicate data causing it to differ significantly from the remaining results for that set.

3B   DETERMINATE ERRORS

Determinate errors have a definite source that usually can be identified. They cause all the results from replicate measurements to be either high or low. Because the results are either all high or all low, determinate errors are also called *systematic errors*.

> Determinate, or systematic, errors affect the accuracy of results.

For example, the last two data sets in Figure 3-3 reveal a negative determinate error that can be traced to the chemical nature of the sample, nicotinic acid. The analytical method used involves decomposing samples with hot concentrated sulfuric acid, which converts the nitrogen to ammonium sulfate. The amount of ammonia in the ammonium sulfate is then determined in the measurement step. Experiments have shown that compounds containing a pyridine ring (such as nicotinic acid) are incompletely decomposed by sulfuric acid unless special precautions are taken. Without these precautions, low results are obtained. It is very likely that the negative errors $(\bar{x}_3 - x_t)$ and $(\bar{x}_4 - x_t)$ in Figure 3-3 are determinate errors that can be blamed on incomplete decomposition.

3B-1   Sources of Determinate Errors

There are three types of determinate error. (1) *Instrument errors* are caused by imperfections in measuring devices and instabilities in their power supplies. (2) *Method errors* arise from nonideal chemical or physical behavior in analytical systems. (3) *Personal errors* result from the carelessness, inattention, or personal limitations of the experimenter.

Instrument Errors

All measuring devices are sources of determinate errors. For example, pipets, burets, and volumetric flasks may have volumes slightly different from those indicated by their graduations. These differences arise from using glassware at a temperature that differs significantly from the calibration temperature, from distortions in container walls due to heating while drying, from errors in the original calibration, or from contaminants on the inner surfaces of the containers. Calibration eliminates most determinate errors of this type.

Electronic instruments are subject to determinate errors. These uncertainties have many sources. For example, some are caused by decreases in the voltage of a battery-operated power supply with use. Errors also result from increased resistance in circuits because of dirty electrical contacts. Temperature changes cause variation in resistors and standard potential sources. Currents induced from 110-V power lines affect electronic instruments. Errors from these and other sources are detectable and correctable.

Method Errors

The nonideal chemical or physical behavior of the reagents and reactions upon which an analysis is based often introduces determinate method errors. Sources of nonideality include the slowness of some reactions, the incompleteness of others, the instability of some species, the nonspecificity of most reagents, and the possible occurrence of side reactions that interfere with the measurement process. For example, a common method error in volumetric methods results from the small excess of reagent required to cause an indicator to undergo the color change that signals completion of the reaction. The accuracy of such an analysis is thus limited by the very phenomenon that makes the titration possible.

Another example of method error was described earlier in connection with the data of analysts 3 and 4 in Figure 3-3. In that case, the source was the incompleteness of the reaction between the pyridine ring in nicotinic acid and the sulfuric acid.

Errors inherent in a method are difficult to detect and are thus the most serious of the three types of determinate error.

Personal Errors

Many measurements require personal judgments. Examples include estimating the position of a pointer between two scale divisions, the color of a solution at the end point in a titration, or the level of a liquid with respect to a graduation in a pipet or buret. Judgments of this type are often subject to systematic, unidirectional errors. For example, one person may read a pointer consistently high, another may be slightly slow in activating a timer, and a third may be less sensitive to color changes. An analyst who is insensitive to color changes tends to use excess reagent in a volumetric analysis. Physical handicaps are often sources of personal determinate errors.

A universal source of personal error is prejudice, or *bias*. Most of us, no matter how honest, have a natural tendency to estimate scale readings in a direction that improves the precision in a set of results. Or we may have a preconceived notion of the true value for the measurement. We then subconsciously cause the results to fall close to this value.

Number bias is another source of personal error that varies considerably from person to person. The most common bias encountered in estimating the position of a needle on a scale involves a preference for the digits 0 and 5. Also prevalent is a prejudice favoring small digits over large and even numbers over odd.

Digital readouts on pH meters, laboratory balances, and other electronic instruments eliminate bias because no judgment is involved in taking a reading.

> Color blindness is a good example of a handicap that amplifies personal errors in volumetric analysis. A famous color-blind analytical chemist enlisted his wife to come to the laboratory to help him detect color changes at titration end points.

> Persons who make measurements must guard against bias to preserve the integrity of the collected data.

3B-2 The Effect of Determinate Errors Upon Analytical Results

Determinate errors may be either *constant* or *proportional*. The magnitude of a constant error does not depend on the size of the quantity measured. Proportional errors increase or decrease in proportion to the size of the sample taken for analysis.

Constant Errors

Constant errors become more serious as the size of the quantity measured decreases. The effect of solubility losses on the results of a gravimetric analysis illustrates this behavior.

Example 3-3

Suppose that 0.50 mg of precipitate is lost as a result of being washed with 200 mL of wash liquid. If the precipitate weighs 500 mg, the relative error due to solubility loss is $-(0.50/500) \times 100\% = -0.1\%$. Loss of the same quantity from 50 mg of precipitate results in a relative error of -1.0%.

The excess of reagent required to bring about a color change during titration is another example of constant error. This volume, usually small, remains the same regardless of the total volume of reagent required for the titration. Again, the relative error from this source becomes more serious as the total volume decreases. One way of minimizing the effect of constant error is to use as large a sample as possible.

Proportional Errors

A common cause of proportional errors is the presence of interfering contaminants in the sample. For example, a widely used method for the determination of copper is based upon the reaction of copper(II) ion with potassium iodide to give iodine. The quantity of iodine produced is then measured and is proportional to the amount of copper. Iron(III), if present, also liberates iodine from potassium iodide. Unless steps are taken to prevent this interference, high results are observed for the percentage of copper because the iodine produced will be a measure of the copper(II) *and* iron(III) in the sample. The size of this error is determined by the *fraction* of iron contamination, which is independent of the size of sample taken. If the sample size is doubled, for example, the amount of iodine liberated by both the copper and the iron contaminant is also doubled. Thus, the magnitude of the reported percentage of copper is independent of sample size.

3B-3 Detection of Determinate Instrument and Personal Errors

Determinate instrument errors are usually found and corrected by calibration. Periodic calibration of equipment is always desirable because the response of most instruments changes with time as a result of wear, corrosion, or mistreatment.

Most personal errors can be minimized by care and self-discipline. It is a good habit to check instrument readings, notebook entries, and calculations systematically. Errors that result from a known physical handicap can usually be avoided by a careful choice of method.

3B-4 Detection of Determinate Method Errors

Determinate method errors are particularly difficult to detect. We may take one or more of the following steps to recognize and adjust for systematic errors of this type.

Analysis of Standard Samples

The best way of estimating the determinate error of an analytical method is by the analysis of *standard reference materials*—materials that contain one or more analytes at exactly known concentration levels. Standard reference materials are obtained in several ways.

Standard materials can sometimes be prepared by synthesis. Here, carefully measured quantities of the pure components of a material are measured out and mixed in such a way as to produce a homogeneous sample whose composition is known from the quantities taken. The overall composition of a synthetic standard material must approximate closely the composition of the samples to be analyzed. Great care must be taken to ensure that the concentration of analyte is known exactly. Unfortunately, the synthesis of such standard samples is often impossible or so difficult and time consuming that this approach is not practical.

Standard reference materials can be purchased from a number of governmental and industrial sources. For example, the National Institute for Standards and Technology (formerly the National Bureau of Standards) offers over 900 standard reference materials, which include such things as rocks and minerals, gas mixtures, glasses, hydrocarbon mixtures, polymers, urban dusts, rainwaters, and river sediments.[3] The concentration of one or more of the components in these materials has been determined in one of three ways: (1) by analysis by a previously validated reference method of analysis, (2) by analysis by two or more independent, reliable measurement methods, or (3) by analysis by a network of cooperating laboratories, technically competent and thoroughly knowledgeable about the material being tested.

Several commercial supply houses also offer analyzed materials for method testing.[4]

Independent Analysis

If standard samples are not available, a second independent and reliable analytical method can be used parallel to the method being evaluated. The independent method should differ as much as possible from the one under study. This minimizes the possibility that some common factor in the sample has the same effect on both methods.

> Standard reference materials (SRM) for detecting determinate errors in analytical methods are materials containing one or more species in exactly known concentration.

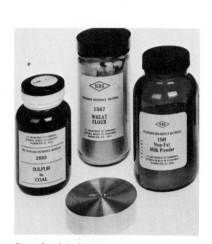

Standard reference materials from NIST. (Photo courtesy of the National Institute for Standards and Technology.)

---

[3]See U.S. Department of Commerce, *NBS Standard Reference Materials,* 1988–89 ed., NBS Special Publication 260. Washington, D.C.: Government Printing Office, 1988. For a description of the reference material programs of the NBS, see R. A. Alvarez, S. D. Rasberry, and F. A. Uriano, *Anal. Chem.,* **1982,** *54,* 1226A; and F. A. Uriano, *ASTM Standardization News,* **1979,** *7,* 8.

[4]For sources of biological and environmental reference materials containing various elements, see C. Veillon, *Anal. Chem.,* **1986,** *58,* 851A.

Blank Determinations

Blank determinations are useful for detecting certain types of constant errors. In a blank determination, or *blank,* all steps of the analysis are performed in the absence of a sample. For instance, if a procedure calls for dissolving samples in 25 mL of hydrochloric acid and titrating, the blank would be 25 mL of hydrochloric acid, which would then be titrated in exactly the same way as the samples. The results from the blank are then applied as a correction to the sample measurements.

Blank determinations reveal errors due to interfering contaminants from the reagents and vessels employed in analysis. Blanks also allow the analyst to correct titration data for the volume of reagent needed to cause an indicator to change color at the end point.

> A blank is a solution that contains the solvent and all the reagents used in an analysis but no sample.

Variation in Sample Size

Example 3-3 demonstrates that as the size of a measurement increases, the effect of a constant error decreases. Thus, constant errors can often be detected by varying the sample size.

3C GROSS ERRORS

Most gross errors are personal and arise from carelessness, laziness, or ineptitude. Gross errors can be random but occur so infrequently that they are generally not considered as indeterminate errors. Sources of gross errors include arithmetic mistakes, transposition of numbers in recording data, reading a scale backward, reversing a sign, using a wrong scale, spilling a solution, or just bad luck. Some gross errors affect only a single result. Others, such as using the wrong scale of an instrument, affect an entire set of replicate measurements.

Because carelessness causes most gross personal errors, they can usually be eliminated through self-discipline. Many scientists follow the practice of re-reading an instrument after a first reading has been recorded and then comparing the two readings to reveal errors.

Gross errors are also encountered as a result of momentary interruptions in power or water supplies and other unexpected events.

3D INDETERMINATE ERRORS

Indeterminate, or *random, errors* occur when a system of measurement is extended to its maximum sensitivity. They are caused by the many uncontrollable variables that are an inevitable part of every physical and chemical measurement.

There are many sources of this type of error, but none of them can be positively identified or measured because most are so small that they cannot be detected individually. The accumulated effect of the individual indeterminate errors, however, causes the data from a set of replicate measurements to fluctuate randomly around the mean of the set. For example, the scatter of data in Figures 3-1 and 3-3 is a direct result of the accumulation of small indeterminate errors. Notice that in Figure 3-3 the indeterminate errors in the results of analysts 2 and 4 are greater than in those of analysts 1 and 3.

Table 3-2

POSSIBLE COMBINATIONS OF FOUR EQUAL-SIZED UNCERTAINTIES

Combinations of Uncertainties	Magnitude of Indeterminate Error	Number of Combinations	Relative Frequency
$+U_1 + U_2 + U_3 + U_4$	$+4U$	1	$1/16 = 0.0625$
$-U_1 + U_2 + U_3 + U_4$ $+U_1 - U_2 + U_3 + U_4$ $+U_1 + U_2 - U_3 + U_4$ $+U_1 + U_2 + U_3 - U_4$	$+2U$	4	$4/16 = 0.250$
$-U_1 - U_2 + U_3 + U_4$ $+U_1 + U_2 - U_3 - U_4$ $+U_1 - U_2 + U_3 - U_4$ $-U_1 + U_2 - U_3 + U_4$ $-U_1 + U_2 + U_3 - U_4$ $+U_1 - U_2 - U_3 + U_4$	0	6	$6/16 = 0.375$
$+U_1 - U_2 - U_3 - U_4$ $-U_1 + U_2 - U_3 - U_4$ $-U_1 - U_2 + U_3 - U_4$ $-U_1 - U_2 - U_3 + U_4$	$-2U$	4	$4/16 = 0.250$
$-U_1 - U_2 - U_3 - U_4$	$-4U$	1	$1/16 = 0.0625$

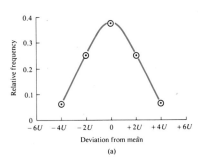

(a)

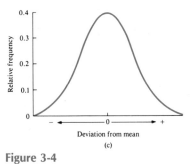

(b)

(c)

Figure 3-4

Frequency distribution for measurements containing (a) four indeterminate uncertainties; (b) ten indeterminate uncertainties; (c) a very large number of indeterminate uncertainties.

Table 3-3

REPLICATE DATA ON THE CALIBRATION OF A 10-mL PIPET*

Trial	Volume, mL	Trial	Volume, mL	Trial	Volume, mL
1	9.988	18	9.975	35	9.976
2	9.973	19	9.980	36	9.990
3	9.986	20	9.994‡	37	9.988
4	9.980	21	9.992	38	9.971
5	9.975	22	9.984	39	9.986
6	9.982	23	9.981	40	9.978
7	9.986	24	9.987	41	9.986
8	9.982	25	9.978	42	9.982
9	9.981	26	9.983	43	9.977
10	9.990	27	9.982	44	9.977
11	9.980	28	9.991	45	9.986
12	9.989	29	9.981	46	9.978
13	9.978	30	9.969†	47	9.983
14	9.971	31	9.985	48	9.980
15	9.982	32	9.977	49	9.983
16	9.983	33	9.976	50	9.979
17	9.988	34	9.983		

Mean volume = 9.982 mL

Median volume = 9.982 mL

Spread = 0.025 mL

Standard deviation = 0.0056 mL

*Data listed in the order obtained.

†Minimum value.

‡Maximum value.

3D-1 Sources of Indeterminate Error

We can get a qualitative idea of the way small errors produce an overall uncertainty in the following way. Imagine a situation in which just four small random errors combine to give an overall error. We will assume that each error has an equal probability of occurring and that each can cause the final result to be high or low by a fixed amount $\pm U$.

Table 3-2 shows all the possible ways the four errors can combine to give the indicated deviations from the mean value. Note that only one combination of errors leads to a deviation of $+4U$, four combinations give a deviation of $+2U$, and six give a deviation of $0U$. The negative errors have the same relationship. This ratio of $1:4:6:4:1$ is a measure of the probability of a deviation of each magnitude. If we make a sufficiently large number of measurements, therefore, we can expect a frequency distribution like that shown in Figure 3-4a. Note that the ordinate in the plot is the relative frequency of occurrence of the five possible combinations.

Figure 3-4b shows the theoretical distribution for ten equal-sized uncertainties. Again we see that the most frequent occurrence is zero deviation from the mean. At the other extreme, a maximum deviation of $10U$ occurs only about once in 500 measurements.

When the same procedure is applied to a very large number of individual errors, a curve like that shown in Figure 3-4c results. This bell-shaped curve is called a *Gaussian curve* or a *normal error curve*.

In our example, all the uncertainties have the same magnitude. This restriction is not necessary to derive the equation for a Gaussian curve.

3D-2 Distribution of Experimental Data

We find empirically that the distribution of replicate data from most quantitative analytical experiments approaches that of the Gaussian curve shown in Figure 3-4c. As an example, consider the data in Table 3-3 for the calibration of a 10-mL pipet. In this experiment a small flask and stopper are weighed, and a 10-mL portion of water is transferred to the flask with the pipet, and the flask is stoppered. Finally, the flask, stopper, and water are weighed again. The temperature of the water is also measured to establish its density. The weight of the water is then calculated by taking the difference between the two weights; this difference is divided by the density of the water to find the volume delivered by the pipet. The experiment was performed 50 times.

The data in Table 3-3 are typical of those obtained by an experienced worker weighing to the nearest milligram (which corresponds to 0.001 mL) on a top-loading balance and making every effort to avoid determinate error. Even so, the standard deviation of the 50 measurements is 0.0056 mL, and the spread is 0.025 mL. This distribution of data about the mean results directly from the many indeterminate errors in the experiment.

The information in Table 3-3 is easier to visualize when the data are rearranged into frequency distribution groups, as in Table 3-4. Here we tabulate the number of data falling into a series of adjacent 0.003-mL *cells* and calculate the percentage of measurements falling into each cell. Note that 26% of the data reside in the cell containing the mean and median

Table 3-4

FREQUENCY DISTRIBUTION OF DATA FROM TABLE 3-3

Volume Range, mL	Number in Range	% in Range
9.969 to 9.971	3	6
9.972 to 9.974	1	2
9.975 to 9.977	7	14
9.978 to 9.980	9	18
9.981 to 9.983	13	26
9.984 to 9.986	7	14
9.987 to 9.989	5	10
9.990 to 9.992	4	8
9.993 to 9.995	1	2

value of 9.982 mL and that more than half the data are within ±0.004 mL of this mean. Note also that 72% of the data are within ±0.0056 mL, or one standard deviation, of the mean. As we will show in Chapter 4, the theoretical percentage in this range is 68.

The frequency-distribution data in Table 3-4 are plotted as a bar graph, or *histogram,* in Figure 3-5, which also shows a theoretical Gaussian curve derived for an infinite set of data. The data used for the Gaussian curve have the same mean (9.982 mL), the same standard deviation (0.0056 mL), and the same area under the curve as the histogram data. As the number of calibration experiments increases and the cell size decreases, the shape of the histogram approaches the shape of the continuous curve.

Variations in replicate results such as those in Table 3-3 result from numerous small and individually undetectable instrument, method, and personal indeterminate errors. These errors are attributable to uncontrollable variables in the experiment, and the net effect of such errors is also indeterminate. Ordinarily, the small errors tend to cancel one another and thus have a minimal effect. Occasionally, however, they occur in the same direction, and then the result is a large positive or negative error.

Sources of uncertainty in the calibration of a pipet include such visual judgments as the level of the water with respect to the marking on the pipet and the mercury level in the thermometer (both personal indeterminate errors). Other sources are variation in the drainage time and in the angle of the pipet as it drains (both method errors). Instrument errors arise from temperature fluctuations, which affect (1) the volume of the pipet, (2) the viscosity of the liquid, and (3) the performance of the balance. Other sources of instrument error are vibrations and drafts that cause small variations in the balance reading. There are numerous other sources of indeterminate error. Thus, many small and uncontrollable variables affect even as simple a process as calibrating a pipet. It is very difficult to determine the influence of any one of these several indetermi-

> A histogram is a bar graph such as that shown in Figure 3-5.

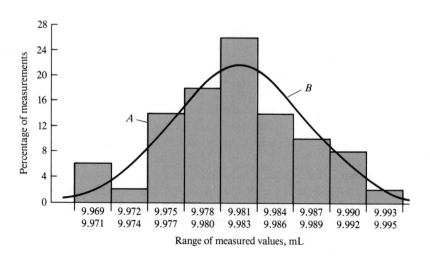

Figure 3-5

A histogram (A) showing distribution of the 50 results in Table 3-3 and a Gaussian curve (B) for data having the same mean and same standard deviation as the data in the histogram.

nate errors, but their cumulative effect is responsible for the scatter of data around the mean.[5]

3E THE STANDARD DEVIATION OF COMPUTED RESULTS

We often need to estimate the standard deviation of a result that has been computed from two or more experimental data, each of which has a known standard deviation. The way such estimates are made depends upon the type of arithmetic operation used.

3E-1 The Standard Deviation of Sums and Differences

Consider the summation:

$$
\begin{array}{ll}
+0.50 & (\pm 0.02) \\
+4.10 & (\pm 0.03) \\
\underline{-1.97} & (\pm 0.05) \\
2.63 &
\end{array}
$$

where the numbers in parentheses are absolute standard deviations. If the three individual standard deviations happen by chance to have the same sign, the standard deviation of the sum could be as large as $+0.02 + 0.03 + 0.05 = +0.10$ or $-0.02 - 0.03 - 0.05 = -0.10$. On the other hand, it is possible that the three could combine to give an accumulated value of zero: $-0.02 - 0.03 + 0.05 = 0$ or $+0.02 + 0.03 - 0.05 = 0$. More likely, however, the sum will lie between these two extremes. It can be shown from statistical theory[6] that the most probable value for the standard deviation of a sum or difference can be found by taking the square root of the sum of the squares of the individual absolute standard deviations. Thus for the computation

$$
y = a(\pm s_a) + b(\pm s_b) - c(\pm s_c)
$$

the standard deviation of the result s_y is given by

$$
s_y = \sqrt{s_a^2 + s_b^2 + s_c^2} \tag{3-8}
$$

where s_a, s_b, and s_c are the standard deviations of the three terms in the sum. Substituting the standard deviations from the example gives

$$
s_y = \sqrt{(\pm 0.02)^2 + (\pm 0.03)^2 + (\pm 0.05)^2} = \pm 0.06
$$

and the sum should be reported as 2.63 (± 0.06).

For a sum or a difference, the *absolute standard deviation of the answer* is the square root of the sum of the squares of the *absolute standard deviations* of the numbers used to calculate the sum or difference.

---

[5]Sources of error can be separated through the statistical procedure known as analysis of variance (ANOVA). See R. L. Anderson, *Practical Statistics for Analytical Chemists*, p. 107. New York: Van Nostrand–Reinhold, 1987.

[6]For a derivation of the various relationships shown in this section, see D. A. Skoog, *Principles of Instrumental Analysis*, 3rd ed., pp. 14–18. Philadelphia: Saunders College Publishing, 1985.

3E-2 The Standard Deviation of Products and Quotients

Consider the following computation, where the numbers in parentheses are again absolute standard deviations:

$$\frac{4.10(\pm0.02) \times 0.0050(\pm0.0001)}{1.97(\pm0.04)} = 0.010406(\pm?)$$

In this situation, the standard deviations of two of the numbers in the calculation are larger than the result. Evidently, the approach we use to calculate standard deviation in multiplication and division cannot be the same as the one we used for addition and subtraction. The relative standard deviation of a product or quotient is determined by the *relative standard deviations* of the numbers forming the computed result. In the case of

$$y = \frac{a \times b}{c} \tag{3-9}$$

we obtain the relative standard deviation s_y/y of the result y by summing the squares of the relative standard deviations of a, b, and c and then extracting the square root of the sum:

For multiplication or division, the *relative standard deviation of the answer* is the square root of the sum of the squares of the *relative standard deviations* of the numbers that are multiplied or divided.

$$\frac{s_y}{y} = \sqrt{\left(\frac{s_a}{a}\right)^2 + \left(\frac{s_b}{b}\right)^2 + \left(\frac{s_c}{c}\right)^2} \tag{3-10}$$

Applying this equation to the numerical example gives

$$\frac{s_y}{y} = \sqrt{\left(\frac{\pm0.02}{4.10}\right)^2 + \left(\frac{\pm0.0001}{0.005}\right)^2 + \left(\frac{\pm0.04}{1.97}\right)^2}$$

$$= \sqrt{(0.0049)^2 + (0.0200)^2 + (0.0203)^2} = \pm0.0289$$

In order to complete the calculation, we must find the absolute standard deviation of the result,

To find the absolute standard deviation in a product or quotient, first find the relative standard deviation in the result and then multiply it by the result.

$$s_y = y \times (\pm0.0289) = 0.0104 \times (\pm0.0289) = \pm0.000301$$

and we can write the answer and its uncertainty as 0.0104 (±0.0003).

The following example demonstrates the calculation of the standard deviation of the result for a more complex calculation.

Example 3-4

Calculate the standard deviation of the result of

$$\frac{[14.3(\pm0.2) - 11.6(\pm0.2)] \times 0.050(\pm0.001)}{[820(\pm10) + 1030(\pm5)] \times 42.3(\pm0.4)} = 1.725(\pm?) \times 10^{-6}$$

First we must calculate the standard deviation of the sum and the difference. For the difference in the numerator,

$$s_a = \sqrt{(\pm 0.2)^2 + (\pm 0.2)^2} = \pm 0.283$$

and for the sum in the denominator,

$$s_b = \sqrt{(\pm 10)^2 + (\pm 5)^2} = 11.2$$

We may then rewrite the equation as

$$\frac{2.7(\pm 0.283) \times 0.050(\pm 0.001)}{1850(\pm 11.2) \times 42.3(\pm 0.4)} = 1.725 \times 10^{-6}$$

The equation now contains only products and quotients, and Equation 3-10 applies. Thus,

$$\frac{s_y}{y} = \sqrt{\left(\pm \frac{0.283}{2.7}\right)^2 + \left(\pm \frac{0.001}{0.050}\right)^2 + \left(\pm \frac{11.2}{1850}\right)^2 + \left(\pm \frac{0.4}{42.3}\right)^2} = 0.107$$

To obtain the absolute standard deviation, we write

$$s_y = y \times 0.107 = 1.725 \times 10^{-6} \times (\pm 0.107) = \pm 0.185 \times 10^{-6}$$

and round the answer to $1.7(\pm 0.2) \times 10^{-6}$.

> The number of significant figures is all of the certain digits plus the first uncertain digit.

3F METHODS FOR REPORTING ANALYTICAL DATA

Because a numerical result is worthless unless something is known about its accuracy, it is always important to indicate the reliability of your data. One of the best ways of indicating reliability is to give a confidence limit at the 90 or 95% confidence level, as we describe in Section 4B-1. Another method is to report the absolute standard deviation or the coefficient of variation of the data. In this case, it is a good idea to indicate the number of data used to obtain the standard deviation. A less satisfactory but more common indicator of the quality of data is the *significant-figure convention*.

3F-1 The Significant-Figure Convention

A simple way of indicating the probable uncertainty associated with an experimental measurement is to round the result so that it contains only *significant figures*. By definition, the significant figures in a number are all of the certain digits *and the first uncertain digit*. For example, when you read a 50-mL buret that has graduations every 0.1 mL, you can easily tell that the liquid level is greater than 30.2 mL and less than 30.3 mL. You can also estimate the position of the liquid between the graduations to about ± 0.02 mL (see Figure 3-6). So, using the significant-figure conven-

Figure 3-6

Buret section showing the liquid level and meniscus.

tion, you should report the volume delivered as, say, 30.24 mL, which is four significant figures. The first three digits here are certain, and the last digit (4) is uncertain.

A zero may or may not be significant depending upon its location in a number. A zero that is surrounded by other digits is always significant (such as in 30.24 mL) because it is read directly and with certainty from a scale or instrument readout. On the other hand, zeros that only locate the decimal point for us are not significant. If we write 30.24 mL as 0.03024 L, there are still only four significant figures. The only function of the 0 before the 3 is to locate the decimal point, and so this 0 is not significant. Terminal zeros may or may not be significant. For example, if the volume of a beaker is expressed as 2.0 L, the presence of the 0 tells us that the volume is known to a few tenths of a liter, and so both the 2 and the 0 are significant figures. If this same volume is reported as 2000 mL, the situation becomes confused. The last two zeros are not significant because the uncertainty is still a few tenths of a liter, or a few hundred milliliters. In order to follow the significant-figure convention here, use scientific notation and report the volume as 2.0×10^3 mL.

An obvious limitation of the significant-figure convention as an indicator of data reliability is its ambiguity. For example, when a result is reported as 61.6, the uncertainty could range from a high of 0.5 to a low of 0.05.

3F-2 Significant Figures in Numerical Computations

Care is required to determine the appropriate number of significant figures in the result of an arithmetic combination of two or more numbers.[7]

Sums and Differences

For addition and subtraction, the number of significant figures can be found by visual inspection. For example, in the expression

$$3.4 + 0.020 + 7.31 = 10.73 = 10.7$$

the second and third decimal places in the answer cannot be significant because 3.4 is uncertain in the first decimal place. Note that the result contains three significant digits even though two of the numbers involved have only two significant digits.

Products and Quotients

A rule of thumb sometimes suggested for multiplication and division is that the answer should be rounded so that it contains the same number of significant digits as the original number with the smallest number of significant digits. Unfortunately, this procedure often leads to incorrect rounding. For example, consider the two calculations

$$\frac{24 \times 4.52}{100.0} = 1.08 \qquad \text{and} \qquad \frac{24 \times 4.02}{100.0} = 0.965$$

To avoid confusion in determining whether terminal zeros are significant, express data in scientific notation.

Rules for significant figures:
1. Disregard all initial zeros.
2. Disregard all final zeros *unless they follow a decimal point*.
3. All remaining digits, including zeros between nonzero digits, are significant.

You have probably heard it said that a chain is only as strong as its weakest link. For addition and subtraction, the weak link is the number with the *fewest* decimal places.

When adding and subtracting numbers in scientific notation, express the numbers to the same power of ten.

The weak link for multiplication and division is the number with the *fewest* significant figures. *Use this rule of thumb with caution.*

---

[7]For an extensive discussion of propagation of significant figures, see L. M. Schwartz, *J. Chem. Educ.*, **1985,** *62,* 693.

By the rule just described, the first answer would be rounded to 1.1 and the second to 0.96. If, however, the last digit of each number making up the first quotient is uncertain by 1, the relative uncertainties associated with these numbers are 1/24, 1/452, and 1/1000. Because the first relative uncertainty is much larger than the other two, the relative uncertainty in the result is also 1/24. The absolute uncertainty is then

$$1.08 \times 1/24 = 0.045 = 0.04$$

By the same argument, the absolute uncertainty of the second answer is

$$0.965 \times 1/24 = 0.040 = 0.04$$

Therefore, the first result should be rounded to three significant figures, or 1.08, but the second should be rounded to only two, that is, 0.96.

Logarithms and Antilogarithms

Be especially careful in rounding the results of calculations involving logarithms. The following rules apply to most situations:[8]

1. In a logarithm of a number, keep as many digits to the right of the decimal point as there are significant figures in the original number.
2. In an antilogarithm of a number, keep as many digits as there are digits to the right of the decimal point in the original number.

The mantissa of a logarithm is the digits to the right of the decimal point. The characteristic of a logarithm is the digits to the left of the decimal point.

The number of significant figures in the *mantissa* is the same as the number of significant figures in the original number. For example,

$$\log(9.57) \times 10^4 = 4.981$$

Example 3-5

Round the following answers so that only significant digits are retained:
(a) $\log 4.000 \times 10^{-5} = -4.3979400$; (b) antilog $12.5 = 3.162278 \times 10^{12}$.

(a) Following rule 1, we retain four digits to the right of the decimal point:

$$\log 4.000 \times 10^{-5} = -4.3979$$

(b) Following rule 2, we retain only one digit:

$$\text{antilog } 12.5 = 3 \times 10^{12}$$

3F-3 Rounding Data

Always round the computed results of a chemical analysis in an appropriate way. For example, consider the replicate results 61.60, 61.46, 61.55, and 61.61. The mean of these data is 61.555, and the standard deviation is 0.069. When we round the mean, do we take 61.55 or 61.56? A good guide to follow when rounding a 5 is always to round to the nearest even number. In this way, we eliminate any tendency to round in a set direction. In

In rounding a number ending in 5, always round so that the result ends with an even number.

---

[8]D. E. Jones, *J. Chem. Educ.*, **1971**, *49*, 753.

other words, there is an equal likelihood that the nearest even number will be the higher or the lower in any given situation. Accordingly, we might choose to report the result as 61.56 ± 0.07. If we had reason to doubt the reliability of the estimated standard deviation, we might report the result as 61.6 ± 0.1.

3F-4 Rounding Results from Chemical Computations

Throughout this text and others, the reader is asked to perform calculations with data whose precision is indicated only by the significant-figure convention. In these circumstances, common-sense assumptions must be made as to the uncertainty in each number. The uncertainty of the result is then estimated using the techniques presented in Section 3E. Finally, the result is rounded so that it contains only significant digits. *It is especially important to postpone rounding until the calculation is completed.* At least one extra digit beyond the significant digits should be carried through the entire computation in order to avoid a rounding error. This extra digit is sometimes called a *guard digit*. Modern calculators generally retain several digits that are not significant, and the user must be careful to round final results properly so that only significant figures are included. The following example illustrates this procedure.

Example 3-6

A 3.4842-g sample of a solid mixture containing benzoic acid (C_6H_5COOH, fw = 122.1247 g) was dissolved and titrated with base to a phenolphthalein end point. The acid consumed 41.36 mL of 0.2328 M NaOH. Calculate the percent benzoic acid (HBz) in the sample.

As will be shown in Section 6C-3, the computation takes the following form:

$$\%HBz = \frac{41.36 \times 0.2328 \times \dfrac{122.125}{1000}}{3.4842} \times 100\%$$

$$= 33.749\%$$

Since all operations are either multiplication or division, the relative uncertainty of the answer is determined by the relative uncertainties of the experimental data. Let us estimate what these uncertainties are.

(a) The position of the liquid level in a buret can be estimated to ± 0.02 mL. Initial and final readings must be made, however, so that the standard deviation of the volume will be (Equation 3-8)

$$\sqrt{(0.02)^2 + (0.02)^2} = \pm 0.028 \text{ mL}$$

The relative uncertainty is then

$$\frac{\pm 0.028}{41.36} \times 1000 \text{ ppt} = \pm 0.68 \text{ ppt}$$

(b) Generally, the absolute uncertainty of a weight obtained with an analytical balance is on the order of ±0.0001 g. Thus the relative uncertainty of the denominator is

$$\frac{0.0001}{3.4842} \times 1000 \text{ ppt} = 0.029 \text{ ppt}$$

(c) Usually we can assume that the absolute uncertainty in the molarity of a reagent solution is 0.0001, and so

$$\frac{0.0001}{0.2328} \times 1000 \text{ ppt} = 0.43 \text{ ppt}$$

(d) The relative uncertainty in the formula weight of HBz is several orders of magnitude smaller than that of the three experimental data and is therefore of no consequence. Note, however, that we should retain enough digits in the calculation so that the formula weight is given to at least one more digit (the guard digit) than any of the experimental data. Thus, in the calculation, we use 122.125 for the formula weight (here we are carrying two extra digits).
(e) No uncertainty is associated with 100% and with the 1000 since these are exact numbers.

It is sufficient in cases like this to round the answer so that its relative uncertainty is of the *same order of magnitude* as the relative uncertainty of the number having the largest relative uncertainty. In practice, this means that the answer is rounded so that its relative uncertainty lies between 0.2 and 2 times the largest relative uncertainty of the input data.
We see that the largest relative uncertainty of the three input data is 0.68 ppt. The answer should then be rounded to the same order of magnitude as 0.68, or about 0.7 ppt. If the answer is rounded to 33.7, the suggested relative uncertainty is (0.1/33.7) × 1000 ppt = 3 ppt, which is well over 2 × 0.7 ppt = 1.4 ppt. Rounding to 33.75 implies a relative uncertainty of (0.01/33.75) × 1000 ppt = 0.3 ppt, which lies between 0.2 × 1 ppt and 2 × 1 ppt. Thus the answer is written as % HBz = 33.75.

With a little practice, you can do in your head the type of rounding decision shown in Example 3-5. For example, looking again at the first equation in this example, we see that the relative uncertainty of the volume measurement is somewhat less than 4 parts in 4000 or 1 part in 1000 (actually 2.8 parts in 4136). Similarly, the uncertainty in the molarity is roughly 1 part in 2000, and the uncertainty in the denominator is somewhat less than 1 in 34,000. Therefore the uncertainty in the result is determined by the uncertainty in the volumetric measurement, which, as we have said, is roughly 1 part in 1000. The uncertainty in the computed result is then 1/1000th of 33.7%, or about 0.03%. Therefore we round to 33.75%.
It is important for you to remember that rounding decisions are a necessary part of *every calculation* and that such decisions *cannot* be based on the number of digits displayed on the readout of a calculator.

3G QUESTIONS AND PROBLEMS

3-1. Explain the difference between
   *(a) accuracy and precision.
   (b) indeterminate and determinate error.
   *(c) mean and median.
   (d) absolute and relative error.
   *(e) constant and proportional error.
   (f) variance and standard deviation.

3-2. Define
   *(a) range.
   (b) coefficient of variation.
   *(c) histogram.
   (d) Gaussian distribution.
   *(e) standard error of a mean.
   (f) significant figures.

*3-3. Name three types of determinate errors.

3-4. How are determinate method errors detected?

*3-5. What kind of determinate errors are detected by varying the sample size?

3-6. Suggest some sources of indeterminate error in measuring the width of a 3-m table with a 1-m metal rule.

3-7. Consider the following sets of replicate measurements:

*A	B	*C	D	*E	F
2.4	69.94	0.0902	2.3	69.65	0.624
2.1	69.92	0.0884	2.6	69.63	0.613
2.1	69.80	0.0886	2.2	69.64	0.596
2.3		0.1000	2.4	69.21	0.607
1.5			2.9		0.582

For each set, calculate the **(a)** mean, **(b)** median, **(c)** spread, **(d)** standard deviation, and **(e)** coefficient of variation.

3-8. The accepted values for the sets of data in Problem 3-7 are: *A, 2.0; B, 69.75; *C, 0.0930; D, 3.0; *E, 69.05; F, 0.635. For each set, calculate **(a)** the absolute error and **(b)** the relative error in ppt.

3-9. A method of analysis yields weights for gold that are low by 0.3 mg. Calculate the percent relative error caused by this uncertainty if the weight of gold in the sample is
   *(a) 800 mg. (b) 500 mg. *(c) 100 mg. (d) 25 mg.

3-10. The method described in Problem 3-9 is to be used for the analysis of ores that assay about 1.2% gold. What minimum sample weight should be taken if the relative error resulting from a 0.3-mg loss is not to exceed

*(a) −0.2%? (b) −0.5%? *(c) −0.8%? (d) −1.2%?

*3-11. Estimate the absolute deviation and the coefficient of variation for the results of the following calculations. Round each result so that it contains only significant digits. The numbers in parentheses are absolute standard deviations.

(a) $y = 6.75(\pm0.03) + 0.843(\pm0.001) - 7.021(\pm0.001) = 0.572$

(b) $y = 19.97(\pm0.04) + 0.0030(\pm0.0001) + 1.29(\pm0.08) = 21.263$

(c) $y = 67.1(\pm0.3) \times 1.03(\pm0.02) \times 10^{-17} = 6.9113 \times 10^{-16}$

(d) $y = 243(\pm1) \times \dfrac{760(\pm2)}{1.006(\pm0.006)} = 183578.5$

(e) $y = \dfrac{143(\pm6) - 64(\pm3)}{1249(\pm1) + 77(\pm8)} = 5.9578 \times 10^{-2}$

(f) $y = \dfrac{1.97(\pm0.01)}{243(\pm3)} = 8.106996 \times 10^{-3}$

3-12. Estimate the absolute standard deviation and the coefficient of variation for the results of the following calculations. Round each result to include only significant figures. The numbers in parentheses are absolute standard deviations.

(a) $y = -1.02(\pm0.02) \times 10^{-7} - 3.54(\pm0.2) \times 10^{-8} = -1.374 \times 10^{-7}$

(b) $y = 100.20(\pm0.08) - 99.62(\pm0.06) + 0.200(\pm0.004) = 0.780$

(c) $y = 0.0010(\pm0.0005) \times 18.10(\pm0.02) \times 200(\pm1) = 3.62$

(d) $y = \dfrac{1.73(\pm0.03) \times 10^{-14}}{1.63(\pm0.04) \times 10^{-16}} = 106.1349693$

(e) $y = \dfrac{100(\pm1)}{2(\pm1)} = 50$

(f) $y = \dfrac{1.43(\pm0.02) \times 10^{-2} - 4.76(\pm0.06) \times 10^{-3}}{24.3(\pm0.7) + 8.06(\pm0.08)} = 2.948 \times 10^{-4}$

3-13. Round each of the following results to include only significant figures.
   *(a) $y = \log 1.73 = 0.238046$
   (b) $y = \log 0.0432 = -1.364516$
   *(c) $y = \log 6.022 \times 10^{23} = 23.77960$
   (d) $y = \log 4.213 \times 10^{-21} = -20.375409$
   *(e) $y = \text{antilog}\,(-3.47) = 3.38844 \times 10^{-4}$
   (f) $y = \text{antilog}\,5.7 = 5.01187 \times 10^{5}$
   *(g) $y = \text{antilog}\,0.99 = 9.77237$
   (h) $y = \text{antilog}\,(-27.2424) = 5.722687 \times 10^{28}$

Chapter 4

Statistical Evaluation of Analytical Data

As we showed in Chapter 3, the magnitude of the indeterminate error associated with an individual measurement is determined by a chance combination of a host of tiny individual errors, each of which may be positive or negative. Because chance is involved in this type of error, we can use the laws of statistics to extract reliability information from experimental data.[1] In this chapter we describe several important statistical procedures and show how they are used to estimate the magnitude of the indeterminate error in an analysis.

4A THE STATISTICAL TREATMENT OF INDETERMINATE ERRORS

Statistics is the mathematical science that deals with chance variations. We must emphasize at the outset that statistics reveals only information that is already present in a set of data. That is, *no new information is created* by statistics. Statistical treatment of a set of data does, however, allow us to make objective judgments concerning the validity of results that are otherwise difficult to make.

Statistics allows us to look at our data in different ways and make objective and intelligent decisions regarding their quality and use.

[1] References on statistical applications include R. L. Anderson, *Practical Statistics for Analytical Chemists*. New York: Van Nostrand–Reinhold, 1987; R. Calcutt and R. Boddy, *Statistics for Analytical Chemists*. New York: Chapman and Hall, 1983; J. Mandel, *Treatise on Analytical Chemistry*, 2nd ed., I. M. Kolthoff and P. J. Elving, eds., Part I, Vol. 1, Chapter 5. New York: Wiley, 1978.

4A-1 The Population and the Sample

In order to use statistics to treat our data, we must assume that the few replicate experimental results gathered in the laboratory are representative of the infinite number of results that could be collected if we had infinite time. Statisticians call this small set of data a *sample* and view it as a subset of a *population*, or a *universe*, of data that in principle exists. For example, the data in Table 3-3 make up a statistical sample of an infinite population of pipet-calibration experiments that can be imagined (but not performed).

The laws of statistics apply strictly to a population of data only. To use these laws, we must assume that the handful of data that make up the typical sample truly represent the infinite population of results. Unfortunately, there is no guarantee that this assumption is valid. As a result, statistical estimates about the magnitude of indeterminate errors are themselves subject to uncertainty and therefore can be made only in terms of probabilities.

The Population Mean (μ) and the Sample Mean ($\bar{x}$)

We will find it useful to differentiate between the *sample mean* and the *population mean*. The sample mean is the mean of a limited sample drawn from a population of data. It is defined by Equation 3-1 when N is a small number. The population mean, in contrast, is the true mean for the population. It is also defined by Equation 3-1 when N is so large that it approaches infinity. If the data are free of determinate error, the population mean is also the true value. To emphasize the difference between the two means, the sample mean is symbolized by $\bar{x}$ and the population mean by μ. More often than not, particularly when N is small, $\bar{x}$ differs from μ because a small sample of data does not exactly represent its population.

The Sample Standard Deviation (s) and Population Standard Deviation (σ)

We must also differentiate between the *sample standard deviation* and the *population standard deviation*. The sample standard deviation was defined as s in Equation 3-2, that is,

$$s = \sqrt{\frac{\displaystyle\sum_{i=1}^{N} (x_i - \bar{x})^2}{N - 1}} \tag{4-1}$$

In contrast, the population standard deviation σ, which is the true standard deviation, is given by

$$\sigma = \sqrt{\frac{\displaystyle\sum_{i=1}^{N} (x_i - \mu)^2}{N}} \tag{4-2}$$

Do not confuse the *statistical sample* with the *analytical sample*. Four analytical samples analyzed in the laboratory represent a single statistical sample. This duplication of the term "sample" is unfortunate but should cause no trouble once you are aware of the two meanings.

Sample mean = $\bar{x}$, where

$$\bar{x} = \frac{\displaystyle\sum_{i=1}^{N} x_i}{N} \quad \text{when } N \text{ is small.}$$

Population mean = μ, where

$$\mu = \frac{\displaystyle\sum_{i=1}^{N} x_i}{N} \quad \text{when } N \to \infty.$$

In the absence of determinate error, the population standard deviation σ is the true value of a measured quantity.

Note that this equation differs from Equation 4-1 in two ways. First, the population mean μ appears in the numerator of Equation 4-2 in place of the sample mean $\bar{x}$. Second, N replaces the number of degrees of freedom $(N - 1)$ that appears in Equation 4-1.

The reason the number of degrees of freedom must be used when N is small is as follows. When σ is unknown, two quantities must be extracted from a set of data: $\bar{x}$ and s. One degree of freedom is used to establish $\bar{x}$ because, with their signs retained, the sum of the individual deviations must add up to zero. Thus, when $N - 1$ deviations have been computed, the final one is known. Consequently, only $N - 1$ deviations provide an *independent* measure of the precision of the set.

When $N \rightarrow \infty$, $\bar{x} \rightarrow \mu$ and $s \rightarrow \sigma$.

4A-2 Properties of the Normal Error Curve

Figure 4-1a shows two Gaussian curves in which the relative frequency of occurrence of various deviations from the mean is plotted as a function of deviation from the mean $(x - \mu)$. The two curves are for two populations of data that differ only in standard deviation. The standard deviation for the population yielding the broader but lower curve (B) is twice that for the population yielding curve A.

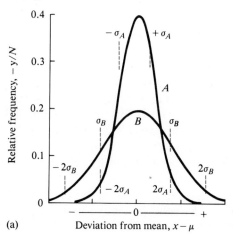

(a)

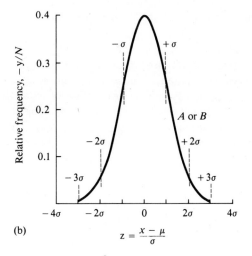

(b)

$$z = \frac{x - \mu}{\sigma}$$

Figure 4-1

Normal error curves. The standard deviation for curve B is twice that for curve A, that is, $\sigma_B = 2\sigma_A$. (a) The abscissa is the deviation from the mean in the units of measurement. (b) The abscissa is the deviation from the mean in units of σ. Thus, the two curves A and B are identical here.

Figure 4-1b shows another type of normal error curve, one in which the abscissa is now a new variable z, which is defined as

$$z = \frac{(x - \mu)}{\sigma} \qquad (4\text{-}3)$$

Note that z is the deviation from the mean stated *in units of standard deviation*. That is, when $x - \mu = \sigma$, z is equal to one standard deviation; when $x - \mu = 2\sigma$, z is equal to two standard deviations; and so forth. Since z is the deviation from the mean in standard deviation units, a plot of relative frequency versus this parameter yields a single Gaussian curve that describes all populations of data regardless of standard deviation. Thus, Figure 4-1b is the normal error curve for the two sets of data used to plot curves A and B in Figure 4-1a.

The normal error curve has several general properties. (1) The mean occurs at the central point of maximum frequency. (2) There is a symmetrical distribution of positive and negative deviations about the maximum. (3) There is an exponential decrease in frequency as the magnitude of the deviations increases. Thus, small indeterminate uncertainties are observed much more often than very large ones.

Areas Under a Normal Error Curve

It can be shown that, regardless of its width, 68.3% of the area beneath a normal error curve lies within one standard deviation ($\pm 1\sigma$) of the mean μ. Thus, 68.3% of the data making up the population lie within these bounds. Furthermore, approximately 95.5% of all data are within $\pm 2\sigma$ of the mean, and 99.7% within $\pm 3\sigma$. Dashed lines show the areas bounded by $\pm 1\sigma$ and $\pm 2\sigma$ in Figure 4-1a and by $\pm 1\sigma$, $\pm 2\sigma$, and $\pm 3\sigma$ in Figure 4-1b.

Because of area relationships such as these, the standard deviation of a population of data is a useful predictive tool. For example, we can say that the chances are 68.3 in 100 that the indeterminate uncertainty of any single measurement is no more than $\pm 1\sigma$. Similarly, the chances are 95.5 in 100 that the error is less than $\pm 2\sigma$, and so forth.

Standard Error of a Mean

The figures on percentage distribution just quoted refer to the probable error for a *single* measurement. If a series of sample sets, each containing N data, are taken randomly from a population of data, the mean of each set will show less and less scatter as N increases. The standard deviation of the mean of each set is known as the *standard error* of the mean and is given the symbol σ_m. It can be shown that the standard error is inversely proportional to the square root of the number of data N used to calculate the mean:

$$\sigma_m = \sigma/\sqrt{N} \qquad (4\text{-}4)$$

where σ is defined by Equation 4-2. An analogous equation can be written for a sample standard deviation:

$$s_m = s/\sqrt{N} \qquad (4\text{-}5)$$

4A-3 Properties of the Standard Deviation

Effect of N on the Reliability of s

Uncertainty in the calculated value of s decreases as N in Equation 4-1 increases. Figure 4-2 shows the error in s as a function of N. When N is greater than about 20, s and σ can be assumed to be identical for all practical purposes. For example, if the 50 measurements in Table 3-3 are divided into ten subgroups of five measurements each, the value of s varies widely from one subgroup to another (0.0023 to 0.0079 mL) even though the average of the computed values of s is that of the entire set (0.0056 mL). In contrast, the computed values of s for two subsets of 25 measurements each are nearly identical (0.0054 and 0.0058 mL).

When $N > 20$, $s \approx \sigma$.

The rapid improvement in the reliability of s as N increases makes it feasible to obtain a good approximation of σ when the method of measurement is not excessively time-consuming and when an adequate supply of sample is available. For example, if the pH of numerous solutions is to be measured in the course of an investigation, it is useful to evaluate s in a series of preliminary experiments. This measurement is simple, requiring only that a pair of rinsed and dried electrodes be immersed in the test solution. The voltage between the electrodes is proportional to pH. To determine s, 20 to 30 portions of a buffer solution of fixed pH can be measured with all steps of the procedure being followed exactly. Normally, it is safe to assume that the indeterminate error in these test measurements is the same as that in subsequent measurements. The value of s calculated from Equation 4-2 is a valid and accurate measure of the theoretical σ.

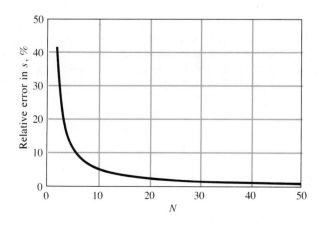

Figure 4-2

Relative error in σ as a function of N.

Pooling Data to Improve the Reliability of s

For analyses that are time-consuming, the foregoing procedure may not always be practical. In this situation, data from a series of samples accumulated over time can often be pooled to provide an estimate of s that is superior to the value for any individual subset. Again, we must assume the same sources of indeterminate error in all the samples. This assumption is usually valid if the samples have similar compositions and have been analyzed in exactly the same way.

To obtain a pooled estimate of the standard deviation s_{pooled}, deviations from the mean for each subset are squared; the squares of all subsets are then summed and divided by an appropriate number of degrees of freedom, as shown in Equation 4-6. The pooled s is obtained by extracting the square root of the quotient. One degree of freedom is lost for each subset. Thus, the number of degrees of freedom for the pooled s is equal to the total number of measurements minus the number of subsets.

$$s_{pooled} = \sqrt{\frac{\sum_{i=1}^{N_1}(x_i - \bar{x}_1) + \sum_{j=1}^{N_2}(x_j - \bar{x}_2) + \sum_{k=1}^{N_3}(x_k - \bar{x}_3) + \cdots}{N_1 + N_2 + N_3 + \cdots - N_s}} \qquad (4\text{-}6)$$

where N_1 is the number of data in set 1, N_2 is the number in set 2, and so forth. The term N_s is the number of data sets that are being pooled.

Example 4-1

The amount of mercury in samples of seven fish taken from the Sacramento River was determined by a method based upon the absorption of radiation by gaseous elemental mercury. Calculate a pooled estimate of the standard deviation for the method, based upon the first three columns of data:

Specimen	Number of Samples Measured	Hg Content, ppm	Mean, ppm Hg	Sum of Squares of Deviations from Mean
1	3	1.80, 1.58, 1.64	1.673	0.0258
2	4	0.96, 0.98, 1.02, 1.10	1.015	0.0115
3	2	3.13, 3.35	3.240	0.0242
4	6	2.06, 1.93, 2.12, 2.16, 1.89, 1.95	2.018	0.0611
5	4	0.57, 0.58, 0.64, 0.49	0.570	0.0114
6	5	2.35, 2.44, 2.70, 2.48, 2.44	2.482	0.0685
7	4	1.11, 1.15, 1.22, 1.04	1.130	0.0170

Number of measurements = 28

Sum of squares = 0.2196

The values in the last two columns for sample 1 were computed as follows:

x_i	$|(x_i - \bar{x})|$	$(x_i - \bar{x})^2$
1.80	0.127	0.0161
1.58	0.093	0.0086
1.64	0.033	0.0011
5.02	Sum of squares =	0.0258

$$\bar{x} = \frac{5.02}{3} = 1.673$$

The other data in columns 4 and 5 were obtained similarly. Then

$$s_{pooled} = \sqrt{\frac{0.0258 + 0.0115 + 0.0242 + 0.0611 + 0.0114 + 0.0685 + 0.0170}{28 - 7}}$$

$$= 0.10 \text{ ppm Hg}$$

Note that one degree of freedom is lost for each of the seven samples. Because more than 20 degrees of freedom remain, however, the computed value of s can be considered a good approximation of σ; that is, $s \rightarrow \sigma = 0.10$ ppm Hg.

4B   THE USES OF STATISTICS

Experimentalists use statistical calculations to sharpen their judgment concerning the effects of indeterminate errors. The most common applications of statistics to analytical chemistry are:

1. Defining the interval around a set mean within which the population mean can be expected at a given probability.
2. Determining the number of replicate measurements required to assure (at a given probability) that an experimental mean falls within a predetermined interval around the population mean.
3. Deciding whether an outlying value in a set of replicate results should be retained or rejected in calculating the mean for the set.
4. Treating calibration data.

We will examine each of these applications in the sections that follow.

4B-1   Confidence Limits

The exact value of the mean μ for a population of data can never be determined exactly because such a determination requires that an infinite number of measurements be made. Statistical theory allows us to set limits around an experimentally determined mean $\bar{x}$, however, and the true mean μ then lies within these limits with a given degree of probabil-

Confidence limits define an interval around $\bar{x}$ that probably contains μ.

ity. These limits are called *confidence limits,* and the interval they define is known as the *confidence interval.*

The size of the confidence interval, which is derived from the sample standard deviation, depends on the certainty with which *s* is known. If there is reason to believe that *s* is a good approximation of σ, then the confidence interval can be significantly narrower than if the estimate of *s* is based upon only two or three measurements.

The Confidence Interval When *s* Is a Good Approximation of σ

Figure 4-3 shows a series of five normal error curves. In each, the relative frequency is plotted as a function of the quantity *z* (Equation 4-3), which is the deviation from the mean *in units of the population standard deviation.* The shaded area in each plot lies between the values of $-z$ and $+z$ indicated to the left and right of the curves. The number within each shaded area is the percentage of the total area under the curve that is included within the *z* values. For example, as shown in the top curve, 50% of the area under any Gaussian curve is located between -0.67σ and $+0.67\sigma$. Proceeding on down, we see that 80% of the total area lies between -1.29σ and $+1.29\sigma$ and 90% lies between -1.64σ and $+1.64\sigma$. Relationships such as these allow us to define a range of values around a measurement within which the true mean is likely to lie with a certain probability. For example, we may assume that 90 times out of 100, the true mean μ will be within $\pm1.64\sigma$ of any measurement that we make. Here, the *confidence level* is 90% and the *confidence interval* is $\pm z\sigma = \pm1.64\sigma$.

We find a general expression for the confidence limits (CL) of a single measurement by rearranging Equation 4-3. Remember that *z* can take positive or negative values. Thus,

$$\text{CL for} \quad \boxed{\mu = x \pm z\sigma} \tag{4-7}$$

Values for *z* at various confidence levels are found in Table 4-1.

> The confidence level is the probability expressed as a percent.

> The confidence limits are the values above and below a measurement that define its confidence interval.

Table 4-1
CONFIDENCE LEVELS FOR VARIOUS VALUES OF z

Confidence Level, %	z
50	0.67
68	1.00
80	1.29
90	1.64
95	1.96
96	2.00
99	2.58
99.7	3.00
99.9	3.29

Example 4-2

Calculate the 50% and 95% confidence limits for the first entry (1.80 ppm Hg) in Example 4-1.

In that example, we calculated *s* to be 0.10 ppm Hg and had sufficient data to assume $s \rightarrow \sigma$. From Table 4-1, we see that $z = 0.67$ and 1.96 for the two confidence levels. Thus, from Equation 4-7

$$50\% \text{ CL for } \mu = 1.80 \pm 0.67 \times 0.10 = 1.80 \pm 0.07$$

$$95\% \text{ CL for } \mu = 1.80 \pm 1.96 \times 0.10 = 1.80 \pm 0.20$$

From these calculations, we conclude that the chances are 50 in 100 that μ, the population mean (and, *in the absence of determinate error,* the true value), lies in the interval between 1.73 and 1.87 ppm Hg. Furthermore, there is a 95% chance that it lies in the interval between 1.60 and 2.00 ppm Hg.

Equation 4-7 applies to the result of a *single measurement*. Application of Equation 4-4 shows that the confidence interval is decreased by $\sqrt{N}$ for the average of N replicate measurements. Thus, a more general form of Equation 4-7 is

$$\text{CL for} \quad \mu = \bar{x} \pm \frac{z\sigma}{\sqrt{N}} \tag{4-8}$$

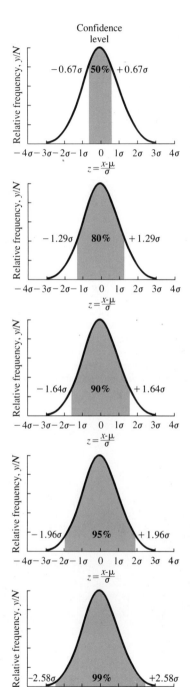

Confidence level

$z = \frac{x-\mu}{\sigma}$

Figure 4-3

Areas under a Gaussian curve for various values of $\pm z$.

Example 4-3

Calculate the 50% and 95% confidence limits for the mean value (1.67 ppm Hg) for specimen 1 in Example 4-1. Again, $s \rightarrow \sigma = 0.10$.

For the three measurements,

$$50\% \text{ CL} = 1.67 \pm \frac{0.67 \times 0.10}{\sqrt{3}} = 1.67 \pm 0.04$$

$$95\% \text{ CL} = 1.67 \pm \frac{1.96 \times 0.10}{\sqrt{3}} = 1.67 \pm 0.11$$

Thus, the chances are 50 in 100 that the population mean is located in the interval from 1.63 to 1.71 ppm Hg and 95 in 100 that it lies between 1.56 and 1.78 ppm.

Example 4-4

How many replicate measurements of specimen 1 in Example 4-1 are needed to decrease the 95% confidence interval to ±0.07 ppm Hg?

The pooled value is a good estimate of σ. For a confidence interval of ±0.07 ppm Hg, then, substitution into Equation 4-8 leads to

$$0.07 = \pm \frac{zs}{\sqrt{N}} = \pm \frac{1.96 \times 0.10}{\sqrt{N}}$$

$$\sqrt{N} = \pm \frac{1.96 \times 0.10}{0.07} = \pm 2.80$$

$$N = (\pm 2.8)^2 = 7.8$$

We conclude that eight measurements would provide a slightly better than 95% chance that the population mean lies within ±0.07 ppm of the experimental mean.

Equation 4-8 tells us that the confidence interval for an analysis can be halved by carrying out four measurements, that 16 measurements will narrow the interval by a factor of 4, and so on. We rapidly reach a point of diminishing returns in acquiring additional data. Ordinarily, we take advantage of the relatively large gain attained by averaging two to four measurements but can seldom afford the time required for additional increases in confidence.

It is essential to keep in mind at all times that confidence intervals based on Equation 4-8 apply only *in the absence of determinate errors and only if we can assume that s ≈ σ.*

The Confidence Limits When σ Is Unknown

Often we are faced with limitations in time or amount of available sample that prevent us from accurately estimating σ. Here, a single set of replicate measurements must provide not only a mean but also an estimate of precision. As indicated earlier, s calculated from a small set of data may be quite uncertain. Thus, confidence limits are necessarily broader when a good estimate of σ is not available.

The *t* statistic is often called *Student's t* because Student was the name used by W. S. Gossett when he wrote the classic paper on *t* that appeared in *Biometrika,* **1908,** *6,* 1. Gossett was employed by the Guinness Brewery to statistically analyze the results of analyses of its products. As a result of this work, he discovered the now-famous statistical treatment of small sets of data. To avoid disclosure of any of his employer's trade secrets, Gossett published the paper under the name Student.

To account for the variability of s, we use the important statistical parameter t, which is defined in exactly the same way as z (Equation 4-3) except that here s is substituted for σ:

$$t = \frac{x - \mu}{s} \qquad (4\text{-}9)$$

Like z in Equation 4-3, t depends on the desired confidence level. It also depends on the number of degrees of freedom in the calculation of s. Table 4-2 provides values for t for a few degrees of freedom. More extensive tables are found in various mathematical and statistical handbooks. Note that $t \rightarrow z$ (Table 4-1) as the number of degrees of freedom approaches infinity.

Table 4-2

VALUES OF *t* FOR VARIOUS LEVELS OF PROBABILITY

Degrees of Freedom	Factor for Confidence Interval				
	80%	90%	95%	99%	99.9%
1	3.08	6.31	12.7	63.7	637
2	1.89	2.92	4.30	9.92	31.6
3	1.64	2.35	3.18	5.84	12.9
4	1.53	2.13	2.78	4.60	8.60
5	1.48	2.02	2.57	4.03	6.86
6	1.44	1.94	2.45	3.71	5.96
7	1.42	1.90	2.36	3.50	5.40
8	1.40	1.86	2.31	3.36	5.04
9	1.38	1.83	2.26	3.25	4.78
10	1.37	1.81	2.23	3.17	4.59
11	1.36	1.80	2.20	3.11	4.44
12	1.36	1.78	2.18	3.06	4.32
13	1.35	1.77	2.16	3.01	4.22
14	1.34	1.76	2.14	2.98	4.14
∞	1.29	1.64	1.96	2.58	3.29

The confidence limits for the mean $\bar{x}$ of N replicate measurements can be derived from t by an equation similar to Equation 4-8:

$$\text{CL for } \quad \mu = \bar{x} \pm \frac{ts}{\sqrt{N}} \qquad \text{(4-10)}$$

Example 4-5

A chemist obtained the following data for the alcohol content of a sample of blood: % C_2H_5OH: 0.084, 0.089, and 0.079. Calculate the 95% confidence limits for the mean assuming (a) no additional knowledge about the precision of the method and (b) that on the basis of previous experience, it is known that $s \rightarrow \sigma = 0.005\%$ C_2H_5OH.

(a)
$$\Sigma\, x_i = 0.084 + 0.089 + 0.079 = 0.252$$
$$\Sigma\, x_i^2 = 0.007056 + 0.007921 + 0.006241 = 0.021218$$
$$s = \sqrt{\frac{0.021218 - (0.252)^2/3}{3 - 1}} = 0.0050\% \ C_2H_5OH$$

Here, $\bar{x} = 0.252/3 = 0.084$. Table 4-2 indicates that $t = 4.30$ for two degrees of freedom and 95% confidence. Thus,

$$95\% \ \text{CL} = \bar{x} \pm \frac{ts}{\sqrt{N}} = 0.084 \pm \frac{4.30 \times 0.0050}{\sqrt{3}} = 0.084 \pm 0.012\% \ C_2H_5OH$$

(b) Because a good value of σ is available,

$$95\% \ \text{CL} = \bar{x} \pm \frac{z\sigma}{\sqrt{N}} = 0.084 \pm \frac{1.96 \times 0.0050}{\sqrt{3}} = 0.084 \pm 0.006\% \ C_2H_5OH$$

Note that a sure knowledge of σ decreases the confidence interval by a significant amount.

4B-2 Rejection of Outliers

When a set of data contains an outlying result that appears to differ excessively from the average, the decision must be made whether to retain or reject the outlier.[2] The choice of criterion for rejecting a suspected result has its perils. If we set a stringent standard that makes rejection difficult, we run the risk of retaining results that are spurious and have an inordinate effect on the average. If we set lenient limits on precision and thereby make rejection easy, we are likely to discard measurements that rightfully belong in the set, thus introducing a bias to the data.

---

[2]J. Mandel, in *Treatise on Analytical Chemistry*, 2nd ed., I. M. Kolthoff and P. J. Elving, Eds., Part I, Vol. 1. pp. 282–289. New York: Wiley, 1978.

Table 4-3

CRITICAL VALUES FOR REJECTION QUOTIENT Q*

Number of Observations	Q_{crit} (Reject if $Q_{exp} > Q_{crit}$)		
	90% Confidence	96% Confidence	99% Confidence
3	0.94	0.98	0.99
4	0.76	0.85	0.93
5	0.64	0.73	0.82
6	0.56	0.64	0.74
7	0.51	0.59	0.68
8	0.47	0.54	0.63
9	0.44	0.51	0.60
10	0.41	0.48	0.57

*Reproduced from W. J. Dixon, *Ann. Math. Stat.*, **1951**, *22*, 68.

It is an unfortunate fact that no universal rule can be invoked to settle the question of retention or rejection.

The Q Test

The Q test is a simple, widely used statistical test.[3] In this test the absolute value (without regard to sign) of the difference between the questionable result x_q and its nearest neighbor x_n is divided by the spread w of the entire set to give the quantity Q_{exp}:

$$Q_{exp} = \frac{|x_q - x_n|}{w} \qquad (4\text{-}11)$$

If $Q_{exp} > Q_{crit}$, reject the outlier.

If $Q_{exp} < Q_{crit}$, keep the result.

This ratio is then compared with rejection values Q_{crit} found in Table 4-3. If Q_{exp} is greater than Q_{crit}, the questionable result can be rejected with the indicated degree of confidence.

Example 4-6

Remember to arrange your data in increasing order when you apply the Q test.

The analysis of a calcite sample yielded CaO percentages of 55.95, 56.00, 56.04, 56.08, and 56.23. The last value appears anomalous; should it be retained or rejected?

The difference between 56.23 and 56.08 is 0.15%. The spread (56.23 − 55.95) is 0.28%. Thus,

$$Q_{exp} = \frac{0.15}{0.28} = 0.54$$

For five measurements, Q_{crit} at the 90% confidence level is 0.64. Because 0.54 < 0.64, we must retain 56.08.

---

[3]R. B. Dean and W. J. Dixon, *Anal. Chem.*, **1951**, *23*, 636.

Other Statistical Tests

Several other statistical tests have been developed to provide criteria for rejecting or retaining outliers. Like the Q test, these others also assume that the distribution of the population data is normal, or Gaussian. Unfortunately, this condition cannot be proved or disproved for samples that have many fewer than 50 results. Consequently, statistical rules, which are perfectly reliable for normal distributions of data, should be *used with extreme caution* when applied to samples containing only a few data. J. Mandel, in discussing the treatment of small sets of data, writes, "Those who believe that they can discard observations with statistical sanction by using statistical rules for the rejection of outliers are simply deluding themselves."[4] Thus, statistical tests for rejection should be used only as aids to common sense when small samples are involved.

The blind application of statistical tests to determine whether you should retain or reject a suspect measurement in a small set of data is not likely to be much more fruitful than an arbitrary decision. The application of good judgment based on broad experience with an analytical method is usually a sounder approach. In the end, the only valid reason for rejecting a result from a small set of data is the sure knowledge that a mistake was made in the measurement process. Without this knowledge, *a cautious approach to rejection of an outlier is wise*.

Use caution when rejecting data for any reason.

Recommendations for the Treatment of Outliers

Recommendations for the treatment of a small set of results that contains a suspect value follow:

1. Reexamine carefully all data relating to the outlying result to see if a gross error could have affected its value. This recommendation demands *a properly kept laboratory notebook containing careful notations of all observations*.
2. If possible, estimate the precision that can be reasonably expected from the procedure to be sure that the outlying result actually is questionable.
3. Repeat the analysis if sufficient sample and time are available. Agreement between the newly acquired data and those data of the original set that appear valid will lend weight to the notion that the outlying result should be rejected. Furthermore, if retention is still indicated after repeating the analysis, the questionable result will have a smaller effect on the mean of the larger data set.
4. If more data cannot be secured, apply the Q test to the existing set to see if the doubtful result should be retained or rejected on statistical grounds.
5. If the statistical test indicates retention, consider reporting the median of the set rather than the mean. The median has the great virtue of allowing inclusion of all data in a set without undue influence from an outlying value. In addition, the median of a normally distributed set

---

[4]J. Mandel, in *Treatise on Analytical Chemistry*, 2nd ed., I. M. Kolthoff and P. J. Elving, Eds., Part I, Vol. 1. p. 282. New York: Wiley, 1978.

containing three measurements provides a better estimate of the correct value than the mean of the remaining two results after the outlying value has been discarded.

4B-3 The Least-Squares Method for Deriving Calibration Plots

Most analytical methods are based on a calibration curve in which a measured quantity y is plotted as a function of the known concentration x of a series of standards. The typical calibration curve shown in Figure 4-4 was derived for the determination of isooctane in a hydrocarbon sample. The ordinate is the area under the chromatographic peak for isooctane, and the abscissa is the mole percent of isooctane. As is typical (and desirable), the plot approximates a straight line. Note, however, that not all the data fall exactly on the line because of the indeterminate errors in the measuring process. Thus, the investigator must try to derive a "best" straight line from the points. A statistical technique called *regression analysis* provides the means for objectively obtaining such a line and also for specifying the uncertainties associated with its subsequent use. We consider here only the simplest regression procedure: the *method of least squares*.

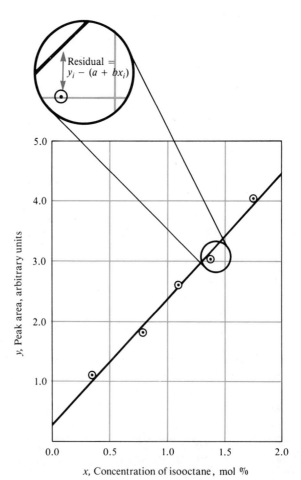

Figure 4-4

Calibration curve for the determination of isooctane in a hydrocarbon mixture.

Assumptions

When the method of least squares is used to generate a calibration curve, two assumptions are required. The first is that there is a linear relationship between the measured variable (y) and the analyte concentration (x). This relationship is stated mathematically as

$$y = a + bx$$

where a is the y intercept (the point at which the straight line drawn through the data points intercepts the y axis) and b is the slope of that line (see Figure 4-5). We also assume that any deviation of individual points from the straight line results from error in the *measurement*. That is, we assume there is no error in the x values of the points, an assumption we can make because concentrations of the standards used to generate a calibration curve are usually known exactly. Both of these assumptions are appropriate for most analytical methods.

Linear least-squares analysis gives you the equation for the best straight line among a set of x, y data pairs.

The Derivation of a Least-Squares Line

As illustrated in Figure 4-4, the vertical deviation of each point from the straight line is called a *residual*. The line generated by the least-squares method is the one that minimizes the sum of the squares of the residuals of all the points. In addition to providing the best fit among the experimental points and the straight line, the method gives the standard deviations for a and b.[5]

For convenience in calculation, we define three quantities S_{xx}, S_{yy}, and S_{xy} as follows:

$$S_{xx} = \Sigma\,(x_i - \bar{x})^2 = \Sigma\,x_i^2 - \frac{(\Sigma\,x_i)^2}{N} \tag{4-12}$$

$$S_{yy} = \Sigma\,(y_i - \bar{y})^2 = \Sigma\,y_i^2 - \frac{(\Sigma\,y_i)^2}{N} \tag{4-13}$$

$$S_{xy} = \Sigma\,(x_i - \bar{x})(y_i - \bar{y}) = \Sigma\,x_iy_i - \frac{\Sigma\,x_i\,\Sigma\,y_i}{N} \tag{4-14}$$

The terms S_{xx} and S_{yy} are similar in form to the numerator in the equation for standard deviation.

where x_i and y_i are individual pairs of data for x and y, N is the number of pairs of data used in preparing the calibration curve, and $\bar{x}$ and $\bar{y}$ are the average values for the variables, that is,

$$\bar{x} = \frac{\Sigma\,x_i}{N} \qquad \text{and} \qquad \bar{y} = \frac{\Sigma\,y_i}{N}$$

Note that S_{xx} and S_{yy} are the sums of the squares of the deviations from the mean for the individual values of x and y. The equivalent expressions shown on the far right in Equations 4-12 and 4-13 are more convenient when a handheld calculator is being used.

---

[5]R. L. Anderson, *Practical Statistics for Analytical Chemists*, pp. 89–121. New York: Van Nostrand–Reinhold, 1987.

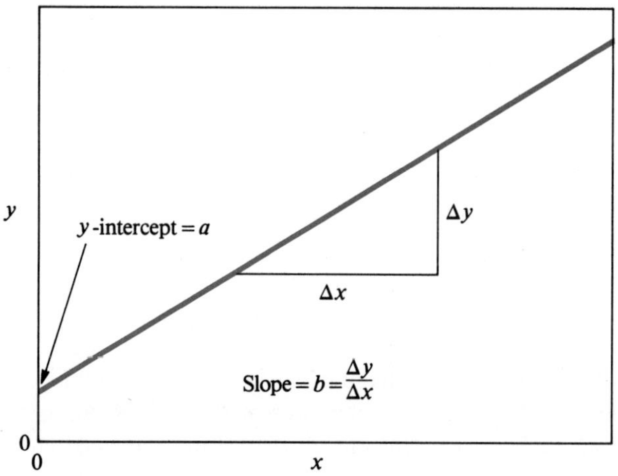

Figure 4-5
The slope-intercept form of a
straight line.

Five useful quantities can be derived from S_{xx}, S_{yy}, and S_{xy}:

1. The slope of the line b:

$$b = \frac{S_{xy}}{S_{xx}}$$ (4-15)

2. The intercept a:

$$a = \bar{y} - b\bar{x}$$ (4-16)

3. The standard deviation about regression s_r:

$$s_r = \sqrt{\frac{S_{yy} - b^2 S_{xx}}{N - 2}}$$ (4-17)

4. The standard deviation of the slope b:

$$s_b = \sqrt{s_r^2/S_{xx}}$$ (4-18)

5. The standard deviation for results obtained from the calibration curve s_c:

$$s_c = \frac{s_r}{b} \sqrt{\frac{1}{M} + \frac{1}{N} + \frac{(\bar{y}_c - \bar{y})^2}{b^2 S_{xx}}}$$ (4-19)

Equation 4-19 gives us a way to calculate the standard deviation from the mean $\bar{y}_c$ of a set of M replicate analyses when a calibration curve containing N points is used; recall that $\bar{y}$ is the mean value of y for the N calibration data.

The standard deviation about regression s_r (Equation 4-17) is the standard deviation for y when the deviations are measured not from the mean of y (as is usually the case) but from the derived straight line:

$$s_r = \sqrt{\frac{\sum\limits_{i=1}^{N} [y_i - (a + bx_i)]^2}{N - 2}} \qquad (4\text{-}20)$$

In this equation, the number of degrees of freedom is $N - 2$ since one degree is lost in the calculation of b and one in determining a.

Example 4-7

Carry out a least-squares analysis of the experimental data provided in the first two columns in Table 4-4 and plotted in Figure 4-4.

Columns 3, 4, and 5 of the table contain computed values for x_i^2, y_i^2, and x_iy_i, with their sums appearing as the last entry in each column. Note that the number of digits carried in the computed values should be the *maximum allowed by the calculator or computer;* that is, *rounding should not be performed until the calculation is complete.*

We now substitute into Equations 4-12, 4-13, and 4-14 and obtain

The standard deviation about regression is analogous to the standard deviation for one-dimensional data.

$$S_{xx} = \Sigma x_i^2 - \frac{(\Sigma x_i)^2}{N} = 6.90201 - (5.365)^2/5 = 1.14537$$

$$S_{yy} = \Sigma y_i^2 - \frac{(\Sigma y_i)^2}{N} = 36.3775 - (12.51)^2/5 = 5.07748$$

$$S_{xy} = \Sigma x_iy_i - \frac{\Sigma x_i \Sigma y_i}{N} = 15.81992 - 5.365 \times 12.51/5 = 2.39669$$

Substitution of these quantities into Equations 4-15 and 4-16 yields

$$b = \frac{2.39669}{1.14537} = 2.0925 = 2.09$$

$$a = \frac{12.51}{5} - 2.0925 \times \frac{5.365}{5} = 0.2567 = 0.26$$

Thus, the equation for the least-squares line is

$$y = 0.26 + 2.09x$$

Substitution into Equation 4-17 yields the standard deviation for the residuals:

$$s_r = \sqrt{\frac{S_{yy} - b^2 S_{xx}}{N - 2}} = \sqrt{\frac{5.07748 - (2.0925)^2 \times 1.14537}{5 - 2}} = 0.144 = 0.14$$

and substitution into Equation 4-18 gives the standard deviation of the slope:

$$s_b = \sqrt{s_r^2/S_{xx}} = \sqrt{(0.144)^2/1.14537} = 0.13$$

Table 4-4

CALIBRATION DATA FOR A CHROMATOGRAPHIC METHOD FOR THE DETERMINATION OF ISOOCTANE IN A HYDROCARBON MIXTURE

Mole Percent Isooctane, x_i	Peak Area, y_i	x_i^2	y_i^2	$x_i y_i$
0.352	1.09	0.12390	1.1881	0.38368
0.803	1.78	0.64481	3.1684	1.42934
1.08	2.60	1.16640	6.7600	2.80800
1.38	3.03	1.90440	9.1809	4.18140
1.75	4.01	3.06250	16.0801	7.01750
5.365	12.51	6.90201	36.3775	15.81992

Example 4-8

The calibration curve derived in Example 4-7 was used for the chromatographic determination of isooctane in a hydrocarbon mixture. A peak area of 2.65 was obtained. Calculate the mole percent of isooctane and the standard deviation for the result if the area was (a) the result of a single measurement and (b) the mean of four measurements.

In either case,

$$x = \frac{y - 0.26}{2.09} = \frac{2.65 - 0.26}{2.09} = 1.14 \text{ mol } \%$$

(a) Substituting into Equation 4-19, we obtain

$$s_c = \frac{0.14}{2.09} \sqrt{\frac{1}{1} + \frac{1}{5} + \frac{(2.65 - 12.51/5)^2}{(2.09)^2 \times 1.145}} = 0.074 \text{ mol } \%$$

(b) For the mean of four measurements,

$$s_c = \frac{0.14}{2.09} \sqrt{\frac{1}{4} + \frac{1}{5} + \frac{(2.65 - 12.51/5)^2}{(2.09)^2 \times 1.145}} = 0.046 \text{ mol } \%$$

Computer Application 4-1
LEAST-SQUARES ANALYSIS IN BASIC

You can do a linear least-squares analysis with a few lines of BASIC programming. The following program accomplishes this task using Equations 4-12 to 4-18. We have added a few new tricks to this program. To preserve the precision of our calculations, we have declared most of our variables to be double precision in statement 100. The DEFDBL A-H,M-Z defines all variables beginning with the letters A through H and M through Z to be double precision. This saves us the trouble of typing # after all variable names and avoids cluttering our program.

Statement 120 declares that all variables beginning with I through L will be integers. Statement 140 sets aside 200 memory locations for the storage of the *x-y* pairs of data points in our analysis. The variable X(J) is an array variable, which in many computer languages is a way of indicating subscripts, so that X(J) is analogous to x_j.

```
100 DEFDBL A-H,M-Z
120 DEFINT I-L
140 DIM X(100),Y(100)
200 INPUT"Enter # points:";I
220 SUMX = 0
240 SUMX2 = 0
260 SUMY = 0
270 SUMY2 = 0
280 SUMXY = 0
300 FOR J = 1 TO I
350 PRINT "Point number "J
400 INPUT"X:";X(J)
500 INPUT"Y:";Y(J)
520 SUMX = SUMX+X(J)
530 SUMX2 = SUMX2+X(J)^2
540 SUMY = SUMY+Y(J)
550 SUMY2 = SUMY2+Y(J)^2
560 SUMXY = SUMXY+X(J)*Y(J)
600 NEXT J
800 A2 = SUMX2-((SUMX)^2/I)
900 B2 = SUMY2-((SUMY)^2/I)
1000 AB = SUMXY-(SUMX*SUMY/I)
1050 XBAR = SUMX/I
1060 YBAR = SUMY/I
1100 B = AB/A2
1200 A = YBAR-B*XBAR
1300 SR = SQR(ABS((B2-(B^2*A2))/(I-2)))
1400 SB = SR/SQR(A2)
2000 PRINT using"_S_L_O_P_E_ = #.####^^^^";B
2010 PRINT using"_ _ _I_N_T_ = #.####^^^^";A
2020 PRINT using"_ _s_(_B_)_ = #.####^^^^";SB
2030 PRINT using"_ _s_(_r_)_ = #.####^^^^";SR
2040 END
```

Run the program, and enter the data from Table 4-3. Compare your results with those in Example 4-7. Add statements at the end of the program to accept a value for Y from the keyboard, and calculate a mole percent from the slope and intercept.

4C   QUESTIONS AND PROBLEMS

4-1. Calculate the 95% confidence interval for each set of data in Problem 3-7. What do these confidence intervals mean?

4-2. Calculate the 95% confidence interval for each set of data in Problem 3-7 if $s \rightarrow \sigma$ and has a value of: *set A, 0.20; set B, 0.050; *set C, 0.0070; set D, 0.50; *set E, 0.15; set F, 0.015.

4-3. The last result in each set of data in Problem 3-7 may be an outlier. Apply the Q test (96% confidence level) to determine whether there is a statistical basis for rejection.

*4-4. Analysis of several plant-food preparations for potassium ion yielded the following data:

Sample	Mean Percent K⁺	Number of Observations	Deviation of Individual Results from Mean
1	4.80	5	0.13, 0.09, 0.07, 0.05, 0.06
2	8.04	3	0.09, 0.08, 0.12
3	3.77	4	0.02, 0.15, 0.07, 0.10
4	4.07	4	0.12, 0.06, 0.05, 0.11
5	6.84	5	0.06, 0.07, 0.13, 0.10, 0.09

(a) Evaluate the standard deviation s for each sample.

(b) Obtain a pooled estimate for s.

4-5. Six bottles of wine were analyzed for residual sugar, with the following results:

Bottle	Percent (w/v) Residual Sugar	Number of Observations	Deviation of Individual Results from Mean
1	0.94	3	0.050, 0.10, 0.08
2	1.08	4	0.060, 0.050, 0.090, 0.060
3	1.20	5	0.05, 0.12, 0.07, 0.00, 0.08
4	0.67	4	0.05, 0.10, 0.06, 0.09
5	0.83	3	0.07, 0.09, 0.10
6	0.76	4	0.06, 0.12, 0.04, 0.03

(a) Evaluate the standard deviation s for each set of data.

(b) Pool the data to establish an absolute standard deviation for the method.

*4-6. Nine samples of illicit heroin preparations were analyzed in duplicate by a gas chromatographic technique. Pool the following data to establish an absolute standard deviation for the procedure:

Sample	Heroin, %	Sample	Heroin, %
1	2.24, 2.27	6	1.07, 1.02
2	8.4, 8.7	7	14.4, 14.8
3	7.6, 7.5	8	21.9, 21.1
4	11.9, 12.6	9	8.8, 8.4
5	4.3, 4.2		

4-7. Calculate a pooled estimate of s from the following spectrophotometric analysis for NTA (nitrilotriacetic acid) in water from the Ohio River:

Sample	NTA, ppb
1	13, 16, 14, 9
2	38, 37, 38
3	25, 29, 23, 29, 26

*4-8. An atomic absorption method for the determination of the amount of iron present in used jet engine oil was found, from pooling 30 triplicate analyses, to have a standard deviation $s \rightarrow \sigma = 2.4$ μg/mL. Calculate the 80 and 95% confidence intervals for the result, 18.5 μg Fe/mL, if it was based upon **(a)** a single analysis, **(b)** the mean of two analyses, **(c)** the mean of four analyses.

4-9. An atomic absorption method for determination of copper in fuels yielded a pooled standard deviation of $s \rightarrow \sigma = 0.32$ μg Cu/mL. The analysis of an oil from a reciprocating aircraft engine showed a copper content of 8.53 μg Cu/mL. Calculate the 80 and 99% confidence interval for the result if it was based upon **(a)** a single analysis, **(b)** the mean of 4 analyses, **(c)** the mean of 16 analyses.

*4-10. How many replicate measurements are needed to decrease the 95 and 99% confidence limits for the analysis described in Problem 4-8 to ±1.5 μg Fe/mL?

4-11. How many replicate measurements are necessary to decrease the 95 and 99% confidence limits for the analysis described in Problem 4-9 to ±0.2 μg Cu/mL?

*4-12. A volumetric calcium analysis on triplicate samples of the blood serum of a patient believed to be suffering from a hyperparathyroid condition produced the following data: meq Ca/L = 3.15, 3.25, 3.26. What is the 95% confidence interval for the mean of the data, assuming

(a) no prior information about the precision of the analysis?

(b) $s \rightarrow \sigma = 0.05$ meq Ca/L?

4-13. A chemist obtained the following data for percent lindane in the triplicate analysis of an insecticide preparation: 7.47, 6.98, 7.27. Calculate the 90% confidence interval for the mean of the three data, assuming that

(a) the only information about the precision of the method is the precision for the three data.

(b) on the basis of long experience with the method, it is believed that $s \rightarrow \sigma = 0.28\%$ lindane.

4-14. A standard method for determining tetraethyl lead (TEL) in gasoline is reported to have a standard deviation of 0.040 mL TEL per gallon. If $s \rightarrow \sigma = 0.040$, how many replicate analyses should be made in order for the mean for the analysis of a sample to be within

*(a) ±0.03 mL/gal of the true mean 99% of the time?

(b) ±0.03 mL/gal of the true mean 95% of the time?

(c) ±0.02 mL/gal of the true mean 90% of the time?

*4-15. Apply the Q test to the following data sets to determine whether the outlying result should be retained or rejected at the 95 to 96% confidence level.

(a) 41.27, 41.61, 41.84, 41.70

(b) 7.295, 7.284, 7.388, 7.292

4-16. Apply the Q test to the following data sets to determine whether the outlying result should be retained or rejected at the 95 to 96% confidence level.

(a) 85.10, 84.62, 84.70

(b) 85.10, 84.62, 84.65, 84.70

4-17. The sulfate ion concentration in natural water can be determined by measuring the turbidity that results when an excess of $BaCl_2$ is added to a measured quantity of the sample. A turbidimeter, the instrument used for this analysis, was calibrated with a series of standard Na_2SO_4 solutions. The following data were obtained in the calibration:

mg SO_4^{2-}/L, C_x	Turbidimeter Reading, R
0.00	0.06
5.00	1.48
10.00	2.28
15.0	3.98
20.0	4.61

Assume that a linear relationship exists between the instrument reading and concentration.

(a) Plot the data and draw a straight line through the points by eye.

(b) Derive a least-squares equation for the relationship between the variables.

(c) Compare the straight line from the relationship derived in (b) with that in (a).

(d) Calculate the standard deviation for the slope and the standard deviation about regression of the least-squares line.

(e) Calculate the concentration of sulfate in a sample yielding a turbidimeter reading of 3.67. Calculate the absolute standard deviation of the result and the coefficient of variation.

(f) Repeat the calculations in (e) assuming that the 3.67 was a mean of six turbidimeter readings.

4-18. The following data were obtained in calibrating a calcium ion electrode for the determination of pCa. A linear relationship between the potential E and pCa is known to exist.

pCa	E, mV
5.00	−53.8
4.00	−27.7
3.00	+ 2.7
2.00	+31.9
1.00	+65.1

(a) Plot the data and draw a line through the points by eye.

(b) Derive a least-squares expression for the best straight line through the points. Plot this line.

(c) Calculate the standard deviation for the slope of the least-squares line.

(d) Calculate the standard deviation about regression from the least-squares line.

(e) Calculate the pCa of a serum solution in which the electrode potential was 20.3 mV. Calculate the absolute and relative standard deviations for pCa if the result was from a single voltage measurement.

(f) Calculate the absolute and relative standard deviations for pCa if the millivolt reading in (e) was the mean of two replicate measurements. Repeat the calculation based upon the mean of eight measurements.

(g) Calculate the molar calcium ion concentration for the sample described in (e).

(h) Calculate the absolute and relative standard deviations in the calcium ion concentration if the measurement was performed as described in (f).

4-19. The following are relative peak areas for chromatograms of standard solutions of methyl vinyl ketone (MVK).

Concn MVK, mmol/L	Relative Peak Area
0.500	3.76
1.50	9.16
2.50	15.03
3.50	20.42
4.50	25.33
5.50	31.97

(a) Derive a least-squares expression assuming the variables bear a linear relationship to one another.

(b) Plot the least-squares line as well as the experimental points.

(c) Calculate the uncertainty of the points around the line as well as the uncertainty of the slope.

(d) Two samples containing MVK yielded relative peak areas of 6.3 and 27.5. Calculate the concentration of MVK in each solution.

(e) Assume that the results in (d) represent a single measurement as well as the mean of four measurements. Calculate the respective absolute and relative standard deviations.

CHAPTER 5

GRAVIMETRIC METHODS of ANALYSIS

Gravimetric methods of analysis, which are based upon the measurement of mass, are of two major types.[1] In *precipitation methods,* the analyte is converted to a sparingly soluble precipitate that is filtered, washed free of impurities, and converted to a product of known composition by suitable heat treatment. This product is then weighed. For example, in a precipitation method for determining calcium in natural water recommended by the Association of Official Analytical Chemists, an excess of oxalic acid, $H_2C_2O_4$, is added to a carefully measured volume of the sample. The addition of ammonia then causes essentially all the calcium in the sample to precipitate as calcium oxalate. The reaction is

$$Ca^{2+}(aq) + C_2O_4^{2-}(aq) \rightarrow CaC_2O_4(s)$$

The precipitate is collected in a weighed filtering crucible, dried, and then ignited at red heat. This process converts the precipitate entirely to calcium oxide:

$$CaC_2O_4(s) \rightarrow CaO(s) + CO(g) + CO_2(g)$$

Gravimetric methods of analysis are based upon mass measurements with an analytical balance, an instrument that yields highly accurate and precise data. In fact, if you perform a gravimetric chloride analysis in the laboratory, you may make some of the most accurate and precise measurements of your life.

---

[1]For an extensive treatment of gravimetric methods, see C. L. Rulfs, in *Treatise on Analytical Chemistry,* I. M. Kolthoff and P. J. Elving, Eds., Part I, Vol. 11, Chapter 13. New York: Wiley, 1975.

The crucible and precipitate are cooled, weighed, and the weight of calcium oxide is determined by subtraction of the known weight of the crucible. The calcium content of the sample is then computed as shown in Example 5-1, Section 5C.

In *volatilization methods,* the analyte or its decomposition products are volatilized at a suitable temperature. The volatile product is then collected and weighed, or alternatively the weight of the product is determined indirectly from the weight loss of the sample. An example of a gravimetric volatilization procedure is the determination of the sodium hydrogen carbonate content of antacid tablets. Here, a weighed quantity of the finely ground sample is treated with dilute sulfuric acid to convert the sodium hydrogen carbonate to carbon dioxide:

$$NaHCO_3(aq) + H_2SO_4(aq) \rightarrow CO_2(g) + H_2O(l) + NaHSO_4(aq)$$

This reaction is carried out in a flask that is connected to a weighed absorption tube that contains an absorbent that retains the carbon dioxide selectively as it is evolved from the solution by heating.[2] The difference in weight of the tube before and after absorption is used to calculate the amount of sodium hydrogen carbonate.

5A   PROPERTIES OF PRECIPITATES AND PRECIPITATING REAGENTS

Dimethylglyoxime (Section 5D-3) precipitates only Ni^{2+} and is thus a specific reagent. Silver nitrate, in contrast, is selective. The only common ions it precipitates from acidic solution are Cl^-, Br^-, I^-, and SCN^-.

Ideally, a gravimetric precipitating agent (the *precipitant*) should react either *specifically* or, if not that, at least *selectively* with the analyte. Specific reagents, which are rare, react only with a single chemical species. Selective reagents, which are more common, react with a limited number of species. In addition to being either specific or selective, the ideal precipitating reagent should react with the analyte to give a product that is

1. Readily filtered and washed free of contaminants.
2. Of sufficiently low solubility so that no significant loss of the analyte occurs during filtration and washing.
3. Unreactive with constituents of the atmosphere.
4. Of known composition after it is dried or, if necessary, ignited.

Few if any reagents produce precipitates that possess all these desirable properties.

Variables that influence solubility are discussed in Chapters 7, 8, and 9. In this section we consider methods for obtaining pure and easily filtered solids of known composition.[3]

---

[2]See Section 5D-5 for a description of the absorbent.

[3]For a more detailed treatment of precipitates, see H. A. Laitinen and W. E. Harris, *Chemical Analysis,* 2nd ed., Chapters 8 and 9. New York: McGraw-Hill, 1975; A. E. Nielsen, in *Treatise on Analytical Chemistry,* 2nd ed., I. M. Kolthoff and P. J. Elving, Eds., Part I, Vol. 3, Chapter 27. New York: Wiley, 1983.

5A-1 Particle Size and Filterability of Precipitates

Precipitates made up of large particles are generally desirable in gravimetric work because large particles are easy to filter and wash free of impurities. In addition, such precipitates are usually purer than are precipitates made up of fine particles.

Factors That Determine the Particle Size of Precipitates

The particle size of solids formed by precipitation varies enormously. At one extreme are *colloidal suspensions,* whose tiny particles are invisible to the naked eye (10^{-7} to 10^{-4} cm in diameter). Colloidal particles show no tendency to settle from solution, nor are they easily filtered. At the other extreme are particles with dimensions on the order of tenths of a millimeter or greater. The temporary dispersion of such particles in a liquid phase is called a *crystalline suspension*. The particles of a crystalline suspension tend to settle spontaneously and are readily filtered.

Scientists have studied precipitate formation for many years, but the mechanism of the process is still not fully understood. It is certain, however, that precipitate particle size is influenced by such experimental variables as precipitate solubility, temperature, reactant concentrations, and rate at which reactants are mixed. The net effect of these variables can be accounted for, at least qualitatively, by assuming that the particle size is related to a single property of the system called its *relative supersaturation,* where

$$\text{relative supersaturation} = \frac{Q - S}{S} \qquad (5\text{-}1)$$

In this equation, Q is the concentration of the solute at any instant and S is its equilibrium solubility.

Precipitation reactions are generally slow, so that even when a precipitating reagent is added drop by drop to a solution of an analyte, some supersaturation is likely. Experimental evidence indicates that the particle size of a precipitate varies inversely with the average relative supersaturation during the time the reagent is being introduced. Thus, when $(Q - S)/S$ is large, the precipitate tends to be colloidal; when $(Q - S)/S$ is small, a crystalline solid is more likely.

Mechanism of Precipitate Formation

The effect of relative supersaturation on particle size can be explained if we assume that precipitates form in two ways: by *nucleation* and by *particle growth*. The particle size of a freshly formed precipitate is determined by which of these two precipitation routes is the faster.

In nucleation, a few ions, atoms, or molecules (perhaps as few as four or five) come together to form a stable solid. These nuclei often form on the surface of suspended solid contaminants, such as dust particles. Further precipitation then involves a competition between additional nucleation and growth on existing nuclei (particle growth). If nucleation predom-

In diffuse light, colloidal suspensions may be perfectly clear and appear to contain no solid. The presence of the second phase can be detected, however, by shining the beam of a flashlight into the solution. Because particles of colloidal dimensions scatter visible radiation, the path of the beam through the solution can be seen by the eye. This phenomenon is called the *Tyndall effect*.

The particles of colloidal suspension are not easily filtered. To trap these particles, the pore size of the filtering medium must be so small that filtrations take too long. With suitable treatment, however, the individual colloidal particles can be made to stick together and thus give a filterable mass.

Equation 5-1 is known as the Von Weimarn equation in recognition of the scientist who proposed it in 1925.

To increase the particle size of a precipitate, minimize the relative supersaturation while the precipitate is being formed.

Nucleation is a process in which a minimum number of atoms, ions, or molecules join together to give a stable solid.

Precipitates form by nucleation and by particle growth. When nucleation predominates, the result is a large number of very fine particles; when particle growth predominates, the result is a smaller number of larger particles.

inates, the precipitate is made up of a large number of small particles; if growth predominates, the precipitate consists of a smaller number of larger particles.

The rate of nucleation is believed to increase enormously with increasing relative supersaturation. In contrast, the rate of particle growth is only moderately enhanced by high relative supersaturation. Thus, when the relative supersaturation is high, nucleation is the major precipitation mechanism, and a large number of small particles are formed. When the relative supersaturation is low, the rate of particle growth tends to predominate, and deposition of solid on existing particles occurs to the exclusion of further nucleation; a crystalline suspension results.

Experimental Control of Particle Size

Experimental variables that minimize supersaturation and thus lead to crystalline precipitates include elevated temperatures to increase the solubility of the precipitate (S in Equation 5-1), dilute solutions (to minimize Q), and slow addition of the precipitating agent with good stirring. The last two measures also minimize the concentration of the solute (Q) at any given instant.

Larger particles can also be obtained by pH control, provided the solubility of the precipitate depends upon pH. For example, large, easily filtered crystals of calcium oxalate are obtained by forming the bulk of the precipitate in a mildly acidic environment, in which the salt is moderately soluble. The precipitation is then completed by slowly adding aqueous ammonia until the acidity is sufficiently low that essentially all the calcium oxalate is precipitated. The additional precipitate produced during this step forms on the solid particles formed in the first step.

Unfortunately, many precipitates cannot be formed as crystals under practical laboratory conditions. A colloidal solid is generally encountered when a precipitate has such a low solubility that S in Equation 5-1 always remains negligible relative to Q. The relative supersaturation thus remains enormous throughout precipitate formation, and a colloidal suspension results. For example, under conditions feasible for an analysis, the hydrous oxides of iron(III), aluminum, and chromium(III) and the sulfides of most heavy-metal ions form only as colloids because of their very low solubilities.[4]

Precipitates with very low solubilities, such as many sulfides and hydrous oxides, generally form as colloids.

5A-2 Colloidal Precipitates

Colloidal suspensions are often stable for indefinite periods and are not usable for gravimetric analysis because their particles are too small to be readily filtered. Fortunately, the stability of most suspensions of this kind can be decreased by heating, by stirring, and by adding an electrolyte. These measures cause the individual colloidal particles to bind together to form an amorphous mass that settles out of solution and is filterable. The

---

[4]Silver chloride illustrates that the relative-supersaturation concept is imperfect. This compound ordinarily forms as a colloid, and yet its molar solubility is not significantly different from that of other compounds, such as $BaSO_4$, which generally form as crystals.

process of converting a colloidal suspension into a filterable solid is called *coagulation* or *agglomeration*.

Coagulation of Colloids

Colloidal suspensions are stable because the particles are either all positively charged or all negatively charged and thus repel one another. This charge results from cations or anions that are bound to the surface of the particles. The process by which ions are retained *on the surface of a solid* is known as *adsorption*. We can readily demonstrate that colloidal particles are charged by observing their migration when placed in an electrical field.

The tendency of ions to be adsorbed on an ionic solid surface originates in the normal bonding forces that are responsible for crystal growth. For example, a silver ion at the surface of a silver chloride particle has a partially unsatisfied bonding capacity for anions because of its surface location. Negative ions are attracted to this site by the same forces that hold chloride ions in the silver chloride lattice. In the same way, chloride ions at the surface of the solid attract cations dissolved in the solvent.

The kind and number of ions retained on the surface of a colloidal particle depend, in a complex way, on several variables. For a suspension produced in the course of a gravimetric analysis, however, the species adsorbed, and hence the charge on the particles, is readily predicted because lattice ions are generally more strongly held than any other ions. For example, when sodium chloride is first added to a solution containing silver nitrate, the colloidal particles of silver chloride formed are positively charged as shown in Figure 5-1. This charge is due to adsorption of

> Adsorption is a process in which a substance (gas, liquid, or solid) is held *on the surface* of a solid. In contrast, absorption, involves retention of a substance *within* the pores of a solid.

> The charge on a colloidal particle formed in a precipitation is determined by the charge of the lattice ion that is in excess when the precipitation is complete.

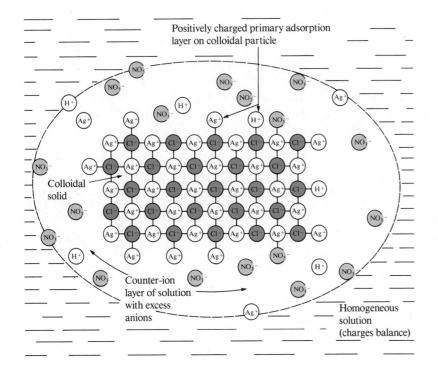

Figure 5-1

A colloidal silver chloride particle suspended in a solution of silver nitrate.

Theodore W. Richards (1868–1928) and his graduate students at Harvard University developed or refined many of the techniques of gravimetric analysis involving silver and chlorine. These techniques were used to determine the atomic weights of 25 of the elements by preparing pure samples of the chlorides of the elements, decomposing known weights of these compounds, and determining their chloride content by gravimetric methods. For this work, Richards became the first American to receive the Nobel Prize in Chemistry (1914).

The electric double layer of a colloid consists of a layer of charge adsorbed on the surface of the particles and a layer of opposite charge in the solution surrounding the particles.

some of the excess silver ions in the medium. The charge on the particles becomes negative, however, when enough sodium chloride has been added to provide an excess of chloride ions. Now, the adsorbed species is primarily chloride ions. The surface charge is at a minimum when the supernatant liquid contains an excess of neither ion.

The extent of adsorption, and thus the charge on a given particle, increase rapidly as the concentration of the lattice ion becomes greater. Eventually, however, the surface of the particles becomes covered with the adsorbed ions, and the charge then becomes constant and independent of concentration.

Figure 5-1 shows a colloidal silver chloride particle in a solution that contains an excess of silver nitrate. Attached directly to the solid surface is the *primary adsorption layer,* which consists mainly of adsorbed silver ions. Surrounding the charged particle is a layer of solution, called *the counter-ion layer,* which contains an excess of negative ions (principally nitrate). The primarily adsorbed silver ions and the negative counter-ion layer constitute an *electric double layer* that stabilizes a colloidal suspension by preventing individual particles from coming close enough to one another to agglomerate.

In Figure 5-2a, the effective charge of a colloidal silver chloride particle in a silver nitrate solution is plotted as a function of distance from its surface. The effective charge can be thought of as a measure of the repulsive force the particle exerts on like particles in the solution. Note that the effective charge falls off rapidly as the distance from the surface increases and approaches zero at the point d_1. This decrease in effective charge (in this case positive) is caused by the negative charge of the excess counter ions in the double layer. At point d_1, the number of counter ions in the layer is approximately equal to the number of primarily adsorbed ions on the surface of the particle so that the effective charge of the particle approaches zero.

Let us now consider two particles described by the upper curve in Figure 5-2a. This system consists of silver chloride particles suspended in a relatively concentrated silver nitrate solution so the particles have a large positive charge. The effective charge on the particles prevents them from approaching one another more closely than about $2d_1$—a distance that is too great for coagulation to occur.

Figure 5-2

Effect of $AgNO_3$ and electrolyte concentration on the thickness of the double layer surrounding a colloidal AgCl particle in a solution containing excesses of $AgNO_3$.

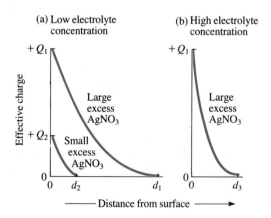

As you add more sodium chloride to this solution, the silver ion concentration becomes smaller as does the number of silver ions adsorbed on each particle. As shown in the lower of the two curves, the effective charge is now much less and extends a shorter distance into the solution. Now the particles can approach within $2d_2$ of one another. With further additions of chloride ion, this distance becomes small enough so that the forces of agglomeration can take over, and a coagulated precipitate begins to appear.

Coagulation of a colloidal suspension can often be brought about by a short period of heating, particularly if accompanied by stirring. Heating decreases the number of adsorbed ions and thus the thickness d_1 of the double layer. The particles may also gain enough kinetic energy at the higher temperature to overcome the barrier to close approach posed by the double layer.

> Colloidal suspensions can often be coagulated by heating, by stirring, and by adding an electrolyte.

An even more effective way to coagulate a colloid is to increase the electrolyte concentration of the solution. If we add a suitable ionic compound to a suspension, the concentration of counter ions increases in the vicinity of each particle. As a result, the volume of solution that contains sufficient counter ions to balance the charge of the primary adsorption layer decreases. The net effect of adding an electrolyte is thus a shrinkage of the counter-ion layer as shown in Figure 5-2b. The particles can then approach one another more closely and agglomerate.

Peptization of Colloids

Peptization refers to the process by which a coagulated colloid reverts to its original dispersed state. When a coagulated colloid is washed, some of the electrolyte responsible for its coagulation is leached from the internal liquid in contact with the solid particles. Removal of this electrolyte has the effect of increasing the volume of the counter-ion layer. The repulsive forces responsible for the original colloidal state are then reestablished, and particles detach themselves from the coagulated mass. The washings become cloudy as the freshly dispersed particles pass through the filter.

> Peptization is a process by which a coagulated colloid is reconverted to a colloidal suspension.

The chemist is thus faced with a dilemma in working with coagulated colloids. On the one hand, washing is needed to minimize contamination; on the other, there is the risk of losses resulting from peptization if pure water is used as the wash liquid. The problem is commonly resolved by washing with a solution containing an electrolyte that volatilizes when the precipitate is dried during the subsequent drying or ignition step. For example, silver chloride is ordinarily washed with a dilute solution of nitric acid. While the precipitate undoubtedly becomes contaminated with the acid, no harm results since the nitric acid is volatilized during the ensuing drying step.

> Peptization is prevented by washing precipitates with a solution of a volatile electrolyte.

Practical Treatment of Colloidal Precipitates

Colloids are best precipitated from hot, stirred solutions containing sufficient electrolyte to ensure coagulation. The filterability of a coagulated colloid frequently improves if the solid is allowed to stand for an hour or more in contact with the hot solution from which it was formed. During this process, which is known as *digestion*, weakly bound water appears to

> Digestion is a process in which a precipitate is heated for an hour or more in the solution from which it was formed (the *mother liquor*).

be lost from the precipitate; the result is a denser mass that is easier to filter.

5A-3 Crystalline Precipitates

Crystalline precipitates are generally more easily filtered and purified than are coagulated colloids. In addition, the size of individual crystalline particles, and thus their filterability, can be controlled to a degree.

Methods of Improving Particle Size and Filterability

The particle size of crystalline solids can often be improved significantly by minimizing Q and/or maximizing S in Equation 5-1. Minimization of Q is generally accomplished by using dilute solutions and adding the precipitating reagent slowly and with good mixing. Often S is increased by precipitating from hot solution or by adjusting the pH of the precipitation medium.

Digesting crystalline precipitates (without stirring) for some time after formation frequently yields a purer, more filterable product. The improvement in filterability undoubtedly results from the dissolution and recrystallization that occur continuously and at an enhanced rate at elevated temperatures. Recrystallization apparently results in bridging between adjacent particles, a process that yields larger and more easily filtered crystalline aggregates. This view is supported by the observation that little improvement in filtering characteristics occurs if the mixture is stirred during digestion.

Digestion improves the purity and filterability of both colloidal and crystalline precipitates.

5A-4 Coprecipitation

Coprecipitation is a phenomenon in which *otherwise soluble* compounds are removed from solution during precipitate formation. It is important to understand that contamination of a precipitate by a second substance whose solubility product has been exceeded *does not constitute coprecipitation*.

There are four types of coprecipitation: *surface adsorption, mixed-crystal formation, occlusion,* and *mechanical entrapment.*[5] Surface adsorption and mixed-crystal formation are equilibrium processes, whereas occlusion and mechanical entrapment arise from the kinetics of crystal growth.

Coprecipitation is a process whereby normally soluble compounds are carried out of solution by a precipitate.

Surface Adsorption

Adsorption is a common source of coprecipitation that is likely to cause significant contamination of precipitates with large specific surface areas—that is, coagulated colloids (see Feature 5-1 for the definition of specific area). Although adsorption does occur in crystalline solids, its

Adsorption is often the major source of contamination in coagulated colloids but of no significance in crystalline precipitates.

---

[5]Several classification systems have been suggested for coprecipitation phenomena. We follow the simple system proposed by A. E. Nielsen, in *Treatise on Analytical Chemistry,* 2nd ed., I. M. Kolthoff and P. J. Elving, Eds., Part I, Vol. 3, p. 333. New York: Wiley, 1983.

effects on purity are usually undetectable because of the relatively small specific surface area of these solids.

Feature 5-1
THE SPECIFIC SURFACE AREA OF COLLOIDS

Specific surface area is defined as the surface area per unit mass of solid and is ordinarily expressed in units of square centimeters per gram.

For a given mass of solid, specific surface area increases dramatically as particle size decreases, and therefore is enormous for colloids. For example, the solid cube shown in Figure 5-3 is 1 cm on a side and so has a surface area of 6 cm². If this cube weighs 2 g, its specific surface area is 6 cm²/2 g = 3 cm²/g. As shown in the lower part of Figure 5-3, this cube can be divided into 1000 cubes, each being 0.1 cm on a side. The surface area of each face of these cubes is now 0.1 cm × 0.1 cm = 0.01 cm², and the total surface area for the six faces of one of these 0.1-cm cubes is 0.06 cm². Because there are 1000 of these cubes, the total surface area for the 2 g of solid is now 60 cm²; the specific surface area is therefore 30 cm²/g.

Continuing in this way, we find that the specific surface area becomes 300 cm²/g when we have 10⁶ cubes that are 0.01 cm on a side. The particle size of a typical crystalline suspension lies in the region between 0.1 and 0.01 cm, and so a typical crystalline precipitate has a specific surface area between 30 and 300 cm²/g. Contrast these figures with that for 2 g of a colloid made up of 10¹⁸ particles having dimensions of 10⁻⁶ cm. Here, the specific surface area is 3 × 10⁶ cm²/g, which converts to something over 3000 ft²/g. Thus 1 g of a colloidal suspension has a surface area that is equivalent to the floor area of a good-sized home.

> The specific surface area of a solid is the ratio of its surface area (cm²) to its mass (g).

Figure 5-3
Increase in surface area per unit mass with decrease in particle size.

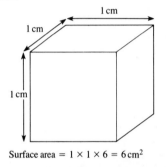

Surface area = 1 × 1 × 6 = 6 cm²

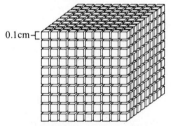

1000 cubes each 0.1 cm on a side
Surface area = 1000 × 0.1 × 0.1 × 6 = 60 cm²

Coagulation of a colloid does not significantly decrease the extent of adsorption because the coagulated solid still contains large internal surface areas that remain exposed to the solvent (Figure 5-4). The coprecipitated contaminant on the coagulated colloid consists of the lattice ion originally adsorbed on the surface before coagulation and the counter ion of opposite charge held in the film of solution immediately adjacent to the particle. *The net effect of surface adsorption is therefore the carrying down of an otherwise soluble compound as a surface contaminant.* For example, the coagulated silver chloride formed in the gravimetric determination of chloride ion is contaminated with primarily adsorbed silver ions and with nitrate or other anions in the counter-ion layer. As a consequence, silver nitrate, a normally soluble compound, is coprecipitated with the silver chloride.

In adsorption, a normally soluble compound is carried out of solution on the surface of a coagulated colloid. This compound consists of the primarily adsorbed ion and an ion of opposite charge from the counter-ion layer.

Methods for Minimizing Adsorbed Impurities on Colloids. The purity of many coagulated colloids is improved by digestion. During this process, water is expelled from the solid to give a denser mass that has a smaller specific surface area for adsorption.

Washing a coagulated colloid will not remove many primarily adsorbed ions because attraction between these ions and the solid surface is too strong. Exchange does occur between existing *counter ions* and ions of like charge in the wash solution, however. It is thus advantageous to select a volatile electrolyte for this solution. For example, in the determination of silver by precipitation with chloride ion, the primarily adsorbed species is chloride. Washing with an acidic solution converts the counter-ion layer largely to hydrogen ions so that both chloride and hydrogen ions are retained by the solid and its counter-ion layer. Volatile HCl is then given off when the precipitate is dried.

Regardless of the method of treatment, a coagulated colloid is always contaminated to some degree, even after extensive washing. The error introduced into the analysis from this source can be as low as 1 to 2 ppt, as in the coprecipitation of silver nitrate on silver chloride. The coprecipitation of heavy-metal hydroxides on the hydrous oxides of trivalent iron or aluminum, on the other hand, may cause errors amounting to several percent, which is generally intolerable.

Reprecipitation. A drastic but effective way to minimize the effects of adsorption is *reprecipitation,* or *double precipitation.* Here, the filtered solid is redissolved and reprecipitated. The first precipitate ordinarily carries down only a fraction of the contaminant present in the original solvent. Thus, the solution containing the redissolved precipitate has a significantly lower contaminant concentration than the original, and even less adsorption occurs during the second precipitation. Reprecipitation adds substantially to the time required for an analysis but is often necessary for such precipitates as the hydrous oxides of iron(III) and aluminum, which have extraordinary tendencies to adsorb the hydroxides of heavy-metal cations, such as zinc, cadmium, and manganese.

Calcium oxalate is often freed from coprecipitated magnesium oxalate. Here the precipitated calcium oxalate is dissolved in dilute sulfuric acid. Ammonium oxalate is added to the acid solution followed by ammonia until a second precipitate of calcium oxalate forms. This precipitate is largely free of magnesium oxalate.

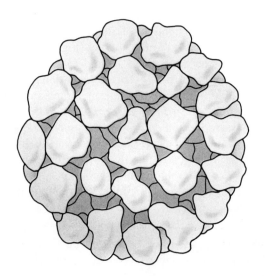

Figure 5-4
A coagulated colloid. Note the extensive *internal* surface area exposed to solvent.

Mixed-Crystal Formation

In mixed-crystal formation, one of the ions in the crystal lattice of a solid is replaced by an ion of another element. For this exchange to occur, it is necessary that the two ions have the same charge and that their sizes differ by no more than about 5%. Furthermore, the two salts must belong to the same crystal class. Thus, for example, barium sulfate formed by adding barium chloride to a solution containing sulfate, lead, and acetate ions is found to be severely contaminated by lead sulfate even though acetate ions normally prevent precipitation of lead sulfate by complexing the lead. Here, lead ions replace some of the barium ions in the barium sulfate crystals. Other examples of coprecipitation by mixed-crystal formation include $MgKPO_4$ in $MgNH_4PO_4$, $SrSO_4$ in $BaSO_4$, and MnS in CdS.

The extent of mixed-crystal contamination is governed by the law of mass action and increases as the ratio of contaminant concentration to analyte concentration increases. Mixed-crystal formation is a particularly troublesome type of coprecipitation because little can be done about it when certain combinations of ions are present in a sample matrix. This problem is encountered with both colloidal suspensions and crystalline precipitates. When mixed-crystal formation occurs, separation of the interfering ion may have to be carried out before the final precipitation step. Alternatively, a different precipitating reagent, one that does not give mixed crystals with the ions in question, may be used.

Occlusion and Mechanical Entrapment

When a crystal is growing rapidly during precipitate formation, foreign ions in the counter-ion layer may become trapped, or *occluded,* within the growing crystal. The amount of occluded material is greatest in that part of a crystal that forms first because supersaturation, and thus growth rate, decrease as a precipitation progresses.

Mechanical entrapment occurs when crystals lie close together during growth. Here, several crystals grow together and in so doing trap a portion of the solution in a tiny pocket.

Both occlusion and mechanical entrapment are at a minimum when the rate of precipitate formation is low—that is, under conditions of low supersaturation. In addition, digestion is often remarkably helpful in decreasing these types of coprecipitation. Undoubtedly, the rapid solution and reprecipitation that goes on at the elevated temperature of digestion open up the pockets and allow the impurities to escape into the solution.

Coprecipitation Errors

Coprecipitated impurities may cause either negative or positive errors in an analysis. If the contaminant is not a compound of the analyte ion, positive errors always result. Thus, a positive error is observed whenever colloidal silver chloride adsorbs silver nitrate during a chloride analysis.

In contrast, when the contaminant does contain the ion being determined, either positive or negative errors may be observed. For example, in the determination of barium by precipitation as barium sulfate, occlusion of other barium salts occurs. If the occluded contaminant is barium nitrate, a positive error is observed because this compound has a larger

Mixed crystal formation is a type of coprecipitation in which a contaminant ion replaces the analyte ion in the lattice of a crystalline precipitate.

Mixed-crystal formation may occur in both colloidal and crystalline precipitates, whereas occlusion and mechanical entrapment are confined to crystalline precipitates.

Occlusion is a type of coprecipitation in which an ion and its counter ion are trapped in a pocket of a rapidly growing crystalline precipitate.

Coprecipitation can cause either negative or positive errors in a gravimetric analysis.

formula weight than the barium sulfate that would have formed had no coprecipitation occurred. If barium chloride is the contaminant, the error is negative because its formula weight is less than that of the sulfate salt.

5A-5 Precipitation from Homogeneous Solution

In a homogeneous precipitation, a precipitating agent is generated homogeneously throughout a solution by a slow chemical reaction.

Precipitation from homogeneous solution is a technique in which a precipitating agent is generated in a solution of the analyte by a slow chemical reaction.[6] Local reagent excesses do not occur because the precipitating agent appears gradually and homogeneously throughout the solution and reacts immediately with the analyte. As a result, the relative supersaturation is kept low during the entire precipitation. In general, homogeneously formed precipitates, both colloidal and crystalline, are better suited to analysis than are solids formed by direct addition of a precipitating reagent.

Urea is often used for the homogeneous generation of hydroxide ion. The reaction can be expressed by the equation

$$(H_2N)_2CO + 3\ H_2O \rightarrow CO_2 + 2\ NH_4^+ + 2\ OH^-$$

Solids formed by homogeneous precipitation are generally purer and more easily filtered than precipitates generated by direct addition of a reagent to the analyte solution.

This reaction proceeds slowly just below 100°C, and a 1 to 2 h heating period is needed to complete a typical precipitation. Urea is particularly valuable for the precipitation of hydrous oxides or basic salts. For example, hydrous oxides of iron(III) and aluminum formed by direct addition of base are bulky and gelatinous masses that are heavily contaminated and difficult to filter. In contrast, when these same products are produced by homogeneous generation of hydroxide ion, they are dense and readily filtered and have considerably higher purity. Figure 5-5 shows hydrous oxide precipitates of aluminum formed by direct addition of base and by homogeneous precipitation with urea. Similar pictures for hydrous oxides of iron(III) are shown in the color plates. Homogeneous precipitation of crystalline precipitates also results in marked increases in crystal size as well as improvements in purity.

Representative methods based upon precipitation by homogeneously generated reagents are given in Table 5-1.

Jöns Jacob Berzelius (1779–1848), considered the leading chemist of his time, developed much of the apparatus and many of the techniques of nineteenth-century analytical chemistry. Examples include the use of ashless filter paper in gravimetry, the use of hydrofluoric acid to decompose silicates, and the use of the metric system in weight determinations. He performed thousands of analyses of pure compounds to determine the atomic weights of most of the elements known then. Berzelius also develop our present system of symbols for elements and compounds.

5B   Drying and Ignition of Precipitates

After filtration, a gravimetric precipitate is heated until its weight becomes constant. Heating removes the solvent and any volatile species carried down with the precipitate. Some precipitates must be ignited to decompose the solid and form a compound of known composition. This new compound is often called the *weighing form*.

The temperature required to produce a suitable weighing form varies from precipitate to precipitate. Figure 5-6 shows weight loss as a function of temperature for several common analytical precipitates. These data

---

[6]For a general reference on this technique, see L. Gordon, M. L. Salutsky, and H. H. Willard, *Precipitation from Homogeneous Solution*. New York: Wiley, 1959.

Figure 5-5
Hydrous aluminum oxide formed (a) by homogeneous generation of hydroxide ion and (b) by direct addition of base.

were obtained with an automatic thermobalance,[7] an instrument that records the weight of a substance continuously as the temperature of the substance is increased at a constant rate in a furnace (Figure 5-7). Heating three of the precipitates—silver chloride, barium sulfate, and aluminum oxide—simply causes removal of water and perhaps volatile electrolytes. Note, however, the vastly different temperatures required to produce an anhydrous precipitate of constant weight. Moisture is completely removed from silver chloride above 110°C, but dehydration of aluminum oxide is not complete until a temperature greater than 1000°C is achieved. It is of interest to note that aluminum oxide formed homogeneously with urea can be completely dehydrated at about 650°C.

The thermal curve for calcium oxalate is considerably more complex than the others shown in Figure 5-6. Below about 135°C, unbound water is eliminated to give the monohydrate $CaC_2O_4 \cdot H_2O$. This compound is then converted to the anhydrous oxalate CaC_2O_4 at 225°C. The abrupt change in weight at about 450°C signals the decomposition of calcium oxalate to calcium carbonate and carbon monoxide. The final step in the curve depicts the conversion of the calcium carbonate to calcium oxide and carbon dioxide. It is evident that the compound finally weighed in a gravimetric calcium determination based upon precipitation as oxalate is highly dependent upon the ignition temperature.

The temperature required to dehydrate a precipitate completely may be as low as 100°C or as high as 1000°C.

---

[7]For descriptions of thermobalances, see W. W. Wendlandt, *Thermal Methods of Analysis,* 2nd ed. New York: Wiley, 1974.

Table 5-1

METHODS FOR THE HOMOGENEOUS GENERATION OF PRECIPITATING AGENTS

Precipitating Agent	Reagent	Generation Reaction	Elements Precipitated
OH^-	Urea	$(NH_2)_2CO + 3\ H_2O \rightarrow$ $CO_2 + 2\ NH_4^+ + 2\ OH^-$	Al, Ga, Th, Bi, Fe, Sn
PO_4^{3-}	Trimethyl phosphate	$(CH_3O)_3PO + 3\ H_2O \rightarrow$ $3CH_3OH + H_3PO_4$	Zr, Hf
$C_2O_4^{2-}$	Ethyl oxalate	$(C_2H_5)_2C_2O_4 + 2\ H_2O \rightarrow$ $2\ C_2H_5OH + H_2C_2O_4$	Mg, Zn, Ca
SO_4^{2-}	Dimethyl sulfate	$(CH_3O)_2SO_2 + 4\ H_2O \rightarrow$ $2\ CH_3OH + SO_4^{2-} + 2\ H_3O^+$	Ba, Ca, Sr, Pb
CO_3^{2-}	Trichloroacetic acid	$Cl_3CCOOH + 2\ OH^- \rightarrow$ $CHCl_3 + CO_3^{2-} + H_2O$	La, Ba, Ra
H_2S	Thioacetamide*	$CH_3CSNH_2 + H_2O \rightarrow$ $CH_3CONH_2 + H_2S$	Sb, Mo, Cu, Cd
DMG†	Biacetyl + hydroxylamine	$CH_3COCOCH_3 +$ $2\ H_2NOH \rightarrow DMG + 2\ H_2O$	Ni
HOQ¶	8-Acetoxyquinoline§	$CH_3COOQ + H_2O \rightarrow$ $CH_3COOH + HOQ$	Al, U, Mg, Zn

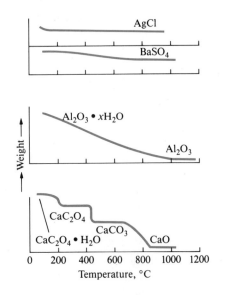

Figure 5-6

Effect of temperature on precipitate weight.

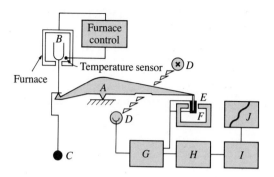

Figure 5-7
Schematic of a thermobalance: *A*, beam; *B*, sample cup and holder; *C*, counterweight; *D*, lamp and photodiodes; *E*, coil; *F*, magnet; *G*, control amplifier; *H*, tare calculator; *I*, amplifier; and *J*, recorder. (Courtesy of Mettler Instrument Corp., Hightstown, NJ.)

5C Calculation of Results from Gravimetric Data

The results of a gravimetric analysis are generally computed from two experimental measurements: the weight of the sample and the weight of a product of known composition. The examples that follow illustrate how such computations are carried out.

Example 5-1

The calcium in a 200.0-mL sample of a natural water was determined by precipitation as CaC_2O_4. The precipitate was filtered, washed, and ignited in a crucible having an empty weight of 26.6002 g. The weight of the crucible plus CaO (fw = 56.08 g) was 26.7134 g. Calculate the grams of Ca (fw = 40.08 g) per 100 mL of the water.

The weight of CaO is

$$\text{wt CaO} = 26.7134 \text{ g} - 26.6002 \text{ g} = 0.1132 \text{ g}$$

The number of moles of Ca in the sample is equal to the number of moles of CaO, or

Note that we carry a guard digit for the amount of CaO and then round at the end.

$$\text{amount Ca} = 0.1132 \text{ g CaO} \times \frac{1 \text{ mol CaO}}{56.08 \text{ g CaO}} \times \frac{1 \text{ mol Ca}}{\text{mol CaO}}$$

$$= 2.0185 \times 10^{-3} \text{ mol Ca}$$

$$\text{concn Ca} = \frac{2.0185 \times 10^{-3} \text{ mol Ca} \times 40.08 \text{ g Ca/mol Ca}}{200.0 \text{ mL}} \times \frac{100 \text{ mL}}{100 \text{ mL}}$$

$$= 0.04045 \text{ g Ca/100 mL}$$

Example 5-2

An iron ore was analyzed by dissolving a 1.1324-g sample in concentrated HCl. The resulting solution was diluted with water, and the iron(III) was precipitated as the hydrous oxide $Fe_2O_3 \cdot xH_2O$ by the addition of NH_3. After filtration and washing, the residue was ignited at a high temperature to give 0.5394 g of pure Fe_2O_3 (fw = 159.69 g). Calculate (a) the % Fe (fw = 55.847 g) and (b) the % Fe_3O_4 (fw = 231.54 g) in the sample.

For both parts of this problem, we will need to calculate the number of moles of Fe_2O_3:

$$\text{amount } Fe_2O_3 = 0.5394 \text{ g } Fe_2O_3 \times \frac{1 \text{ mol } Fe_2O_3}{159.69 \text{ g } Fe_2O_3}$$

$$= 3.3778 \times 10^{-3} \text{ mol } Fe_2O_3$$

(a) The number of moles of Fe is twice the number of moles of Fe_2O_3, and

$$\text{wt Fe} = 3.3778 \times 10^{-3} \text{ mol } Fe_2O_3 \times \frac{2 \text{ mol Fe}}{\text{mol } Fe_2O_3} \times 55.847 \frac{\text{g Fe}}{\text{mol Fe}}$$

$$= 0.37728 \text{ g Fe}$$

$$\text{% Fe} = \frac{0.37728 \text{ g Fe}}{1.1324 \text{ g sample}} \times 100\% = 33.317\% = 33.32\%$$

(b) As shown by the following equation, 3 mol of Fe_2O_3 is chemically equivalent to 2 mol of Fe_3O_4:

$$3 Fe_2O_3 \rightarrow 2 Fe_3O_4 + \tfrac{1}{2}O_2$$

$$\text{wt } Fe_3O_4 = 3.3778 \times 10^{-3} \text{ mol } Fe_2O_3 \times \frac{2 \text{ mol } Fe_3O_4}{3 \text{ mol } Fe_2O_3} \times \frac{231.54 \text{ g } Fe_3O_4}{\text{mol } Fe_3O_4}$$

$$= 0.52140 \text{ g } Fe_3O_4$$

$$\text{% } Fe_3O_4 = \frac{0.5214 \text{ g } Fe_3O_4}{1.1324 \text{ g sample}} \times 100\% = 46.043\% = 46.04\%$$

Example 5-3

A 0.2356-g sample containing only NaCl (fw = 58.44 g) and $BaCl_2$ (fw = 208.25 g) yielded 0.4637 g of dried AgCl (fw = 143.32 g). Calculate the percent of each halogen compound in the sample.

If we let x be the weight of NaCl in grams and y the weight of $BaCl_2$ in grams, we can write as a first equation

$$x + y = 0.2356 \text{ g sample} \tag{1}$$

To obtain the weight of AgCl from the NaCl, we write an expression for the amount of AgCl formed from the NaCl:

$$\text{amount AgCl from NaCl} = x \text{ g NaCl} \times \frac{1 \text{ mol NaCl}}{58.44 \text{ g NaCl}} \times \frac{1 \text{ mol AgCl}}{\text{mol NaCl}}$$

$$= 0.017112x \text{ mol AgCl}$$

The weight of AgCl from this source is

$$\text{wt AgCl from NaCl} = 0.017112x \ \cancel{\text{mol AgCl}} \times 143.32 \ \frac{\text{g AgCl}}{\cancel{\text{mol AgCl}}}$$

$$= 2.4524x \text{ g AgCl}$$

Note again that we carried five figures throughout the calculation and then rounded it to four at the end.

Proceeding in the same way, we can write that the amount of AgCl from the $BaCl_2$ is

$$\text{amount AgCl from } BaCl_2 = y \ \cancel{\text{g } BaCl_2} \times \frac{1 \ \cancel{\text{mol } BaCl_2}}{208.25 \ \cancel{\text{g } BaCl_2}} \times \frac{2 \text{ mol AgCl}}{\cancel{\text{mol } BaCl_2}}$$

$$= 9.6038 \times 10^{-3} \ y \text{ mol AgCl}$$

$$\text{wt AgCl from } BaCl_2 = 9.6038 \times 10^{-3} \ y \ \cancel{\text{mol AgCl}} \times 143.32 \ \frac{\text{g AgCl}}{\cancel{\text{mol AgCl}}}$$

$$= 1.3764 \ y \text{ g AgCl}$$

Because 0.4637 g of AgCl comes from the two compounds, we can write

$$2.4524x + 1.3764y = 0.4637 \qquad (2)$$

Equation 1 can be rewritten as

$$y = 0.2356 - x$$

Substituting into Equation 2 gives

$$2.4524x + 1.3764(0.2356 - x) = 0.4637$$

which rearranges to

$$1.0760x = 0.13942$$

$$x = \text{wt NaCl} = 0.12957 \text{ g NaCl}$$

$$\% \text{ NaCl} = \frac{0.12957 \text{ g NaCl}}{0.2356 \text{ g sample}} \times 100\% = 54.996\% = 55.00\%$$

$$\% \ BaCl_2 = 100.00\% - 55.00\% = 45.00\%$$

5D   APPLICATIONS OF GRAVIMETRIC METHODS

Gravimetric methods have been developed for most inorganic anions and cations, as well as for such neutral species as water, sulfur dioxide, carbon dioxide, and iodine. A variety of organic substances can also be readily determined gravimetrically. Examples include lactose in milk products, salicylates in drug preparations, phenolphthalein in laxatives, nicotine in pesticides, cholesterol in cereals, and benzaldehyde in almond extracts. Indeed, gravimetric methods are among the most widely applicable of all analytical procedures.

Gravimetric methods do not require a calibration or standardization step (as do all other analytical procedures except coulometry) because the results are calculated directly from the experimental data and atomic weights. Thus, when only one or two samples are to be analyzed, a gravimetric approach may be the method of choice because it requires less time and effort than a procedure that requires preparation of standards and calibration.

Computer Application 5-1
EUREKA: THE SOLVER

One by-product of the age of personal computers is the wide variety of programs available to students and practicing scientists alike. In addition to computer languages (such as BASIC), word processors, grammar checkers, spreadsheets, desktop organizers, database management systems, and other productivity tools have appeared in large numbers.

One tool of special interest to chemists is the equation solver, a program that quickly solves complex algebraic equations. Most equation-solver programs are capable of solving many types of equations, from simple linear equations in one variable to differential equations. A word of caution is in order here. Although computers are fast, accurate, and precise, they are generally pretty dumb. They do exactly what you tell them to do—no more and no less. They have sets of rules and statements that you must learn how to manipulate in order to get them to do the things you want to do.

Several equation solvers are available. Among these are TK Solver Plus®, MathCad®, and Eureka: The Solver®. All of these programs have been rated highly, and so the choice of which to use depends upon the tasks at hand and the resources available.[8] Because of its wide availability, low cost to educational institutions, and ease of use, we have chosen to use Eureka for several computer applications. Eureka has the added advantage that it is available for both IBM PC and Macintosh computers.

You can invoke Eureka from the keyboard of your computer by typing EUREKA [Return]. The following screen then appears.

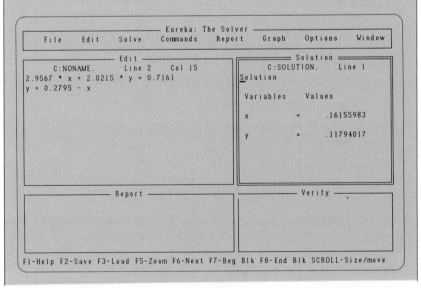

---

[8]F. T. Chau and Andy S. W. Chik, *J. Chem. Educ.* **1989,** *62*(2), A61; K. R. Foster, *Science* **1988,** *240,* 1353–1358.

Eureka is what is called a *menu-driven* program. The window at the top of the screen contains a menu of eight words representing the eight selectable options. You can select one of the options by depressing the first letter of the option or by using the arrow keys, ⬅ or ➡, to move the cursor to the chosen option and pressing **Return**. For our first example, we shall solve the two simultaneous equations of Example 5-3. By depressing **E**, we move the cursor to the edit window. We then type the equations to be solved into the edit window. As you can see, we have typed the equations exactly as they appear in Example 5-3. Note that in Eureka arithmetic operations—that is, +, −, *, /, ^, and ( )—have the same meaning that they have in BASIC. The relational operators <, >, and = are also the same. We will discuss other functions as they become necessary for other computer applications.

Once the equations have been entered, we press **Esc** to move the cursor back to the main menu at the top of the screen. We then type **S** to instruct Eureka to solve the two equations. When Eureka has finished, the solutions for x and y appear as shown in the solution window.

If you have access to Eureka: The Solver, run the program, enter the equations shown in the example, and verify their solution. You can also use Eureka to solve the equations you will set up in Problems 5-27, 5-28, and 5-31.

5D-1 Inorganic Precipitating Agents

Table 5-2 lists common inorganic precipitating agents. These reagents typically form slightly soluble salts or hydrous oxides with the analyte. As you can see from the many entries for each reagent, most inorganic precipitants are not very selective.

5D-2 Reducing Agents

Table 5-3 lists several reagents that convert an analyte to its elemental form for weighing.

5D-3 Organic Precipitating Agents

Numerous organic reagents have been developed for the gravimetric determination of inorganic species. Some of these reagents are significantly more selective in their reactions than are most of the inorganic reagents listed in Table 5-2.

We encounter two types of organic reagents. One forms slightly soluble, nonionic products called *coordination compounds;* the other forms products in which the bonding between the inorganic species and the reagent is largely ionic.

Table 5-2

SOME INORGANIC PRECIPITATING AGENTS†

Precipitating Agent	Element Precipitated‡
$NH_3(aq)$	**Be** (BeO), **Al** (Al_2O_3), **Sc** (Sc_2O_3), Cr (Cr_2O_3),* **Fe** (Fe_2O_3), Ga (Ga_2O_3), Zr (ZrO_2), **In** (In_2O_3), Sn (SnO_2), U (U_3O_8)
H_2S	Cu (CuO),* **Zn** (ZnO, or $ZnSO_4$), **Ge** (GeO_2), As (As_2O_3, or As_2O_5), Mo (MoO_3), Sn (SnO_2),* Sb ($\underline{Sb_2O_3}$, or Sb_2O_5), Bi (Bi_2S_3)
$(NH_4)_2S$	Hg ($\underline{HgS}$), Co (Co_3O_4)
$(NH_4)_2HPO_4$	**Mg** ($Mg_2P_2O_7$), Al ($AlPO_4$), Mn ($Mn_2P_2O_7$), Zn ($Zn_2P_2O_7$), Zr ($Zr_2P_2O_7$), Cd ($Cd_2P_2O_7$), Bi ($BiPO_4$)
H_2SO_4	Li, Mn, **Sr, Cd, Pb, Ba** (all as sulfates)
H_2PtCl_6	K (K_2PtCl_6, or Pt), Rb (Rb_2PtCl_6), Cs ($\underline{Cs_2PtCl_6}$)
$H_2C_2O_4$	Ca (CaO), Sr (SrO), **Th** (ThO_2)
$(NH_4)_2MoO_4$	Cd ($CdMoO_4$),* Pb ($\underline{PbMoO_4}$)
HCl	**Ag** (AgCl), Hg ($Hg_2\underline{Cl_2}$), Na (as NaCl from butyl alcohol), Si (SiO_2)
$AgNO_3$	**Cl** (AgCl), Br ($\underline{AgBr}$), $\underline{I(AgI)}$
$(NH_4)_2CO_3$	**Bi** (Bi_2O_3)
NH_4SCN	Cu [$Cu_2(SCN)_2$]
$NaHCO_3$	Ru, Os, Ir (precipitated as hydrous oxides; reduced with H_2 to metallic state)
HNO_3	Sn (SnO_2)
H_5IO_6	Hg [$Hg_5(IO_6)_2$]
NaCl, $Pb(NO_3)_2$	F (PbClF)
$BaCl_2$	$\mathbf{SO_4^{2-}}$ ($BaSO_4$)
$MgCl_2$, NH_4Cl	$\mathbf{PO_4^{3-}}$ ($Mg_2P_2O_7$)

†From W. F. Hillebrand, G. E. F. Lundell, H. A. Bright, and J. I. Hoffman, *Applied Inorganic Analysis.* New York: Wiley, 1953. By permission of John Wiley & Sons, Inc.

‡Boldface type indicates that gravimetric analysis is the preferred method for the element or ion. The weighed form is indicated in parentheses. An asterisk indicates that the gravimetric method is seldom used. An underscore indicates the most reliable gravimetric method.

Table 5-3

SOME REDUCING AGENTS EMPLOYED IN GRAVIMETRIC METHODS

Reducing Agent	Analyte
SO_2	Se, Au
$SO_2 + H_2NOH$	Te
H_2NOH	Se
$H_2C_2O_4$	Au
H_2	Re, Ir
HCOOH	Pt
$NaNO_2$	Au
$SnCl_2$	Hg
Electrolytic reduction	Co, Ni, Cu, Zn, Ag, In, Sn, Sb, Cd, Re, Bi

Chelates are cyclical metal-organic compounds in which the metal is a part of one or more five- or six-membered rings. The chelate pictured below is heme, which is a part of hemoglobin, the oxygen-carrying molecule in human blood.

Notice the four six-membered rings that are formed with Fe^{2+}.

Organic reagents that yield sparingly soluble coordination compounds typically contain at least two functional groups. Each of these groups is capable of bonding with a cation by sharing a pair of electrons. The functional groups are located in the molecule such that a five- or six-membered ring results from the reaction. Reagents that form compounds of this type are called *chelating agents,* and their products are called *chelates.*

Metal chelates are relatively nonpolar and, as a consequence, have solubilities that are low in water but high in organic liquids. Usually these compounds possess low densities and are often intensely colored. Because they are not wetted by water, coordination compounds are readily freed of moisture at low temperatures. Two widely used chelating reagents are described in the paragraphs that follow.

8-Hydroxyquinoline

Approximately two dozen cations form sparingly soluble chelates with 8-hydroxyquinoline. The structure of magnesium 8-hydroxyquinolate is typical of these chelates:

8-Hydroxyquinoline is sometimes called oxine.

The solubilities of metal 8-hydroxyquinolates vary widely from cation to cation and are pH-dependent because 8-hydroxyquinoline is always deprotonated during chelation. Therefore, we can achieve a considerable degree of selectivity in the use of 8-hydroxyquinoline by controlling pH.

Dimethylglyoxime

Dimethylglyoxime is an organic precipitating agent of unparalleled specificity. Only nickel(II) is precipitated from a weakly alkaline solution. The reaction is

Nickel dimethylglyoxime is spectacular in appearance. It has a beautiful and vivid red color.

This precipitate is so bulky that only small amounts of nickel can be handled conveniently. It also has an exasperating tendency to creep up the sides of the container as it is filtered and washed. The solid is readily dried at 110°C and has the composition indicated by its formula.

Creeping is the process by which a precipitate (usually a metal-organic chelate) moves up the sides of a wetted surface of a glass container or a filter paper.

Sodium Tetraphenylboron

Sodium tetraphenylboron, $(C_6H_5)_4B^-Na^+$, is an important example of an organic precipitating reagent that forms saltlike precipitates. In cold mineral-acid solutions, it is a near-specific precipitating agent for potassium and ammonium ions. The composition of the precipitates is stoichiometric and contains one mole of potassium or ammonium ion for each mole of tetraphenylboron ion; these ionic compounds are readily filtered and can be brought to constant weight at 105 to 120°C. Only mercury(II), rubidium, and cesium interfere and must be removed by prior treatment.

Potassium tetraphenyl boron is a salt rather than a chelate.

5D-4 Organic Functional-Group Analysis

Several reagents react selectively with certain organic functional groups and thus can be used for the determination of most compounds containing these groups. A list of gravimetric functional-group reagents is given in Table 5-4. Many of the reactions shown can also be used for volumetric and spectrophotometric determinations.

5D-5 Volatilization Methods

The two most common gravimetric methods based on volatilization are those for water and carbon dioxide.

Water is quantitatively eliminated from many inorganic samples by ignition. In its direct determination, water is collected on any of several solid desiccants, and its mass is then determined from the weight gain of the desiccant.

The indirect method, in which the amount of water is determined by the loss of weight of the sample during heating, is less satisfactory because it must be assumed that water is the only component volatilized. This as-

Table 5-4
GRAVIMETRIC METHODS FOR ORGANIC FUNCTIONAL GROUPS

Functional Group	Basis for Method	Reaction and Product Weighed*
Carbonyl	Weight of precipitate with 2,4-dinitrophenyl-hydrazine	$RCHO + H_2NNHC_6H_3(NO_2)_2 \rightarrow$ $\underline{R{-}CH = NNHC_6H_3(NO_2)_2}(s) + H_2O$ (RCOR′ reacts similarly)
Aromatic carbonyl	Weight of CO_2 formed at 230°C in quinoline; CO_2 distilled, absorbed, and weighed	$ArCHO \xrightarrow[CuCO_3]{230°C} Ar + \underline{CO_2}(g)$
Methoxyl and ethoxyl	Weight of AgI formed after distillation and decomposition of CH_3I or C_2H_5I	$ROCH_3 \ \ + HI \rightarrow ROH \ \ \ + CH_3I$ $RCOOCH_3 + HI \rightarrow RCOOH + CH_3I$ $\left.\right\}$ $CH_3I + Ag^+ + H_2O \rightarrow$ $ROC_2H_5 \ \ + HI \rightarrow ROH \ \ \ + C_2H_5I$ $\quad\underline{AgI}(s) + CH_3OH$
Aromatic nitro	Weight loss of Sn	$RNO_2 + \frac{3}{2}\underline{Sn}(s) + 6\,H^+ \rightarrow RNH_2 + \frac{3}{2}Sn^{4+} + 2\,H_2O$
Azo	Weight loss of Cu	$RN = NR' + 2\,\underline{Cu}(s) + 4\,H^+ \rightarrow RNH_2 + R'NH_2 + 2\,Cu^{2+}$
Phosphate	Weight of Ba salt	$\overset{\text{O}}{\overset{\|}{ROP(OH)_2}} + Ba^{2+} \rightarrow \overset{\text{O}}{\overset{\|}{\underline{ROPO_2Ba}}}(s) + 2\,H^+$
Sulfamic acid	Weight of $BaSO_4$ after oxidation with HNO_2	$RNHSO_3H + HNO_2 + Ba^{2+} \rightarrow ROH + \underline{BaSO_4}(s) + N_2 + 2\,H^+$
Sulfinic acid	Weight of Fe_2O_3 after ignition of Fe^{3+} sulfinate	$3\,ROSOH + Fe^{3+} \rightarrow (ROSO)_3Fe(s) + 3\,H^+$ $(ROSO)_3Fe \xrightarrow{O_2} CO_2 + H_2O + SO_2 + \underline{Fe_2O_3}(s)$

*The substance weighed is underlined.

sumption is frequently unjustified because heating of many substances results in their decomposition and a consequent change in weight, irrespective of the presence of water. Nevertheless, the indirect method has found wide use for the determination of water in items of commerce. For example, a semiautomated instrument for the determination of moisture in cereal grains can be purchased. It consists of a platform balance upon which a 10-g sample is heated with an infrared lamp. The percent residue is read directly.

Carbonates are ordinarily decomposed by acids to give carbon dioxide, which is readily evolved from solution by heat. As in the direct analysis for water, the weight of carbon dioxide is established from the increase in the weight of a solid absorbent. Ascarite II,[9] which consists of sodium hydroxide on a nonfibrous silicate, retains carbon dioxide by the reaction

$$2 \text{ NaOH} + CO_2 \rightarrow Na_2CO_3 + H_2O$$

The absorption tube must also contain a desiccant to prevent loss of the evolved water.

Sulfides and sulfites can also be determined by volatilization. Hydrogen sulfide or sulfur dioxide evolved from the sample after treatment with acid is collected in a suitable absorbent.

Finally, the classical method for the determination of carbon and hydrogen in organic compounds is a gravimetric procedure in which the combustion products (H_2O and CO_2) are collected selectively on weighed absorbents. The increase in weight serves as the analytical parameter.

Automatic instruments for the routine determination of water in various products of agriculture and commerce are marketed by several instrument manufacturers.

5E QUESTIONS AND PROBLEMS

5-1. Explain the difference between
    *(a) a colloidal and a crystalline precipitate.
    (b) specific and selective reagents.
    *(c) precipitation and coprecipitation.
    (d) peptization and coagulation.
    *(e) occlusion and mixed-crystal formation.
    (f) nucleation and particle growth.
5-2. Define
    *(a) digestion.
    (b) adsorption.
    *(c) reprecipitation.
    (d) precipitation from homogeneous solution.
    *(e) counter-ion layer.
    (f) mother liquor.
    *(g) relative supersaturation.
*5-3. What are the structural characteristics of a chelating agent?
5-4. How can the relative supersaturation during precipitate formation be varied?
*5-5. An aqueous solution contains $NaNO_3$ and KSCN. The thiocyanate ion is precipitated as AgSCN by ad-

dition of $AgNO_3$. After an excess of the precipitating reagent has been added,
    (a) what is the charge on the surface of the coagulated colloidal particles?
    (b) what is the source of the charge?
    (c) what ions make up the counter-ion layer?
5-6. Suggest a method by which Cu^{2+} can be precipitated homogeneously as CuS.
*5-7. What is peptization and how is it avoided?
5-8. Suggest a precipitation method for separation of K^+ from Na^+ and Li^+.
5-9. Write an equation showing how the weight of the substance on the left can be obtained from the weight of the substance on the right.

Sought	Weighed
*(a) SO_3	$BaSO_4$
(b) Zn	$Zn_2P_2O_7$
*(c) In	In_2O_3
(d) K	K_2PtCl_6

---

[9]®Thomas Scientific, Swedesboro, NJ.

Sought	Weighed
*(e) CuO	$Cu_2(SCN)_2$
(f) Mn_2O_3	Mn_3O_4
*(g) Pb_3O_4	PbO_2
(h) $U_2P_2O_{11}$	P_2O_5
*(i) $Na_2B_4O_7 \cdot 10\ H_2O$	B_2O_3
(j) Na_2O	$NaZn(UO_2)_3(C_2H_3O_2)_9 \cdot 6\ H_2O$

*5-10. Treatment of a 0.4000-g sample of impure potassium chloride with an excess of $AgNO_3$ resulted in the formation of 0.7332 g of AgCl. Calculate the percentage of KCl in the sample.

5-11. The aluminum in a 1.200-g sample of impure ammonium aluminum sulfate was precipitated with aqueous ammonia as the hydrous $Al_2O_3 \cdot xH_2O$. The precipitate was filtered and ignited at 1000°C to give anhydrous Al_2O_3, which weighed 0.1798 g. Express the result of this analysis in terms of
(a) % $NH_4Al(SO_4)_2$.   (b) % Al_2O_3.   (c) % Al.

*5-12. What weight of $Cu(IO_3)_2$ can be formed from 0.400 g of $CuSO_4 \cdot 5H_2O$?

5-13. What weight of KIO_3 is needed to convert the copper in 0.4000 g of $CuSO_4 \cdot 5H_2O$ to $Cu(IO_3)_2$?

*5-14. What weight of AgI can be produced from a 0.240-g sample that assays 30.6% MgI_2?

5-15. Precipitates used in the gravimetric determination of uranium include $Na_2U_2O_7$ (fw = 634.0 g), $(UO_2)_2P_2O_7$ (fw = 714.0 g), and $V_2O_5 \cdot 2UO_3$ (fw = 753.9 g). Which of these weighing forms provides the greatest weight of precipitate from a given quantity of uranium?

*5-16. A 0.7406-g sample of impure magnesite, $MgCO_3$, was decomposed with HCl; the liberated CO_2 was collected on calcium oxide and found to weigh 0.1881 g. Calculate the percentage of magnesium in the sample.

5-17. The hydrogen sulfide in a 50.0-g sample of crude petroleum was removed by distillation and collected in a solution of $CdCl_2$. The precipitated CdS was then filtered, washed, and ignited to $CdSO_4$. Calculate the percentage of H_2S in the sample if 0.108 g of $CdSO_4$ was recovered.

*5-18. A 0.1799-g sample of an organic compound was burned in a stream of oxygen, and the CO_2 produced was collected in a solution of barium hydroxide. Calculate the percentage of carbon in the sample if 0.5613 g of $BaCO_3$ was formed.

5-19. A 5.000-g sample of a pesticide was decomposed with metallic sodium in alcohol, and the liberated chloride ion was precipitated as AgCl. Express the results of this analysis in terms of the percentage of DDT ($C_{14}H_9Cl_5$) based upon the recovery of 0.1606 g of AgCl.

*5-20. The mercury in a 0.7152-g sample was precipitated with an excess of paraperiodic acid, H_5IO_6:

$$5\ Hg^{2+} + 2\ H_5IO_6 \rightarrow Hg_5(IO_6)_2(s) + 10\ H^+$$

The precipitate was filtered, washed free of precipitating agent, dried, and found to weigh 0.3408 g. Calculate the percentage of Hg_2Cl_2 in the sample.

5-21. The iodide in a sample that also contained chloride was converted to iodate by treatment with an excess of bromine:

$$3\ H_2O + 3\ Br_2 + I^- \rightarrow 6\ Br^- + IO_3^- + 6\ H^+$$

The unused bromine was removed by boiling; an excess of barium ion was then added to precipitate the iodate:

$$Ba^{2+} + 2\ IO_3^- \rightarrow Ba(IO_3)_2(s)$$

In the analysis of a 2.72-g sample, 0.0720 g of barium iodate was recovered. Express the results of this analysis as percent potassium iodide.

*5-22. Ammoniacal nitrogen can be determined by treating the sample with chloroplatinic acid; the product is slightly soluble ammonium chloroplatinate:

$$H_2PtCl_6 + 2\ NH_4^+ \rightarrow (NH_4)_2PtCl_6(s) + 2\ H^+$$

The precipitate decomposes upon ignition, yielding metallic platinum and gaseous products:

$$(NH_4)_2PtCl_6 \rightarrow Pt(s) + 2\ Cl_2(g) + 2\ NH_3(g) + 2\ HCl(g)$$

Calculate the percentage of ammonia in a 0.2213-g sample if 0.5881 g of platinum was produced.

5-23. A 0.6447-g portion of manganese dioxide was added to an acidic solution in which 1.1402 g of a chloride-containing sample was dissolved. Evolution of chlorine took place as a consequence of the following reaction:

$$MnO_2(s) + 2\ Cl^- + 4\ H^+ \rightarrow Mn^{2+} + Cl_2(g) + 2\ H_2O$$

After the reaction was complete, the excess MnO_2 was collected by filtration, washed, and found to weigh 0.3521 g. Express the results of this analysis in terms of percentage of aluminum chloride.

*5-24. A series of sulfate samples are to be analyzed by precipitation as $BaSO_4$. If it is known that the sulfate content in these samples ranges between 20 and 55%, what minimum sample weight should be taken to ensure that a precipitate weight no smaller than 0.300 g is produced? What is the maximum precipitate weight to be expected if this quantity of sample is taken?

5-25. The addition of dimethylglyoxime, $H_2C_4H_6O_2N_2$, to a solution containing nickel(II) ion gives rise to a precipitate:

$$Ni^{2+} + 2\ H_2C_4H_6O_2N_2 \rightarrow 2\ H^+ + Ni(HC_4H_6O_2N_2)_2(s)$$

Nickel dimethylglyoxime is a bulky precipitate that is inconvenient to manipulate in amounts greater than 175 mg. The amount of nickel in a type of permanent-magnet alloy ranges between 24 and 35%. Calculate the sample size that should not be exceeded when analyzing this alloy for nickel.

*5-26. The success of a particular catalyst is highly dependent upon its zirconium content. The starting material for this preparation is received in batches that assay between 68 and 84% $ZrCl_4$. Because it has been established that there are no sources of chloride ion other than the $ZrCl_4$ in the sample, routine analysis based upon precipitation of AgCl is feasible.

(a) What sample weight should be taken to ensure a AgCl precipitate that weighs at least 0.400 g?

(b) If this sample weight is used, what is the maximum weight of AgCl that can be expected in this analysis?

(c) To simplify calculations, what sample weight should be taken in order to have the percentage of $ZrCl_4$ exceed the weight of AgCl produced by a factor of 100?

5-27. A 0.8720-g sample of a mixture consisting solely of sodium bromide and potassium bromide yields 1.505 g of silver bromide. What are the percentages of the two salts in the sample?

*5-28. A 0.6407-g sample containing chloride and iodide ions gave a silver halide precipitate weighing 0.4430 g. This precipitate was then strongly heated in a stream of Cl_2 gas to convert the AgI to AgCl; upon completion of this treatment, the precipitate weighed 0.3181 g. Calculate the percentage of chloride and iodide in the sample.

5-29. The phosphorus in a 0.2374-g sample was precipitated as the slightly soluble $(NH_4)_3PO_4 \cdot 12MoO_3$. This precipitate was filtered, washed, and then redissolved in acid. Treatment of the resulting solution with an excess of Pb^{2+} resulted in the formation of 0.2752 g of $PbMoO_4$. Express the results of this analysis in terms of percentage of P_2O_5.

*5-30. What weight of CO_2 is evolved from a 1.204-g sample that is 36.0% $MgCO_3$ and 44.0% K_2CO_3 by weight?

5-31. A 6.881-g sample containing magnesium chloride and sodium chloride was dissolved in sufficient water to give 500 mL of solution. Analysis for the chloride content of a 50.0-mL aliquot resulted in the formation of 0.5923 g of AgCl. The magnesium in a second 50.0-mL aliquot was precipitated as $MgNH_4PO_4$; upon ignition, 0.1796 g of $Mg_2P_2O_7$ was found. Calculate the percentage of $MgCl_2 \cdot 6H_2O$ and of NaCl in the sample.

*5-32. A 50.0-mL portion of a solution containing 0.200 g of $BaCl_2 \cdot 2H_2O$ is mixed with 50.0 mL of a solution containing 0.300 g of $NaIO_3$. Assume that the solubility of $Ba(IO_3)_2$ in water is negligibly small and calculate

(a) the weight of the precipitated $Ba(IO_3)_2$.

(b) the weight of the unreacted component that remains in solution.

5-33. When a 100.0-mL portion of a solution containing 0.500 g of $AgNO_3$ is mixed with 100.0 mL of a solution containing 0.300 g of K_2CrO_4, a bright red precipitate of Ag_2CrO_4 forms.

(a) Assuming the solubility of Ag_2CrO_4 is negligible, calculate the weight of the precipitate.

(b) Calculate the weight of the unreacted component that remains in solution.

Chapter 6

Volumetric Methods of Analysis

Volumetric methods constitute a large and powerful group of quantitative procedures that find widespread use in analytical chemistry. Volumetric methods are one of three types of *titrimetry* in which analyses are based upon measuring the amount of a reagent of known concentration that is consumed by the analyte. The other types of titrimetric methods are *weight* (or *gravimetric*) *titrimetry* and *coulometric titrimetry*. In volumetric titrimetry, the volume of a solution of known concentration that is needed to react essentially completely with the analyte is determined. Gravimetric titrimetry differs only in that the weight of the reagent is measured instead of its volume. In coulometric titrimetry, the "reagent" is a constant direct electrical current of known magnitude that reacts with the analyte; here, the time required to complete the electrochemical reaction is measured.

This chapter provides an introduction to volumetric methods. Chapters 10 through 17 deal with the theory and applications of various types of volumetric titrimetry, although much of the material also applies to weight titrimetry. Coulometric titrimetry is considered in Section 19C-5. All these titrimetric methods are used for routine analyses because they are generally rapid, convenient, accurate, and readily automated.

The three types of quantitative titrimetry are: volumetric, gravimetric, and coulometric. Volumetric is by far the most widely used.

Titrimetric methods are analytical procedures in which the amount of analyte is determined from the amount of a standard reagent required to react with the analyte completely.

92

6A SOME GENERAL ASPECTS OF VOLUMETRIC TITRIMETRY[1]

6A-1 Definition of Some Terms

A *standard solution* (or *standard titrant*) is a reagent of known concentration that is used to carry out a volumetric analysis. A *titration* is performed by slowly adding a standard solution from a buret or other volumetric measuring device to a solution of the analyte until the reaction between the two is complete. The volume needed to complete the titration is determined from the difference between the initial and final buret readings.

The *equivalence point* in a titration is reached when the amount of added titrant is chemically equivalent to the amount of analyte in the sample. For example, the equivalence point in the titration of sodium chloride with silver nitrate occurs after exactly one mole of silver ion has been added for each mole of chloride ion in the sample. The equivalence point in the titration of sulfuric acid with sodium hydroxide is reached after introduction of two moles of base for each mole of acid.

It is sometimes necessary to add an excess of the standard titrant and then determine the excess amount by *back-titration* with a second standard titrant. Here, the equivalence point corresponds to the point where the amount of initial titrant is chemically equivalent to the amount of analyte plus the amount of back-titrant.

> The equivalence point is the point in a titration when the amount of added standard reagent exactly equals the amount of analyte.

> Back-titrations are often required when the rate of reaction between the analyte and reagent is slow or when the reagent lacks stability.

6A-2 Equivalence Points and End Points

The equivalence point of a titration is a theoretical point that cannot be determined experimentally. Instead, we can only estimate it by observing some physical change associated with the condition of equivalence. This change is called the *end point* for the titration. Every effort is made to ensure that any volume difference between the equivalence point and the end point is small. Such differences do exist, however, as a result of inadequacies in the physical changes and in our ability to observe them. The difference in volume between the equivalence point and the end point is the *titration error*.

An *indicator* is often added to the analyte solution in order to give an observable physical change (the end point) at or near the equivalence point. We shall see that large changes in the relative concentration of analyte or titrant occur in the equivalence-point region. These concentration changes cause the indicator to change in appearance. Typical indicator changes are the appearance or disappearance of a color, a change in color, and the appearance or disappearance of turbidity.

We often use instruments to detect end points. These instruments respond to certain properties of the solution that change in a characteristic way during the titration. Among such instruments are voltmeters, ammeters, and ohmmeters; colorimeters; temperature recorders; and refractometers.

> The end point is the point in a titration when a physical change occurs that is associated with the condition of chemical equivalence.

> All volumetric methods are based upon a primary standard whose chemical composition and purity are known exactly.

In volumetric methods, the titration error E_t is given by

$$E_t = V_{ep} - V_{eq}$$

where V_{ep} is the actual volume used to arrive at the end point, and V_{eq} is the theoretical volume of reagent required to reach the equivalence point.

---

[1]For a detailed discussion of volumetric methods, see J. I. Watters, in *Treatise on Analytical Chemistry,* I. M. Kolthoff and P. J. Elving, Eds., Part I, Vol. 11, Chapter 114. New York: Wiley, 1975.

6A-3 Primary Standards

A *primary standard* is a highly purified compound that serves as a reference material in all volumetric titrimetric methods. The accuracy of such methods is critically dependent on the properties of this compound. Important requirements for a primary standard are:

1. High purity. Established methods for confirming purity should be available.
2. Stability in air.
3. Absence of hydrate water so that the composition does not change with variations in relative humidity.
4. Ready availability at modest cost.
5. Reasonable solubility in the titration medium.
6. Reasonably large formula weight so that the relative error associated with weighing is minimized.

The number of compounds that meet or even approach these criteria is so small that only a limited number of primary standard substances are available to the chemist. As a consequence, less pure compounds must sometimes be employed in lieu of a primary standard. The purity of such a *secondary standard* must be established by careful analysis.

6B   STANDARD SOLUTIONS

Standard solutions play a central role in all volumetric methods of analysis.

6B-1 Desirable Properties of Standard Solutions

The ideal standard solution for a volumetric method will

1. Be sufficiently stable so that it is necessary to determine its concentration only once.
2. React rapidly with the analyte so that the time required between additions of titrant is minimized.
3. React more or less completely with the analyte so that satisfactory end points are realized.
4. Undergo a selective reaction with the analyte that can be described by a simple balanced equation.

Relatively few reagents meet all these ideals perfectly.

6B-2 Methods for Establishing the Concentration of Standard Solutions

The accuracy of a volumetric method can be no better than the accuracy of the concentration of the standard solution used in the titration. Two basic methods are used to establish the concentration of standard solutions. In the *direct method*, a carefully weighed quantity of a primary standard is dissolved and diluted to an exactly known volume in a volumetric flask. In the second, the solution is *standardized* by titrating (1) a weighed quantity of a primary standard, (2) a weighed quantity of a secondary standard, or (3) a measured volume of another standard solution. A titrant that is standardized against a secondary standard or against another standard solution is sometimes referred to as a *secondary standard solution*. A secondary standard solution is less desirable than a primary standard solution because the concentration of the former is subject to greater uncertainty. The best standard solutions are those prepared by the direct method. This procedure cannot be used when the reagent does not possess the properties required for a primary standard.

> Standardization is a process in which the concentration of a solution is determined by using the solution to titrate a known amount of another reagent.

6B-3 Methods for Expressing the Concentration of Standard Solutions

The concentrations of standard solutions are generally expressed in units of either *molarity C* or *normality* C_N. The first gives the number of moles of reagent contained in 1 L of solution, and the second gives the number of equivalents of reagent in the same volume.

> The molarity of a solution is the number of moles of a reagent contained in one liter of the solution.

Many chemists believe that the terms "normality" and "equivalent" offer so few real advantages that they can be abandoned without serious loss.[2] We are sympathetic to this view and base volumetric calculations in this text on molarities. We also recognize, however, that the use of normalities and equivalents abound in the chemical and biochemical literature of the last century and are also still found in various monographs and compilations of analytical methods of considerable value to present-day chemists. For this reason, we have included a section in Appendix 6 that demonstrates how volumetric computations are performed with normalities and equivalents. You may find this appendix useful when consulting the analytical literature.

6C VOLUMETRIC CALCULATIONS

6C-1 Some Useful Algebraic Relationships

Most volumetric calculations are based on two pairs of simple equations that are derived from the definitions of millimole, mole, and molar concentration. For the chemical species A, we may write

$$\text{amount A} = \text{no. mmol A} = \frac{\text{wt A (g)}}{\text{mfw A (g/mmol)}} \tag{6-1}$$

$$\text{amount A} = \text{no. mol A} = \frac{\text{wt A (g)}}{\text{fw A (g/mol)}} \tag{6-2}$$

> The units for the amount of a chemical species are millimoles and moles.

---

[2]For example, *Analytical Chemistry*, a leading scientific journal of analytical chemistry, no longer permits the use of either "normality" or "titer" in the papers it accepts for publication.

The second pair are derived from the definition of molar concentration:

$$\text{amount A} = \text{no. mmol A} = V(\text{mL}) \times C_A(\text{mmol A/mL}) \qquad \text{(6-3)}$$
$$\text{amount A} = \text{no. mol A} = V(\text{L}) \times C_A(\text{mol A/L}) \qquad \text{(6-4)}$$

where V is the volume of the solution.

You should use Equations 6-1 and 6-3 when volumes are measured in milliliters and Equations 6-2 and 6-4 when volumes are measured in liters.

Feature 6-1
UNITS IN USING EQUATIONS 6-1 THROUGH 6-4

It is useful to know that any combination of grams, moles, and liters can be replaced with any analogous combination expressed in milligrams, millimoles, and milliliters. For example, a 0.1 M solution contains 0.1 mol of a species per liter or 0.1 mmol per milliliter. Similarly, the number of moles of a compound is equal to either the weight in grams of that compound divided by its formula weight in grams or the weight in milligrams divided by its milliformula weight in milligrams.

6C-2 Calculation of the Molarity of Standard Solutions

The following four examples illustrate how volumetric reagents are prepared.

Example 6-1

Describe the preparation of 5.000 L of 0.1000 M Na_2CO_3 (fw = 105.99 g) from the primary-standard solid.

Since the volume is in liters, we base our calculations on the mole rather than the millimole. Thus to obtain the amount of Na_2CO_3 needed, we write

$$\text{amount } Na_2CO_3 = V_{\text{soln}}(\text{L}) \times C_{Na_2CO_3}(\text{mol/L})$$
$$= 5.000 \ \text{L} \times \frac{0.1000 \ \text{mol } Na_2CO_3}{\text{L}} = 0.5000 \ \text{mol } Na_2CO_3$$

To obtain the weight of Na_2CO_3, we rearrange Equation 6-2 to give

$$\text{wt } Na_2CO_3 = 0.5000 \ \text{mol } Na_2CO_3 \times \frac{105.99 \ \text{g } Na_2CO_3}{\text{mol } Na_2CO_3} = 53.00 \ \text{g } Na_2CO_3$$

Therefore the solution is prepared by dissolving 53.00 g of Na_2CO_3 in water and diluting to exactly 5.000 L.

Example 6-2

A standard 0.0100 M solution of Na^+ is required to calibrate a flame-photometric method for determining the element. Describe how 500 mL of this solution can be prepared from primary-standard Na_2CO_3.

We wish to compute the weight of reagent required to give a species molarity of 0.0100. Here, we will use millimoles since the volume is in milliliters. Because Na_2CO_3 dissociates to give two Na^+ ions, we can write that the number of millimoles of Na_2CO_3 needed is

$$\text{amount } Na_2CO_3 = 500 \text{ mL} \times \frac{0.0100 \text{ mmol Na}}{\text{mL}} \times \frac{1 \text{ mmol } Na_2CO_3}{2 \text{ mmol Na}}$$

$$= 2.50 \text{ mmol } Na_2CO_3$$

From the definition of millimole, we write

$$\text{wt } Na_2CO_3 = 2.50 \text{ mmol } Na_2CO_3 \times 0.10599 \frac{g \ Na_2CO_3}{\text{mmol } Na_2CO_3}$$

$$= 0.265 \text{ g } Na_2CO_3$$

The solution is therefore prepared by dissolving 0.265 g of Na_2CO_3 in water and diluting to 500 mL.

Example 6-3

How would you prepare 50.0-mL portions of standard solutions that are 0.00500 M, 0.00200 M, and 0.00100 M in Na^+ from the solution in Example 6-2?

The amount of Na^+ taken from the concentrated solution must equal the amount in the diluted solutions. Thus,

$$\text{no. mol } Na^+ \text{ from concd soln} = \text{no. mmol } Na^+ \text{ in dil soln}$$

Recall that the number of millimoles is equal to the number of millimoles per milliliter times number of milliliters:

$$V_{concd} \times C_{concd} = V_{dil} \times C_{dil}$$

A useful algebraic relationship is $V_{concd} \times C_{concd} = V_{dil} \times C_{dil}$.

where V_{concd} and V_{dil} are the volumes in milliliters of the concentrated and diluted solutions, respectively, and C_{concd} and C_{dil} are their molar Na^+ concentrations. This equation rearranges to

$$V_{concd} = \frac{V_{dil} \times C_{dil}}{C_{concd}} = \frac{50.0 \text{ mL} \times 0.00500 \text{ mmol } Na^+/\text{mL}}{0.0100 \text{ mmol } Na^+/\text{mL}} = 25.0 \text{ mL}$$

Thus, to produce 50.0 mL of 0.00500 M Na^+, 25.0 mL of the concentrated solution should be diluted to exactly 50.0 mL.

Repeat the calculation for the other two molarities to confirm that diluting 10.0 and 5.00 mL of the concentrated solution to 50.0 mL produces the desired solutions.

Example 6-4

Describe the preparation of 100 mL of approximately 6.0 M HCl from the commercial concentrated reagent. The label on the bottle states that the reagent is 37% HCl and has a specific gravity of 1.18.

Generally, the percentages employed in describing commercial reagents are weight/weight; that is, the reagent in question contains about 37 g of HCl per 100 g of the concentrated reagent. Furthermore, we assume (Section 2B-2) that the reagent has a density of 1.18 g/mL. Therefore,

$$\frac{\text{wt HCl}}{\text{vol concd soln}} = \frac{1.18 \text{ g concd soln}}{\text{mL concd soln}} \times \frac{37 \text{ g HCl}}{100 \text{ g concd soln}}$$

$$= 0.437 \frac{\text{g HCl}}{\text{mL concd soln}}$$

To obtain the weight of HCl required, we calculate the number of millimoles of the acid that are required and multiply this figure by the formula weight:

$$\text{wt HCl required} = 100 \text{ mL dil soln} \times \frac{6.00 \text{ mmol HCl}}{\text{mL dil soln}} \times \frac{0.0365 \text{ g HCl}}{\text{mmol HCl}}$$

$$= 21.9 \text{ g HCl}$$

The volume of concentrated reagent required is

$$V = 21.9 \text{ g HCl} \times \frac{1 \text{ mL concd soln}}{0.437 \text{ g HCl}} = 50.2 \text{ mL concd soln}$$

The solution is prepared by diluting about 50 mL of the concentrated reagent to a volume of about 100 mL.

Remember: The percent on commercial bottles of acids and bases is a weight/weight percent.

Remember: At room temperature, specific gravity is approximately equal to density in grams per liter.

6C-3 Treatment of Titration Data

In this section, we describe two types of volumetric calculations. The first involves computing the molarity of solutions that have been standardized against either a primary standard or another standard solution. The second involves calculating the amount of analyte in a sample from titration data. Both types are based on three algebraic relationships. Two of these are Equations 6-1 and 6-3, both of which are based on millimoles and milliliters. The third relationship is the stoichiometric ratio of the number of millimoles of analyte and the number of millimoles of titrant.

Calculation of Molarities from Standardization Data
Examples 6-5 and 6-6 illustrate how standardization data are treated.

Example 6-5

Exactly 50.00 mL of an HCl solution required 29.71 mL of 0.01963 M $Ba(OH)_2$ to reach an end point with bromocresol green indicator. Calculate the molarity of the HCl.

In the titration, 1 mmol of $Ba(OH)_2$ reacts with 2 mmol of HCl, and thus the stoichiometric ratio is

$$\text{stoichiometric ratio} = \frac{2 \text{ mmol HCl}}{1 \text{ mmol Ba(OH)}_2}$$

The number of millimoles of the standard is obtained by substituting into Equation 6-3:

$$\text{amount Ba(OH)}_2 = 29.71 \; \cancel{\text{mL Ba(OH)}_2} \times 0.01963 \; \frac{\text{mmol Ba(OH)}_2}{\cancel{\text{mL Ba(OH)}_2}}$$

To obtain the number of millimoles of HCl, we multiply this result by the stoichiometric ratio:

$$\text{amount HCl} = (29.71 \times 0.01963) \; \cancel{\text{mmol Ba(OH)}_2} \times \frac{2 \text{ mmol HCl}}{1 \; \cancel{\text{mmol Ba(OH)}_2}}$$

To obtain the number of millimoles of HCl per milliliter, we divide by the volume of acid:

$$c_{HCl} = \frac{(29.71 \times 0.01963 \times 2) \text{ mmol HCl}}{50.00 \text{ mL HCl}}$$

$$= 0.023328 \; \frac{\text{mmol HCl}}{\text{mL HCl}} = 0.02333 \text{ M}$$

Example 6-6

Titration of 0.2121 g of pure $Na_2C_2O_4$ (fw = 134.00 g) required 43.31 mL of $KMnO_4$. What is the molarity of the $KMnO_4$ solution? The chemical reaction is

$$2 \text{ MnO}_4^- + 5 \text{ C}_2\text{O}_4^{2-} + 16 \text{ H}^+ \rightarrow 2 \text{ Mn}^{2+} + 10 \text{ CO}_2 + 8 \text{ H}_2\text{O}$$

From this equation, we see that the stoichiometric ratio is

$$\text{stoichiometric ratio} = \frac{2 \text{ mmol KMnO}_4}{5 \text{ mmol Na}_2\text{C}_2\text{O}_4}$$

The amount of primary-standard $Na_2C_2O_4$ is given by Equation 6-1:

$$\text{amount Na}_2\text{C}_2\text{O}_4 = 0.2121 \; \cancel{\text{g Na}_2\text{C}_2\text{O}_4} \times \frac{1 \text{ mmol Na}_2\text{C}_2\text{O}_4}{0.13400 \; \cancel{\text{g Na}_2\text{C}_2\text{O}_4}}$$

To obtain the number of millimoles of $KMnO_4$, we multiply this result by the stoichiometric factor:

$$\text{amount } KMnO_4 = \frac{0.2121}{0.13400} \text{ mmol Na}_2\text{C}_2\text{O}_4 \times \frac{2 \text{ mmol } KMnO_4}{5 \text{ mmol Na}_2\text{C}_2\text{O}_4}$$

The molarity is then obtained by dividing by the volume of $KMnO_4$ consumed:

$$C_{KMnO_4} = \frac{\left(\frac{0.2121}{0.13400} \times \frac{2}{5}\right) \text{ mmol } KMnO_4}{43.31 \text{ mL } KMnO_4} = 0.01462 \text{ M}$$

Note that units are carried through all calculations as a check on the correctness of the relationships used in Examples 6-5 and 6-6.

Calculation of Quantity of Analyte from Titration Data
As shown by the examples that follow, the same general approach is also used to compute analyte concentrations from titration data.

Example 6-7

A 0.8040-g sample of an iron ore is dissolved in acid. The iron is then reduced to Fe^{2+} and titrated with 47.22 mL of 0.02242 M $KMnO_4$ solution. Calculate the results of this analysis in terms of (a) percent Fe (fw = 55.847 g) and (b) percent Fe_3O_4 (fw = 231.54 g). The reaction of the analyte with the reagent is described by the equation

$$MnO_4^- + 5 Fe^{2+} + 8 H^+ \rightarrow Mn^{2+} + 5 Fe^{3+} + 4 H_2O$$

(a) stoichiometric ratio $= \dfrac{5 \text{ mmol } Fe^{2+}}{1 \text{ mmol } KMnO_4}$

$$\text{amount } KMnO_4 = 47.22 \text{ mL } KMnO_4 \times \frac{0.02242 \text{ mmol } KMnO_4}{\text{mL } KMnO_4}$$

$$\text{amount } Fe^{2+} = (47.22 \times 0.02242) \text{ mmol } KMnO_4 \times \frac{5 \text{ mmol } Fe^{2+}}{1 \text{ mmol } KMnO_4}$$

The weight of Fe^{2+} is then given by

$$\text{wt } Fe^{2+} = (47.22 \times 0.02242 \times 5) \text{ mmol } Fe^{2+} \times 0.055847 \frac{\text{g } Fe^{2+}}{\text{mmol } Fe^{2+}}$$

The percent Fe^{2+} is

$$\% Fe^{2+} = \frac{(47.22 \times 0.02242 \times 5 \times 0.055847) \text{ g } Fe^{2+}}{0.8040 \text{ g sample}} \times 100\% = 36.77\%$$

(b) In order to derive a stoichiometric ratio, we note that

$$5 \; Fe^{2+} \equiv 1 \; MnO_4^-$$

The symbol ≡ means chemically equivalent to.

Therefore,

$$5 \; mmol \; Fe_3O_4 \equiv 15 \; mmol \; Fe^{2+} \equiv 3 \; mmol \; MnO_4^-$$

and

$$\text{stoichiometric ratio} = \frac{5 \; mmol \; Fe_3O_4}{3 \; mmol \; KMnO_4}$$

As in part (a),

$$\text{amount } KMnO_4 = 47.22 \; \cancel{mL \; KMnO_4} \times \frac{0.02242 \; mmol \; KMnO_4}{\cancel{mL \; KMnO_4}}$$

$$\text{amount } Fe_3O_4 = (47.22 \times 0.02242) \; \cancel{mmol \; KMnO_4} \times \frac{5 \; mmol \; Fe_3O_4}{3 \; \cancel{mmol \; KMnO_4}}$$

$$\text{wt } Fe_3O_4 = \left(47.22 \times 0.02242 \times \frac{5}{3}\right) \; \cancel{mmol \; Fe_3O_4} \times 0.23154 \; \frac{g \; Fe_3O_4}{\cancel{mmol \; Fe_3O_4}}$$

$$\% \; Fe_3O_4 = \frac{\left(47.22 \times 0.02242 \times \frac{5}{3}\right) \times 0.23154 \; g \; Fe_3O_4}{0.8040 \; g \; sample} \times 100\% = 50.81\%$$

Feature 6-2
ANOTHER APPROACH TO EXAMPLE 6-7a

Some people find it easier to write out the solution to a problem in such a way that the units in the denominator of each term cancel the units in the numerator of the preceding term until the units of the answer are obtained. For example, the solution to part (a) of Example 6-7 can be written

$$47.22 \; \cancel{mL \; KMnO_4} \times \frac{0.02242 \; \cancel{mmol \; KMnO_4}}{\cancel{mL \; KMnO_4}} \times \frac{5 \; \cancel{mmol \; Fe}}{1 \; \cancel{mmol \; KMnO_4}}$$

$$\times \frac{0.055847 \; g \; Fe}{\cancel{mmol \; Fe}} \times \frac{1}{0.8040 \; g \; sample} \times 100\%$$

$$= 36.77\% \; Fe$$

Example 6-8

The organic matter in a 3.776-g sample of a mercuric ointment is decomposed with HNO_3. After dilution, the Hg^{2+} is titrated with 21.30 mL of a

0.1144 M solution of NH_4SCN. Calculate the percent Hg (fw = 200.59 g) in the ointment. This titration involves the formation of a stable neutral complex, $Hg(SCN)_2$:

$$Hg^{2+} + 2\ SCN^- \rightarrow Hg(SCN)_2(aq)$$

At the equivalence point,

$$\text{stoichiometric ratio} = \frac{1\ \text{mmol Hg}^{2+}}{2\ \text{mmol NH}_4\text{SCN}}$$

$$\text{amount NH}_4\text{SCN} = 21.30\ \cancel{\text{mL NH}_4\text{SCN}} \times 0.1144\ \frac{\text{mmol NH}_4\text{SCN}}{\cancel{\text{mL NH}_4\text{SCN}}}$$

$$\text{amount Hg}^{2+} = (21.30 \times 0.1144)\ \cancel{\text{mmol NH}_4\text{SCN}} \times \frac{1\ \text{mmol Hg}^{2+}}{2\ \cancel{\text{mmol NH}_4\text{SCN}}}$$

$$\text{wt Hg}^{2+} = (21.30 \times 0.1144 \times \tfrac{1}{2})\ \cancel{\text{mmol Hg}^{2+}} \times \frac{0.20059\ \text{g Hg}^{2+}}{\cancel{\text{mmol Hg}^{2+}}}$$

$$\%\ \text{Hg} = \frac{(21.30 \times 0.1144 \times \tfrac{1}{2} \times 0.20059)\ \text{g Hg}^{2+}}{3.776\ \text{g sample}} \times 100\%$$

$$= 6.472\% = 6.47\%$$

Feature 6-3
ROUNDING THE ANSWER IN EXAMPLE 6-8

You should note that the input data for Example 6-8 all contain either four or five significant figures but the answer is rounded to three. Why is this?

   Let us proceed as we did on page 42 and make the rounding decision by doing a couple of rough calculations in our heads. We will assume that the input data are uncertain to one part in the last significant figure. The largest *relative* error will then be associated with the molarity of the reagent. Here, the relative uncertainty is 0.0001/ 0.1144. But we really do not need to know the error this accurately, and we can simply figure that the uncertainty is about 1 part in 1000 (compared with about 1 part in 2000 for the volume and 1 part in 3800 for the weight). We then assume the calculated result is uncertain to about the same amount as the least accurate measurement, or 1 part in 1000. The absolute uncertainty of the final result is then 6.472% × 1/1000 = 0.0065 = 0.01%, and we round to the second figure to the right of the decimal point. Thus, we report 6.47%.

   You should practice making this rough type of rounding decision whenever you make a computation in this course and others.

Example 6-9

A 0.4755-g sample containing $(NH_4)_2C_2O_4$ and inert materials was dissolved in water and made strongly alkaline with KOH, which converted NH_4^+ to NH_3. The liberated NH_3 was distilled into exactly 50.00 mL of 0.05035 M H_2SO_4. The excess H_2SO_4 was back-titrated with 11.13 mL of 0.1214 M NaOH. Calculate (a) the percent N (fw = 14.007 g) and (b) the percent $(NH_4)_2C_2O_4$ (fw = 124.10 g) in the sample.

(a) The H_2SO_4 reacts with both NH_3 and NaOH, and so two stoichiometric ratios are needed:

$$\frac{2 \text{ mmol NH}_3}{1 \text{ mmol H}_2SO_4} \quad \text{and} \quad \frac{1 \text{ mmol H}_2SO_4}{2 \text{ mmol NaOH}}$$

$$\text{total amount H}_2SO_4 = 50.00 \text{ mL H}_2SO_4 \times 0.05035 \frac{\text{mmol H}_2SO_4}{\text{mL H}_2SO_4}$$

$$= 2.5175 \text{ mmol H}_2SO_4$$

The amount of H_2SO_4 consumed by the NaOH in the back-titration is

$$\text{amount H}_2SO_4 = (11.13 \times 0.1214) \text{ mmol NaOH} \times \frac{1 \text{ mmol H}_2SO_4}{2 \text{ mmol NaOH}}$$

$$= 0.6756 \text{ mmol H}_2SO_4$$

The amount of H_2SO_4 that reacted with NH_3 is then

$$\text{amount H}_2SO_4 = (2.5175 - 0.6756) \text{ mmol H}_2SO_4 = 1.8419 \text{ mmol H}_2SO_4$$

The amount of NH_3, which is equal to the amount of N, is

$$\text{amount N} = \text{amount NH}_3 = 1.8419 \text{ mmol H}_2SO_4 \times \frac{2 \text{ mmol N}}{1 \text{ mmol H}_2SO_4}$$

$$= 3.6838 \text{ mmol N}$$

$$\%N = \frac{3.6838 \text{ mmol N} \times 0.014007 \text{ g N/mmol N}}{0.4755 \text{ g sample}} \times 100\% = 10.85$$

(b) Since each millimole of $(NH_4)_2C_2O_4$ produces 2 mmol of NH_3, which reacts with 1 mmol of H_2SO_4,

$$\text{stoichiometric ratio} = \frac{1 \text{ mmol }(NH_4)_2C_2O_4}{1 \text{ mmol H}_2SO_4}$$

$$\text{amount }(NH_4)_2C_2O_4 = 1.8419 \text{ mmol H}_2SO_4 \times \frac{1 \text{ mmol }(NH_4)_2C_2O_4}{1 \text{ mmol H}_2SO_4}$$

$$\text{wt }(NH_4)_2C_2O_4 = 1.8419 \text{ mmol }(NH_4)_2C_2O_4 \times \frac{0.12410 \text{ g }(NH_4)_2C_2O_4}{\text{mmol }(NH_4)_2C_2O_4}$$

$$= 0.22858 \text{ g}$$

$$\% (NH_4)_2C_2O_4 = \frac{0.22858 \text{ g }(NH_4)_2C_2O_4}{0.4755 \text{ g sample}} \times 100\% = 48.07\%$$

Example 6-10

The CO in a 20.3-L sample of gas was converted to CO_2 by passing the gas over iodine pentoxide heated to 150°C:

$$I_2O_5(s) + 5\ CO(g) \rightarrow 5\ CO_2(g) + I_2(g)$$

The iodine distilled at this temperature and was collected in an absorber containing 8.25 mL of 0.01101 M $Na_2S_2O_3$:

$$I_2(aq) + 2\ S_2O_3^{2-}(aq) \rightarrow 2\ I^-(aq) + S_4O_6^{2-}(aq)$$

The excess $Na_2S_2O_3$ was back-titrated with 2.16 mL of 0.00947 M I_2 solution. Calculate the number of milligrams of CO (fw = 28.01 g) per liter of sample.

Based on the two reactions, the stoichiometric ratios are

$$\frac{5\ \text{mmol CO}}{1\ \text{mmol } I_2} \qquad \text{and} \qquad \frac{2\ \text{mmol } Na_2S_2O_3}{1\ \text{mmol } I_2}$$

We divide the first ratio by the second to get a third useful ratio:

$$\frac{5\ \text{mmol CO}}{2\ \text{mmol } Na_2S_2O_3}$$

This relationship reveals that 5 mmol of CO is responsible for the consumption of 2 mmol of $Na_2S_2O_3$. The total amount of $Na_2S_2O_3$ is

$$\text{amount } Na_2S_2O_3 = 8.25\ \cancel{\text{mL } Na_2S_2O_3} \times 0.01101\ \frac{\text{mmol } Na_2S_2O_3}{\cancel{\text{mL } Na_2S_2O_3}}$$

$$= 0.09083\ \text{mmol } Na_2S_2O_3$$

The amount of $Na_2S_2O_3$ consumed in the back-titration is

$$\text{amount } Na_2S_2O_3 = 2.16\ \cancel{\text{mL } I_2} \times 0.00947\ \frac{\cancel{\text{mmol } I_2}}{\cancel{\text{mL } I_2}} \times \frac{2\ \text{mmol } Na_2S_2O_3}{\cancel{\text{mmol } I_2}}$$

$$= 0.04091\ \text{mmol } Na_2S_2O_3$$

The amount of CO can then be obtained by employing the third stoichiometric ratio:

$$\text{amount CO} = (0.09083 - 0.04091)\ \cancel{\text{mmol } Na_2S_2O_3} \times \frac{5\ \text{mmol CO}}{2\ \cancel{\text{mmol } Na_2S_2O_3}}$$

$$= 0.1248\ \text{mmol CO}$$

$$\text{wt CO} = 0.1248\ \cancel{\text{mmol CO}} \times \frac{28.01\ \text{mg CO}}{\cancel{\text{mmol CO}}} = 3.4958\ \text{mg}$$

$$\frac{\text{wt CO}}{\text{vol sample}} = \frac{3.4958\ \text{mg CO}}{20.3\ \text{L sample}} = 0.172\ \frac{\text{mg CO}}{\text{L}}$$

6D QUESTIONS AND PROBLEMS

*6-1. Distinguish between the end point and the equivalence point in a titration,

6-2. What is a primary standard?

*6-3. What is a secondary standard? How does it differ from a primary standard?

*6-4. Calculate the molar concentration of a solution that is 25.0% H_2SO_4 (w/w) and has a specific gravity of 1.19.

6-5. Calculate the molar concentration of a 12.0% solution (w/w) of $CuSO_4$ that has a specific gravity of 1.13.

*6-6. Describe the preparation of 3.00 L of 0.0800 M H_2SO_4 from
(a) 4.00 M H_2SO_4.
(b) 13.0% (w/w) H_2SO_4.
(c) the concentrated reagent (sp gr = 1.84; % H_2SO_4 = 95).

6-7. Describe the preparation of 800 mL of 0.0500 M KOH from
(a) a 6.00-M KOH solution.
(b) a 3.61% (w/w) KOH solution.
(c) a concentrated reagent (sp gr = 1.505; % KOH = 50.0).

*6-8. A solution of $HClO_4$ was standardized by dissolving 0.3745 g of primary-standard HgO in a solution of KBr:

$$HgO(s) + 4\ Br^- + H_2O \rightarrow HgBr_4^{2-} + 2\ OH^-$$

The liberated OH^- required 37.79 mL of the acid. Calculate the molarity of the $HClO_4$.

*6-9. A 0.3367-g sample of primary-standard Na_2CO_3 required 28.66 mL of an H_2SO_4 solution to reach the end point in the reaction

$$CO_3^{2-} + 2\ H^+ \rightarrow H_2O + CO_2(g).$$

Calculate the molarity of the H_2SO_4.

6-10. A 0.3396-g sample that assayed 96.4% Na_2SO_4 was titrated with a solution of $BaCl_2$:

$$Ba^{2+} + SO_4^{2-} \rightarrow BaSO_4(s)$$

Calculate the molarity of the $BaCl_2$ solution if the end point was observed when 37.70 mL of the reagent was added.

*6-11. A 40.00-mL aliquot of a solution of $HClO_4$ was added to a solution containing 0.4793 g of primary-standard Na_2CO_3. The solution was boiled to remove CO_2, and the excess $HClO_4$ was back-titrated with 8.70 mL of an NaOH solution. In a separate experiment, 25.00 mL of the NaOH neutralized 27.43 mL of $HClO_4$. Calculate the molarity of the $HClO_4$ and that of the NaOH.

*6-12. The sulfur in a petroleum product was determined by burning a 4.476-g sample in a tube furnace and bubbling the combustion products through a 3% solution of H_2O_2. The SO_2 was converted to H_2SO_4:

$$SO_2(g) + H_2O_2 \rightarrow H_2SO_4$$

A 25.00-mL portion of 0.00923 M NaOH was added to the solution, and the excess base was back-titrated with 13.33 mL of 0.01007 M HCl. Calculate the parts per million of sulfur in the sample.

6-13. A 100.0-mL sample of a spring water was analyzed for its iron content by acidifying and reducing all the iron present to Fe^{2+}. A 25.00-mL aliquot of a 0.002107 M solution of $K_2Cr_2O_7$ was added, which resulted in the reaction

$$6\ Fe^{2+} + Cr_2O_7^{2-} + 14\ H^+ \rightarrow 6\ Fe^{3+} + 2\ Cr^{3+} + 7\ H_2O$$

The excess $K_2Cr_2O_7$ was back-titrated with 7.47 mL of 0.00979 M Fe^{2+}. Calculate the parts per million of iron in the sample.

*6-14. The thiourea in a 1.455-g sample of organic material was extracted into a dilute H_2SO_4 solution and titrated with 37.31 mL of 0.009372 M Hg^{2+} via the reaction

$$4\ (NH_2)_2CS + Hg^{2+} \rightarrow [(NH_2)_2CS]_4Hg^{2+}$$

Calculate the percent $(NH_2)_2CS$ (fw = 76.12 g) in the sample.

6-15. The ethyl acetate concentration in an alcoholic solution was determined by diluting a 10.00-mL sample to exactly 100 mL. A 20.00-mL portion of the diluted solution was refluxed with 40.00 mL of 0.04672 M KOH:

$$CH_3COOC_2H_5 + OH^- \rightarrow CH_3COO^- + C_2H_5OH$$

After cooling, the excess OH^- was back-titrated with 3.41 mL of 0.05042 M H_2SO_4. Calculate the number of grams of ethyl acetate (fw = 88.11 g) per 100 mL of the original sample.

*6-16. The arsenic in a 1.223-g sample of a pesticide was converted to H_3AsO_4 by suitable treatment. The acid was then neutralized, and exactly 40.00 mL of 0.07891 M $AgNO_3$ added to precipitate the arsenic quantitatively as Ag_3AsO_4. The excess Ag^+ in the filtrate and washings from the precipitation was titrated with 11.27 mL of 0.1000 M KSCN; the reaction was

$$Ag^+ + SCN^- \rightarrow AgSCN(s)$$

Calculate the percent As_2O_3 in the sample.

*6-17. A solution of $Ba(OH)_2$ was standardized against 0.1016 g of primary-standard benzoic acid, C_6H_5COOH (fw = 122.12 g). An end point was observed after addition of 44.42 mL of base.

(a) Calculate the molarity of the base.

(b) Calculate the standard deviation of the molarity if the standard deviation for weighing was ±0.2 mg and that for the volume measurement was ±0.03 mL.

(c) Assuming an error of −0.3 mg in the weighing, calculate the absolute and relative determinate error in the molarity.

6-18. A 0.1475 M solution of $Ba(OH)_2$ was used to titrate the acetic acid (fw = 60.05 g) in a dilute aqueous solution. The following results were obtained.

Sample	Sample Volume, mL	$Ba(OH)_2$ Volume, mL
1	50.00	43.17
2	49.50	42.68
3	25.00	21.47
4	50.00	43.33

(a) Calculate the mean w/v percentage of acetic acid in the sample.

(b) Calculate the standard deviation of the results.

(c) At the 90% confidence level, could any of the results be discarded?

(d) Assume that the buret used to measure out the acetic acid had a determinate error of −0.05 mL at all volumes delivered. Calculate the determinate error in the mean result.

CHAPTER 7

AQUEOUS-SOLUTION CHEMISTRY

This chapter provides a review of aqueous-solution chemistry, including chemical equilibrium and simple equilibrium-constant calculations. The material is a prerequisite to understanding the contents of the chapters that follow.

7A THE CHEMICAL COMPOSITION OF AQUEOUS SOLUTIONS

Water is the most plentiful solvent available on earth and finds widespread use as a medium for carrying out chemical analyses.

7A-1 Solutions of Electrolytes

Most of the solutes we will discuss are *electrolytes,* which are substances that form ions when dissolved in water or other solvents and thus produce solutions that conduct electricity. *Strong electrolytes* ionize essentially completely in a solvent, whereas *weak electrolytes* ionize only partially. Therefore, weak electrolytes impart less conductivity to a solvent than do strong electrolytes. Table 7-1 is a compilation of solutes that act as strong and weak electrolytes in water. Among the strong electrolytes listed are acids, bases, and salts.

A salt is the product formed in the reaction of an acid with a base. Examples include $NaCl$, Na_2SO_4, and CH_3COONa (sodium acetate).

Svante Arrhenius (1859–1927), Swedish chemist, formulated many of the early ideas regarding ionic dissociation in solution. His ideas were not accepted at first; in fact, he was given the lowest possible passing grade for his Ph.D. examination. In 1903 Arrhenius was awarded the Nobel Prize in Chemistry for these revolutionary ideas. He was one of the first scientists to suggest the relationship between the amount of carbon dioxide in the atmosphere and global temperature, a phenomenon that has come to be known as the greenhouse effect.

An acid is a substance that donates protons. A base is a substance that accepts protons.

An acid donates protons only in the presence of a proton acceptor (a base). Likewise, a base accepts protons only in the presence of a proton donor (an acid).

A conjugate base is the species formed when an acid loses a proton.

For example, acetate ion is the conjugate base of acetic acid, and ammonium ion is the conjugate acid of ammonia.

A conjugate acid is the species formed when a base accepts a proton.

Table 7-1
CLASSIFICATION OF ELECTROLYTES

Strong	Weak
1. The inorganic acids HNO_3, $HClO_4$, H_2SO_4*, HCl, HI, HBr, $HClO_3$, $HBrO_3$	1. Many inorganic acids, including H_2CO_3, H_3BO_3, H_3PO_4, H_2S, H_2SO_3
2. Alkali and alkaline-earth hydroxides	2. Most organic acids
3. Most salts	3. Ammonia and most organic bases
	4. Halides, cyanides, and thiocyanates of Hg, Zn, and Cd

*H_2SO_4 is completely dissociated into HSO_4^- and H_3O^+ ions and for this reason is classified as a strong electrolyte. However, it should be noted that the HSO_4^- ion is a weak electrolyte, being only partially dissociated.

7A-2 Acids and Bases

In 1923, two chemists, J. N. Brønsted in Denmark and J. M. Lowry in England, proposed independently a theory of acid/base behavior that is particularly useful in analytical chemistry.[1] According to the Brønsted-Lowry theory, *an acid is a proton donor* and *a base is a proton acceptor*. In order for a species to behave as an acid, a proton acceptor (or base) must be present. The reverse is also true.

Conjugate Acids and Bases

An important feature of the Brønsted-Lowry concept is that when an acid gives up a proton, a *conjugate base* is formed that is capable of accepting a proton. For example, when the species acid₁ gives up a proton, the species base₁ is formed, as shown by the reaction

$$acid_1 \rightleftarrows base_1 + proton$$

Here, acid₁ and base₁ are a conjugate acid/base pair.

Similarly, every base produces its *conjugate acid* as a result of accepting a proton. That is,

$$base_2 + proton \rightleftarrows acid_2$$

When these two processes are combined, the result is an acid/base, or neutralization, reaction:

$$acid_1 + base_2 \rightleftarrows base_1 + acid_2$$

The extent to which this reaction proceeds depends upon the relative tendencies of the two bases to accept a proton (or the two acids to donate

[1]For a thorough treatment of the various acid/base concepts, see I. M. Kolthoff, in *Treatise on Analytical Chemistry,* 2nd ed., I. M. Kolthoff and P. J. Elving, Eds,, Part I. Vol. 2, Chapter 17. New York: Wiley, 1979.

a proton). Examples of conjugate acid/base relationships are shown in Equations 7-1 through 7-4.

Many solvents are proton donors or proton acceptors and can thus induce basic or acidic behavior in solutes dissolved in them. For example, in an aqueous solution of ammonia, water donates a proton and thus acts as an acid with respect to the solute:

$$NH_3 + H_2O \rightleftarrows NH_4^+ + OH^- \qquad (7\text{-}1)$$

Base₁   Acid₂   Conjugate   Conjugate
                 Acid₁   Base₂

Water can act as either an acid or a base.

In this reaction, ammonia (base₁) reacts with water, which is labeled acid₂, to give the conjugate acid ammonium ion (acid₁) and hydroxide ion, which is the conjugate base (base₂) of the acid water. In contrast, water acts as a proton acceptor, or base, in an aqueous solution of nitrous acid:

$$H_2O + HNO_2 \rightleftarrows H_3O^+ + NO_2^- \qquad (7\text{-}2)$$

Base₁   Acid₂   Conjugate   Conjugate
                 Acid₁   Base₂

Nitrite ion is the conjugate base of the acid HNO_2; H_3O^+ is the conjugate acid of the base H_2O. Neither NH_3 nor HNO_2 reacts completely with H_2O; therefore, ammonia and nitrous acid are both termed weak electrolytes.

Feature 7-1
AMPHIPROTIC SPECIES

Some compounds behave as both an acid and a base. An example is dihydrogen phosphate ion ($H_2PO_4^-$), which behaves as a base in the presence of a proton donor, such as H_3O^+:

$$H_2PO_4^- + H_3O^+ \rightleftarrows H_3PO_4 + H_2O$$

Base₁   Acid₂   Acid₁   Base₂

Here, H_3PO_4 is the conjugate acid of the original base. In the presence of a proton acceptor, such as water, however, $H_2PO_4^-$ behaves as an acid and forms the conjugate base HPO_4^{2-}:

$$H_2PO_4^- + H_2O \rightleftarrows HPO_4^{2-} + H_3O^+$$

Acid₁   Base₂   Base₁   Acid₂

Amphiprotic Solvents

Water is the classic example of an *amphiprotic* solvent—that is, a solvent that can act either as an acid (Equation 7-1) or as a base (Equation 7-2), depending upon the solute. Other common amphiprotic solvents are

Amphiprotic solvents behave as acids in the presence of basic solutes and as bases in the presence of acidic solutes.

methanol, ethanol, and anhydrous acetic acid. In methanol, for example, the equilibria analogous to those shown in Equations 7-1 and 7-2 are

$$NH_3 + CH_3OH \rightleftharpoons NH_4^+ + CH_3O^- \quad (7\text{-}3)$$

$$CH_3OH + HNO_2 \rightleftharpoons CH_3OH_2^+ + NO_2^- \quad (7\text{-}4)$$

Base₁ — $Base_1$ Acid₂ — $Acid_2$ Conjugate Acid₁ — $Conjugate\ Acid_1$ Conjugate Base₂ — $Conjugate\ Base_2$

It is important to emphasize that an acid that has donated a proton becomes a conjugate base capable of accepting a proton to reform the original acid; the converse holds equally well. Thus, nitrite ion, the species produced by the loss of a proton from nitrous acid, is a potential acceptor of a proton from a suitable donor. It is this proton-accepting reaction that causes an aqueous solution of sodium nitrite to be slightly basic:

$$NO_2^- + H_2O \rightleftharpoons HNO_2 + OH^-$$

$Base_1$ $Acid_2$ $Conjugate\ Acid_1$ $Conjugate\ Base_2$

7A-3 Autoprotolysis

Amphiprotic solvents undergo self-ionization, or *autoprotolysis,* to form a pair of ionic species. Autoprotolysis is yet another example of acid/base behavior, as illustrated by the following equations:

$$base_1 + acid_2 \rightleftharpoons acid_1 + base_2$$
$$H_2O + H_2O \rightleftharpoons H_3O^+ + OH^-$$
$$CH_3OH + CH_3OH \rightleftharpoons CH_3OH_2^+ + CH_3O^-$$
$$HCOOH + HCOOH \rightleftharpoons HCOOH_2^+ + HCOO^-$$
$$NH_3 + NH_3 \rightleftharpoons NH_4^+ + NH_2^-$$

The hydronium ion H_3O^+ is the reaction product of water with an acid.

The product H_3O^+ formed by the autoprotolysis of water is called the *hydronium ion* and consists of a proton covalently bonded to a water molecule by one of the unshared electron pairs of the oxygen. Higher hydrates, such as $H_5O_2^+$ and $H_9O_4^+$, also exist, but they are orders of magnitude less stable than H_3O^+. Essentially no unhydrated protons exist in aqueous solutions.[2]

To emphasize the extraordinary stability of the singly hydrated proton, many chemists use the notation H_3O^+ when writing equations for aqueous solutions containing the proton. Others use H^+ because this notation simplifies the balancing of equations in which the proton is a participant. Ordinarily, we shall use H_3O^+ in acid/base equilibrium calculations but the simpler H^+ notation otherwise.

In this text we shall use the symbol H_3O^+ in those chapters that deal with acid/base equilibria and acid/base equilibrium calculations. In the remaining chapters we simplify to the more convenient H^+, it being understood that this symbol represents the hydronium ion.

[2]See P. A. Giguere, *J. Chem. Educ.,* **1979,** *56,* 571.

Strongest acid

Weakest acid

$$HClO_4 + H_2O \rightleftharpoons H_3O^+ + ClO_4^-$$
$$HCl + H_2O \rightleftharpoons H_3O^+ + Cl^-$$
$$H_3PO_4 + H_2O \rightleftharpoons H_3O^+ + H_2PO_4^-$$
$$Al(H_2O)_6^{3+} + H_2O \rightleftharpoons H_3O^+ + AlOH(H_2O)_5^{2+}$$
$$HC_2H_3O_2 + H_2O \rightleftharpoons H_3O^+ + C_2H_3O_2^-$$
$$H_2PO_4^- + H_2O \rightleftharpoons H_3O^+ + HPO_4^{2-}$$
$$NH_4^+ + H_2O \rightleftharpoons H_3O^+ + NH_3$$

Weakest base

Strongest base

Figure 7-1
Dissociation reactions and relative strengths of some common acids and their conjugate bases.

The extent to which water undergoes autoprotolysis is slight at room temperature. Thus, the hydronium and hydroxide ion concentrations in pure water are only about 10^{-7} M. Nevertheless, this dissociation reaction is of utmost importance in understanding the behavior of aqueous solutions.

7A-4 Strengths of Acids and Bases

Figure 7-1 shows the dissociation reactions of a few common acids in water. The first two are *strong acids* because reaction with the solvent is sufficiently complete as to leave no undissociated solute molecules in aqueous solution. The remainder are *weak acids,* which react incompletely with water to give solutions that contain significant quantities of both the parent acid and its conjugate base. Note that acids can be cationic, anionic, or electrically neutral.

The acids in Figure 7-1 become progressively weaker from top to bottom. Perchloric acid and hydrochloric acid are completely dissociated. In contrast, only about 1% of acetic acid ($HC_2H_3O_2$) is dissociated. Ammonium ion is an even weaker acid; only about 0.01% of this ion is dissociated into hydronium ions and ammonia molecules. Another generality illustrated in Figure 7-1 is that the weakest acid forms the strongest conjugate base; that is, ammonia has a much stronger affinity for protons than any base above it. Perchlorate and chloride ions have no affinity for protons in water.

The tendency of a solvent to accept or donate protons determines the strength of a solute acid or base dissolved in it. For example, perchloric and hydrochloric acids are strong acids in water. If anhydrous acetic acid, a poorer proton acceptor than water, is substituted *as the solvent,* neither perchloric nor hydrochloric acid undergoes complete dissociation; instead, equilibria such as the following are established:

$$CH_3COOH + HClO_4 \rightleftharpoons CH_3COOH_2^+ + ClO_4^-$$

Base₁   Acid₂   Acid₁   Base₂

Perchloric acid is, however, considerably stronger than hydrochloric acid in this solvent; its dissociation is about 5000 times greater than that of the hydrochloric acid. Acetic acid thus acts as a *differentiating* solvent toward these two acids in the sense that its use reveals inherent differences in their acidities. Water, on the other hand, is a *leveling* solvent for perchloric, hydrochloric, nitric, and sulfuric acids in that all are completely ionized in it and thus exhibit no differences in strength.

The common strong acids are HCl, $HClO_3$, HNO_3, the first hydrogen in H_2SO_4, HBr, HI, and the organic sulfonic acids (RSO_3H).

The common strong bases are NaOH, KOH, $Ba(OH)_2$, and the quaternary ammonium hydroxides (R_4NOH, where R is an alkyl group such as CH_3 or C_2H_5).

Of all the acids listed in Figure 7-1, only perchloric acid is a strong acid in methanol and ethanol. Thus, these two alcohols are also differentiating solvents.

In a differentiating solvent, various acids dissociate to different degrees and are thus of different strengths. In a leveling solvent, several acids are completely dissociated and are thus of the same strength.

7B   CHEMICAL EQUILIBRIUM

Generally, the reactions used in analytical chemistry are never complete. Instead, they proceed to a state of *chemical equilibrium* in which the ratio of concentrations of reactants and products is constant. *Equilibrium-constant expressions* are *algebraic* equations that describe the concentration relationships among reactants and products at chemical equilibrium. Such relationships permit calculation of the quantity of analyte that remains unreacted when a steady state has been reached. From these data, the error resulting from incompleteness of the reaction upon which an analysis is based can be computed.

The discussion that follows deals with use of equilibrium-constant expressions to gain information about analytical systems in which no more than one or two equilibria are important. Chapter 9 extends these methods to systems containing several simultaneous equilibria. Such complex systems are often encountered in analytical chemistry.

7B-1   The Equilibrium State

Consider the chemical equilibrium

$$H_3AsO_4 + 3 I^- + 2 H^+ \rightleftharpoons H_3AsO_3 + I_3^- + H_2O \qquad (7\text{-}5)$$

The rate of this reaction and the extent to which it proceeds to the right can be readily judged by observing the orange-red color of the triiodide ion (I_3^-) since all the other participants in the reaction are colorless. If, for example, 1 mmol of arsenic acid (H_3AsO_4) is added to 100 mL of a solution containing 3 mmol of potassium iodide, the red color of the triiodide ion appears almost immediately, and within a few seconds the intensity of the color becomes constant, which shows that the triiodide concentration has become constant.

A solution of identical color intensity (and hence identical triiodide concentration) can also be produced by adding 1 mmol of arsenous acid (H_3AsO_3) to 100 mL of a solution containing 1 mmol of triiodide ion. Here, the color intensity is initially greater than in the first solution but rapidly decreases as a result of the reaction

$$H_3AsO_3 + I_3^- + H_2O \rightleftharpoons H_3AsO_4 + 3 I^- + 2 H^+$$

Ultimately the color of the two solutions is identical. Many other combinations of the four reactants can be employed to yield solutions that are indistinguishable from the two just described.

The foregoing discussion illustrates that the concentration relationship at chemical equilibrium (that is, the *position of equilibrium*) is independent of the route by which the equilibrium state is achieved. Moreover, a system that is in equilibrium will not spontaneously depart from this condition unless a stress is applied to the system. Such stresses include changes in temperature, in pressure (if one of the reactants or products is a gas), or in total concentration of a reactant or a product. These effects

This reaction is shown in the color plates.

The position of a chemical equilibrium is independent of the route by which equilibrium is reached.

can be predicted qualitatively from the *principle of Le Châtelier,* which states that the position of chemical equilibrium always shifts in the direction that tends to relieve the effect of an applied stress. Thus, an increase in temperature alters the concentration relationship in the direction that tends to absorb heat, and an increase in pressure favors those participants that occupy a smaller total volume.

In an analysis, the effect of introducing an additional amount of a participating species to the reaction mixture is particularly important. Here, the resulting stress is relieved by a shift in equilibrium in the direction that partially uses up the added substance. Thus, for the equilibrium we have been considering (Equation 7-5), the addition of either arsenic acid (H_3AsO_4) or hydrogen ions causes an increase in color as more triiodide ion and arsenous acid are formed; the addition of arsenous acid has the reverse effect. An equilibrium shift brought about by changing the amount of one of the participating species is called a *mass-action effect.*

If it were possible to examine the system under discussion at the molecular level, we would find that interactions among the participating species continue unabated even after equilibrium is achieved. The constant concentration ratio of reactants and products results from the equality in the rates of the forward and reverse reactions. In other words, chemical equilibrium is a dynamic state in which the rates of the forward and reverse reactions are identical.

7B-2 Equilibrium-Constant Expressions

The influence of concentration (or pressure if the species are gases) on the position of a chemical equilibrium is conveniently described in quantitative terms by means of an equilibrium-constant expression. Such expressions are readily derived from thermodynamic theory. They are of great practical importance because they permit the chemist to predict the direction and completeness of a chemical reaction. We must emphasize, however, that an equilibrium-constant expression yields no information concerning the *rate* at which equilibrium is approached. In fact, we sometimes encounter reactions that have highly favorable equilibrium constants but are of little analytical use because their rates are low. This limitation can often be overcome by the use of a catalyst, which speeds the attainment of equilibrium without changing its position.

Let us consider a generalized equation for a chemical equilibrium:

$$wW + xX \rightleftharpoons yY + zZ \qquad (7\text{-}6)$$

where the capital letters represent the formulas of participating chemical species and the lowercase italic letters are the small whole numbers required to balance the equation. Thus, the equation states that w mol of W reacts with x mol of X to form y mol of Y and z mol of Z. The equilibrium-constant expression for this reaction is

$$\frac{[Y]^y[Z]^z}{[W]^w[X]^x} = K \qquad (7\text{-}7)$$

The Le Châtelier principle states that the position of an equilibrium always shifts in such a direction as to relieve any stress applied to the system.

The mass action effect is a shift in the position of an equilibrium caused by adding one of the reactants or products to a system.

Chemical reactions do not cease at equilibrium. Instead, the amounts of reactants and products appear to be constant because the rates of the forward and reverse processes are identical.

Thermodynamics is a branch of chemical science that deals with heat or energy flow in chemical reactions. The position of a chemical equilibrium can be related to the heat generated or dissipated in a chemical reaction.

Equilibrium-constant expressions provide *no* information as to whether a chemical reaction is fast enough to be used for an analysis.

Cato Guldberg (1836–1902) and Peter Waage (1833–1900) were Norwegian chemists whose primary interests were in the field of thermodynamics. In 1864, these workers were the first to propose the law of mass action, which is expressed in Equation 7-7.

where the square-bracketed terms have the following meanings:

1. Molar concentration if the species is a dissolved solute.
2. Partial pressure in atmospheres if the species is a gas; in fact, we will often replace the square-bracketed term (say [Z] in Equation 7-7) with the symbol p_Z, which stands for the partial pressure of the gas Z in atmospheres.
3. Unity if the species is (a) a pure liquid, (b) a pure solid, or (c) the solvent in a dilute solution.

The term $[Z]^z$ in Equation 7-7 is replaced with

1. p_z in atmospheres if Z is a gas.
2. 1.00 if Z is a pure solid present in excess.
3. 1.00 if Z is a pure liquid present in excess.
4. 1.00 if Z is the solvent H_2O.

The rationale for these substitutions is explained in the description of the various types of equilibrium constants found in Section 7B-3.

The constant K in Equation 7-7 is a temperature-dependent numerical quantity called the *equilibrium constant*. By convention, the concentrations of the products, *as the equation is written,* are always placed in the numerator and the concentrations of the reactants in the denominator.

Equation 7-7 is only an approximate form of a thermodynamic equilibrium-constant expression. The exact form is given by Equation 8-3. Generally we will use the approximate form of this equation throughout this text because it is less tedious and time-consuming to use. In Section 8B, we show when the use of Equation 7-7 is likely to lead to serious errors in equilibrium calculations and how Equation 8-3 is applied in these cases.

7B-3 Types of Equilibrium Constants Encountered in Analytical Chemistry

Table 7-2 summarizes the types of chemical equilibria and equilibrium constants important in analytical chemistry. Applications of some of these constants are illustrated in the paragraphs that follow.

The Ion-Product Constant for Water

Aqueous solutions contain small amounts of hydronium and hydroxide ions as a consequence of the dissociation reaction

$$2\ H_2O \rightleftarrows H_3O^+ + OH^- \tag{7-8}$$

An equilibrium constant for this reaction can be formulated as shown in Equation 7-7:

Note that assuming $[H_2O]$ to be constant and including this concentration in the equilibrium constant is simply another way of stating the convention that the concentration term for a solvent in a dilute solution is 1.00.

$$\frac{[H_3O^+][OH^-]}{[H_2O]^2} = K \tag{7-9}$$

The concentration of water in dilute aqueous solutions is enormous, however, when compared with the concentration of hydrogen and hydroxide ions. As a consequence, $[H_2O]$ in Equation 7-9 can be taken as constant, and we write

A useful relationship is obtained by taking the negative logarithm of Equation 7-10

$$-\log K_w = -\log[H_3O^+] - \log[OH^-]$$

By definition of p-functions,

$$pK_w = pH + pOH$$

At 25°C, $pK_w = 14.00$.

$$K[H_2O]^2 = K_w = \boxed{[H_3O^+][OH^-]} \tag{7-10}$$

Table 7-2
EQUILIBRIA AND EQUILIBRIUM CONSTANTS OF IMPORTANCE TO ANALYTICAL CHEMISTRY

Type of Equilibrium	Name and Symbol of Equilibrium Constant	Typical Example	Equilibrium-Constant Expression
Dissociation of water	Ion-product constant, K_w	$2\,H_2O \rightleftharpoons H_3O^+ + OH^-$	$K_w = [H_3O^+][OH^-]$
Heterogeneous equilibrium between a slightly soluble substance and its ions in a saturated solution	Solubility product, K_{sp}	$BaSO_4(s) \rightleftharpoons Ba^{2+} + SO_4^{2-}$	$K_{sp} = [Ba^{2+}][SO_4^{2-}]$
Dissociation of a weak acid or base	Dissociation constant, K_a or K_b	$CH_3COOH + H_2O \rightleftharpoons$ $H_3O^+ + CH_3COO^-$ $CH_3COO^- + H_2O \rightleftharpoons$ $OH^- + CH_3COOH$	$K_a = \dfrac{[H_3O^+][CH_3COO^-]}{[CH_3COOH]}$ $K_b = \dfrac{[OH^-][CH_3COOH]}{[CH_3COO^-]}$
Formation of a complex ion	Formation constant, β_n	$Ni^{2+} + 4\,CN^- \rightleftharpoons Ni(CN)_4^{2-}$	$\beta_4 = \dfrac{[Ni(CN)_4^{2-}]}{[Ni^{2+}][CN^-]^4}$
Oxidation/reduction equilibrium	K_{redox}	$MnO_4^- + 5\,Fe^{2+} + 8\,H^+ \rightleftharpoons$ $Mn^{2+} + 5\,Fe^{3+} + 4\,H_2O$	$K_{redox} = \dfrac{[Mn^{2+}][Fe^{3+}]^5}{[MnO_4^-][Fe^{2+}]^5[H^+]^8}$
Distribution equilibrium between immiscible solvents	K_d	$I_2(aq) \rightleftharpoons I_2(org)$	$K_d = \dfrac{[I_2]_{org}}{[I_2]_{aq}}$

Feature 7-2
STEPWISE AND OVERALL FORMATION CONSTANTS FOR COMPLEX IONS

Stepwise-formation constants are symbolized by K_i. For example,

$$Ni^{2+} + CN^- \rightleftharpoons Ni(CN)^+ \qquad K_1 = \frac{[Ni(CN)^+]}{[Ni^{2+}][CN^-]}$$

$$Ni(CN)^+ + CN^- \rightleftharpoons Ni(CN)_2 \qquad K_2 = \frac{[Ni(CN)_2]}{[Ni(CN)^+][CN^-]}$$

$$Ni(CN)_2 + CN^- \rightleftharpoons Ni(CN)_3^- \qquad K_3 = \frac{[Ni(CN)_3^-]}{[Ni(CN)_2][CN^-]}$$

$$Ni(CN)_3^- + CN^- \rightleftharpoons Ni(CN)_4^{2-} \qquad K_4 = \frac{[Ni(CN)_4^{2-}]}{[Ni(CN)_3^-][CN^-]}$$

Overall constants are designated by the symbol β_n. Thus,

$$Ni^{2+} + 2\,CN^- \rightleftharpoons Ni(CN)_2 \qquad \beta_2 = K_1 K_2 = \frac{[Ni(CN)_2]}{[Ni^{2+}][CN^-]^2}$$

$$Ni^{2+} + 3\,CN^- \rightleftharpoons Ni(CN)_3^- \qquad \beta_3 = K_1 K_2 K_3 = \frac{[Ni(CN)_3^-]}{[Ni^{2+}][CN^-]^3}$$

$$Ni^{2+} + 4\,CN^- \rightleftharpoons Ni(CN)_4^{2-} \qquad \beta_4 = K_1 K_2 K_3 K_4 = \frac{[Ni(CN)_4^{2-}]}{[Ni^{2+}][CN^-]^4}$$

Feature 7-3

WHY [H_2O] DOES NOT APPEAR IN EQUILIBRIUM-CONSTANT
EXPRESSIONS FOR AQUEOUS SOLUTIONS

In a dilute aqueous solution, the molar concentration of water is

$$[H_2O] = \frac{1000 \text{ g } H_2O}{L\ H_2O} \times \frac{1 \text{ mol } H_2O}{18.0 \text{ g } H_2O} = 55.6 \text{ M}$$

Let us suppose we have 0.1 mol of HCl present in 1 L of water. The
presence of this acid will shift the equilibrium shown in Equation 7-8
to the left by the mass-action effect. Originally, however, there was
only 10^{-7} mol/L of OH^- to consume the added protons. Thus, even if
all the OH^- ions are converted to H_2O, the water concentration will
increase only to

$$[H_2O] = 55.6 \frac{\text{mol } H_2O}{L\ H_2O} + 1 \times 10^{-7} \frac{\text{mol } OH^-}{L\ H_2O} \times \frac{1 \text{ mol } H_2O}{\text{mol } OH^-}$$

$$= 55.6 \text{ M}$$

The percent change in water concentration is

$$\frac{10^{-7} \text{ M}}{55.6 \text{ M}} \times 100\% = 2 \times 10^{-7}\%$$

which is certainly negligible. Thus, $K\,[H_2O]^2$ in Equation 7-9 is, for all
practical purposes, a constant. That is,

$$K(55.6)^2 = K_w = 1.00 \times 10^{-14}$$

where the new constant K_w is given a special name, the *ion-product
constant for water.*

At 25°C, the ion-product constant for water is 1.008×10^{-14}. For convenience, we will use the approximation throughout this text that $K_w \approx 1.00 \times 10^{-14}$. Table 7-3 shows the dependence of this constant upon temperature.

The ion-product constant for water allows us to calculate the hydronium and hydroxide ion concentrations of aqueous solutions.

Table 7-3

VARIATION OF K_w WITH
TEMPERATURE

Temperature, °C	K_w
0	0.114×10^{-14}
25	1.01×10^{-14}
50	5.47×10^{-14}
100	49×10^{-14}

Example 7-1

Calculate the hydronium and hydroxide ion concentrations of pure water
at 25 and 100°C.

Because OH^- and H_3O^+ are formed only from the dissociation of
water, their concentrations must be equal:

$$[H_3O^+] = [OH^-]$$

Substitution into Equation 7-10 gives

$$[H_3O^+]^2 = [OH^-]^2 = K_w$$
$$[H_3O^+] = [OH^-] = \sqrt{K_w}$$

At 25°C,

$$[H_3O^+] = [OH^-] = \sqrt{1.00 \times 10^{-14}} = 1.00 \times 10^{-7} \text{ M}$$

At 100°C, from Table 7-3,

$$[H_3O^+] = [OH^-] = \sqrt{49 \times 10^{-14}} = 7.0 \times 10^{-7} \text{ M}$$

Example 7-2

Calculate the hydronium and hydroxide ion concentrations in 0.200 M aqueous NaOH.

Sodium hydroxide is a strong electrolyte, and so its contribution to the hydroxide ion concentration in this solution is 0.200 mol/L. As in Example 7-1, hydroxide ions and hydronium ions are formed *in equal amounts* from the dissociation of water. Therefore, we write

$$[OH^-] = 0.200 + [H_3O^+]$$

where $[H_3O^+]$ accounts for the hydroxide ions contributed by the solvent. The concentration of OH^- from the water is insignificant, however, when compared with 0.200, and so we can write

$$[OH^-] \approx 0.200$$

Equation 7-10 is then used to calculate the hydronium ion concentration:

$$[H_3O^+] = \frac{K_w}{[OH^-]} = \frac{1.00 \times 10^{-14}}{0.200} = 5.00 \times 10^{-14}$$

Note that the approximation

$$[OH^-] = 0.200 + 5.00 \times 10^{-14} \approx 0.200$$

causes no significant error.

Solubility-Product Constants

Most sparingly soluble salts are essentially completely dissociated in saturated aqueous solution. For example, when an excess of barium iodate is equilibrated with water, the dissociation process is adequately described by the equation

$$Ba(IO_3)_2(s) \rightleftarrows Ba^{2+} + 2 IO_3^-$$

When we say that a sparingly soluble salt is completely dissociated, *we do not imply* that all of the salt dissolves. Only a small amount goes into solution, but what does dissociates completely.

Application of Equation 7-7 leads to

$$\frac{[Ba^{2+}][IO_3^-]^2}{[Ba(IO_3)_2(s)]} = K$$

The denominator represents the molar concentration of $Ba(IO_3)_2$ *in the solid,* which is a second phase in contact with the saturated solution. The concentration of a compound in its solid state is, however, constant. In other words, the number of moles of $Ba(IO_3)_2$ divided by the *volume* of the $Ba(IO_3)_2$ is constant no matter how much excess solid is present. Therefore, the foregoing equation can be rewritten in the form

$$[Ba^{2+}][IO_3^-]^2 = K[Ba(IO_3)_2(s)] = K_{sp} \qquad (7\text{-}11)$$

Inclusion of $[Ba(IO_3)_2]$ in K_{sp} is another way of saying that this bracketed term has a numerical value of 1.00.

For Equation 7-11 to apply, it is necessary only that *some solid be present. You should always keep in mind that if no Ba(IO₃)(s) is present, Equation 7-11 is not valid.*

where the new constant is called the *solubility-product constant* or the *solubility product.* It is important to appreciate that Equation 7-11 shows that the position of this equilibrium is independent of the *amount* of $Ba(IO_3)_2$ present; that is, it does not matter whether there is a few milligrams or several grams of solid present just as long as some solid is present to keep the solution saturated.

A table of solubility products for numerous inorganic salts is found in Appendix 2. The examples that follow demonstrate some typical uses of solubility-product expressions. Further applications are considered in Chapters 9 and 10.

The Solubility of a Precipitate in Water

The solubility-product expression permits us to calculate the solubility of a sparingly soluble substance that ionizes in water.

Example 7-3

How many grams of $Ba(IO_3)_2$ (fw = 487 g) can be dissolved in 500 mL of water at 25°C?

The solubility-product constant for $Ba(IO_3)_2$ is 1.57×10^{-9} (Appendix 2). The equilibrium between the solid and its ions in solution is described by the equation

$$Ba(IO_3)_2(s) \rightleftarrows Ba^{2+} + 2\,IO_3^-$$

and so

$$[Ba^{2+}][IO_3^-]^2 = 1.57 \times 10^{-9} = K_{sp}$$

The equation describing the equilibrium reveals that 1 mol of Ba^{2+} is formed for each mole of $Ba(IO_3)_2$ that dissolves. Therefore,

The molar solubility is equal to $[Ba^+]$ or to $\frac{1}{2}[IO_3^-]$.

$$\text{molar solubility of } Ba(IO_3)_2 = [Ba^{2+}]$$

The iodate concentration is clearly twice that for barium ion:

$$[IO_3^-] = 2[Ba^{2+}]$$

Substituting this equality into the equilibrium-constant expression gives

$$[Ba^{2+}](2[Ba^{2+}])^2 = 4[Ba^{2+}]^3 = 1.57 \times 10^{-9}$$
$$[Ba^{2+}] = \left(\frac{1.57 \times 10^{-9}}{4}\right)^{1/3} = 7.32 \times 10^{-4} \text{ M}$$

Since 1 mol Ba^{2+} is produced for every mole of $Ba(IO_3)_2$

$$\text{solubility} = 7.32 \times 10^{-4} \text{ M}$$

To compute the number of millimoles of $Ba(IO_3)_2$ dissolved in 500 mL of solution, we write

$$\text{amount } Ba(IO_3)_2 = 7.32 \times 10^{-4} \frac{\text{mmol } Ba(IO_3)_2}{\text{mL}} \times 500 \text{ mL}$$

The weight of $Ba(IO_3)_2$ in 500 mL is given by

$$\text{wt } Ba(IO_3)_2 = (7.32 \times 10^{-4} \times 500) \text{ mmol } Ba(IO_3)_2$$
$$\times 0.487 \frac{\text{g } Ba(IO_3)_2}{\text{mmol } Ba(IO_3)_2}$$
$$= 0.178 \text{ g}$$

The Effect of a Common Ion on Solubility. The common-ion effect predicted from the Le Châtelier principle is demonstrated by the following examples.

Example 7-4

Calculate the molar solubility of $Ba(IO_3)_2$ in a solution that is 0.0200 M in $Ba(NO_3)_2$.

The solubility is no longer equal to $[Ba^{2+}]$ because $Ba(NO_3)_2$ is also a source of barium ions. We know, however, that solubility is related to $[IO_3^-]$:

$$\text{molar solubility of } Ba(IO_3)_2 = \tfrac{1}{2}[IO_3^-]$$

There are two sources of barium ions: $Ba(NO_3)_2$ and $Ba(IO_3)_2$. The contribution from the former is 0.0200 M, and that from the latter is equal to the molar solubility, or $\tfrac{1}{2}[IO_3^-]$. Thus,

$$[Ba^{2+}] = 0.0200 + \tfrac{1}{2}[IO_3^-]$$

Substitution of these quantities into the solubility-product expression yields

$$(0.0200 + \tfrac{1}{2}[IO_3^-])[IO_3^-]^2 = 1.57 \times 10^{-9}$$

Since the exact solution for $[IO_3^-]$ requires solving a cubic equation, we seek an approximation that simplifies the algebra. The small numerical value of K_{sp} suggests that the solubility of $Ba(IO_3)_2$ is not large, and this is confirmed by the result obtained in Example 7-3. So let us suppose that the barium ion concentration resulting from the solubility of $Ba(IO_3)_2$ is small relative to that from the $Ba(NO_3)_2$. That is, $\tfrac{1}{2}[IO_3^-] \ll 0.0200$, and

$$[Ba^{2+}] = 0.0200 + \tfrac{1}{2}[IO_3^-] \approx 0.0200$$

The original equation then simplifies to

$$0.0200[IO_3^-]^2 = 1.57 \times 10^{-9}$$
$$[IO_3^-] = \sqrt{7.85 \times 10^{-8}} = 2.80 \times 10^{-4} \text{ M}$$

The assumption that $(0.0200 + \tfrac{1}{2} \times 2.80 \times 10^{-4}) \approx 0.0200$ does not appear to cause serious error because the second term, representing the amount of Ba^{2+} arising from the dissociation of $Ba(IO_3)_2$, is only about 0.7% of 0.0200. Ordinarily, we consider an assumption of this type to be satisfactory if the discrepancy is less than 10%. Finally, then,

$$\text{solubility of } Ba(IO_3)_2 = \tfrac{1}{2}[IO_3^-] = \tfrac{1}{2} \times 2.80 \times 10^{-4} = 1.40 \times 10^{-4} \text{ M}$$

If we compare this result with the solubility of barium iodate in pure water (Example 7-3), we see that the presence of a small excess of the common ion has lowered the molar solubility of $Ba(IO_3)_2$ by a factor of about five.

Example 7-5

Calculate the solubility of $Ba(IO_3)_2$ in a solution prepared by mixing 200 mL of 0.0100 M $Ba(NO_3)_2$ with 100 mL of 0.100 M $NaIO_3$.

We must first establish whether either reactant is present in excess at equilibrium. The amounts taken are

$$\text{amount } Ba^{2+} = 200 \text{ mL} \times 0.0100 \text{ mmol/mL} = 2.00 \text{ mmol}$$
$$\text{amount } IO_3^- = 100 \text{ mL} \times 0.100 \text{ mmol/mL} = 10.0 \text{ mmol}$$

If formation of $Ba(IO_3)_2$ is complete,

$$\text{amount excess } NaIO_3 = 10.0 \text{ mmol} - 2 \times 2.00 \text{ mmol} = 6.0 \text{ mmol}$$

Thus,

$$[IO_3^-] = \frac{6.0 \text{ mmol}}{300 \text{ mL}} = 0.0200 \text{ M}$$

As in Example 7-3,

$$\text{molar solubility of Ba(IO}_3)_2 = [Ba^{2+}]$$

Here, however,

$$[IO_3^-] = 0.0200 + 2[Ba^{2+}]$$

where $2[Ba^{2+}]$ represents the iodate contributed by the sparingly soluble $Ba(IO_3)_2$. We can obtain a provisional answer after making the assumption that $[IO_3^-] \approx 0.0200$; thus

$$\text{solubility of Ba(IO}_3)_2 = [Ba^{2+}] = \frac{K_{sp}}{[IO_3^-]^2} = \frac{1.57 \times 10^{-9}}{(0.0200)^2}$$
$$= 3.93 \times 10^{-6} \text{ mol/L}$$

The approximation appears to be reasonable.

The uncertainty in $[IO_3^-]$ is 0.1 part in 6.0 or 1 part in 60. Thus $0.0200 \times (1/60) = 0.0003$ and we round to 0.0200 M.

Note that the results from the last two examples demonstrate that an excess of iodate ion is more effective in decreasing the solubility of $Ba(IO_3)_2$ than is the same excess of barium ion.

A 0.02 M excess of Ba^{2+} lowers the solubility of $Ba(IO_3)_2$ by a factor of about five; this same excess of IO_3^- lowers the solubility by roughly 200.

Acid and Base Dissociation Constants

When a weak acid or a weak base is dissolved in water, partial dissociation occurs. Thus, for nitrous acid, we can write

$$HNO_2 + H_2O \rightleftharpoons H_3O^+ + NO_2^- \qquad \frac{[H_3O^+][NO_2^-]}{[HNO_2]} = K_a$$

where K_a is the *acid dissociation constant* for nitrous acid. In an analogous way, the *base dissociation constant* for ammonia is

$$NH_3 + H_2O \rightleftharpoons NH_4^+ + OH^- \qquad \frac{[NH_4^+][OH^-]}{[NH_3]} = K_b$$

Note that $[H_2O]$ does not appear in the denominator of either equation because its concentration is so large relative to the concentration of the weak acid or base that the dissociation does not alter $[H_2O]$ appreciably (see Feature 7-3). Just as in the derivation of the ion-product constant for water, $[H_2O]$ is incorporated in the equilibrium constants K_a and K_b. Dissociation constants for weak acids and weak bases are found in Appendixes 3 and 4.

Dissociation Constants for Conjugate Acid/Base Pairs
Consider the dissociation-constant expressions for ammonia and its conjugate acid, ammonium ion:

$$NH_3 + H_2O \rightleftharpoons NH_4^+ + OH^- \qquad \frac{[NH_4^+][OH^-]}{[NH_3]} = K_b$$

$$NH_4^+ + H_2O \rightleftharpoons NH_3 + H_3O^+ \qquad \frac{[NH_3][H_3O^+]}{[NH_4^+]} = K_a$$

Multiplication of one equilibrium-constant expression by the other gives

$$\frac{[NH_3][H_3O^+]}{[NH_4^+]} \times \frac{[NH_4^+][OH^-]}{[NH_3]} = [H_3O^+][OH^-] = K_a K_b$$

but

$$[H_3O^+][OH^-] = K_w$$

and therefore

$$K_a K_b = K_w \qquad (7\text{-}12)$$

This relationship is general for all conjugate acid/base pairs. Most tables of dissociation constants do not list both the acid and the base dissociation constant for conjugate pairs since it is so easy to calculate one from the other with Equation 7-12.

Feature 7-4
RELATIVE STRENGTHS OF CONJUGATE ACID/BASE PAIRS

Equation 7-12 confirms the observation in Figure 7-1 that, as the acid of a conjugate acid/base pair becomes weaker, the conjugate base becomes stronger and vice versa. Thus the conjugate base of an acid with a dissociation constant of 10^{-2} has a basic dissociation constant of 10^{-12}, whereas an acid with a dissociation constant of 10^{-9} has a conjugate base with a dissociation constant of 10^{-5}.

Example 7-6

What is K_b for the equilibrium

$$CN^- + H_2O \rightleftharpoons HCN + OH^-$$

Examination of Appendix 4 (dissociation constants for bases) reveals no entry for CN^-. Appendix 3, however, lists a K_a value of 2.1×10^{-9} for

HCN. Thus,

$$\frac{[HCN][OH^-]}{[CN^-]} = K_b = \frac{K_w}{K_{HCN}}$$

$$K_b = \frac{1.00 \times 10^{-14}}{2.1 \times 10^{-9}} = 4.8 \times 10^{-6}$$

Hydronium Ion Concentration in Solutions of Weak Acids

When the weak acid HA is dissolved in water, two equilibria are established that yield hydronium ions:

$$HA + H_2O \rightleftharpoons H_3O^+ + A^- \qquad \frac{[H_3O^+][A^-]}{[HA]} = K_a$$

$$2\ H_2O \rightleftharpoons H_3O^+ + OH^- \qquad [H_3O^+][OH^-] = K_w$$

Ordinarily, the hydronium ions produced from the first reaction suppress the dissociation of water to such an extent that the contribution of hydronium ions from the second equilibrium is negligible. Under these circumstances, one H_3O^+ ion is formed for each A^- ion, and we write

$$[A^-] \approx [H_3O^+] \qquad (7\text{-}13)$$

Furthermore, the sum of the molar concentrations of the weak acid and its conjugate base must equal the analytical concentration of the acid C_{HA} because the solution contains no other source of A^- ions. Thus,

$$C_{HA} = [A^-] + [HA] \qquad (7\text{-}14)$$

Substituting $[H_3O^+]$ for $[A^-]$ (Equation 7-13) in Equation 7-14 yields

$$C_{HA} = [H_3O^+] + [HA]$$

which rearranges to

$$[HA] = C_{HA} - [H_3O^+] \qquad (7\text{-}15)$$

When $[A^-]$ and $[HA]$ are replaced by their equivalent terms from Equations 7-13 and 7-15, the equilibrium-constant expression becomes

$$\frac{[H_3O^+]^2}{C_{HA} - [H_3O^+]} = K_a \qquad (7\text{-}16)$$

which rearranges to

$$[H_3O^+]^2 + K_a[H_3O^+] - K_a C_{HA} = 0 \qquad (7\text{-}17)$$

The positive solution to this quadratic equation is

$$[H_3O^+] = \frac{-K_a + \sqrt{K_a^2 + 4K_aC_{HA}}}{2} \tag{7-18}$$

As an alternative to using Equation 7-18, Equation 7-17 may be solved by successive approximations as shown in Feature 7-5.

Equation 7-15 can frequently be simplified by making the assumption that dissociation does not appreciably decrease the molar concentration of HA. Thus, provided $[H_3O^+] \ll C_{HA}$, $C_{HA} - [H_3O^+] \approx C_{HA}$, and Equation 7-16 reduces to

$$\frac{[H_3O^+]^2}{C_{HA}} = K_a$$

$$[H_3O^+] = \sqrt{K_aC_{HA}} \tag{7-19}$$

The magnitude of the error introduced by the assumption that $[H_3O^+] \ll C_{HA}$ increases as the molar concentration of acid becomes smaller and as the dissociation constant for the acid becomes larger. This statement is supported by the data in Table 7-4. Note that the error introduced by the assumption is 0.5% when the ratio C_{HA}/K_a is 10^4. The error increases to 1.6% when the ratio is 10^3, to 5.3% when it is 10^2, and to 17% when it is 10. Figure 7-2 illustrates the effect graphically. Note that the hydronium ion

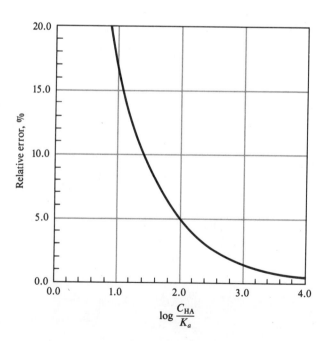

Figure 7-2

Relative error resulting from the assumption that $[H_3O^+] \ll C_{HA}$ in Equation 7-15.

Table 7-4

ERROR INTRODUCED BY ASSUMING H_3O^+ CONCENTRATION IS SMALL RELATIVE TO C_{HA} IN EQUATION 7-15

K_a	C_{HA}	$[H_3O^+]$ Using Assumption	$[H_3O^+]$ Using More Exact Equation	Percent Error
1.00×10^{-2}	1.00×10^{-3}	3.16×10^{-3}	0.92×10^{-3}	244
	1.00×10^{-2}	1.00×10^{-2}	0.62×10^{-2}	61
	1.00×10^{-1}	3.16×10^{-2}	2.70×10^{-2}	17
1.00×10^{-4}	1.00×10^{-4}	1.00×10^{-4}	0.62×10^{-4}	61
	1.00×10^{-3}	3.16×10^{-4}	2.70×10^{-4}	17
	1.00×10^{-2}	1.00×10^{-3}	0.95×10^{-3}	5.3
	1.00×10^{-1}	3.16×10^{-3}	3.11×10^{-3}	1.6
1.00×10^{-6}	1.00×10^{-5}	3.16×10^{-6}	2.70×10^{-6}	17
	1.00×10^{-4}	1.00×10^{-5}	0.95×10^{-5}	5.3
	1.00×10^{-3}	3.16×10^{-5}	3.11×10^{-5}	1.6
	1.00×10^{-2}	1.00×10^{-4}	9.95×10^{-5}	0.5
	1.00×10^{-1}	3.16×10^{-4}	3.16×10^{-4}	0.0

concentration computed with the approximation becomes equal to or greater than the molar concentration of the acid when the ratio is unity or smaller, which is clearly a meaningless result.

In general, it is good practice to make the simplifying assumption and obtain a trial value for $[H_3O^+]$ that can be compared with C_{HA} in Equation 7-15. If the trial value alters [HA] by an amount smaller than the allowable error in the calculation, the solution may be considered satisfactory. Otherwise, a better value for $[H_3O^+]$ must be obtained by solving the quadratic equation rigorously or by the method of successive approximations (Feature 7-5). Alternatively, Computer Application 7-1 can be used to give an exact result.

Example 7-7

Calculate the hydronium ion concentration in 0.120 M nitrous acid. The principal equilibrium is

$$HNO_2 + H_2O \rightleftarrows H_3O^+ + NO_2^-$$

for which (Appendix 3)

$$\frac{[H_3O^+][NO_2^-]}{[HNO_2]} = K_a = 5.1 \times 10^{-4}$$

Substitution into Equations 7-13 and 7-15 gives

$$[NO_2^-] = [H_3O^+]$$
$$[HNO_2] = 0.120 - [H_3O^+]$$

When these relationships are introduced into the expression for K_a, we obtain

$$\frac{[H_3O^+]^2}{0.120 - [H_3O^+]} = 5.1 \times 10^{-4}$$

If we now assume $[H_3O^+] \ll 0.120$, we find

$$[H_3O^+] = \sqrt{0.120 \times 5.1 \times 10^{-4}} = 7.8 \times 10^{-3} \text{ M}$$

We now examine the assumption that $0.120 - 0.0078 \approx 0.120$ and see that the error is about 7%. The relative error in $[H_3O^+]$ is smaller than this, however, as we can see by calculating $\log(C_{HA}/K_a) = 2.4$, which, from Figure 7-2, suggests an error of about 3%. If a more accurate value is needed, solution of the quadratic equation yields 7.6×10^{-3} M for the hydronium ion concentration.

Example 7-8

Calculate the hydronium ion concentration in a solution that is 2.0×10^{-4} M in aniline hydrochloride ($C_6H_5NH_3Cl$).

In aqueous solution, dissociation of the salt to Cl^- and $C_6H_5NH_3^+$ is complete. The weak acid $C_6H_5NH_3^+$ dissociates as follows:

$$C_6H_5NH_3^+ + H_2O \rightleftarrows C_6H_5NH_2 + H_3O^+ \qquad \frac{[H_3O^+][C_6H_5NH_2]}{[C_6H_5NH_3^+]} = K_a$$

Inspection of Appendix 3 reveals no entry for $C_6H_5NH_3^+$, but Appendix 4 gives a base constant for aniline ($C_6H_5NH_2$):

$$C_6H_5NH_2 + H_2O \rightleftarrows C_6H_5NH_3^+ + OH^- \qquad K_b = 3.94 \times 10^{-10}$$

The acid dissociation constant is then obtained from Equation 7-12:

$$K_a = \frac{1.00 \times 10^{-14}}{3.94 \times 10^{-10}} = 2.54 \times 10^{-5}$$

Proceeding as in Example 7-7, we have

$$[H_3O^+] = [C_6H_5NH_2]$$
$$[C_6H_5NH_3^+] = 2.0 \times 10^{-4} - [H_3O^+]$$

Let us assume that $[H_3O^+] \ll 2.0 \times 10^{-4}$ and substitute the simplified value for $[C_6H_5NH_3^+]$ into the dissociation-constant expression:

$$\frac{[H_3O^+]^2}{2.0 \times 10^{-4}} = 2.54 \times 10^{-5}$$
$$[H_3O^+] = \sqrt{5.08 \times 10^{-9}} = 7.1 \times 10^{-5} \text{ M}$$

Comparison of 7.1×10^{-5} with 2.0×10^{-4} suggests that a significant error has been introduced by the assumption that $[H_3O^+] \ll C_{C_6H_5NH_3^+}$ (Figure 7-2 indicates that this error is about 20%). Thus, unless only an approximate value for $[H_3O^+]$ is needed, it is necessary to use the expression

$$\frac{[H_3O^+]^2}{2.0 \times 10^{-4} - [H_3O^+]} = 2.54 \times 10^{-5}$$

which rearranges to

$$[H_3O^+]^2 + 2.54 \times 10^{-5}[H_3O^+] - 5.08 \times 10^{-9} = 0$$

$$[H_3O^+] = \frac{-2.54 \times 10^{-5} + \sqrt{(2.54 \times 10^{-5})^2 + 4 \times 5.08 \times 10^{-9}}}{2}$$

$$= 6.0 \times 10^{-5}$$

The quadratic equation can also be solved by the iterative method shown in Feature 7-5.

Feature 7-5
THE METHOD OF SUCCESSIVE APPROXIMATIONS

For convenience, let us write the quadratic equation in Example 7-8 in the form

$$x^2 + 2.54 \times 10^{-5}x - 5.08 \times 10^{-9} = 0$$

where $x = [H_3O^+]$.
    As a first step, let us rearrange the equation to the form

$$x = \sqrt{5.08 \times 10^{-9} - 2.54 \times 10^{-5}x}$$

We then assume that x on the right-hand side of the equation is zero and calculate a first value, x_1.

$$x_1 = \sqrt{5.08 \times 10^{-9} - 2.54 \times 10^{-5} \times 0} = 7.13 \times 10^{-5}$$

We then substitute this value into the original equation and compute a second value x_2:

$$x_2 = \sqrt{5.08 \times 10^{-9} - 2.54 \times 10^{-5} \times 7.13 \times 10^{-5}} = 5.72 \times 10^{-5}$$

Repeating this calculation gives

$$x_3 = \sqrt{5.08 \times 10^{-9} - 2.54 \times 10^{-5} \times 5.72 \times 10^{-5}} = 6.02 \times 10^{-5}$$

Continuing in the same way we obtain

$$x_4 = 5.96 \times 10^{-5}$$
$$x_5 = 5.97 \times 10^{-5}$$
$$x_6 = 5.97 \times 10^{-5}$$

Note that after three iterations, x_3 is 6.02×10^{-5}, which is within about 0.8% of the final value of 5.97×10^{-5} M.

Occasionally, the final result will oscillate between a high and a low value. In this case, you may have to solve the quadratic equation.

The method of successive approximations is particularly useful when cubic or higher-power equations need to be solved.

Computer Application 7-1
EUREKA AND HIGH ORDER EQUATIONS

Eureka: The Solver makes short work of high order equations in one variable. The equation discussed in Feature 7-5 is solved by typing the equation directly into the Edit window of Eureka, as shown below. Then type $\boxed{\text{ESC}}$ $\boxed{\text{S}}$ and the solution will appear in the window as shown.

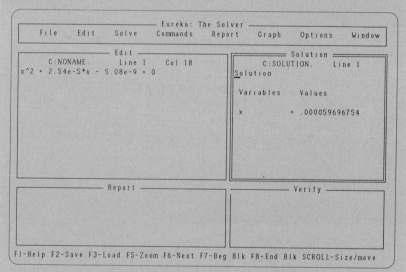

Try this equation and verify the solution obtained in Feature 7-5. Enter and solve the following equation:

$$x^4 + 8x^3 + 19.25x^2 - 17.5x - 122 = 0$$

Finally, use Eureka to solve the cubic equation in Example 7-4 and verify that the approximations that we made were valid. You may want to use the iterate command ($\boxed{\text{ESC}}$ $\boxed{\text{C}}$$\boxed{\text{I}}$) to reach an acceptable solution.

Hydronium-Ion Concentration in Solutions of Weak Bases

The techniques discussed in previous sections are readily adapted to the calculation of the hydroxide or hydronium ion concentration in solutions of weak bases.

Aqueous ammonia is basic by virtue of the reaction

$$NH_3 + H_2O \rightleftarrows NH_4^+ + OH^-$$

The predominant species in such solutions has been clearly demonstrated to be NH_3. Nevertheless, solutions of ammonia are sometimes called ammonium hydroxide because at one time chemists thought that NH_4OH rather than NH_3 was the undissociated form of the base. Application of the mass law to the equilibrium as written yields

$$\frac{[NH_4^+][OH^-]}{[NH_3]} = K_b$$

Example 7-9

Calculate the hydronium ion concentration of a 0.075 M NH_3 solution.
The predominant equilibrium is

$$NH_3 + H_2O \rightleftarrows NH_4^+ + OH^-$$

From Appendix 3

$$\frac{[NH_4^+][OH^-]}{[NH_3]} = 1.76 \times 10^{-5} = K_b$$

The concentration of OH^- from water dissociation is negligible. Therefore,

$$[NH_4^+] = [OH^-]$$

Both NH_4^+ and NH_3 come from the 0.075 M solution. Thus

$$[NH_4^+] + [NH_3] = C_{NH_3} = 0.075 \text{ M} \tag{1}$$

If we substitute $[OH^-]$ for $[NH_4^+]$ in Equation 1 and rearrange, we find that

$$[NH_3] = 0.075 - [OH^-]$$

Substituting these quantities into the dissociation-constant expression yields

$$\frac{[OH^-]^2}{7.5 \times 10^{-2} - [OH^-]} = 1.76 \times 10^{-5}$$

Provided that $[OH^-] \ll 7.5 \times 10^{-2}$, this equation simplifies to

$$[OH^-]^2 \approx 7.5 \times 10^{-2} \times 1.76 \times 10^{-5}$$
$$[OH^-] = 1.15 \times 10^{-3} \text{ M}$$

Upon comparing the calculated value for $[OH^-]$ with 7.5×10^{-2}, we see that the error in $[OH^-]$ is less than 2%. If needed, a better value for $[OH^-]$ can be obtained by solving the quadratic equation.

Finally, then,

$$[H_3O^+] = \frac{K_w}{[OH^-]} = \frac{1.00 \times 10^{-14}}{1.15 \times 10^{-3}} = 8.7 \times 10^{-12} \text{ M}$$

Example 7-10

Calculate the hydronium ion concentration in a 0.010 M sodium hypochlorite solution.

The equilibrium between OCl^- and water is

$$OCl^- + H_2O \rightleftarrows HOCl + OH^-$$

for which

$$\frac{[HOCl][OH^-]}{[OCl^-]} = K_b$$

Appendix 4 does not contain a value for K_b, but Appendix 3 reveals that the acid dissociation constant for HOCl is 3.0×10^{-8}. Therefore, we rearrange Equation 7-12 and write

$$K_b = \frac{K_w}{K_a} = \frac{1.00 \times 10^{-14}}{3.0 \times 10^{-8}} = 3.3 \times 10^{-7}$$

Proceeding as in Example 7-9, we have

$$[OH^-] = [HOCl]$$
$$[OCl^-] + [HOCl] = 0.010$$
$$[OCl^-] = 0.010 - [OH^-] \approx 0.010$$

Here we have assumed that $[OH^-] \ll 0.010$. Substitution into the equilibrium-constant expression gives

$$\frac{[OH^-]^2}{0.010} = 3.3 \times 10^{-7}$$
$$[OH^-] = 5.74 \times 10^{-5} \text{ M}$$

Note that the error resulting from the approximation is small. To obtain the hydronium ion concentration, we write

$$[H_3O^+] = \frac{1.00 \times 10^{-14}}{5.74 \times 10^{-5}} = 1.7 \times 10^{-10} \text{ M}$$

7C QUESTIONS AND PROBLEMS

7-1. Define
  *(a) salt.
  (b) electrolyte.
  *(c) hydronium ion.
  (d) conjugate base.
  *(e) base.
  (f) mass law.
  *(g) Le Châtelier principle.
  (h) autoprotolysis.
  *(i) amphiprotic solvent.
  (j) common-ion effect.

*7-2. Explain why water is termed a leveling solvent toward mineral acids, whereas ethanol is a differentiating solvent.

*7-3. The solubility product for Ag_2CO_3 is 8.1×10^{-12} and that for AgCl is 1.8×10^{-10}. Why is it wrong to conclude that Ag_2CO_3 is less soluble than silver chloride because its K_{sp} is smaller?

7-4. Write the equilibrium-constant expressions, and give the numerical values for each constant in
  *(a) the basic dissociation of ethylamine ($C_2H_5NH_2$).
  (b) the acidic dissociation of hydrogen cyanide (HCN).
  *(c) the acidic dissociation of pyridine hydrochloride (C_5H_5NHCl).
  (d) the basic dissociation of NaCN.
  *(e) the dissociation of H_3AsO_4 to H_3O^+ and AsO_4^{3-}.
  (f) the reaction of CO_3^{2-} with H_2O to give H_2CO_3 and OH^-.

7-5. Calculate the solubility-product constant for each of the following substances, given that the molar concentrations of their saturated solutions are as indicated:
  *(a) AgSeCN (2.0×10^{-8} mol/L; products are Ag^+ and $SeCN^-$).
  (b) $RaSO_4$ (6.6×10^{-6} mol/L).
  *(c) $Pb(BrO_3)_2$ (1.7×10^{-1} mol/L).
  (d) $PbBr_2$ (2.1×10^{-2} mol/L).
  *(e) $Ce(IO_3)_3$ (1.9×10^{-3} mol/L).
  (f) BiI_3 (1.3×10^{-5} mol/L).

*7-6. Calculate the weight (in grams) of PbI_2 that dissolves in 100 mL of
  (a) H_2O.
  (b) 2.00×10^{-2} M KI.
  (c) 2.00×10^{-2} M $Pb(NO_3)_2$.

7-7. Calculate the weight (in grams) of $La(IO_3)_3$ that dissolves in 100 mL of
  (a) H_2O.
  (b) 2.00×10^{-2} M $NaIO_3$.
  (c) 2.00×10^{-2} $La(NO_3)_3$.

*7-8. The solubility product for Tl_2CrO_4 is 9.8×10^{-13}. What CrO_4^{2-} concentration is required to
  (a) initiate precipitation of Tl_2CrO_4 from a solution that is 2.12×10^{-3} M in Tl^+?
  (b) lower the concentration of Tl^+ in a solution to 1.00×10^{-6} M?

7-9. What hydroxide ion concentration is required to
  (a) initiate precipitation of $Mg(OH)_2$ from a 1.00×10^{-3} M solution of $MgSO_4$?
  (b) lower the Mg^{2+} concentration in the foregoing solution to 1.00×10^{-6} M?

*7-10. The solubility-product constant for $Ce(IO_3)_3$ is 3.2×10^{-10}. What is the Ce^{3+} concentration in a solution prepared by mixing 50.0 mL of 0.0500 M Ce^{3+} with 50.00 mL of
  (a) water?
  (b) 0.050 M IO_3^-?
  (c) 0.150 M IO_3^-?
  (d) 0.300 M IO_3^-?

7-11. The solubility-product constant for K_2PtCl_6 is 1.1×10^{-5} ($K_2PtCl_6 \rightleftarrows 2 K^+ + PtCl_6^{2-}$). What is the K^+ concentration of a solution prepared by mixing 50.0 mL of 0.400 M KCl with 50.0 mL of
  (a) 0.100 M $PtCl_6^{2-}$?
  (b) 0.200 M $PtCl_6^{2-}$?
  (c) 0.400 M $PtCl_6^{2-}$?

7-12. At 25°C, what are the molar H_3O^+ and OH^- concentrations in
  *(a) 0.0200 M HOCl?
  (b) 0.0800 M propanoic acid?
  *(c) 0.200 M methylamine?
  (d) 0.100 M trimethylamine?
  *(e) 0.120 M NaOCl?
  (f) 0.0860 M CH_3COONa?
  *(g) 0.100 M hydroxylamine hydrochloride?
  (h) 0.0500 M ethanolamine hydrochloride?

*7-13. What is the hydronium ion concentration in water at 0°C?

7-14. At 25°C, what is the hydronium ion concentration in
  *(a) 0.100 M chloroacetic acid?
  *(b) 0.100 M sodium chloroacetate?
  (c) 0.0100 M methylamine?
  (d) 0.0100 M methylamine hydrochloride?
  *(e) 1.00×10^{-3} M aniline hydrochloride?
  (f) 0.200 M HIO_3?

*7-15. The solubility products for a series of iodides are

TlI	$K_{sp} = 6.5 \times 10^{-8}$
AgI	$K_{sp} = 8.3 \times 10^{-17}$
PbI$_2$	$K_{sp} = 7.1 \times 10^{-9}$
BiI$_3$	$K_{sp} = 8.1 \times 10^{-19}$

List these four compounds in order of decreasing molar solubility in

(a) water.

(b) 0.10 M NaI.

(c) a 0.010 M solution of the solute cation.

7-16. The solubility products for a series of iodates are

AgIO$_3$	$K_{sp} = 3.0 \times 10^{-8}$
Sr(IO$_3$)$_2$	$K_{sp} = 3.3 \times 10^{-7}$
La(IO$_3$)$_3$	$K_{sp} = 6.2 \times 10^{-12}$
Ce(IO$_3$)$_4$	$K_{sp} = 4.7 \times 10^{-17}$

List these four compounds in order of decreasing molar solubility in

(a) water.

(b) 0.10 M NaIO$_3$.

(c) a 0.10 M solution of the solute cation.

CHAPTER 8

ACTIVITIES
AND ACTIVITY COEFFICIENTS

We have noted that Equation 7-7 (page 113) is an approximate form of the thermodynamic equilibrium-constant expression derived from theory. In this chapter we describe when the use of this approximate form is likely to lead to error. We also show how the equation is modified to give a more exact description of a system at chemical equilibrium.

8A THE EFFECT OF ELECTROLYTES ON CHEMICAL EQUILIBRIA

Experimentally we find that the position of most solution equilibria depends upon the electrolyte concentration of the medium. This happens even when the added electrolyte contains no ion in common with those involved in the equilibrium. For example, consider again the oxidation of iodide ion by arsenic acid:

$$H_3AsO_4 + 3\,I^- + 2\,H^+ \rightleftharpoons H_3AsO_3 + I_3^- + H_2O$$

If an electrolyte, such as barium nitrate, potassium sulfate, or sodium perchlorate, is added to this solution, the color becomes less intense. This decrease in color intensity indicates that the concentration of I_3^- has decreased and that the equilibrium has been shifted to the left by the added electrolyte.

Figure 8-1 further illustrates the effect of electrolytes. Curve A is a plot of the product of the molar hydronium and hydroxide ion *concentrations*

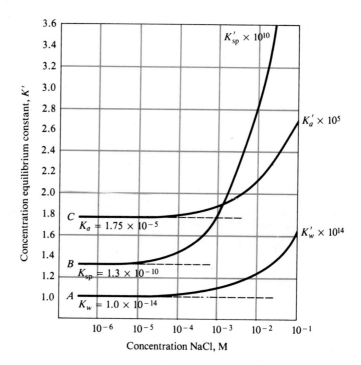

Figure 8-1
Effect of electrolyte concentration on concentration-based equilibrium constants.

Concentration-based equilibrium constants are commonly indicated by adding a prime mark. For example: K'_w, K'_{sp}, K'_a.

($\times 10^{14}$) as a function of the concentration of sodium chloride. This *concentration*-based ion product is designated K'_w. At low sodium chloride concentrations, K'_w becomes independent of the electrolyte concentration and is equal to 1.00×10^{-14}, which is the *thermodynamic* ion-product constant for water, K_w. This relationship, and others like it, where the numerical value for the relationship reaches a constant value as some concentration parameter approaches zero, is called a *limiting law;* the numerical constant obtained at the limit is referred to as a *limiting value*.

The vertical axis for curve B in Figure 8-1 is the product of the molar concentrations of barium and sulfate ions ($\times 10^{10}$) in saturated solutions of barium sulfate. This concentration-based solubility product is designated as K'_{sp}. At low electrolyte concentrations, K'_{sp} has a limiting value of 1.3×10^{-10}, which is the accepted thermodynamic value of K_{sp} for barium sulfate.

Curve C is a plot of K'_a ($\times 10^5$), the concentration quotient for the equilibrium involving the dissociation of acetic acid, as a function of electrolyte concentration. Here again, the ordinate function approaches a limiting value K_a, which is the thermodynamic acid dissociation constant for acetic acid.

As the electrolyte concentration becomes very small, concentration-based equilibrium constants approach their thermodynamic values: K_w, K_{sp}, K_a.

The dashed lines in Figure 8-1 represent ideal behavior of the solutes. Note that departures from ideality can be significant. For example, the product of the molar concentrations of hydrogen and hydroxide ion increases from 1.0×10^{-14} in pure water to about 1.7×10^{-14} in a solution that is 0.1 M in sodium chloride. The effect is even more pronounced with barium sulfate; here, K'_{sp} in 0.1 M sodium chloride is more than double that of its limiting value.

The electrolyte effect shown in Figure 8-1 is not peculiar to sodium chloride. Indeed, we would see identical curves when potassium nitrate or sodium perchlorate is substituted for sodium chloride. In each case, the effect has as its origin the electrostatic attraction between the ions of the electrolyte and the ions of reacting species of opposite charge. Since the electrostatic forces associated with all singly charged ions are approximately the same, the three salts exhibit essentially identical effects on equilibria.

We must now consider how we can take into account the electrolyte effect when we wish to make more accurate equilibrium calculations.

8A-1 The Effect of Reactant and Product Charge

Extensive studies have revealed that the magnitude of the electrolyte effect is highly dependent upon the charges of the participants in an equilibrium. When only neutral species are involved, the position of equilibrium is essentially independent of electrolyte concentration. With ionic participants, the magnitude of the electrolyte effect increases with charge. This generality is demonstrated by the three solubility curves in Figure 8-2. Note, for example, that in a 0.02 M solution of potassium nitrate, the solubility of barium sulfate, with its pair of doubly charged ions, is larger than it is in pure water by a factor of 2. This same change in electrolyte concentration increases the solubility of barium iodate by a factor of only 1.25 and that of silver chloride by 1.2. The enhanced effect due to doubly charged ions is also reflected in the greater slope of curve B in Figure 8-1.

8A-2 The Effect of Ionic Strength

Systematic studies have shown that the effect of added electrolyte on equilibria is independent of the chemical nature of the electrolyte but depends upon a property of the solution called the *ionic strength*. This quantity is defined as

$$\text{ionic strength} = \mu = \tfrac{1}{2}([A]Z_A^2 + [B]Z_B^2 + [C]Z_C^2 + \cdots) \qquad \text{(8-1)}$$

where [A], [B], [C], . . . represent the molar concentrations of ions A, B, C, . . . and Z_A, Z_B, Z_C, . . . are their charges.

Figure 8-2

Effect of electrolyte concentration on the solubility of some salts.

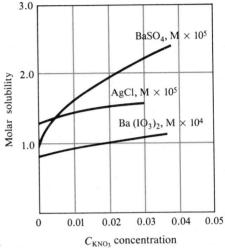

Example 8-1

Calculate the ionic strength of (a) a 0.1 M solution of KNO_3 and (b) a 0.1 M solution of Na_2SO_4.

(a) For the KNO_3 solution, $[K^+]$ and $[NO_3^-]$ are 0.1 M and

$$\mu = \tfrac{1}{2}(0.1 \times 1^2 + 0.1 \times 1^2) = 0.1$$

(b) For the Na_2SO_4 solution, $[Na^+] = 0.2$ and $[SO_4^{2-}] = 0.1$. Therefore,

$$\mu = \tfrac{1}{2}(0.2 \times 1^2 + 0.1 \times 2^2) = 0.3$$

Example 8-2

What is the ionic strength of a solution that is 0.05 M in KNO_3 and 0.1 M in Na_2SO_4?

$$\mu = \tfrac{1}{2}(0.05 \times 1^2 + 0.05 \times 1^2 + 0.2 \times 1^2 + 0.1 \times 2^2) = 0.35 \text{ M}$$

EFFECT OF CHARGE ON
IONIC STRENGTH

Type of Electrolyte	Example	Ionic Strength*
1:1	NaCl	C
1:2	$Ba(NO_3)_2$, Na_2SO_4	$3C$
1:3	$Al(NO_3)_3$, Na_3PO_4	$6C$
2:2	$MgSO_4$	$4C$

*C = molarity of the salt.

The activity of a species is a measure of its effective concentration as determined by the lowering of the freezing point of water, by electrical conductivity, and by the mass action effect.

As these examples show, the ionic strength of a solution of a strong electrolyte consisting solely of singly charged ions is identical with the total molar salt concentration. If the solution contains ions with multiple charges, however, the ionic strength is greater than the molar concentration.

For solutions with ionic strengths of 0.1 M or less, the electrolyte effect is independent of the *kind* of ions and dependent *only upon the ionic strength*. Thus, the degree of dissociation of acetic acid is the same in aqueous sodium chloride, potassium nitrate, or barium iodide, provided the concentrations of these species are such that their ionic strengths are identical. Note that this independence with respect to electrolyte species disappears at high ionic strengths.

8B   ACTIVITY COEFFICIENTS

In order to describe quantitatively the effect of ionic strength on equilibria, chemists use a term called *activity,* which for the species X is defined as

$$a_X = [X]f_X \tag{8-2}$$

where a_X is the activity of the species X, [X] is its molar concentration, and f_X is a dimensionless quantity called the *activity coefficient*. The activity coefficient and thus the activity of X vary with ionic strength such that substitution of a_X for [X] in any equilibrium-constant expression frees the numerical value of the constant from dependence on the ionic strength. To illustrate, for the generalized equilibrium we considered in Chapter 7 (page 113),

$$wW + xX \rightleftarrows yY + zZ$$

The thermodynamic equilibrium constant K is defined by the equation

$$K = \frac{a_Y^y \cdot a_Z^z}{a_W^w \cdot a_X^x} \tag{8-3}$$

Applying Equation 8-2 gives

$$K = \frac{[Y]^y[Z]^z}{[W]^w[X]^x} \times \frac{f_Y^y \cdot f_Z^z}{f_W^w \cdot f_X^x} = K' \frac{f_Y^y \cdot f_Z^z}{f_W^w \cdot f_X^x} \tag{8-4}$$

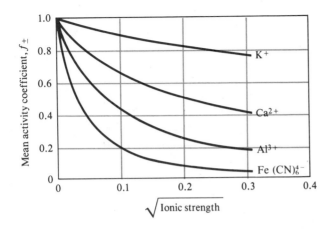

Figure 8-3
Effect of ionic strength on activity coefficients.

where

$$K' = \frac{[Y]^y [Z]^z}{[W]^w [X]^x}$$

Here K' is the *concentration equilibrium constant* and K is the thermodynamic equilibrium constant.[1] The activity coefficients f_W, f_X, f_Y, and f_Z vary with ionic strength in such a way as to keep K numerically constant and independent of ionic strength (in contrast to the concentration constants K' shown in Figure 8-1).

8B-1 Properties of Activity Coefficients

Activity coefficients have the following properties:

1. The activity coefficient of a species is a measure of the effectiveness with which that species influences an equilibrium in which it is a participant. In very dilute solutions, where the ionic strength is minimal, this effectiveness becomes constant, and the activity coefficient is unity. Under such circumstances, the activity and the molar concentration are identical (as are the thermodynamic and concentration equilibrium constants). As the ionic strength increases, however, an ion loses some of its effectiveness, and its activity coefficient decreases. We may summarize this behavior in terms of Equation 8-2. At moderate ionic strengths $f_X < 1$; as the solution approaches infinite dilution, however, $f_X \to 1$ and thus $a_X \to [X]$ and $K' \to K$. At high ionic strengths ($\mu > 0.1$ M), activity coefficients often increase and may even become greater than unity. Because interpretation of the behavior of solutions in this region is difficult, we shall confine our discussion to regions of low or moderate ionic strength (that is, where $\mu < 0.1$ M). The variation of typical activity coefficients as a function of ionic strength is shown in Figure 8-3.

As $\mu \to 0$, $f_X \to 1$, $a_X \to [X]$, and $K' \to K$.

---

[1] In the chapters that follow, we use the prime notation only when it is necessary to distinguish between thermodynamic and concentration equilibrium constants.

2. In solutions that are not too concentrated, the activity coefficient for a given species is independent of the nature of the electrolyte and dependent only upon the ionic strength.

3. For a given ionic strength, the activity coefficient of an ion departs farther from unity as the charge carried by the species increases. This effect is shown in Figure 8-3. The activity coefficient of an uncharged molecule is approximately unity, regardless of ionic strength.

4. At any given ionic strength, the activity coefficients of ions of the same charge are approximately equal. The small variations that do exist can be correlated with the effective diameter of the hydrated ions.

5. The activity coefficient of a given ion describes the ion's effective behavior in all equilibria in which it participates. For example, at a given ionic strength, a single activity coefficient for cyanide ion describes the influence of that species upon any of the following equilibria:

$$HCN + H_2O \rightleftharpoons H_3O^+ + CN^-$$

$$Ag^+ + CN^- \rightleftharpoons AgCN(s)$$

$$Ni^{2+} + 4\ CN^- \rightleftharpoons Ni(CN)_4^{2-}$$

8B-2 The Debye-Hückel Equation

As we have noted, the electrolyte effect results from the electrostatic attractive and repulsive forces that exist between the ions of an electrolyte and the ions involved in an equilibrium. These forces cause each ion from the dissociated reactant to be surrounded by a sheath of solution that contains a slight excess of electrolyte ions of opposite charge. For example, when a barium sulfate precipitate is equilibrated with a sodium chloride solution, each dissolved barium ion is surrounded by an ionic atmosphere that carries a net negative charge on the average due to repulsion of sodium ions and attraction of chloride ions. Similarly, each sulfate ion is surrounded by an ionic atmosphere that tends to be slightly positive. These charged layers make the barium ions somewhat less positive and the sulfate ions somewhat less negative than they would be in the absence of electrolyte. The consequence of this effect is a decrease in overall attraction between barium and sulfate ions and an increase in solubility, which becomes greater as the number of electrolyte ions in the solution becomes larger.

In 1923, P. Debye and E. Hückel used this model to derive a theoretical expression that permits the calculation of activity coefficients of ions from their charge and average size.[2] This equation, which has become known as the *Debye-Hückel equation,* takes the form

$$-\log f_X = \frac{0.51 Z_X^2 \sqrt{\mu}}{1 + 0.33 \alpha_X \sqrt{\mu}} \tag{8-5}$$

---

[2] P. Debye and E. Hückel, *Physik. Z.,* **1923,** *24,* 185.

Table 8-1

ACTIVITY COEFFICIENTS FOR IONS AT 25°C*

Ion	α_X, Å	Activity Coefficient at Indicated Ionic Strength				
		0.001	0.005	0.01	0.05	0.1
H_3O^+	9	0.967	0.933	0.914	0.86	0.83
Li^+, $C_6H_5COO^-$	6	0.965	0.929	0.907	0.84	0.80
Na^+, IO_3^-, HSO_3^-, HCO_3^-, $H_2PO_4^-$, $H_2AsO_4^-$, OAc^-	4–4.5	0.964	0.928	0.902	0.82	0.78
OH^-, F^-, SCN^-, HS^-, ClO_3^-, ClO_4^-, BrO_3^-, IO_4^-, MnO_4^-	3.5	0.964	0.926	0.900	0.81	0.76
K^+, Cl^-, Br^-, I^-, CN^-, NO_2^-, NO_3^-, $HCOO^-$	3	0.964	0.925	0.899	0.80	0.76
Rb^+, Cs^+, Tl^+, Ag^+, NH_4^+	2.5	0.964	0.924	0.898	0.80	0.75
Mg^{2+}, Be^{2+}	8	0.872	0.755	0.69	0.52	0.45
Ca^{2+}, Cu^{2+}, Zn^{2+}, Sn^{2+}, Mn^{2+}, Fe^{2+}, Ni^{2+}, Co^{2+}, $Phthalate^{2-}$	6	0.870	0.749	0.675	0.48	0.40
Sr^{2+}, Ba^{2+}, Cd^{2+}, Hg^{2+}, S^{2-}	5	0.868	0.744	0.67	0.46	0.38
Pb^{2+}, CO_3^{2-}, SO_3^{2-}, $C_2O_4^{2-}$	4.5	0.868	0.742	0.665	0.46	0.37
Hg_2^{2+}, SO_4^{2-}, $S_2O_3^{2-}$, CrO_4^{2-}, HPO_4^{2-}	4.0	0.867	0.740	0.660	0.44	0.36
Al^{3+}, Fe^{3+}, Cr^{3+}, La^{3+}, Ce^{3+}	9	0.738	0.54	0.44	0.24	0.18
PO_4^{3-}, $Fe(CN)_6^{3-}$	4	0.725	0.50	0.40	0.16	0.095
Th^{4+}, Zr^{4+}, Ce^{4+}, Sn^{4+}	11	0.588	0.35	0.255	0.10	0.065
$Fe(CN)_6^{4-}$	5	0.57	0.31	0.20	0.048	0.021

*From J. Kielland, *J. Am. Chem. Soc.*, **1937**, *59*, 1675. By courtesy of the American Chemical Society.

where

f_X = activity coefficient of the species X

Z_X = charge on the species X

μ = ionic strength of the solution

α_X = effective diameter of the hydrated ion X in angstrom units (1 Å = 10^{-8} cm)

The constants 0.51 and 0.33 are applicable to aqueous solutions at 25°C; other values must be employed at other temperatures.

Unfortunately, considerable uncertainty exists regarding the magnitude of α_X in Equation 8-5. Its value appears to be approximately 3 Å for most singly charged ions; for these species, then, the denominator of the Debye-Hückel equation simplifies to approximately $1 + \sqrt{\mu}$. For ions with higher charge, α_X may be as large as 10 Å. This increase in size with increase in charge makes good chemical sense. The larger the charge on an ion, the larger the number of polar water molecules that will be held in the solvation shell about the ion. It should be noted that the second term of the denominator in Equation 8-5 is small with respect to the first when the ionic strength is less than 0.01, so that at these ionic strengths, uncertainties in α_X are of little significance in calculating activity coefficients.

Kielland[3] has derived values of α_X for numerous ions from a variety of experimental data. His best values for effective diameters are given in Table 8-1. Also presented are activity coefficients calculated from Equation 8-5 using these values for the size parameter.

---

[3] J. Kielland, *J. Amer. Chem. Soc.*, **1937**, *59*, 1675.

Experimental determination of single-ion activity coefficients such as those shown in Table 8-1 is unfortunately impossible because all experimental methods give only a mean activity coefficient for the positively and negatively charged ions in a solution. In other words, it is impossible to measure the properties of individual ions in the presence of counter ions of opposite charge and solvent molecules. It should be pointed out, however, that mean activity coefficients calculated from the data in Table 8-1 agree satisfactorily with the experimental values.

Example 8-3

Use Equation 8-5 to calculate the activity coefficient for Hg^{2+} in a solution that has an ionic strength of 0.085. Use 5 Å for the effective diameter of the ion. Compare the calculated value with $f_{Hg^{2+}}$ obtained by interpolation of data from Table 8-1.

$$-\log f_{Hg^{2+}} = \frac{(0.51)(2)^2 \sqrt{0.085}}{1 + (0.33)(5) \sqrt{0.085}} = 0.4016$$

$$\frac{1}{f_{Hg^{2+}}} = 2.52$$

$$f_{Hg^{2+}} = 0.397 = 0.40$$

Table 8-1 indicates that $f_{Hg^{2+}} = 0.46$ when $\mu = 0.05$ and 0.38 when $\mu = 0.1$. Thus, for $\mu = 0.085$

$$f_{Hg^{2+}} = 0.38 + \frac{(0.10 - 0.085)}{(0.10 - 0.05)} \times (0.46 - 0.38) = 0.404 = 0.40$$

Values for activity coefficients at ionic strengths not listed in Table 8-1 can be approximated by interpolation as shown here.

The Debye-Hückel relationship and the data in Table 8-1 give satisfactory activity coefficients for ionic strengths up to about 0.1. Beyond this value, the equation fails, and experimentally determined mean activity coefficients must be used.

Feature 8-1
MEAN ACTIVITY COEFFICIENTS

The mean activity coefficient of the electrolyte $A_m B_n$ is defined as

$$f_{\pm} = \text{mean activity coefficient} = (f_A^m \cdot f_B^n)^{1/(m+n)}$$

The mean activity coefficient can be measured in any of several ways, but it is impossible experimentally to resolve this term into the individual activity coefficients for f_A and f_B. For example, if $A_m B_n$ is a precipitate, we can write

$$K_{sp} = [A]^m[B]^n \cdot f_A^m \cdot f_B^n = [A]^m[B]^n \cdot f_{\pm}^{m+n}$$

By measuring the solubility of A_mB_n in a solution in which the electrolyte concentration approaches zero (that is, where both f_A and $f_B \to$ 1), we can obtain K_{sp}. A second solubility measurement at some ionic strength μ_1 gives values for [A] and [B]. These data then permit the calculation of $f_A^m \cdot f_B^n = f_\pm^{m+n}$ for ionic strength μ_1.

It is important to understand that this procedure does not provide enough experimental data to permit the calculation of the *individual* quantities f_A and f_B and that there appears to be no additional experimental information that would permit evaluation of these quantities. This situation is general, and the *experimental* determination of individual activity coefficients is impossible.

8B-3 Equilibrium Calculations Employing Activity Coefficients

Equilibrium calculations with activities yield more correct results than those obtained with molar concentrations. Unless otherwise specified, equilibrium constants found in tables are generally based upon activities and are thus thermodynamic equilibrium constants. The example that follows illustrates how activity coefficients from Table 8-1 are applied to such data.

Example 8-4

Use activities to calculate the solubility of $Ba(IO_3)_2$ in a 0.033 M solution of $Mg(IO_3)_2$. The thermodynamic solubility product for $Ba(IO_3)_2$ is 1.57×10^{-9} (Appendix 2).

At the outset, we write the solubility-product expression in terms of activities:

$$a_{Ba^{2+}} \times a_{IO_3^-}^2 = K_{sp} = 1.57 \times 10^{-9}$$

where $a_{Ba^{2+}}$ and $a_{IO_3^-}$ are the activities of barium and iodate ions. Replacing activities in this equation by activity coefficients and concentrations from Equation 8-2 yields

$$[Ba^{2+}]f_{Ba^{2+}} \times [IO_3^-]^2 f_{IO_3^-}^2 = K_{sp}$$

where $f_{Ba^{2+}}$ and $f_{IO_3^-}$ are the activity coefficients for the two ions. Rearranging this expression gives

$$[Ba^{2+}][IO_3^-]^2 = \frac{K_{sp}}{f_{Ba^{2+}} \times f_{IO_3^-}^2} = K_{sp}' \tag{8-6}$$

where K_{sp}' is the *concentration-based solubility product*.

The ionic strength of the solution is obtained by substituting into Equation 8-1:

$$\mu = \tfrac{1}{2}([Mg^{2+}] \times 2^2 + [IO_3^-] \times 1^2)$$
$$= \tfrac{1}{2}(0.033 \times 4 + 0.066 \times 1) = 0.099 \approx 0.1$$

In calculating μ, we have assumed that the Ba^{2+} and IO_3^- ions from the precipitate do not significantly affect the ionic strength of the solution. This simplification seems justified, considering the low solubility of barium iodate and the relatively high concentration of $Mg(IO_3)_2$. In situations where it is not possible to make such an assumption, the concentrations of the two ions can be approximated by solubility calculations in which activities and concentrations are assumed to be identical (as in Examples 7-3 and 7-4). These concentrations can then be introduced to give a better value for μ.

Turning now to Table 8-1, we find that at an ionic strength of 0.1,

$$f_{Ba^{2+}} = 0.38 \qquad f_{IO_3^-} = 0.78$$

If the calculated ionic strength did not match that of one of the columns in the table, $f_{Ba^{2+}}$ and $f_{IO_3^-}$ could be calculated from Equation 8-5 or by interpolation.

Substituting into the thermodynamic solubility-product expression (Equation 8-6) gives

$$\frac{1.57 \times 10^{-9}}{(0.38)(0.78)^2} = 6.8 \times 10^{-9} = K'_{sp}$$

$$[Ba^{2+}][IO_3^-]^2 = 6.8 \times 10^{-9}$$

Proceeding now as in earlier solubility calculations,

$$\text{solubility} = [Ba^{2+}]$$
$$[IO_3^-] = 0.066$$
$$[Ba^{2+}](0.066)^2 = 6.8 \times 10^{-9}$$
$$[Ba^{2+}] = \text{solubility} = 1.56 \times 10^{-6} \text{ M}$$

It is interesting to note that the calculated solubility if we neglect the effects of ionic strength is 3.6×10^{-7} M, which is less than one fourth as large as the value obtained here.

8B-4 Omission of Activity Coefficients in Equilibrium Calculations

We shall ordinarily neglect activity coefficients and simply use molar concentrations in applications of the equilibrium law. This recourse simplifies the calculations and greatly decreases the amount of data needed. For most purposes, the error introduced by the assumption of unity for the activity coefficient is not large enough to lead to false conclusions. It should be apparent from the preceding examples, however, that disregard of activity coefficients may introduce a significant numerical error in calculations of this kind. Note, for example, that an error greater than -75% was incurred in Example 8-4. The reader should be alert to the conditions under which the substitution of concentration for activity is likely to lead to the largest error. Significant discrepancies occur when the ionic strength is large (0.01 or larger) and when the ions involved have multiple charges (Table 8-1). With dilute solutions (ionic strength < 0.01) of nonelectrolytes or of singly charged ions, the use of concentrations in a mass-law calculation often provides reasonably accurate results.

It is also important to note that the decrease in solubility resulting from the presence of an ion common to the precipitate is in part counteracted by the larger electrolyte concentration associated with the presence of the salt containing the common ion. This effect is illustrated in Example 8-4.

8C QUESTIONS AND PROBLEMS

*8-1. What is a thermodynamic equilibrium constant?

8-2. What is the numerical value of the activity coefficient of acetic acid (CH_3COOH)?

*8-3. Why is the slope of curve B in Figure 8-1 greater than that of curve A?

8-4. What is a mean activity coefficient?

8-5. Calculate the ionic strength of a solution that is
*(a) 0.040 M in $MgSO_4$.
(b) 0.20 M in Na_2SO_4.
*(c) 0.10 M in $Al(NO_3)_3$ and 0.20 M in $Mg(NO_3)_2$.
(d) 0.06 M in $LaCl_3$ and 0.030 M in Na_2SO_4.

8-6. Using the data in Table 8-1, calculate the mean activity coefficient for each of the following at an ionic strength of 0.050:
*(a) Na_2SO_4 *(c) $Na_4Fe(CN)_6$
(b) $Ba(NO_3)_2$ (d) $AlCl_3$

8-7. Use Equation 8-5 to calculate the activity coefficient of
*(a) Fe^{3+} at $\mu = 0.075$.   *(c) Ce^{4+} at $\mu = 0.080$.
(b) Pb^{2+} at $\mu = 0.012$.   (d) Sn^{4+} at $\mu = 0.060$.

8-8. Calculate activity coefficients for the species in Problem 8-7 by linear interpolation of the data in Table 8-1.

8-9. For a solution in which μ is 5.0×10^{-2}, calculate K'_{sp} for
*(a) AgSCN.   *(c) $La(IO_3)_3$.
(b) PbI_2.   (d) $MgNH_4PO_4$.

8-10. Use activities to calculate the molar solubility of $Mn(OH)_2$ in
(a) 0.0100 M $NaNO_3$.
*(b) 0.0167 M K_2SO_4.
(c) the solution that results when 60.0 mL of 0.0333 M $MnCl_2$ is mixed with 40.0 mL of 0.0500 M KOH.

*8-11. Use (1) activities and (2) molar concentrations to calculate the solubilities of the following compounds in a 0.0333 M solution of $Mg(ClO_4)_2$:
(a) AgSCN.
(b) PbI_2.
(c) $BaSO_4$.
(d) $Cd_2Fe(CN)_6$
$[Cd_2Fe(CN)_6(s) \rightleftarrows 2Cd^{2+} + Fe(CN)_6^{4-}$;
$K_{sp} = 3.2 \times 10^{-17}]$.

8-12. Use (1) activities and (2) molar concentrations to calculate the solubilities of the following compounds in a 0.0167 M solution of $Ba(NO_3)_2$:
(a) $AgIO_3$.   (c) $BaSO_4$.
(b) $Mg(OH)_2$.   (d) $La(IO_3)_2$.

8-13. Use (1) activities and (2) molar concentrations to calculate the hydronium ion concentration in the following solutions:
*(a) 0.0200 M HNO_2 that is 0.0010 M in $NaNO_3$.
*(b) 0.0200 M HNO_2 that is 0.10 M in $NaNO_3$.
*(c) 0.0500 M NH_3 that is 0.0010 M in NaCl.
(d) 0.0500 M NH_3 that is 0.100 M in NaCl.
(e) 0.0500 M NH_4^+ that is 0.001 M in NaCl.
(f) 0.0500 M NH_4^+ that is 0.100 M in NaCl.

8-14. Mercury(II) forms a soluble neutral complex with Cl^-:

$$Hg^{2+} + 2 Cl^- \rightleftarrows HgCl_2 \qquad K_f = 1.6 \times 10^{13}$$

Use (1) activities and (2) molarities to calculate the Hg^{2+} concentration of the following solutions:
*(a) a solution prepared by dissolving 0.0100 mol of $HgCl_2$ in 1.00 L of water.
(b) a solution prepared by dissolving 0.0100 mol of $HgCl_2$ in 1.00 L of 0.0500 M $NaNO_3$.

*(c) a solution prepared by dissolving 0.0100 mol of HgCl$_2$ in 1.00 L of 0.0500 M NaCl.

(d) a solution prepared by dissolving 0.0100 mol of HgCl$_2$ in 1.00 L of 0.0333 M Hg(NO$_3$)$_2$.

*(e) a solution prepared by mixing 50.0 mL of a solution that is 0.0100 M in Hg(NO$_3$)$_2$ with 50.0 mL of a solution that is 0.0400 M in NaCl and 0.0600 M in NaNO$_3$.

(f) a solution prepared by mixing 50.0 mL of a solution that is 0.0100 M in Hg(NO$_3$)$_2$ with 50.0 mL of a solution that is 0.0400 M in BaCl$_2$ and 0.0500 M in NaNO$_3$.

Chapter 9

A Systematic Method
for Performing
Equilibrium Calculations

Aqueous solutions encountered in the laboratory often contain several species that interact with one another and with water to yield two or more equilibria. For example, when 1.0 mmol of sodium hydrogen carbonate is dissolved in 100 mL of water, the species sodium ion concentration is 0.010 M. The species concentration of hydrogen carbonate ion is less than 0.010 M, however, because this anion reacts with water to form two products:

$$HCO_3^- + H_2O \rightleftharpoons CO_3^{2-} + H_3O^+ \tag{9-1}$$

$$HCO_3^- + H_2O \rightleftharpoons H_2CO_3 + OH^- \tag{9-2}$$

All equilibria involving H_3O^+ or OH^- are necessarily influenced by the autoprotolysis of water. Therefore, in order to describe a sodium hydrogen carbonate system in quantitative terms, we must consider a third equilibrium:

$$2\ H_2O \rightleftharpoons H_3O^+ + OH^- \tag{9-3}$$

We see from these three equations that a 0.010 M solution of sodium hydrogen carbonate contains five species in addition to Na^+: HCO_3^-, CO_3^{2-}, H_2CO_3, H_3O^+, and OH^-. To calculate the concentration of one or

More often than not, developing the necessary algebraic expressions is easier than solving them unless a suitable computer program is available.

more of these species from equilibrium-constant expressions, we must generate five independent algebraic expressions and solve them simultaneously. As we demonstrate in this chapter, generating such algebraic equations is relatively easy if the problem is approached systematically.

Throughout this discussion, you should bear in mind that *the validity and form of a particular equilibrium-constant expression are unaffected by other equilibria in the solution.* For example, if barium chloride is added to the sodium hydrogen carbonate solution just described, barium carbonate precipitates and a new equilibrium is established:

$$BaCO_3(s) \rightleftarrows Ba^{2+} + CO_3^{2-} \qquad (9\text{-}4)$$

This additional equilibrium does not affect the equilibrium relationships among the other carbonate species and the hydronium and hydroxide ion. That is, the following two equilibrium-constant expressions, which describe the equilibria shown in Equations 9-1 and 9-2, are valid *regardless of whether barium chloride has been added.*

$$\frac{[H_3O^+][CO_3^{2-}]}{[HCO_3^-]} = K_a \qquad (9\text{-}5)$$

$$\frac{[OH^-][H_2CO_3]}{[HCO_3^-]} = K_b \qquad (9\text{-}6)$$

The introduction of a new equilibrium system into a solution containing equilibria does not change the equilibrium constants for the original equilibria.

To be sure, except for the sodium ion, the concentrations of all species in the original solution are markedly altered by the added barium ion. Nevertheless, Equations 9-5 and 9-6 still describe the relationships among the species involved.

In this chapter, we first outline the steps in a systematic approach to solving multiple-equilibrium problems. We then illustrate this approach with several examples that involve calculating precipitate solubilities in the presence of species that interact with the ions of the precipitates and thus enhance their solubilities. In subsequent chapters, we will use this same systematic approach for the solution of problems involving other types of equilibria.

9A   A SYSTEMATIC METHOD FOR SOLVING MULTIPLE-EQUILIBRIUM PROBLEMS

In order to solve a multiple-equilibrium problem, we need as many independent algebraic equations as there are participants in the system being studied. Thus, five such equations are required for the sodium hydrogen carbonate system just described. A sixth will be needed if we add barium ion to the solution.

Three types of algebraic equations are used in solving multiple-equilibrium problems: (1) equilibrium-constant expressions, (2) *mass-balance* equations, and (3) a single *charge-balance* equation. We have already shown how equilibrium-constant expressions are written; methods for obtaining the other two types of equations are given in the next two sections.

9A-1 Mass-Balance Equations

Mass-balance equations relate the equilibrium concentrations of various species in a solution to one another and to the analytical concentrations of the various solutes. They are derived from information about how the solution was prepared and from a knowledge of the kinds of equilibria established in the solution.

The term "mass-balance equation," although widely used, is misleading because such equations are really based upon balancing *concentrations* rather than *masses*.

Example 9-1

Write mass-balance expressions for a solution that is 0.050 M in acetic acid (HOAc).

First we write chemical equations for all important equilibria. Here,

For a slightly soluble salt with a 1 : 1 stoichiometry, the equilibrium molarity of the cations is equal to equilibrium molarity of the anion. This equality is the mass-balance expression.

$$HOAc + H_2O \rightleftarrows H_3O^+ + OAc^-$$

$$2\,H_2O \rightleftarrows H_3O^+ + OH^-$$

Because there is only one source of HOAc and OAc^-, we may write that the analytical concentration of acetic acid C_{HOAc} is equal to the sum of their concentrations:

$$C_{HOAc} = 0.050 = [HOAc] + [OAc^-]$$

A second mass-balance expression can be obtained by examining the two chemical equations. Here we see that one mole of H_3O^+ forms for each mole of OAc^- and for each mole of OH^-. Therefore, the total concentration of H_3O^+ is

$$[H_3O^+] = [OAc^-] + [OH^-]$$

For slightly soluble salts with stoichiometries other than 1 : 1, the mass-balance expression is obtained by multiplying the concentration of one of the ions by the stoichiometric ratio. For example, the mass-balance expression for a saturated solution of Ag_2CrO_4 is

$$2\,[CrO_4^{2-}] = [Ag^+]$$

Example 9-2

Write mass-balance expressions for the system formed when a 0.010 M NH_3 solution is saturated with AgBr.

Here, equations for the pertinent equilibria are

$$AgBr(s) \rightleftarrows Ag^+ + Br^-$$
$$Ag^+ + 2\,NH_3 \rightleftarrows Ag(NH_3)_2^+$$
$$NH_3 + H_2O \rightleftarrows NH_4^+ + OH^-$$

Because the only source of Br^-, Ag^+, and $Ag(NH_3)_2^+$ is AgBr and because silver and bromide ions are present in a 1 : 1 ratio in the starting material, one mass-balance equation is

$$[Ag^+] + [Ag(NH_3)_2^+] = [Br^-]$$

where the bracketed terms are molar species concentrations. Also, we know that the only source of ammonia-containing species is the 0.010 M

NH_3. Therefore, a second mass-balance expression is

$$C_{NH_3} = [NH_3] + [NH_4^+] + 2[Ag(NH_3)_2^+] = 0.010$$

9A-2 Charge-Balance Equation

We know that electrolyte solutions are electrically neutral even though they may contain many millions of charged ions. Solutions are neutral because the *molar concentration of positive charge* in an electrolyte solution always equals *the molar concentration of negative charge*. That is, for any solution containing electrolytes, we may write

$$\text{no. mol positive charge/L} = \text{no. mol negative charge/L}\qquad(9\text{-}7)$$

This equation represents the charge-balance condition and is called the *charge-balance equation*. To make this equation useful for equilibrium-constant calculations, we must express the two sides of the equation in terms of molar concentrations of negative and positive ions in the solution.

How much charge is contributed to a solution by 1 mol of Na^+? Or, how about 1 mol of Mg^{2+} or 1 mol of PO_4^{3-}? The concentration of charge contributed to a solution by an ion is equal to the molar concentration of that ion multiplied by its charge. Thus the molar concentration of positive charge in a solution due to the presence of sodium ions is

> Always remember that a charge-balance equation is based upon the equality in *molar charge concentrations*. To obtain the charge concentration of an ion, you must multiply the molar concentration of the ion by its charge.

$$\text{no. mol positive charge/L} = \frac{\text{mol } Na^+}{L} \times \frac{1 \text{ mol positive charge}}{\text{mol } Na^+}$$
$$= 1 \times [Na^+]$$

The concentration of positive charge due to magnesium ions is

$$\text{no. mol positive charge/L} = \frac{\text{mol } Mg^{2+}}{L} \times \frac{2 \text{ mol positive charge}}{\text{mol } Mg^{2+}}$$
$$= 2 \times [Mg^{2+}]$$

since each mole of magnesium ion contributes 2 mol of positive charge to the solution. Similarly, we may write for phosphate ion

$$\text{no. mol negative charge/L} = \frac{\text{mol } PO_4^{3-}}{L} \times \frac{3 \text{ mol negative charge}}{\text{mol } PO_4^{3-}}$$
$$= 3 \times [PO_4^{3-}]$$

Now, consider how we would write a charge-balance equation for a 0.100 M solution of sodium chloride. Positive charges in this solution are supplied by Na^+ and H_3O^+ (from dissociation of water). Negative charges come from Cl^- and OH^-. The molarity of positive and negative charges are

$$mol \text{ positive charge/L} = [Na^+] + [H_3O^+] = 0.100 + 1 \times 10^{-7}$$
$$mol \text{ negative charge/L} = [Cl^-] + [OH^-] = 0.100 + 1 \times 10^{-7}$$

We write the charge-balance equation by equating the concentrations of positive and negative charges. That is,

$$[Na^+] + [H_3O^+] = [Cl^-] + [OH^-] = 0.100 + 1 \times 10^{-7}$$

Let us now consider a solution that has an analytical concentration of magnesium chloride of 0.100 M. Here, the molarities of positive and negative charge are given by

$$mol \text{ positive charge/L} = 2[Mg^{2+}] + [H_3O^+] = 2 \times 0.100 + 1 \times 10^{-7}$$
$$mol \text{ negative charge/L} = [Cl^-] + [OH^-] = 2 \times 0.100 + 1 \times 10^{-7}$$

In the first equation, the molar concentration of magnesium ion is multiplied by two (2×0.100) because 1 mol of that ion contributes 2 mol of positive charge to the solution. In the second equation the molar chloride ion concentration is twice that of the magnesium chloride concentration or 2×0.100. To obtain the charge-balance equation, we equate the concentration of positive charge with the concentration of negative charge to obtain

$$2[Mg^{2+}] + [H_3O^+] = [Cl^-] + [OH^-] = 0.200 + 1 \times 10^{-7}$$

For a neutral solution, $[H_3O^+]$ and $[OH^-]$ are very small and equal so that we can simplify the charge-balance equation to

$$2[Mg^{2+}] = [Cl^-] = 0.200$$

Example 9-3

Write a charge-balance equation for the system in Example 9-2.

$$[Ag^+] + [Ag(NH_3)_2^+] + [H_3O^+] + [NH_4^+] = [OH^-] + [Br^-]$$

Example 9-4

Neglecting the dissociation of water, write a charge-balance equation for a solution that contains NaCl, $Mg(NO_3)_2$, and $Al_2(SO_4)_3$.

$$[Na^+] + 2[Mg^{2+}] + 3[Al^{3+}] = [Cl^-] + [NO_3^-] + 2[SO_4^{2-}]$$

9A-3 Steps for Solving Problems Involving Several Equilibria

1. Write a set of balanced chemical equations for all pertinent equilibria.
2. Write an equation that expresses the quantity being sought in terms of the equilibrium concentrations of other species in the system.

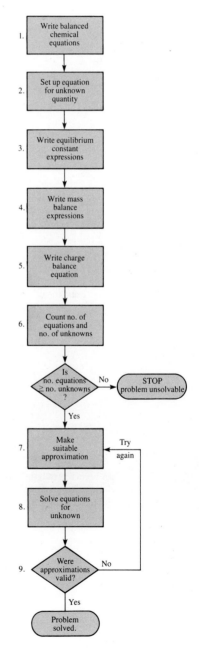

3. Write equilibrium-constant expressions for all equilibria developed in step 1, and find numerical values for the constants in tables of equilibrium constants.
4. Write mass-balance expressions for the system.
5. If possible, write a charge-balance expression for the system.
6. Count the number of unknown concentrations in the equations developed in steps 3, 4, and 5, and compare this number with the number of independent equations. If the number of equations is equal to the number of unknowns, proceed to step 7. If the number of equations is smaller than the number of unknowns, seek additional equations. If enough equations cannot be developed, try to eliminate unknowns by suitable approximations regarding the concentration of one or more of the unknowns. If such approximations cannot be found, the problem cannot be solved.
7. Make suitable approximations to simplify the algebra.
8. Solve the algebraic equations for the equilibrium concentrations needed to give a provisional answer to the equation developed in step 2.
9. Check the validity of the approximations made in steps 6 and 7 using the provisional concentrations computed in step 8.

> In some systems, charge-balance equations either cannot be written or are of no use. Often the charge-balance equation is identical to one of the mass-balance expressions.

Step 6 is particularly important because it shows whether or not an exact solution to the problem is possible. If the number of unknowns is identical to the number of equations, the problem has been reduced to one of *algebra* alone. That is, answers can be obtained with sufficient perseverance. On the other hand, if there are not enough equations even after approximations are made, the problem should be abandoned.

> Do not waste time starting the algebra in an equilibrium calculation until you are absolutely sure you have enough independent equations to make solution feasible.

9A-4 The Use of Approximations in Equilibrium Calculations

When step 6 of the systematic approach is complete, we have a *mathematical* problem of solving several nonlinear simultaneous equations. This job is often formidable, tedious, and time-consuming unless a suitable computer program is available or unless we can find approximations that decrease the number of unknowns and equations. In this section, we consider in general terms how equations describing equilibrium relationships can be simplified by suitable approximations.

Bear in mind that *only* the mass-balance and charge-balance equations can be simplified because only in these equations do the concentration terms appear as sums or differences rather than as products or quotients. It is always possible to assume that one (or more) of the terms in a sum or difference is so much smaller than the others that it can be assumed to have a value of zero without significantly affecting the equality. However, if a term in an equilibrium-constant expression is assumed to be zero the product or quotient becomes equal to zero or infinity and is thus meaningless.

> Approximations can be made only in charge-balance and mass-balance equations—never in equilibrium-constant expressions.

Many students find step 7 to be the most troublesome because they fear that making invalid approximations will lead to serious errors in their computed results. Such fears are groundless. Experienced scientists are often as puzzled as beginners when making an approximation that simplifies an equilibrium calculation. Nonetheless, they make such approximations without fear because they know that the effects of an invalid assumption will become obvious by the time a computation is completed. Generally, questionable assumptions should be tried at the outset and provisional answers computed. If the assumption leads to an intolerable error (which is easily recognized), a recalculation without the faulty approximation is then performed. Usually, it is more efficient to try a questionable assumption at the outset than to make a more time-consuming and tedious calculation without the assumption.

Never be afraid to make an assumption in attempting to solve an equilibrium problem. If the assumption is not valid, you will know it as soon as you have a provisional answer.

9B   THE CALCULATION OF SOLUBILITY BY THE SYSTEMATIC METHOD

The use of the systematic method is illustrated in this section with examples involving the solubility of precipitates under various conditions. In later chapters we apply this method to other types of equilibria.

9B-1 Metal Hydroxides

Examples 9-5 and 9-6 involve calculating the solubilities of two metal hydroxides. One of these compounds is relatively soluble compared with the second. These examples illustrate the importance of making approximations and provisional calculations in solving mass-law problems.

Example 9-5

Calculate the molar solubility of $Mg(OH)_2$ in water.

Step 1. Pertinent Equilibria

Two equilibria that need to be considered are

$$Mg(OH)_2(s) \rightleftharpoons Mg^{2+} + 2\ OH^-$$

$$2\ H_2O \rightleftharpoons H_3O^+ + OH^-$$

Step 2. Definition of Unknown

Since 1 mol of Mg^{2+} is formed for each mole of $Mg(OH)_2$ dissolved,

$$\text{solubility } Mg(OH)_2 = [Mg^{2+}]$$

Step 3. Equilibrium-Constant Expressions

$$[Mg^{2+}][OH^-]^2 = K_{sp} = 1.8 \times 10^{-11} \qquad (9\text{-}8)$$

$$[H_3O^+][OH^-] = K_w = 1.00 \times 10^{-14} \qquad (9\text{-}9)$$

Step 4. Mass-Balance Expression

$$[OH^-] = 2[Mg^{2+}] + [H_3O^+] \qquad (9\text{-}10)$$

If $[OH^-]_{H_2O}$ and $[OH^-]_{Mg(OH)_2}$ are the concentrations of OH^- produced from H_2O and $Mg(OH)_2$, respectively, then

$$[OH^-]_{H_2O} = [H_3O^+]$$

$$[OH^-]_{Mg(OH)_2} = 2 [Mg^{2+}]$$

$$[OH^-]_{total} = [OH^-]_{H_2O} + [OH^-]_{Mg(OH)_2}$$

$$= [H_3O^+] + 2 [Mg^{2+}]$$

The first term on the right-hand side of Equation 9-10 represents the hydroxide ion concentration resulting from dissolved $Mg(OH)_2$, and the second term is the hydroxide ion concentration resulting from the dissociation of water.

Step 5. Charge-Balance Expression

$$[OH^-] = 2[Mg^{2+}] + [H_3O^+]$$

Note that this equation is identical to Equation 9-10. Often a mass-balance equation and a charge-balance equation are the same.

Step 6. Number of Independent Equations and Unknowns

We have developed three independent algebraic equations (Equations 9-8, 9-9, and 9-10) and have three unknowns ($[Mg^{2+}]$, $[OH^-]$, and $[H_3O^+]$). Therefore, the problem can be solved rigorously.

Step 7. Approximations

We can make approximations only in Equation 9-10. Since the solubility-product constant for $Mg(OH)_2$ is relatively large, the solution will be somewhat basic. Therefore, it is reasonable to assume that $[H_3O^+] \ll [Mg^{2+}]$. Equation 9-10 then simplifies to

$$2[Mg^{2+}] \approx [OH^-] \qquad (9\text{-}11)$$

Step 8. Solution to Equations

Substitution of Equation 9-11 into Equation 9-8 gives

$$[Mg^{2+}](2[Mg^{2+}])^2 = 1.8 \times 10^{-11}$$

$$[Mg^{2+}]^3 = \frac{1.8 \times 10^{-11}}{4} = 4.50 \times 10^{-12}$$

$$[Mg^{2+}] = \text{solubility} = 1.651 \times 10^{-4} = 1.7 \times 10^{-4} \text{ mol/L}$$

Step 9. Check of Assumptions

Substitution into Equation 9-11 yields

$$[OH^-] = 2 \times 1.651 \times 10^{-4} = 3.30 \times 10^{-4} = 3.3 \times 10^{-4} \text{ mol/L}$$

$$[H_3O^+] = \frac{1.00 \times 10^{-14}}{3.33 \times 10^{-4}} = 3.0 \times 10^{-11} \text{ mol/L}$$

Thus our assumption that $3.0 \times 10^{-11} \ll 1.7 \times 10^{-4}$ is certainly valid.

Example 9-6

Calculate the solubility of $Fe(OH)_3$ in water. Proceeding by the systematic approach used in Example 9-5, we write.

Step 1. Pertinent Equilibria

$$Fe(OH)_3(s) \rightleftharpoons Fe^{3+} + 3\ OH^-$$

$$2\ H_2O \rightleftharpoons H_3O^+ + OH^-$$

Step 2. Definition of Unknown

$$\text{solubility} = [Fe^{3+}]$$

Step 3. Equilibrium-Constant Expressions

$$[Fe^{3+}][OH^-]^3 = 4 \times 10^{-38}$$
$$[H_3O^+][OH^-] = 1.00 \times 10^{-14}$$

Steps 4 and 5. Mass-Balance and Charge-Balance Equations
As in Example 9-5, the mass-balance equation and the charge-balance equations are identical. That is,

$$[OH^-] = 3[Fe^{3+}] + [H_3O^+]$$

Step 6. Number of Independent Equations and Unknowns
We see that we have three equations and three unknowns.
Step 7. Approximations
As in Example 9-5, let us assume $[H_3O^+] \ll 3[Fe^{3+}]$, so that

$$3[Fe^{3+}] \approx [OH^-]$$

Step 8. Solution to Equations
Substituting this equation into the solubility-product expression gives

$$[Fe^{3+}](3[Fe^{3+}])^3 = 4 \times 10^{-38}$$

$$[Fe^{3+}] = \left(\frac{4 \times 10^{-38}}{27}\right)^{1/4} = 2 \times 10^{-10}$$

$$\text{solubility} = [Fe^{3+}] = 2 \times 10^{-10} \text{ mol/L}$$

Step 9. Check of Assumptions
From the assumptions made in step 7, we can calculate a provisional value of $[OH^-]$. That is,

$$[OH^-] \approx 3[Fe^{3+}] = 3 \times 2 \times 10^{-10} = 6 \times 10^{-10}$$

Let us use this value of $[OH^-]$ to compute a *provisional* value for $[H_3O^+]$:

$$[H_3O^+] = \frac{1.00 \times 10^{-14}}{6 \times 10^{-10}} = 1.7 \times 10^{-5}$$

But 1.7×10^{-5} is not much smaller than three times our provisional value of $[Fe^{3+}]$. This discrepancy means that our assumption was invalid and the provisional values for $[Fe^{3+}]$, $[OH^-]$, and $[H_3O^+]$ are all significantly in error. Therefore, let us go back to step 7 and assume that

$$3[Fe^{3+}] \ll [H_3O^+]$$

Now the mass-balance expression becomes

$$[H_3O^+] = [OH^-]$$

Substituting this equality into the expression for K_w gives

$$[H_3O^+] = [OH^-] = 1.00 \times 10^{-7}$$

Substituting this number into the solubility-product expression developed in step 3 gives

$$[Fe^{3+}] = \frac{4 \times 10^{-38}}{(1.00 \times 10^{-7})^3} = 4 \times 10^{-17} \text{ mol/L}$$

In this case we have assumed that $3[Fe^{3+}] \ll [OH^-]$ or $3 \times 4 \times 10^{-17} \ll 10^{-7}$. Clearly, the assumption is valid and we may write

$$\text{solubility} = 4 \times 10^{-17} \text{ mol/L}$$

The faulty assumption in Example 9-6 was readily recognized.

Note the very large error introduced by the invalid assumption.

Example 9-6 illustrates how readily the effects of an invalid assumption are detected. It may bother you, that we used a faulty value for $[OH^-]$ to obtain $[H_3O^+]$, which we then compared with $3[Fe^{3+}]$. But the point is that the faulty value of $[OH^-]$ also arose because of the invalid assumption. Had the assumption been valid we would have had an internally consistent set of calculated concentrations. The presence of even one internal inconsistency is a clear indication of an invalid assumption. A valid assumption leads to concentrations that satisfy all of the algebraic equalities developed in steps 3, 4, 5.

9B-2 The Effect of pH on Solubility

All precipitates that contain an anion that is the conjugate base of a weak acid are more soluble at low pH than at high pH.

The solubility of precipitates containing an anion with basic properties, a cation with acidic properties, or both is dependent upon pH. For example, when barium sulfate is equilibrated with a solution containing hydrochloric acid, the following equilibria are established:

$$BaSO_4(s) \rightleftarrows Ba^{2+} + SO_4^{2-}$$
$$SO_4^{2-} + H_3O^+ \rightleftarrows HSO_4^- + H_2O$$

If more hydronium ion is added to this system, the second equilibrium is shifted to the right by the common-ion effect. The resulting decrease in the sulfate ion concentration causes the first equilibrium to shift to the right, thus increasing the solubility. The example that follows illustrates how the effect of pH on solubility can be treated in quantitative terms.

Solubility Calculations When the pH Is Fixed and Known
Analytical precipitations are frequently performed in solutions in which the pH is fixed at some predetermined and known value. The calculation of solubility under this circumstance is illustrated by the following example.

Example 9-7

Calculate the molar solubility of calcium oxalate in a solution that has been buffered so that its pH is constant and equal to 4.00.
Step 1. Pertinent Equilibria

$$CaC_2O_4 \rightleftharpoons Ca^{2+} + C_2O_4^{2-} \qquad (9\text{-}12)$$

Oxalate ions react with water to form $HC_2O_4^-$ and $H_2C_2O_4$. Thus, two other equilibria in this solution are

$$H_2C_2O_4 + H_2O \rightleftharpoons H_3O^+ + HC_2O_4^- \qquad (9\text{-}13)$$
$$HC_2O_4^- + H_2O \rightleftharpoons H_3O^+ + C_2O_4^{2-} \qquad (9\text{-}14)$$

Step 2. Definition of the Unknown
Calcium oxalate is a strong electrolyte, and so its molar analytical concentration is equal to the equilibrium calcium ion concentration:

$$\text{solubility} = [Ca^{2+}] \qquad (9\text{-}15)$$

Step 3. Equilibrium-Constant Expressions

$$[Ca^{2+}][C_2O_4^{2-}] = K_{sp} = 2.3 \times 10^{-9} \qquad (9\text{-}16)$$

$$\frac{[H_3O^+][HC_2O_4^-]}{[H_2C_2O_4]} = K_1 = 5.36 \times 10^{-2} \qquad (9\text{-}17)$$

$$\frac{[H_3O^+][C_2O_4^{2-}]}{[HC_2O_4^-]} = K_2 = 5.42 \times 10^{-5} \qquad (9\text{-}18)$$

Step 4. Mass-Balance Expressions
Because CaC_2O_4 is the only source of Ca^{2+} and of the three oxalate species,

$$[Ca^{2+}] = [C_2O_4^{2-}] + [HC_2O_4^-] + [H_2C_2O_4] \qquad (9\text{-}19)$$

Moreover, the problem states that the pH is 4.00. Thus,

$$[H_3O^+] = 1.00 \times 10^{-4}$$

Step 5. Charge-Balance Expressions
A buffer is required to maintain the pH at 4.00. The buffer most likely consists of some weak acid HA and its conjugate base A⁻ (Section 11C).

A buffer keeps the pH of a solution constant.

The nature of the three species and their concentrations have no specified, however, and so we do not have enough information to charge-balance equation.

Step 6. Number of Independent Equations and Unknowns

We have four unknowns ($[Ca^{2+}]$, $[C_2O_4^{2-}]$, $[HC_2O_4^-]$, and $[H_2C_2O_4]$) as well as four independent algebraic relationships (Equations 9-16, 9-17, 9-18, and 9-19). Therefore, an exact solution can be obtained, and the problem becomes one of algebra.

Step 7. Approximations

An exact solution is so readily obtained that we will not bother with approximations.

Step 8. Solution of the Equations

A convenient way to solve the problem is to substitute Equations 9-17 and 9-18 into 9-19 in such a way as to develop a relationship between $[Ca^{2+}]$, $[C_2O_4^{2-}]$, and $[H_3O^+]$. Thus, we rearrange Equation 9-18 to give

$$[HC_2O_4^-] = \frac{[H_3O^+][C_2O_4^{2-}]}{K_2}$$

Substituting numerical values gives

$$[HC_2O_4^-] = \frac{1.00 \times 10^{-4}[C_2O_4^{2-}]}{5.42 \times 10^{-5}} = 1.85[C_2O_4^{2-}]$$

Substituting this relationship into Equation 9-17 gives, upon rearranging,

$$[H_2C_2O_4] = \frac{[H_3O^+][C_2O_4^{2-}] \times 1.85}{K_2}$$
$$= \frac{1.85 \times 10^{-4}[C_2O_4^{2-}]}{5.36 \times 10^{-2}} = 3.45 \times 10^{-3}[C_2O_4^{2-}]$$

Substituting this equation and the earlier equation for $[HC_2O_4^-]$ into Equation 9-19 gives

$$[Ca^{2+}] = [C_2O_4^{2-}] + 1.85[C_2O_4^{2-}] + 3.45 \times 10^{-3}[C_2O_4^{2-}]$$
$$= 2.85[C_2O_4^{2-}]$$

or

$$[C_2O_4^{2-}] = \frac{[Ca^{2+}]}{2.85}$$

Substituting into Equation 9-16 gives

$$\frac{[Ca^{2+}][Ca^{2+}]}{2.85} = 2.3 \times 10^{-9}$$

$$[Ca^{2+}] = \text{solubility} = \sqrt{2.85 \times 2.3 \times 10^{-9}} = 8.1 \times 10^{-5} \text{ mol/L}$$

Computer Application 9-1
USING EUREKA TO SOLVE COMPLEX EQUILIBRIUM PROBLEMS

You can use Eureka: The Solver to solve sets of equations such as Equations 9-16 through 9-19 of Example 9-7. Run Eureka, press E to select the edit window, press F5 to expand the window to fill the entire screen, and then enter the equations as shown in the following screen. Notice that semicolons indicate comments in the program. Eureka ignores the comments, but they are helpful, particularly when you plan to store the program and use it later. The symbols for the four unknown species are entered as they appear in the equations of Example 9-7, except that subscripts and charges are omitted. Eureka and BASIC both use a common method of expressing numbers in scientific notation. For example, $1.0 \times 10^{-4} = 1.0e-4$.

The first few lines of the program set limits on the variables. We know that concentrations must be greater than zero, even though our equations may have negative solutions. So we can ensure that Eureka will ignore the negative solutions by declaring that each of the variables must be greater than zero.

```
    C:EX97.EKA      Line 13    Col 11   Insert Indent Tab
;Program to find solutions to the equations of Example 9-7

;Constraints

Ca > 0                          ;All concentrations are greater than zero.

H2C2O4 > 0

HC2O4 > 0

C2O4 > 0

;Equations

H3O = 1.0e-4                    ;pH = 4.00

Ca * C2O4 = 2.3e-9             ;Equation 9-16

H3O * HC2O4 / H2C2O4 = 5.36e-2  ;Equation 9-17

H3O * C2O4 / HC2O4 = 5.42e-5    ;Equation 9-18

Ca = H2C2O4 + C2O4 + HC2O4      ;Equation 9-19
```

After you have entered the program, press Esc S and compare the displayed solution to the result of Example 9-7. Notice that Eureka evaluates all four variables without making any approximations. As an exercise, find the solubility of calcium oxalate as a function of pH over the range 1 to 14 and make a plot of log $[Ca^{2+}]$ vs. pH. To calculate $[Ca^{2+}]$ at a different pH, you need only change the value for the hydronium ion concentration in the program and give Eureka the command to solve the system of equations. If you accept the challenge in the margin note on p. 158 to write the system of equations to determine the solubility of barium carbonate, write a program to solve the equations.

Solubility Calculations When the pH Is Variable

Saturating an unbuffered solution with a sparingly soluble salt made up of a basic anion or an acidic cation causes the pH of the solution to change. For example, pure water saturated with barium carbonate is basic as a consequence of the reactions

$$BaCO_3(s) \rightleftarrows Ba^{2+} + CO_3^{2-}$$

$$CO_3^{2-} + H_2O \rightleftarrows HCO_3^- + OH^-$$

$$HCO_3^- + H_2O \rightleftarrows H_2CO_3 + OH^-$$

In contrast to Example 9-6, the hydroxide ion concentration now becomes an unknown, and an additional algebraic equation must therefore be developed if the solubility of barium carbonate is to be calculated.

It is not difficult to write the algebraic relationships needed to calculate the solubility of such a precipitate. Solving the equations, however, is tedious.

Challenge: Write enough algebraic equations to determine the solubility of $BaCO_3$ in pure water.

9B-3 The Solubility of Precipitates in the Presence of Complexing Agents

The solubility of a precipitate may increase dramatically in the presence of reagents that form complexes with the anion or the cation of the precipitate. For example, fluoride ions prevent the quantitative precipitation of aluminum hydroxide even though the solubility product of this precipitate is remarkably small (2×10^{-32}). The cause of the increase in solubility is shown by the equations

The solubility of a precipitate always increases in the presence of a complexing agent that reacts with the cation of the precipitate.

$$Al(OH)_3(s) \rightleftarrows Al^{3+} + 3\ OH^-$$
$$+$$
$$6\ F^-$$
$$\uparrow\downarrow$$
$$AlF_6^{3-}$$

The fluoride complex is sufficiently stable to permit fluoride ions to compete successfully with hydroxide ions for aluminum ions.

Complex Formation with an Ion Common to the Precipitate

Many precipitates react with the precipitating reagent to form soluble complexes. This tendency may have the unfortunate effect of reducing the recovery of analytes if too large an excess of reagent is used in a gravimetric analysis. For example, in solutions containing high concentrations of chloride, silver chloride forms chloro complexes such as $AgCl_2^-$ and $AgCl_3^{2-}$. The effect of these complexes is illustrated in Figure 9-1, in which the experimentally determined solubility of silver chloride is plotted against the logarithm of the potassium chloride concentration. For low anion concentrations, the experimental solubilities do not differ greatly from those calculated with the solubility-product constant for silver chloride; beyond a chloride ion concentration of about 10^{-3} M, however, the calculated solubilities approach zero while the measured values

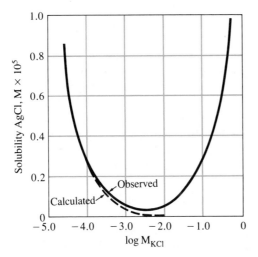

Figure 9-1
Solubility of silver chloride in potassium chloride solutions. The dashed curve is calculated from K_{sp}; the solid curve is plotted from experimental data of A. Pinkus and A. M. Timmermans, *Bull. Soc. Belges*, **1937**, *46*, 46–73.

rise steeply. Note that the solubility of silver chloride in 0.3 M KCl is about the same as in pure water; in a 1 M KCl solution, however, the silver chloride solubility increases to almost eight times its value in pure water. We can describe these effects quantitatively if the composition of the complexes and their formation constants are known.

Increases in solubility caused by large excesses of a common ion are not unusual. Of particular interest are amphoteric hydroxides, which are sparingly soluble in dilute base but are redissolved by excess hydroxide ion. The hydroxides of zinc and aluminum, for example, are converted to the soluble zincate and aluminate ions upon treatment with excess base. For zinc, the equilibria can be represented as

$$Zn^{2+} + 2\,OH^- \rightleftarrows Zn(OH)_2(s)$$

$$Zn(OH)_2(s) + 2\,OH^- \rightleftarrows Zn(OH)_4^{2-}$$

As with silver chloride, the solubilities of amphoteric hydroxides pass through minima and then increase rapidly with increasing concentration of base. The hydroxide ion concentration at which the solubility is a minimum can be calculated, provided equilibrium constants for the reactions are available.

Quantitative Treatment of the Effect of Complex Formation on Solubility

Solubility calculations for a precipitate in the presence of a complexing reagent are similar in principle to those discussed in the previous section. Formation constants for the complexes involved must be available.[1]

In gravimetric procedures, a small excess of precipitating agent minimizes solubility losses, but a large excess often causes increased losses due to complex formation.

---

[1]For an example of such a calculation, see D. A. Skoog, D. M. West, and F. J. Holler, *Fundamentals of Analytical Chemistry*, 5th ed., pp. 141–145. Philadelphia: Saunders College Publishing, 1988.

9C   SEPARATION OF IONS BASED UPON SOLUBILITY DIFFERENCES; SULFIDE SEPARATIONS

Several precipitating agents are useful for separating ions based upon solubility differences. Such separations require close control of the active reagent concentration at a suitable and predetermined level. Most often, such control is achieved by controlling the pH of the solution with appropriate buffers. This technique is applicable to reagents in which the anion is the conjugate base of a weak acid. Examples include sulfide ion (the conjugate base of hydrogen sulfide), hydroxide ion (the conjugate base of water), and the anions of several organic weak acids.

Sulfide ion forms precipitates with heavy-metal cations that have solubility products from 10^{-10} to 10^{-50} or smaller. In addition, the concentration of S^{2-} can be varied over a range of about 0.1 M to 10^{-22} M by controlling the pH of a saturated solution of hydrogen sulfide. These two properties make possible a number of useful cation separations. To illustrate the use of hydrogen sulfide to separate cations based upon pH control, let us consider the precipitation of the divalent cation M^{2+} from a solution that is kept saturated with hydrogen sulfide by bubbling the gas continuously through the solution. The important equilibria in this solution are

$$MS(s) \rightleftharpoons M^{2+} + S^{2-} \qquad\qquad [M^{2+}][S^{2-}] = K_{sp}$$

$$H_2S + H_2O \rightleftharpoons H_3O^+ + HS^- \qquad\qquad \frac{[H_3O^+][HS^-]}{[H_2S]} = K_1 = 5.7 \times 10^{-8}$$

$$HS^- + H_2O \rightleftharpoons H_3O^+ + S^{2-} \qquad\qquad \frac{[H_3O^+][S^{2-}]}{[HS^-]} = K_2 = 1.2 \times 10^{-15}$$

We may also write

$$\text{solubility} = [M^{2+}]$$

The concentration of hydrogen sulfide in a saturated solution of the gas is approximately 0.1 M. Thus, we may write as a mass-balance expression

$$0.1 = [S^{2-}] + [HS^-] + [H_2S]$$

Because we know the hydronium ion concentration, we have four unknowns: the concentrations of the metal ion and the three sulfide species.

We can simplify the calculation greatly by assuming that $([S^{2-}] + [HS^-]) \ll [H_2S]$, so that

The equilibrium hydrogen sulfide concentration ($[H_2S]$) of a solution through which H_2S is continually bubbled is about 0.1 M.

$$[H_2S] \approx 0.10 \text{ mol/L}$$

The two dissociation-constant expressions for hydrogen sulfide may be multiplied together to give an expression for the overall dissociation of hydrogen sulfide to sulfide ion:

$$H_2S + 2 H_2O \rightleftharpoons 2 H_3O^+ + S^{2-} \qquad\qquad \frac{[H_3O^+]^2[S^{2-}]}{[H_2S]} = K_1K_2 = 6.8 \times 10^{-23}$$

The constant for this overall reaction is just the product of K_1 and K_2. Substituting the numerical value for $[H_2S]$ into this equation gives

$$\frac{[H_3O^+]^2[S^{2-}]}{0.10} = 6.8 \times 10^{-23}$$

Upon rearranging this equation, we obtain

$$[S^{2-}] = \frac{6.8 \times 10^{-24}}{[H_3O^+]^2} \tag{9-20}$$

Thus, we see that the sulfide ion concentration of a saturated hydrogen sulfide solution varies inversely as the square of the hydronium ion concentration. Figure 9-2, which was obtained with Equation 9-20, reveals that the sulfide ion concentration of an aqueous solution can be varied by over 20 orders of magnitude by varying the pH from 1 to 11.

Substituting Equation 9-20 into the solubility-product expression gives

$$\frac{[M^{2+}] \times 6.8 \times 10^{-24}}{[H_3O^+]^2} = K_{sp}$$

$$[M^{2+}] = \text{solubility} = \frac{[H_3O^+]^2 K_{sp}}{6.8 \times 10^{-24}}$$

Thus, the solubility of a divalent metal sulfide increases as the square of the hydronium ion concentration.

Figure 9-2

Sulfide ion concentration as a function of pH in a saturated H_2S solution.

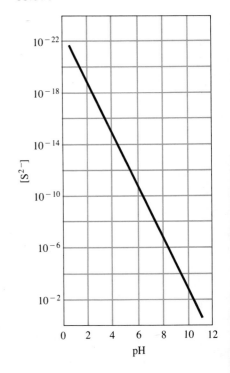

Example 9-8

Cadmium sulfide is less soluble than thallium(I) sulfide. Find the conditions under which Cd^{2+} and Tl^+ can, in theory, be separated quantitatively with H_2S from a solution that is 0.1 M in each cation.

The constants for the two solubility equilibria are

$$CdS(s) \rightleftarrows Cd^{2+} + S^{2-} \qquad [Cd^{2+}][S^{2-}] = 2 \times 10^{-28}$$
$$Tl_2S(s) \rightleftarrows 2\,Tl^+ + S^{2-} \qquad [Tl^+]^2[S^{2-}] = 1 \times 10^{-22}$$

Since CdS precipitates at a lower sulfide ion concentration than does Tl_2S, we first compute the sulfide ion concentration necessary for quantitative removal of Cd^{2+} from solution. In order to make such a calculation, we must first specify what constitutes a quantitative removal. The decision here is arbitrary and depends upon the purpose of the separation. In this example, we consider a separation to be quantitative when all but 1 part in 1000 of the Cd^{2+} has been removed, or, in other words, when the concentration of the cation has been reduced to 1.00×10^{-4} M. Substituting this value into the solubility-product expression gives

Ordinarily we shall assume a precipitation is quantitative if less than one part per thousand of the analyte remains in the solution.

$$10^{-4}[S^{2-}] = 2 \times 10^{-28}$$
$$[S^{2-}] = 2 \times 10^{-24}$$

Thus, if we maintain the sulfide ion concentration at this level or greater, we may assume that quantitative removal of the cadmium will take place.

Next, we compute the sulfide ion concentration needed to initiate precipitation of Tl_2S from a 0.1 M solution. Precipitation will begin when the solubility product is just exceeded. Since the solution is 0.1 M in Tl^+,

$$(0.1)^2[S^{2-}] = 1 \times 10^{-22}$$

$$[S^{2-}] = 1 \times 10^{-20}$$

These two calculations show that quantitative precipitation of Cd^{2+} takes place when $[S^{2-}]$ is made greater than 2×10^{-24} M. No precipitation of Tl^+ occurs, however, until $[S^{2-}]$ becomes greater than 1×10^{-20} M.

Substituting these two values for $[S^{2-}]$ into Equation 9-20 permits calculation of the $[H_3O^+]$ range required for the separation:

$$[H_3O^+]^2 = \frac{6.8 \times 10^{-24}}{2 \times 10^{-24}} = 3.4$$

$$[H_3O^+] = 1.8 \text{ M}$$

and

$$[H_3O^+]^2 = \frac{6.8 \times 10^{-24}}{1 \times 10^{-20}} = 6.8 \times 10^{-4}$$

$$[H_3O^+] \approx 0.026 = 0.03 \text{ M}$$

By maintaining $[H_3O^+]$ between 0.03 and 1.8 M, we can in theory separate CdS quantitatively from Tl_2S.

9D   QUESTIONS AND PROBLEMS

Unless directed otherwise, base all calculations in this section on molar concentrations rather than activities.

9-1. Write mass-balance expressions for a solution that is
*(a) 0.10 M in H_3PO_4.
(b) 0.10 M in Na_2HPO_4.
*(c) 0.100 M in HNO_2 and 0.0500 M in $NaNO_2$.
(d) 0.025 M in NaF and saturated with CaF_2.
*(e) 0.100 M in NaOH and saturated with $Zn(OH)_2$ (which undergoes the reaction $Zn(OH)_2 + 2 OH^- \rightleftarrows Zn(OH)_4^{2-}$).
(f) saturated with $MgCO_3$.
*(g) saturated with CaF_2.

9-2. Write a charge-balance equation for each solution in Problem 9-1.

9-3. Calculate the molar solubility of Ag_2CO_3 in a solution that has a H_3O^+ concentration of
*(a) 1.0×10^{-6} M.   *(c) 1.0×10^{-9} M.
(b) 1.0×10^{-7} M.   (d) 1.0×10^{-11} M.

9-4. Calculate the molar solubility of $BaSO_4$ in a solution that has a $[H_3O^+]$ of
*(a) 2.0 M. (b) 1.0 M. *(c) 0.50 M. (d) 0.10 M.

*9-5. The solubility-product constant for $CdCO_3$ is 2.5×10^{-14}. Calculate the equilibrium solubility of $CdCO_3$ in
(a) a solution in which the equilibrium concentration of CO_3^{2-} is 0.10 M.
(b) a solution in which $[H_3O^+]$ is 1.0×10^{-8} M.

9-6. Calculate the molar solubility of $MgCO_3$ in
(a) a solution in which $[H_3O^+] = 1.0 \times 10^{-8}$ M.
(b) 0.10 M Na_2CO_3.

*9-7. Calculate the molar solubility of CuS in a solution in which $[H_3O^+]$ is held constant at
(a) 1.0×10^{-1} M.   (b) 1.0×10^{-4} M.

9-8. Calculate the concentration of CdS in a solution in which $[H_3O^+]$ is held constant at
(a) 1.0×10^{-1} M.   (b) 1.0×10^{-4} M.

*9-9. What is the equilibrium solubility of MnS in a solution in which $[H_3O^+]$ is held constant at
(a) 1.0×10^{-5} M?   (b) 1.0×10^{-8} M?

9-10. What weight of AgBr dissolves in 200 mL of 0.100 M NaCN? $(Ag^+ + 2\ CN^- \rightleftarrows Ag(CN)_2^-$;
$K_f = 1.3 \times 10^{21})$

*9-11. The equilibrium constant for formation of $CuCl_2^-$ is given by

$$Cu^+ + 2\ Cl^- \rightleftarrows CuCl_2^-$$

$$\beta_2 = \frac{[CuCl_2^-]}{[Cu^+][Cl^-]^2} = 7.9 \times 10^4$$

What is the solubility of CuCl in solutions having the following analytical NaCl concentrations:
(a) 1.0 M?   (d) 1.0×10^{-3} M?
(b) 1.0×10^{-1} M?   (e) 1.0×10^{-4} M?
(c) 1.0×10^{-2} M?

9-12. The equilibrium constant for the formation of $Al(OH)_4^-$ is given by

$$Al(OH)_3(s) + OH^- \rightleftarrows Al(OH)_4^- \qquad \beta_2 = 10$$

How many milliliters of 1.0 M NaOH is required to completely redissolve 1.00 g of $Al(OH)_3$ suspended in 100.0 mL of water?

*9-13. What concentration of OH^- must be maintained in order to dissolve 0.200 g of $Pb(OH)_2$ in 200 mL of solution? $[Pb(OH)_2(s) + OH^- \rightleftarrows Pb(OH)_3^-$;
$K_f = 5.0 \times 10^{-2}]$.

9-14. Formation constants for the reaction of Ag^+ with $S_2O_3^{2-}$ are

$$Ag^+ + S_2O_3^{2-} \rightleftarrows AgS_2O_3^- \qquad K_1 = 6.6 \times 10^8$$
$$AgS_2O_3^- + S_2O_3^{2-} \rightleftarrows Ag(S_2O_3)_2^{3-} \qquad K_2 = 4.4 \times 10^3$$

Calculate the solubility of AgI in 0.200 M $Na_2S_2O_3$ (assume that $S_2O_3^{2-}$ does not combine with H_3O^+).

*9-15. Calculate the solubility of $CdCO_3$ $(K_{sp} = 2.5 \times 10^{-14})$ in pure water. In order to simplify the calculation, assume that the solution is significantly basic as a consequence of the reaction $CO_3^{2-} + H_2O \rightleftarrows HCO_3^- + OH^-$, which makes $[H_3O^+]$ negligible with respect to the concentrations of all other ions. Furthermore, assume that $[H_2CO_3] \ll [HCO_3^-]$.

9-16. Calculate the solubility of CdS $(K_{sp} = 2 \times 10^{-28})$ in pure water. Because of the low solubility, it is per-

missible to assume that the presence of CdS does not alter the pH of the solution, and therefore $[H_3O^+] = [OH^-] = 1.0 \times 10^{-7}$ M.

*9-17. Silver ion is being considered as a reagent for separating I^- from SCN^- in a solution that is 0.060 M in KI and 0.070 M in NaSCN.
(a) What Ag^+ concentration is needed to lower the I^- concentration to 1.0×10^{-6} M?
(b) What is the Ag^+ concentration of the solution when AgSCN begins to precipitate?
(c) What is the ratio of SCN^- to I^- ion concentrations when AgSCN begins to precipitate?
(d) What is the ratio of SCN^- to I^- concentrations when the Ag^+ concentration is 1.0×10^{-3} M?

9-18. A solution is 0.040 M in Na_2SO_4 and 0.050 M in NaOH. To this is added a dilute solution containing Pb^{2+}.
(a) Which compound precipitates first, $PbSO_4$ or $Pb(OH)_2$?
(b) What is the Pb^{2+} concentration as the first precipitate appears?
(c) What Pb^{2+} concentration is required to initiate precipitation of the more soluble substance?
(d) What is the concentration of the first anion when the more soluble precipitate begins to form?

*9-19. Using 1.0×10^{-6} M as the criterion for quantitative removal, determine whether it is feasible to use
(a) SO_4^{2-} to separate Ba^{2+} from Sr^{2+} in a solution that is initially 0.10 M in Sr^{2+} and 0.25 M in Ba^{2+}.
(b) SO_4^{2-} to separate Ba^{2+} from Ag^+ in a solution that is initially 0.040 M in each cation. For Ag_2SO_4, $K_{sp} = 1.6 \times 10^{-5}$.
(c) OH^- to separate Be^{2+} from Hf^{4+} in a solution that is initially 0.020 M in Be^{2+} and 0.010 M in Hf^{4+}. For $Be(OH)_2$, $K_{sp} = 7.0 \times 10^{-22}$; for $Hf(OH)_4$, $K_{sp} = 4.0 \times 10^{-26}$.
(d) IO_3^- to separate In^{3+} from Tl^+ in a solution that is initially 0.11 M in In^{3+} and 0.060 M in Tl^+. For $In(IO_3)_3$, $K_{sp} = 3.3 \times 10^{-11}$; for $TlIO_3$, $K_{sp} = 3.1 \times 10^{-6}$.

9-20. Dilute NaOH is introduced into a solution that is 0.050 M in Cu^{2+} and 0.040 M in Mn^{2+}.
(a) Which hydroxide precipitates first?
(b) What OH^- concentration is needed to initiate precipitation of the first hydroxide?
(c) What is the concentration of the cation forming the less soluble hydroxide when the more soluble hydroxide begins to form?

CHAPTER 10

PRECIPITATION TITRATIONS WITH SILVER NITRATE

Silver nitrate is one of the most important and widely used titrimetric reagents. It is employed for the determination of anions that precipitate as silver salts. Among such species are the halides, several divalent anions, mercaptans, and certain fatty acids. Titrimetric methods based upon silver nitrate are sometimes termed *argentometric methods*.

With a few minor exceptions, no other precipitating agents react rapidly enough to be useful as titrimetric reagents. Thus, most precipitation titrations make use of silver nitrate.

"Argentometric" is derived from the Latin noun *argentum,* which means silver.

10A TITRATION CURVES IN TITRIMETRIC METHODS

As noted in Section 6A-2, an end point is an observable physical change that occurs near the equivalence point of a titration. The two most widely used end points involve (1) a change in color due to the reagent, the analyte, or an indicator and (2) a change in potential of an electrode that responds to the concentration of one of the reactants.

To help us understand the theoretical basis of end points and the sources of titration errors, we will derive a *titration curve* for the system under consideration. Titration curves consist of a plot of reagent volume as the horizontal axis and some function of the analyte or reagent concentration as the vertical axis.

Titration curves are plots of a concentration-related variable as a function of reagent volume.

10A-1 Types of Titration Curves

Two general types of titration curves (and thus two general types of end points) are encountered in titrimetric methods. In the first type, called a *sigmoidal curve,* important observations are confined to a small region (typically ±0.1 to ±0.5 mL) surrounding the equivalence point. A sigmoidal curve is shown in Figure 10-1a. In the second type, called a *linear-segment curve,* measurements are made on both sides of, but well away from, the equivalence point. Measurements near equivalence are avoided. A typical linear-segment curve is shown in Figure 10-1b. The sigmoidal type offers the advantages of speed and convenience. The linear-segment type is advantageous for reactions that are complete only in the presence of a considerable excess of reagent or analyte.

Here and in the several chapters that follow, we will be dealing exclusively with sigmoidal titration curves. Linear-segment curves are considered in Chapters 19 and 21.

10A-2 Concentration Changes During Titrations

The equivalence point in a titration is characterized by major changes in the *relative* concentrations of reagent and analyte. Table 10-1 illustrates this phenomenon. The data in the second column of the table show the changes in concentration of an analyte A as a 50.00-mL aliquot of a 0.1000 M solution of A is titrated with a 0.1000 M solution of a reagent R. The chemical reaction produces the precipitate AR, which has a solubility product of 1.0×10^{-10}. In order to emphasize the changes in *relative* concentration that occur in the equivalence region, the volume increments selected are those required to cause tenfold decreases in the concentration of A. Thus, we see in the third column that an addition of 40.9 mL of R is needed to decrease the concentration of the analyte by one order of magnitude, from 0.10 M to 0.010 M. An addition of only 8.1 mL is required to lower the concentration by another factor of 10, to 0.0010 M;

> Sigmoidal means s-shaped.

> The vertical axis in a sigmoidal titration curve is either the p-function of the analyte or reagent or the potential of an analyte- or reagent-sensitive electrode.

> The vertical axis in a linear-segment titration is an instrumental signal that is proportional to the concentration of the analyte or the reagent.

> At the beginning of the titration described in Table 10-1, about 41 mL of reagent brings about a tenfold decrease in the concentration of A; only 0.1 mL is required to cause this same change at the equivalence point.

Table 10-1
CONCENTRATION CHANGES DURING A TITRATION*

Volume of 0.1000 M R, mL	[A], mol/L	Volume of R That Causes a Tenfold Decrease in [A], mL	pA	pR
0.00	1.0×10^{-1}	—	1.00	—
40.90	1.0×10^{-2}	40.9	2.00	8.00
49.00	1.0×10^{-3}	8.1	3.00	7.00
49.90	1.0×10^{-4}	0.9	4.00	6.00
50.00	1.0×10^{-5}	0.1	5.00	5.00
50.10	1.0×10^{-6}	0.1	6.00	4.00
51.00	1.0×10^{-7}	0.9	7.00	3.00
61.10	1.0×10^{-8}	10.1	8.00	2.00

*Reaction: $A(aq) + R(aq) \rightarrow AR(s)$
For AR(s), $K_{sp} = 1.00 \times 10^{-10}$
Volume of 0.1000 M A = 50.00 mL

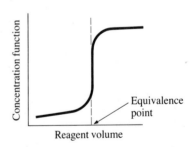

(a) Sigmoidal curve

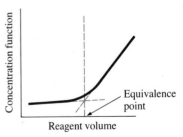

(b) Linear-segment curve
Figure 10-1

Two types of titration curves.

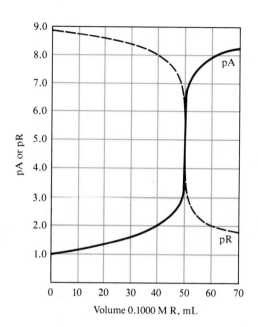

Figure 10-2

Precipitation titration curve for 50.00 mL of 0.1000 M A with 0.1000 M R.

0.9 mL causes yet another tenfold decrease. Corresponding increases in [R] occur at the same time. End-point detection, then, is based upon this large change in *relative* concentration of analyte (or reagent) that occurs at the equivalence point for every type of titration.

The large relative-concentration changes that occur in the region of chemical equivalence are shown by plotting the negative logarithm of analyte or reagent concentration (the p-function) against reagent volume, as has been done in Figure 10-2. The data for these plots are found in the fourth and fifth columns of Table 10-1. Titration curves for reactions involving precipitation, neutralization, complex formation, and oxidation/reduction all exhibit the same sharp increase or decrease in p-function shown in Figure 10-2. Titration curves define the properties required of an indicator and allow us to estimate the error associated with titration methods.

10B   TITRATION CURVES FOR ARGENTOMETRIC METHODS

The derivation of curves for an argentometric titration requires three types of calculation, each of which applies to a particular stage in the titration: (1) preequivalence points, (2) equivalence point, and (3) postequivalence points.

In the preequivalence stage, we compute the concentration of the analyte from its starting concentration and the volumetric data. The silver ion concentration is then obtained by substituting the analyte concentration into the expression for the solubility-product constant. At the equivalence point, silver ion and analyte ion exist in stoichiometric proportion, and the silver ion concentration is derived directly from the solubility-product constant. In the postequivalence stage, the analytical concentration of the excess silver nitrate is computed and assumed to be identical to its equilibrium concentration.

Example 10-1

Derive a titration curve for the titration of 50.00 mL of 0.00500 M NaBr with 0.01000 M $AgNO_3$ (for AgBr, $K_{sp} = 5.2 \times 10^{-13}$).

Initial Point

At the outset, the solution is 0.000 M in Ag^+, and pAg is indeterminate.

We cannot determine pAg initially because no $AgNO_3$ has been added.

After Addition of 5.00 mL of Reagent

The bromide ion concentration is decreased as a result of both precipitate formation and dilution. Thus, the analytical concentration of NaBr is

$$C_{NaBr} = \frac{\text{no. mmol NaBr after addition of } AgNO_3}{\text{total volume soln}}$$

$$= \frac{\text{original no. mmol NaBr} - \text{no. mmol } AgNO_3 \text{ added}}{\text{total volume soln}}$$

$$= \frac{(50.00 \text{ mL} \times 0.00500 \text{ M}) - (5.00 \text{ mL} \times 0.01000 \text{ M})}{50.00 \text{ mL} + 5.00 \text{ mL}}$$

$$= \frac{(0.2500 \text{ mmol} - 0.0500 \text{ mmol})}{55.00 \text{ mL}} = 3.64 \times 10^{-3} \text{ M}$$

Expressions for calculating concentrations during a titration can be derived from the charge-balance equation for the system:

$$[Na^+] + [Ag^+] = [Br^-] + [NO_3^-]$$

The first term in the numerator of these equations is the number of millimoles of NaBr originally in the sample, and the second term is the number of millimoles of $AgNO_3$ added, which is equal to the number of millimoles of Br^- that have reacted. The denominator takes into account the dilution of the solution by the added reagent.

Both the unreacted NaBr and the slightly soluble AgBr contribute to the species concentration of bromide ion. Thus, the equilibrium concentration of Br^- is larger than the analytical concentration of NaBr by an amount equal to the molar solubility of the precipitate:

$$[Br^-] = 3.64 \times 10^{-3} + [Ag^+]$$

From the charge-balance expression,

$$[Br^-] = ([Na^+] - [NO_3^-]) + [Ag^+]$$
$$= 3.64 \times 10^{-3} + [Ag^+]$$

The contribution of the silver bromide to the equilibrium bromide ion concentration is equal to $[Ag^+]$ because one silver ion is formed for each bromide ion from this source. Unless the concentration of NaBr is very small, this term can be neglected. That is, $[Ag^+] \ll 3.64 \times 10^{-3}$, and so

$$[Br^-] \approx C_{NaBr} = 3.64 \times 10^{-3}$$

The silver ion concentration is given by the relationship

$$[Ag^+] = \frac{K_{sp}}{[Br^-]}$$

To obtain an expression for pAg, we take the negative logarithm of both sides of this equation. Thus,

$$-\log[Ag^+] = -\log K_{sp} - (-\log[Br^-])$$

pAg + pBr = pK_{sp}
     = −log 5.2 × 10^{-13}
     = 12.28

From the definition of p-functions, we may write

$$pAg = pK_{sp} - pBr$$
$$= 12.28 - (-\log 3.64 \times 10^{-3})$$
$$= 12.28 - 2.44 = 9.84$$

This relationship applies to solutions containing silver and bromide ions in contact with solid silver bromide. Note that the calculated pAg corresponds to $[Ag^+] = 1.4 \times 10^{-10}$, which, as we assumed at the outset, is certainly much smaller than 3.4×10^{-4}.

Other data points in the region before chemical equivalence can be derived in this same way. Data for several such points are found in column 3 of Table 10-2.

At equivalence, $[Na^+] = [NO_3]$ and the charge-balance equation becomes

[N̶a̶⁺] + $[Ag^+]$ = $[Br^-]$ + [N̶O̶₃̶]
$[Ag^+]$ = $[Br^-]$

Equivalence Point

At the equivalence point, neither NaBr nor AgNO$_3$ is in excess, and so the concentrations of silver and bromide ions must be equal. Substituting this equality into the solubility-product expression yields

$$[Ag^+] = [Br^-] = \sqrt{5.2 \times 10^{-13}} = 7.21 \times 10^{-7}$$
$$pAg = pBr = -\log(7.21 \times 10^{-7}) = 6.14$$

After Addition of 25.10 mL of Reagent

The solution now contains an excess of AgNO$_3$, and we can write

$$C_{AgNO_3} = \frac{\text{total no. mmol AgNO}_3 - \text{original no. mmol NaBr}}{\text{total volume of solution}}$$

$$= \frac{(25.10 \text{ mL} \times 0.01000 \text{ M}) - (50.00 \text{ mL} \times 0.00500 \text{ M})}{(50.00 + 25.10) \text{ mL}}$$

$$= 1.33 \times 10^{-5} \text{ M}$$

and the equilibrium concentration of silver ion is

From charge-balance:

$[Ag^+]$ = ($[NO_3^-]$ − $[Na^+]$) + $[Br^-]$
     = 1.33 × 10^{-5} + $[Br^-]$

$$[Ag^+] = 1.33 \times 10^{-5} + [Br^-] \approx 1.33 \times 10^{-5}$$

In this equation, $[Br^-]$ is a measure of the Ag^+ concentration resulting from the slight solubility of AgBr; it can ordinarily be neglected. Thus,

$$pAg = -\log(1.33 \times 10^{-5}) = 4.876 = 4.88$$

Additional points defining the titration curve beyond the equivalence point can be obtained in an analogous way and are found in Table 10-2. The data are plotted as curve B in Figure 10-3.

10B-1 Significant Figures in Titration-Curve Calculations

The concentration data associated with the equivalence-point region of a titration curve are usually of low precision because they are based upon small differences between large numbers. For example, in the calculation

Table 10-2

CHANGES IN pAg DURING TITRATION WITH SOLUTIONS OF DIFFERENT CONCENTRATIONS

Volume AgNO₃, mL	pAg		
	50.00 mL 0.0500 M Br⁻ with 0.1000 M AgNO₃	50.00 mL 0.00500 M Br⁻ with 0.01000 M AgNO₃	50.00 mL 0.000500 M Br⁻ with 0.001000 M AgNO₃
0.00	—	—	—
10.00	10.68	9.68	8.68
20.00	10.13	9.13	8.13
23.00	9.72	8.72	7.72
24.90	8.41	7.41	6.50*
24.95	8.10	7.10	6.33*
25.00	6.14	6.14	6.14
25.05	4.18	5.18	5.95†
25.10	3.88	4.88	5.78†
27.00	2.58	3.58	4.58
30.00	2.20	3.20	4.20

*Approximation $[Ag^+] \ll C_{NaBr}$ not valid.
†Approximation $[Br^-] \ll C_{AgNO_3}$ not valid.

of C_{AgNO_3} following the introduction of 25.10 mL of 0.01000 M AgNO₃, the numerator $(0.2510 - 0.2500 = 0.0010)$ contains only two significant figures; at best then, C_{AgNO_3} is known to two significant figures. To minimize the rounding error, however, three digits (1.33×10^{-5}) were retained in the calculation, and rounding was postponed until after pAg was computed.

In rounding the calculated value for pAg, it is important to recall (page 41) that it is the *mantissa of a logarithm (that is, the number to the right*

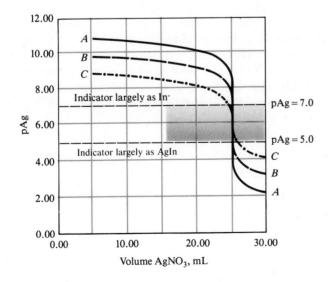

Figure 10-3

Effect of titrant concentration on titration curves: *A*, 50.00 mL of 0.0500 M NaBr with 0.1000 M AgNO₃; *B*, 50.00 mL of 0.00500 M NaBr with 0.01000 M AgNO₃; *C*, 50.00 mL of 0.000500 M NaBr with 0.001000 M AgNO₃.

pAg = 4.88 has only *two* significant figures and not three because the 4 merely sets the decimal point.

of the decimal point) that should be rounded to include only significant figures because the characteristic (the number to the left of the decimal point) serves merely to locate the decimal point. Thus, in Example 10-1, we rounded pAg to two figures to the right of the decimal point; that is, pAg = 4.88. Fortunately, the large changes in p-functions characteristic of most equivalence points are not obscured by the limited precision of the calculated data.

10B-2 Factors Influencing End-Point Sharpness

A sharp and easily located end point is observed when small additions of titrant cause large changes in p-function. It is therefore of interest to examine the variables that influence the magnitude of such changes during a titration.

As a rule of thumb, satisfactory end points require a change of 2 in p-function within ±0.1 mL of the equivalence point.

Reagent Concentration

Table 10-2 is a compilation of data computed by the method shown in Example 10-1. Three concentrations of titrant and analyte, each differing by a factor of 10, are used. The data are plotted in Figure 10-3. It is apparent that increases in analyte and reagent concentrations enhance the change in pAg in the equivalence-point region (an analogous effect is observed when pBr is plotted rather than pAg).

End points improve as solutions become more concentrated.

These effects have practical significance for the titration of bromide ion. If the analyte concentration is sufficient to permit the use of a silver nitrate solution that is 0.1 M or stronger, easily detected end points are observed, and the titration error is minimal. On the other hand, with standard titrant solutions that are 0.001 M or less, the change in pAg or pBr is so small that end-point detection is difficult and a large titration error must be expected. We shall show that titrant concentration also affects end-point sharpness in neutralization and complex-formation titrations.

Reaction Completeness

End points improve as the analytical reaction becomes more complete.

Figure 10-4 demonstrates the effect of product solubility on the sharpness of the end point in titrations using 0.1 M silver nitrate. Clearly, the greatest change in pAg occurs in the titration of iodide ion, which, of all the anions considered, forms the least soluble silver salt and hence reacts most completely with silver ion. The smallest change in pAg is observed for the reaction product that is most soluble—that is, in the titration of bromate ion.

10B-3 Chemical Indicators for Precipitation Titrations

Indicators compete with the analyte for reagent or with the reagent for analyte.

The end point produced by a chemical indicator usually consists of a color change or, occasionally, the appearance or disappearance of turbidity in the solution being titrated. Chemical indicators function by reacting competitively with one of the reactants of the titration. For example, consider

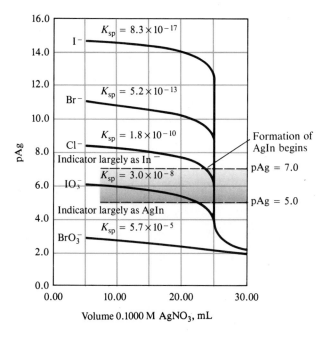

Figure 10-4
Effect of reaction completeness on titration curves. For each curve, 50.00 mL of a 0.0500 M solution of the anion was titrated with 0.1000 M $AgNO_3$.

Feature 10-1

TITRATION CURVES OF MIXTURES

Figure 10-5 is an argentometric titration curve for a mixture of iodide and chloride ions. Initial additions of silver nitrate result in formation of silver iodide exclusively because the solubility of that salt is only about 5×10^{-7} that of silver chloride.

It can be shown that this solubility difference is great enough so that formation of silver chloride is delayed until all but $7 \times 10^{-5}\%$ of the iodide has precipitated. Thus, short of the equivalence point, the curve is essentially indistinguishable from that for iodide alone. Just beyond the iodide equivalence point, the silver ion concentration is determined by the concentration of chloride ion in the solution, and the titration curve becomes essentially identical to the curve shown in Figure 10-4 for chloride ion by itself.

Figure 10-5 shows that it is possible in principle to determine both chloride and iodide in a mixture because the pAg change is greater than 2 in both equivalence-point regions. As will be shown in Chapter 18, curves resembling Figure 10-5 can be obtained experimentally by measuring the potential of a silver electrode immersed in the solution. Hence, a chloride/iodide mixture can be analyzed for each of its components. This technique is not as applicable to analyzing iodide/bromide or bromide/chloride mixtures, however, because the solubility differences between the silver salts are not great enough. Thus, the more soluble salt begins to form in significant amounts before precipitation of the less soluble salt is complete.

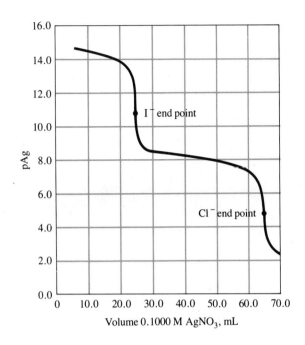

Figure 10-5
Titration curve for 50.00 mL of a solution that is 0.0500 M in I⁻ and 0.0800 M in Cl⁻.

the titration of an analyte A with a titrant R in the presence of an indicator In that reacts with R. A chemical description of this system throughout the titration is

$$A + R \rightleftharpoons AR(s)$$

$$In + R \rightleftharpoons InR$$

For satisfactory indicator action, a very low concentration of InR must radically change the appearance of the titration solution. In addition, the equilibrium constant for the indicator reaction must be such that the ratio [InR]/[In] increases very quickly as a consequence of the change in [R] (or pR) that occurs in the equivalence-point region. For most indicators, the ratio [InR]/[In] must shift by one to two orders of magnitude in order to produce a color change detectable to the human eye. Such a change corresponds to a change of 1 or 2 in pR.

To illustrate, consider the application of a hypothetical indicator In⁻ to the three bromide titrations described in Table 10-2 and Figure 10-3. We will assume that the equilibrium constant for the reaction of the indicator with silver ion to give AgIn is such that a full color change occurs when pAg shifts from 7 to 5. It is clear that each of the three titrations requires a different volume of titrant to encompass this range. For example, the data in the second column of Table 10-2 indicate that pAg changes from 8.10 to 4.18 as the volume of 0.1000 M silver nitrate is increased from 24.95 mL to 25.05 mL. Thus, an indicator with a pAg range of 7 to 5 would exhibit its color change within less than ±0.05 mL of the equivalence point at 25.00 mL. An abrupt change in color and a minimal titration error can be ex-

For a color change to be seen, [InR]/[In] must change by a factor of 10 to 100.

pected. In contrast, with 0.001 M $AgNO_3$, it can be shown that the color starts to change at about 24.5 mL and is complete at 25.8 mL. Exact location of the end point in this titration is impossible. The pAg change for the titration with 0.01 M reagent is such that somewhat less than 0.2 mL is required to bring about a color change; here, the indicator is usable, but the titration error is significant.

Consider the applicability and effectiveness of this same indicator for the titrations represented by the curves in Figure 10-4. Here, the indicator is largely AgIn throughout the titration of bromate and iodate ions, and so no color change is observed after the very first drops of silver nitrate are added. In contrast, the solubilities of silver bromide and silver iodide are small enough to prevent formation of significant amounts of AgIn until the equivalence-point region is reached. Note that pAg remains well above 7 until just before the equivalence point for the bromide titration and until just beyond the equivalence point for the iodide titration. In both cases, the color change occurs over a range of less than 0.01 mL of reagent, and titration errors are therefore negligible.

Turning to the titration curve for chloride ion in Figure 10-4, we see that an indicator with a pAg range of 5 to 7 is not satisfactory because appreciable amounts of AgIn will form approximately 1 mL short of the equivalence point and AgIn formation continues over a range of about 1 mL, making exact location of the end point impossible. In contrast, an indicator with a pAg range of 4 to 6 would be perfectly satisfactory. No chemical indicator exists for the iodate and bromate titrations because the changes in pAg in the equivalence-point region are far too small.

The Volhard End Point

In the Volhard method silver ions are titrated with a standard solution of thiocyanate ion:

$$Ag^+ + SCN^- \rightleftarrows AgSCN(s)$$

Iron(III) serves as the indicator. The solution turns red with the first slight excess of thiocyanate ion:

$$Fe^{3+} + SCN^- \rightleftarrows FeSCN^{2+} \qquad K_f = 1.4 \times 10^2 = \frac{[Fe(SCN)^{2+}]}{[Fe^{3+}][SCN^-]}$$
$$\text{Red}$$

The titration must be carried out in acidic solution to prevent precipitation of iron(III) as the hydrated oxide.

Example 10-2

It has been found from experiment that the average observer can just detect the red color of $Fe(SCN)^{2+}$ when its concentration is 6.4×10^{-6} M. In the titration of 50.0 mL of 0.050 M Ag^+ with 0.100 M KSCN, what concentration of Fe^{3+} should be used to reduce the titration error to zero?

For a zero titration error, the $FeSCN^{2+}$ color should appear when the concentration of Ag^+ remaining in the solution is identical to the sum of the two thiocyanate species. That is, at the equivalence point,

$$[Ag^+] = [SCN^-] + [Fe(SCN)^{2+}]$$
$$= [SCN^-] + 6.4 \times 10^{-6}$$

or $\qquad \dfrac{K_{sp}}{[SCN^-]} = \dfrac{1.1 \times 10^{-12}}{[SCN^-]} = [SCN^-] + 6.4 \times 10^{-6}$

which rearranges to

$$[SCN^-]^2 + 6.4 \times 10^{-6}[SCN^-] - 1.1 \times 10^{-12} = 0$$
$$[SCN^-] = 1.7 \times 10^{-7}$$

The formation constant for $FeSCN^{2+}$ is

$$K_f = 1.4 \times 10^2 = \frac{[Fe(SCN)^{2+}]}{[Fe^{3+}][SCN^-]}$$

If we now substitute the $[SCN^-]$ necessary to give a detectable concentration of $FeSCN^{2+}$ at the equivalence point, we obtain

$$1.4 \times 10^2 \doteq \frac{6.4 \times 10^{-6}}{[Fe^{3+}]1.7 \times 10^{-7}}$$
$$[Fe^{3+}] = 0.27$$

The indicator concentration is not critical in the Volhard titration. In fact, calculations similar to those in Example 10-2 demonstrate that a titration error of one part in a thousand or less is, in theory, possible if the iron(III) concentration is held between 0.002 and 1.6 M. In practice, it is found that an indicator concentration greater than 0.2 M imparts sufficient color to the solution to make detection of the thiocyanate complex difficult. Therefore, lower concentrations (usually about 0.01 M) of iron(III) ion are employed.

The most important application of the Volhard method is for the indirect determination of halide ions. A measured excess of standard silver nitrate solution is added to the sample, and the excess silver ion is determined by back-titration with a standard thiocyanate solution. The strongly acidic environment required for the Volhard procedure represents a distinct advantage over other methods of halide analysis because such ions as carbonate, oxalate, and arsenate (which form slightly soluble silver salts in neutral media) do not interfere.

Silver chloride is more soluble than silver thiocyanate. As a consequence, in chloride determinations by the Volhard method, the reaction

> The Volhard procedure requires that the analyte solution be distinctly acidic.

$$AgCl(s) + SCN^- \rightleftarrows AgSCN(s) + Cl^-$$

occurs to a significant extent near the end of the back-titration of the excess silver ion. This reaction causes the end point to fade and results in an overconsumption of thiocyanate ion, which in turn leads to low values for the chloride analysis. This error can be circumvented by filtering out the silver chloride before undertaking the back-titration. Filtration is not required in the determination of other halides because they all form silver salts that are less soluble than silver thiocyanate.

Adsorption Indicators

An *adsorption indicator* is an organic compound that tends to be adsorbed onto the surface of the solid in a precipitation titration. Ideally, the adsorption (or desorption) occurs near the equivalence point and results not only in a color change but also in a transfer of color from the solution to the solid (or the reverse).

Fluorescein is a typical adsorption indicator that is useful for the titration of chloride ion with silver nitrate. In aqueous solution, fluorescein partially dissociates into hydronium ions and negatively charged fluoresceinate ions that are yellow-green. The fluoresceinate ion forms a highly colored reddish silver salt. Whenever this dye is used as an indicator, however, *its concentration is never large enough to precipitate as silver fluoresceinate.*

In the early stages of the titration of chloride ion with silver nitrate, the colloidal silver chloride particles are negatively charged because of adsorption of excess chloride ions (Section 5A-2). The dye anions are repelled from this surface by electrostatic repulsion and impart a yellow-green color to the solution. Beyond the equivalence point, however, the silver chloride particles strongly adsorb silver ions and thereby acquire a positive charge. Fluoresceinate anions are now attracted *into the counter-ion layer* that surrounds each colloidal silver chloride particle. The net result is the appearance of the red color of silver fluoresceinate *in the surface layer of the solution surrounding the solid.* It is important to emphasize that the color change is an *adsorption* (not a precipitation) process since the solubility product of the silver fluoresceinate is never exceeded. The adsorption is reversible, the dye being desorbed upon back-titration with chloride ion.

Titrations involving adsorption indicators are rapid, accurate, and reliable, but their application is limited to the relatively few precipitation reactions in which a colloidal precipitate is formed rapidly.

Adsorption indicators were introduced in 1935 by a Polish chemist, K. Fajans. Titrations with these indicators are sometimes called Fajans titrations.

10B-4 Other Methods of End-Point Detection

Potentiometric methods, which can be applied to the detection of end points in some precipitation reactions, are described in Chapter 18.

10C APPLICATIONS OF STANDARD SILVER NITRATE SOLUTIONS

Table 10-3 lists some typical applications of precipitation titrations in which silver nitrate is the standard solution. In most of these methods, the analyte is precipitated with a measured excess of silver nitrate and the

Table 10-3

TYPICAL ARGENTOMETRIC PRECIPITATION METHODS

Substance Being Determined	End Point	Remarks
AsO_4^{3-}, Br^-, I^-, CNO^-, SCN^-	Volhard	Removal of silver salt not required
CO_3^{2-}, CrO_4^{2-}, CN^-, Cl^-, $C_2O_4^{2-}$, PO_4^{3-}, S^{2-}, NCN^{2-}	Volhard	Removal of silver salt required before back-titration of excess Ag^+
BH_4^-	Modified Volhard	Titration of excess Ag^+ following BH_4^- + 8 Ag^+ + 8 $OH^- \rightarrow$ 8 Ag(s) + $H_2BO_3^-$ + 5 H_2O
Epoxide	Volhard	Titration of excess Cl^- following hydrohalogenation
K^+	Modified Volhard	Precipitation of K^+ with known excess of $B(C_6H_5)_4^-$, addition of excess Ag^+ giving $AgB(C_6H_5)_4(s)$, and back-titration of the excess
Br^-, Cl^-	2 Ag^+ + $CrO_4^{2-} \rightarrow$ $Ag_2CrO_4(s)$ red	In neutral solution
Br^-, Cl^-, I^-, SeO_3^{2-}	Adsorption indicator	
$V(OH)_4^+$, fatty acids, mercaptans	Electroanalytical	Direct titration with Ag^+
Zn^{2+}	Modified Volhard	Precipitation as $ZnHg(SCN)_4$, filtration, dissolution in acid, addition of excess Ag^+, back-titration of excess Ag^+
F^-	Modified Volhard	Precipitation as PbClF, filtration, dissolution in acid, addition of excess Ag^+, back-titration of excess Ag^+

excess determined by a Volhard back-titration with standard potassium thiocyanate.

Both silver nitrate and potassium thiocyanate are obtainable in primary-standard quality. The latter is somewhat hygroscopic, however, and so thiocyanate solutions are ordinarily standardized against silver nitrate. Both standard solutions are stable indefinitely.

10D QUESTIONS AND PROBLEMS

*10-1. Why does a Volhard determination of chloride ion require more steps than a Volhard determination of bromide ion?

10-2. Give equations for the volumetric determination of $C_2O_4^{2-}$ by the Volhard method. Would it be necessary to isolate the $Ag_2C_2O_4$ before back-titration with KSCN? Why or why not?

*10-3. Outline a method for the determination of K^+ based on argentometry. Write balanced equations for the reactions.

10-4. Suggest an argentometric method for the determination of F^-. Write balanced equations for the reactions.

10-5. Calculate the molar concentration of a $AgNO_3$ solution if a 26.12-mL portion is needed to react with
*(a) 0.2124 g of KSCN.
(b) 0.6120 g of $BaCl_2 \cdot 2\ H_2O$.
*(c) 222.4 mg of Na_3PO_4.
(d) 50.00 mL of 0.01375 M H_2S.
*(e) 50.00 mL of 0.05451 M H_3AsO_4.
(f) 50.00 mL of 0.02756 M K_2CrO_4.

10-6. The As in a 9.13-g sample of pesticide was converted to AsO_4^{3-} and precipitated as Ag_3AsO_4 with 50.00 mL of 0.02105 M $AgNO_3$. The excess Ag^+ was then back-titrated with 4.75 mL of 0.04321 M KSCN. Calculate the percentage of As_2O_3 in the sample.

*10-7. Lead(II) can be titrated with standard K_2CrO_4 (product: $PbCrO_4$); an adsorption indicator signals the end point. A 1.1622-g mineral sample consisting mainly of Pb_3O_4 was decomposed by treatment with acid and subsequently titrated with 34.47 mL of 0.04176 M K_2CrO_4. Calculate the percentage of Pb_3O_4 in the sample.

10-8. A 50.00-mL aliquot of 0.1011 M $AgNO_3$ was added to an ammoniacal solution containing 0.2005 g of a sample containing an unknown amount of propargyl alcohol. The reaction is

$$2\ Ag^+ + NO_3^- + HC\equiv CCH_2OH \rightarrow$$
$$AgC\equiv CCH_2OH \cdot AgNO_3(s) + H^+$$

When reaction was complete, the excess Ag^+ was back-titrated with 6.80 mL of 0.08143 M KSCN. Calculate the percentage of propargyl alcohol (fw = 56.065 g) in the sample.

*10-9. The monochloroacetic acid ($ClCH_2COOH$) preservative in 100.0 mL of a carbonated beverage was extracted into diethyl ether and then returned to aqueous solution as $ClCH_2COO^-$ by extraction with 1 M NaOH. This aqueous extract was acidified and treated with 50.00 mL of 0.04521 M $AgNO_3$. The reaction is

$$ClCH_2COOH + Ag^+ + H_2O \rightarrow$$
$$HOCH_2COOH + H^+ + AgCl(s)$$

After filtration of the AgCl, titration of the filtrate and washings required 10.43 mL of an NH_4SCN solution. Titration of a blank taken through the entire process used 22.98 mL of the NH_4SCN. Calculate the weight (in milligrams) of $ClCH_2COOH$ in the sample.

*10-10. A 0.1064-g sample of a pesticide was decomposed by the action of sodium biphenyl in toluene. The liberated Cl^- was extracted with water and titrated with 23.28 mL of 0.03337 M $AgNO_3$ using an adsorption indicator. Express the results of this analysis in terms of percent aldrin, $C_{12}H_8Cl_6$ (fw = 364.92 g).

10-11. A carbonate fusion was needed to free the Bi from a 0.6423-g sample containing the mineral eulytite ($2\ Bi_2O_3 \cdot 3\ SiO_2$). The fused mass was dissolved in dilute acid, following which the Bi^{3+} was titrated with 27.36 mL of 0.03369 M NaH_2PO_4 solution. The reaction is

$$Bi^{3+} + H_2PO_4^- \rightarrow BiPO_4(s) + 2\ H^+$$

Calculate the percentage purity of eulytite (fw = 1112 g) in the sample.

*10-12. A 20-tablet sample of soluble saccharin was treated with 20.00 mL of 0.08181 M $AgNO_3$. The reaction is

After removal of the solid, titration of the filtrate and washings required 2.81 mL of 0.04124 M KSCN. Calculate the average number of milligrams of saccharin (fw = 205.17 g) in each tablet.

10-13. An adsorption indicator is to be used in the routine analysis of solids for their chloride content. It is desired that the volume of standard $AgNO_3$ used in these titrations be numerically equal to the percent Cl^- when 0.2500-g samples are used. What should the molarity of the $AgNO_3$ solution be?

*10-14. The Association of Official Analytical Chemists recommends a Volhard titration for analysis of the insecticide heptachlor ($C_{10}H_5Cl_7$). The percentage of heptachlor is given by

$$\% \text{ heptachlor} = \frac{(mL_{Ag} \times C_{Ag} - mL_{SCN} \times C_{SCN}) \times 37.33}{\text{wt sample}}$$

What does this calculation reveal concerning the stoichiometry of this titration?

10-15. An analysis for borohydride ion is based upon its reaction with Ag^+:

$$BH_4^- + 8\ Ag^+ + 8\ OH^- \rightarrow$$
$$H_2BO_3^- + 8\ Ag(s) + 5\ H_2O$$

The purity of a quantity of KBH_4 for use in an organic synthesis was established by diluting 0.3213 g of the material to exactly 500.0 mL with water, treating a 100.0-mL aliquot with 50.00 mL of 0.2221 M $AgNO_3$, and titrating the excess silver ion with 3.36 mL of 0.0397 M KSCN. Calculate the percent purity of the KBH_4 (fw = 53.95 g).

10-16. What volume of 0.04642 M KSCN would be needed if the analysis in Problem 10-15 were completed by filtering off the metallic Ag, dissolving it in acid, diluting to 250.0 mL with water, and titrating a 50.00-mL aliquot?

*10-17. A 100-mL sample of brackish water was made ammoniacal and the sulfide it contained titrated with
8.47 mL of 0.01310 M $AgNO_3$. The net reaction is

$$2\,Ag^+ + S^{2-} \rightarrow Ag_2S(s)$$

Calculate the parts per million of H_2S in the water.

10-18. A 2.000-L water sample was evaporated to a small
volume and treated with an excess of sodium
tetraphenylboron, $NaB(C_6H_5)_4$. The precipitated
$KB(C_6H_5)_4$ was filtered and then redissolved in acetone. The analysis was completed by titration, with
37.90 mL of 0.03981 M $AgNO_3$ being used. The net
reaction is

$$KB(C_6H_5)_4 + Ag^+ \rightarrow AgB(C_6H_5)_4 + K^+$$

Express the results of this analysis in terms of parts
per million of K (that is mg K/L).

*10-19. The action of an alkaline I_2 solution upon the roden-
ticide warfarin ($C_{19}H_{16}O_4$, fw = 308.34 g) results in
the formation of 1 mol of iodoform (CHI_3, fw =
393.73 g) for 1 mol of the parent compound. Analysis for warfarin can then be based upon the reaction
between CHI_3 and Ag^+:

$$CHI_3 + 3\,Ag^+ + H_2O \rightarrow 3\,AgI(s) + 3\,H^+ + CO(g)$$

The CHI_3 produced from a 13.96-g sample was
treated with 25.00 mL of 0.02979 M $AgNO_3$, and the
excess Ag^+ was then titrated with 2.85 mL of
0.05411 M KSCN. Calculate the percentage of warfarin in the sample.

*10-20. A 1.998-g sample containing Cl^- and ClO_4^- was dissolved in sufficient water to give 250.0 mL of solu-
tion. A 50.00-mL aliquot required 13.97 mL of
0.08551 M $AgNO_3$ to titrate the Cl^-. A second
50.00-mL aliquot was treated with $V_2(SO_4)_3$ to reduce the ClO_4^- to Cl^-:

$$ClO_4^- + 4\,V_2(SO_4)_3 + 4\,H_2O \rightarrow$$
$$Cl^- + 12\,SO_4^{2-} + 8\,VO^{2+} + 8\,H^+$$

Titration of the reduced sample required 40.12 mL
of the $AgNO_3$ solution. Calculate the percentages of
Cl^- and ClO_4^- in the sample.

10-21. A 2.4414-g sample containing KCl, K_2SO_4, and inert materials was dissolved in sufficient water to
give 250.0 mL of solution. Titration of a 50.00-mL
aliquot required 41.36 mL of 0.05818 M $AgNO_3$. A
second 50.00-mL aliquot was treated with 40.00 mL
of 0.1083 M $NaB(C_6H_5)_4$. The reaction is

$$NaB(C_6H_5)_4 + K^+ \rightarrow KB(C_6H_5)_4(s) + Na^+$$

The solid was filtered, redissolved in acetone, and
titrated with 49.98 mL of the $AgNO_3$ solution (see
Problem 10-18). Calculate the percentages of KCl
and K_2SO_4 in the sample.

*10-22. A 0.2185-g sample containing only KCl and K_2SO_4
yielded a precipitate of $KB(C_6H_5)_4$ that, after isolation and solution in acetone, required 25.02 mL of
0.1126 M $AgNO_3$ for the reaction

$$KB(C_6H_5)_4 + Ag^+ \rightarrow AgB(C_6H_5)_4(s) + K^+$$

Calculate the percentages of KCl and K_2SO_4 in the
sample.

10-23. For each of the following precipitation titrations,
calculate the cation and anion concentrations at
equivalence as well as at reagent volumes corresponding to ± 20.00 mL, ± 10.00 mL, and ± 1.00 mL
of equivalence. Construct a titration curve from the
data, plotting the p-function of the cation versus
reagent volume.

*(a) 25.00 mL of 0.05000 M $AgNO_3$ with 0.02500 M
NH_4SCN

(b) 20.00 mL of 0.06000 M $AgNO_3$ with 0.03000 M
KI

*(c) 30.00 mL of 0.07500 M $AgNO_3$ with 0.07500 M
NaCl

(d) 35.00 mL of 0.4000 M Na_2SO_4 with 0.2000 M
$Pb(NO_3)_2$

*(e) 40.00 mL of 0.02500 M $BaCl_2$ with 0.05000 M
Na_2SO_4

(f) 50.00 mL of 0.2000 M NaI with 0.4000 M $TlNO_3$
(K_{sp} for TlI = 6.5×10^{-8})

10-24. Calculate the silver ion concentration after the addition of 5.00*, 15.00, 25.00, 30.00, 35.00, 39.00,
40.00*, 41.00, 45.00*, and 50.00 mL of 0.05000 M
$AgNO_3$ to 50.0 mL of 0.0400 M KBr. Construct a
titration curve from these data, plotting pAg as a
function of titrant volume.

10-25. Calculate the Hg_2^{2+} concentration after the addition
of 0, 10.00*, 20.00, 30.00*, 35.00, 39.00, 40.00*,
41.00, 45.00, and 50.00* mL of 0.2000 M NaCl to
80.0 mL of a solution that is 0.0500 M in Hg_2^{2+}. For
the process,

$$Hg_2Cl_2 \rightleftharpoons Hg_2^{2+} + 2\,Cl^- \qquad K_{sp} = 1.3 \times 10^{-18}$$

Construct a titration curve from these data, plotting
pHg_2 as a function of titrant volume. *Note:* No evidence exists for the intermediate species Hg_2Cl^+.

Chapter 11

Theory of Neutralization Titrations

Standard solutions of strong acids and strong bases are used extensively for determining analytes that are themselves acids or bases or that can be converted to such species by chemical treatment. This chapter deals with the theoretical aspects of titrations with such reagents.

We begin our discussion by first describing a class of organic weak acids and bases that have indicator properties. We then show how these substances provide end points in acid/base titrations. Finally, we turn our attention to various types of pH calculations and to the derivation of titration curves for strong and weak acids and bases.[1]

In preparation for this discussion, you may find it helpful to review the material on acids and bases in Sections 7A-2 and 7B-3.

> The standard reagents used in acid/base titrations are always strong acids or strong bases, most commonly HCl, $HClO_4$, H_2SO_4, NaOH, and KOH. Weak acids and bases are never used as standard reagents because they react incompletely with analytes.

11A ACID/BASE INDICATORS

Many substances color solutions to a hue that depends upon pH. Some of these substances, which have been used for centuries to indicate the acidity or alkalinity of water, find application as acid/base indicators.

> For a list of common indicators and their colors, look inside the front cover of this book.

---

[1] For a general reference on these topics, see D. R. Rosenthal and P. Zuman, in *Treatise on Analytical Chemistry*, 2nd ed., I. M. Kolthoff and P. J. Elving, Eds., Part I, Vol. 2, Chapter 18. New York: Wiley, 1979.

11A-1 Theory of Indicator Behavior

An acid/base indicator is a weak organic acid or a weak organic base whose undissociated form differs in color from its conjugate-base or conjugate-acid form. For example, the behavior of an acid-type indicator HIn is described by the equilibrium

$$HIn + H_2O \rightleftharpoons In^- + H_3O^+$$

Acid   Base
Color   Color

Here, internal structural changes accompany dissociation and cause the color change (Figure 11-1). The equilibrium for a base-type indicator In is

$$In + H_2O \rightleftharpoons InH^+ + OH^-$$

Base   Acid
Color   Color

In the paragraphs that follow, we focus on the behavior of acid-type indicators. The discussion, however, can be readily applied to base-type indicators as well.

The equilibrium-constant expression for the dissociation of an acid-type indicator takes the form

$$\frac{[H_3O^+][In^-]}{[HIn]} = K_a \tag{11-1}$$

Rearranging leads to

$$[H_3O^+] = K_a \frac{[HIn]}{[In^-]} \tag{11-2}$$

We see then that the hydronium ion concentration determines the ratio of acid and conjugate-base form of the indicator.

The human eye is not very sensitive to color differences in a solution containing a mixture of In$^-$ and HIn, particularly when the ratio [In$^-$]/[HIn] is greater than about 10 or smaller than about 0.1. To the average observer, consequently, the color of a solution of a typical indicator appears to change rapidly only within the limited concentration ratio of

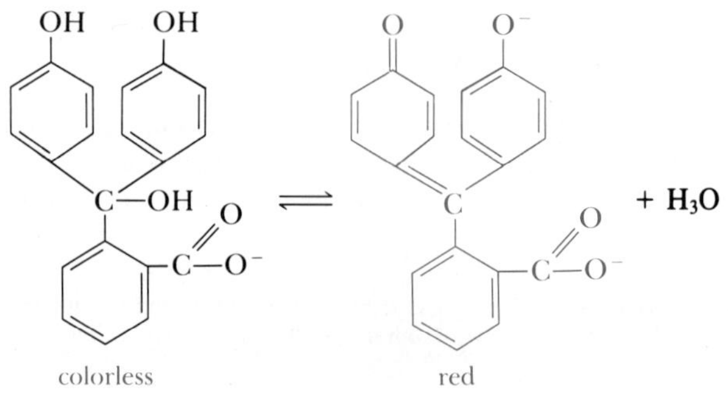

Figure 11-1
Color change for phenolphthalein.

approximately 10 to 0.1. At greater or smaller ratios, the color becomes essentially constant to the human eye and independent of the ratio. Therefore, we can write that the indicator HIn exhibits its pure acid color when

$$\frac{[\text{HIn}]}{[\text{In}^-]} \geq \frac{10}{1}$$

and its pure base color when

$$\frac{[\text{HIn}]}{[\text{In}^-]} \leq \frac{1}{10}$$

The color appears to be intermediate for ratios between these two values. These ratios vary considerably from indicator to indicator, of course. Furthermore, people differ significantly in their ability to distinguish between colors, with a color-blind person representing one extreme.

If the two concentration ratios are substituted into Equation 11-1, the range of hydronium ion concentrations needed to cause the complete indicator color change can be evaluated. Thus, for the full acid color,

$$[\text{H}_3\text{O}^+] \geq K_a \frac{10}{1}$$

and similarly for the full base color,

$$[\text{H}_3\text{O}^+] \leq K_a \frac{1}{10}$$

To obtain the indicator range, we take the negative logarithms of the two expressions:

$$\text{pH (acid color)} = -\log (K_a \cdot 10) = pK_a + 1$$
$$\text{pH (basic color)} = -\log (K_a/10) = pK_a - 1$$

$$\text{indicator range} = pK_a \pm 1 \qquad (11\text{-}3)$$

Thus, the typical indicator with an acid dissociation constant of 1×10^{-5} ($pK_a = 5$) shows a complete color change when the pH of the solution in which it is dissolved changes from 4 to 6 (Figure 11-2). A similar relationship is easily derived for a basic indicator.

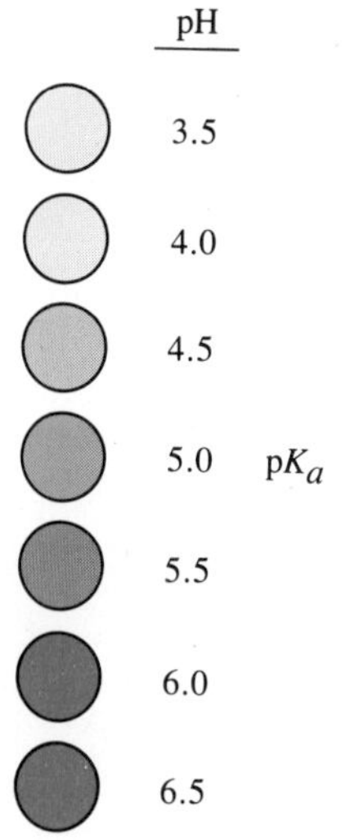

11A-2 Variables That Influence Indicator Behavior

The pH range over which an indicator changes color depends upon temperature, ionic strength, as well as the presence of organic solvents and colloidal particles. Some of these effects, particularly the latter two, can cause the transition range to shift by one or more pH units.[2]

---

[2]For a discussion of these effects, see H. A. Laitinen and W. E. Harris, *Chemical Analysis,* 2nd ed., pp. 48–51. New York: McGraw-Hill, 1975.

Figure 11-2
Indicator color as a function of pH ($pK_a = 5.0$).

	pH
	3.5
	4.0
	4.5
	5.0 pK_a
	5.5
	6.0
	6.5

The approximate pH range of most acid-type indicators is $pK_a \pm 1$.

11A-3 Some Common Indicators

The list of acid/base indicators is large and includes a number of organic structures. Indicators are available for any desired pH range. A few common indicators and their properties are listed in Table 11-1. Note that most of the transition ranges cover something less than two pH units.

11B TITRATION CURVES FOR STRONG ACIDS AND STRONG BASES

When both reagent and analyte are strong electrolytes, the neutralization reaction is described by the equation

$$H_3O^+ + OH^- \rightleftarrows 2\,H_2O$$

Titration curves based on this reaction are derived by methods analogous to those shown in Section 10B with K_w rather than K_{sp} being used as the equilibrium constant. Normally, pH serves as the ordinate in titration curves for strong bases as well as strong acids.

11B-1 The Titration of a Strong Acid with a Strong Base

The hydronium ions in an aqueous solution of a strong acid have two sources: (1) the reaction of the solute with water and (2) the dissociation

Table 11-1

SOME IMPORTANT ACID/BASE INDICATORS*

Common Name	Transition Range, pH	pK_a**	Color Change†	Indicator Type‡
Thymol blue	1.2–2.8	1.65	R–Y	1
	8.0–9.6	8.90	Y–B	
Methyl yellow	2.9–4.0		R–Y	2
Methyl orange	3.1–4.4	3.46§	R–O	2
Bromocresol green	3.8–5.4	4.66	Y–B	1
Methyl red	4.2–6.3	5.00§	R–Y	2
Bromocresol purple	5.2–6.8	6.12	Y–P	1
Bromothymol blue	6.2–7.6	7.10	Y–B	1
Phenol red	6.8–8.4	7.81	Y–R	1
Cresol purple	7.6–9.2		Y–P	1
Phenolphthalein	8.3–10.0		C–R	1
Thymolphthalein	9.3–10.5		C–B	1
Alizarin yellow GG	10–12		C–Y	2

*From C. A. Streuli, in *Handbook of Analytical Chemistry*, L. Meites, Ed., pp. **3**-35 and **3**-36. New York: McGraw-Hill, 1963.

**At ionic strength of 0.1.

†B = blue; C = colorless; O = orange; P = purple; R = red; Y = yellow.

‡(1) Acid type: HIn + H$_2$O $\rightleftarrows$ H$_3$O$^+$ + In$^-$

(2) Base type: In + H$_2$O $\rightleftarrows$ InH$^+$ + OH$^-$

§For the reaction InH$^+$ + H$_2$O $\rightleftarrows$ H$_3$O$^+$ + In

of water. In all but the most dilute solutions, however, the contribution from the solute far exceeds that from the solvent. Thus, for a solution of HCl with a concentration greater than about 1×10^{-6} M, we can write

$$[H_3O^+] = C_{HCl} + [OH^-] \approx C_{HCl} \qquad (11\text{-}4)$$

where $[OH^-]$ represents the contribution of water dissociation to the hydronium ion concentration.

An analogous relationship applies for a solution of a strong base, such as sodium hydroxide

$$[OH^-] = C_{NaOH} + [H_3O^+] \approx C_{NaOH} \qquad (11\text{-}5)$$

As seen in the following example, the derivation of a curve for the titration of a strong acid with a strong base is simplified because the concentration of the hydronium ion or the hydroxide ion is obtained directly from stoichiometric calculations.

In solutions of a strong acid that are more concentrated than about 1×10^{-6} M, it is proper to assume that the equilibrium concentration of $[H_3O^+]$ is equal to the analytical concentration of the acid. The same is true for the hydroxide concentration in solutions of a strong base.

A useful relationship is obtained by taking the negative logarithm of each side of the ion-product constant for water:

$$-\log K_w = -\log [H_3O^+][OH^-]$$
$$= -\log [H_3O^+] - \log[OH^-]$$
$$pK_w = pH + pOH$$

Example 11-1

Derive a titration curve for the reaction of 50.00 mL of 0.05000 M HCl with 0.1000 M NaOH. Round pH data to two places to the right of the decimal point.

Initial Point

The solution is 5.00×10^{-2} M in HCl. Since HCl is completely dissociated,

$$[H_3O^+] = 5.00 \times 10^{-2} \text{ M}$$
$$pH = -\log (5.00 \times 10^{-2}) = -\log 5.00 - \log 10^{-2}$$
$$= -0.699 + 2 = 1.301 = 1.30$$

Remember: pH = pK_w − pOH
$= 14.00 - $ pOH

pH after Addition of 10.00 mL of NaOH

The volume of the solution is now 60.00 mL, and part of the HCl has been neutralized. Thus,

$$[H_3O^+] = \frac{50.00 \times 0.0500 - 10.00 \times 0.1000}{60.00} = 2.50 \times 10^{-2} \text{ M}$$
$$pH = 2 - \log 2.50 = 1.60$$

Before the equivalence point, we calculate the pH from the concentration of unreacted strong acid.

Additional data to define the curve in the region before the equivalence point are calculated in the same way. The results of such calculations are given in column 2 of Table 11-2.

pH after Addition of 25.00 mL of NaOH

At the equivalence point, the solution contains neither an excess of HCl nor an excess of NaOH, and

$$[H_3O^+] = [OH^-] = \sqrt{K_w} = 1.00 \times 10^{-7}$$
$$pH = -\log 1.00 \times 10^{-7} = 7.00$$

At the equivalence point the solution is neutral.

After the equivalence point we first calculate pOH, then pH.

pH after Addition of 25.10 mL of NaOH

The concentration of excess base is

$$C_{NaOH} = \frac{25.10 \times 0.1000 - 50.0 \times 0.0500}{75.10} = 1.33 \times 10^{-4} \text{ M}$$

and so

$$[OH^-] = C_{NaOH} = 1.33 \times 10^{-4} \text{ M}$$
$$pOH = -\log(1.33 \times 10^{-4}) = 3.88$$

Thus,

$$pH = 14.00 - 3.88 = 10.12$$

Additional data for this titration, calculated in the same way, are given in column 2 of Table 11-2.

The Effect of Concentration

The effects of reagent and analyte concentration on the neutralization titration curves for strong acids are shown by the two sets of data in Table 11-2 and the plots in Figure 11-3.

With 0.1 M NaOH as the titrant (curve *A*), the change in pH in the equivalence-point region is large. With 0.001 M NaOH (curve *B*), the change is markedly less but still pronounced.

Indicator Choice

Figure 11-3 shows that the selection of an indicator is not critical when the reagent concentration is approximately 0.1 M. Here, the volume differ-

Table 11-2

CHANGES IN pH DURING THE TITRATION OF A STRONG ACID WITH A STRONG BASE

Volume of NaOH, mL	50.00 mL of 0.0500 M HCl with 0.1000 M NaOH pH	50.00 mL of 0.000500 M HCl with 0.001000 M NaOH pH
0.00	1.30	3.30
10.00	1.60	3.60
20.00	2.15	4.15
24.00	2.87	4.87
24.90	3.87	5.87
25.00	7.00	7.00
25.10	10.12	8.12
26.00	11.12	9.12
30.00	11.80	9.80

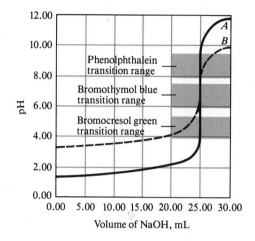

Figure 11-3
Titration curves for HCl with NaOH. *A*: 50.00 mL of 0.0500 M HCl with 0.1000 M NaOH. *B*: 50.00 mL of 0.000500 M HCl with 0.001000 M NaOH.

ences in titrations with the three indicators shown are of the same magnitude as the uncertainties associated with reading the buret and thus negligible. Note, however, that bromocresol green is clearly unsuited for a titration involving the 0.001 M reagent because the color begins to change after addition of about 18 mL of reagent and is complete with addition of approximately 23 mL. Phenolphthalein is equally unsatisfactory. Of the three indicators, then, only bromothymol blue provides a satisfactory end point with a minimal determinate titration error when the more dilute solutions are used.

11B-2 The Titration of a Strong Base With a Strong Acid

Titration curves for a strong base are derived in a way analogous to that for strong acids. Short of the equivalence point, the solution is highly basic, and the hydroxide ion concentration is equal to or a multiple of the analytical molarity of the base. The solution is neutral at the equivalence point. Finally, the solution is acidic in the region beyond the equivalence point; here, the hydronium ion concentration is equal to the analytical concentration of the excess strong acid. A curve for the titration of a strong base with 0.1 M hydrochloric acid is shown in Figure 11-8. Indicator selection is based upon the considerations described for the titration of a strong acid with a strong base.

11C BUFFER SOLUTIONS

By definition, a *buffer solution* is one that resists changes in pH. Most buffers consist of either a weak acid and its conjugate base or a weak base and its conjugate acid. Chemists use buffers whenever they need to maintain the pH of a solution at a constant and predetermined level. You will find many references to the use of buffers throughout this text.

Buffers are used in all types of chemistry whenever it is desirable to maintain the pH of a solution at a constant and predetermined level.

Buffered aspirin contains buffers to help maintain stomach contents at a relatively high pH.

A buffer solution is formed whenever a weak acid is titrated with a strong base or a weak base is titrated with a strong acid. As a consequence of the presence of a buffer, titration curves for weak acids and weak bases are significantly different from the titration curves we have thus far encountered.

11C-1 The Calculation of Buffer-Solution pH

Weak Acid/Conjugate Base Buffers

A solution containing a weak acid HA and its conjugate base A^- may be acidic, neutral, or basic, depending upon the position of two competitive equilibria:

$$HA + H_2O \rightleftharpoons H_3O^+ + A^- \qquad \frac{[H_3O^+][A^-]}{[HA]} = K_a \qquad (11\text{-}6)$$

$$A^- + H_2O \rightleftharpoons OH^- + HA \qquad \frac{[OH^-][HA]}{[A^-]} = K_b = \frac{K_w}{K_a} \qquad (11\text{-}7)$$

If the first equilibrium lies farther to the right than the second, the solution is acidic. If the second equilibrium is dominant, the solution is basic. These two equilibrium-constant expressions show that the relative concentrations of the hydronium and hydroxide ions depend not only upon the magnitudes of K_a and K_b but also upon the ratio of the concentrations of the acid and its conjugate base.

In order to compute the pH of a solution containing both an acid HA and its salt NaA, we need to express the equilibrium concentrations of HA and NaA in terms of their analytical concentrations C_{HA} and C_{NaA}. An examination of the two equilibria reveals that the first reaction decreases the concentration of HA by an amount equal to $[H_3O^+]$, whereas the second increases the HA concentration by an amount equal to $[OH^-]$. Thus, the species concentration of HA is related to its analytical concentration by

$$[HA] = C_{HA} - [H_3O^+] + [OH^-] \qquad (11\text{-}8)$$

Similarly, the first equilibrium increases the concentration of A^- by an amount equal to $[H_3O^+]$, and the second decreases this concentration by the amount $[OH^-]$. Thus the equilibrium concentration is given by

$$[A^-] = C_{NaA} + [H_3O^+] - [OH^-] \qquad (11\text{-}9)$$

Because of the inverse relationship between $[H_3O^+]$ and $[OH^-]$, it is *always* possible to eliminate one or the other from Equations 11-8 and

Feature 11-1
APPLICATION OF THE SYSTEMATIC METHOD TO BUFFER
CALCULATIONS

Equations 11-8 and 11-9 can also be derived from mass- and charge-balance expressions. Thus, mass-balance considerations require that

$$C_{HA} + C_{NaA} = [HA] + [A^-] \qquad (1)$$

Electrical-neutrality considerations require that

$$[Na^+] + [H_3O^+] = [A^-] + [OH^-]$$

but

$$[Na^+] = C_{NaA}$$

Therefore, the charge-balance equation is

$$C_{NaA} + [H_3O^+] = [A^-] + [OH^-] \qquad (2)$$

which rearranges to Equation 11-9:

$$[A^-] = C_{NaA} + [H_3O^+] - [OH^-]$$

Subtraction of Equation 2 from Equation 1 and rearrangement give

$$[HA] = C_{HA} - [H_3O^+] + [OH^-]$$

which corresponds to Equation 11-8.

The difference between the hydronium and hydroxide ion concentrations in a buffer is usually small relative to either C_{HA} or C_{NaA}.

11-9. Moreover, the *difference* in concentration between these two species is often so small relative to the molar concentrations of acid and conjugate base that Equations 11-8 and 11-9 simplify to

$$[HA] = C_{HA} \qquad (11-10)$$
$$[A^-] = C_{NaA} \qquad (11-11)$$

Substitution of Equations 11-10 and 11-11 into the dissociation-constant expression yields, upon rearrangement,

$$[H_3O^+] = K_a \frac{C_{HA}}{C_{NaA}} \qquad (11\text{-}12)$$

The assumption leading to Equations 11-10 and 11-11 sometimes breaks down with acids or bases that have dissociation constants greater than about 10^{-3} or when the molar concentration of either the acid or its conjugate base (or both) is very small. In these circumstances, either $[OH^-]$ or $[H_3O^+]$ must be retained in Equations 11-8 and 11-9, depending upon whether the solution is acidic or basic. In any case, Equations 11-10 and 11-11 should always be used initially. The provisional values for $[H_3O^+]$ and $[OH^-]$ can then be used to test the assumptions.

Within the limits imposed by the assumptions made in its derivation, Equation 11-12 says that the hydronium ion concentration of a solution containing a weak acid and its conjugate base is dependent only upon the *ratio* of the molar concentrations of these two solutes. Furthermore, this ratio is *independent of dilution* because the concentration of each component changes proportionately when the volume changes.

Feature 11-2
THE HENDERSON-HASSELBALCH EQUATION

The Henderson-Hasselbalch equation is an alternative form of Equation 11-12 that is encountered frequently in the biological literature and biochemical texts. It is obtained by expressing each term in the equation in the form of its negative logarithm and inverting the concentration ratio to keep all signs positive:

$$-\log [H_3O^+] = -\log K_a - \log \frac{C_{HA}}{C_{NaA}}$$

Therefore,

$$pH = pK_a + \log \frac{C_{NaA}}{C_{HA}} \qquad (11\text{-}13)$$

Example 11-2

What is the pH of a solution that is 0.400 M in formic acid and 1.00 M in sodium formate?

The equilibrium governing the hydronium ion concentration in this solution is

$$H_2O + HCOOH \rightleftarrows H_3O^+ + HCOO^-$$

for which

$$K_a = \frac{[H_3O^+][HCOO^-]}{[HCOOH]} = 1.77 \times 10^{-4}$$

$$[HCOO^-] \approx C_{HCOO^-} = 1.00$$

$$[HCOOH] \approx C_{HCOOH} = 0.400$$

Substitution into Equation 11-12 gives, with rearrangement,

$$[H_3O^+] = 1.77 \times 10^{-4} \times \frac{0.400}{1.00} = 7.08 \times 10^{-5}$$

Note that the assumption that $[H_3O^+] \ll C_{HCOOH}$ and C_{HCOO^-} is valid. Thus,

$$pH = -\log (7.08 \times 10^{-5}) = 4.15$$

Weak Base/Conjugate Acid Buffers

The calculation of the hydroxide ion concentration for a solution consisting of a weak base and its conjugate acid is entirely analogous to that developed in the preceding section.

Example 11-3

Calculate the pH of a solution that is 0.200 M in NH_3 and 0.300 M in NH_4Cl. The base dissociation constant for NH_3 is 1.76×10^{-5}.

The equilibria of interest are

$$NH_3 + H_2O \rightleftarrows NH_4^+ + OH^- \qquad K_b = 1.76 \times 10^{-5}$$

$$NH_4^+ + H_2O \rightleftarrows NH_3 + H_3O^+ \qquad K_a = \frac{1.00 \times 10^{-14}}{1.76 \times 10^{-5}} = 5.68 \times 10^{-10}$$

Since $K_b > K_a$, the solution is alkaline. Using the arguments in the previous example, the molar concentrations of NH_3 and NH_4^+ are

$$[NH_3] = 0.200 - [\cancel{OH^-}] + [\cancel{H_3O^+}] \approx 0.200$$

$$[NH_4^+] = 0.300 + [\cancel{OH^-}] - [\cancel{H_3O^+}] \approx 0.300$$

A provisional value for $[OH^-]$ is obtained by substituting the approximate values for $[NH_3]$ and $[NH_4^+]$ into the rearranged form of the equilibrium-constant expression for NH_3:

$$[OH^-] = K_b \times \frac{[NH_3]}{[NH_4^+]} = 1.76 \times 10^{-5} \times \frac{0.200}{0.300} = 1.173 \times 10^{-5}$$

Figure 11-4
The effect of dilution on the pH of buffered and unbuffered solutions. The dissociation constant for HA is 1.00×10^{-4}. Initial solute concentrations are 1.00 M.

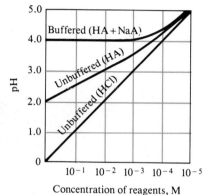

Concentration of reagents, M

The approximations are clearly justified; thus,

$$pOH = 5.000 - \log 1.173 = 4.931$$
$$pH = 14.00 - 4.931 = 9.07$$

11C-2 Properties of Buffer Solutions

In this section we illustrate the resistance of buffers to changes in pH brought about either by dilution or by addition of strong acids or bases.

The Effect of Dilution
The pH of a buffer solution remains essentially independent of dilution until the concentrations of the species it contains are decreased to the point where the approximations used to develop Equations 11-10 and 11-11 become invalid. Figure 11-4 contrasts the behavior of buffered and unbuffered solutions with dilution. For each, the initial solute concentration is 1.00 M. The resistance of the buffered solution to changes in pH during dilution is clear.

The Effect of Added Acids and Bases
Example 11-4 illustrates a second property of buffer solutions: their resistance to pH change after addition of small amounts of strong acids or bases.

Example 11-4

Calculate the pH change that takes place when a 100-mL portion of (a) 0.0500 M NaOH and (b) 0.0500 M HCl is added to 400 mL of the buffer solution described in Example 11-3.
(a) Addition of NaOH converts part of the NH_4^+ in the buffer to NH_3:

$$NH_4^+ + OH^- \rightarrow NH_3 + H_2O$$

The analytical concentrations of NH_3 and NH_4Cl then become

$$c_{NH_3} = \frac{400 \times 0.200 + 100 \times 0.0500}{500} = \frac{85.0}{500} = 0.170 \text{ M}$$

$$c_{NH_4Cl} = \frac{400 \times 0.300 - 100 \times 0.0500}{500} = \frac{115}{500} = 0.230 \text{ M}$$

When substituted into the dissociation-constant expression, these values yield

Buffers do not maintain pH at an absolutely constant level, but changes in pH are relatively small when small amounts of acid or base are added.

$$[OH^-] = \frac{1.76 \times 10^{-5} \times 0.170}{0.230} = 1.30 \times 10^{-5}$$

$$pH = 14.00 - (-\log 1.30 \times 10^{-5}) = 9.11$$

In Example 11-3, we found the original pH to be 9.07. Thus, the change in pH is

$$\Delta pH = 9.11 - 9.07 = 0.04$$

(b) Addition of HCl converts part of the NH_3 to NH_4^+; thus,

$$NH_3 + H_3O^+ \rightarrow NH_4^+ + H_2O$$

$$C_{NH_3} = \frac{400 \times 0.200 - 100 \times 0.0500}{500} = \frac{75}{500} = 0.150 \text{ M}$$

$$C_{NH_4^+} = \frac{400 \times 0.300 + 100 \times 0.0500}{500} = \frac{125}{500} = 0.250 \text{ M}$$

$$[OH^-] = 1.76 \times 10^{-5} \times \frac{0.150}{0.250} = 1.06 \times 10^{-5}$$

$$pH = 14.00 - (-\log 1.06 \times 10^{-5}) = 9.02$$

$$\Delta pH = 9.02 - 9.07 = -0.05$$

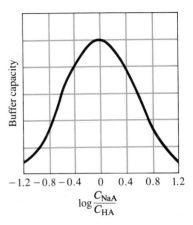

Figure 11-5
Buffer capacity as a function of the ratio C_{NaA}/C_{HA}.

It is of interest to contrast the behavior of an unbuffered solution with a pH of 9.07 to that of the buffer in Example 11-3. You can show that adding the same quantity of base to the unbuffered solution would increase the pH to 12.00—a pH change of 2.93 units. Adding the acid would decrease the pH by slightly more than 7 units.

Buffer Capacity
Figure 11-4 and Example 11-4 demonstrate that a solution containing a conjugate acid/base pair possesses remarkable resistance to changes in pH. The ability of a buffer to prevent a significant change in pH is directly related to the total concentration of the buffering species as well as to their concentration ratio. For example, if we diluted the buffer described in Example 11-3 by a factor of 10, the pH would still be 9.07. However, if we treated the diluted buffer in the same way as the more concentrated buffer was treated in Example 11-4, the pH change would be 0.4 to 0.5 unit instead of 0.04 to 0.05.

The *buffer capacity* of a solution is the number of moles of a strong acid or a strong base that causes 1.00 L of the buffer to undergo a 1.00-unit change in pH. The capacity of a buffer depends not only on the total concentrations of the two buffer components but also on their concentration ratio. Buffer capacity falls off moderately rapidly as the concentration ratio of acid to conjugate base becomes larger or smaller than unity (Figure 11-5). For this reason, the pK_a of the acid chosen for a given application should lie within ±1 unit of the desired pH in order for the buffer to have a reasonable capacity.

The Preparation of Buffers
In principle, a buffer solution of any desired pH can be prepared by combining calculated quantities of a suitable conjugate acid/base pair. In practice, however, the pH values of buffers prepared from theoretically

> The buffer capacity is the number of moles of strong acid or strong base that 1 L of the buffer can absorb while changing the pH by 1 unit.

generated recipes differ from the predicted values (1) because of uncertainties in the numerical values of many dissociation constants and (2) because of the simplifications used in calculations. Most important is the fact that the ionic strength of a buffer is usually so high that good values for the activity coefficients of the ions in the solution cannot be obtained from the Debye-Hückel relationship. Because of these uncertainties, we prepare buffers by making up a solution of approximately the desired pH and then adjust by adding acid or conjugate base until the required pH is indicated by a pH meter.

Alternatively, empirically derived recipes for preparing buffer solutions of known pH are available in chemical handbooks and reference works.[3]

Buffers are of tremendous importance in biological and biochemical studies where a low but constant concentration of hydronium ion (10^{-6} to 10^{-10} M) must be maintained. Several biological supply houses offer a variety of such buffers.

11D TITRATION CURVES FOR WEAK ACIDS

Four distinctly different types of calculations are needed to derive a titration curve for a weak acid (or weak base):

1. At the beginning, the solution contains only a weak acid or a weak base, and the pH is calculated from the concentration of that solute and the corresponding dissociation constant.
2. After various increments of titrant have been added (in volumes up to, but not including, the equivalence point), the solution consists of a series of buffers. The pH of each buffer is calculated from the analytical concentrations of the conjugate base or acid and the residual concentrations of the weak acid or base.
3. At the equivalence point, the solution contains only the conjugate of the weak acid or base being titrated (that is, a salt), and the pH is calculated from the concentration of this product.
4. Beyond the equivalence point, the excess of strong acid or base titrant represses the acidic or basic character of the reaction product to such an extent that the pH is governed largely by the concentration of the excess titrant.

Example 11-5

Derive a titration curve for the reaction of 50.00 mL of 0.1000 M acetic acid ($K_a = 1.75 \times 10^{-5}$) with 0.1000 M sodium hydroxide.

Initial pH

Initially, we must calculate the pH of a 0.1000 M solution of HOAc using Equation 7-19:

$$[H_3O^+] = \sqrt{K_a C_{HOAc}} = \sqrt{1.75 \times 10^{-5} \times 0.1000} = 1.32 \times 10^{-3}$$
$$pH = -\log 1.32 \times 10^{-3} = 2.88$$

---

[3]See, for example, L. Meites, Ed., *Handbook of Analytical Chemistry*, pp. **5**-112 and **11**-3 to **11**-8. New York: McGraw-Hill, 1963.

pH after Addition of 10.00 mL of Reagent

A buffer solution consisting of NaOAc and HOAc has now been produced. The analytical concentrations of the two constituents are

$$C_{HOAc} = \frac{50.00 \text{ mL} \times 0.1000 \text{ M} - 10.00 \text{ mL} \times 0.1000 \text{ M}}{60.00 \text{ mL}} = \frac{4.000}{60.00} \text{ M}$$

$$C_{NaOAc} = \frac{10.00 \text{ mL} \times 0.1000 \text{ M}}{60.00 \text{ mL}} = \frac{1.000}{60.00} \text{ M}$$

We substitute these concentrations into the dissociation-constant expression for acetic acid and obtain

$$\frac{[H_3O^+](1.000/60.00)}{4.000/60.00} = K_a = 1.75 \times 10^{-5}$$

$$[H_3O^+] = 7.00 \times 10^{-5}$$

$$pH = 4.16$$

Calculations similar to this provide points on the curve throughout the buffer region. Data from such calculations are given in column 2 of Table 11-3.

Equivalence-Point pH

At the equivalence point, all the acetic acid has been converted to sodium acetate. The solution is therefore similar to one formed by dissolving this conjugate base in water, and the pH calculation is identical to that shown in Example 7-10 for a weak base. In the present example, the NaOAc concentration is 0.0500 M. Thus,

$$OAc^- + H_2O \rightleftharpoons HOAc + OH^-$$

$$[OH^-] = [HOAc]$$

$$[OAc^-] = 0.0500 - [OH^-] \approx 0.0500$$

Table 11-3

CHANGES IN pH DURING THE TITRATION OF A WEAK ACID WITH A STRONG BASE

Volume of NaOH, mL	pH	
	50.00 mL of 0.1000 M HOAc Titrated with 0.1000 M NaOH	50.00 mL of 0.001000 M HOAc Titrated with 0.001000 M NaOH
0.00	2.88	3.91
10.00	4.16	4.30
25.00	4.76	4.80
40.00	5.36	5.38
49.00	6.45	6.46
49.90	7.46	7.47
50.00	8.73	7.73
50.10	10.00	8.09
51.00	11.00	9.00
60.00	11.96	9.96
75.00	12.30	10.30

Substituting in the base dissociation-constant expression for OAc^- gives

$$\frac{[OH^-]^2}{0.0500} = \frac{K_w}{K_a} = \frac{1.00 \times 10^{-14}}{1.75 \times 10^{-5}} = 5.714 \times 10^{-10}$$

$$[OH^-] = \sqrt{0.0500 \times 5.714 \times 10^{-10}} = 5.35 \times 10^{-6}$$

$$pH = 14.00 - (-\log 5.35 \times 10^{-6}) = 8.73$$

pH after Addition of 50.10 mL of Base

After the addition of 50.10 mL of NaOH, both the excess base and the acetate ion are sources of hydroxide ion. The contribution of the latter is small, however, because the strong base represses the reaction of acetate ion with water. This fact becomes evident when we consider that the hydroxide ion concentration is only 5.34×10^{-6} M at the equivalence point; once an excess of strong base is added, the contribution from the reaction of the acetate is even smaller. Thus,

$$[OH^-] \approx C_{NaOH} = \frac{50.10 \text{ mL} \times 0.1000 \text{ M} - 50.00 \text{ mL} \times 0.1000 \text{ M}}{100.1 \text{ mL}}$$

$$= 1.00 \times 10^{-4} \text{ M}$$

$$pH = 14.00 - (-\log 1.00 \times 10^{-4}) = 10.00$$

Titration curves for strong and weak acids are identical at all points just slightly beyond the equivalence point. The same is true for strong and weak bases.

Note that, in the region slightly beyond the equivalence point, the titration curve for a weak acid with a strong base is identical to that for a strong acid with a strong base (Figure 11-3).

At the half neutralization point in the titration of a weak acid, $[H_3O^+] = K_a$ or $pH = pK_a$.

Note that the analytical concentrations of acid and conjugate base are identical when an acid has been half neutralized (in Example 11-5, after the addition of exactly 25.00 mL of base). Thus, these terms cancel in the equilibrium-constant expression, and the hydronium ion concentration is numerically equal to the dissociation constant. Likewise, in the titration of a weak base, the hydroxide ion concentration is numerically equal to the dissociation constant of the base at the midpoint in the titration curve. In addition, the buffer capacity of both solutions is at a maximum at this point.

At the half neutralization point in the titration of a weak base, $[OH^-] = K_b$ or $pOH = pK_b$.

Feature 11-3
DETERMINATION OF DISSOCIATION CONSTANTS FOR WEAK ACIDS AND BASES

The dissociation constants of weak acids or weak bases are often determined by monitoring the pH of the solution while the acid or base is being titrated. A pH meter and a glass electrode are used for the measurements. For an acid, the measured pH when the acid is exactly half neutralized is numerically equal to pK_a. For a weak base, the pH at half neutralization must be converted to pOH, which is then equal to pK_b.

11D-1 The Effect of Concentration

The second and third columns of Table 11-3 contain pH data for the titration of 0.1000 M and 0.001000 M acetic acid with sodium hydroxide solutions of the same two concentrations. In deriving the data for the more dilute acid, none of the approximations shown in Example 11-5 were valid, and solution of a quadratic equation was necessary throughout.

Figure 11-6 is a plot of the data in Table 11-3. Note that the initial pH values are higher and the equivalence-point pH is lower for the more dilute solution. At intermediate titrant volumes, however, the pH values of the two solutions differ from each other only slightly because of the buffering action of the acetic acid/sodium acetate system present in this region. Figure 11-6 is graphic confirmation of the fact that the pH of buffers is largely independent of dilution.

Challenge: Derive the data in the third column of Table 11-3.

The pH of a buffer is relatively independent of dilution.

11D-2 The Effect of Reaction Completeness

Titration curves for 0.1000 M solutions of acids with different dissociation constants are shown in Figure 11-7. Note that the pH change in the equivalence-point region becomes smaller as the acid becomes weaker—that is, as the reaction between the acid and the base becomes less complete. The effects of reactant concentration and reaction completeness illustrated by Figures 11-6 and 11-7 are analogous to these effects on precipitation titration curves (Section 10B-2).

11D-3 Indicator Choice; The Feasibility of Titration

Figures 11-6 and 11-7 show clearly that the choice of indicator for the titration of a weak acid is more limited than that for a strong acid. For example, from Figure 11-6, it is obvious that bromocresol green is totally unsuited for the titration of 0.1000 M acetic acid. Bromothymol blue is also unsatisfactory because its full color change occurs over a range from about 46 to 50 mL of 0.1000 M base. An indicator exhibiting a color change in the basic region, such as phenolphthalein, provides a sharp end point with a minimal titration error.

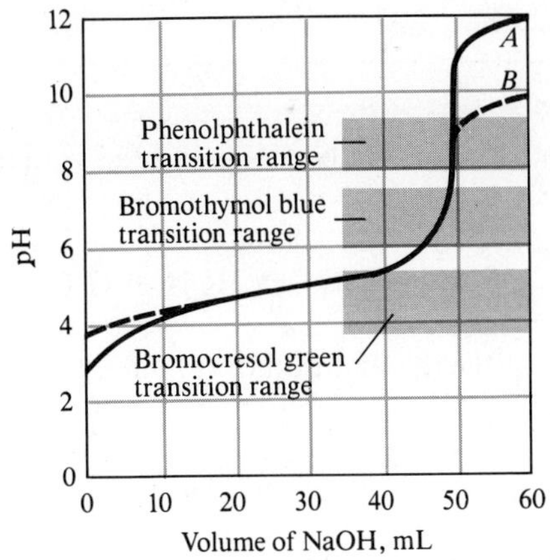

Figure 11-6

Curve for the titration of acetic acid with sodium hydroxide: A: 0.1000 M acid with 0.1000 M base. B: 0.001000 M acid with 0.001000 M base.

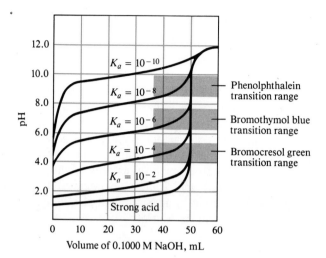

Figure 11-7

The effect of acid strength on titration curves. Each curve represents the titration of 50.0 mL of 0.1000 M acid with 0.1000 M NaOH.

The end-point pH change associated with the titration of 0.001000 M acetic acid (curve B, Figure 11-6) is so small that a significant titration error is likely to be introduced regardless of indicator. However, use of an indicator with a transition range between that of phenolphthalein and that of bromothymol blue in conjunction with a suitable color-comparison standard makes it possible to establish the end point in this titration with a reproducibility of a few percent relative.

Figure 11-7 illustrates that similar problems exist as the strength of the acid being titrated decreases. Precision on the order of ±2 ppt can be achieved in the titration of a 0.1000 M solution of an acid with a dissociation constant of 10^{-8} provided a suitable color-comparison standard is available. With more concentrated solutions, somewhat weaker acids can be titrated with reasonable precision.

11E TITRATION CURVES FOR WEAK BASES

The derivation of a curve for the titration of a weak base is analogous to that of a weak acid.

Example 11-6

A 50.00-mL aliquot of 0.0500 M NaCN is titrated with 0.1000 M HCl. The reaction is

$$CN^- + H_3O^+ \rightleftarrows HCN + H_2O$$

Calculate the pH after the addition of (a) 0.00, (b) 10.00, (c) 25.00, and (d) 26.00 mL of acid.

(a) Initial pH

The pH of a solution of NaCN can be derived by the method shown in Example 7-10:

$$CN^- + H_2O \rightleftharpoons HCN + OH^-$$

$$\frac{[OH^-][HCN]}{[CN^-]} = K_b = \frac{K_w}{K_a} = \frac{1.00 \times 10^{-14}}{2.1 \times 10^{-9}} = 4.76 \times 10^{-6} \qquad (1)$$

$$[OH^-] = [HCN]$$

$$[CN^-] = C_{NaCN} - [OH^-] \approx C_{NaCN} = 0.0500$$

Substitution into the dissociation-constant expression gives, after rearrangement,

$$[OH^-] = \sqrt{K_b C_{NaCN}} = \sqrt{4.76 \times 10^{-6} \times 0.0500} = 4.88 \times 10^{-4}$$

$$pH = 14.00 - (-\log 4.88 \times 10^{-4}) = 10.69$$

(b) pH after Addition of 10.00 mL of Titrant
Addition of acid produces a buffer with a composition given by

$$C_{NaCN} = \frac{50.00 \times 0.0500 - 10.00 \times 0.1000}{60.00} = \frac{1.500}{60.00} \text{ M}$$

$$C_{HCN} = \frac{10.00 \times 0.1000}{60.00} = \frac{1.000}{60.00} \text{ M}$$

These values are then substituted into the expression for the acid dissociation constant of HCN to give

$$[H_3O^+] = \frac{2.1 \times 10^{-9} \times (1.000/60.00)}{1.500/60.00} = 1.4 \times 10^{-9}$$

$$pH = 8.85$$

The pH of the buffer can be calculated with K_a for HCN, as was done here, or equally well with K_b (Equation 1). We used K_a because it gives $[H_3O^+]$ directly; K_b gives $[OH^-]$.

(c) pH after Addition of 25.00 mL of Titrant
This volume corresponds to the equivalence point, where the principal solute species is the weak acid HCN. Thus,

$$C_{HCN} = \frac{25.00 \times 0.1000}{75.00} = 0.03333 \text{ M}$$

Applying Equation 7-19 gives

$$[H_3O^+] = \sqrt{K_a C_{HCN}} = \sqrt{2.1 \times 10^{-9} \times 0.03333} = 8.37 \times 10^{-6}$$

$$pH = 5.08$$

(d) pH after Addition of 26.00 mL of Titrant
The excess of strong acid now present represses the dissociation of the HCN to the point where its contribution to the pH is negligible. Thus,

$$[H_3O^+] = C_{HCl} = \frac{26.00 \times 0.1000 - 50.00 \times 0.0500}{76.00} = 1.32 \times 10^{-3}$$

$$pH = 2.88$$

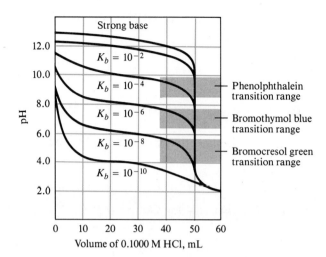

Figure 11-8

The effect of base strength on titration curves. Each curve represents the titration of 50.0 mL of 0.1000 M base with 0.1000 M HCl.

Figure 11-8 shows theoretical curves for a series of weak bases of different strengths. Clearly, indicators with *acidic* transition ranges must be employed for weak bases.

11F   TITRATION CURVES FOR MIXTURES OF ACIDS OR BASES

Figure 11-9 shows titration curves for mixtures consisting of 0.100 M HCl and a series of weak acids having dissociation constants that vary from 10^{-2} to 10^{-8}. In each case, the concentration of the weak acid is 0.0800 M. Note that the rise in pH at the first equivalence point is either small or essentially nonexistent when the weak acid has a relatively large dissociation constant (curve *A*). For mixtures of this kind, only the total millimoles of acid can be determined. Conversely, when the weak acid has a dissociation constant somewhat less than 10^{-8}, only the strong acid can be determined. For weak acids of intermediate strength (K_a between 10^{-4} and 10^{-8}), two useful end points can be obtained.

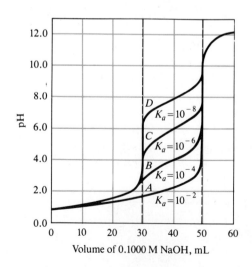

Figure 11-9

Curves for the titration of strong acid/weak acid mixtures with 0.1000 M NaOH. Each titration is on 25.00 mL of a solution that is 0.1200 M in HCl and 0.0800 M in HA.

Curves such as *B*, *C*, and *D* in Figure 11-9 are readily derived by the techniques discussed earlier in this chapter. Up to a few tenths of a milliliter of the first equivalence point, the strong acid represses the dissociation of the weak acid to the extent that the latter contributes little to the pH of the solution. That is, the first part of the titration is identical to that of a strong acid in the absence of a weak one. At the first equivalence point and beyond, the curve is identical to that of a weak acid alone.

> The amount of each component in a mixture of a weak and a strong acid can be determined by titration, provided that the dissociation constant of the weak acid is in the range of about 10^{-4} to 10^{-8}.

11G   THE COMPOSITION OF BUFFER SOLUTIONS AS A FUNCTION OF pH

The changes in composition that occur while a solution of a weak acid or a weak base is being titrated are sometimes of interest and can be visualized by plotting the *relative* concentration of the weak acid as well as the relative concentration of the conjugate base as a function of pH. These relative concentrations are called *alpha values*. For example, if we let C_T be the sum of the analytical concentrations of acetic acid and sodium acetate at any point in the titration discussed in Example 11-5, we may write

$$C_T = C_{\text{HOAc}} + C_{\text{NaOAc}} \qquad (11\text{-}14)$$

We then define α_0 as

$$\alpha_0 = \frac{[\text{HOAc}]}{C_T} \qquad (11\text{-}15)$$

and α_1 as

$$\alpha_1 = \frac{[\text{OAc}^-]}{C_T} \qquad (11\text{-}16)$$

Alpha values are unitless ratios whose sum must equal unity. That is,

> Alpha values do not depend on C_T.

$$\alpha_0 + \alpha_1 = 1$$

Alpha values are determined by $[\text{H}_3\text{O}^+]$ and K_a alone and are independent of C_T. To obtain an expression for α_0, we rearrange the dissociation-constant expression to

$$[\text{OAc}^-] = \frac{K_a[\text{HOAc}]}{[\text{H}_3\text{O}^+]} \qquad (11\text{-}17)$$

Mass balance requires that

$$C_T = [\text{HOAc}] + [\text{OAc}^-] \tag{11-18}$$

Substituting Equation 11-17 into Equation 11-18 gives

$$C_T = [\text{HOAc}] + \frac{K_a[\text{HOAc}]}{[\text{H}_3\text{O}^+]} = [\text{HOAc}] \left(\frac{[\text{H}_3\text{O}^+] + K_a}{[\text{H}_3\text{O}^+]} \right)$$

Upon rearrangement, we obtain the expression for α_0 as given by Equation 11-15.

$$\alpha_0 = \frac{[\text{HOAc}]}{C_T} = \frac{[\text{H}_3\text{O}^+]}{[\text{H}_3\text{O}^+] + K_a} \tag{11-19}$$

To obtain an expression for α_1, we rearrange the dissociation-constant expression to

$$[\text{HOAc}] = \frac{[\text{H}_3\text{O}^+][\text{OAc}^-]}{K_a}$$

and substitute into Equation 11-18

$$C_T = \frac{[\text{H}_3\text{O}^+][\text{OAc}^-]}{K_a} + [\text{OAc}^-] = [\text{OAc}^-] \left(\frac{[\text{H}_3\text{O}^+] + K_a}{K_a} \right)$$

Rearranging this equation gives α_1 as defined by Equation 11-16

$$\alpha_1 = \frac{[\text{OAc}^-]}{C_T} = \frac{K_a}{[\text{H}_3\text{O}^+] + K_a} \tag{11-20}$$

Note that the denominator is the same for both α_0 and α_1.

Figure 11-10
Solid lines show the change in relative amounts of HOAc (α_0) and OAc$^-$ (α_1) during the titration of 50.00 mL of 0.1000 M acetic acid. The curved line is the titration curve for the system.

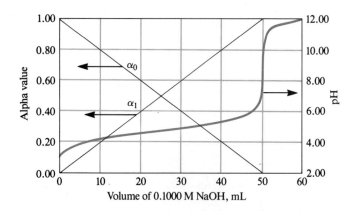

The solid lines labeled α_0 and α_1 in Figure 11-10 were derived with Equations 11-19 and 11-20 using values for $[H_3O^+]$ shown in column 2 of Table 11-3. The titration curve is shown as the curved line in Figure 11-10. Note that α_0 is nearly 1 (0.987) at the outset of the titration, meaning that 98.7% of the acetate-containing species is present as HOAc and only 1.3% is present as OAc^-. At the equivalence point, α_0 has decreased to 1.1×10^{-4} and α_1 approaches 1. Thus, only about 0.011% of the acetate-containing species is HOAc. Note that when the acid is half neutralized (25.00 mL), α_0 and α_1 are each 0.5.

11H   QUESTIONS AND PROBLEMS

*11-1. Explain qualitatively how the titration curves for 0.10 M NaOH and 0.1 M NH_3 differ from each other. Account for the difference.

11-2. What is a buffer solution and what are its properties?

*11-3. Why are buffers important in physiological fluids?

11-4. Define buffer capacity.

*11-5. Which has the greater buffer capacity, (a) a mixture containing 0.100 mol of NH_3 and 0.200 mol of NH_4Cl or (b) a mixture containing 0.0500 mol of NH_3 and 0.100 mol of NH_4Cl?

11-6. Would a buffer prepared by mixing 1 mol of acetic acid and 0.5 mol of NaOH differ from one prepared by mixing 1 mol of sodium acetate and 0.5 mol of HCl?

*11-7. Explain the fundamental difference between the equivalence point in a titration and the end point.

11-8. What variables can cause the pH range of an indicator to shift?

*11-9. Why are the standard reagents used in neutralization titrations generally strong acids and bases rather than weak acids and bases?

11-10. Before glass electrodes and pH meters became so widely used, pH was often determined by measuring the concentration of the acid and base forms of the indicator colorimetrically. If bromothymol blue is introduced into a solution and the concentration ratio of acid to base form is found to be 1.43, what is the pH of the solution?

*11-11. The procedure described in Question 11-10 was used to determine pH with methyl orange as the indicator. The concentration ratio of the acid to base form of the indicator was 1.64. Calculate the pH of the solution.

11-12. Which is expected to yield a sharper end point in a titration with 0.10 M HCl or 0.1 M NaOH
 *(a) 0.10 M NaOCl or 0.10 M hydroxylamine?
 (b) 0.10 M anilinium hydrochloride ($C_6H_5NH_3^+Cl^-$) or benzoic acid?
 *(c) 0.10 M NaOCl or 0.10 M hydrazine?
 (d) 0.10 M sodium phenolate ($NaOC_6H_5$) or 0.100 M NH_3?

11-13. Consult Appendixes 3 and 4 and pick out a suitable acid/base pair to prepare a buffer of pH

*(a) 3.5.   (b) 7.6.   *(c) 9.3.   (d) 5.1.

*11-14. Values for K_w at 0, 50, and 100°C are 1.14×10^{-15}, 5.47×10^{-14}, and 4.9×10^{-13}, respectively. Calculate the pH of a neutral solution at each of these temperatures.

11-15. What is the pH of a 1.00×10^{-2} M NaOH solution at 0°C?

*11-16. What is the pH of an aqueous solution that is 39.9% HCl by weight and has a density of 1.200 g/mL?

11-17. Calculate the pH of a solution that contains 1.00% (w/w) NaOH and has a density of 1.011 g/mL.

*11-18. Calculate the pH of the solution that results upon mixing 20.0 mL of 0.2000 M HCl with 25.0 mL of
 (a) distilled water.
 (b) 0.132 M $AgNO_3$.
 (c) 0.132 M NaOH.
 (d) 0.132 M NH_3.
 (e) 0.232 M NaOH.

11-19. What is the pH of the solution that results when 0.102 g of $Mg(OH)_2$ is mixed with
 (a) 75.0 mL of 0.0600 M HCl?
 (b) 15.0 mL of 0.0600 M HCl?
 (c) 30.0 mL of 0.0600 M HCl?
 (d) 30.0 mL of 0.0600 M $MgCl_2$?

*11-20. What is the pH of a 2.0×10^{-7} M HCl solution?

11-21. What is the pH of a solution that is 0.0100 M in H_2SO_4?

*11-22. Calculate the hydronium ion concentration and pH of a solution that is 0.0500 M in HCl
 (a) neglecting activities.
 (b) using activities.

11-23. A solution is 0.0500 M in NH_4Cl and 0.0300 M in NH_3. Calculate its OH^- concentration and its pH
 (a) neglecting activities.
 (b) taking activities into account.

*11-24. Calculate the pH of a solution prepared by
 (a) dissolving 43.0 g of lactic acid in water and diluting to 500 mL.
 (b) diluting 25.0 mL of the solution in (a) to 250 mL.
 (c) diluting 10.0 mL of the solution in (b) to 1.00 L.

11-25. Calculate the pH of an iodic acid solution that is (a) 1.00×10^{-1} M, (b) 1.00×10^{-2} M, (c) 1.00×10^{-4} M.

*11-26. Calculate the pH of an NH_4Cl solution that is **(a)** 1.00×10^{-1} M, **(b)** 1.00×10^{-2} M, **(c)** 1.00×10^{-4} M.

11-27. Calculate the pH of a HOCl solution that is **(a)** 1.00×10^{-1} M, **(b)** 1.00×10^{-2} M, **(c)** 1.00×10^{-4} M.

*11-28. Calculate the pH of a NaOCl solution that is **(a)** 1.00×10^{-1} M, **(b)** 1.00×10^{-2} M, **(c)** 1.00×10^{-4} M.

11-29. Calculate the pH of a solution in which the concentration of piperidine is **(a)** 1.00×10^{-1}M, **(b)** 1.00×10^{-2} M, **(c)** 1.00×10^{-4} M.

*11-30. Calculate the pH of the solution that results when 20.0 mL of 0.200 M formic acid is
(a) diluted to 45.0 mL with distilled water.
(b) mixed with 25.0 mL of 0.160 M NaOH solution.
(c) mixed with 25.0 mL of 0.200 M NaOH solution.
(d) mixed with 25.0 mL of 0.200 sodium formate solution.

11-31. Calculate the pH of the solution that results when 40.0 mL of 0.100 M NH_3 is
(a) diluted to 60 mL with distilled water.
(b) mixed with 20.0 mL of 0.200 M HCl solution.
(c) mixed with 20.0 mL of 0.250 M HCl solution.
(d) mixed with 20.0 mL of 0.200 M NH_4Cl solution.
(e) mixed with 20.0 mL of 0.100 M HCl solution.

*11-32. What is the pH of a solution that is
(a) prepared by dissolving 9.20 g of lactic acid (fw = 90.08 g) and 11.15 g of sodium lactate (fw = 112.06 g) in water and diluting to 1.00 L?
(b) 0.055 M in acetic acid and 0.011 M in sodium acetate?
(c) prepared by dissolving 3.00 g of salicylic acid $[C_6H_4(OH)COOH]$ (fw = 138.12 g) in 50.0 mL of 0.1130 M NaOH and diluting to 500.0 mL?
(d) 0.010 M in picric acid and 0.100 M in sodium picrate?

11-33. What is the pH of a solution that is
(a) prepared by dissolving 3.30 g of $(NH_4)_2SO_4$ in water, adding 125.0 mL of 0.1011 M NaOH, and diluting to 500.0 mL?
(b) 0.120 M in piperidine and 0.080 M in its chloride salt?
(c) 0.050 M in ethylamine and 0.167 M in its chloride salt?
(d) prepared by dissolving 2.32 g of aniline (fw = 93.12 g) in 100 mL of 0.0200 M HCl and diluting to 250.0 mL?

*11-34. Calculate the change in pH that occurs in each solution as a result of a tenfold dilution with water:
(a) H_2O.
(b) 0.0500 M HCl.
(c) 0.0500 M NaOH.
(d) 0.0500 M NH_3.
(e) 0.0500 M NH_4Cl.
(f) 0.0500 M NH_3 + 0.0500 M NH_4Cl.
(g) 0.500 M NH_3 + 0.500 M NH_4Cl.

*11-35. Calculate the change in pH that occurs when 1.00 mmol of a strong acid is added to 100 mL of the solutions listed in Problem 11-34.

*11-36. Calculate the change in pH that occurs when 1.00 mmol of a strong base is added to 100 mL of the solutions listed in Problem 11-34.

11-37. Calculate the change in pH that occurs when 0.50 mmol of a strong acid is added to 100 mL of a solution that is
(a) 0.0200 M in lactic acid and 0.0800 M in sodium lactate.
(b) 0.0800 M in lactic acid and 0.0200 M in sodium lactate.
(c) 0.0500 M in lactic acid and 0.0500 M in sodium lactate.

*11-38. What weight of sodium formate must be added to 400 mL of 1.00 M formic acid to produce a buffer solution that has a pH of 3.50?

11-39. What weight of sodium glycolate (fw = 76.05 g) should be added to 300 mL of 1.00 M glycolic acid to produce a buffer solution with a pH of 4.00?

*11-40. What volume of 0.200 M HCl must be added to 250 mL of 0.300 M sodium mandelate to produce a buffer solution with a pH of 3.37?

11-41. What volume of 2.00 M NaOH must be added to 300 mL of 1.00 M glycolic acid to produce a buffer solution having a pH of 4.00?

*11-42. In a titration of 50.00 mL of 0.05000 M formic acid with 0.1000 M KOH, the titration error must be smaller than ±0.05 mL. What indicator can be chosen to realize this goal?

11-43. In a titration of 50.00 mL of 0.1000 M ethylamine with 0.1000 M $HClO_4$, the titration error must be no more than ±0.05 mL. What indicator can be chosen to realize this goal?

*11-44. A 50.00-mL aliquot of 0.1000 M NaOH is titrated with 0.1000 M HCl. Calculate the pH of the solution after the addition of 0.00, 10.00, 25.00, 40.00, 45.00, 49.00, 50.00, 51.00, 55.00, and 60.00 mL of acid. Plot a titration curve from the data.

11-45. Calculate the pH after addition of 0.00, 5.00, 15.00, 25.00, 40.00, 45.00, 49.00, 50.00, 51.00, 55.00, and 60.00 mL of 0.1000 M NaOH in the titration of 50.00 mL of
*(a) 0.1000 M HNO_2.
(b) 0.1000 M lactic acid.
*(c) 0.1000 M pyridinium chloride.

11-46. Calculate the pH after addition of 0.00, 5.00, 15.00, 25.00, 40.00, 45.00, 49.00, 50.00, 51.00, 55.00, and 60.00 mL of 0.1000 M HCl in the titration of 50.00 mL of
*(a) 0.1000 M ammonia.
(b) 0.1000 M hydrazine.
(c) 0.1000 M sodium cyanide.

11-47. Calculate the pH after addition of 0.00, 5.00, 15.00, 25.00, 40.00, 49.00, 50.00, 51.00, 55.00, and 60.00 mL of reagent in the titration of

*(a) 0.1000 M anilinium chloride with 0.1000 M NaOH.

(b) 0.01000 M picric acid with 0.01000 M NaOH.

*(c) 0.1000 M hypochlorous acid with 0.1000 M NaOH.

(d) 0.1000 M hydroxylamine with 0.1000 M HCl. Plot titration curves from the data.

11-48. Calculate α_0 and α_1 for

*(a) lactic acid species in a solution with a pH of 4.61.

(b) iodic acid species in a solution with a pH of 1.39.

*(c) butanoic acid species in a solution with a pH of 4.86.

(d) hypochlorous acid species in a solution with a pH of 7.97.

11-49. What is the equilibrium concentration of undissociated lactic acid in a solution in which the total analytical concentration of lactate-containing species is 0.1500 M and the pH is 4.25?

11-50. Supply the missing data:

Acid	$C_T =$ $C_{HA} + C_{A^-}$	pH	[HA]	[A$^-$]	α_0	α_1
*Lactic	0.120				0.640	
Iodic	0.200					0.765
*Butanoic		5.00	0.0644			
Hypochlorous	0.280	7.00				
*Nitrous					0.105	0.413
Hydrogen cyanide			0.145	0.221		
Sulfamic	0.250	1.20				

Chapter 12

Titration Curves for Polyfunctional Acids and Bases

An examination of Appendix 3 reveals that many acids have more than one functional group capable of donating a proton. Similarly, numerous compounds have more than one basic functional group. We have chosen to call such species *polyfunctional* acids or bases. In this chapter, we examine how polyfunctional compounds behave during neutralization titrations.

12A POLYFUNCTIONAL ACIDS

Phosphoric acid is a typical polyfunctional acid. In aqueous solution it undergoes the following three dissociation reactions:

Generally, $K_1 > K_2$, often by a factor of 10^4 to 10^5, because of electrostatic forces. That is, the first dissociation involves separating a single positively charged proton from a singly charged anion. In the second step, a proton is separated from a doubly charged anion, a process that requires considerably more energy.

$$H_3PO_4 + H_2O \rightleftharpoons H_2PO_4^- + H_3O^+$$

$$K_1 = \frac{[H_3O^+][H_2PO_4^-]}{[H_3PO_4]} = 7.11 \times 10^{-3}$$

$$H_2PO_4^- + H_2O \rightleftharpoons HPO_4^{2-} + H_3O^+$$

$$K_2 = \frac{[H_3O^+][HPO_4^{2-}]}{[H_2PO_4^-]} = 6.34 \times 10^{-8}$$

$$HPO_4^{2-} + H_2O \rightleftharpoons PO_4^{3-} + H_3O^+$$

$$K_3 = \frac{[H_3O^+][PO_4^{3-}]}{[HPO_4^{2-}]} = 4.2 \times 10^{-13}$$

With this acid, as with others, $K_1 > K_2 > K_3$.

Feature 12-1

COMBINING EQUILIBRIUM-CONSTANT EXPRESSIONS

When two adjacent stepwise equilibria are *added*, the equilibrium constant for the resulting overall reaction is the *product* of the two constants. Thus for the first two dissociation equilibria for H_3PO_4, we may write

$$H_3PO_4 + H_2O \rightleftharpoons H_2PO_4^- + H_3O^+$$
$$\underline{H_2PO_4^- + H_2O \rightleftharpoons HPO_4^- + H_3O^+}$$
$$H_3PO_4 + 2\,H_2O \rightleftharpoons HPO_4^- + 2\,[H_3O^+]$$

and

$$K_1 K_2 = \frac{[H_3O^+]^2[HPO_4^-]}{H_3PO_4}$$

$$= 7.11 \times 10^{-3} \times 6.34 \times 10^{-8} = 4.51 \times 10^{-10}$$

Similarly,

$$K_1 K_2 K_3 = \frac{[H_3O^+]^3[PO_4^{3-}]}{H_3PO_4}$$

$$= 7.11 \times 10^{-3} \times 6.34 \times 10^{-8} \times 4.2 \times 10^{-13} = 1.9 \times 10^{-22}$$

12B POLYFUNCTIONAL BASES

Polyfunctional bases are also common, an example being sodium carbonate. Carbonate ion, the conjugate base of the hydrogen carbonate ion, is involved in the stepwise equilibria

$$CO_3^{2-} + H_2O \rightleftharpoons HCO_3^- + OH^-$$

$$K_{b1} = \frac{[HCO_3^-][OH^-]}{[CO_3^{2-}]} = \frac{K_w}{K_2} = \frac{1.00 \times 10^{-14}}{4.7 \times 10^{-11}} = 2.1 \times 10^{-4}$$

$$HCO_3^- + H_2O \rightleftharpoons H_2CO_3 + OH^-$$

$$K_{b2} = \frac{[H_2CO_3][OH^-]}{[HCO_3^-]} = \frac{K_w}{K_1} = \frac{1.00 \times 10^{-14}}{4.45 \times 10^{-7}} = 2.25 \times 10^{-8}$$

where K_1 and K_2 are the first and second dissociation constants for carbonic acid and K_{b1} and K_{b2} are the first and second basic dissociation constants for carbonate ion.

The overall basic dissociation reaction of sodium carbonate is described by the equations

$$CO_3^{2-} + 2\,H_2O \rightleftharpoons H_2CO_3 + 2\,OH^-$$

$$K_{b1}K_{b2} = \frac{[H_2CO_3][OH^-]^2}{[CO_3^{2-}]} = 2.1 \times 10^{-4} \times 2.25 \times 10^{-8} = 4.8 \times 10^{-12}$$

The pH of polyfunctional systems, such as phosphoric acid or sodium carbonate, can be computed rigorously through use of the systematic approach to multiple-equilibrium problems described in Chapter 9. Solution of the several simultaneous equations involved is difficult and time-consuming, however, unless a simplifying assumption is made.

12C   CALCULATION OF THE pH OF NaHA SOLUTIONS

Thus far, we have not considered how to calculate the pH of solutions of salts that have both acidic and basic properties—that is, salts that are *amphiprotic*. Such salts are formed during the neutralization titration of polyfunctional acids and bases. For example, when 1 mol of NaOH is added to a solution containing 1 mol of the acid H_2A, 1 mol of NaHA is formed. The pH of this solution is determined by the equilibria

> Amphiprotic salts have both acidic and basic properties.

$$HA^- + H_2O \rightleftharpoons A^{2-} + H_3O^+$$

$$HA^- + H_2O \rightleftharpoons H_2A + OH^-$$

One of these reactions produces hydronium ions and the other hydroxide ions. A solution of NaHA will be either acidic or basic, depending upon the relative magnitude of the equilibrium constants for these processes:

$$K_2 = \frac{[H_3O^+][A^{2-}]}{[HA^-]} \tag{12-1}$$

$$K_{b2} = \frac{K_w}{K_1} = \frac{[H_2A][OH^-]}{[HA^-]} \tag{12-2}$$

If K_{b2} is greater than K_2, the solution is basic; otherwise, it is acidic.

A solution of NaHA can be described in terms of mass balance:

$$C_{NaHA} = [HA^-] + [H_2A] + [A^{2-}] \tag{12-3}$$

and charge balance:

$$[Na^+] + [H_3O^+] = [HA^-] + 2\,[A^{2-}] + [OH^-]$$

Since the sodium ion concentration is equal to the molar analytical concentration of the salt, the last equation can be rewritten as

$$C_{NaHA} + [H_3O^+] = [HA^-] + 2\,[A^{2-}] + [OH^-] \tag{12-4}$$

One additional algebraic equation is needed to solve for the five unknowns. The ion-product constant for water serves this purpose:

$$K_w = [H_3O^+][OH^-]$$

The rigorous computation of the hydronium ion concentration from these five equations is difficult. However, a reasonable approximation applicable to solutions of most amphiprotic salts can be obtained as follows.

When we subtract the mass-balance equation from the charge-balance equation, we obtain

$$C_{NaHA} + [H_3O^+] = [HA^-] + 2\,[A^{2-}] + [OH^-]$$
$$\underline{C_{NaHA} \qquad\qquad = [H_2A] + [HA^-] + [A^{2-}]}$$
$$[H_3O^+] = [A^{2-}] + [OH^-] - [H_2A]$$

We then express this last equation in terms of the principal solute species $[HA^-]$, $[H_3O^+]$, the dissociation constants for carbonic acid, and K_w:

$$[H_3O^+] = \frac{K_2[HA^-]}{[H_3O^+]} + \frac{K_w}{[H_3O^+]} - \frac{[H_3O^+][HA^-]}{K_1}$$

Multiplication by $[H_3O^+]$ gives

$$[H_3O^+]^2 = K_2[HA^-] + K_w - \frac{[H_3O^+]^2[HA^-]}{K_1}$$

We collect terms to obtain

$$[H_3O^+]^2 \left(\frac{[HA^-]}{K_1} + 1\right) = K_2[HA^-] + K_w$$

Finally, this equation rearranges to

$$[H_3O^+] = \sqrt{\frac{K_2[HA^-] + K_w}{1 + [HA^-]/K_1}} \qquad (12\text{-}5)$$

Under most circumstances, it can be assumed that

$$[HA^-] \approx C_{NaHA} \qquad (12\text{-}6)$$

Substitution of this equality into Equation 12-5 gives

$$[H_3O^+] = \sqrt{\frac{K_2 C_{NaHA} + K_w}{1 + C_{NaHA}/K_1}} \qquad (12\text{-}7)$$

It is important to understand that the approximation shown as Equation 12-6 requires that $[HA^-]$ be much larger than any of the other equilibrium concentrations in Equations 12-3 and 12-4. This assumption is not valid for very dilute solutions of NaHA or when K_2 or K_w/K_1 is relatively large.

Frequently, the ratio C_{NaHA}/K_1 is much larger than unity and $K_2 C_{NaHA}$ is considerably greater than K_w. With these assumptions, Equation 12-7

Note that

$$[H_2A] = \frac{[H_3O^+][HA^-]}{K_1}$$

simplifies to

Make sure that you always check the assumptions that are in Equation 12-8.

$$[H_3O^+] \approx \sqrt{K_1 K_2} \qquad (12\text{-}8)$$

Note that Equation 12-8 does not contain C_{NaHA}, which implies that the pH of solutions of this type remains constant over a considerable range of solute concentrations.

Example 12-1

Calculate the hydronium ion concentration of a 0.100 M $NaHCO_3$ solution.

We first examine the assumptions leading to Equation 12-8. The dissociation constants for H_2CO_3 are $K_1 = 4.45 \times 10^{-7}$ and $K_2 = 4.7 \times 10^{-11}$. Clearly, C_{NaHA}/K_1 is much larger than unity; in addition, $K_2 C_{NaHA}$ has a value of 4.7×10^{-12}, which is substantially greater than K_w. Thus Equation 12-8 applies and

$$[H_3O^+] = \sqrt{4.45 \times 10^{-7} \times 4.7 \times 10^{-11}} = 4.6 \times 10^{-9}$$

Example 12-2

Calculate the hydronium ion concentration of a 1.0×10^{-3} M Na_2HPO_4 solution.

The pertinent dissociation constants are K_2 and K_3, which both contain $[HPO_4^{2-}]$. Their values are $K_2 = 6.34 \times 10^{-8}$ and $K_3 = 4.2 \times 10^{-13}$. Considering again the assumptions that led to Equation 12-8, we find that $(1.0 \times 10^{-3})/(6.34 \times 10^{-8})$ is again large enough so that the denominator can be simplified. The product $K_2 C_{Na_2HPO_4}$ is by no means much larger than K_w, however. We therefore use a partially simplified version of Equation 12-7:

$$[H_3O^+] = \sqrt{\frac{4.2 \times 10^{-13} \times 1.0 \times 10^{-3} + 1.0 \times 10^{-14}}{(1.0 \times 10^{-3})/(6.34 \times 10^{-8})}} = 8.1 \times 10^{-10}$$

Use of Equation 12-8 yields a value of 1.6×10^{-10} M.

Example 12-3

Find the hydronium ion concentration of a 0.0100 M NaH_2PO_4 solution.

The two dissociation constants of importance (those containing $[H_2PO_4^-]$) are $K_1 = 7.11 \times 10^{-3}$ and $K_2 = 6.34 \times 10^{-8}$. We see that the denominator of Equation 12-7 cannot be simplified, but the numerator reduces to $K_2 C_{NaH_2PO_4}$. Thus, Equation 12-7 becomes

$$[H_3O^+] = \sqrt{\frac{6.34 \times 10^{-8} \times 1.0 \times 10^{-2}}{1.00 + (1.0 \times 10^{-2})/(7.11 \times 10^{-3})}} = 1.6 \times 10^{-5}$$

12D TITRATION CURVES FOR POLYFUNCTIONAL ACIDS

Compounds with two or more acidic functional groups yield multiple end points in a titration, provided the functional groups differ sufficiently in their strengths as acids. The computational techniques described in Chapter 11 permit derivation of reasonably accurate theoretical titration curves for polyprotic acids if the ratio K_1/K_2 is somewhat greater than 10^3. If this ratio is smaller, the error, particularly in the region of the first equivalence point, becomes excessive, and a more rigorous treatment of the equilibrium relationships is required.

Figure 12-1 shows the titration curve for a diprotic acid H_2A with dissociation constants of 1.00×10^{-3} and 1.00×10^{-7}. Because K_1/K_2 is significantly greater than 10^3, we can derive this curve (except for the first equivalence point) using the techniques developed in Chapter 11 for monoprotic weak acids. Thus, to obtain the initial pH (point A), we treat the system as if it contained a single monoprotic acid with a dissociation constant of 1.00×10^{-3}. In region B we have the equivalent of a simple buffer solution consisting of the weak acid H_2A and its conjugate base NaHA. That is, we assume that the concentration of A^{2-} is negligible with respect to that of the other two A-containing species. At the first equivalence point (C), we are dealing with a solution of an amphiprotic salt and so use Equation 12-7 or 12-8 to compute the hydronium ion concentration. In the region labeled D, we have a second buffer, this one consisting of a weak acid HA^- and its conjugate base Na_2A, and we calculate the pH using the second dissociation constant, 1.00×10^{-7}. At point E, the solu-

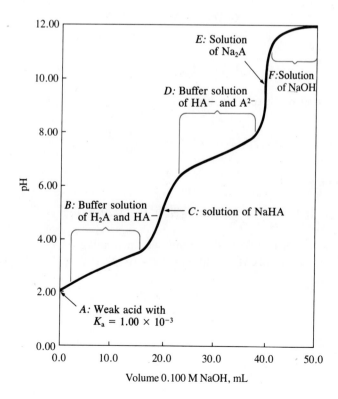

Figure 12-1

Titration of 20.0 mL of 0.100 M H_2A with 0.100 M NaOH. For H_2A, $K_1 = 1.00 \times 10^{-3}$ and $K_2 = 1.00 \times 10^{-7}$. Method of pH calculation is shown for several points and regions on the titration curve.

tion contains the conjugate base of a weak acid with a dissociation constant of 1.00×10^{-7}. That is, we assume that the hydroxide ion concentration of the solution is determined solely by the reaction of A^{2-} with water to form HA^- and OH^-. Finally, in the region labeled F, we compute the hydroxide ion concentration from the molarity of the NaOH and find the pH from this quantity.

Example 12-4

Derive a curve for the titration of 25.00 mL of 0.1000 M maleic acid, $C_2H_2(COOH)_2$, with 0.1000 M NaOH.

The two dissociation equilibria can be written as

$$H_2M + H_2O \rightleftharpoons H_3O^+ + HM^- \qquad K_1 = 1.20 \times 10^{-2}$$
$$HM^- + H_2O \rightleftharpoons H_3O^+ + M^{2-} \qquad K_2 = 5.96 \times 10^{-7}$$

where H_2M symbolizes the free acid. Because the ratio K_1/K_2 is large (2×10^4), we proceed as just described.

Initial pH

Only the first dissociation makes an appreciable contribution to $[H_3O^+]$; thus,

$$[H_3O^+] = [HM^-]$$

Mass balance requires that

$$[H_2M] + [HM^-] = 0.1000$$

or

$$[H_2M] = 0.1000 - [HM^-] = 0.1000 - [H_3O^+]$$

Substituting these relationships into the expression for K_1 gives

$$K_1 = 1.20 \times 10^{-2} = \frac{[H_3O^+]^2}{0.1000 - [H_3O^+]}$$

Rearranging yields

$$[H_3O^+]^2 + 1.20 \times 10^{-2}[H_3O^+] - 1.20 \times 10^{-3} = 0$$

Because K_1 for maleic acid is large, we must solve the quadratic equation exactly or by successive approximations. When we do so, we obtain

$$[H_3O^+] = 2.92 \times 10^{-2}$$
$$pH = 2 - \log 2.92 = 1.54$$

First Buffer Region

The addition of 5.00 mL of base results in the formation of a buffer consisting of the weak acid H_2M and its conjugate base HM^-. To the extent that dissociation of HM^- to give M^{2-} is negligible, the solution can be treated as a simple buffer system. Thus applying Equations 11-10 and 11-11 gives

$$C_{NaHM} \approx [HM^-] = \frac{5.00 \times 0.1000}{30.00} = 1.67 \times 10^{-2} \text{ M}$$

$$C_{H_2M} \approx [H_2M] = \frac{25.00 \times 0.1000 - 5.00 \times 0.1000}{30.00} = 6.67 \times 10^{-2} \text{ M}$$

Substitution of these values into the equilibrium-constant expression for K_1 yields a provisional value of 4.8×10^{-2} M for $[H_3O^+]$. It is clear, however, that the approximation $[H_3O^+] \ll C_{H_2M}$ or C_{HM^-} is not valid; therefore Equations 11-8 and 11-9 must be used:

$$[HM^-] = 1.67 \times 10^{-2} + [H_3O^+] - [OH^-]$$
$$[H_2M] = 6.67 \times 10^{-2} - [H_3O^+] + [OH^-]$$

Because the solution is quite acidic, the approximation that $[OH^-]$ is very small is surely justified. Substitution of these last two expressions into the dissociation-constant relationship gives

$$\frac{[H_3O^+](1.67 \times 10^{-2} + [H_3O^+])}{6.67 \times 10^{-2} - [H_3O^+]} = 1.20 \times 10^{-2} = K_1$$

$$[H_3O^+]^2 + (2.87 \times 10^{-2})[H_3O^+] - 8.00 \times 10^{-4} = 0$$

$$[H_3O^+] = 1.74 \times 10^{-2}$$

$$pH = 1.76$$

Additional points in the first buffer region can be computed in a similar way.

First Equivalence Point

At the first equivalence point,

$$[HM^-] \approx C_{NaHM} = \frac{2.500}{50.00} = 5.00 \times 10^{-2}$$

Simplification of the numerator in Equation 12-7 is clearly justified. On the other hand, the second term in the denominator is not $\ll 1$. Hence,

$$[H_3O^+] = \sqrt{\frac{K_2 C_{HM^-}}{1 + C_{HM^-}/K_1}} = \sqrt{\frac{5.96 \times 10^{-7} \times 5.00 \times 10^{-2}}{1 + (5.00 \times 10^{-2})/(1.20 \times 10^{-2})}}$$

$$= 7.60 \times 10^{-5}$$

$$pH = 4.12$$

Second Buffer Region

Further additions of base to the solution create a new buffer system consisting of HM^- and M^{2-}. When enough base has been added so that the reaction of HM^- with water to give OH^- can be neglected (a few tenths of a milliliter beyond the first equivalence point), the pH of the mixture is readily obtained from K_2. With the introduction of 25.50 mL of NaOH, for example,

$$C_{Na_2M} = \frac{(25.50 - 25.00)(0.1000)}{50.50} = \frac{0.050}{50.50}\ M$$

and the molar concentration of NaHM is

$$C_{NaHM} = \frac{(25.00 \times 0.1000) - (25.50 - 25.00)(0.1000)}{50.50} = \frac{2.45}{50.50}\ M$$

Substituting these values into the expression for K_2 gives

$$\frac{[H_3O^+](0.050/50.50)}{2.45/50.50} = 5.96 \times 10^{-7}$$

$$[H_3O^+] = 2.92 \times 10^{-5}$$

$$pH = 4.54$$

The assumption that $[H_3O^+]$ is small relative to C_{HM^-} and $C_{M^{2-}}$ is valid.

Second Equivalence Point

After the addition of 50.00 mL of 0.1000 M sodium hydroxide, the solution is 0.0333 M in Na_2M. Reaction of the base M^{2-} with water is the predominant equilibrium in the system and the only one that must be taken into account. Thus,

$$M^{2-} + H_2O \rightleftarrows OH^- + HM^-$$

$$\frac{[OH^-][HM^-]}{[M^{2-}]} = \frac{K_w}{K_2} = \frac{1.00 \times 10^{-14}}{5.96 \times 10^{-7}} = 1.68 \times 10^{-8}$$

$$[OH^-] = [HM^-]$$

$$[M^{2-}] = 0.0333 - [OH^-] \approx 0.0333$$

$$\frac{[OH^-]^2}{0.0333} = \frac{1.00 \times 10^{-14}}{5.96 \times 10^{-7}}$$

$$[OH^-] = 2.36 \times 10^{-5}$$

$$pH = 14.00 - (-\log 2.36 \times 10^{-5}) = 9.37$$

Beyond the Second Equivalence Point

Further additions of sodium hydroxide suppress the basic dissociation of M^{2-}. The pH is calculated from the concentration of NaOH added in excess of that required for the complete neutralization of H_2M. Thus after addition of 51.00 mL of NaOH, we have 1.00 mL excess of 0.1000 M NaOH, and

$$[OH^-] = \frac{1.00 \times 0.1000}{76.00} = 1.32 \times 10^{-3}$$

$$pH = 14.00 - (-\log 1.32 \times 10^{-3}) = 11.12$$

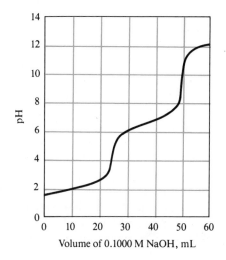

Figure 12-2
Titration curve for 25.00 mL of 0.1000 M maleic acid, H_2M, with 0.1000 M NaOH.

Figure 12-2 is the titration curve for 0.1000 M maleic acid derived as shown in Example 12-4. Two end points are apparent, either of which could in principle be used as a measure of the concentration of the acid. The second end point is clearly more satisfactory, however, because the pH change is more pronounced.

Figure 12-3 shows titration curves for three other polyprotic acids. These curves illustrate that a well-defined end point corresponding to the first equivalence point is observed only when the degree of dissociation of the first proton is sufficiently different from that of the second. The ratio of K_1 to K_2 for oxalic acid (curve B) is approximately 1000. The curve for this titration shows an inflection corresponding to the first equivalence point. However, the magnitude of the pH change is too small to permit precise location of equivalence with an indicator. The second end point, however, provides a means for the accurate determination of oxalic acid concentration.

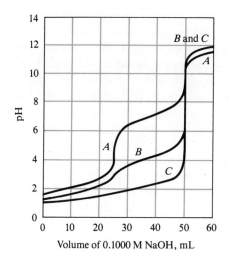

Figure 12-3
Curves for the titration of polyprotic acids. A 0.1000 M NaOH solution is used to titrate 25.00 mL of 0.1000 M H_3PO_4 (A), 0.1000 M oxalic acid (B), and 0.1000 M H_2SO_4 (C).

In titrating a polyprotic acid or base, two usable end points are obtained if the ratio of dissociation constants is greater than 10^4 and if the weaker acid or base has a dissociation constant greater than 10^{-8}.

Curve A in Figure 12-3 is the theoretical titration curve for triprotic phosphoric acid. Here, the ratio K_1/K_2 is approximately 10^5, as is K_2/K_3. This results in two well-defined end points, either of which is satisfactory for analytical purposes. An acid-range indicator will provide a color change when 1 mol of base has been introduced for 1 mol of acid; a base-range indicator will require 2 mol of base for 1 mol of acid. The third hydrogen of phosphoric acid is so slightly dissociated ($K_3 = 4.2 \times 10^{-13}$) that no practical end point is associated with its neutralization. The buffering effect of the third dissociation is noticeable, however, and causes the pH for curve A to be lower than the pH for the other two curves in the region beyond the second equivalence point.

Curve C is the titration curve for sulfuric acid, a substance that has one fully dissociated proton and one that is dissociated to a relatively large extent ($K_2 = 1.2 \times 10^{-2}$). Because of the similarity in strengths of the two acids, only a single end point, corresponding to the titration of both protons, is observed.

In general, the titration of acids or bases that have two reactive groups yields two end points of practical value only when the ratio between the two dissociation constants is at least 10^4. If the ratio is much smaller than this, the pH change at the first equivalence point will prove unsatisfactory for analysis.

12E TITRATION CURVES FOR POLYFUNCTIONAL BASES

The derivation of a titration curve for a polyfunctional base involves no new principles. To illustrate, consider the titration of a sodium carbonate solution with standard hydrochloric acid. The important equilibrium constants are

$$CO_3^{2-} + H_2O \rightleftharpoons OH^- + HCO_3^- \qquad K_{b1} = \frac{K_w}{K_2} = 2.13 \times 10^{-4}$$

$$HCO_3^- + H_2O \rightleftharpoons OH^- + H_2CO_3 \qquad K_{b2} = \frac{K_w}{K_1} = 2.25 \times 10^{-8}$$

The reaction of carbonate ion with water governs the initial pH of the solution, which can be computed by the method shown for the second equivalence point in Example 12-4. With the first additions of acid, a carbonate/hydrogen carbonate buffer is established. In this region, the pH can be derived from *either* the hydroxide ion concentration calculated from K_{b1} or the hydronium ion concentration calculated from K_2.

Sodium hydrogen carbonate is the principal solute species at the first equivalence point, and Equation 12-8 is used to compute the hydronium ion concentration. With the addition of more acid, a new buffer consisting of sodium hydrogen carbonate and carbonic acid is formed. The pH of this buffer is readily obtained from either K_{b2} or K_1.

At the second equivalence point, the solution consists of carbonic acid and sodium chloride. The carbonic acid can be treated as a simple weak acid having a dissociation constant K_1. Finally, after excess hydrochloric acid has been introduced, the dissociation of the weak acid is suppressed

Feature 12-2

THE DISSOCIATION OF SULFURIC ACID

Sulfuric acid is unusual in that one of its protons behaves as a strong acid and the other as a weak acid ($K_2 = 1.20 \times 10^{-2}$). Let us see how the hydronium ion concentration of sulfuric acid solutions is computed, using as an example a solution that has an analytical concentration of 0.0400 M.

If the dissociation of HSO_4^- is assumed to be negligible, it follows that

$$[H_3O^+] = [HSO_4^-] = 0.0400$$

However, an estimate of $[SO_4^{2-}]$ based upon this approximation and the expression for K_2 reveals that

$$\frac{\cancel{0.0400}[SO_4^{2-}]}{\cancel{0.0400}} = 1.20 \times 10^{-2}$$

Clearly, $[SO_4^{2-}]$ is *not* small relative to $[HSO_4^-]$, and so a more rigorous solution is required.

Stoichiometric considerations require that

$$[H_3O^+] = 0.0400 + [SO_4^{2-}]$$

The first term on the right is the concentration of H_3O^+ from dissociation of the H_2SO_4 to HSO_4^-. The second term is the contribution of the dissociation of HSO_4^-. Rearrangement yields

$$[SO_4^{2-}] = [H_3O^+] - 0.0400$$

Mass-balance considerations require that

$$C_{H_2SO_4} = 0.0400 = [HSO_4^-] + [SO_4^{2-}]$$

Combining the last two equations and rearranging yield

$$[HSO_4^-] = 0.0800 - [H_3O^+]$$

Introduction of these equations for $[SO_4^{2-}]$ and $[HSO_4^-]$ into the expression for K_2 yields

$$\frac{[H_3O^+]([H_3O^+] - 0.0400)}{0.0800 - [H_3O^+]} = 1.20 \times 10^{-2}$$

$$[H_3O^+]^2 - (0.0280)[H_3O^+] - 9.60 \times 10^{-4} = 0$$

$$[H_3O^+] = 0.0480$$

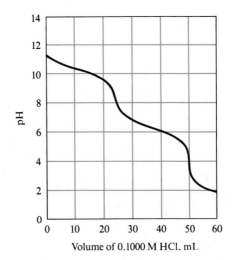

Figure 12-4

Curve for the titration of 25.00 mL of 0.1000 M Na_2CO_3 with 0.1000 M HCl.

to a point where the hydronium ion concentration is essentially that of the molar concentration of the strong acid.

Figure 12-4 illustrates that two end points are observed in the titration of sodium carbonate, the second being appreciably sharper than the first. It is apparent that the individual components in mixtures of sodium carbonate and sodium hydrogen carbonate can be determined by neutralization methods.

12F   THE COMPOSITION OF POLYPROTIC-ACID SOLUTIONS AS A FUNCTION OF pH

In Section 11G, we showed how alpha values are useful in visualizing the various concentration changes that occur in a titration of a simple weak acid. Alpha values can also be derived for polyfunctional acids and bases. For example, if we let C_T be the sum of the molar concentrations of maleate-containing species throughout the titration described in Example 12-4, the alpha value for the free acid is

$$\alpha_0 = \frac{[H_2M]}{C_T}$$

where

$$C_T = [H_2M] + [HM^-] + [M^{2-}] \tag{12-9}$$

The alpha values for HM^- and M^{2-} are given by similar equations:

$$\alpha_1 = \frac{[HM^-]}{C_T}$$

$$\alpha_2 = \frac{[M^{2-}]}{C_T}$$

As noted earlier, the sum of the alpha values for a system must equal unity:

$$\alpha_0 + \alpha_1 + \alpha_2 = 1$$

The alpha values for the maleic acid system are readily expressed in terms of $[H_3O^+]$, K_1, and K_2. To obtain such expressions, we follow the method used to derive Equations 11-19 and 11-20 in Section 11G and obtain

$$\alpha_0 = \frac{[H_3O^+]^2}{[H_3O^+]^2 + K_1[H_3O^+] + K_1K_2} \qquad (12\text{-}10)$$

$$\alpha_1 = \frac{K_1[H_3O^+]}{[H_3O^+]^2 + K_1[H_3O^+] + K_1K_2} \qquad (12\text{-}11)$$

$$\alpha_2 = \frac{K_1K_2}{[H_3O^+]^2 + K_1[H_3O^+] + K_1K_2} \qquad (12\text{-}12)$$

Note that the denominator is the same for each expression. Note also that the fractional amount of each species is fixed at any pH and is *independent* of the total concentration C_T.

Feature 12-3

A GENERAL EXPRESSION FOR ALPHA VALUES

For the weak acid H_nA, the denominator in all alpha-value expressions takes the form:

$$[H_3O^+]^n + K_1[H_3O^+]^{(n-1)} + K_1K_2[H_3O^+]^{(n-2)} + \cdots K_1K_2 \cdots K_n$$

The numerator for α_0 is the first term in the denominator, the numerator for α_1 is the second term, and so forth. Thus, if we let D be the denominator, $\alpha_0 = [H_3O^+]^n/D$ and $\alpha_1 = K_1[H_3O^+]^{(n-1)}/D$.

Alpha values for polyfunctional bases are generated in an analogous way, with the equations being written in terms of base dissociation constants and $[OH^-]$.

The three curves plotted in Figure 12-5 show the alpha value for each maleate-containing species as a function of pH. The solid curves in Figure 12-6 depict the same alpha values but now plotted as a function of sodium hydroxide volume as the acid is titrated. The titration curve is shown by the dashed line. Consideration of these curves gives a clear picture of all concentration changes that occur during the titration. For example, Figure 12-6 reveals that, before the addition of any base, α_0 for H_2M is roughly 0.7, α_1 for HM^- is approximately 0.3, and for all practical purposes, α_2 is zero. Thus, approximately 70% of the maleic acid exists as H_2M and 30% as HM^-. With addition of base, the pH rises, as does the

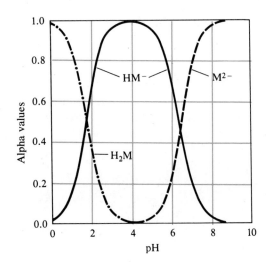

Figure 12-5
Composition of H_2M solutions as a function of pH.

fraction of HM^-. At the first equivalence point (pH = 4.12), essentially all of the maleate is present as HM^- ($\alpha_1 \rightarrow 1$). Beyond the first equivalence point, HM^- decreases and M^{2-} increases. At the second equivalence point (pH = 9.37) and beyond, essentially all of the maleate exists as M^{2-}.

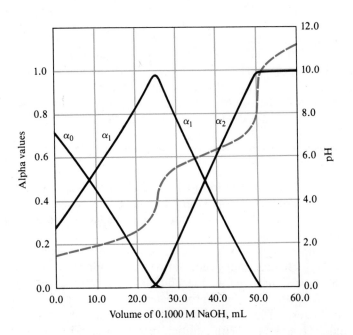

Figure 12-6
Titration of 25.00 mL of 0.1000 M maleic acid with 0.1000 M NaOH. The solid curves are plots of alpha values as a function of volume. The broken curve is a plot of pH as a function of volume.

12G   QUESTIONS AND PROBLEMS

*12-1. What is an amphiprotic solute?

12-2. Indicate whether an aqueous solution of the following compounds is acidic, neutral, or basic:
   *(a) NH_4OAc.
   (b) NH_4NO_2.
   *(c) $(NH_4)_2C_2O_4$.
   (d) $NaHC_2O_4$.
   *(e) $Na_2C_2O_4$.
   (f) Na_2HCit (where Cit^{3-} is the citrate anion).
   *(g) NaH_2Cit.
   (h) Na_3Cit.

12-3. Suggest an indicator that could be used to provide an end point for the titration of the first proton in H_3AsO_4.

*12-4. Suggest an indicator that would give an end point when two of the protons in H_3AsO_4 have been titrated.

12-5. Suggest a method for the determination of the amounts of H_3PO_4 and NaH_2PO_4 in an aqueous solution.

12-6. Suggest a suitable indicator for a titration based upon the following reactions; use 0.05 M if an equivalence-point concentration is needed.
   *(a) $H_2CO_3 + NaOH \rightarrow NaHCO_3 + H_2O$
   (b) $H_2P + 2\ NaOH \rightarrow Na_2P + 2\ H_2O$
       ($H_2P = o$-phthalic acid)
   *(c) $H_2T + 2\ NaOH \rightarrow Na_2T + 2\ H_2O$
       (H_2T = tartaric acid)
   (d) $NH_2C_2H_4NH_2 + HCl \rightarrow NH_2C_2H_4NH_3Cl$
   *(e) $NH_2C_2H_4NH_2 + 2\ HCl \rightarrow ClNH_3C_2H_4NH_3Cl$
   (f) $H_2SO_3 + NaOH \rightarrow NaHSO_3 + H_2O$
   *(g) $H_2SO_3 + 2\ NaOH \rightarrow Na_2SO_3 + 2\ H_2O$

*12-7. Calculate the pH of a solution that is 0.0600 M in
   (a) hydrogen sulfide.
   (b) sulfuric acid.
   (c) malonic acid.
   (d) sodium sulfide.
   (e) ethylenediamine.
   (f) sodium oxalate.

12-8. Calculate the pH of a solution that is 0.0600 M in
   (a) phosphoric acid.
   (b) oxalic acid.
   (c) phosphorous acid.
   (d) sodium sulfite.
   (e) trisodium phosphate.
   (f) sodium sulfate.

*12-9. Calculate the pH of a solution that is 0.0400 M in
   (a) sodium hydrogen sulfide.
   (b) sodium hydrogen oxalate.
   (c) sodium hydrogen sulfite.
   (d) ethylenediamine hydrochloride ($NH_2C_2H_4NH_3Cl$).

12-10. Calculate the pH of a solution that is 0.0400 M in
   (a) sodium hydrogen fumarate.
   (b) sodium hydrogen sulfate.
   (c) disodium hydrogen arsenate.
   (d) sodium dihydrogen arsenate.

*12-11. Calculate the pH of a solution that is
   (a) 0.0100 M in HCl and 0.0200 M in picric acid.
   (b) 0.0100 M in HCl and 0.0200 M in benzoic acid.
   (c) 0.0100 M in NaOH and 0.100 M in Na_2CO_3.
   (d) 0.0100 M in NaOH and 0.100 M in NH_3.

12-12. Calculate the pH of a solution that is
   (a) 0.0100 M in $HClO_4$ and 0.0300 M in monochloroacetic acid.

(b) 0.0100 M in HCl and 0.0150 M in H_2SO_4.
(c) 0.0100 M in NaOH and 0.0300 M in Na_2S.
(d) 0.0100 M in NaOH and 0.0300 M in sodium acetate.

*12-13. Identify the principal conjugate acid/base pair and calculate the ratio between them in a solution that is buffered to pH 6.00 and contains species derived from
   (a) H_2SO_3.
   (b) citric acid.
   (c) malonic acid.
   (d) tartaric acid.

12-14. Identify the principal conjugate acid/base pair and calculate the ratio between them in a solution that is buffered to pH 9.00 and contains species derived from
   (a) H_2S.
   (b) ethylenediamine dihydrochloride.
   (c) H_3AsO_4.
   (d) H_2CO_3.

*12-15. Calculate the pH of a solution made up to contain the following analytical concentrations:
   (a) 0.0500 M in H_3AsO_4 and 0.0200 M in NaH_2AsO_4.
   (b) 0.0300 M in NaH_2AsO_4 and 0.0500 M in Na_2HAsO_4.
   (c) 0.0600 M in Na_2CO_3 and 0.0300 M in $NaHCO_3$.
   (d) 0.0400 M in H_3PO_4 and 0.0200 M in Na_2HPO_4.
   (e) 0.0500 M in $NaHSO_4$ and 0.0400 M in Na_2SO_4.

12-16. Calculate the pH of a solution made up to contain the following analytical concentrations:
   (a) 0.240 M in H_3PO_3 and 0.480 M in NaH_2PO_3.
   (b) 0.0670 M in Na_2SO_3 and 0.0315 M in $NaHSO_3$.
   (c) 0.640 M in $HOC_2H_4NH_2$ and 0.750 M in $HOC_2H_4NH_3Cl$.
   (d) 0.240 M in $H_2C_2O_4$ and 0.360 M in $Na_2C_2O_4$.
   (e) 0.0100 M in $Na_2C_2O_4$ and 0.0400 M in $NaHC_2O_4$.

*12-17. What is the pH of the buffer formed upon mixing 50 mL of 0.200 M NaH_2PO_4 with
   (a) 50.0 mL of 0.120 M HCl?
   (b) 50.0 mL of 0.120 M NaOH?

12-18. What is the pH of the buffer formed by adding 100 mL of 0.150 M potassium hydrogen phthalate to
   (a) 100 mL of 0.0800 M NaOH?
   (b) 100 mL of 0.0800 M HCl?

*12-19. Describe the preparation of 1.00 L of a buffer of pH 9.60 from 0.300 M Na_2CO_3 and 0.200 M HCl.

12-20. Describe how to prepare 1.00 L of a buffer of pH 7.00 from 0.200 M H_3PO_4 and 0.160 M NaOH.

12-21. Describe the preparation of 1.00 L of a buffer of pH 6.00 from 0.500 M Na_3AsO_4 and 0.400 M HCl.

*12-22. How many grams of $Na_2HPO_4 \cdot 2\ H_2O$ must be added to 400 mL of 0.200 M H_3PO_4 to give a buffer of pH 7.30?

12-23. How many grams of dipotassium phthalate must be added to 750 mL of 0.0500 M phthalic acid to give a buffer of pH 5.75?

12-24. Derive a curve for the titration of 50.00 mL of a 0.1000 M solution of compound A with a 0.2000 M solution of compound B in the following list. For each titration, calculate the pH after the addition of 0.00, 12.50, 20.00, 24.00, 25.00, 26.00, 37.50, 45.00, 49.00, 50.00, 51.00, and 60.00 mL of compound B:

	A	B
*(a)	Na_2CO_3	HCl
(b)	ethylenediamine	HCl
*(c)	H_2SO_4	NaOH
(d)	$H_2C_2O_4$	NaOH

*12-25. Generate a titration curve for 50.00 mL of a solution in which the analytical concentration of NaOH is 0.1000 M and that for hydrazine is 0.08000 M. Calculate the pH after addition of 0.00, 10.00, 20.00, 24.00, 25.00, 26.00, 35.00, 44.00, 45.00, 46.00, and 50.00 mL of 0.2000 M $HClO_4$.

12-26. Generate a titration curve for 50.00 mL of a solution in which the analytical concentration of $HClO_4$ is 0.1000 M and that for formic acid is 0.08000 M. Calculate the pH after addition of 0.00, 10.00, 20.00, 24.00, 25.00, 26.00, 35.00, 44.00, 45.00, 46.00, and 50.00 mL of 0.2000 M KOH.

*12-27. Give equations that define α_0, α_1, α_2, and α_3 for H_3AsO_4.

12-28. Formulate equilibrium constants for the following equilibria, and calculate numerical values for the constants:

*(a) $H_2AsO_4^- + H_2AsO_4^- \rightleftarrows H_3AsO_4 + HAsO_4^{2-}$
(b) $HAsO_4^{2-} + HAsO_4^{2-} \rightleftarrows AsO_4^{3-} + H_2AsO_4^-$

*12-29. Derive a numerical value for the equilibrium constant for the reaction
$$NH_4^+ + OAc^- \rightleftarrows NH_3 + HOAc.$$

12-30. For pH values of 2.00, 6.00, and 10.00, calculate the alpha value for each species in an aqueous solution of

*(a) phthalic acid.   (d) arsenic acid.
(b) phosphoric acid.   *(e) phosphorous acid.
*(c) citric acid.   (f) oxalic acid.

CHAPTER 13

Applications of Neutralization Titrations

Neutralization titrations are widely used for determining the concentration of analytes that either are acids or bases or else are convertible to such species by suitable treatment.[1] Water is the usual solvent for neutralization titrations because it is readily available, inexpensive, and nontoxic. Its low temperature coefficient of expansion is an added virtue. Some analytes, however, are not titratable in aqueous media because their solubilities are too low or because their strengths as acids or bases are not sufficiently great to provide satisfactory end points. The concentration of such substances can often be determined by titration in a solvent other than water.[2]

In this text, we shall restrict our discussions to aqueous systems.

13A   REAGENTS FOR NEUTRALIZATION REACTIONS

Standard solutions for neutralization titrations are always prepared from strong acids or strong bases because this type of reagent provides the sharpest end point.

---

[1]For a review of applications of neutralization titrations, see D. Rosenthal and P. Zuman, in *Treatise on Analytical Chemistry,* 2nd ed., I. M. Kolthoff and P. J. Elving, Eds., Part I, Vol. 2, Chapter 18. New York: Wiley, 1979.

[2]For a review of nonaqueous acid/base titrimetry, see *Treatise on Analytical Chemistry,* 2nd ed., I. M. Kolthoff and P. J. Elving, Eds., Part I, Vol. 2, Chapters 19A–19E. New York: Wiley, 1979.

13A-1 Preparation of Standard Acid Solutions

Hydrochloric acid is widely used for the titration of bases. Dilute solutions of the reagent are stable indefinitely and do not cause troublesome precipitation reactions with most cations. It is reported that 0.1 M solutions of HCl can be boiled for as long as 1 h without loss of acid, provided that the water lost by evaporation is periodically replaced; 0.5 M solutions can be boiled for at least 10 min without significant loss.

Solutions of perchloric acid and sulfuric acid are also stable and are useful for titrations where chloride ion interferes by forming precipitates. Standard solutions of nitric acid are seldom used because of their oxidizing properties.

Standard acid solutions are ordinarily prepared by diluting an approximate volume of the concentrated reagent and subsequently standardizing the diluted solution against a primary-standard base.

Solutions of HCl, HClO$_4$, and H$_2$SO$_4$ are stable indefinitely. Restandardization is never required.

13A-2 The Standardization of Acids

Sodium Carbonate

Acids are frequently standardized by titration of weighed quantities of sodium carbonate. Primary-standard-grade sodium carbonate is available commercially or can be prepared by heating purified sodium hydrogen carbonate between 270 to 300°C for 1 h:

$$2\ NaHCO_3(s) \rightarrow Na_2CO_3(s) + H_2O(g) + CO_2(g)$$

Two end points are observed in the titration of sodium carbonate. The first, at about pH 8.3, corresponds to the conversion of carbonate to hydrogen carbonate; the second, at about pH 3.8, involves the formation of carbonic acid. The second end point is always used for standardization because the change in pH here is greater than at the first end point.

An even sharper end point can be achieved by boiling the solution briefly to eliminate the carbonic acid produced by the reaction. The sample is titrated to the first appearance of the acidic color of the indicator (such as bromocresol green or methyl orange). At this point, the solution contains a large amount of carbonic acid and a small amount of unreacted hydrogen carbonate. Boiling effectively destroys this buffer by eliminating the carbonic acid:

$$H_2CO_3(aq) \rightarrow CO_2(g) + H_2O(l)$$

The solution then becomes alkaline again due to the residual hydrogen carbonate ion. The titration is completed after the solution has cooled. Now, however, a substantially larger decrease in pH occurs during the final additions of acid, thus giving a more abrupt color change.

As an alternative, the acid solution can be introduced in an amount slightly in excess of that needed to convert the sodium carbonate to carbonic acid. The solution is boiled as before to remove carbon dioxide and cooled; the excess acid is then back-titrated with a dilute solution of base. Any indicator suitable for a strong acid/strong base titration is satis-

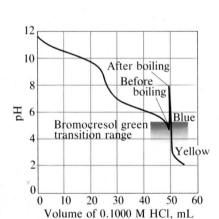

Titration of 25.00 mL of 0.1000 M Na$_2$CO$_3$ with 0.1000 M HCl. After about 49 mL of HCl have been added, the solution is boiled, causing the increase in pH shown. The change in pH on further addition of HCl is much larger.

factory. The volume ratio of acid to base must of course be established by an independent titration.

Other Primary Standards for Acids

Tris-(hydroxymethyl)aminomethane, $(HOCH_2)_3CNH_2$, known also as TRIS or THAM, is available in primary-standard purity from commercial sources. It has the advantage of a substantially greater formula weight (121.1) than sodium carbonate (53.0).

A high formula weight is desirable in a primary standard because a large weight of reagent must be used, thus reducing the relative weighing error.

Sodium tetraborate decahydrate and mercury(II) oxide have also been recommended as primary standards. The reaction of an acid with the tetraborate is

$$B_4O_7^{2-} + 2\ H_3O^+ + 3\ H_2O \rightarrow 4\ H_3BO_3$$

Feature 13-1
EQUIVALENT WEIGHTS OF ACIDS AND BASES

The equivalent weight of a participant in a neutralization reaction is the weight that reacts with or supplies one mole of protons *in a particular reaction*. For example, the equivalent weight of H_2SO_4 is one half of its formula weight. The equivalent weight of Na_2CO_3 is usually one half of its formula weight because in most applications its reaction is

$$Na_2CO_3 + 2\ H_3O^+ \rightarrow H_2O + H_2CO_3 + 2\ Na^+$$

When titrated with some indicators, however, it consumes but a single proton:

$$Na_2CO_3 + H_3O^+ \rightarrow NaHCO_3 + Na^+$$

Here, the equivalent weight and the formula weight of Na_2CO_3 are identical. These observations demonstrate that the equivalent weight of a compound cannot be defined without having a particular reaction in mind (Appendix 6).

13A-3 Preparations of Standard Base Solutions

Sodium hydroxide is the most common base for preparing standard solutions, although potassium hydroxide and barium hydroxide are also encountered. None of these is obtainable in primary-standard purity, and so standardization is required after preparation.

The Effect of Carbon Dioxide upon Standard Base Solutions

In solution as well as in the solid state, the hydroxides of sodium, potassium, and barium react avidly with atmospheric carbon dioxide to produce the corresponding carbonate:

$$CO_2(g) + 2\ OH^- \rightarrow CO_3^{2-} + H_2O$$

A negative determinate error results from the absorption of carbon dioxide by a standardized solution of sodium or potassium hydroxide in titrations that require a basic range indicator. No determinate error is incurred with an acidic range indicator.

Generally, carbonate ion in standard solutions of bases is undesirable because it decreases the sharpness of end points.

Although production of each carbonate ion uses up two hydroxide ions, the uptake of carbon dioxide by a solution of base does not necessarily alter its combining capacity for hydronium ions. Thus, at the end point of a titration that requires an acid-range indicator (such as bromocresol green), each carbonate ion produced from sodium or potassium hydroxide will have reacted with two hydronium ions of the analyte (Figure 12-4):

$$CO_3^{2-} + 2\,H_3O^+ \rightarrow H_2CO_3 + 2\,H_2O$$

Because the amount of hydronium ion consumed by this reaction is identical to the amount of hydroxide lost during formation of the carbonate ion, no error is incurred.

Unfortunately, most applications that use standard base require an indicator with a basic transition range (phenolphthalein, for example). Here, each carbonate ion has reacted with only one hydronium ion when the color change of the indicator is observed:

$$CO_3^{2-} + H_3O^+ \rightarrow HCO_3^- + H_2O$$

The effective concentration of the base is thus diminished by absorption of carbon dioxide, and a determinate error (called a *carbonate error*) results.

The solid reagents used to prepare standard solutions of base are always contaminated by significant amounts of carbonate ion. Thus, freshly prepared solutions of base contain substantial quantities of carbonate ion unless steps are taken to remove this species. The presence of this contaminant does not cause a carbonate error if the same indicator is used for both standardization and analysis. It does, however, lead to less sharp end points. Consequently, steps are usually taken to remove carbonate ion before a solution of a base is standardized.

The best method for preparing carbonate-free sodium hydroxide solutions takes advantage of the very low solubility of sodium carbonate in concentrated solutions of the reagent. An approximately 50% aqueous solution of sodium hydroxide is prepared (or purchased from commercial sources). The solid sodium carbonate is allowed to settle, and then the clear liquid is decanted and diluted to give the desired concentration (alternatively, the solid is removed by vacuum filtration).

Water for preparing carbonate-free solutions of base must be free of carbon dioxide. Distilled water, which is sometimes supersaturated with carbon dioxide, should be boiled briefly to eliminate the gas. The water is then allowed to cool to room temperature before the introduction of base because hot alkali solutions rapidly absorb carbon dioxide. Deionized water ordinarily does not contain significant amounts of carbon dioxide.

Standard solutions of base are reasonably stable as long as they are protected from contact with the atmosphere. Figure 13-1 shows an arrangement for preventing the uptake of atmospheric carbon dioxide during storage and when the reagent is dispensed. Air entering the vessel is passed over a solid absorbent for CO_2, such as soda lime or Ascarite II®.[3]

[3]Thomas Scientific, Swedesboro, NJ. Ascarite II® consists of sodium hydroxide deposited on a nonfibrous silicate structure.

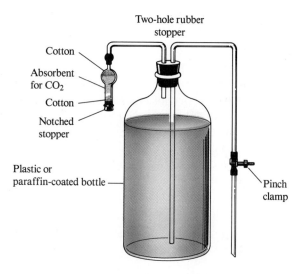

Figure 13-1
Arrangement for the storage of standard base solutions.

The contamination that occurs as the solution is transferred from this storage bottle to the buret is ordinarily negligible.

A tightly capped polyethylene bottle usually provides sufficient short-term protection against the uptake of atmospheric carbon dioxide. Before capping, the bottle is squeezed to minimize the interior air space. Care should also be taken to keep the bottle closed except during the brief periods when the contents are being transferred to a buret. Sodium hydroxide solutions will ultimately cause a polyethylene bottle to become brittle.

The concentration of sodium hydroxide solutions decreases slowly (0.1 to 0.3% per week) when the base is stored in glass bottles. The loss in strength is caused by the reaction of the base with the glass to form sodium silicates. For this reason, standard solutions of base should not be stored for extended periods (longer than one or two weeks) in glass containers. In addition, bases should never be kept in glass-stoppered containers because the reaction between the base and the stopper may cause the latter to "freeze" after a brief period. Finally, to avoid the same type of freezing, burets with glass stopcocks should be promptly drained and thoroughly rinsed with water after use with standard base solutions. This problem is avoided with burets equipped with Teflon stopcocks.

Solutions of bases should be stored in polyethylene bottles. Such solutions should never be stored in glass-stoppered bottles because removal of the stopper often becomes impossible after a brief period.

13A-4 The Standardization of Bases

Several excellent primary standards are available for the standardization of bases. Most are weak organic acids that require the use of an indicator with a basic transition range.

Standard solutions of bases cannot be prepared directly by weight and must always be standardized against a primary standard or a standard acid solution.

Potassium Hydrogen Phthalate, $KHC_8H_4O_4$
Potassium hydrogen phthalate is an ideal primary standard. It is a nonhygroscopic crystalline solid with a high equivalent weight (204.2). For most purposes, the commercial analytical-grade salt can be used without further purification. For the most exacting work, potassium hydrogen phtha-

late of certified purity is available from the National Institute for Standards and Technology.

Other Primary Standards for Bases

Benzoic acid is obtainable in primary-standard purity and can be used for the standardization of bases. Because its solubility in water is limited, this reagent is ordinarily dissolved in ethanol prior to dilution with water and titration. A blank should always be carried through this standardization procedure because commercial alcohol is sometimes slightly acidic.

Potassium hydrogen iodate, $KH(IO_3)_2$, is an excellent primary standard with a high equivalent weight. It is also a strong acid that can be titrated using virtually any indicator with a transition range between pH 4 and 10.

13B   TYPICAL APPLICATIONS OF NEUTRALIZATION TITRATIONS

Neutralization titrations are among the most widely used analytical methods.

Neutralization titrations are used to determine the innumerable inorganic, organic, and biological species that have inherent acidic or basic properties. Equally important, however, are the many applications that involve conversion of an analyte to an acid or base by suitable chemical treatment followed by titration with a standard strong base or acid.

Two major types of end points find widespread use in neutralization titrations. The first is a visual end point based on indicators such as those described in Section 11A. The second is a *potentiometric* end point, in which the potential of a glass/calomel electrode system is determined with a voltage-measuring device. The measured potential is directly proportional to pH. Potentiometric end points are described in Chapter 18.

13B-1 Elemental Analysis

Several important elements that occur in organic and biological systems are conveniently determined by methods that involve an acid/base titration as the final step. Generally, the elements susceptible to this type of analysis are nonmetallic and include carbon, nitrogen, sulfur, chlorine, bromine, and fluorine as well as a few other less common species. In each instance, the element is converted to an inorganic acid or base that is then titrated. A few examples follow.

Nitrogen

Nitrogen is found in a wide variety of substances of interest in research, industry, and agriculture. For example it is found in amino acids, proteins, synthetic drugs, fertilizers, explosives, soils, potable water supplies, and dyes. Thus analytical methods for the determination of nitrogen, particularly in organic substrates, are very important.

Hundreds of thousands of Kjeldahl nitrogen determinations are performed each year, primarily to provide a measure of the protein content of meats, grains, and animal feeds.

The most common method for determining organic nitrogen is the *Kjeldahl method,* which is based on a neutralization titration. The procedure is straightforward, requires no special equipment, and is readily adapted to the routine analysis of large numbers of samples. It is the standard means for determining the protein content of grains, meats, and other biological materials. Since most proteins contain approximately the same percentage of nitrogen, multiplication of this percentage by a suitable factor (6.25 for meats, 6.38 for dairy products, and 5.7 for cereals) gives the percentage of protein in a sample.

The Kjeldahl method was developed by a Danish chemist who first described it in 1883: J. Kjeldahl, *Z. Anal. Chem.* **1883,** *22,* 366.

Feature 13-2
OTHER METHODS FOR DETERMINING ORGANIC NITROGEN

Two other methods are used to determine the nitrogen content of organic materials. In the *Dumas method,* the sample is mixed with powdered copper(II) oxide and ignited in a combustion tube to give carbon dioxide, water, nitrogen, and small amounts of nitrogen oxides. A stream of carbon dioxide carries these products through a packing of hot copper, which reduces any oxides of nitrogen to elemental nitrogen. The mixture is then passed into a gas buret filled with concentrated potassium hydroxide. The only component not absorbed by the base is nitrogen, and its volume is measured directly.

The newest method for determining organic nitrogen involves combusting the sample at 1100°C for a few minutes to convert the nitrogen to nitric oxide, NO. Ozone is then introduced into the gaseous mixture, which oxidizes the nitric oxide to nitrogen dioxide. This reaction gives off visible radiation (*chemiluminescence*), the intensity of which is proportional to the nitrogen content of the sample. An instrument for this procedure is available from commercial sources.

In the Kjeldahl method, the sample is decomposed in hot, concentrated sulfuric acid to convert the bound nitrogen to ammonium ion. The resulting solution is then cooled, diluted, and made basic. The liberated ammonia is distilled, collected in an acidic solution, and determined by a neutralization titration.

The critical step in the Kjeldahl method is the decomposition with sulfuric acid, which oxidizes the carbon and hydrogen in the sample to carbon dioxide and water. The fate of the nitrogen, however, depends upon its state of combination in the original sample. Amine and amide nitrogen is quantitatively converted to ammonium ion. In contrast, nitro, azo, and azoxy groups are likely to yield the element or various nitrogen oxides, all of which are lost from the hot acidic medium. This loss can be avoided by first treating the sample with a reducing agent to form reduced products that behave as amide or amine nitrogen. In one such reduction scheme, salicylic acid and sodium thiosulfate are added to the concentrated sulfuric acid solution containing the sample. After a brief period, the digestion is performed in the usual way.

Certain aromatic heterocyclic compounds, such as pyridine and its derivatives, are particularly resistant to complete decomposition by sulfuric acid. Such compounds yield low results as a consequence (Figure 3-3) unless special precautions are taken.

The decomposition step is frequently the most time-consuming aspect of a Kjeldahl determination. Some samples may require heating periods in

$-NO_2$ $-N{=}N-$ $\underset{O^-}{-N_+{=}N-}$

nitro group   azo group

azoxy group

excess of 1 h. Numerous modifications of the original procedure have been proposed with the aim of shortening the digestion time. In the most widely used modification, a neutral salt, such as potassium sulfate, is added to increase the boiling point of the sulfuric acid solution and thus the temperature at which the decomposition occurs.

Many substances catalyze the decomposition of organic compounds by sulfuric acid. Mercury, copper, and selenium, either combined or in the elemental state, are effective. Mercury(II), if present, must be precipitated with hydrogen sulfide prior to distillation to prevent retention of ammonia as a mercury(II) ammine complex.

Figure 13-2 illustrates typical equipment for a Kjeldahl distillation. The long-necked container, which is used for both digestion and distillation, is called a *Kjeldahl flask*. After the decomposition is judged complete, the digested mixture is cooled, diluted with water, and made basic to liberate the ammonia. The equilibria are:

$$NH_4^+ + OH^- \rightleftharpoons NH_3(aq) + H_2O$$

$$NH_3(aq) \rightleftharpoons NH_3(g)$$

In the apparatus shown in Figure 13-2a, the base is added slowly by partially opening the stopcock from the NaOH storage vessel; the liberated ammonia is then carried to the receiving flask by steam distillation.

In an alternative method (Figure 13-2b), a dense, concentrated sodium hydroxide solution is carefully poured down the side of the Kjeldahl flask to form a second, lower layer. The flask is then quickly connected to a spray trap and an ordinary condenser before loss of ammonia can occur. Only then are the two layers mixed by gentle swirling of the flask.

Quantitative collection of ammonia requires the tip of the condenser to extend into the liquid in the receiving flask throughout the distillation

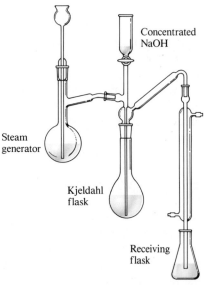

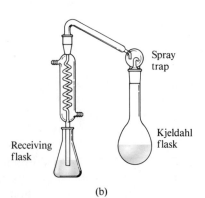

Figure 13-2
Kjeldahl distillation apparatus.   (a)   (b)

step. The tip must be removed before heating is discontinued, however. Otherwise, the liquid will be drawn back into the apparatus.

Two methods are commonly used for collecting and determining the ammonia liberated from the sample. In one, the ammonia is distilled into a measured volume of standard acid. After the distillation is complete, the excess acid is back-titrated with standard base. An indicator with an acidic transition range is required because of the acidity of the ammonium ions present at equivalence. A convenient alternative, which requires only one standard solution, involves the collection of the ammonia in an unmeasured excess of boric acid, which retains the ammonia by the reaction

$$H_3BO_3 + NH_3 \rightarrow NH_4^+ + H_2BO_3^-$$

The dihydrogen borate ion produced is a reasonably strong base that can be titrated with a standard solution of hydrochloric acid:

$$H_2BO_3^- + H_3O^+ \rightarrow H_3BO_3 + H_2O$$

At the equivalence point, the solution contains boric acid and ammonium ions; an indicator with an acidic transition interval (such as bromocresol green) is again required.

Sulfur

Sulfur in organic and biological materials is conveniently determined by burning the sample in a stream of oxygen. The sulfur dioxide (as well as the sulfur trioxide) formed during the oxidation is collected by distillation into a dilute solution of hydrogen peroxide:

$$SO_2(g) + H_2O_2 \rightarrow H_2SO_4$$

The sulfuric acid is then titrated with standard base.

Other Elements

Table 13-1 lists other elements that can be determined by neutralization methods.

Table 13-1
ELEMENTAL ANALYSES BASED ON NEUTRALIZATION TITRATIONS

Element	Converted to	Absorption or Precipitation Products	Titration
N	NH_3	$NH_3(g) + H_3O^+ \rightarrow NH_4^+ + H_2O$	Excess HCl with NaOH
S	SO_2	$SO_2(g) + H_2O_2 \rightarrow H_2SO_4$	NaOH
C	CO_2	$CO_2(g) + Ba(OH)_2 \rightarrow BaCO_3(s) + H_2O$	Excess $Ba(OH)_2$ with HCl
Cl (Br)	HCl	$HCl(g) + H_2O \rightarrow Cl^- + H_3O^+$	NaOH
F	SiF_4	$SiF_4(g) + H_2O \rightarrow H_2SiF_6$	NaOH
P	H_3PO_4	$12\ H_2MoO_4 + 3\ NH_4^+ + H_3PO_4 \rightarrow$ $(NH_4)_3PO_4 \cdot 12\ MoO_3(s) + 12\ H_2O + 3\ H^+$ $(NH_4)_3PO_4 \cdot 12\ MoO_3(s) + 26\ OH^- \rightarrow$ $HPO_4^{2-} + 12\ MoO_4^{2-} + 14\ H_2O + 3\ NH_3(g)$	Excess NaOH with HCl

13B-2 The Determination of Inorganic Substances

Numerous inorganic species can be determined by titration with strong acids or bases. A few examples follow.

Ammonium Salts

Ammonium salts are conveniently determined by conversion to ammonia with strong base followed by distillation in the Kjeldahl apparatus shown in Figure 13-2. The ammonia is collected and titrated as in the Kjeldahl method.

Nitrates and Nitrites

The method just described for ammonium salts can be extended to the determination of inorganic nitrate or nitrite. These ions are first reduced to ammonium ion by Devarda's alloy (50% Cu, 45% Al, 5% Zn). Granules of the alloy are introduced into a strongly alkaline solution of the sample in a Kjeldahl flask. The ammonia is distilled after reaction is complete. Arnd's alloy (60% Cu, 40% Mg) has also been used as the reducing agent.

Carbonate and Carbonate Mixtures

The qualitative and quantitative determination of the constituents in a solution containing sodium carbonate, sodium hydrogen carbonate, and sodium hydroxide, either alone or admixed, provides interesting examples of how neutralization titrations can be employed to analyze mixtures. No more than two of these three constituents can exist in appreciable amount in any solution because reaction eliminates the third. Thus, mixing sodium hydroxide with sodium hydrogen carbonate results in the formation of sodium carbonate until one or the other (or both) of the original reactants is exhausted. If the sodium hydroxide is used up, the solution will contain sodium carbonate and sodium hydrogen carbonate; if the sodium hydrogen carbonate is depleted, sodium carbonate and sodium hydroxide will remain; if equimolar amounts of sodium hydrogen carbonate and sodium hydroxide are mixed, the principal solute species will be sodium carbonate.

The analysis of such mixtures requires two titrations; one with an alkaline-range indicator, such as phenolphthalein, and the other with an acid-range indicator, such as bromocresol green. The composition of the solution can then be deduced from the relative volumes of acid needed to titrate equal volumes of the sample (Table 13-2 and Figure 12-4). Once the composition of the solution has been established, the volume data can be used to determine the concentration of each component in the sample.

Example 13-1

A solution contains $NaHCO_3$, Na_2CO_3, and/or NaOH, either alone or in permissible combination. Titration of a 50.0-mL portion to a phenolphthalein end point requires 22.1 mL of 0.100 M HCl. A second 50.0-mL aliquot requires 48.4 mL of the HCl when titrated to a bromocresol green end point. Deduce the composition, and calculate the molar solute concentrations of the original solution.

Table 13-2

VOLUME RELATIONSHIPS IN THE ANALYSIS OF MIXTURES
CONTAINING HYDROXIDE, CARBONATE, AND HYDROGEN
CARBONATE IONS

Constituent(s) in Sample	Relationship between V_{phth} and V_{bcg} in the Titration of an Equal Volume of Sample*
NaOH	$V_{phth} = V_{bcg}$
Na_2CO_3	$V_{phth} = \frac{1}{2}V_{bcg}$
$NaHCO_3$	$V_{phth} = 0; V_{bcg} > 0$
NaOH, Na_2CO_3	$V_{phth} > \frac{1}{2}V_{bcg}$
Na_2CO_3, $NaHCO_3$	$V_{phth} < \frac{1}{2}V_{bcg}$

*V_{phth} = volume of acid needed for a phenolphthalein end point; V_{bcg} = volume of acid needed for a bromocresol green end point.

If the solution contained only NaOH, the volume of acid required would be the same regardless of indicator (that is, $V_{phth} = V_{bcg}$). Similarly, we can rule out the presence of Na_2CO_3 alone because titration of this compound to a bromocresol green end point would require just twice the volume of acid required to reach the phenolphthalein end point. In fact, however, the second titration requires 48.4 mL. Because less than half of this amount is involved in the first titration, the solution must contain some $NaHCO_3$ in addition to Na_2CO_3. We can now calculate the concentration of the two constituents.

Compatible mixtures containing two of the following can also be analyzed in a similar way: HCl, H_3PO_4, NaH_2PO_4, Na_2HPO_4, Na_3PO_4, and NaOH.

When the phenolphthalein end point is reached, the CO_3^{2-} originally present is converted to HCO_3^-. Thus,

$$\text{amount } Na_2CO_3 = 22.1 \text{ mL} \times 0.100 \text{ mmol/mL} = 2.21 \text{ mmol}$$

The titration from the phenolphthalein end point to the bromocresol green end point (48.4 − 22.1 = 26.3 mL) involves both the hydrogen carbonate originally present and that formed by titration of the carbonate. Thus,

$$\text{amount } NaHCO_3 + \text{amount } Na_2CO_3 = 26.3 \times 0.100 = 2.63 \text{ mmol}$$

Hence,

$$\text{amount } NaHCO_3 = 2.63 - 2.21 = 0.42 \text{ mmol}$$

The molar concentrations are readily calculated from these data:

$$C_{Na_2CO_3} = \frac{2.21 \text{ mmol}}{50.0 \text{ mL}} = 0.0442 \text{ M}$$

$$C_{NaHCO_3} = \frac{0.42 \text{ mmol}}{50.0 \text{ mL}} = 0.0084 \text{ M}$$

The method described in Example 13-1 is not entirely satisfactory because the pH change corresponding to the hydrogen carbonate equivalence point is not sufficient to give a sharp color change with a chemical indicator (Figure 12-4). Relative errors of 1% or more must be expected as a consequence.

How could you analyze a mixture of HCl and H_3PO_4? A mixture of Na_3PO_4 and Na_2HPO_4? See Figure 12-3 A.

13C QUESTIONS AND PROBLEMS

*13-1. The boiling points of HCl and CO_2 are nearly the same (-85 and $-78°C$). Explain why CO_2 can be removed from an aqueous solution by boiling briefly while essentially no HCl is lost even after boiling for 1 h or more.

13-2. Why is HNO_3 seldom used to prepare standard acid solutions?

*13-3. Explain how primary-standard grade Na_2CO_3 can be prepared from primary standard $NaHCO_3$.

13-4. Suggest two advantages of $KH(IO_3)_2$ as a primary standard for bases.

*13-5. What is the carbonate error in acid/base titrations? How can it be avoided?

13-6. A sample of alfalfa was found to contain 2.83% nitrogen when analyzed by the Kjeldahl procedure. What was the protein content?

*13-7. Outline a method for the determination of $NaNO_2$ in a mixture of inorganic halides.

13-8. Outline a method for the determination of SO_2 in the effluent gases from a paper mill.

*13-9. If 1.000 L of 0.1500 M NaOH is unprotected from the air after standardization and absorbs 11.2 mmol of CO_2, what is its new molarity when it is standardized against a standard solution of HCl using
  (a) phenolphthalein?
  (b) bromocresol green?

13-10. A NaOH solution was 0.1019 M immediately after standardization. Exactly 500.0 mL of the reagent was left exposed to air for several days and absorbed 0.652 g of CO_2. Calculate the relative carbonate error in the determination of acetic acid with this solution if phenolphthalein is used as an indicator.

*13-11. Describe the preparation of 2.00 L of
  (a) 0.15 M KOH from the solid.
  (b) 0.015 M $Ba(OH)_2 \cdot 8H_2O$ from the solid.
  (c) 0.200 M HCl from a reagent that has a density of 1.0579 g/mL and is 11.50% HCl (w/w).

13-12. Describe the preparation of 500 mL of
  (a) 0.250 M H_2SO_4 from a reagent that has a density of 1.1539 g/mL and is 21.8% H_2SO_4 (w/w).
  (b) 0.30 M NaOH from the solid.
  (c) 0.0800 M Na_2CO_3 from the pure solid.

*13-13. The following data were obtained from the standardization of HCl against samples of sodium tetraborate, $Na_2B_4O_7 \cdot 10H_2O$:

$$B_4O_7^{2-} + 2 H_3O^+ + 3 H_2O \rightarrow 4 H_3BO_3$$

$Na_2B_4O_7 \cdot 10H_2O$, g	HCl, mL
0.6442	33.74
0.7102	37.56
0.5934	31.26

  (a) Calculate the mean molarity for the set.

  (b) Calculate the standard deviation for the molarity.

13-14. (a) Calculate the mean molarity of a $Ba(OH)_2$ solution from the following standardization data:

$KH(IO_3)_2$, g	$Ba(OH)_2$, mL
0.2574	26.77
0.2733	28.45
0.2885	30.11

  (b) Calculate the standard deviation for the molarity.

13-15. Suggest a range of sample weights for the indicated primary standard if it is desired to use between 35 and 45 mL of titrant:
  *(a) 0.150 M $HClO_4$ titrated against Na_2CO_3 (CO_2 product).
  (b) 0.075 M HCl titrated against $Na_2C_2O_4$:

$$Na_2C_2O_4 \rightarrow Na_2CO_3 + CO$$
$$CO_3^{2-} + 2 H^+ \rightarrow H_2O + CO_2$$

  *(c) 0.20 M NaOH titrated against benzoic acid.
  (d) 0.030 M $Ba(OH)_2$ titrated against $KH(IO_3)_2$.
  *(e) 0.040 M $HClO_4$ titrated against THAM.
  (f) 0.080 M H_2SO_4 titrated against $Na_2B_4O_7 \cdot 10 H_2O$ (Problem 13-13).

*13-16. A 25.00-mL aliquot of dilute H_2SO_4 yielded 0.3472 g of $BaSO_4$. Calculate the molarity of the acid.

13-17. A 50.00-mL aliquot of dilute HCl yielded 0.4771 g of AgCl. What was the molarity of the acid?

*13-18. A 50.00-mL sample of a white dinner wine required 21.48 mL of 0.03776 M NaOH to achieve a phenolphthalein end point. Express the acidity of the wine in terms of grams of tartaric acid, $H_2C_4H_4O_6$ (fw $= 150.09$ g), per 100 mL. (Assume that two hydrogens of the acid are titrated.)

13-19. A 25.00-mL sample of a household cleaning solution was diluted to 250.0 mL in a volumetric flask. A 50.00-mL aliquot of this solution required 40.38 mL of 0.2506 M HCl to reach a bromocresol green end point. Calculate the weight/volume percentage of NH_3 in the sample. (Assume that all the alkalinity results from the ammonia.)

*13-20. A 0.2296-g sample of a recrystallized organic acid required a 29.83-mL titration with 0.1000 M NaOH to reach a phenolphthalein end point. What is the equivalent weight of the acid (that is, the number of grams of acid per mole of titratable protons)?

13-21. In order to establish the identity of the cation in a pure carbonate, a 0.1401-g sample was dissolved in 50.00 mL of 0.1140 M HCl and boiled to remove CO_2. The excess HCl was back-titrated with 24.21 mL of 0.09802 M NaOH. Identify the carbonate.

*13-22. The active ingredient in Antabuse, a drug used for

the treatment of chronic alcoholism, is tetraethyl-thiuram disulfide,

$$
\begin{array}{c c}
S & S \\
\parallel & \parallel
\end{array}
$$
$$(C_2H_5)_2NCSSCN(C_2H_5)_2$$

(fw = 296.54 g). The sulfur in a 0.4329-g sample of an Antabuse preparation was oxidized to SO_2, which was absorbed in H_2O_2 to give H_2SO_4. The acid was titrated with 22.13 mL of 0.03736 M base. Calculate the percentage of active ingredient in the preparation.

13-23. Neohetramine, $C_{16}H_{21}ON_4$ (fw = 285.37 g), is a common antihistamine. A 0.1247-g sample containing this compound was analyzed by the Kjeldahl method. The ammonia produced was collected in H_3BO_3; the resulting $H_2BO_3^-$ was titrated with 26.13 mL of 0.01477 M HCl. Calculate the percentage of neohetramine in the sample.

*13-24. A 0.9092-g sample of a wheat flour was analyzed by the Kjeldahl procedure. The ammonia formed was distilled into 50.00 mL of 0.05063 M HCl; a 7.46-mL back-titration with 0.04917 M base was required. Calculate the percentage of protein in the flour using 5.70 as the multiplier (page 227).

13-25. The formaldehyde content of a pesticide preparation was determined by weighing 0.3124 g of the liquid sample into a flask containing 50.0 mL of 0.0996 M NaOH and 50 mL of 3% H_2O_2. Upon heating, the following reaction took place:

$$OH^- + HCHO + H_2O_2 \rightarrow HCOO^- + 2 H_2O$$

After cooling, the excess base was titrated with 23.3 mL of 0.05250 M H_2SO_4. Calculate the percentage of HCHO (fw = 30.03 g) in the sample.

*13-26. A 3.00-L sample of urban air was bubbled through a solution containing 50.0 mL of 0.0116 M $Ba(OH)_2$, and $BaCO_3$ precipitated:

$$Ba^{2+} + 2 OH^- + CO_2 \rightarrow BaCO_3(s) + H_2O$$

The excess base was back-titrated to a phenolphthalein end point with 23.6 mL of 0.0108 M HCl. Calculate the parts per million of CO_2 in the air (that is, mL $CO_2/10^6$ mL air); use 1.98 g/L for the density of CO_2.

13-27. Air was bubbled at a rate of 30.0 L/min through a trap containing 75 mL of 1% H_2O_2 ($H_2O_2 + SO_2 \rightarrow H_2SO_4$). After 10.0 min, the H_2SO_4 was titrated with 11.1 mL of 0.00204 M NaOH. Calculate the parts per million of SO_2 (that is, mL $SO_2/10^6$ mL air) if the density of SO_2 is 0.00285 g/mL.

*13-28. A 0.8160-g sample containing dimethylphthalate, $C_6H_4(COOCH_3)_2$ (fw = 194.19 g), and unreactive species was refluxed with 50.00 mL of 0.1031 M NaOH to hydrolyze the ester groups (this process is called *saponification*):

$$C_6H_4(COOCH_3)_2 + 2 OH^- \rightarrow$$
$$C_6H_4(COO)_2^{2-} + 2 CH_3OH$$

After the reaction was complete, the excess NaOH was back-titrated with 24.27 mL of 0.1644 M HCl. Calculate the percentage of dimethylphthalate in the sample.

13-29. A 1.219-g sample containing $(NH_4)_2SO_4$, NH_4NO_3, and nonreactive substances was diluted to 200 mL in a volumetric flask. A 50.00-mL aliquot was made basic with strong alkali, and the liberated NH_3 was distilled into 30.00 mL of 0.08421 M HCl. The excess HCl required 10.17 mL of 0.08802 M NaOH. A 25.00-mL aliquot of the sample was made alkaline after the addition of Devarda's alloy, and the NO_3^- was reduced to NH_3. The NH_3 from both NH_4^+ and NO_3^- was then distilled into 30.00 mL of the standard acid and back-titrated with 14.16 mL of the base. Calculate the percentage of $(NH_4)_2SO_4$ and NH_4NO_3 in the sample.

*13-30. A 50.00-mL sample containing acetaldehyde, ethyl acetate, and unreactive substances was diluted to 250.0 mL with water. A 50.00-mL aliquot of the diluted sample was mixed with 40.00 mL of 0.05454 M NaOH and 20 mL of 3% H_2O_2. Upon refluxing for 30 min, quantitative hydrolysis of the ester occurred; at the same time, the acetaldehyde was oxidized to sodium acetate. The two reactions were:

$$C_2H_5COOC_2H_5 + OH^- \rightarrow C_2H_5COO^- + C_2H_5OH$$
$$H_2O_2 + CH_3CHO + OH^- \rightarrow CH_3COO^- + 2 H_2O$$

The unreacted NaOH consumed 10.05 mL of 0.02513 M HCl. A 25.00-mL aliquot of the sample solution was treated with $NH_2OH \cdot HCl$ to convert the acetaldehyde to the corresponding oxime:

$$CH_3CHO + NH_2OH \cdot HCl \rightarrow$$
$$CH_3CH{=}NOH + HCl + H_2O$$

The liberated HCl consumed 10.97 mL of the standard base. Calculate the weight/volume percentage of acetaldehyde and ethyl acetate in the sample.

13-31. The digestion of 0.1417 g of a phosphorus-containing sample in a mixture of HNO_3 and H_2SO_4 resulted in the formation of CO_2, H_2O, and H_3PO_4. Addition of ammonium molybdate yielded a solid with the composition $(NH_4)_3PO_4 \cdot 12 MoO_3$ (fw = 1876.3 g). This precipitate was filtered, washed, and dissolved in 50.00 mL of 0.2000 M NaOH:

$$(NH_4)_3PO_4 \cdot 12 MoO_3(s) + 26 OH^- \rightarrow$$
$$HPO_4^{2-} + 12 MoO_4^{2-} + 14 H_2O + 3 NH_3(g)$$

After the solution was boiled to remove the NH_3, the excess NaOH was titrated with 14.17 mL of 0.1741 M HCl to a phenolphthalein end point. Calculate the percentage of phosphorus in the sample.

*13-32. A 1.217-g sample of commercial KOH contaminated by K_2CO_3 and H_2O was dissolved in water, and the resulting solution was diluted to 500.0 mL. A 50.00-mL aliquot of this solution was treated with 40.00 mL of 0.05304 M HCl and boiled to remove CO_2. The excess acid consumed 4.74 mL of 0.04983 M NaOH (phenolphthalein indicator). An excess of $BaCl_2$ was added to another 50.00-mL aliquot to precipitate the carbonate as $BaCO_3$. The solution was then titrated with 28.56 mL of the acid to a phenolphthalein end point. Calculate the percentages of KOH, K_2CO_3, and H_2O in the sample, assuming that these are the only compounds present.

13-33. A 0.5000-g sample containing $NaHCO_3$, Na_2CO_3, and H_2O was dissolved and diluted to 250.0 mL. A 25.00-mL aliquot was then boiled with 50.00 mL of 0.01255 M HCl. After cooling, the excess acid in the solution required 2.34 mL of 0.01063 M NaOH when titrated to a phenolphthalein end point. A second 25.00-mL aliquot was then treated with an excess of $BaCl_2$ and 25.00 mL of the base; precipitation of all the carbonate resulted, and the excess base was titrated with 7.63 mL of the HCl. Calculate the composition of the mixture.

*13-34. Calculate the volume of 0.06122 M HCl needed to titrate
  (a) 20.00 mL of 0.05555 M Na_3PO_4 to a thymolphthalein (transition pH = 9.3–10.5) end point.
  (b) 25.00 mL of 0.05555 M Na_3PO_4 to a bromocresol green end point.
  (c) 40.00 mL of a solution that is 0.02102 M in Na_3PO_4 and 0.01655 M in Na_2HPO_4 to a bromocresol green end point.
  (d) 20.00 mL of a solution that is 0.02102 M in Na_3PO_4 and 0.01655 M in NaOH to a thymolphthalein end point.
  (See Figure 12-3 A for a titration curve for H_3PO_4.)

13-35. Calculate the volume of 0.07731 M NaOH needed to titrate
  (a) 25.00 mL of a solution that is 0.03000 M in HCl and 0.01000 M in H_3PO_4 to a bromocresol green end point.
  (b) the solution in (a) to a thymolphthalein (transition pH range = 9.3–10.5) end point.
  (c) 30.00 mL of 0.06407 M NaH_2PO_4 to a thymolphthalein end point.
  (d) 25.00 mL of a solution that is 0.02000 M in H_3PO_4 and 0.03000 M in NaH_2PO_4 to a thymolphthalein end point.
  (See Figure 12-3 A for a titration curve for H_3PO_4.)

*13-36. Several solutions containing NaOH, Na_2CO_3, and $NaHCO_3$, alone or in compatible combination, were titrated with 0.1202 M HCl. Tabulated below are the volumes of acid needed to titrate 25.00-mL portions of each solution to a (1) phenolphthalein and (2) bromocresol green end point. Use this information to deduce the composition of the solutions. In addition, calculate the number of milligrams of each solute per milliliter of solution.

	(1), mL	(2), mL
(a)	22.42	22.44
(b)	15.67	42.13
(c)	29.64	36.42
(d)	16.12	32.23
(e)	0.00	33.33

13-37. Solutions containing NaOH, Na_3AsO_4, and Na_2HAsO_4, alone or in compatible combination, were titrated with 0.08601 M HCl. Tabulated below are the volumes of acid needed to titrate 25.00-mL portions of each solution to a (1) phenolphthalein and (2) bromocresol green end point. Use this information to deduce the composition of the solutions. In addition, calculate the number of milligrams of each solute per milliliter of solution.

	(1), mL	(2), mL
(a)	0.00	18.15
(b)	21.00	28.15
(c)	19.80	39.61
(d)	18.04	18.03
(e)	16.00	37.37

Chapter 14

Complex-Formation Titrations

Complex-formation reagents are widely used to titrate cations. The most useful of these reagents are organic compounds with several electron-donor groups that form multiple covalent bonds with metal ions.

14A COMPLEX-FORMATION REACTIONS

Most metal ions react with electron-pair donors to form coordination compounds, or complex ions. The donor species, or *ligand,* must have at least one pair of unshared electrons available for bond formation. Water, ammonia, and halide ions are common inorganic ligands.

The number of covalent bonds a cation tends to form with electron donors is its *coordination number*. Typical values for coordination numbers are two, four, and six. The species formed as a result of coordination can be electrically positive, neutral, or negative. For example, copper(II), which has a coordination number of four, forms a cationic complex with ammonia, $Cu(NH_3)_4^{2+}$; a neutral complex with glycine, $Cu(NH_2CH_2COO)_2$; and an anionic complex with chloride ion, $CuCl_4^{2-}$.

Titrimetric methods based upon complex formation (sometimes called *complexometric methods*) have been used for more than a century. The truly remarkable growth in their analytical application began in the 1940s, however, and is based upon a particular class of coordination compounds called *chelates*. A chelate is produced when a metal ion coordinates with two (or more) donor groups of a single ligand to form a five- or six-

> A ligand is an ion or a molecule that forms a covalent bond with a cation by donating a pair of electrons, which are then shared by the ligand and the cation.

> The coordination number of a cation is the number of covalent bonds it tends to form with electron donor species.

> Complexometric methods are titrimetric methods that are based upon complex-formation reactions.

> A chelate is a cyclic complex formed when a cation is bonded by two or more donor groups contained in a single ligand.

235

membered heterocyclic ring. The copper complex of glycine, mentioned in the previous paragraph, is an example. Here, copper bonds to both the oxygen of the carboxyl group and the nitrogen of the amine group:

"Chelate" is pronounced kee′late.

$$Cu^{2+} + 2\ \underset{\substack{|\\H}}{\overset{\substack{NH_2\ O\\|\ \ \ \|}}{H-C-C-OH}} \longrightarrow \underset{\substack{|\\H_2C-N\\ \ \ \ H_2}}{\overset{\substack{O=C-O\\ \ \ \ \ }}{\diagdown}}Cu\underset{\substack{|\\N-CH_2\\H_2}}{\overset{\substack{O-Cu=O\\ \diagup}}{\diagup}} + 2\ H^+$$

A ligand that has a single donor group, such as ammonia, is called *unidentate* (single-toothed), whereas one such as glycine, which has two groups available for covalent bonding, is *bidentate. Tridentate, tetradentate, pentadentate,* and *hexadentate* chelating agents are also known.

As titrants, multidentate ligands, particularly those having four or six donor groups, have two advantages over their unidentate counterparts. First, multidentate ligands generally react more completely with cations and thus provide sharper end points. Second, they ordinarily react with metal ions in a single-step process, whereas complex formation with unidentate ligands usually involves two or more intermediate species.

The advantage of single-step reaction is illustrated by the titration curves shown in Figure 14-1. Each titration involves a reaction that has an overall equilibrium constant of 10^{20}. Curve *A* is derived for a reaction in which a metal ion M having a coordination number of four reacts with a tetradentate ligand D to form the complex MD (here we have omitted the charges on the two reactants for convenience). Curve *B* is for the reaction of M with a hypothetical bidentate ligand B to give MB_2 in two steps. The formation constant for the first step is 10^{12}, and that for the second is 10^{8}. Curve *C* involves a unidentate ligand A that forms MA_4 in four steps with successive formation constants of 10^{8}, 10^{6}, 10^{4}, and 10^{2}. These curves demonstrate that a much sharper end point is obtained with a reaction that takes place in a single step. For this reason, polydentate ligands are ordinarily preferred for complexometric titrations.

"Dentate" means having toothlike projections.

Tetradentate and hexadentate ligands are more satisfactory as titrants than ligands with a lesser number of donor groups for two reasons. First, they react more completely, with cations and second they tend to form 1 : 1 complexes.

Figure 14-1
Curves for complex-formation titrations. Titration of 60.0 mL of a solution that is 0.020 M in M with (*A*) a 0.020 M solution of the tetradentate ligand D to give MD as product, (*B*) a 0.040 M solution of the bidentate ligand B to give MB_2, and (*C*) a 0.080 M solution of the unidentate ligand A to give MA_4. The overall formation constant for each product is 1.0 × 10^{20}.

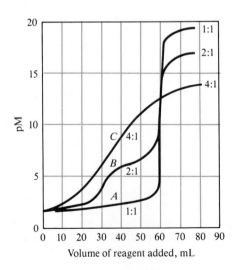

14B TITRATIONS WITH AMINOPOLYCARBOXYLIC ACIDS

Tertiary amines that also contain carboxylic acid groups form remarkably stable chelates with many metal ions.[1] Schwarzenbach first recognized their potential as analytical reagents in 1945. Since this original work, investigators throughout the world have described applications of these compounds to the volumetric determination of most of the metals in the periodic table.

14B-1 Ethylenediaminetetraacetic Acid (EDTA)

Ethylenediaminetetraacetic acid, also called (ethylenedinitrilo)tetraacetic acid and commonly shortened to EDTA, has the structure

EDTA stands for ethylenediaminetetraacetic acid.

$$\text{HOOC—CH}_2 \diagdown \qquad\qquad \diagup \text{CH}_2\text{—COOH}$$
$$:\text{N—CH}_2\text{—CH}_2\text{—N}:$$
$$\text{HOOC—CH}_2 \diagup \qquad\qquad \diagdown \text{CH}_2\text{—COOH}$$

The EDTA molecule has six potential sites for bonding with a metal ion: the four carboxyl groups and the two amino groups, each of the latter with an unshared pair of electrons. Thus, EDTA is a hexadentate ligand.

Acidic Properties of EDTA

The dissociation constants for the four carboxylic acid groups in EDTA are $K_1 = 1.02 \times 10^{-2}$, $K_2 = 2.14 \times 10^{-3}$, $K_3 = 6.92 \times 10^{-7}$, and $K_4 = 5.50 \times 10^{-11}$. It is of interest that the first two constants are of the same order of magnitude, which suggests that the two protons involved dissociate from opposite ends of the rather long molecule. As a consequence of the physical separation of these protons, the negative charge created by the first dissociation does not greatly inhibit the removal of the second proton. The same cannot be said for the dissociation of the other two protons, however, which are much closer to the negatively charged carboxylate ions created by the initial dissociations.

EDTA, a hexadentate ligand, is among the most important and widely used reagents in titrimetry.

The various EDTA species are often abbreviated H_4Y, H_3Y^-, H_2Y^{2-}, HY^{3-}, and Y^{4-}. Figure 14-2 illustrates how the relative amounts of these five species vary as a function of pH. It is apparent that H_2Y^{2-} predominates in a moderately acidic medium (pH 3 to 6). Only at pH values greater than 10 does Y^{4-} become a major component of EDTA solutions.

Reagents

The free acid of EDTA, H_4Y, and the dihydrate of the sodium salt, $Na_2H_2Y \cdot 2 H_2O$, are available in reagent quality. The former can serve as a primary standard after it has been dried for several hours at 130 to

---

[1]These reagents are the subject of several excellent monographs. See, for example, R. Pribil, *Applied Complexometry*. New York: Pergamon, 1982; A. Ringbom and E. Wanninen, in *Treatise on Analytical Chemistry,* 2nd ed., I. M. Kolthoff and P. J. Elving, Eds., Part I, Vol. 2, Chapter 11. New York: Wiley, 1979.

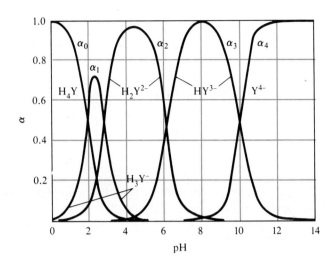

Figure 14-2

Composition of EDTA solutions as a function of pH.

145°C. It is then dissolved in the minimum amount of base required for complete solution.

Under normal atmospheric conditions, the dihydrate contains 0.3% moisture in excess of the stoichiometric amount. For all but the most exacting work, this excess is sufficiently reproducible to permit use of a corrected weight of the salt in the direct preparation of a standard solution. If necessary, the pure dihydrate can be prepared by drying at 80°C for several days in an atmosphere of 50% relative humidity.

A number of compounds related chemically to EDTA have also been investigated but do not appear to offer advantages over the original species. We shall thus confine our discussion to the properties and applications of EDTA.

Standard EDTA solutions are ordinarily prepared by dissolving weighed quantities of $Na_2H_2Y \cdot 2\ H_2O$ and diluting to the mark in a volumetric flask.

Nitrilotriacetic acid (NTA) is the second most common aminopolycarboxylic acid used for titrimetry. Its structure is

$$HOOC{-}CH_2 \qquad CH_2{-}COOH$$
$$\diagdown \qquad \diagup$$
$$N$$
$$\diagdown$$
$$CH_2{-}COOH$$

14B-2 Complexes of EDTA and Metal Ions

EDTA combines with metal ions in a 1 : 1 ratio regardless of the charge on the cation. For example, formation of the silver and aluminum complexes is described by the equations

$$Ag^+ + Y^{4-} \rightleftarrows AgY^{3-}$$
$$Al^{3+} + Y^{4-} \rightleftarrows AlY^-$$

The reaction of the EDTA anion with a metal ion M^{n+} is described by the equation

$$M^{n+} + Y^{4-} \rightleftarrows MY^{(n-4)+}$$

The reagent is remarkable not only because it forms chelates with all cations but also because most of these chelates are sufficiently stable to form the basis for a titrimetric method. This great stability undoubtedly results from the several complexing sites within the molecule that give rise to a cagelike structure in which the cation is effectively surrounded and isolated from solvent molecules. One form of the complex is depicted in Figure 14-3. Note that all six ligand groups are involved in bonding the divalent metal ion.

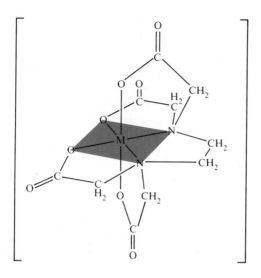

Figure 14-3
Structure of a metal/EDTA chelate.

Table 14-1 lists formation constants K_{MY} for common EDTA complexes. Note that the constant refers to the equilibrium involving the species Y^{4-} with the metal ion:

$$M^{n+} + Y^{4-} \rightleftarrows MY^{(n-4)+} \qquad \frac{[MY^{(n-4)+}]}{[M^{n+}][Y^{4-}]} = K_{MY} \qquad (14\text{-}1)$$

14B-3   Equilibrium Calculations Involving EDTA

A titration curve for the reaction of a cation with EDTA consists of a plot of pM as a function of reagent volume. Values for pM are readily computed in the early stage of a titration by assuming that the equilibrium concentration of M^{n+} is equal to its analytical concentration, which in turn is readily derived from stoichiometric data.

Table 14-1
FORMATION CONSTANTS FOR EDTA COMPLEXES*

Cation	K_{MY}	log K_{MY}	Cation	K_{MY}	log K_{MY}
Ag^+	2.1×10^7	7.32	Cu^{2+}	6.3×10^{18}	18.80
Mg^{2+}	4.9×10^8	8.69	Zn^{2+}	3.2×10^{16}	16.50
Ca^{2+}	5.0×10^{10}	10.70	Cd^{2+}	2.9×10^{16}	16.46
Sr^{2+}	4.3×10^8	8.63	Hg^{2+}	6.3×10^{21}	21.80
Ba^{2+}	5.8×10^7	7.76	Pb^{2+}	1.1×10^{18}	18.04
Mn^{2+}	6.2×10^{13}	13.79	Al^{3+}	1.3×10^{16}	16.13
Fe^{2+}	2.1×10^{14}	14.33	Fe^{3+}	1.3×10^{25}	25.1
Co^{2+}	2.0×10^{16}	16.31	V^{3+}	7.9×10^{25}	25.9
Ni^{2+}	4.2×10^{18}	18.62	Th^{4+}	1.6×10^{23}	23.2

*Data from G. Schwarzenbach, *Complexometric Titrations*, p. 8. London: Chapman and Hall, 1957. With permission. Constants valid at 20°C and an ionic strength of 0.1.

Calculation of M^{n+} at and beyond the equivalence point requires the use of Equation 14-1. The computation in this region is troublesome and time-consuming if the pH is unknown and variable because both $[MY^{(n-4)+}]$ and $[M^{n+}]$ are pH-dependent. Fortunately, EDTA titrations are always performed in solutions that are buffered to a known pH to avoid interference by other cations and to ensure satisfactory indicator behavior. Calculating $[M^{n+}]$ in a buffered solution containing EDTA is a relatively straightforward procedure provided the pH is known. In this computation, use is made of the alpha value for H_4Y. Recall (Section 12F) that α_4 for H_4Y is defined as

$[Y^{4-}] = \alpha_4 C_T$

$$\alpha_4 = \frac{[Y^{4-}]}{C_T} \qquad (14\text{-}2)$$

where C_T is the total molar concentration of *uncomplexed* EDTA:

$$C_T = [Y^{4-}] + [HY^{3-}] + [H_2Y^{2-}] + [H_3Y^-] + [H_4Y]$$

Conditional Formation Constants

Conditional, or *effective, formation constants* are pH-dependent equilibrium constants that apply *at a single pH only*. To obtain the conditional constant for the equilibrium shown in Equation 14-1, we substitute $\alpha_4 C_T$ from Equation 14-2 for $[Y^{4-}]$ in the formation-constant expression (Equation 14-1):

Conditional formation constants are pH-dependent.

$$M^{n+} + Y^{4-} \rightleftarrows MY^{(n-4)+} \qquad \frac{[MY^{(n-4)+}]}{[M^{n+}]\alpha_4 C_T} = K_{MY} \qquad (14\text{-}3)$$

Solving this equation for the product of the two constants K_{MY} and α_4 yields a new constant:

$$\frac{[MY^{(n-4)+}]}{[M^{n+}]C_T} = \alpha_4 K_{MY} = K'_{MY} \qquad (14\text{-}4)$$

where K'_{MY}, the conditional formation constant, describes equilibrium relationships *only at the pH for which α_4 is applicable.*

Conditional constants are readily computed and provide a simple means by which the equilibrium concentrations of the metal ion and the complex can be calculated at any point in a titration curve. Note that replacement of $[Y^{4-}]$ with C_T in the equilibrium-constant expression greatly simplifies calculations because C_T is readily determined from the reaction stoichiometry, whereas $[Y^{4-}]$ is not.

The Computation of α_4 Values for EDTA Solutions

An expression for calculating α_4 at a given hydrogen ion concentration is readily derived by the method demonstrated in Section 12F. Thus, α_4 for EDTA is

$$\alpha_4 = \frac{K_1 K_2 K_3 K_4}{[H^+]^4 + K_1[H^+]^3 + K_1 K_2[H^+]^2 + K_1 K_2 K_3[H^+] + K_1 K_2 K_3 K_4} \quad (14\text{-}5)$$

$$\alpha_4 = \frac{K_1 K_2 K_3 K_4}{D} \quad (14\text{-}6)$$

where K_1, K_2, K_3, and K_4 are the four dissociation constants for H_4Y and D is the denominator of Equation 14-5.

Table 14-2 lists α_4 at selected pH values. Note that only about $4 \times 10^{-12}\%$ of the EDTA exists as Y^{4-} at pH 2.00. Example 14-1 illustrates how $[Y^{4-}]$ is calculated for a solution of known pH.

Alpha values for the other EDTA species are readily obtained in a similar way and are found to be

$$\alpha_0 = [H^+]^4/D$$
$$\alpha_1 = K_1[H^+]^3/D$$
$$\alpha_2 = K_1 K_2[H^+]^2/D$$
$$\alpha_3 = K_1 K_2 K_3[H^+]/D$$

Only α_4 is needed, however, in deriving titration curves.

In this chapter and those that follow, we shall revert to the use of H^+ as a convenient shorthand notation for H_3O^+; from time to time, we shall refer to the species represented by H^+ as the hydrogen ion.

Example 14-1

Calculate the molar Y^{4-} concentration in a 0.0200 M EDTA solution that has been buffered to a pH of 10.00.

At pH 10.00, α_4 is 0.35 (Table 14-2). Thus,

$$[Y^{4-}] = \alpha_4 C_T = (0.35)(0.0200) = 7.0 \times 10^{-3} \text{ M}$$

The Calculation of the Cation Concentrations in EDTA Solutions
Example 14-2 demonstrates how the cation concentration in a solution of an EDTA complex is computed. Example 14-3 illustrates this calculation for a solution that contains an excess of EDTA.

Example 14-2

Calculate the equilibrium concentration of Ni^{2+} in a solution with an analytical NiY^{2-} concentration of 0.0150 M at pH (a) 3.0 and (b) 8.0.

From Table 14-1,

$$Ni^{2+} + Y^{4-} \rightleftarrows NiY^{2-} \qquad \frac{[NiY^{2-}]}{[Ni^{2+}][Y^{4-}]} = K_{MY} = 4.2 \times 10^{18}$$

The equilibrium concentration of NiY^{2-} is equal to the analytical concentration of the complex minus the concentration lost by dissociation. The latter is given by the equilibrium nickel ion concentration. Thus,

$$[NiY^{2-}] = 0.0150 - [Ni^{2+}]$$

If we assume $[Ni^{2+}] \ll 0.0150$, an assumption that is almost certainly valid in light of the large formation constant of the complex, the foregoing equation simplifies to

$$[NiY^{4-}] \approx 0.0150$$

Assume:

$$[NiY^{2-}] \approx C_{Ni} = 0.0150 \gg [Ni^{2+}]$$

Table 14-2

VALUES FOR α_4 FOR EDTA AT
SELECTED pH VALUES

pH	α_4	pH	α_4
2.0	3.7×10^{-14}	7.0	4.8×10^{-4}
3.0	2.5×10^{-11}	8.0	5.4×10^{-3}
4.0	3.6×10^{-9}	9.0	5.2×10^{-2}
5.0	3.5×10^{-7}	10.0	3.5×10^{-1}
6.0	2.2×10^{-5}	11.0	8.5×10^{-1}
		12.0	9.8×10^{-1}

Since the complex is the only source of both Ni^{2+} and the EDTA species,

$$[Ni^{2+}] = [Y^{4-}] + [HY^{3-}] + [H_2Y^{2-}] + [H_3Y^-] + [H_4Y] = C_T$$

Substitution of this equality into Equation 14-4 gives

$$\frac{[NiY^{2-}]}{[Ni^{2+}]C_T} = \frac{[NiY^{2-}]}{[Ni^{2+}]^2} = \alpha_4 K_{MY} = K'_{MY}$$

(a) Table 14-2 indicates that α_4 is 2.5×10^{-11} at pH 3.0. Substitution of this value and the concentration of NiY^{2-} into the equation for K'_{MY} gives

$$\frac{0.0150}{[Ni^{2+}]^2} = 2.5 \times 10^{-11} \times 4.2 \times 10^{18} = 1.05 \times 10^8$$

$$[Ni^{2+}] = \sqrt{1.43 \times 10^{-10}} = 1.2 \times 10^{-5}\ M$$

Our assumption is valid.

Note that $[Ni^{2+}] \ll 0.0150$, as assumed.

(b) At pH 8.0, the conditional constant is much larger. Thus,

$$K'_{MY} = 5.4 \times 10^{-3} \times 4.2 \times 10^{18} = 2.27 \times 10^{16}$$

and substitution into the equation for K'_{MY} followed by rearrangement gives

$$[Ni^{2+}] = \sqrt{0.0150/(2.27 \times 10^{16})} = 8.1 \times 10^{-10}\ M$$

Example 14-3

Calculate the concentration of Ni^{2+} in a solution prepared by mixing 50.0 mL of 0.0300 M Ni^{2+} with 50.0 mL of 0.0500 M EDTA. The mixture is buffered to a pH of 3.00.

Here, the solution has an excess of EDTA, and so the analytical concentration of the complex is determined by the amount of Ni^{2+} originally present. Thus,

$$C_{NiY^{2-}} = 50.0 \times \frac{0.03000}{100} = 0.0150\ M$$

$$C_{EDTA} = \frac{50.0 \times 0.0500 - 50.0 \times 0.0300}{100} = 0.0100\ M$$

Again let us assume that $[Ni^{2+}] \ll [NiY^{2-}]$, so that

$$[NiY^{2-}] = 0.0150 - [Ni^{2+}] \approx 0.0150$$

At this point, the total concentration of uncomplexed EDTA species is:

$$C_T = 0.0100\ M$$

Substitution into Equation 14-4 gives

$$\frac{0.0150}{[Ni^{2+}]0.0100} = \alpha_4 K_{MY} = K'_{MY}$$

$$= 2.5 \times 10^{-11} \times 4.2 \times 10^{18} = 1.05 \times 10^8$$

$$[Ni^{2+}] = \frac{0.0150}{0.0100 \times 1.05 \times 10^8} = 1.4 \times 10^{-8} \text{ M}$$

Again our assumption is valid.

Computer Application 14-1
USING EUREKA TO CALCULATE ALPHAS

Eureka can be used to make repetitive calculations and plot the results on the screen of your computer. You can either list the results on the screen or print them out. To use Eureka in this way, you must express equations in functional form. For example, the equation $y = x^2$ should be written as $f(x) = x^2$.

Let us use Eureka to check the α values shown in Table 14-2. These data were calculated by substitution of $[H^+]$ values into Equation 14-5, which is reproduced here.

$$\alpha_4 = \frac{K_1 K_2 K_3 K_4}{[H^+]^4 + K_1[H^+]^3 + K_1 K_2[H^+]^2 + K_1 K_2 K_3[H^+] + K_1 K_2 K_3 K_4}$$

However, we want to enter our data as pH rather than $[H^+]$. Since pH = $-\log[H^+]$, $[H^+]$ = antilog($-$pH) = 10^{-pH}. We then write alpha in the above equation (Equation 14-5) as a function of x in which x = pH and $[H^+] = 10^{-pH} = 10^{-x}$. Substituting these relationships into the equation for alpha gives

$$\alpha_4(x) =$$

$$\frac{K_1 K_2 K_3 K_4}{(10^{-x})^4 + K_1(10^{-x})^3 + K_1 K_2(10^{-x})^2 + K_1 K_2 K_3(10^{-x}) + K_1 K_2 K_3 K_4}$$

Note that we have used the Greek alpha in this equation. Eureka permits use of the IBM extended character set so that some Greek letters and other special characters can be shown on the screen. To produce α on the screen, simultaneously depress [Shift] and [Alt] and type 224 on the numeric keypad of your computer. When you release [Shift] and [Alt], α should appear on the screen.

To calculate values for α, run Eureka, depress [E] to edit and [F5] to expand the edit window to full screen. Enter the following equations.

```
;
;Dissociation constants for EDTA
;
K1 = 1.02 e − 2
```

```
K2 = 2.14 e - 3
K3 = 6.92 e - 7
K4 = 5.50 e - 11
;
;Overall constants for each step
;
B1 = K1
B2 = K1 * K2
B3 = K1 * K2 * K3
B4 = K1 * K2 * K3 * K4


;
;Alpha values
;
α4(x) = B4/(10^(-x)^4 + B1*10^(-x)^3 + B2*10^(-x)^2 +
    B3*10^(-x) + B4)
;
```

Notice that for convenience, we have represented K_1, K_1K_2, $K_1K_2K_3$, and $K_1K_2K_3K_4$ as B1, B2, B3, and B4, respectively.

After you have typed the program, type **ESC** **S** so that Eureka will assign numerical values to B1–B4. Then type **ESC** , **G** for graph, and **L** to indicate that you wish to produce a list of values for the function ($\alpha_4(x)$ in this case). Eureka will then prompt you to specify the starting value of x, the increment between values of x, and the number of values of x for which to calculate α. Type **2** **Return** , **1** **Return** , **11** **Return** . These entries tell Eureka to calculate the eleven values of α4(x) for values of x from 2 to 12. Eureka then produces the list of values shown on the screen below. Check these values against the values listed in Table 14-2. You can page through the values by typing **PgUp** or **PgDn** . You can also expand the list window to fill the screen by typing **F5** as before.

```
       C:LIST.         Line 1    Col 1    Insert Indent Tab

List of function values.
α4

        x            α4(x)
    2.0000000     3.7116360e-14
    3.0000000     2.5142119e-11
    4.0000000     3.6106388e-09
    5.0000000     3.5441647e-07
    6.0000000     .000022487373
    7.0000000     .00048032190
    8.0000000     .0053924163
    9.0000000     .052061390
    10.000000     .35480563
    11.000000     .84615196
    12.000000     .98214283
```

You can plot $\alpha 4(x)$ vs. x on the screen by typing **ESC** **G** **P** (for plot). Eureka will then prompt you to specify the first and last values of x to be plotted. Type **2** **Return** **12** **Return**. Again, you may expand the plot to fill the whole screen by typing **F5**.

Edit the program to calculate values of $\alpha 2(x)$. List the values as a function of x and plot them on the screen. If you have a printer, print the list and the plot. Repeat this exercise for all α values of EDTA.

14B-4 The Derivation of EDTA Titration Curves

Example 14-4 demonstrates how an EDTA titration curve for a metal ion is derived for a solution of fixed pH.

___

Example 14-4

Derive a curve (pCa as a function of EDTA volume) for the titration of 50.0 mL of 0.00500 M Ca^{2+} with 0.0100 M EDTA in a solution buffered to pH 10.0.

Calculation of a Conditional Constant

The conditional formation constant for the calcium/EDTA complex at pH 10 is obtained from the formation constant of the complex (Table 14-1) and the α_4 value for EDTA at pH 10 (Table 14-2). Thus, substitution into Equation 14-4 gives

$$\frac{[CaY^{2-}]}{[Ca^{2+}]C_T} = \alpha_4 K_{CaY} = K'_{CaY}$$

$$= 0.35 \times 5.0 \times 10^{10} = 1.75 \times 10^{10}$$

Preequivalence-Point Values for pCa

Before the equivalence point is reached, the equilibrium concentration of Ca^{2+} is equal to the sum of the contributions from the untitrated excess of the cation and from the dissociation of the complex, the latter being numerically equal to C_T. It is ordinarily reasonable to assume that C_T is small relative to the analytical concentration of the uncomplexed calcium ion. Thus, for example, after the addition of 10.0 mL of reagent,

$$[Ca^{2+}] = \frac{50.0 \times 0.00500 - 10.0 \times 0.0100}{60.0} + C_T \approx 2.50 \times 10^{-3}\ M$$

$$pCa = -\log 2.50 \times 10^{-3} = 2.60$$

Other preequivalence-point data are derived in this same way.

The Equivalence-Point pCa

Following the method shown in Example 14-2, we first compute the analytical concentration of CaY^{2-}:

$$C_{CaY^{2-}} = \frac{50.0 \times 0.00500}{50.0 + 25.0} = 3.33 \times 10^{-3}\ M$$

The only source of Ca^{2+} ions is the dissociation of this complex. It also follows that the calcium ion concentration must be identical to the sum of the concentrations of the uncomplexed EDTA, C_T. Thus,

$$[Ca^{2+}] = C_T$$

$$[CaY^{2-}] = 0.00333 - [Ca^{2+}] \approx 0.00333 \text{ M}$$

Substituting into the conditional formation-constant expression gives

$$\frac{0.00333}{[Ca^{2+}]^2} = 1.75 \times 10^{10}$$

$$[Ca^{2+}] = \sqrt{\frac{0.00333}{1.75 \times 10^{10}}} = 4.36 \times 10^{-7} \text{ M}$$

$$pCa = -\log 4.36 \times 10^{-7} = 6.36$$

Postequivalence-Point pCa

Beyond the equivalence point, analytical concentrations of CaY^{2-} and EDTA are obtained directly from the stoichiometric data. Then a calculation similar to that in Example 14-3 is performed. Thus, after the addition of 35.0 mL of reagent,

$$C_{CaY^{2-}} = \frac{50.0 \times 0.00500}{85.0} = 2.94 \times 10^{-3} \text{ M}$$

$$C_{EDTA} = \frac{35.0 \times 0.0100 - 50.0 \times 0.00500}{85.0} = 1.18 \times 10^{-3} \text{ M}$$

As an approximation, we can write

$$[CaY^{2-}] = 2.94 \times 10^{-3} - [Ca^{2+}] \approx 2.94 \times 10^{-3}$$

$$C_T = 1.18 \times 10^{-3} + [Ca^{2+}] \approx 1.18 \times 10^{-3} \text{ M}$$

and substitution into the conditional formation-constant expression gives

$$\frac{2.94 \times 10^{-3}}{[Ca^{2+}] \times 1.18 \times 10^{-3}} = 1.75 \times 10^{10} = K'_{CaY}$$

$$[Ca^{2+}] = \frac{2.94 \times 10^{-3}}{1.18 \times 10^{-3} \times 1.75 \times 10^{10}} = 1.42 \times 10^{-10}$$

$$pCa = -\log 1.42 \times 10^{-10} = 9.85$$

The approximation was clearly valid.

Other postequivalence-point data are computed in this same way.

Curve A in Figure 14-4 is a plot of data for the titration in Example 14-4. Curve B is the titration curve for a solution of magnesium ion under identical conditions. The formation constant for the EDTA complex of magnesium is smaller than that for the calcium complex. Consequently,

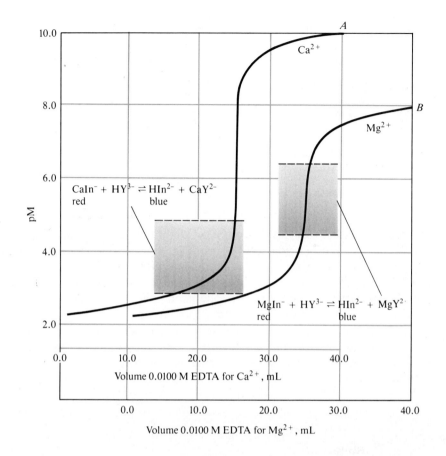

Figure 14-4

EDTA titration curves for 50.00 mL of 0.00500 M Ca^{2+} (K' for CaY^{2-} = 1.75 × 10^{10}) and Mg^{2+} (K' for MgY^{2-} = 1.72 × 10^{8}) at pH 10.0. The shaded areas show the transition range for Eriochrome Black T.

the reaction of calcium ion with the EDTA is more complete, and a larger change in p-function occurs in the equivalence region. The effect here is analogous to the effects seen earlier for precipitation and neutralization titrations.

Figure 14-5 shows titration curves for calcium ion in solutions buffered to various pH levels. Recall that α_4, and hence K'_{CaY}, become smaller as

End points for EDTA titrations become less sharp with pH decreases because complex formation is less complete under these circumstances.

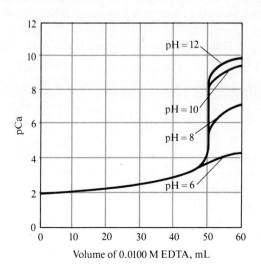

Figure 14-5

Influence of pH on the titration of 0.0100 M Ca^{2+} with 0.0100 M EDTA.

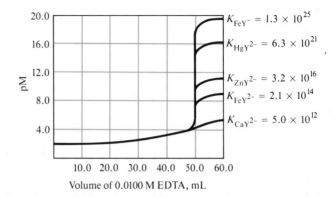

Figure 14-6
Titration curves for 50.0 mL of 0.0100 M cation solutions at pH 6.0.

the pH decreases. The less favorable equilibrium constant leads to a smaller change in pCa in the equivalence-point region. It is apparent from Figure 14-5 that an adequate end point in the titration of calcium requires a pH of about 8 or greater. As shown in Figure 14-6, however, cations with larger formation constants provide good end points even in acidic media. Figure 14-7 shows the minimum permissible pH for a satisfactory end point in the titration of various metal ions in the absence of competing complexing agents. Note that a moderately acidic environment is satisfactory for many divalent heavy-metal cations and that a strongly acidic medium can be tolerated in the titration of such ions as iron(III) and indium(III).

Figure 14-7 shows that most cations having a +3 or +4 charge can be titrated in a distinctly acidic solution.

14B-5 The Effect of Other Complexing Agents on EDTA Titration Curves

Many cations form hydrous oxide precipitates when the pH is raised to the level required for their titration with EDTA. When this problem is encountered, an auxiliary complexing agent is needed to keep the cation in solution. For example, zinc(II) is ordinarily titrated in a medium that has fairly high concentrations of ammonia and ammonium chloride. These species buffer the solution to a pH that ensures complete reaction between cation and titrant; in addition, ammonia forms ammine complexes with zinc(II) and prevents formation of the sparingly soluble zinc hydroxide, particularly in the early stages of the titration. A realistic description of the reaction is then

Often, auxiliary complexing agents must be used in EDTA titrations to prevent precipitation of the analyte as a hydrous oxide. Such reagents cause end points to be less sharp.

$$Zn(NH_3)_4^{2+} + HY^{3-} \rightarrow ZnY^{2-} + 3\ NH_3 + NH_4^+$$

The solution also contains such other zinc/ammonia species as $Zn(NH_3)_3^{2+}$, $Zn(NH_3)_2^{2+}$, and $Zn(NH_3)^{2+}$. Calculation of pZn in a solution that contains ammonia must take these species into account.[2] Qualitatively, complexation of a cation by an auxiliary complexing reagent causes preequivalence pM values to be larger than in a comparable solution with no such reagent.

[2]See, for example, D. A. Skoog, D. M. West, and F. J. Holler, *Fundamentals of Analytical Chemistry*, 5th ed., pp. 269–271. Philadelphia: Saunders College Publishing, 1988.

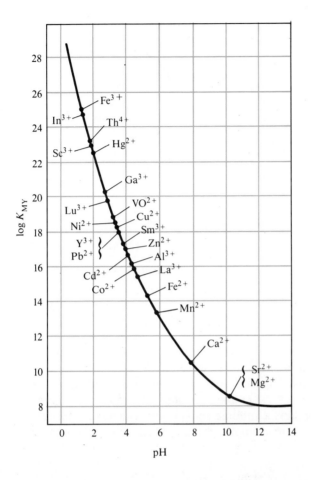

Figure 14-7

Minimum pH needed for satisfactory titration of various cations with EDTA. (From C. N. Reilley and R. W. Schmid, *Anal. Chem.,* **1958,** *30,* 947. With permission of the American Chemical Society.)

Figure 14-8 shows two theoretical curves for the titration of zinc(II) with EDTA at pH 9.00. The equilibrium concentration of ammonia was 0.100 M for one titration and 0.0100 M for the other. Note that the presence of ammonia decreases the change in pZn near the equivalence point. For this reason, the concentration of auxiliary complexing reagents should always be kept to the minimum required to prevent analyte precipitation. Note that the auxiliary complexing agent does not affect pZn beyond the equivalence point. Keep in mind, on the other hand, that α_4, and thus pH, play an important role in defining this part of the titration curve (Figure 14-5).

14B-6 Indicators for EDTA Titrations

Reilley and Barnard[3] have listed nearly 200 organic compounds that have been suggested as indicators for metal ions in EDTA titrations. In general, these indicators are organic dyes that form colored chelates with metal ions in a pM range that is characteristic of the particular cation and dye.

[3]C. N. Reilley and A. J. Barnard Jr., in *Handbook of Analytical Chemistry,* L. Meites, Ed., p. **3**-77. New York: McGraw-Hill, 1963.

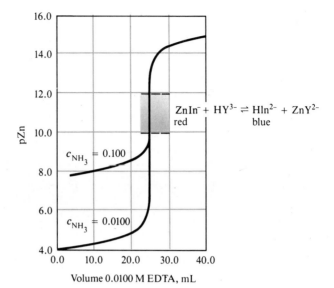

Figure 14-8

Influence of ammonia concentration on the end point for the titration of 50.0 mL of 0.00500 M Zn^{2+}. Solutions buffered to pH 9.00. The shaded area shows the transition range for Eriochrome Black T.

The complexes are often intensely colored and are discernible to the eye at concentrations in the range of 10^{-6} to 10^{-7} M.

Eriochrome Black T is a typical metal ion indicator that is used in the titration of several common cations. As shown in Figure 14-9, this compound contains a sulfonic acid group, which is completely dissociated in water, and two phenolic groups, which dissociate only partially. Its behavior as a weak acid is described by the equations

$$H_2O + H_2In^- \rightleftharpoons HIn^{2-} + H_3O^+ \qquad K_1 = 5 \times 10^{-7}$$

Red   Blue

$$H_2O + HIn^{2-} \rightleftharpoons In^{3-} + H_3O^+ \qquad K_2 = 2.8 \times 10^{-12}$$

Blue   Orange

Note that the acids and their conjugate bases have different colors. Thus, Eriochrome Black T behaves as an acid/base indicator as well as a metal ion indicator.

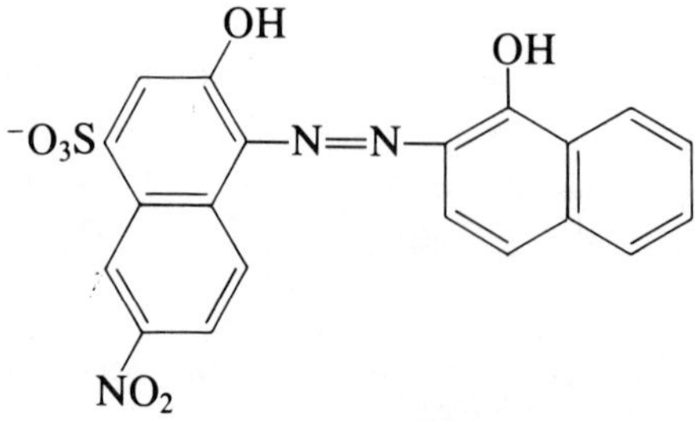

Figure 14-9

Eriochrome Black T.

The metal complexes of Eriochrome Black T are generally red, as is H_2In^-. Thus, for metal ion detection, it is necessary to adjust the pH to 7 or above so that the blue form of the species, HIn^{2-}, predominates in the absence of a metal ion. Until the equivalence point in a titration, the indicator complexes the excess metal ion, and so the solution is red. When EDTA becomes present in slight excess, the solution turns blue as a consequence of the reaction

$$MIn^- + HY^{3-} \rightleftarrows HIn^{2-} + MY^{2-}$$

Red   Blue

Eriochrome Black T forms red complexes with more than two dozen metal ions, but the formation constants of only a few are appropriate for end-point detection. Transition ranges for magnesium and calcium are indicated on the titration curves in Figure 14-4. The indicator is clearly ideal for magnesium titration but totally unsatisfactory for calcium. Eriochrome Black T is also well suited for the titration of zinc with EDTA (Figure 14-8).

A limitation of Eriochrome Black T is that its solutions decompose slowly upon standing; refrigeration slows this process. It is claimed that solutions of Calmagite, an indicator that for all practical purposes is identical in behavior to Eriochrome Black T, do not suffer this disadvantage.

Many other metal indicators have been developed for EDTA titrations.[4] In contrast to Eriochrome Black T, some of these indicators are suitable for use in strongly acidic media.

14B-7 Types of EDTA Titrations

The paragraphs that follow describe several ways in which EDTA is employed for the determination of cations.

Direct-Titration Methods

Approximately 40 cations can be determined by direct titration with a standard EDTA solution using a metal ion indicator. Often, cations can be determined by direct titration even when no satisfactory indicator for them is available. For example, calcium ion indicators are not nearly as satisfactory as those for magnesium ion. Calcium can be determined by direct titration, however, if a small amount of magnesium chloride is introduced into the EDTA solution. The titration is then carried out with Eriochrome Black T indicator. During the early stages of the titration, calcium ions displace magnesium ions from the EDTA because the calcium complex is considerably more stable than the magnesium complex. After the calcium ions have been fully complexed, the EDTA then recombines with magnesium ions, ultimately giving an Eriochrome Black T end point. To use this approach, it is necessary to standardize the magnesium-containing EDTA against primary-standard calcium carbonate.

> Direct-titration procedures with a metal ion indicator are the easiest and the most convenient to use.

---

[4]See, for example, L. Meites, *Handbook of Analytical Chemistry,* pp. **3**-101 to **3**-165. New York: McGraw-Hill, 1963.

Direct titration is also used to determine those metal ions for which specific ion electrodes are available. Electrodes of this type are described in Section 18D.

Back-Titration Methods

Back-titration procedures are used when no suitable indicator is available, when the reaction between analyte and EDTA is slow, or when the analyte forms precipitates at the pH required for its titration.

Back-titration procedures are useful when a satisfactory indicator for a cation is unavailable or when a cation reacts only slowly with EDTA. A measured excess of standard EDTA solution is added to the analyte solution. After the reaction is judged complete, the excess EDTA is back-titrated with a standard magnesium or zinc ion solution to an Eriochrome Black T or Calmagite end point.[5] For this procedure to be successful, it is necessary that the magnesium or zinc ions form an EDTA complex that is less stable than the corresponding analyte complex.

Back-titration is also useful for analyzing samples that contain anions that would otherwise form sparingly soluble precipitates with the analyte under the conditions needed for satisfactory complexation. Here, the excess EDTA prevents precipitate formation.

Displacement Methods

In a displacement titration, an unmeasured excess of a solution containing the magnesium or zinc complex of EDTA is introduced into the analyte solution. If the analyte forms a more stable complex than that of magnesium or zinc, the following displacement reaction occurs:

$$MgY^{2-} + M^{2+} \rightarrow MY^{2-} + Mg^{2+}$$

where M^{2+} represents the analyte cation. The liberated magnesium or zinc is then titrated with a standard EDTA solution.

Displacement titrations are used when no indicator for an analyte is available.

Displacement methods are useful where no satisfactory indicator is available for the metal ion being determined.

14B-8 The Scope of EDTA Titrations

Complexometric titrations with EDTA have been applied to the determination of virtually every metal cation with the exception of the alkali metal ions. Because EDTA complexes most cations, the reagent might appear at first glance to be totally lacking in selectivity. In fact, however, considerable control over interferences can be realized by pH regulation. For example, trivalent cations can usually be titrated without interference from divalent species by maintaining the solution at a pH of about 1 (Figure 14-7). At this pH, the less stable divalent chelates do not form to any significant extent, but the trivalent ions are quantitatively complexed.

Similarly, ions such as cadmium and zinc, which form more stable EDTA chelates than does magnesium, can be determined in the presence of the latter ion by buffering the mixture to pH 7 before titration. Erio-

---

[5]For an analysis of the back-titration procedure, see C. Macca and M. Fiorana, *J. Chem. Educ.*, **1986,** *63,* 121.

chrome Black T serves as an indicator for the cadmium or zinc end point without interference from magnesium because the indicator chelate with magnesium is not formed at this pH.

Finally, interference from a particular cation can sometimes be eliminated by adding a suitable *masking agent,* an auxiliary ligand that preferentially forms highly stable complexes with the potential interference.[6] For example, cyanide ion is often employed as a masking agent to permit the titration of magnesium and calcium ions in the presence of ions such as cadmium, cobalt, copper, nickel, zinc, and palladium. All of the latter form sufficiently stable cyanide complexes to prevent reaction with EDTA.

> A masking agent is a complexing agent that reacts selectively with a component in a solution and in so doing prevents that component from interfering in an analysis.

14B-9   The Determination of Water Hardness

Historically, water "hardness" was defined in terms of the capacity of cations in the water to replace the sodium or potassium ions in soaps and form sparingly soluble products. Most multiply charged cations share this undesirable property. In natural waters, however, the concentrations of calcium and magnesium ions generally far exceed that of any other metal ion. Consequently, hardness is now expressed in terms of the concentration of calcium carbonate equivalent to the total concentration of all multivalent cations in the sample.

> Hard water contains calcium, magnesium, and heavy-metal ions that form precipitates with soap (but not detergents).

The determination of hardness is a useful analytical test that provides a measure of water quality for household and industrial uses. The test is important to industry because hard water, upon being heated, precipitates calcium carbonate, which then clogs boilers and pipes.

Water hardness is ordinarily determined by an EDTA titration after the sample has been buffered to pH 10. Magnesium, which forms the least stable EDTA complex of all the common multivalent cations in typical water samples, is not titrated until enough reagent has been added to complex all the other cations in the sample. Therefore, a magnesium ion indicator, such as Calmagite or Eriochrome Black T, can serve as indicator in water-hardness titrations. Often, a small amount of the magnesium/EDTA complex is incorporated into the buffer or into the titrant to ensure the presence of sufficient magnesium ions for satisfactory indicator action.

Test kits for determining the hardness of household water are available commercially. They consist of a vessel calibrated to contain a known volume of water, a measuring scoop to deliver an appropriate amount of a solid buffer mixture, an indicator solution, and a bottle of standard EDTA, which is equipped with a medicine dropper. The drops of standard reagent needed to cause a color change are counted. The concentration of the EDTA solution is ordinarily such that one drop corresponds to one grain (about 0.065 g) of calcium carbonate per gallon of water.

---

[6]For further information, see D. D. Perrin, *Masking and Demasking of Chemical Reactions.* New York: Wiley-Interscience, 1970; and C. N. Reilley and A. J. Barnard Jr., in *Handbook of Analytical Chemistry,* L. Meites, Ed., pp. 3-208 to 3-225. New York: McGraw-Hill, 1963.

Table 14-3

TYPICAL INORGANIC COMPLEX-FORMATION TITRATIONS*

Titrant	Analyte	Remarks
$Hg(NO_3)_2$	Br^-, Cl^-, SCN^-, CN^-, thiourea	Products are neutral mercury(II) complexes; various indicators used
$AgNO_3$	CN^-	Product is $Ag(CN)_2^-$; indicator is I^-; titrate to first turbidity of AgI
$NiSO_4$	CN^-	Product is $Ni(CN)_4^{2-}$; indicator is AgI; titrate to first turbidity of AgI
KCN	Cu^{2+}, Hg^{2+}, Ni^{2+}	Products are $Cu(CN)_4^{2-}$, $Hg(CN)_2$, $Ni(CN)_4^{2-}$; various indicators used

*For further applications and selected references, see L. Meites, *Handbook of Analytical Chemistry*, pp. **3**-226. New York: McGraw-Hill, 1963.

14C   TITRATIONS WITH INORGANIC COMPLEXING AGENTS

Table 14-3 lists the common inorganic complexing agents as well as some of their applications.[7]

14D   QUESTIONS AND PROBLEMS

14-1. Define
  *(a) chelate.
   (b) tetradentate chelating agent.
  *(c) ligand.
   (d) coordination number.
  *(e) conditional formation constant.
   (f) NTA.
  *(g) water hardness.
   (h) EDTA displacement titration.
*14-2. Describe three general methods for performing EDTA titrations. What are the advantages of each?
14-3. Propose a complexometric method for the determination of the individual components in a solution containing In^{3+}, Zn^{2+}, and Mg^{2+}.
*14-4. Why are multidentate ligands preferable to unidentate ligands for complexometric titrations?
14-5. Write chemical equations and equilibrium-constant expressions for the stepwise formation of
  *(a) $Ag(S_2O_3)_2^{3-}$.   *(c) $Cd(NH_3)_4^{2+}$.
   (b) AlF_6^{3-}.   (d) $Ni(SCN)_3^-$.
14-6. Explain how stepwise and overall formation constants are related.
14-7. Why is a small amount of MgY^{2-} often added to a solution that is to be titrated for hardness?
*14-8. An EDTA solution was prepared by dissolving 3.853 g of purified and dried $Na_2H_2Y \cdot 2H_2O$ in suffi-

cient water to give 1.000 L. Calculate the molar concentration, given that the solute contained 0.3% excess moisture (p. 238).
14-9. An EDTA solution was prepared by dissolving about 3.0 g of $Na_2H_2Y \cdot 2H_2O$ in approximately 1 L of water. Calculate the molar concentration of this solution if an average of 32.22 mL was required to titrate 50.00-mL aliquots of 0.004517 M Mg^{2+}.
*14-10. Calculate the molar concentration of an EDTA solution from the accompanying titration data:

	EDTA Soln	0.01470 M Mg^{2+} Soln
Final volume, mL	46.39	31.69
Initial volume, mL	0.04	0.00

14-11. A 50.00-mL aliquot of a solution containing iron(II) and iron(III) required 13.73 mL of 0.01200 M EDTA when titrated at pH 2.0 and 29.62 mL when titrated at pH 6.0. Express the concentration of the solution in terms of the parts per million of each solute.
*14-12. A 24-h urine specimen was diluted to 2.000 L. After the solution was buffered to pH 10, a 10.00-mL aliquot was titrated with 26.81 mL of 0.003474 M EDTA. The calcium in a second 10.00-mL aliquot

---

[7]For further information, see I. M. Kolthoff and V. A. Stenger, *Volumetric Analysis*, Vol. 2, pp. 282–331. New York: Interscience, 1947.

was isolated as $CaC_2O_4(s)$, redissolved in acid, and titrated with 11.63 mL of the EDTA solution. Assuming that 15 to 300 mg of magnesium and 50 to 400 mg of calcium per day are normal, did this specimen fall within these ranges?

14-13. An EDTA solution was prepared by dissolving approximately 4 g of the disodium salt in approximately 1 L of water. An average of 42.35 mL of this solution was required to titrate 50.00-mL aliquots of a standard that contained 0.7682 g of $MgCO_3$ per liter. Titration of a 25.00-mL sample of mineral water at pH 10 required 18.81 mL of the EDTA solution. A 50.00-mL aliquot of the mineral water was rendered strongly alkaline to precipitate the magnesium as $Mg(OH)_2$. Titration with a calcium-specific indicator required 31.54 mL of the EDTA solution. Calculate
(a) the molarity of the EDTA solution.
(b) the ppm of $CaCO_3$ in the mineral water.
(c) the ppm of $MgCO_3$ in the mineral water.

*14-14. The sulfate in a 1.515-g sample was homogeneously precipitated as $BaSO_4$ by adding an excess of BaY^{2-} solution and slowly increasing the acid concentration to liberate Ba^{2+}. The precipitate was filtered and washed, and the filtrate and washings were collected in a 250-mL volumetric flask. A pH-10.00 buffer was added and the solution diluted to the mark. A 25.00-mL aliquot required a 28.73-mL titration with 0.01545 M Mg^{2+} solution. Express the results of this analysis in terms of percent $Na_2SO_4 \cdot 10H_2O$ (fw = 322.2 g).

14-15. Calamine, which is used for relief of skin irritations, is a mixture of zinc and iron oxides. A 1.022-g sample of dried calamine was dissolved in acid and diluted to 250.0 mL. Potassium fluoride was added to a 10.00-mL aliquot of the diluted solution to mask the iron; after suitable adjustment of the pH, the Zn^{2+} consumed 38.78 mL of 0.01294 M EDTA. A second 50.00-mL aliquot was suitably buffered and titrated with 2.40 mL of 0.002727 M ZnY^{2-} solution:

$$Fe^{3+} + ZnY^{2-} \rightarrow FeY^- + Zn^{2+}$$

Calculate the percentages of ZnO and Fe_2O_3 in the sample.

*14-16. A 3.650-g sample containing bromate and bromide was dissolved in sufficient water to give 250.0 mL. After acidification, silver nitrate was introduced to a 25.00-mL aliquot to precipitate AgBr, which was filtered, washed, and then redissolved in an ammoniacal solution of potassium tetracyanonickelate(II):

$$Ni(CN)_4^{2-} + 2\,AgBr(s) \rightarrow$$
$$2\,Ag(CN)_2^- + Ni^{2+} + 2\,Br^-$$

The liberated nickel ion required 26.73 mL of 0.02089 M EDTA. The bromate in a 10.00-mL aliquot was reduced to bromide with arsenic(III) prior to the addition of silver nitrate. The same procedure was followed, and the released nickel ion required 21.94 mL of the EDTA solution. Calculate the percentages of NaBr and $NaBrO_3$ in the sample.

14-17. The potassium ion in a 250.0-mL sample of mineral water was precipitated with sodium tetraphenylboron:

$$K^+ + B(C_6H_4)_4^- \rightarrow KB(C_6H_5)_4(s)$$

The precipitate was filtered, washed, and redissolved in an organic solvent. An excess of the mercury(II)/EDTA chelate was added:

$$4\,HgY^{2-} + B(C_6H_4)_4^- + 4\,H_2O \rightarrow$$
$$H_3BO_3 + 4\,C_6H_5Hg^+ + 4\,HY^{3-} + OH^-$$

The liberated EDTA was titrated with 29.64 mL of 0.05581 M Mg^{2+}. Calculate the potassium ion concentration in parts per million.

*14-18. A 0.4085-g sample containing lead, magnesium, and zinc was dissolved and treated with cyanide to complex and mask the zinc:

$$Zn^{2+} + 4\,CN^- \rightarrow Zn(CN)_4^{2-}$$

Titration of the lead and magnesium required 42.22 mL of 0.02064 M EDTA. The lead was next masked with BAL (2,3-dimercaptopropanol), and the released EDTA was titrated with 19.35 mL of a 0.007657 M magnesium solution. Finally, formaldehyde was introduced to demask the zinc:

$$Zn(CN)_4^{2-} + 4\,HCHO + 4\,H_2O \rightarrow$$
$$Zn^{2+} + 4\,HOCH_2CN + 4\,OH^-$$

The zinc was then titrated with 28.63 mL of 0.02064 M EDTA. Calculate the percentages of the three metals in the sample.

14-19. Chromel is an alloy composed of nickel, iron, and chromium. A 0.6472-g sample was dissolved and diluted to 250.0 mL. When a 50.00-mL aliquot of 0.05180 M EDTA was mixed with an equal volume of the diluted sample, all three ions were chelated, and a 5.11-mL back-titration with 0.06241 M copper(II) was required. The chromium in a second 50.0-mL aliquot was masked through the addition of hexamethylenetetramine; titration of the Fe and Ni required 36.28 mL of 0.05182 M EDTA. Iron and chromium were masked with pyrophosphate in a third 50.0-mL aliquot, and the nickel was titrated with 25.91 mL of the EDTA solution. Calculate the percentages of nickel, chromium, and iron in the alloy.

*14-20. A 0.3284-g sample of brass (containing lead, zinc, copper, and tin) was dissolved in nitric acid. The sparingly soluble $SnO_2 \cdot 4\,H_2O$ was removed by filtration, and the combined filtrate and washings were then diluted to 500.0 mL. A 10.00-mL aliquot

was suitably buffered; titration of the lead, zinc, and copper in this aliquot required 37.56 mL of 0.002500 M EDTA. The copper in a 25.00-mL aliquot was masked with thiosulfate; the lead and zinc were then titrated with 27.67 mL of the EDTA solution. Cyanide ion was used to mask the copper and zinc in a 100-mL aliquot; lead ion was then titrated with 10.80 mL of the EDTA solution. Determine the composition of the brass sample; evaluate the percentage of tin by difference.

*14-21. Calculate conditional constants for the formation of the EDTA complex of Mn^{2+} at pH

(a) 6.0.   (b) 8.0.   (c) 10.0.

14-22. Calculate conditional constants for the formation of the EDTA complex of Sr^{2+} at pH

(a) 7.0.   (b) 9.0.   (c) 11.0.

*14-23. Derive a titration curve for 50.00 mL of 0.01000 M Sr^{2+} with 0.02000 M EDTA in a solution buffered to pH 11.0. Calculate pSr values after the addition of 0.00, 10.00, 24.00, 24.90, 25.00, 25.10, 26.00, and 30.00 mL of titrant.

14-24. Derive a titration curve for 50.00 mL of 0.0150 M Fe^{2+} with 0.0300 M EDTA in a solution buffered to pH 7.0. Calculate pFe values after the addition of 0.00, 10.00, 24.00, 24.90, 25.00, 25.10, 26.00, and 30.00 mL of titrant.

Chapter 15

An Introduction to Electrochemistry

We now turn our attention to several analytical methods that are based upon oxidation/reduction reactions. These methods, which are described in Chapters 16 through 19, include oxidation/reduction titrimetry, potentiometry, coulometry, and electrogravimetry. This chapter deals with certain fundamentals of electroanalytical chemistry that are necessary for understanding the theory of these methods.

15A OXIDATION/REDUCTION REACTIONS

An oxidation/reduction reaction is one in which electrons are transferred from one reactant to another. An example is the oxidation of iron(II) by cerium(IV). The reaction is described by the equation

$$Ce^{4+} + Fe^{2+} \rightleftarrows Ce^{3+} + Fe^{3+} \qquad (15\text{-}1)$$

Here, Ce^{4+} ion extracts an electron from Fe^{2+} to form Ce^{3+} and Fe^{3+} ions. A substance that, like Ce^{4+}, has a strong affinity for electrons and thus tends to remove them from other species is called an *oxidizing agent* or an *oxidant*. A *reducing agent*, or *reductant*, is a reagent that, like Fe^{2+}, readily donates electrons to another species. With regard to Equation 15-1, we say that Fe^{2+} is *oxidized* by Ce^{4+}, and Ce^{4+} is *reduced* by Fe^{2+}.

Oxidation/reduction reactions are sometimes called redox reactions.

A reducing agent is an electron donor. An oxidizing agent is an electron acceptor.

257

We can split any oxidation/reduction equation into two *half-reactions* that show clearly which species gains electrons and which loses them. For example, Equation 15-1 is made up of the two half-reactions

$$Ce^{4+} + e^- \rightleftarrows Ce^{3+} \qquad \text{(reduction of } Ce^{4+}\text{)}$$

$$Fe^{2+} \rightleftarrows Fe^{3+} + e^- \qquad \text{(oxidation of } Fe^{2+}\text{)}$$

> It is important to understand that a half-reaction is a theoretical concept. A half-reaction *cannot* occur alone in a solution. It is always accompanied by a second half-reaction.

The rules for balancing half-reactions are the same as those for balancing ordinary reactions; that is, the number of atoms of each element as well as the net charge must be the same on the two sides of the equation. Thus, for the oxidation of Fe^{2+} by MnO_4^-, the half-reactions are

$$MnO_4^- + 5\ e^- + 8\ H^+ \rightleftarrows Mn^{2+} + 4\ H_2O$$

$$5\ Fe^{2+} \rightleftarrows 5\ Fe^{3+} + 5\ e^-$$

In the first half-reaction, the net charge on the left side is $(-1 - 5 + 8) = +2$, which is the same as that on the right. Note also that we have multiplied the Fe^{2+}/Fe^{3+} half-reaction by 5 so that the number of electrons gained by MnO_4^- equals the number lost by Fe^{2+}. A balanced net ionic equation for the overall reaction is then obtained by adding the two half-reactions:

$$MnO_4^- + 5\ Fe^{2+} + 8\ H^+ \rightleftarrows Mn^{2+} + 5\ Fe^{3+} + 4\ H_2O$$

15A-1 Comparison of Oxidation/Reduction Reactions with Acid/Base Reactions

Oxidation/reduction reactions can be viewed in a way that is analogous to the Brønsted-Lowry concept of acid/base reactions (Section 7A-2). Both involve the transfer of one or more charged particles from a donor to an acceptor—the particles being electrons in oxidation/reduction and protons in neutralization. When an acid donates a proton, that acid then becomes a conjugate base capable of accepting a proton. By analogy, when a reducing agent donates an electron, that reducing agent becomes an oxidizing agent that can accept an electron. This product could be called a conjugate oxidant (but seldom, if ever, is). We may then write a generalized equation for a redox reaction as

> Recall that in the Brønsted-Lowry concept an acid/base reaction is described by the equation
>
> $$acid_1 + base_2 \rightleftarrows base_1 + acid_2$$

$$Red_1 + Ox_2 \rightleftarrows Ox_1 + Red_2 \qquad (15\text{-}2)$$

Here, oxidant Ox_2 accepts electrons from Red_1 to form the new reductant, Red_2. At the same time, reductant Red_1, having given up electrons, becomes an oxidizing agent, Ox_1. If we know from chemical evidence that the equilibrium in Equation 15-2 lies to the right, we can state that Ox_2 is a better electron acceptor (stronger oxidant) than Ox_1. Furthermore, Red_1 is a more effective electron donor (better reductant) than Red_2.

Feature 15-1
BALANCING REDOX EQUATIONS

A knowledge of how to balance oxidation/reduction reactions is essential to understanding all the concepts covered in this chapter. Although you probably remember this technique from your general chemistry course, we present a quick review here to remind you of how the process works. As practice, let us complete and balance the following equation after adding H^+, OH^-, or H_2O as needed:

$$MnO_4^- + NO_2^- \rightleftharpoons Mn^{2+} + NO_3^-$$

First we must write and balance the two half-reactions involved. For MnO_4^- we write

$$MnO_4^- \rightleftharpoons Mn^{2+}$$

To take up the four oxygens on the left, we add eight H^+ to give

$$MnO_4^- + 8\ H^+ \rightleftharpoons Mn^{2+} + 4\ H_2O$$

To balance the charge, we need to add five electrons to the left side:

$$MnO_4^- + 8\ H^+ + 5\ e^- \rightleftharpoons Mn^{2+} + 4\ H_2O$$

For the nitrite half-reaction, we add one H_2O to the left side to supply the needed oxygen:

$$NO_2^- + H_2O \rightleftharpoons NO_3^- + 2\ H^+$$

Then we add two electrons to the right side to balance the charge:

$$NO_2^- + H_2O \rightleftharpoons NO_3^- + 2\ H^+ + 2\ e^-$$

Before combining the two equations, we must multiply the first by 2 and the second by 5 so that the electrons will cancel. We then add the two equations and obtain

$$2\ MnO_4^- + 16\ H^+ + \cancel{10\ e^-} + 5\ NO_2^- + 5\ H_2O \rightleftharpoons$$
$$2\ Mn^{2+} + 8\ H_2O + 5\ NO_3^- + 10\ H^+ + \cancel{10\ e^-}$$

which rearranges to the balanced equation

$$2\ MnO_4^- + 6\ H^+ + 5\ NO_2^- \rightleftharpoons 2\ Mn^{2+} + 5\ NO_3^- + 3\ H_2O$$

Example 15-1

As written, the following reactions proceed to the right:

$$Sn^{4+} + H_2(g) \rightleftarrows Sn^{2+} + 2\,H^+$$

$$2\,H^+ + Sn(s) \rightleftarrows H_2(g) + Sn^{2+}$$

$$2\,Fe^{3+} + Sn^{2+} \rightleftarrows 2\,Fe^{2+} + Sn^{4+}$$

What can we deduce regarding the strengths of Sn^{4+}, Sn^{2+}, H^+, and Fe^{3+} as electron acceptors, or oxidizing agents?

The first reaction shows that Sn^{4+} is a more effective electron acceptor than H^+ because it accepts electrons from H_2 to give H^+. The second reaction shows that H^+ is more effective than Sn^{2+} in accepting electrons, and the third reaction reveals that Fe^{3+} is more effective than Sn^{4+}. Thus, the order of oxidizing strength is $Fe^{3+} > Sn^{4+} > H^+ > Sn^{2+}$.

15A-2 Oxidation/Reduction Reactions in Electrochemical Cells

Many oxidation/reduction reactions can be carried out in two ways that are physically quite different. For example, if we immerse a strip of copper in a solution containing silver nitrate, silver ions migrate to the metal and are reduced:

$$Ag^+ + e^- \rightarrow Ag(s)$$

At the same time, an equivalent quantity of copper is oxidized:

$$Cu(s) \rightleftarrows Cu^{2+} + 2\,e^-$$

Multiplication of the silver half-reaction by 2 followed by addition of the half-reactions yields a net ionic equation for the overall process:

$$2\,Ag^+ + Cu(s) \rightleftarrows 2\,Ag(s) + Cu^{2+} \tag{15-3}$$

A unique aspect of oxidation/reduction reactions is that the transfer of electrons—and thus the identical net reaction—can often be brought about in an *electrochemical cell* in which the oxidizing agent and the reducing agent are physically separated from one another. Figure 15-1a shows such an arrangement. Note that a *salt bridge* isolates the reactants but maintains electrical contact between the two halves of the cell. The bridge is necessary to prevent Ag^+ from reacting directly with the copper metal. An external metallic conductor connects the two metal electrodes. In this cell, metallic copper is oxidized, silver ions are reduced, and electrons flow through the external circuit to the silver electrode. The voltmeter measures the potential difference between the two metals at any instant and is a measure of the tendency of the cell reaction to proceed toward equilibrium. As the reaction proceeds, this tendency, and thus the potential, decrease continuously and approach zero as the equilibrium state for the overall reaction is approached.

When the $CuSO_4$ and $AgNO_3$ solutions are 1 M, the cell develops a potential of 0.462 V as shown in Figure 15-1(a).

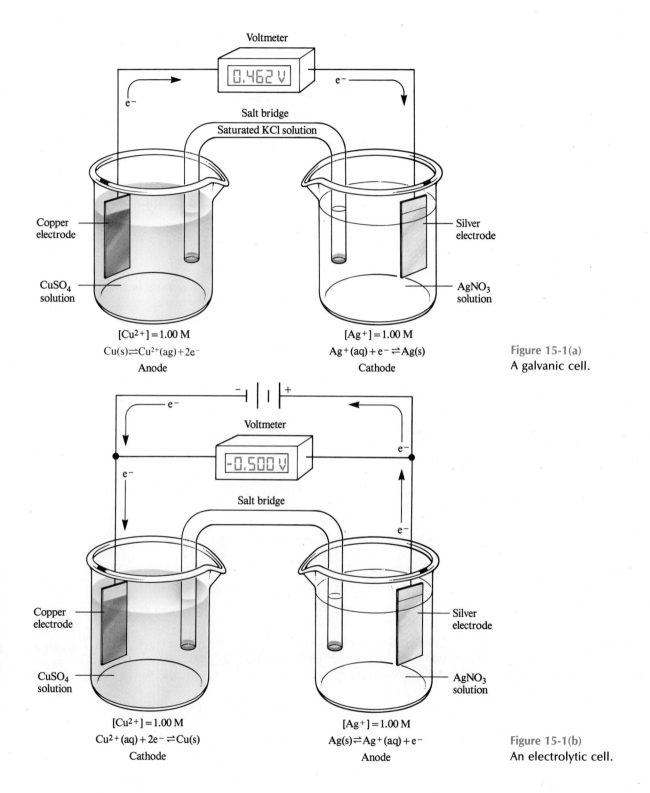

Figure 15-1(a)
A galvanic cell.

Figure 15-1(b)
An electrolytic cell.

The equilibrium-constant expression for the reaction shown in Figure 15-1 is

$$K_{eq} = \frac{[Cu^{2+}]}{[Ag^+]^2} = 4.1 \times 10^{15} \quad (15\text{-}4)$$

This expression applies regardless of whether the reaction occurs directly between reactants or within an electrochemical cell.

When zero voltage is reached, the concentrations of Cu(II) and Ag(I) ions will have values that satisfy the equilibrium-constant expression in Equation 15-4 (margin). At this point, no further net flow of electrons will occur. *It is important to realize that the overall reaction and its equilibrium position are totally independent of the way the reaction is carried out,* whether by direct reaction in a solution or by indirect action in an electrochemical cell.

15B   ELECTROCHEMICAL CELLS

Any oxidation/reduction equilibrium is conveniently studied by measuring the potential of an electrochemical cell in which the two half-reactions making up the equilibrium are participants. For this reason, we need to consider some of the characteristics of cells.

An electrochemical cell consists of two conductors called *electrodes,* each of which is immersed in an electrolyte solution. In most of the cells of interest to us, the solutions surrounding the two electrodes are different and must be separated to prevent direct reaction between the reactants. The most common way of avoiding mixing is to insert a salt bridge, such as that shown in Figure 15-1a, between the solutions. Conduction of electricity from one electrolyte solution to the other then occurs by migration of potassium ions from the bridge in one direction and chloride ions in the other. In this way, direct contact between copper metal and silver ions is prevented.

> A salt bridge is a conducting solution that is used to separate the two compartments of an electrochemical cell.

> In some cells, known as cells without liquid junction, the electrodes share a common electrolyte.

15B-1 Cathodes and Anodes

The *cathode* in an electrochemical cell is the electrode at which a reduction reaction occurs. The *anode* is the electrode at which an oxidation takes place.

Examples of typical cathodic reactions are

Reduction occurs at the cathode. Oxidation occurs at the anode.

$$Ag^+ + e^- \rightleftarrows Ag(s)$$
$$Fe^{3+} + e^- \rightleftarrows Fe^{2+}$$
$$NO_3^- + 10\ H^+ + 8\ e^- \rightleftarrows NH_4^+ + 3\ H_2O$$

The first reaction takes place at the surface of a silver electrode. The other two occur at an inert electrode, such as platinum.

Typical anodic reactions are

The reaction

$$2\ H^+ + 2\ e^- \rightleftarrows H_2(g)$$

occurs at a cathode when a solution contains no more easily reducible species.

$$Cu(s) \rightleftarrows Cu^{2+} + 2\ e^-$$
$$2\ Cl^- \rightleftarrows Cl_2(g) + 2\ e^-$$
$$Fe^{2+} \rightleftarrows Fe^{3+} + e^-$$

The reaction

$$2\ H_2O \rightleftarrows O_2(g) + 4\ H^+ + 4\ e^-$$

occurs at an anode when a solution contains no more easily oxidizible species.

The Cu/Cu^{2+} reaction requires a copper anode, but the other two reactions can be carried out at the surface of an inert platinum electrode.

15B-2 Types of Electrochemical Cells

Electrochemical cells are either galvanic or electrolytic. They can also be classified as reversible or irreversible.

Galvanic, or *voltaic,* cells are batteries that store electrical energy. The reactions at the two electrodes in such cells tend to proceed spontaneously and produce a flow of electrons from anode to cathode via an external conductor. The cell shown in Figure 15-1a is a galvanic cell that develops a potential of about 0.46 V when electrons move through the external circuit from the copper anode to the silver cathode.

An *electrolytic* cell, in contrast, requires an external source of electrical energy for operation. As shown in Figure 15-1b, the cell in Figure 15-1a can be operated as an electrolytic cell by inserting a 0.5 V dc source into the circuit. The silver electrode is connected to the positive terminal of the source and becomes the anode of the cell. The copper electrode is connected to the negative terminal and is now the cathode. The cell reaction of the electrolytic cell is the reverse of that of the galvanic cell. That is,

$$2 \, Ag(s) + Cu^{2+} \rightleftarrows 2 \, Ag^+ + Cu(s) \qquad (15\text{-}5)$$

The cell in Figure 15-1 is an example of a *reversible cell,* that is, one in which the direction of the electrochemical reaction is reversed when the direction of electron flow is changed. In an *irreversible* cell, changing the direction of current causes entirely different half-reactions to occur at one or both electrodes.

> The Fe^{2+} half-reaction may seem somewhat unusual because a cation rather than an anion migrates to the electrode and gives up an electron. As we shall see, oxidation of a cation at an anode or reduction of an anion at a cathode is not uncommon.

> Galvanic cells store electricity; electrolytic cells consume electricity.

For both galvanic and electrolytic cells, remember that

1. Reduction always takes place at the cathode.
2. Oxidation always takes place at the anode.

> In a reversible cell, reversing the current reverses the cell reaction. In an irreversible cell, reversing the current causes a different half-reaction to occur at one or both electrodes.

Feature 15-2
THE DANIELL GRAVITY CELL

The Daniell gravity cell was one of the earliest galvanic cells that found widespread practical application. It was used in the mid-1800s as a battery to power telegraphic communication systems. As shown in Figure 15-2, the cathode was a piece of copper immersed in a saturated solution of copper sulfate. A much less dense solution of dilute zinc sulfate was layered on top of the copper sulfate, and a massive zinc electrode was located in this solution. The two half-cells were separated only by the boundary between the two solutions of different densities. The electrode reactions are

$$Zn(s) \rightleftarrows Zn^{2+} + 2 \, e^- \qquad \text{anode}$$
$$Cu^{2+} + 2 \, e^- \rightleftarrows Cu(s) \qquad \text{cathode}$$

This cell develops an initial voltage of 1.18 V. In common with all galvanic cells, this potential gradually decreases with use.

Allessandro Volta (1745–1827), Italian physicist, was the inventor of the first battery, the so-called voltaic pile. It consisted of alternating disks of copper and zinc separated by disks of cardboard soaked with salt solution. In honor of his many contributions to electrical science, the unit of potential difference, the volt, is named for Volta.

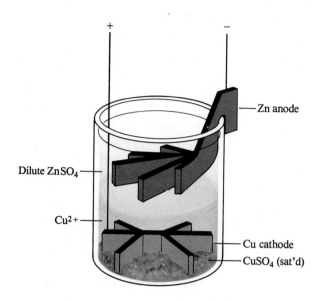

Figure 15-2
A Daniell gravity cell.

15B-3 Currents in Electrochemical Cells

As shown in Figure 15-3, electricity is transported through an electrochemical cell by three mechanisms:

1. Electrons carry electricity within the electrodes as well as through the external conductor.
2. Anions and cations carry electricity within the cell. As shown in Figure 15-3, copper and silver ions move toward the silver cathode, whereas anions, such as sulfate, nitrate, and hydrogen sulfate ions, are attracted to the copper anode. Within the salt bridge, chloride ions migrate toward and into the copper compartment, whereas potassium ions move in the opposite direction.
3. The ionic conduction of the solution is coupled to the electronic conduction in the electrodes by the reduction reaction at the cathode and the oxidation reaction at the anode.

In a cell, electricity is carried by movement of anions toward the anode and cations toward the cathode.

The phase boundary between an electrode and its solution is called an interface.

15C ELECTRODE POTENTIALS

The potential difference that develops between the cathode and the anode of the cell in Figure 15-1a is a measure of the tendency for the reaction

$$2 \, Ag^+ + Cu(s) \rightleftarrows 2 \, Ag(s) + Cu^{2+}$$

to proceed from a nonequilibrium state to an equilibrium state. Thus, as shown in Figure 15-4a, when the copper and silver ion activities in the cell are 1.00 M, a potential of 0.462 V develops, which shows that this reaction is far from equilibrium. As the reaction proceeds, this potential becomes smaller and smaller until at equilibrium the meter reads 0.000 V. As shown in Figure 15-4b the copper ion equilibrium concentration is then just slightly less than 1.500 M and the silver ion concentration is 1.9×10^{-8} M.

To be able to calculate the cell potential exactly, it is necessary to know the *activities* of the reactants rather than the molar concentrations.

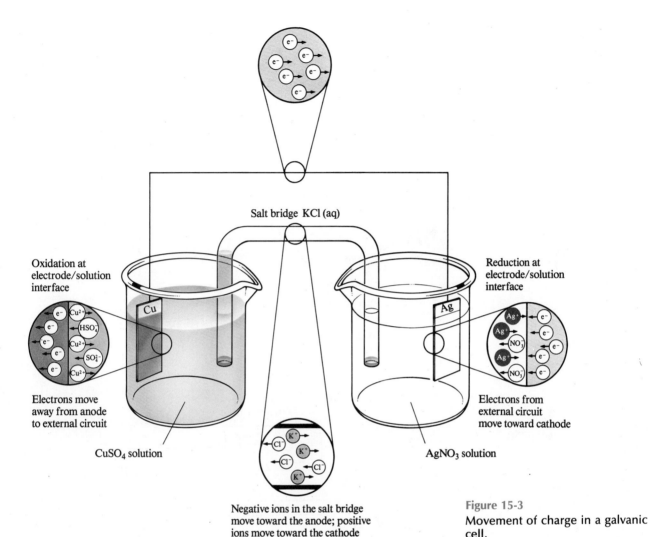

Oxidation at
electrode/solution
interface

Electrons move
away from anode
to external circuit

$CuSO_4$ solution

Salt bridge KCl (aq)

Reduction at
electrode/solution
interface

Electrons from
external circuit
move toward cathode

$AgNO_3$ solution

Negative ions in the salt bridge
move toward the anode; positive
ions move toward the cathode

Figure 15-3
Movement of charge in a galvanic
cell.

The potential of cells such as those shown in Figure 15-4 is the difference between two *half-cell,* or *single-electrode, potentials,* one associated with the half-reaction at the silver cathode ($E_{cathode}$) and the other
associated with the half-reaction at the copper anode (E_{anode}).

Although we cannot determine absolute potentials of electrodes such as
these (see Feature 15-3), we can readily determine *relative* electrode potentials. For example, if we replace the copper anode in Figure 15-1a with
a cadmium electrode immersed in a cadmium sulfate solution, the voltmeter reads about 0.7 V greater than with the original cell. Since the cathode
compartment remains unaltered, we conclude that the half-cell potential
for the oxidation of cadmium is about 0.7 V greater than that for copper
(that is, cadmium is a stronger reductant than is copper). The tendency for
other ions to donate electrons can also be compared by substituting other
anode half-cells while keeping the cathode half-cell unchanged.

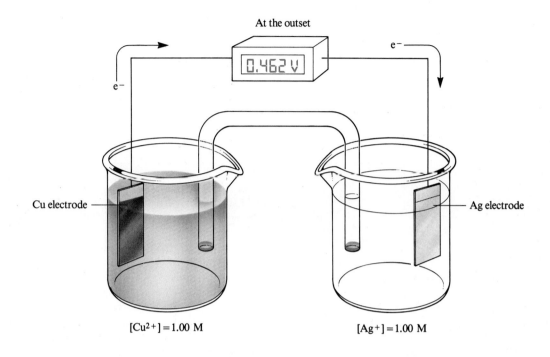

Figure 15-4
Change in cell potential after passage of current.

Feature 15-3
WHY ABSOLUTE ELECTRODE POTENTIALS CANNOT BE MEASURED

Although it is not difficult to measure *relative* half-cell potentials, it is impossible to determine absolute half-cell potentials because all voltage-measuring devices measure only *differences* in potential.

To measure the potential of an electrode, one contact of a voltmeter is connected to the electrode in question. The other contact from the meter must then be brought into electrical contact, via another conductor, with the solution in the electrode compartment. This second contact, however, inevitably involves a solid/solution interface that acts as a second half-cell at which chemical change *must occur* if charge is to flow and the potential is to be measured. A potential is associated with this second reaction. Thus what is obtained is not an absolute half-cell potential but rather the difference between the half-cell potential of interest and the potential of the half-cell made up of the second contact and the solution.

Our inability to measure absolute half-cell potentials presents no obstacle because relative half-cell potentials are just as useful, provided they are all measured against a common reference half-cell. Relative potentials can be combined to give cell potentials. We can also use them to calculate equilibrium constants and generate titration curves.

15C-1 The Standard Hydrogen Reference Electrode

For relative electrode potential data to be widely applicable and useful, we must have a generally agreed-upon reference half-cell against which all others are compared. Such an electrode must be easy to construct, be reversible, and be highly reproducible in its behavior. The *standard hydrogen electrode* (SHE) meets these specifications and has been used throughout the world for many years as a universal reference electrode. It is a typical *gas electrode*.

Figure 15-5 shows how a hydrogen electrode is constructed. The metal conductor is a piece of platinum that has been coated, or *platinized,* with finely divided platinum (*platinum black*) to increase its specific surface area. This electrode is immersed in an aqueous acidic solution of known, constant hydrogen ion activity. The solution is kept saturated with hydrogen by bubbling hydrogen at constant pressure over the surface of the electrode. The platinum does not take part in the electrochemical reaction and serves only as the site where electrons are transferred. The half-reaction responsible for the potential that develops at this electrode is

$$2\ H^+(aq) + 2\ e^- \rightleftarrows H_2(g) \tag{15-6}$$

The hydrogen electrode is reversible and acts either as an anode or as a cathode, depending upon the half-cell with which it is coupled. Hydrogen is oxidized to hydrogen ion when this electrode serves as the anode.

Platinum black catalyzes the reaction shown in Equation 15-6. Remember that catalysts do not change the position of equilibrium but simply shorten the time needed to reach equilibrium.

The reaction shown as Equation 15-6 involves two equilibria:

$$2\ H^+ + 2\ e^- \rightleftarrows H_2(aq)$$
$$H_2(aq) \rightleftarrows H_2(g)$$

The continuous stream of gas at constant pressure provides the solution with a constant molecular hydrogen concentration.

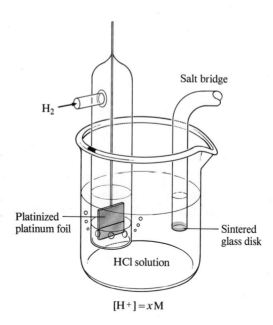

Salt bridge

H_2

Platinized platinum foil

Sintered glass disk

HCl solution

$[H^+] = x M$

Figure 15-5
A hydrogen gas electrode.

Hydrogen ion is reduced to molecular hydrogen when the hydrogen electrode acts as a cathode.

The potential of a hydrogen electrode depends upon temperature and upon the activities of hydrogen ion and molecular hydrogen in the solution. The latter activity, in turn, is proportional to the pressure of the gas used to keep the solution saturated with hydrogen. For the *standard* hydrogen electrode, the activity of hydrogen ions is specified as unity and the partial pressure of the gas is specified as one atmosphere. *By convention, the potential of the standard hydrogen electrode is assigned a value of zero at all temperatures.* As a consequence of this definition, any potential developed in a galvanic cell consisting of a standard hydrogen electrode and some other electrode is attributed entirely to the other electrode.

Several other reference electrodes that are more convenient for routine measurements have been developed. Some of these are described in Section 18B.

> At $p_{H_2} = 1.00$ atm and $a_{H^+} = 1.00$, the potential of the hydrogen electrode is 0.000 V at all temperatures.

15C-2 The Definition of Electrode Potential

An *electrode potential* is defined as the potential of a cell consisting of the electrode in question *acting as a cathode* and the standard hydrogen electrode *acting as an anode*. It should be emphasized that, despite its name, *an electrode potential is in fact the potential of an electrochemical cell involving a carefully defined reference electrode*. It could be more properly called a "relative electrode potential" (but seldom is).

> An electrode potential is the potential of a cell in which a standard hydrogen electrode is the anode.

The cell in Figure 15-6 illustrates the definition of electrode potential for the half-reaction

$$Ag^+ + e^- \rightleftarrows Ag(s)$$

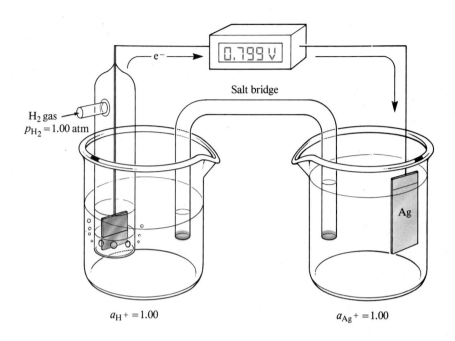

Figure 15-6
Definition of the standard electrode potential for $Ag^+ + e^- \rightarrow$ $Ag(s)$.

Here, the half-cell on the right consists of a strip of pure silver in contact with a solution that has a silver ion activity of 1.00; the electrode on the left is the standard hydrogen electrode. This galvanic cell develops a potential of 0.799 V with the silver electrode functioning *as the cathode;* that is, the spontaneous cell reaction is

$$2 \, Ag^+ + H_2(g) \rightleftarrows 2 \, Ag(s) + 2 \, H^+$$

Because the silver electrode serves as the cathode, the measured potential is, *by definition,* the electrode potential for the silver half-reaction (or the silver *couple*). Note that the silver electrode is positive with respect to the hydrogen anode (that is, electrons flow from the negative hydrogen anode to the silver cathode). Therefore, the electrode potential is given a positive sign, and we write

$$Ag^+ + e^- \rightleftarrows Ag(s) \qquad E_{Ag^+} = +0.799 \text{ V}$$

> A half-cell is sometimes called a couple.

Figure 15-7 illustrates the definition of the electrode potential for the half-reaction

$$Cd^{2+} + 2 \, e^- \rightleftarrows Cd(s)$$

In contrast to the silver electrode, the cadmium electrode acts as the anode of the galvanic cell. That is, the spontaneous cell reaction is

$$Cd(s) + 2 \, H^+ \rightleftarrows H_2(g) + Cd^{2+}$$

Here, the cadmium electrode bears a negative charge with respect to the standard hydrogen electrode. In order to reverse this reaction so that the

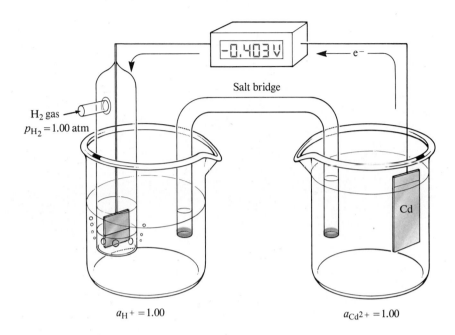

Figure 15-7

Measurement of the standard electrode potential for $Cd^{2+} + 2 e^- \rightarrow Cd(s)$.

cadmium electrode acts as a cathode, a potential more negative than -0.403 V must be applied to the cell. Consequently, the electrode potential of the Cd/Cd^{2+} couple is *by convention* given a negative sign and is equal to -0.403 V.

A zinc electrode immersed in a solution with a zinc ion activity of unity develops a potential of -0.763 V when paired with a standard hydrogen electrode. Because the zinc electrode also behaves as an anode in the galvanic cell, its electrode potential is also negative.

The electrode potentials for the four half-cells just described can be arranged in the order

Half-Reaction	Electrode Potential, V
$Ag^+ + e^- \rightleftarrows Ag(s)$	$+0.799$
$2 H^+ + 2 e^- \rightleftarrows H_2(g)$	0.000
$Cd^{2+} + 2 e^- \rightleftarrows Cd(s)$	-0.403
$Zn^{2+} + 2 e^- \rightleftarrows Zn(s)$	-0.763

The magnitudes of these electrode potentials indicate the relative strength of the four ionic species as electron acceptors (oxidizing agents); that is, in decreasing strength, $Ag^+ > H^+ > Cd^{2+} > Zn^{2+}$.

15C-3 The Sign Convention for Electrode Potentials

Historically, electrochemists have not always used the sign convention just described. Indeed, disagreements regarding the conventions to be used in specifying signs for half-cell processes caused much controversy and confusion in the development of electrochemistry. The International Union of Pure and Applied Chemistry (IUPAC) addressed itself to this problem at its 1953 meeting in Stockholm. The usages adopted at that meeting are collectively referred to as either the *Stockholm Convention* or the *IUPAC Convention* and are now generally accepted. The sign convention described in the previous section and in the paragraphs that follow is based upon the IUPAC recommendations.

Any sign convention must be based upon expressing half-cell processes in a single way—that is, either as oxidations or as reductions. According to the IUPAC convention, the term "electrode potential" (or, more exactly, "relative electrode potential") *is reserved exclusively to describe half-reactions written as reductions.* There is no objection to the use of the term "oxidation potential" to indicate a process written in the opposite sense, but it is not proper to refer to such a potential as an electrode potential.

The sign of an electrode potential is determined by the sign of the half-cell in question when it is coupled to a standard hydrogen electrode. When the half-cell of interest behaves spontaneously as a cathode, it is the positive electrode of the galvanic cell. Thus, its electrode potential is positive. When the half-cell of interest behaves as an anode, the electrode is negative and so is its electrode potential.

It is important to emphasize that the electrode potential refers to a half-cell process written *as a reduction.* For the zinc and cadmium electrodes we have been considering, the spontaneous reactions are oxidations. *It is evident, then, that the sign of an electrode potential indicates whether the reduction is spontaneous with respect to the standard hydrogen electrode.* The positive sign associated with the electrode potential for silver indicates that the process

> An electrode potential is, by definition, a reduction potential. An oxidation potential is the potential for the half-reaction written in the opposite way. The sign of an oxidation potential is therefore opposite that for a reduction potential, but the magnitude is the same.

$$2\ Ag^+ + H_2(g) \rightleftarrows 2\ Ag(s) + 2\ H^+$$

favors the products under ordinary conditions. Similarly, the negative sign of the electrode potential for zinc means that the analogous reaction

> The IUPAC sign convention is based upon the sign of the half-cell of interest when it is connected with the standard hydrogen electrode.

$$Zn^{2+} + H_2(g) \rightleftarrows Zn(s) + 2\ H^+$$

is not spontaneous. That is, the reaction tends to go in the other direction,

> Zinc reacts with acid, but hydrogen gas will not react with Zn^{2+}.

$$Zn(s) + 2\ H^+ \rightleftarrows Zn^{2+} + H_2(g)$$

15C-4 The Effect of Concentration on Electrode Potentials: The Nernst Equation

An electrode potential is a measure of the extent to which the existing concentrations in a half-cell differ from their equilibrium values. Thus, for example, there is a greater tendency for the process

$$Ag^+ + e^- \rightleftarrows Ag(s)$$

to occur in a concentrated solution of silver(I) than in a dilute solution of that ion. It follows that the magnitude of the electrode potential for this process must also become larger (more positive) as the silver ion concentration of the solution is increased. We now examine the quantitative relationship between concentration and electrode potential.

Consider the reversible half-reaction

$$a\text{A} + b\text{B} + n\text{e}^- \rightleftarrows c\text{C} + d\text{D}$$

where the capital letters represent formulas for the participating species (atoms, molecules, or ions), n represents the number of moles of electrons, and the lower case italic letters indicate the number of moles of each species appearing in the half-reaction as it has been written. The electrode potential for this process is described by the equation

$$E = E^0 - \frac{RT}{nF} \ln \frac{[\text{C}]^c[\text{D}]^d \cdots}{[\text{A}]^a[\text{B}]^b \cdots} \tag{15-7}$$

where

E^0 = a constant called the *standard electrode potential,* which is characteristic for each half-reaction

R = the gas constant, 8.314 J K^{-1} mol^{-1}

T = temperature in K (kelvins)

n = number of moles of electrons that appear in the half-reaction for the electrode process as it has been written

F = the faraday = 96,485 C (coulombs)

ln = the natural logarithm = 2.303 × log

Substituting numerical values for the constants, converting to base 10 logarithms, and specifying 25°C for the temperature give

The meanings of the bracketed terms in Equations 15-7 and 15-8 are

for a solute A, [A] = molar concentration

for a gas B, [B] = p_B = partial pressure in atmospheres

for a pure solid or pure liquid in excess C, [C] = 1.00

for a solvent D, [D] = 1.00

$$E = E^0 - \frac{0.0592}{n} \log \frac{[\text{C}]^c[\text{D}]^d \cdots}{[\text{A}]^a[\text{B}]^b \cdots} \tag{15-8}$$

The letters in brackets strictly represent activities, but we shall ordinarily follow our practice of substituting molar concentrations for activities in most calculations. Thus, if some participating species A is a solute, [A] is the concentration of A in moles per liter. If A is a gas, [A] in Equation 15-8 is replaced by p_A, the partial pressure of A in atmospheres. If A is a pure liquid or a pure solid present in excess as a second phase, [A] has a value of 1.00. Similarly, [A] is assigned a value of unity if A is the solvent. The rationale for these assumptions is the same as that described in Section 7B, which deals with equilibrium-constant expressions.

Equation 15-8 is known as the *Nernst equation* in honor of the German chemist responsible for its development.

Example 15-2

Typical half-cell reactions and their corresponding Nernst expressions follow.

(1) $Zn^{2+} + 2\,e^- \rightleftarrows Zn(s)$ $E = E^0 - \dfrac{0.0592}{2} \log \dfrac{1}{[Zn^{2+}]}$

The activity of elemental zinc is unity, by definition, because it is a pure second phase. Thus, the electrode potential varies linearly with the logarithm of the reciprocal of the zinc ion concentration.

(2) $Fe^{3+} + e^- \rightleftarrows Fe^{2+}$ $E = E^0 - \dfrac{0.0592}{1} \log \dfrac{[Fe^{2+}]}{[Fe^{3+}]}$

The potential for this couple can be measured with an inert metallic electrode immersed in a solution containing both iron species. The potential depends upon the logarithm of the ratio of the molar concentrations of these ions.

(3) $2\,H^+ + 2\,e^- \rightleftarrows H_2(g)$ $E = E^0 - \dfrac{0.0592}{2} \log \dfrac{p_{H_2}}{[H^+]^2}$

In this example, p_{H_2} is the partial pressure of hydrogen (in atmospheres) at the surface of the electrode. Ordinarily, its value will be the same as the atmospheric pressure.

(4) $MnO_4^- + 5\,e^- + 8\,H^+ \rightleftarrows Mn^{2+} + 4\,H_2O$

$$E = E^0 - \dfrac{0.0592}{5} \log \dfrac{[Mn^{2+}]}{[MnO_4^-][H^+]^8}$$

Here, the potential depends not only upon the concentration of manganese species but also on the pH of the solution.

(5) $AgCl(s) + e^- \rightleftarrows Ag(s) + Cl^-$ $E = E^0 - \dfrac{0.0592}{1} \log [Cl^-]$

This half-reaction describes the behavior of a silver electrode immersed in a chloride solution that is *saturated* with AgCl. To ensure this condition, an excess of the solid must always be present. Note that this electrode reaction is the sum of two reactions, namely,

$$AgCl(s) \rightleftarrows Ag^+ + Cl^-$$
$$Ag^+ + e^- \rightleftarrows Ag(s)$$

The activities of metallic Ag and of AgCl are equal to unity as long as some of each is in contact with the solution. Therefore, the logarithmic term in the Nernst equation contains the Cl^- concentration only.

Walther Hermann Nernst (1864–1941) was a German physical chemist who made many contributions to our understanding of electrochemistry. He is probably most famous for the equation that bears his name (Equation 15-7), but he was also known for discoveries and inventions in other fields. He invented the Nernst glower, a source of infrared radiation that is shown on the stamp. Nernst received the Nobel Prize in Chemistry in 1920 for his numerous contributions to the field of chemical thermodynamics.

The Nernst expression in part 5 of Example 15-2 applies only if there is present an excess of solid silver chloride *so that the solution is saturated* with AgCl at all times.

Feature 15-4
THE ACTIVITY QUOTIENT IN THE NERNST EQUATION IS UNITLESS

Although it is not obvious from our presentation, the quotient in the logarithmic term of the Nernst equation is unitless. Each individual term is a ratio of the activity of the species to its activity in the standard state, which has been arbitrarily assigned a value of unity (see Section 7B-2, p. 113). Thus, for the equation shown in part 3 of Example 15-2,

$$E = E^0 - \frac{0.0592}{2} \log \frac{p_{H_2}/(p_{H_2})_0}{[H^+]^2/([H^+]_0)^2}$$

$$= E^0 - \frac{0.0592}{2} \log \frac{p_{H_2}(\text{atm})/1.00(\text{atm})}{[H^+]^2(\text{mol/L})^2/1.00^2(\text{mol/L})^2}$$

Hydrogen in its standard state $(p_{H_2})_0$ is 1.00 atm. Similarly, $[H^+]_0$ represents the concentration of hydrogen ion in its standard state, 1.00 mol/L. Thus, the units cancel.

15C-5 The Standard Electrode Potential, E^0

The standard electrode potential, E^0, is defined as the electrode potential when all reactants and products that appear in a half-reaction have unit activity.

Examination of Equations 15-7 and 15-8 reveals that the constant E^0 is the electrode potential whenever the concentration quotient has a value of unity. This constant is called the *standard electrode potential* for the half-reaction. Note that the quotient is always unity when the activities of the reactants and products of a half-reaction are unity.

The standard electrode potential is an important physical constant that provides quantitative information regarding the driving force for a half-cell reaction.[1] The important characteristics of this constant are

1. The standard electrode potential is a relative quantity in the sense that it is the potential of an electrochemical cell in which the anode is the standard hydrogen electrode, whose potential has been arbitrarily set at 0.000 V.
2. The standard electrode potential for a half-reaction refers exclusively to a reduction reaction; that is, it is a relative reduction potential.
3. The standard electrode potential measures the relative force tending to drive the half-reaction from a state in which the reactants and products are at unit activity to a state in which the reactants and products are at their equilibrium activities relative to the standard hydrogen electrode.
4. The standard electrode potential is independent of number of moles of reactant and products shown in the balanced half-reaction. Thus, the standard electrode potential for the half-reaction

$$\text{Fe}^{3+} + \text{e}^- \rightleftarrows \text{Fe}^{2+} \qquad E^0 = +0.771$$

---

[1]For further reading on standard electrode potentials, see R. G. Bates, in *Treatise on Analytical Chemistry*, 2nd ed., I. M. Kolthoff and P. J. Elving, Eds., Part I, Vol. 1, Chapter 13. New York: Wiley, 1978.

does not change if we choose to write the reaction as

$$5\ Fe^{3+} + 5\ e^- \rightleftharpoons 5\ Fe^{2+} \qquad E^0 = +0.771$$

Note, however, that the Nernst equation must be consistent with the half-reaction as written. For the first case, it is

$$E = 0.771 - \frac{0.0592}{1} \log \frac{[Fe^{2+}]}{[Fe^{3+}]}$$

and for the second

$$E = 0.771 - \frac{0.0592}{5} \log \frac{[Fe^{2+}]^5}{[Fe^{3+}]^5}$$

Note that the two logarithmic terms have identical values:

$$\frac{0.0592}{1} \log \frac{[Fe^{2+}]}{[Fe^{3+}]} = \frac{0.0592}{5} \log \frac{[Fe^{2+}]^5}{[Fe^{3+}]^5}$$

5. A positive electrode potential indicates that the half-reaction in question is spontaneous with respect to the half-reaction for the standard hydrogen electrode. That is, the oxidant in the half-reaction is a stronger oxidant than is hydrogen ion. A negative sign indicates just the opposite.
6. The standard electrode potential for a half-reaction is temperature-dependent.

Standard-electrode-potential data are available for an enormous number of half-reactions. Many have been determined directly from electrochemical measurements. Others have been computed from equilibrium studies of oxidation/reduction systems and from thermochemical data associated with such reactions. Table 15-1 contains standard-electrode

Table 15-1
STANDARD ELECTRODE POTENTIALS*

Reaction	E^0 at 25°C, V
$Cl_2(g) + 2\ e^- \rightleftharpoons 2\ Cl^-$	+1.359
$O_2(g) + 4\ H^+ + 4\ e^- \rightleftharpoons 2\ H_2O$	+1.229
$Br_2(aq) + 2\ e^- \rightleftharpoons 2\ Br^-$	+1.087
$Br_2(l) + 2\ e^- \rightleftharpoons 2\ Br^-$	+1.065
$Ag^+ + e^- \rightleftharpoons Ag(s)$	+0.799
$Fe^{3+} + e^- \rightleftharpoons Fe^{2+}$	+0.771
$I_3^- + 2\ e^- \rightleftharpoons 3\ I^-$	+0.536
$Cu^{2+} + 2\ e^- \rightleftharpoons Cu(s)$	+0.337
$UO_2^{2+} + 4\ H^+ + 2\ e^- \rightleftharpoons U^{4+} + 2\ H_2O$	+0.334
$Hg_2Cl_2(s) + 2\ e^- \rightleftharpoons 2\ Hg(l) + 2\ Cl^-$	+0.268
$AgCl(s) + e^- \rightleftharpoons Ag(s) + Cl^-$	+0.222
$Ag(S_2O_3)_2^{3-} + e^- \rightleftharpoons Ag(s) + 2\ S_2O_3^{2-}$	+0.017
$2\ H^+ + 2\ e^- \rightleftharpoons H_2(g)$	0.000
$AgI(s) + e^- \rightleftharpoons Ag(s) + I^-$	-0.151
$PbSO_4(s) + 2\ e^- \rightleftharpoons Pb(s) + SO_4^{2-}$	-0.350
$Cd^{2+} + 2\ e^- \rightleftharpoons Cd(s)$	-0.403
$Zn^{2+} + 2\ e^- \rightleftharpoons Zn(s)$	-0.763

*See Appendix 1 for a more extensive list.

Feature 15-5

STANDARD ELECTRODE POTENTIALS FOR SYSTEMS INVOLVING
PRECIPITATES OR COMPLEX IONS

In Table 15-1, we find several entries involving Ag(I), including

$$Ag^+ + e^- \rightleftarrows Ag(s) \qquad\qquad E^0_{Ag^+} = +0.799 \text{ V}$$

$$AgCl(s) + e^- \rightleftarrows Ag(s) + Cl^- \qquad\qquad E^0_{AgCl} = +0.222 \text{ V}$$

$$Ag(S_2O_3)_2^{3-} + e^- \rightleftarrows Ag(s) + 2\ S_2O_3^{2-} \qquad\qquad E^0_{Ag(S_2O_3)_2^{3-}} = +0.017 \text{ V}$$

Each gives the potential of a silver electrode in a different environment. Let us see how the three potentials are related.

The Nernst expression for the first half-reaction is

$$E = E^0_{Ag^+} - \frac{0.0592}{1} \log \frac{1}{[Ag^+]}$$

If we replace $[Ag^+]$ with $K_{sp}/[Cl^-]$, we obtain

$$E = E^0_{Ag^+} - \frac{0.0592}{1} \log \frac{[Cl^-]}{K_{sp}}$$

$$= E^0_{Ag^+} + 0.0592 \log K_{sp} - 0.0592 \log [Cl^-] \qquad (1)$$

The standard potentials for Ag^+ and AgCl are related by the expression

$$E^0_{Ag^+} = E^0_{AgCl} - 0.0592 \log K_{sp}$$

where K_{sp} is the solubility-product constant for AgCl.

By definition, the standard potential for the Ag/AgCl half-reaction is the potential where $[Cl^-] = 1.00$. That is, when $[Cl^-] = 1.00$, $E = E^0_{AgCl}$. Taking the K_{sp} value for AgCl from Appendix 2 and substituting in Equation 1, we get

$$E^0_{AgCl} = E^0_{Ag^+} + 0.0592 \log 1.82 \times 10^{-10} - 0.0592 \log (1.00)$$

$$= 0.799 + (-0.577) - 0.000 = 0.222 \text{ V}$$

Figure 15-8 illustrates the definition of the standard electrode potential for the Ag/AgCl electrode.

If we proceed in the same way, we obtain for the third equilibrium

$$E^0_{Ag(S_2O_3)_2^{3-}} = E^0_{Ag^+} - 0.0592 \log K_f$$

where K_f is the formation constant for the complex, that is,

$$K_f = \frac{[Ag(S_2O_3)_2^{3-}]}{[Ag^+][S_2O_3^{2-}]^2}$$

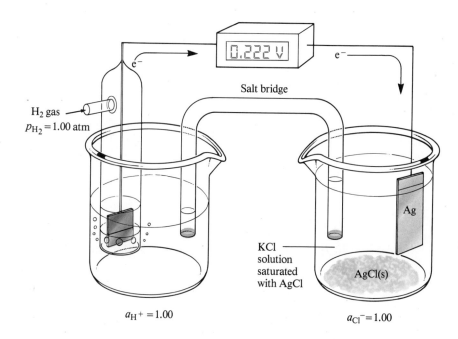

Figure 15-8
Definition of the standard electrode potential for a Ag/AgCl electrode.

data from several half-reactions we will be considering in the pages that follow. A more extensive listing is found in Appendix 1.[2]

Table 15-1 and Appendix 1 illustrate the two common ways for tabulating standard potential data. In Table 15-1, potentials are listed in decreasing numerical order. As a consequence, the species in the upper left part (that is, on the left side of the first few half-reactions) are the most effective electron acceptors, as evidenced by their large positive E^0 values. They are therefore the strongest *oxidizing agents*. As we proceed down the left side of such a table, each succeeding species is a less effective acceptor of electrons than the one above it. The half-cell reactions at the bottom of the table have little tendency to take place as written. On the other hand, they do tend to occur in the opposite sense. The most effective *reducing agents,* then, are those species on the right side of the last few half-reactions shown in the table.

Compilations of electrode-potential data, such as that shown in Table 15-1, provide the user with qualitative insights into the extent and direction of electron-transfer reactions. For example, the standard potential for silver(I) (+0.799 V) is more positive than that for copper(II) (+0.337 V). We therefore conclude that a piece of copper immersed in a silver(I) solution will cause the reduction of that ion and the oxidation of the copper. On the other hand, we would expect no reaction if we place a piece of silver in a copper(II) solution.

Based upon the E^0 values in Table 15-1 for Fe^{3+} and I_3^-, which species would you expect to predominate in a solution produced by mixing iron(III) and iodide ions? See the color plates in the middle of the book.

[2]Comprehensive sources for standard electrode potentials include *Standard Electrode Potentials in Aqueous Solutions,* A. J. Bard, R. Parsons, and J. Jordan, Eds. New York: Marcel Dekker, 1985; G. Milazzo and S. Caroli, *Tables of Standard Electrode Potentials.* New York: Wiley-Interscience, 1977; M. S. Antelman and F. J. Harris, *Chemical Electrode Potentials.* New York: Plenum Press, 1982. Some compilations are arranged alphabetically by element; others are tabulated according to the value of E^0.

In contrast to the data in Table 15-1, Appendix 1 is arranged alphabetically by element to make it easier to locate data for a given electrode reaction.

Example 15-3

Calculate the electrode potential of a silver electrode immersed in a 0.0500 M solution of NaCl using (a) $E^0_{Ag^+}$ = 0.799 V and (b) E^0_{AgCl} = 0.222 V.

(a) $Ag^+ + e^- \rightleftarrows Ag(s)$ $E^0 = 0.799$ V

The Ag^+ concentration of this solution is given by

$$[Ag^+] = K_{sp}/[Cl^-] = 1.82 \times 10^{-10}/0.0500 = 3.64 \times 10^{-9} \text{ M}$$

Substituting into the Nernst expression gives

$$E = 0.799 - 0.0592 \log \frac{1}{3.64 \times 10^{-9}} = 0.299 \text{ V}$$

(b) Here we may write

$$E = 0.222 - 0.0592 \log [Cl^-] = 0.222 - 0.0592 \log 0.0500$$
$$= 0.299 \text{ V}$$

15C-6 Formal Potentials

In many oxidation/reduction systems, electrode-potential calculations are complicated by the presence of other equilibria (such as dissociation, association, complex formation, and solvolysis) that involve one or more of the species involved in the redox half-reaction. The effect of these other equilibria on calculated potentials can be accounted for if equilibrium-constant data are available. Often they are not, however.

Swift[3] proposed substitution of *formal potentials* (also called *conditional potentials*) in place of standard electrode potentials to compensate for activity effects as well as errors due to the existence of competing equilibria. The formal potential of a system is the potential of the half-cell (with respect to the standard hydrogen electrode) when the *concentration* of each solute participating in the half-reaction is exactly 1 M and the concentrations of all other solutes are carefully specified.

Formal potentials for many half-reactions are listed in Appendix 1. Note that large differences exist between the formal and standard potentials for some half-reactions. For example, the standard electrode potential for the reduction of iron(III) to iron(II) is 0.771 V. In 1 M perchloric

A formal potential is the electrode potential when the *concentrations* of reactants and products of a half-reaction are exactly 1 M and the concentrations of any other solutes are specified.

---

[3]E. H. Swift, *A System of Chemical Analysis*, p. 50. San Francisco: Freeman, 1939.

acid, the formal potential for the same half-reaction is 0.732 V. This difference is attributable to the fact that the activity coefficient of iron(III) is considerably smaller than that of iron(II) at the high ionic strength of the 1 M perchloric acid medium. As a consequence, the ratio of activities of the two species is less than unity, a condition that leads to a decrease in the electrode potential. In 1 M hydrochloric acid, the formal potential for this couple is only 0.700 V. Here, the iron(III)/iron(II) activity ratio is even smaller because the chloro complexes of the former are more stable than those of iron(II); a change in potential larger than that seen in perchloric acid is the consequence.

Feature 15-6
WHY THERE ARE TWO ELECTRODE POTENTIALS FOR Br_2 IN TABLE 15-1

In Table 15-1, we find the following data for Br_2:

$$Br_2(aq) + 2\ e^- \rightleftharpoons 2\ Br^- \qquad E^0 = +1.087\ V$$
$$Br_2(l) + 2\ e^- \rightleftharpoons 2\ Br^- \qquad E^0 = +1.065\ V$$

The second standard potential applies *only to a solution that is saturated with Br$_2$* and not to undersaturated solutions of the element. You should therefore use 1.065 V to calculate the electrode potential of a 0.0100 M solution of KBr that is saturated with Br_2 and in contact with an excess of the liquid element. Here,

$$E = 1.065 - \frac{0.0592}{2} \log \frac{[Br^-]^2}{1.00} = 1.065 - \frac{0.0592}{2} \log (0.0100)^2$$

$$= 1.065 - \frac{0.0592}{2} (-4.00) = 1.183\ V$$

In this calculation, the activity of Br_2 in the liquid state is constant and assigned a value of 1.00.

The standard electrode potential shown in the entry for $Br_2(aq)$ is a *hypothetical* standard potential because the solubility of Br_2 at 25°C is only about 0.18 M. Thus, the recorded value of 1.087 V is based upon a system that—in terms of our definition of E^0—cannot be realized experimentally. Nevertheless, the hypothetical potential does permit us to calculate electrode potentials for solutions that are undersaturated in Br_2. For example, if we wish to calculate the electrode potential for a solution that is 0.0100 M in KBr and 0.00100 M in Br_2, we would write

$$E = 1.087 - \frac{0.0592}{2} \log \frac{[Br^-]^2}{[Br_2(aq)]} = 1.087 - \frac{0.0592}{2} \log \frac{(0.0100)^2}{0.00100}$$

$$= 1.087 - \frac{0.0592}{2} \log 0.100 = 1.117\ V$$

Substitution of formal potentials for standard electrode potentials in the Nernst equation yields better agreement between calculated and experimental results—provided, of course, that the electrolyte concentration of the solution approximates that for which the formal potential is applicable. Not surprisingly, attempts to apply formal potentials to systems that differ substantially in type and in electrolyte concentration can result in errors that are larger than those associated with the use of standard electrode potentials. Henceforth we use whichever is the more appropriate.

15D   THE POTENTIAL OF ELECTROCHEMICAL CELLS

We can use standard electrode potentials and the Nernst equation to calculate the potential obtainable from a galvanic cell or the potential required to operate an electrolytic cell. The calculated potentials (sometimes called *thermodynamic potentials*) are theoretical in the sense that they refer to cells in which there is no current. Additional factors must be taken into account if a current is involved.

15D-1 The Schematic Representation of Cells

Chemists frequently use a shorthand notation to describe electrochemical cells. The cell in Figure 15-1a, for example, is described by

$$Cu|CuSO_4(1\ M)\|AgNO_3(1\ M)|Ag$$

where the molar concentrations of $CuSO_4$ and $AgNO_3$ are exactly 1 M. An alternative way of writing the cell is

$$Cu|Cu^{2+}(1\ M)\|Ag^+(1\ M)|Ag \tag{15-9}$$

In this representation, which is more commonly encountered, only the active participants in the cell half-reactions are indicated.

By convention, the anodic process is *always* displayed on the left in these representations. A single vertical line indicates a phase boundary, or interface, at which a potential develops. For example, in the first representation, the first vertical line indicates that a potential develops at the phase boundary between the copper anode and the copper sulfate solution. The double vertical line represents two phase boundaries, one at each end of the salt bridge. A *liquid-junction potential* develops at each of the two salt-bridge interfaces. This potential results from differences in the rates at which the ions in the cell compartments and the salt bridge migrate across the interfaces. A liquid-junction potential can amount to as much as several hundredths of a volt, but the junction potentials at the two ends of a salt bridge tend to cancel each other. The net effect on the overall potential of a cell is thus only a few millivolts or less. For our purposes, we will neglect the contribution of liquid-junction potentials to the total potential of the cell.

The cell shown in Figure 15-6 can be represented as

$$Pt,H_2(p = 1.00\ atm)|H^+(a_{H^+} = 1.00\ M)\|Ag^+(a_{Ag^+} = 1.00)|Ag$$

A liquid-junction potential is a potential that develops across the interface between two solutions with different electrolyte composition.

or alternatively as

$$SHE\|Ag^+(a_{Ag^+} = 1.00)|Ag$$

The cell in Figure 15-7 can be represented by

$$Cd|Cd^{2+}(a_{Cd^{2+}} = 1.00)\|H^+(a_{H^+} = 1.00)|H_2(p = 1.00 \text{ atm}),Pt$$

Note that the standard hydrogen electrode is the anode in the first cell and the cathode in the second.

15D-2 Cell Potential Calculations

The potential of an electrochemical cell E_{cell} is the difference between the *electrode potential* of the cathode and the *electrode potential* of the anode. That is,

$$E_{cell} = E_{cathode} - E_{anode} \qquad (15\text{-}10)$$

where $E_{cathode}$ and E_{anode} are the half-cell potentials of the cathode and anode, respectively.

Gustav Robert Kirchhoff (1824–1877) was a German physicist who made many important contributions to physics and chemistry. In addition to his work in spectroscopy, he is known for Kirchhoff's laws of electrical circuits. These laws are shown in the form of equations on the stamp above.

Example 15-4

Calculate the thermodynamic potential of the following cell

$$Cu|Cu^{2+}(0.100 \text{ M})\|Ag^+(0.200 \text{ M})|Ag$$

Note that this cell is similar to the galvanic cell shown in Figure 15-1a.
  The two half-reactions and standard potentials are

$$
\begin{aligned}
Ag^+ + 2\,e^- &\rightleftarrows Ag(s) &\quad E^0 &= 0.799 \text{ V} \\
Cu^{2+} + 2\,e^- &\rightleftarrows Cu(s) &\quad E^0 &= 0.337 \text{ V}
\end{aligned}
$$

The electrode potentials are

$$E_{Ag^+} = 0.799 - 0.0592 \log \frac{1}{0.200} = 0.758 \text{ V}$$

$$E_{Cu^{2+}} = 0.337 - \frac{0.0592}{2} \log \frac{1}{0.100} = 0.307 \text{ V}$$

We see from the cell diagram that the silver electrode is the cathode and the copper electrode is the anode. Therefore applying Equation 15-10 gives

$$E_{cell} = E_{Ag^+} - E_{Cu^{2+}} = 0.758 - 0.307 = +0.451 \text{ V}$$

Example 15-5

Calculate the thermodynamic potential for the cell

$$Ag|Ag^+(0.200 \text{ M})\|Cu^{2+}(0.100 \text{ M})|Cu$$

Note that this cell is similar to the electrolytic cell shown in Figure 15-1b.
   The electrode potentials for the two half-reactions are identical to the electrode potentials calculated in Example 15-4. That is,

$$E_{Ag^+} = 0.758 \text{ V} \quad \text{and} \quad E_{Cu^{2+}} = 0.307 \text{ V}$$

In contrast to the previous example, however, the silver electrode is the anode and the copper electrode the cathode. Substituting these electrode potentials into Equation 15-10 gives

$$E_{cell} = E_{Cu^{2+}} - E_{Ag^+} = 0.307 - 0.758 = -0.451 \text{ V}$$

A negative cell potential means the cell is electrolytic; a positive potential means the cell is galvanic.

   Examples 15-4 and 15-5 illustrate an important fact. The cell potential computed from Equation 15-10 is positive for a galvanic cell and negative for an electrolytic cell.

Example 15-6

Calculate the potential of the following cell and indicate whether it is galvanic or electrolytic.

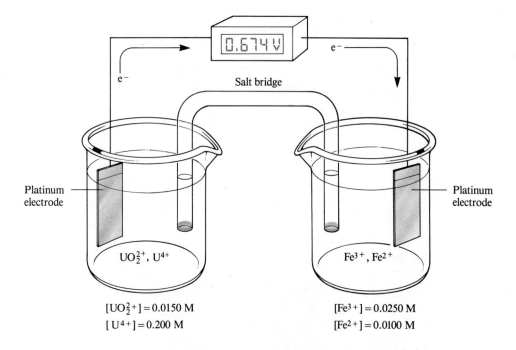

Figure 15-9
Cell for Example 15-4.

Platinum electrode

UO_2^{2+}, U^{4+}

$[UO_2^{2+}] = 0.0150 \text{ M}$
$[U^{4+}] = 0.200 \text{ M}$

Platinum electrode

Fe^{3+}, Fe^{2+}

$[Fe^{3+}] = 0.0250 \text{ M}$
$[Fe^{2+}] = 0.0100 \text{ M}$

Salt bridge

e^- e^-

0.674 V

$Pt|UO_2^{2+}(0.0150\ M),U^{4+}(0.200\ M),H^+(0.0300\ M)\|$

$$Fe^{2+}(0.0100\ M),Fe^{3+}(0.0250\ M)|Pt$$

The two half-reactions are

$$Fe^{3+} + e^- \rightleftarrows Fe^{2+} \qquad\qquad E^0 = +0.771\ V$$

$$UO_2^{2+} + 4\ H^+ + 2\ e^- \rightleftarrows U^{4+} + 2\ H_2O \qquad E^0 = +0.334\ V$$

The electrode potential for the cathode is

$$E_{cathode} = 0.771 - 0.0592\ \log \frac{[Fe^{2+}]}{[Fe^{3+}]}$$

$$= 0.771 - 0.0592\ \log \frac{0.0100}{0.0250} = 0.799 - (-0.0236)$$

$$= 0.7946\ V$$

The electrode potential for the anode is

$$E_{anode} = 0.334 - \frac{0.0592}{2}\ \log \frac{[U^{4+}]}{[UO_2^{2+}][H^+]^4}$$

$$= 0.334 - \frac{0.0592}{2}\ \log \frac{0.200}{(0.0150)(0.0300)^4}$$

$$= 0.334 - 0.2136 = 0.1204\ V$$

and

$$E_{cell} = E_{cathode} - E_{anode} = 0.7946 - 0.1204 = +0.674\ V$$

The positive sign means that the cell is galvanic.

Example 15-7

Calculate the theoretical potential for the cell

$$Ag|AgCl(sat'd),HCl(0.0200\ M)|H_2(0.800\ atm),Pt$$

Note that this cell does not require two compartments (or a salt bridge) because molecular H_2 has little tendency to react directly with the low concentration of Ag^+ in the electrolyte solution. This cell, shown in Figure 15-10, is an example of a cell *without liquid junction*.

The two half-reactions and their corresponding standard electrode potentials are (Table 15-1)

$$2\ H^+ + 2\ e^- \rightleftarrows H_2(g) \qquad\qquad E^0 = 0.000\ V$$

$$AgCl(s) + e^- \rightleftarrows Ag(s) + Cl^- \qquad E^0 = 0.222\ V$$

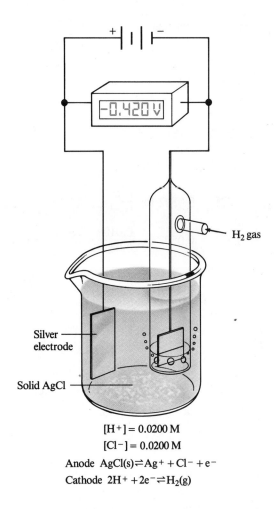

$[H^+] = 0.0200$ M
$[Cl^-] = 0.0200$ M
Anode $AgCl(s) \rightleftharpoons Ag^+ + Cl^- + e^-$
Cathode $2H^+ + 2e^- \rightleftharpoons H_2(g)$

Figure 15-10
Cell for Example 15-7.

The two electrode potentials are

$$E_{\text{cathode}} = 0.000 - \frac{0.0592}{2} \log \frac{0.800}{(0.0200)^2} = -0.0977 \text{ V}$$

$$E_{\text{anode}} = 0.222 - \frac{0.0592}{1} \log 0.0200 = 0.3226 \text{ V}$$

The cell diagram specifies the silver electrode as the anode and the hydrogen electrode as the cathode. Thus,

$$E_{\text{cell}} = -0.0977 - 0.3226 = -0.420 \text{ V}$$

The negative sign indicates that the cell reaction

$$2 \text{ H}^+ + 2 \text{ Ag(s)} + 2 \text{ Cl}^- \rightleftharpoons \text{H}_2(\text{g}) + 2 \text{ AgCl(s)}$$

is nonspontaneous, and thus the cell is electrolytic. It would require an external power source for operation in this way (Figure 15-10).

15E QUESTIONS AND PROBLEMS

15-1. Briefly describe or define
*(a) oxidation.
(b) reducing agent.
*(c) galvanic cell.
(d) electrolytic cell.
*(e) anode.
(f) cathode.
*(g) liquid junction.
(h) irreversible cell.
*(i) standard hydrogen electrode.
(j) electrode potential.
*(k) standard electrode potential.
(l) salt bridge.
*(m) formal potential.
(n) Nernst equation.

15-2. Use activities to calculate the electrode potential of a hydrogen electrode in which the electrolyte is 0.0100 M HCl and the activity of hydrogen gas is 1.00 atm.

*15-3. The standard electrode potential for the reduction of Ni^{2+} to Ni is -0.25 V. Would the potential of a nickel electrode immersed in a 1.00 M NaOH solution saturated with $Ni(OH)_2$ be more negative than $E^0_{Ni^{2+}}$ or less negative? Explain.

15-4. Why is it necessary to bubble hydrogen gas through the electrolyte in a hydrogen gas electrode?

*15-5. The following two entries are found in a table of standard electrode potentials:

$$I_2(s) + 2\,e^- \rightleftarrows 2\,I^- \qquad E^0 = 0.536 \text{ V}$$
$$I_2(aq) + 2\,e^- \rightleftarrows 2\,I^- \qquad E^0 = 0.615 \text{ V}$$

What is the significance of the difference between these two entries?

*15-6. Complete and balance the following equations, adding H^+, OH^-, or H_2O as required:
(a) $Tl^{3+} + Ag(s) + Br^- \rightleftarrows TlBr(s) + AgBr(s)$
(b) $Fe^{2+} + UO_2^{2+} \rightleftarrows Fe^{3+} + U^{4+}$
(c) $N_2(g) + H_2(g) \rightleftarrows N_2H_5^+$
(d) $Cr_2O_7^{2-} + I^- \rightleftarrows Cr^{3+} + I_3^-$
(e) $H_2O_2 + Ce^{4+} \rightleftarrows O_2 + Ce^{3+}$
(f) $IO_3^- + I^- \rightleftarrows I_2(s)$

15-7. Complete and balance the following equations, adding H^+, OH^-, or H_2O as required:
(a) $Cu^{2+} + I^- \rightleftarrows CuI(s) + I_3^-$
(b) $I_3^- + S_2O_3^{2-} \rightleftarrows I^- + S_4O_6^{2-}$
(c) $MnO_4^- + H_2SO_3 \rightleftarrows Mn^{2+} + SO_4^{2-}$
(d) $MnO_4^- + Mn^{2+} \rightleftarrows MnO_2(s)$
(e) $IO_3^- + H_2AsO_3 + Cl^- \rightleftarrows ICl_2^- + H_3AsO_4$
(f) $S_2O_8^{2-} + Mn^{2+} \rightleftarrows SO_4^{2-} + MnO_4^-$

*15-8. Identify the oxidizing agent and the reducing agent on the left side of each equation in Problem 15-6; write a balanced equation for each half-reaction.

15-9. Identify the oxidizing agent and the reducing agent on the left side of each equation in Problem 15-7; write a balanced equation for each half-reaction.

*15-10. Calculate the electrode potential of a mercury electrode immersed in
(a) 0.0400 M $Hg(NO_3)_2$.
(b) 0.0400 M $Hg_2(NO_3)_2$.
(c) 0.0400 M KCl saturated with Hg_2Cl_2.
(d) 0.0400 M $Hg(SCN)_2$.
$(Hg^{2+} + 2\,SCN^- \rightleftarrows Hg(SCN)_2; K_f = 1.8 \times 10^{17})$.

15-11. Calculate the electrode potential for a copper electrode immersed in
(a) 0.0200 M Cu^{2+}.
(b) 0.0200 M Cu^+.
(c) 0.0300 M KI saturated with CuI.
(d) 0.0100 M NaOH saturated with $Cu(OH)_2$.

*15-12. Calculate the electrode potential for a platinum electrode immersed in a solution that is
(a) 0.075 M in $Fe_2(SO_4)_3$ and 0.060 M in $FeSO_4$.
(b) 0.244 M in V^{3+}, 0.414 M in VO^{2+}, and 1.00×10^{-5} M in NaOH.
(c) 0.111 M in KI and 0.200 M in KI_3.
(d) 0.117 M in $K_4Fe(CN)_6$ and 0.333 M in $K_3Fe(CN)_6$.
(e) 0.0731 M in $SbONO_3$, 0.0100 M in HNO_3, and saturated with Sb_2O_5.
(f) 0.0731 M in $SbONO_3$, 1.00×10^{-5} M in HNO_3, and saturated with Sb_2O_5.

15-13. Calculate the electrode potential for a platinum electrode immersed in a solution that is
(a) 0.313 M in $Tl_2(SO_4)_3$ and 0.209 M in Tl_2SO_4.
(b) saturated with hydrogen at 1.00 atm and has a pH of 3.50.
(c) 0.0774 M in UO_2^{2+}, 0.0507 M in U^{4+}, and 1.00×10^{-4} M in $HClO_4$.
(d) 0.0627 M in $S_2O_3^{2-}$ and 0.0714 M in $S_4O_6^{2-}$.
(e) 0.0540 M in $Cr_2O_7^{2-}$, 0.149 M in Cr^{3+}, and 0.100 M in $HClO_4$.

*15-14. Indicate whether each half-cell behaves as anode or cathode when coupled with a standard hydrogen electrode in a galvanic cell, and calculate the cell potential:
(a) $Pb|Pb^{2+}(2.00 \times 10^{-4}$ M$)$
(b) $Pt|Sn^{4+}(0.200$ M$),Sn^{2+}(0.100$ M$)$
(c) $Pt|Sn^{4+}(1.00 \times 10^{-6}$ M$),Sn^{2+}(0.50$ M$)$
(d) $Pt|Ti^{3+}(0.300$ M$),TiO^{2+}(0.100$ M$),H^+(0.200$ M$)$
(e) $Ag|AgBr(sat'd),KBr(1.00 \times 10^{-4}$ M$)$
(f) $Ag|Ag(CN)_2^-(0.200$ M$),CN^-(1.00 \times 10^{-6}$ M$)$

15-15. Indicate whether each half-cell behaves as anode or cathode when coupled with a standard hydrogen electrode in a galvanic cell, and calculate the cell potential:
(a) $Pt|V^{3+}(0.50$ M$),V^{2+}(1.00 \times 10^{-6}$ M$)$
(b) $Ag|AgNO_3(0.0100$ M$),Na_2S_2O_3(0.0800$ M$)$
(c) $Ag|AgNO_3(0.0100$ M$),Na_2S_2O_3(0.200$ M$)$
(d) $Bi|BiCl_4^-(0.0100$ M$),Cl^-(0.500$ M$)$
(e) $Ag|Ag(CN)_2^-(0.400$ M$),CN^-(1.00 \times 10^{-4}$ M$)$

*15-16. Indicate in which direction the reactions in Problem 15-6 proceed if all species are initially at unit activity.

$$Tl^{3+} + Br^- + 2\,e^- \rightleftharpoons TlBr(s) \qquad E^0 = 1.44 \text{ V}$$

15-17. Indicate in which direction the reactions in Problem 15-7 proceed if all species are initially at unit activity.

15-18. Calculate the theoretical potential for the following cells. Indicate whether each cell, as written, is galvanic or electrolytic:

(a) $Pb|PbSO_4(sat'd)SO_4^{2-}(0.200\text{ M})\|$
$$Sn^{2+}(0.150\text{ M}),Sn^{4+}(0.250\text{ M})|Pt$$

(b) $Pt|Fe^{3+}(0.0100\text{ M}),Fe^{2+}(0.00100\text{ M})\|$
$$Ag^+(0.0350\text{ M})|Ag$$

(c) $Cu|CuI(sat'd),KI(0.0100\text{ M})\|$
$$KI(0.200\text{ M}),CuI(sat'd)|Cu$$

(d) $Pt|UO_2^{2+}(0.100\text{ M}),U^{4+}(0.0100\text{ M}),$
$H^+(1.00 \times 10^{-6}\text{ M})\|$
$$AgCl(sat'd),KCl(1.00 \times 10^{-4}\text{ M})|Ag$$

(e) $Hg|Hg_2Cl_2(sat'd),Cl^-(0.0500\text{ M})\|$
$$V^{2+}(0.200\text{ M}),V^{3+}(0.300\text{ M})|Pt$$

(f) $Pt|VO^{2+}(0.250\text{ M}),V^{3+}(0.100\text{ M}),$
$H^+(1.00 \times 10^{-3}\text{ M})\|$
$$Tl^{3+}(0.100\text{ M}),Tl^+(0.0500\text{ M})|Pt$$

15-19. Calculate the theoretical cell potential for each of the following. Is the cell as written galvanic or electrolytic?

(a) $Ag|AgBr(sat'd),Br^-(0.0400\text{ M})\|$
$$H^+(1.00 \times 10^{-4}\text{ M})|H_2(0.90\text{ atm}),Pt$$

(b) $Pt|Cr^{3+}(0.0500\text{ M}),Cr^{2+}(0.0250\text{ M})\|$
$$Ni^{2+}(0.0100\text{ M})|Ni$$

(c) $Ag|Ag(CN)_2^-(0.240\text{ M}),CN^-(0.100\text{ M})\|$
$$Br_2(1.00 \times 10^{-3}\text{ M}),KBr(0.200\text{ M})|Pt$$

(d) $Ag|AgCl(sat'd),HCl(5.00 \times 10^{-3}\text{ M})|$
$$H_2(0.300\text{ atm}),Pt$$

(e) $Hg|Hg_2Cl_2(sat'd),HCl(0.0050\text{ M})\|$
$$HCl(1.50\text{ M}),Hg_2Cl_2(sat'd)|Hg$$

(f) $Pt|TiO^{2+}(0.200\text{ M}),Ti^{3+}(0.100\text{ M}),$
$H^+(2.00 \times 10^{-3}\text{ M})\|$
$$SO_4^{2-}(0.200\text{ M}),PbSO_4(sat'd)|Pb$$

*15-20. The solubility-product constant for Ag_2SO_3 is 1.5×10^{-14}. Calculate E^0 for the process

$$Ag_2SO_3(s) + 2\,e^- \rightleftharpoons 2\,Ag(s) + SO_3^{2-}$$

15-21. The solubility-product constant for $Ni_2P_2O_7$ is 1.7×10^{-13}. Calculate E^0 for the process

$$Ni_2P_2O_7(s) + 4\,e^- \rightleftharpoons 2\,Ni(s) + P_2O_7^{4-}$$

*15-22. Compute E^0 for the process

$$ZnY^{2-} + 2\,e^- \rightleftharpoons Zn(s) + Y^{4-}$$

where Y^{4-} is the completely deprotonated anion of EDTA. The formation constant for ZnY^{2-} is 3.2×10^{16}.

15-23. Calculate E^0 for the process

$$VY^- + e^- \rightarrow VY^{2-}$$

if the formation constant for the EDTA complex of V^{2+} is 5.0×10^{12} and that for the V^{3+} complex is 7.9×10^{25}.

CHAPTER 16

THEORY of Oxidation/Reduction Titrations

This chapter is concerned with titration curves and indicators for oxidation/reduction titrations. Before describing how such curves are derived, we need to know how equilibrium constants for oxidation/reduction reactions are computed from standard electrode potentials.

16A EQUILIBRIUM CONSTANTS FOR OXIDATION/REDUCTION REACTIONS

Let us again consider the equilibrium established when a piece of copper is immersed in a solution containing a dilute solution of silver nitrate:

$$Cu(s) + 2\ Ag^+ \rightleftarrows Cu^{2+} + 2\ Ag(s) \qquad (16\text{-}1)$$

The equilibrium constant for this reaction is

$$K_{eq} = \frac{[Cu^{2+}]}{[Ag^+]^2} \qquad (16\text{-}2)$$

This reaction can also be carried out in the galvanic cell

$$Cu|Cu^{2+}(x\ M)\|Ag^+(y\ M)|Ag$$

Photograph of a "silver tree."   287

Figure 16-1

Potential of a galvanic cell as it discharges.

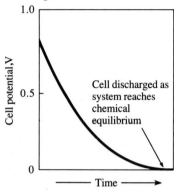

Cell discharged as system reaches chemical equilibrium

A sketch of this cell is shown in Figure 15-1. Its cell potential at any instant is given by

$$E_{cell} = E_{cathode} - E_{anode} = E_{Ag^+} - E_{Cu^{2+}}$$

As the reaction proceeds, the concentration of Cu(II) ions increases and the concentration of Ag(I) ions decreases. These changes make the potential of the copper electrode more positive and that of the silver electrode less positive. As shown in Figure 16-1, the net effect of these changes is a continuous decrease in the potential of the cell as it discharges. Ultimately the concentrations of Cu(II) and Ag(I) attain their equilibrium values as determined by Equation 16-2, and the current ceases. Under these equilibrium conditions, *the potential of the cell becomes zero. Thus, at chemical equilibrium,* we may write

$$E_{cell} = 0 = E_{cathode} - E_{anode} = E_{Ag^+} - E_{Cu^{2+}}$$

or,

$$E_{cathode} = E_{anode} = E_{Ag^+} = E_{Cu^{2+}} \tag{16-3}$$

16A-1 Electrode Potentials in Equilibrum Systems

We can generalize Equation 16-3 by stating that *at equilibrium, the electrode potentials for all half-reactions in an oxidation/reduction system are equal.* This generalization applies regardless of the number of half-reactions present in the system because interactions among *all systems* must take place until their electrode potentials are identical. For example, if we have four oxidation/reduction systems in a solution, interaction among all four takes place until

Remember that *when redox systems are at equilibrium, the electrode potentials of all systems are identical.* This generality applies whether the reactions take place directly in solution or indirectly in a galvanic cell.

$$E_{Ox_1} = E_{Ox_2} = E_{Ox_3} = E_{Ox_4} \tag{16-4}$$

where E_{Ox_1}, E_{Ox_2}, E_{Ox_3}, and E_{Ox_4} are the electrode potentials for the four half-reactions.

16A-2 The Calculation of Equilibrium Constants

Returning to the reaction shown in Equation 16-1, let us substitute Nernst expressions for the two electrode potentials in Equation 16-3, which gives

$$E^0_{Ag^+} - \frac{0.0592}{2} \log \frac{1}{[Ag^+]^2} = E^0_{Cu^{2+}} - \frac{0.0592}{2} \log \frac{1}{[Cu^{2+}]} \tag{16-5}$$

It is important to note that the Nernst equation is applied to the silver half-reaction as it appears in the balanced equation (Equation 16-1):

$$2\ Ag^+ + 2\ e^- \rightleftarrows 2\ Ag(s) \qquad E^0 = 0.799\ V$$

Rearrangement of Equation 16-5 gives

$$E^0_{Ag^+} - E^0_{Cu^{2+}} = \frac{0.0592}{2} \log \frac{1}{[Ag^+]^2} - \frac{0.0592}{2} \log \frac{1}{[Cu^{2+}]}$$

$$= \frac{0.0592}{2} \log \frac{1}{[Ag^+]^2} + \frac{0.0592}{2} \log \frac{[Cu^{2+}]}{1}$$

Finally, then,

$$\frac{2(E^0_{Ag^+} - E^0_{Cu^{2+}})}{0.0592} = \log \frac{[Cu^{2+}]}{[Ag^+]^2} = \log K_{eq} \qquad (16\text{-}6)$$

The concentration terms in Equation 16-6 are *equilibrium concentrations;* *the ratio [Cu²⁺]/[Ag⁺]² in the logarithmic term is therefore the equilibrium* *constant for the reaction.*

Example 16-1

Calculate the equilibrium constant for the reaction shown in Equation 16-1.

Substituting numerical values into Equation 16-6 yields

$$\log K_{eq} = \log \frac{[Cu^{2+}]}{[Ag^+]^2} = \frac{2(0.799 - 0.337)}{0.0592}$$

$$= 15.6$$

$$K_{eq} = \text{antilog } 15.6 = 4.1 \times 10^{15} = 4 \times 10^{15}$$

In making calculations of this sort, you should follow the rounding rule for antilogarithms given on page 41.

Example 16-2

Calculate the equilibrium constant for the reaction

$$2\ Fe^{3+} + 3\ I^- \rightleftarrows 2\ Fe^{2+} + I_3^-$$

In Appendix 1 we find

$$2\ Fe^{3+} + 2\ e^- \rightleftarrows 2\ Fe^{2+} \qquad E^0 = 0.771\ V$$

$$I_3^- + 2\ e^- \rightleftarrows 3\ I^- \qquad E^0 = 0.536\ V$$

We have multiplied the first half-reaction by 2 so that the number of moles of Fe^{3+} and Fe^{2+} are the same as in the balanced overall equation. The Nernst equation for Fe^{3+} is thus based upon the half-reaction for a two-electron transfer:

$$E_{Fe^{3+}} = E^0_{Fe^{3+}} - \frac{0.0592}{2} \log \frac{[Fe^{2+}]^2}{[Fe^{3+}]^2}$$

and

$$E_{I_3^-} = E^0_{I_3^-} - \frac{0.0592}{2} \log \frac{[I^-]^3}{[I_3^-]}$$

At equilibrium, the electrode potentials are equal and

$$E_{Fe^{3+}} = E_{I_3^-}$$

$$E^0_{Fe^{3+}} - \frac{0.0592}{2} \log \frac{[Fe^{2+}]^2}{[Fe^{3+}]^2} = E^0_{I_3^-} - \frac{0.0592}{2} \log \frac{[I^-]^3}{[I_3^-]}$$

This equation rearranges to

$$E^0_{Fe^{3+}} - E^0_{I_3^-} = \frac{0.0592}{2} \log \frac{[Fe^{2+}]^2}{[Fe^{3+}]^2} + \frac{0.0592}{2} \log \frac{[I_3^-]}{[I^-]^3}$$

Notice that we have changed the sign of the second logarithmic term by inverting the fraction. Further rearrangement gives

$$\log \frac{[Fe^{2+}]^2[I_3^-]}{[Fe^{3+}]^2[I^-]^3} = \frac{(E^0_{Fe^{3+}} - E^0_{I_3^-})2}{0.0592}$$

Recall, however, that the concentration terms here are *equilibrium concentrations* and

$$\log K_{eq} = \frac{(E^0_{Fe^{3+}} - E^0_{I_3^-})2}{0.0592} = \frac{(0.771 - 0.536)2}{0.0592} = 7.94$$

$$K_{eq} = \text{antilog } 7.94 = 8.7 \times 10^7$$

We round the answer to two figures because $\log K_{eq}$ contains only two significant figures (the two to the right of the decimal point).

Example 16-3

Calculate the equilibrium constant for the reaction

$$2\ MnO_4^- + 3\ Mn^{2+} + 2\ H_2O \rightleftharpoons 5\ MnO_2(s) + 4\ H^+$$

In Appendix 1 we find

$$2\ MnO_4^- + 8\ H^+ + 6\ e^- \rightleftharpoons 2\ MnO_2(s) + 4\ H_2O \qquad E^0 = +1.695\ V$$
$$3\ MnO_2(s) + 12\ H^+ + 6\ e^- \rightleftharpoons 3\ Mn^{2+} + 6\ H_2O \qquad E^0 = +1.23\ V$$

Again we have multiplied both equations by integers so that the number of electrons are equal. When this system is at equilibrium,

$$E_{MnO_4^-} = E_{MnO_2}$$

$$1.695 - \frac{0.0592}{6} \log \frac{1}{[MnO_4^-]^2[H^+]^8} = 1.23 - \frac{0.0592}{6} \log \frac{[Mn^{2+}]^3}{[H^+]^{12}}$$

Inverting the log term on the right and rearranging lead to

$$\frac{(1.695 - 1.23)6}{0.0592} = \log \frac{[H^+]^{12}}{[MnO_4^-]^2[Mn^{2+}]^3[H^+]^8}$$

Challenge: Explain why the result in Example 16-3 has only one significant figure.

$$47.1 = \log \frac{[H^+]^4}{[MnO_4^-]^2[Mn^{2+}]^3} = \log K_{eq}$$

$$K_{eq} = \text{antilog } 47.1 = 1.3 \times 10^{47} = 1 \times 10^{47}$$

Note that the final result has only one significant figure.

16B REDOX TITRATION CURVES

Because most redox indicators respond to changes in electrode potential, the vertical axis in oxidation/reduction titration curves is an electrode potential instead of the p-functions that were used for precipitation, complex-formation, and neutralization titration curves. Because there is a logarithmic relationship between electrode potential and analyte or titrant concentration, redox titration curves look much like those for the other types of titrations.

16B-1 Electrode Potentials for Redox Titration Systems

In order to demonstrate the nature of the electrode potential used in deriving redox titration curves, let us consider the titration of iron(II) with a standard solution of cerium(IV). The titration is described by the equation

$$Fe^{2+} + Ce^{4+} \rightleftarrows Fe^{3+} + Ce^{3+}$$

This reaction is rapid and reversible, so that the system is at equilibrium at all times throughout the titration. Consequently, the electrode potentials for the two half-reactions are always identical (Section 16A); that is,

$$E_{Ce^{4+}} = E_{Fe^{3+}} = E_{system}$$

where we have termed E_{system} as *the potential of the system*. If a redox indicator is present in this solution, the ratio of the concentrations of its oxidized and reduced forms must adjust so that the electrode potential for the indicator is also equal to the system potential; that is,

$$E_{In} = E_{Ce^{4+}} = E_{Fe^{3+}} = E_{system}$$

The electrode potential of a system is readily derived from standard-potential data. Thus, for the reaction under consideration, the titration mixture is treated as if it were part of the hypothetical cell

$$SHE\|Ce^{4+}, Ce^{3+}, Fe^{3+}, Fe^{2+}|Pt$$

where SHE symbolizes the standard hydrogen electrode. The potential of the platinum electrode with respect to the standard hydrogen electrode is determined by the tendencies of iron(III) and cerium(IV) to accept electrons— that is, by the tendencies of the following half-reactions to occur:

$$Fe^{3+} + e^- \rightleftarrows Fe^{2+}$$

$$Ce^{4+} + e^- \rightleftarrows Ce^{3+}$$

At equilibrium, the concentration ratios of the oxidized and reduced forms of the two species are such that their attraction for electrons (and thus their electrode potentials) are identical. Note that these concentration ratios vary continuously throughout the titration, as must E_{system}. End points are determined from the characteristic variation in E_{system} that occurs during the titration.

> Most end points in oxidation/reduction titrations take advantage of rapid changes in E_{system} that occur at or near chemical equivalence.

Because $E_{system} = E_{Fe^{3+}} = E_{Ce^{4+}}$, data for a titration curve can be obtained by applying the Nernst equation for *either* the cerium(IV) half-reaction or the iron(III) half-reaction. It turns out, however, that one or the other is more convenient, depending upon the stage of the titration. For example, the iron(III) potential is easier to compute in the region short of the equivalence point because here the concentrations of iron(II) and iron(III) are appreciable and are equal to their analytical concentrations. In contrast, the concentration of cerium(IV), which is negligible prior to equivalence because of the large excess of iron(II), can be obtained at this stage only by calculations based upon the equilibrium constant for the reaction. Beyond the equivalence point, the concentrations of cerium(IV) and cerium(III) are readily computed directly from the volumetric data, but that for iron(II) is not. In this region, then, electrode potential for the cerium(IV) couple is the easier to use. Equivalence-point potentials are derived by the method shown in the next section.

> Before equivalence, E_{system} calculations are easier with the Nernst equation for the analyte. Beyond equivalence, the Nernst equation for the reagent is easier to use.

16B-2 Equivalence-Point Potentials

At the equivalence point, the concentrations of cerium(IV) and iron(II) are minute and cannot be obtained from the stoichiometry of the reaction. Fortunately, equivalence-point potentials are readily obtained by taking advantage of the fact that the two reactant species and the two product species have known concentration ratios at chemical equivalence.

At the equivalence point in the titration of iron(II) with cerium(IV), the potential of the system E_{eq} is given by both

$$E_{eq} = E^0_{Ce^{4+}} - \frac{0.0592}{1} \log \frac{[Ce^{3+}]}{[Ce^{4+}]}$$

and

$$E_{eq} = E^0_{Fe^{3+}} - \frac{0.0592}{1} \log \frac{[Fe^{2+}]}{[Fe^{3+}]}$$

Adding these two expressions gives

$$2E_{eq} = E^0_{Ce^{4+}} + E^0_{Fe^{3+}} - \frac{0.0592}{1} \log \frac{[Ce^{3+}][Fe^{2+}]}{[Ce^{4+}][Fe^{3+}]} \qquad (16\text{-}7)$$

The definition of equivalence point requires that

$$[Fe^{3+}] = [Ce^{3+}]$$

$$[Fe^{2+}] = [Ce^{4+}]$$

Substitution of these equalities into Equation 16-7 results in the concentration quotient becoming unity and the logarithmic term becoming zero:

$$2E_{eq} = E^0_{Ce^{4+}} + E^0_{Fe^{3+}} - \frac{0.0592}{1} \log \frac{[\cancel{Ce^{3+}}][\cancel{Ce^{4+}}]}{[\cancel{Ce^{4+}}][\cancel{Ce^{3+}}]} = E^0_{Ce^{4+}} + E^0_{Fe^{3+}}$$

$$E_{eq} = \frac{E^0_{Ce^{4+}} + E^0_{Fe^{3+}}}{2} \qquad (16\text{-}8)$$

Example 16-4 illustrates how the equivalence-point potential is derived for a more complex reaction.

The concentration quotient in Equation 16-7 is *not* the usual ratio of product concentrations and reactant concentrations that appears in equilibirum-constant expressions.

Example 16-4

Derive an expression for the equivalence-point potential in the titration of 0.0500 M U^{4+} with 0.1000 M Ce^{4+}. Assume both solutions are 1.0 M in H_2SO_4.

$$U^{4+} + 2\,Ce^{4+} + 2\,H_2O \rightleftarrows UO_2^{2+} + 2\,Ce^{3+} + 4\,H^+$$

In Appendix 1 we find

$$UO_2^{2+} + 4\,H^+ + 2\,e^- \rightarrow U^{4+} + 2\,H_2O \qquad\qquad E^0 = 0.334\ V$$

$$Ce^{4+} + e^- \rightleftarrows Ce^{3+} \qquad\qquad E^f = 1.44\ V$$

Here we use the formal potential for Ce^{4+} in 1.0 M H_2SO_4.
   Proceeding as in the text above, we write

$$E_{eq} = E^0_{UO_2^{2+}} - \frac{0.0592}{2} \log \frac{[U^{4+}]}{[UO_2^{2+}][H^+]^4}$$

$$E_{eq} = E^f_{Ce^{4+}} - 0.0592 \log \frac{[Ce^{3+}]}{[Ce^{4+}]}$$

In order to combine the logarithmic terms and eliminate the concentration terms, we must multiply the first equation by 2 to give

$$2E_{eq} = 2E^0_{UO_2^{2+}} - 0.0592 \log \frac{[U^{4+}]}{[UO_2^{2+}][H^+]^4}$$

Adding this to the cerium(IV) equation leads to

$$3E_{eq} = 2E^0_{UO_2^{2+}} + E^f_{Ce^{4+}} - 0.0592 \log \frac{[U^{4+}][Ce^{3+}]}{[UO_2^{2+}][Ce^{4+}][H^+]^4}$$

But at equivalence

$$[U^{4+}] = 2[Ce^{4+}]$$

and

$$[UO_2^{2+}] = 2[Ce^{3+}]$$

Substituting these equations gives, upon rearranging,

$$
\begin{aligned}
E_{eq} &= \frac{2E^0_{UO_2^{2+}} + E^f_{Ce^{4+}}}{3} - \frac{0.0592}{3} \log \frac{2[Ce^{4+}][Ce^{3+}]}{2[Ce^{3+}][Ce^{4+}][H^+]^4} \\
&= \frac{2E^0_{UO_2^{2+}} + E^f_{Ce^{4+}}}{3} - \frac{0.0592}{3} \log \frac{1}{[H^+]^4}
\end{aligned}
$$

We see that the equivalence-point potential is pH-dependent in this titration.

16B-3 The Derivation of Titration Curves

We now show how data for titration curves are computed from standard-potential data.

Consider the titration of 50.00 mL of 0.05000 M Fe^{2+} with 0.1000 M Ce^{4+} in a medium that is 1.0 M in H_2SO_4 at all times. Formal potential data for both half-cell processes are available in Appendix 1 and are used for these calculations:

$$Ce^{4+} + e^- \rightleftarrows Ce^{3+} \qquad E^f = 1.44 \text{ V} \quad (1 \text{ M } H_2SO_4)$$
$$Fe^{3+} + e^- \rightleftarrows Fe^{2+} \qquad E^f = 0.68 \text{ V} \quad (1 \text{ M } H_2SO_4)$$

Initial Potential
The solution contains no cerium species at the outset. In all likelihood, a small but unknown amount of Fe^{3+} is present due to air oxidation of Fe^{2+}. In any event, we lack sufficient information to calculate an initial potential.

Remember: The equation for this reaction is

$$Fe^{2+} + Ce^{4+} \rightleftarrows Fe^{3+} + Ce^{3+}$$

The quantity $[Ce^{4+}]$ is equal to the concentration of $[Fe^{2+}]$ that *does not react* with Ce^{4+}. Thus $[Fe^{3+}]$ must be decreased by this amount, and $[Fe^{2+}]$ must be increased by this amount. $[Ce^{4+}]$ is so small that we can neglect it in both cases.

Potential After Addition of 5.00 mL of Cerium(IV)
When oxidant is added, Ce^{3+} and Fe^{3+} are formed, and the solution now contains appreciable and readily calculated concentrations of three of the participants; that of the fourth, Ce^{4+}, is vanishingly small. Therefore, it is more convenient to use the concentrations of the two iron species to calculate the electrode potential of the system.

The concentration of Fe(III) is equal to its molar concentration less the equilibrium concentration of the unreacted Ce(IV):

$$[Fe^{3+}] = \frac{5.00 \times 0.1000}{50.00 + 5.00} - [Ce^{4+}] \approx \frac{0.500}{55.00}$$

Similarly, the Fe^{2+} concentration is given by its molarity plus $[Ce^{4+}]$:

$$[Fe^{2+}] = \frac{50.00 \times 0.05000 - 5.00 \times 0.1000}{55.00} + [Ce^{4+}] \approx \frac{2.000}{55.00}$$

The validity of the indicated assumptions is readily shown by computing the equilibrium constant for the reaction between Fe(II) and Ce(IV) using the technique described in Section 16A-2. The large value for this constant (7×10^{12}) indicates clearly that the concentration of Ce(IV) must indeed be inconsequential relative to the concentrations of the two iron species.

Substitution for $[Fe^{2+}]$ and $[Fe^{3+}]$ in the Nernst equation gives

$$E_{system} = +0.68 - \frac{0.0592}{1} \log \frac{2.00/55.00}{0.500/55.00} = 0.64 \text{ V}$$

Note that the volumes in the numerator and denominator cancel, which indicates that the potential is independent of dilution. This independence persists until the solution becomes so dilute that the two assumptions made in the calculation become invalid.

It is worth emphasizing again that the use of the Nernst equation for the Ce(IV)/Ce(III) system would yield the same value for E_{system}, but to do so would require computing $[Ce^{4+}]$ by means of the equilibrium constant for the reaction.

Additional potentials needed to define the titration curve short of the equivalence point can be obtained similarly. Such data are given in Table 16-1. You may want to confirm one or two of these values.

Equivalence-Point Potential

Substitution of the two formal potentials into Equation 16-8 yields

$$E_{eq} = \frac{E^f_{Ce^{4+}} + E^f_{Fe^{3+}}}{2} = \frac{1.44 + 0.68}{2} = 1.06 \text{ V}$$

Table 16-1
ELECTRODE POTENTIAL VERSUS SHE IN TITRATIONS WITH 0.1000 M Ce^{4+}

Reagent Volume, mL	Potential, V vs. SHE	
	50.00 mL of 0.05000 M Fe^{2+}	50.00 mL of 0.02500 M U^{4+} *
5.00	0.64	0.316
15.00	0.69	0.339
20.00	0.72	0.352
24.00	0.76	0.375
24.90	0.82	0.405
25.00	1.06 ← Equivalence point →	0.703
25.10	1.30	1.30
26.00	1.36	1.36
30.00	1.40	1.40

*H_2SO_4 concentration is such that $[H^+] = 1.0$ throughout.

Potential After Addition of 25.10 mL of Cerium(IV)

The molar concentrations of Ce(III), Ce(IV), and Fe(III) are readily computed at this point, but that for Fe(II) is not. Therefore, E_{system} computations based on the cerium half-reaction are more convenient. The concentrations of the two cerium ion species are

Again, [Fe²⁺] is equal to the amount of Ce⁴⁺ that is left unreacted, so it is added to $C_{Ce^{4+}}$ calculated from the volumes of the two solutions and subtracted from $C_{Ce^{3+}}$.

$$[Ce^{3+}] = \frac{25.00 \times 0.1000}{75.10} - [Fe^{2+}] \approx \frac{2.500}{75.10}$$

$$[Ce^{4+}] = \frac{25.10 \times 0.1000 - 50.00 \times 0.05000}{75.10} + [Fe^{2+}] \approx \frac{0.010}{75.10}$$

The indicated approximations should be reasonable in view of the favorable equilibrium constant. Substitution of these concentrations into the Nernst equation for the cerium couple gives

$$E = +1.44 - \frac{0.0592}{1} \log \frac{[Ce^{3+}]}{[Ce^{4+}]} = +1.44 - \frac{0.0592}{1} \log \frac{2.500/75.10}{0.010/75.10}$$

$$= +1.30 \text{ V}$$

The additional postequivalence potentials in Table 16-1 were derived in a similar fashion.

The titration curve of iron(II) with cerium(IV) appears as A in Figure 16-2. This plot resembles closely the curves encountered in neutralization, precipitation, and complex-formation titrations, with the equivalence point being signaled by a rapid change in the ordinate function. A

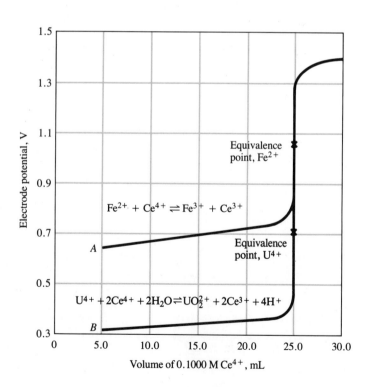

Figure 16-2

Titration curves for 0.1000 M Ce⁴⁺ titration. *A*. Titration of 50.00 mL of 0.05000 M Fe²⁺. *B*. Titration of 50.00 mL of 0.02500 M U⁴⁺.

titration involving 0.00500 M iron(II) and 0.01000 M cerium(IV) yields a curve that is, for all practical purposes, identical to the one derived here, since the electrode potential of the system is independent of dilution.

Example 16-5

Derive a titration curve for the titration of 50.00 mL of 0.02500 M U^{4+} with 0.1000 M Ce^{4+}. Assume that the solution is 1.0 M in H_2SO_4 throughout the titration ($[H^+]$ for such a solution will be about 1.0 M).

The analytical reaction is

$$U^{4+} + 2 H_2O + 2 Ce^{4+} \rightleftharpoons UO_2^{2+} + 2 Ce^{3+} + 4 H^+$$

and in Appendix 1 we find

$$Ce^{4+} + e^- \rightleftharpoons Ce^{3+} \qquad E^f = +1.44 \text{ V}$$
$$UO_2^{2+} + 4 H^+ + 2 e^- \rightleftharpoons U^{4+} + 2 H_2O \qquad E^0 = +0.334 \text{ V}$$

Why is it impossible to calculate the potential before titrant is added?

Potential After Adding 5.00 mL of Ce^{4+}

$$\text{original amount } U^{4+} = 50.00 \text{ mL } U^{4+} \times 0.02500 \, \frac{\text{mmol } U^{4+}}{\text{mL } U^{4+}}$$

$$= 1.250 \text{ mmol } U^{4+}$$

$$\text{amount } Ce^{4+} \text{ added} = 5.00 \text{ mL } Ce^{4+} \times 0.1000 \, \frac{\text{mmol } Ce^{4+}}{\text{mL } Ce^{4+}}$$

$$= 0.5000 \text{ mmol } Ce^{4+}$$

$$\text{amount } UO_2^{2+} \text{ formed} = 0.5000 \text{ mmol } Ce^{4+} \times \frac{1 \text{ mmol } UO_2^{2+}}{2 \text{ mmol } Ce^{4+}}$$

$$= 0.2500 \text{ mmol } UO_2^{2+}$$

$$\text{amount } U^{4+} \text{ remaining} = 1.250 \text{ mmol } U^{4+} - 0.2500 \text{ mmol } UO_2^{2+}$$

$$\times \frac{1 \text{ mmol } U^{4+}}{\text{mmol } UO_2^{2+}} = 1.000 \text{ mmol } U^{4+}$$

$$\text{total volume of solution} = (50.00 + 5.00) \text{ mL} = 55.00 \text{ mL}$$

Applying the Nernst equation for UO_2^{2+}, we obtain

$$E = 0.334 - \frac{0.0592}{2} \log \frac{[U^{4+}]}{[UO_2^{2+}][H^+]^4}$$

$$= 0.334 - \frac{0.0592}{2} \log \frac{[U^{4+}]}{[UO_2^{2+}](1.00)^4}$$

Substituting concentrations of the two uranium species gives

$$E = 0.334 - \frac{0.0592}{2} \log \frac{1.000 \text{ mmol } U^{4+}/55.00 \text{ mL}}{0.2500 \text{ mmol } UO_2^{2+}/55.00 \text{ mL}} = 0.316 \text{ V}$$

Other preequivalence-point data, calculated in the same way, are given in Table 16-1.

Equivalence-Point Potential

Following the procedure shown in Example 16-4, we obtain

$$E_{eq} = \frac{E^f_{Ce^{4+}} + 2E^0_{UO_2^{2+}}}{3} - \frac{0.0592}{3} \log \frac{1}{[H^+]^4}$$

Substituting gives

$$E_{eq} = \frac{1.44 + (2 \times 0.334)}{3} + \frac{0.0592}{3} \log \frac{1}{(1.00)^4} = 0.703 \text{ V}$$

Potential After Adding 25.10 mL of Ce^{4+}

$$\text{original amount } U^{4+} = 50.00 \text{ mL } U^{4+} \times 0.02500 \frac{\text{mmol } U^{4+}}{\text{mL } U^{4+}}$$

$$= 1.250 \text{ mmol } U^{4+}$$

$$\text{amount } Ce^{4+} \text{ added} = 25.10 \text{ mL } Ce^{4+} \times 0.1000 \frac{\text{mmol } Ce^{4+}}{\text{mL } Ce^{4+}}$$

$$= 2.510 \text{ mmol } Ce^{4+}$$

$$\text{amount } Ce^{3+} \text{ formed} = 1.250 \text{ mmol } U^{4+} \times \frac{2 \text{ mmol } Ce^{3+}}{\text{mmol } U^{4+}}$$

$$= 2.500 \text{ mmol } Ce^{3+}$$

$$\text{amount } Ce^{4+} \text{ remaining} = 2.510 \text{ mmol } Ce^{4+} - 2.500 \text{ mmol } Ce^{3+}$$

$$\times \frac{1 \text{ mmol } Ce^{4+}}{\text{mmol } Ce^{3+}} = 0.010 \text{ mmol } Ce^{4+}$$

$$\text{volume of solution} = 75.10 \text{ mL}$$

$$[Ce^{3+}] = 2.500 \text{ mmol } Ce^{3+}/75.10 \text{ mL}$$

$$[Ce^{4+}] = 0.010 \text{ mmol } Ce^{4+}/75.10 \text{ mL}$$

Substituting into the Nernst expression for the Ce^{4+} formal potential gives

$$E = 1.44 - 0.0592 \log \frac{2.500/75.10}{0.010/75.10} = 1.30 \text{ V}$$

Table 16-1 contains other postequivalence-point data obtained in this same way.

Redox titration curves are symmetric when the reactants combine in a 1:1 ratio. Otherwise, they are asymmetric.

The data in the third column of Table 16-1 are plotted as curve B in Figure 16-2. The two curves in the figure are identical for volumes greater than 25.00 mL because the concentrations of the two cerium species are identical in this region. It is also interesting that the curve for iron(II) is symmetric around the equivalence point, while that for uranium(IV) is asymmetric. In general, symmetric curves are always obtained when the analyte and titrant react in a 1:1 molar ratio.

16B-4 The Effect of System Variables on Redox Titration Curves

In earlier chapters, we considered how reactant concentrations and reaction completeness affect titration curves. Here, we describe the effects of these variables on oxidation/reduction titration curves.

Reactant Concentration

As we have just seen, E_{system} for an oxidation/reduction titration is ordinarily independent of dilution. Consequently, titration curves for oxidation/reduction reactions are usually independent of analyte and reagent concentrations. This behavior is in distinct contrast to that observed in the other types of titration curves we have encountered.

Feature 16-1
WHEN REDOX TITRATION CURVES BECOME
CONCENTRATION-DEPENDENT

Electrode potentials become dependent upon dilution when the number of moles of the reactant and product of a half-reaction differ. An example is the reaction

$$I_3^- + 2\,e^- \rightleftarrows 3\,I^-$$

Application of the Nernst equation gives

$$E = E^0 - \frac{0.0592}{2} \log \frac{[I^-]^3}{[I_3^-]}$$

Here, the concentration term in the numerator bears units of $(mol/L)^3$, but the units in the denominator are mol/L. Consequently, the ratio of the two terms is related to the square of concentration. The result is that, in a titration in which the standard reagent is triiodide ion, potentials at and beyond the equivalence point depend upon dilution.

As mentioned earlier, electrode potentials also become concentration-dependent when we cannot assume that the molar concentrations of the various species are equal to the concentrations derived from stoichiometry.

Completeness of the Reaction

The change in E_{system} in the equivalence-point region of an oxidation/reduction titration becomes larger as the reaction becomes more complete. This effect is demonstrated in Figure 16-3, which shows curves for the titration of a hypothetical reductant having a standard electrode potential of 0.20 V with several hypothetical oxidants with standard potentials ranging from 0.40 to 1.20 V; the corresponding equilibrium constants lie between about 2×10^3 and 8×10^{16}. Clearly, the greatest change in

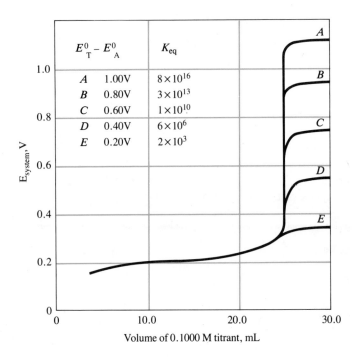

$E_T^0 - E_A^0$		K_{eq}
A	1.00V	8×10^{16}
B	0.80V	3×10^{13}
C	0.60V	1×10^{10}
D	0.40V	6×10^6
E	0.20V	2×10^3

Figure 16-3
Effect of titrant electrode potential upon reaction completeness. The standard electrode potential for the analyte (E_A^0) is 0.200 V; starting with curve A, standard electrode potentials for the titrant (E_T^0) are 1.20, 1.00, 0.80, 0.60, and 0.40 V, respectively. Both analyte and titrant undergo a one-electron change.

potential of the system is associated with the reaction that is most nearly complete. Thus, in this respect, oxidation/reduction titration curves are similar to those involving other types of reactions.

16C   OXIDATION/REDUCTION INDICATORS

Two types of chemical indicators are used in oxidation/reduction titrations. *Specific indicators* react with one of the participants in the titration to produce a color. True *oxidation/reduction indicators,* on the other hand, respond to the potential of the system rather than to the appearance or disappearance of some species during the course of the titration.

16C-1 Specific Indicators

Perhaps the best known specific indicator for an oxidation/reduction titration is starch, which forms a deep blue complex with iodine. The appearance or disappearance of this complex serves as a sensitive indicator for titrations in which iodine is either produced or consumed.

Thiocyanate ion acts as a specific indicator in titrations in which iron(III) is either a reagent or an analyte. For example, the disappearance of the red color of the iron(III)/thiocyanate complex signals the end point in the titration of iron(III) with standard titanium(III).

16C-2 Oxidation/Reduction Indicators

True oxidation/reduction indicators respond to changes in the electrode potential of a system. They are substantially more versatile than specific indicators.

The half-reaction for a typical oxidation/reduction indicator can be written as

$$\text{In}_{ox} + n\text{e}^- \rightleftarrows \text{In}_{red}$$

If the indicator reaction is reversible, we may write

$$E = E^0 - \frac{0.0592}{n} \log \frac{[\text{In}_{red}]}{[\text{In}_{ox}]} \qquad (16\text{-}9)$$

Typically, a color change from the oxidized form of the indicator to the reduced form requires a change in their concentration ratio from

$$\frac{[\text{In}_{red}]}{[\text{In}_{ox}]} \leq \frac{1}{10}$$

to

$$\frac{[\text{In}_{red}]}{[\text{In}_{ox}]} \geq \frac{10}{1}$$

Thus, the color change of an oxidation/reduction indicator requires about a 100-fold change in the concentration ratio of the two forms. The potential change associated with this transition can be established by substituting these boundary values into Equation 16-9, which gives

$$E = E^0 \pm \frac{0.0592}{n} \qquad (16\text{-}10)$$

In order for a typical indicator to undergo a useful transition in color, the titrant must cause a change of $0.118/n$ V in the potential of the system. For many indicators, n is 2, and so a change of 0.059 V is sufficient.

Protons are involved in the reduction of many indicators. For these, the transition range is also pH-dependent.

The potential about which a color transition occurs depends upon the standard potential for the particular indicator system. The indicators shown in Table 16-2 have transition potentials ranging from +1.25 V to about +0.3 V. Turning again to Figure 16-3, we see that all the indicators in Table 16-2 except the first and the last will provide a satisfactory color change for titrations involving titrant A. In contrast, use of titrant D would succeed only if indigo tetrasulfonate (or an indicator with a similar transition range) were used.

End points for many oxidation/reduction titrations can be readily observed by making the analyte solution part of the cell:

reference electrode‖analyte solution|Pt

A plot of the potential of this cell as a function of titrant volume yields curves similar to those in Figures 16-2 and 16-3, and end points can be evaluated graphically. We consider potentiometric titrations in Chapter 18.

Table 16-2
SELECTED OXIDATION/REDUCTION INDICATORS*

Indicator	Color		Transition Potential, V	Conditions
	Oxidized	Reduced		
5-Nitro-1, 10-phenanthroline iron(II) complex	Pale blue	Red-violet	+1.25	1 M H_2SO_4
2,3'-Diphenylamine dicarboxylic acid	Blue-violet	Colorless	+1.12	7–10 M H_2SO_4
1,10-Phenanthroline iron(II) complex	Pale blue	Red	+1.11	1 M H_2SO_4
Erioglaucin A	Blue-red	Yellow-green	+0.98	0.5 M H_2SO_4
Diphenylamine sulfonic acid	Red-violet	Colorless	+0.85	Dilute acid
Diphenylamine	Violet	Colorless	+0.76	Dilute acid
ρ-Ethoxychrysoidine	Yellow	Red	+0.76	Dilute acid
Methylene blue	Blue	Colorless	+0.53	1 M acid
Indigo tetrasulfonate	Blue	Colorless	+0.36	1 M acid
Phenosafranine	Red	Colorless	+0.28	1 M acid

*Data in part from I. M. Kolthoff and V. A. Stenger, *Volumetric Analysis,* 2nd ed., Vol. 1, p. 140. New York: Interscience, 1942.

Feature 16-2
A TYPICAL REDOX INDICATOR

A class of organic compounds known as 1,10-phenanthrolines (or orthophenanthrolines) form stable complexes with iron(II) and certain other ions. The parent compound has a pair of nitrogen atoms, each located in such a position that it can form a covalent bond with the iron(II) ion. Three phenanthroline molecules combine with each iron ion to yield a complex with the structure

This complex, which is sometimes called "ferroin," is conveniently formulated as $(Phen)_3Fe^{2+}$.

The complexed iron in the ferroin undergoes a reversible oxidation/reduction reaction that can be written

$$(Phen)_3Fe^{3+} + e^- \rightleftarrows (Phen)_3Fe^{2+} \qquad E^0 = +1.06 \text{ V}$$

Pale Blue   Red

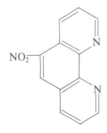

5-Nitro-1,10-phenanthroline

In practice, the blue color of the oxidized form is so slight as to go undetected, and the color change associated with this reduction appears to be from nearly colorless to red. Because of the difference in color intensity, the end point is usually taken when only about 10% of the indicator is in the iron(II) form. The transition potential is thus approximatley +1.11 V in 1 M sulfuric acid.

Of all the oxidation/reduction indicators, ferroin approaches most closely the ideal substance. It reacts rapidly and reversibly, the color change is pronounced, and its solutions are stable and readily prepared. In contrast to many indicators, the oxidized form of ferroin is remarkably inert toward strong oxidizing agents. At temperatures above 60°C, solutions of ferroin decompose.

A number of substituted phenanthrolines have been investigated for their indicator properties, and some have proved to be as useful as the parent compound. Among these, the 5-nitro and 5-methyl derivatives are noteworthy, with transition potentials of +1.25 V and +1.02 V, respectively.

5-Methyl-1,10-phenanthroline

16D   QUESTIONS AND PROBLEMS

*16-1. Why are the transition potentials of many redox indicators pH-dependent?

16-2. Why are titration curves for most redox reactions independent of the concentrations of analyte and standard?

*16-3. Under what conditions are oxidation/reduction titration curves symmetric about the equivalence point?

16-4. Differentiate between a specific and a true redox indicator.

*16-5. Calculate the equilibrium constant for each reaction in Problem 15-6. (For $Tl^{3+} + Br^- + 2 e^- \rightleftarrows TlBr(s)$, $E^0 = 1.44$ V.)

16-6. Calculate the equilibrium constant for each reaction in Problem 15-7.

16-7. Suggest a true redox indicator suitable for the following (assume $[H_3O^+] = 1.00$):

*(a) the titration of quinone, $C_6H_4O_2$, with a standard solution of $V(OH)_4^+$.

(b) the titration of selenous acid, H_2SeO_3, with a standard solution of Ce^{4+}.

*(c) the titration of Fe^{3+} with a standard solution of U^{4+}.

(d) the titration of $Fe(CN)_6^{4-}$ with a standard solution of Tl^{3+}.

*16-8. Calculate the electrode potential of the system at the equivalence point for each of the following reactions. Where necessary, assume that $[H^+] = 1.00$ at equivalence.

(a) $2 Ti^{2+} + Sn^{4+} \rightleftarrows 2 Ti^{3+} + Sn^{2+}$

(b) $2 Ce^{4+} + H_2SeO_3 + H_2O \rightleftarrows$
$SeO_4^{2-} + 2 Ce^{3+} + 4 H^+$ (in 1 M H_2SO_4)

(c) $2 MnO_4^- + 5 HNO_2 + H^+ \rightleftarrows$
$2 Mn^{2+} + 5 NO_3^- + 3 H_2O$

16-9. Calculate the electrode potential of the system at the equivalence point for each of the following reactions. Where necessary, assume that $[H^+] = 0.100$ at equivalence.

(a) $Cr^{2+} + Fe(CN)_6^{3-} \rightleftarrows Cr^{3+} + Fe(CN)_6^{4-}$

(b) $Tl^{3+} + 2 V^{3+} + 2 H_2O \rightleftarrows Tl^+ + 2 VO^{2+} + 4 H^+$

(c) $5 U^{4+} + 2 MnO_4^- + 2 H_2O \rightleftarrows$
$5 UO_2^{2+} + 2 Mn^{2+} + 4 H^+$

*16-10. Calculate equilibrium constants for the reactions in Problem 16-8.

16-11. Calculate equilibrium constants for the reactions in Problem 16-9.

16-12. Construct curves for the following titrations. Calculate potentials after the addition of 10.00, 25.00, 49.00, 49.90, 50.00, 50.10, 51.00, and 60.00 mL of the reagent. Where necessary, assume that $[H^+] = 1.00$ throughout.

*(a) 50.00 mL of 0.1000 M V^{2+} with 0.05000 M Sn^{4+}

(b) 50.00 mL of 0.1000 M $Fe(CN)_6^{3-}$ with 0.1000 M Cr^{2+}

*(c) 50.00 mL of 0.1000 M $Fe(CN)_6^{4-}$ with 0.05000 M Tl^{3+}

(d) 50.00 mL of 0.1000 M Fe^{3+} with 0.05000 M Sn^{2+}

*(e) 50.00 mL of 0.05000 M U^{4+} with 0.02000 M MnO_4^-

(f) 50.00 mL of 0.02000 M Sn^{2+} with 0.02000 M I_3^- (assume that $[I^-] = 0.500$ M throughout the titration)

CHAPTER 17

Applications
of Oxidation/Reduction
Titrations

This chapter is concerned with the preparation of standard solutions of oxidants and reductants and with their applications in analytical chemistry. In addition, auxiliary reagents that convert an analyte to a single oxidation state are described.[1]

17A AUXILIARY OXIDIZING AND REDUCING REAGENTS

The analyte in an oxidation/reduction titration must be in a single oxidation state at the outset. Often, however, the steps that precede the titration (dissolution of the sample and separation of interferences) convert the analyte to a mixture of oxidation states. For example, the solution produced when an iron-containing sample is dissolved usually contains a mixture of iron(II) and iron(III). If we choose to use a standard oxidant for the determination of iron, we must first treat the sample solution with an auxiliary reducing agent; if, on the other hand, we plan to titrate with a

[1]For further reading on redox titrimetry, see J. A. Goldman and V. A. Stenger, in *Treatise on Analytical Chemistry*, I. M. Kolthoff and P. J. Elving, Eds., Part I, Vol. 11, Chapter 119. New York: Wiley, 1975; and I. M. Kolthoff and R. Belcher, *Volumetric Analysis*, Vol. 3. New York: Interscience, 1957.

standard reductant, pretreatment with an auxiliary oxidizing reagent will be needed.[2]

To be useful as a preoxidant or a prereductant, a reagent must react quantitatively with the analyte. In addition, excesses of the reagent must be readily removable because such excesses will ordinarily interfere by consuming standard solution.

17A-1 Auxiliary Reducing Reagents

A number of metals are good reducing agents and have been used for the prereduction of analytes. Included among these are zinc, aluminum, cadmium, lead, nickel, copper, and silver (in the presence of chloride ion). Sometimes, sticks or coils of the metal are immersed directly in the analyte solution. After reduction is judged complete, the solid is removed manually or by filtration.

An alternative to filtration is the use of a *reductor,* such as that shown in Figure 17-1.[3] Here, a finely divided metal is held in a vertical glass tube through which the solution is drawn under a mild vacuum. The metal in a reductor is ordinarily sufficient for hundreds of reductions.

A typical *Jones reductor* has a diameter of about 2 cm and is packed with a 40- to 50-cm column of amalgamated zinc. Amalgamation is accomplished by allowing zinc granules to stand briefly in a solution of mercury(II) chloride:

$$2 \text{ Zn(s)} + \text{Hg}^{2+} \rightarrow \text{Zn}^{2+} + \text{Zn(Hg)(s)}$$

Zinc amalgam is nearly as effective a reducing agent as the pure metal and has the important virtue of inhibiting the reduction of hydrogen ions by zinc, a parasitic reaction that not only needlessly uses up the reducing agent but also heavily contaminates the sample solution with zinc(II) ions. Solutions that are quite acidic can be passed through a Jones reductor without significant hydrogen formation.

Table 17-1 lists the principal applications of the Jones reductor.

Also listed in Table 17-1 are reductions that can be accomplished with a *Walden reductor,* in which granular metallic silver held in a narrow glass column is the reductant. Silver is not a good reducing agent unless chloride or some other ions that forms a silver salt of low solubility is present. For this reason, prereductions with a Walden reductor are generally carried out from hydrochloric acid solutions of the analyte. The coating of silver chloride produced on the metal is removed periodically by dipping a zinc rod into the solution that covers the packing.

Table 17-1 suggests that the Walden reductor is somewhat more selective in its action than is the Jones reductor.

---

[2]For a brief summary of auxiliary reagents, see J. A. Goldman and V. A. Stenger, in *Treatise on Analytical Chemistry,* I. M. Kolthoff and P. J. Elving, Eds., Part I, Vol. 11, pp. 7204–7206. New York: Wiley, 1975.

[3]For a discussion of reductors, see F. Hecht, in *Treatise on Analytical Chemistry,* I. M. Kolthoff and P. J. Elving, Eds., Part I, Vol. 11, pp. 6703–6707. New York: Wiley, 1975.

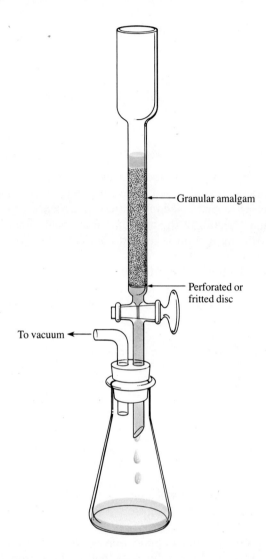

Granular amalgam

Perforated or
fritted disc

To vacuum ◄─

Figure 17-1
A Jones reductor.

Table 17-1
USES OF THE WALDEN REDUCTOR AND THE JONES REDUCTOR*

Walden $Ag(s) + Cl^- \rightarrow AgCl(s) + e^-$	Jones $Zn(Hg)(s) \rightarrow Zn^{2+} + Hg + 2\ e^-$
$Fe^{3+} + e^- \rightarrow Fe^{2+}$	$Fe^{3+} + e^- \rightarrow Fe^{2+}$
$Cu^{2+} + e^- \rightarrow Cu^+$	$Cu^{2+} + 2\ e^- \rightarrow Cu(s)$
$H_2MoO_4 + 2\ H^+ + e^- \rightarrow MoO_2^+ + 2\ H_2O$	$H_2MoO_4 + 6\ H^+ + 3\ e^- \rightarrow Mo^{3+} + 4\ H_2O$
$UO_2^{2+} + 4\ H^+ + 2\ e^- \rightarrow U^{4+} + 2\ H_2O$	$UO_2^{2+} + 4\ H^+ + 2\ e^- \rightarrow U^{4+} + 2\ H_2O$
	$UO_2^{2+} + 4\ H^+ + 3\ e^- \rightarrow U^{3+} + 2\ H_2O$†
$V(OH)_4^+ + 2\ H^+ + e^- \rightarrow VO^{2+} + 3\ H_2O$	$V(OH)_4^+ + 4\ H^+ + 3\ e^- \rightarrow V^{2+} + 4\ H_2O$
TiO^{2+} not reduced	$TiO^{2+} + 2\ H^+ + e^- \rightarrow Ti^{3+} + H_2O$
Cr^{3+} not reduced	$Cr^{3+} + e^- \rightarrow Cr^{2+}$

*From I. M. Kolthoff and R. Belcher, *Volumetric Analysis,* Vol. 3, p. 12. New York:
Interscience, 1957. With permission.

†A mixture of oxidation states is obtained. The Jones reductor may still be used for the
analysis of uranium, however, because any U^{3+} formed can be converted to U^{4+} by shaking
the solution with air for a few minutes.

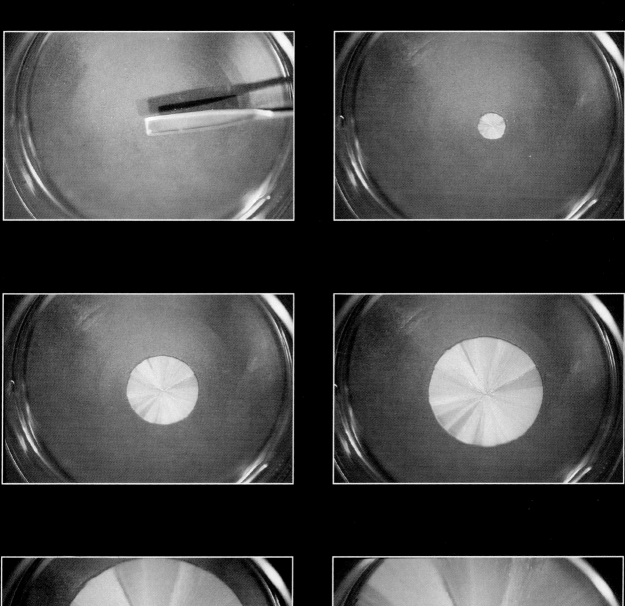
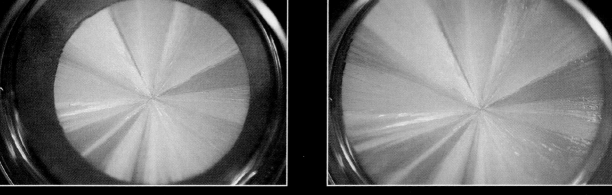

Crystallization of sodium.acetate from a
supersaturated solution (Section 5A-1).

Tyndall effect (see page 69).

Precipitation of 0.1041 g of Fe(III) with NH_3 (left), and homogeneously with urea (right) (Section 5A-5).

Chemical Equilibrium 1: Reaction between iodine and arsenic(III) at pH 1 (Section 7B-1).

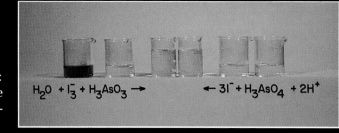

$$H_2O + I_3^- + H_3AsO_3 \longrightarrow \quad \longleftarrow 3I^- + H_3AsO_4 + 2H^+$$

Chemical Equilibrium 2: Reaction between iodine and arsenic(III) at pH 7 (Section 7B-1).

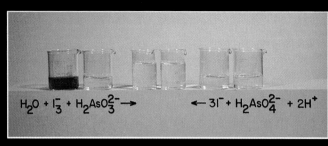

$$H_2O + I_3^- + H_2AsO_3^{2-} \longrightarrow \quad \longleftarrow 3I^- + H_2AsO_4^{2-} + 2H^+$$

Chemical Equilibrium 3: Reaction between iodine and ferrocyanide (Section 7B-1).

$$I_3^- + 2Fe(CN)_6^{4-} \longrightarrow \quad \longleftarrow 3I^- + 2Fe(CN)_6^{3-}$$

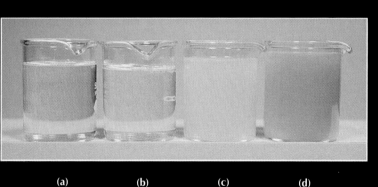

(a) (b) (c) (d)

Argentometric determination of chloride: Fajans method (Section 10B-3). (a) aqueous 2′, 7′-dichlorofluorescein; (b) same, plus 1 mL 0.10 M Ag^+. Note the absence of a precipitate; (c) same, plus AgCl and an excess of Cl^-; (d) same, plus AgCl and the first slight excess of Ag^+.

(a) (b) (c) (d)

In (c) and (d) the AgCl has coagulated. Note that in (d) the dye has been carried down on the precipitate and that the supernatant solution is clear while in (c) the dye remains in solution.

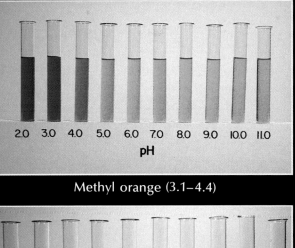

Methyl orange (3.1–4.4)

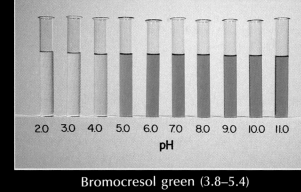

Bromocresol green (3.8–5.4)

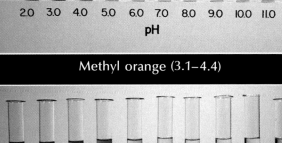
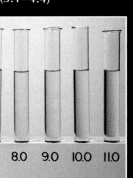

Methyl red (4.2–6.3)

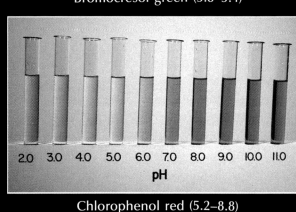

Chlorophenol red (5.2–8.8)

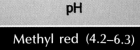
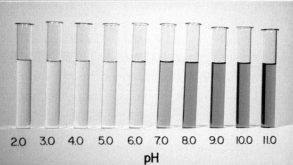

Bromothymol blue (6.0–7.6)

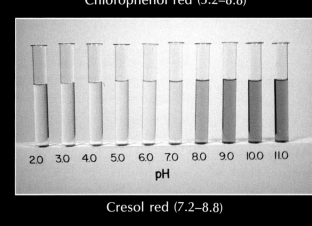

Cresol red (7.2–8.8)

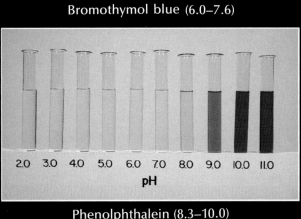

Phenolphthalein (8.3–10.0)

ACID-BASE INDICATORS AND
THEIR TRANSITION pH RANGES
(SECTION 11A).

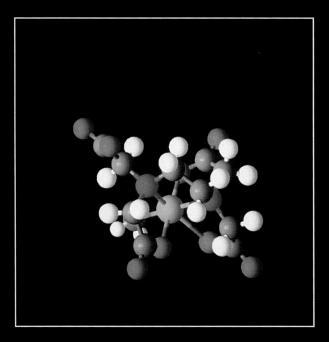

Model of an EDTA/cation chelate (Section 14B-2).

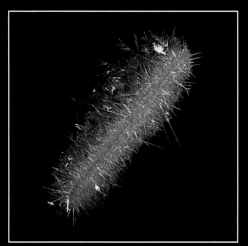

Reduction of silver(I) by direct reaction with copper (Section 15A-2).

Reduction of silver(I) by copper in an electrochemical cell (Section 15A-2).

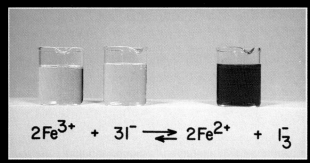

$$2Fe^{3+} + 3I^- \rightleftharpoons 2Fe^{2+} + I_3^-$$

Reaction between Iron(III) and Iodide (margin note page 277).

(a)   (b)   (c)   (d)

Starch/iodine end point (Section 17B-2): (a) iodine solution; (b) same, within a few drops of the equivalence point; (c) same as (b), with starch added; (d) same as (c), at equivalence.

Reaction between permanganate and oxalate (Section 17C-1).

Laminar flow burner for atomic absorption spectroscopy (Section 23B-1). Courtesy of Varian Instruments, Sunnyvale, CA.

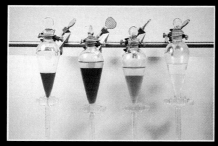

Extraction of iodine with chloroform (Section 24A). Left to right: iodine solution; after single 40-mL extraction; after two 20-mL extractions; after four 10-mL extractions.

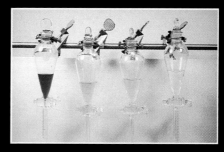

Extraction of iodine with chloroform, aqueous layers only.

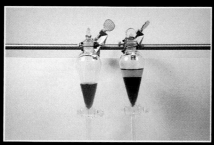

Extraction of iodine with chloroform (Section 24A). Original solution (left); after a single 40-mL extraction (right).

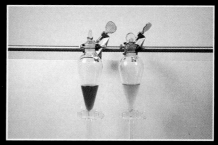

Extraction of iodine with chloroform, aqueous phase only.

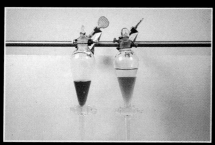

Extraction of iodine with chloroform (Section 24A). Original solution (left); after two 20-mL extractions (right).

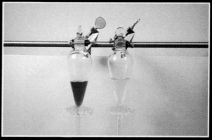

Extraction of iodine with chloroform, aqueous phase only.

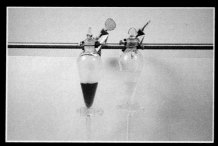

Extraction of Iodine with Chloroform (Section 24A). Original solution (left); after four 10-mL extractions (right).

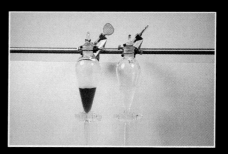

Extraction of iodine with chloroform, aqueous phase only.

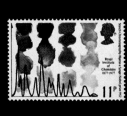

Stamps from various countries celebrating scientists and events of importance to analytical chemistry. The numbers following each citation refer to the Scott Standard Postage Stamp Catalogue. Stamps commemorating (1) the Centennial of the Metric Convention of 1875, France, #1475; (2) T. W. Richards, Sweden, #1104; (3) Svante Arrhenius, Sweden, #549; (4) Cato Guldberg and Peter Waage, Norway, #453; (5) Walther Nernst, Sweden, #655; (6) Allessandro Volta, Italy, #527; (7) André Marie Ampère, Monaco, #1001; (8) Gustav Robert Kirchhoff, Germany, #9N345; (9) spectroscopic analysis, Vatican, #655; (10) Jöns Jacob Berzelius, Sweden, #1293. (11) Stamp honoring Archer J. P. Martin and Richard L. M. Synge for their work in chromatography, Great Britain, #808. (12) Stamp depicting the visible spectrum and atomic lines, Canada, #613. (13) Stamp announcing an international spectroscopy colloquium in Madrid, Spain, #1570. These stamps are from the private collection of Professor C. M. Lang.

17A-2 Auxiliary Oxidizing Reagents

Sodium Bismuthate

Sodium bismuthate is an extremely powerful oxidizing agent capable, for example, of converting manganese(II) quantitatively to permanganate ion. This bismuth salt is a sparingly soluble solid with a formula that is usually written as $NaBiO_3$, although the exact composition is somewhat uncertain. Oxidations are performed by suspending the bismuthate in the analyte solution and boiling for a brief period. The unused reagent is then removed by filtration. The half-reaction for the reduction of sodium bismuthate can be written as

$$NaBiO_3(s) + 4\ H^+ + 2\ e^- \rightleftarrows BiO^+ + Na^+ + 2\ H_2O$$

Ammonium Peroxydisulfate

Ammonium peroxydisulfate, $(NH_4)_2S_2O_8$, is also a powerful oxidizing agent. In acidic solution, it converts chromium(III) to dichromate, cerium(III) to its tetravalent state, and manganese(II) to permanganate. The half-reaction is

$$S_2O_8^{2-} + 2\ e^- \rightleftarrows 2\ SO_4^{2-} \qquad E^0 = 2.01\ V$$

The oxidations are catalyzed by traces of silver ion. The excess reagent is readily decomposed by a brief period of boiling:

$$2\ S_2O_8^{2-} + 2\ H_2O \rightarrow 4\ SO_4^{2-} + O_2(g) + 4\ H^+$$

Sodium Peroxide and Hydrogen Peroxide

Peroxide is a convenient oxidizing agent either as the solid sodium salt or as a dilute solution of the acid. The half-reaction for hydrogen peroxide in acidic solution is

$$H_2O_2 + 2\ H^+ + 2\ e^- \rightleftarrows 2\ H_2O \qquad E^0 = 1.78\ V$$

After oxidation is complete, the solution is freed of excess reagent by boiling:

$$2\ H_2O_2 \rightarrow 2\ H_2O + O_2(g)$$

17B APPLICATIONS OF STANDARD REDUCTANTS

Standard solutions of most reducing agents tend to react with atmospheric oxygen. For this reason, reductants are seldom used for the direct titration of oxidizing analytes; indirect methods are used instead. The two most common indirect methods, discussed in the paragraphs that follow, are based upon standard solutions of iron(II) and standard solutions of sodium thiosulfate.

17B-1 Iron(II) Solutions

Solutions of iron(II) are readily prepared from iron(II) ammonium sulfate, $Fe(NH_4)_2(SO_4)_2 \cdot 6\ H_2O$ (Mohr's salt), or from the closely related iron(II) ethylenediamine sulfate, $FeC_2H_4(NH_3)_2(SO_4)_2 \cdot 4\ H_2O$ (Oesper's salt). Air-oxidation of iron(II) takes place rapidly in neutral solutions but is inhibited in the presence of acids, with the most stable preparations being about 0.5 M in H_2SO_4. Such solutions are stable for no longer than one day, if that long.

Numerous oxidizing agents are conveniently determined by treatment of the sample solution with a measured excess of standard iron(II) followed by immediate titration of the excess with a standard solution of potassium dichromate or cerium(IV) (Sections 17C-1 and 17C-2). Just before or after the samples are titrated, the concentration of the iron(II) solution is established by titrating two or three aliquots of the reagent by itself.

This procedure has been applied to the determination of organic peroxides; chromium(VI); cerium(IV); molybdenum(VI); nitrate, chlorate, and perchlorate ions; and numerous other oxidants.

17B-2 Sodium Thiosulfate

Thiosulfate ion is a moderately strong reducing agent that has been widely used for determining oxidizing agents by an indirect procedure that involves iodine as an intermediate. With iodine, thiosulfate ion is oxidized quantitatively to tetrathionate ion, the half-reaction being

$$2\ S_2O_3^{2-} \rightleftharpoons S_4O_6^{2-} + 2\ e^-$$

In its reaction with iodine, each thiosulfate ion gains one electron.

The quantitative aspect of the reaction with iodine is unique. Other oxidants oxidize the tetrathionate ion, wholly or in part, to sulfate ion.

The scheme used for determining oxidizing agents involves adding an unmeasured excess of potassium iodide to a slightly acidic solution of the analyte. Reduction of the analyte produces an amount of iodine that is stoichiometrically related to the number of moles of analyte. The liberated iodine is then titrated with a standard solution of sodium thiosulfate, $Na_2S_2O_3$, one of the few reducing agents that is stable toward air-oxidation. An example of this procedure is the determination of sodium hypochlorite in bleaches. The reactions are

$$OCl^- + 2\ I^- + 2\ H^+ \rightarrow Cl^- + I_2 + H_2O \qquad \text{(excess KI)}$$
$$I_2 + 2\ S_2O_3^{2-} \rightarrow 2\ I^- + S_4O_6^{2-} \tag{17-1}$$

The quantitative conversion of thiosulfate ion to tetrathionate ion shown in Equation 17-1 requires a pH somewhat lower than 7. If strongly acidic solutions must be titrated, air-oxidation of the excess iodide must be prevented by blanketing the solution with an inert gas, such as carbon dioxide or nitrogen.

End Points in Iodine/Thiosulfate Titrations

The color of iodine can be discerned in a solution that is about 5×10^{-6} M in I_2, a concentration that corresponds to less than one drop of a 0.05 M solution in 100 mL. Thus, provided the analyte solution is colorless, the reagent can serve as its own indicator.

More commonly, titrations involving iodine are performed with a suspension of starch as an indicator. The deep blue color of starch solutions containing iodine is believed to arise from the absorption of iodine into the helical chain of β-amylose, a macromolecular component of most starches (Figure 17-2). The closely related α-amylose forms a red adduct with iodine. This reaction is not readily reversible and is thus undesirable. So-called *soluble starch*, which is available from commercial sources, consists principally of β-amylose, the alpha fraction having been removed; indicator solutions are readily prepared from this product.

Aqueous starch suspensions decompose within a few days, primarily because of bacterial action. The decomposition products tend to interfere with the solution's indicator properties and may also be oxidized by iodine. The rate of decomposition can be inhibited by preparing and storing the indicator under sterile conditions and by adding mercury(II) iodide or chloroform as a bacteriostat. Perhaps the simplest alternative is to prepare a fresh suspension of the indicator, which requires only a few minutes, on the day it is to be used.

Starch decomposes irreversibly in solutions containing large concentrations of iodine. Therefore, in titrating solutions of iodine with sodium thiosulfate, as in the indirect determination of oxidants, addition of the indicator is delayed until the titration is nearly complete (indicated by a change in color from deep red to faint yellow). When thiosulfate solutions are being titrated directly with iodine, the indicator can be introduced at the outset.

> When titrating iodine solutions with thiosulfate ion, addition of the starch must be delayed until most of the iodine has been consumed in order to avoid decomposition of the starch by the high concentration of iodine.

The Stability of Sodium Thiosulfate Solutions

Although sodium thiosulfate solutions are resistant to air-oxidation, they do tend to decompose to give sulfur and hydrogen sulfite ion:

$$S_2O_3^{2-} + H^+ \rightleftarrows HSO_3^- + S(s)$$

Variables that influence the rate of this reaction include pH, presence of microorganisms, solution concentration, presence of copper(II) ions, and exposure to sunlight. These variables may cause the concentration of a thiosulfate solution to change by several percent over a period of a few weeks. On the other hand, proper attention to detail will yield solutions that need only occasional restandardization.

The rate of the decomposition reaction increases markedly with increasing acidity.

The most important single cause for the instability of neutral or slightly basic thiosulfate solutions can be traced to bacteria that metabolize thiosulfate ion to sulfite and sulfate ions as well as elemental sulfur. To minimize this problem, standard solutions of the reagent are prepared under reasonably sterile conditions. Bacterial activity appears to be at a

> When sodium thiosulfate is added to a strongly acidic medium, a cloudiness develops almost immediately as elemental sulfur forms. Even in neutral solution, this reaction proceeds at such a rate that standard sodium thiosulfate must be restandardized periodically.

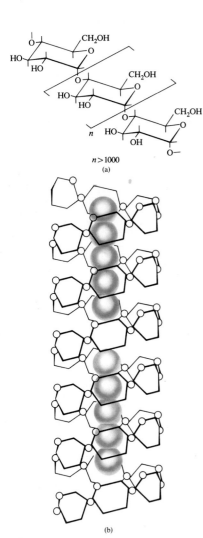

Figure 17-2
Thousands of glucose molecules polymerize to form huge molecules of β-amylose as shown schematically in (a) above. Molecules of β-amylose tend to assume a helical structure. The iodine species I_5^- as shown in (b) is incorporated into the amylose helix. For further details see R. C. Teitelbaum, S. L. Ruby, and T. J. Marks, *J. Amer. Chem. Soc.*, **1980**, *102*, 3322.

minimum at a pH between 9 and 10, which accounts, at least in part, for the reagent's greater stability in slightly basic solutions. The presence of a bactericide, such as chloroform, sodium benzoate, or mercury(II) iodide, also slows decomposition.

The Standardization of Thiosulfate Solutions

Potassium iodate is an excellent primary standard for thiosulfate solutions. In this application, weighed amounts of primary-standard-grade reagent are dissolved in water containing an excess of potassium iodide. When this mixture is acidified with a strong acid, the reaction

$$IO_3^- + 5\ I^- + 6\ H^+ \rightleftarrows 3\ I_2 + 3\ H_2O$$

occurs instantly. The liberated iodine is then titrated with the thiosulfate solution. The stoichiometry of the reactions is

$$1\ mol\ IO_3^- \equiv 3\ mol\ I_2 \equiv 6\ mol\ S_2O_3^{2-}$$

Example 17-1

A solution of sodium thiosulfate was standardized by dissolving 0.1210 g of KIO_3 (fw = 214.00 g) in water, adding a large excess of KI, and acidifying with HCl. The liberated iodine required 41.64 mL of the thiosulfate solution to decolorize the blue starch/iodine complex. Calculate the molarity of the $Na_2S_2O_3$.

$$amount\ NaS_2O_3 = 0.1210\ g\ KIO_3 \times \frac{1\ mmol\ KIO_3}{0.21400\ g\ KIO_3} \times \frac{6\ mmol\ Na_2S_2O_3}{mmol\ KIO_3}$$

$$= 3.3925\ mmol\ Na_2S_2O_3$$

$$C_{Na_2S_2O_3} = \frac{3.3925\ mmol\ Na_2S_2O_3}{41.64\ mL\ Na_2S_2O_3} = 0.08147\ M$$

Table 17-2
SOME APPLICATIONS OF SODIUM THIOSULFATE AS A REDUCTANT

Substance	Half-Reaction	Special Conditions
IO_4^-	$IO_4^- + 8\ H^+ + 7\ e^- \rightarrow \frac{1}{2}I_2 + 4\ H_2O$	Acidic solution
	$IO_4^- + 2\ H^+ + 2\ e^- \rightarrow IO_3^- + H_2O$	Neutral solution
IO_3^-	$IO_3^- + 6\ H^+ + 5\ e^- \rightarrow \frac{1}{2}I_2 + 3\ H_2O$	Strong acid
BrO_3^-, ClO_3^-	$XO_3^- + 6\ H^+ + 6\ e^- \rightarrow X^- + 3\ H_2O$	Strong acid
Br_2, Cl_2	$X_2 + 2\ I^- \rightarrow I_2 + 2\ X^-$	
NO_2^-	$HNO_2 + H^+ + e^- \rightarrow NO(g) + H_2O$	
Cu^{2+}	$Cu^{2+} + I^- + e^- \rightarrow CuI(s)$	
O_2	$O_2 + 4\ Mn(OH)_2(s) + 2\ H_2O \rightarrow 4\ Mn(OH)_3(s)$	Basic solution
	$Mn(OH)_3(s) + 3\ H^+ + e^- \rightarrow Mn^{2+} + 3\ H_2O$	Acidic solution
O_3	$O_3(g) + 2\ H^+ + 2\ e^- \rightarrow O_2(g) + H_2O$	
Organic peroxide	$ROOH + 2\ H^+ + 2\ e^- \rightarrow ROH + H_2O$	

Other primary standards for sodium thiosulfate are potassium dichromate, potassium bromate, potassium hydrogen iodate, potassium ferricyanide, and metallic copper. All these compounds liberate stoichiometric amounts of iodine when treated with excess potassium iodide.

Applications of Sodium Thiosulfate Solutions
Numerous substances can be determined by the indirect method involving titration with sodium thiosulfate; typical applications are summarized in Table 17-2.

17C APPLICATIONS OF STANDARD OXIDANTS

Table 17-3 summarizes the properties of five of the most widely used volumetric oxidizing reagents. Note that the standard potentials for these reagents vary from 0.5 to 1.5 V. The choice among them depends upon analyte strength as a reducing agent, rate of reaction between oxidant and analyte, stability of the standard oxidant solutions, cost, and availability of a satisfactory indicator.

17C-1 The Strong Oxidants: Potassium Permanganate and Cerium(IV)

Solutions of permanganate ion and cerium(IV) ion are strong oxidizing reagents whose applications closely parallel one another. Half-reactions for the two are

$$MnO_4^- + 8 H^+ + 5 e^- \rightleftarrows Mn^{2+} + 4 H_2O \qquad E^0 = 1.51 \text{ V}$$

$$Ce^{4+} + e^- \rightleftarrows Ce^{3+} \qquad E^f = 1.44 \text{ V (1 M } H_2SO_4)$$

Table 17-3
SOME COMMON OXIDANTS USED AS STANDARD SOLUTIONS

Reagent and Formula	Reduction Product	Standard Potential, V	Standardized with	Indicator*	Stability†
Potassium permanganate, $KMnO_4$	Mn^{2+}	1.51	$Na_2C_2O_4$, Fe, As_2O_3	MnO_4^-	(b)
Potassium bromate, $KBrO_3$	Br^-	1.44	$KBrO_3$	(1)	(a)
Cerium(IV), Ce^{4+}	Ce^{3+}	1.44‡	$Na_2C_2O_4$, Fe, As_2O_3	(2)	(a)
Potassium dichromate, $K_2Cr_2O_7$	Cr^{3+}	1.33	$K_2Cr_2O_7$, Fe	(3)	(a)
Iodine, I_2	I^-	0.536	$BaS_2O_3 \cdot H_2O$, $Na_2S_2O_3$	starch	(c)

*(1) α-Naphthoflavone; (2) 1, 10-phenanthroline iron(II) complex (ferroin); (3) diphenylamine sulfonic acid.

†(a) Indefinitely stable; (b) moderately stable, requires periodic standardization; (c) somewhat unstable, requires frequent standardization.

‡E^f in 1 M H_2SO_4.

The formal potential shown for the reduction of cerium(IV) is for solutions that are 1 M in sulfuric acid. In 1 M perchloric acid and 1 M nitric acid, the potentials are 1.70 and 1.61 V, respectively. Solutions of cerium(IV) in the latter two acids are not very stable and thus find limited application.

The half-reaction shown for permanganate ion occurs only in solutions that are 0.1 M or greater in strong acid. The product may be Mn(III), Mn(IV), or Mn(VI) in less acidic media.

Comparison of the Two Reagents

For all practical purposes, the oxidizing strengths of permanganate and cerium(IV) solutions are comparable. Solutions of cerium(IV) in sulfuric acid, however, are stable indefinitely, whereas permanganate solutions decompose slowly and thus require occasional restandardization. Furthermore, cerium(IV) solutions in sulfuric acid do not oxidize chloride ion and thus can be used to titrate hydrochloric acid solutions of analytes. In contrast, permanganate ion cannot be used with hydrochloric acid solutions unless special precautions are taken to prevent the slow oxidation of chloride ion that leads to overconsumption of the standard reagent. A further advantage of cerium(IV) is that a primary-standard-grade salt of the reagent is available, thus making possible the direct preparation of standard solutions.

Despite these several advantages of cerium solutions over permanganate solutions, the latter are more widely used. One reason for the popularity of permanganate solutions is their modest cost. The cost of 1 L of 0.02 M solution is about $0.08, whereas 1 L of cerium(IV) of comparable strength costs about $2.20 ($4.40 if primary-standard-grade reagent is employed). Another disadvantage of cerium(IV) solutions arises from their tendency to form precipitates of basic cerium(IV) salts in solutions that are less than 0.1 M in strong acid.

End Points

A useful property of a potassium permanganate solution is its intense purple color, which is sufficient to serve as an indicator for most titrations. As little as 0.01 mL of a 0.02 M solution imparts a perceptible color to 100 mL of water. If the permanganate solution is very dilute, diphenylamine sulfonic acid or the 1,10-phenanthroline complex of iron(II) (Table 16-2) provides a sharper end point.

The permanganate end point is not permanent because excess permanganate ions react slowly with the relatively large concentration of manganese(II) ions present at the end point:

$$2\ MnO_4^- + 3\ Mn^{2+} + 2\ H_2O \rightarrow 5\ MnO_2(s) + 4\ H^+$$

The equilibrium constant for this reaction is about 10^{47}, which indicates that the *equilibrium* concentration of permanganate ion is vanishingly small even in highly acidic media. Fortunately, the rate at which this equilibrium is approached is so low that the end point fades only gradually over a period of perhaps 30 s.

Solutions of cerium(IV) are yellow-orange, but the color is not intense enough to act as an indicator in titrations. The most widely used indicator for titrations with cerium(IV) is the iron(II) complex of 1,10-phenanthroline or one of its substituted derivatives (Table 16-2).

The Preparation and Stability of Standard Solutions

Potassium Permanganate Solutions. Aqueous solutions of permanganate are not entirely stable because the ion tends to oxidize water:

$$4 \, MnO_4^- + 2 \, H_2O \rightarrow 4 \, MnO_2(s) + 3 \, O_2(g) + 4 \, OH^-$$

Although the equilibrium constant for this reaction indicates that the products are favored, permanganate solutions, when properly prepared, are reasonably stable because the decomposition reaction is slow.

Decomposition is catalyzed by light, heat, acids, bases, manganese(II), and manganese dioxide. Moderately stable solutions of permanganate ion can be prepared if the effects of these catalysts, particularly manganese dioxide, are minimized. Manganese dioxide is a contaminant in even the best grade of solid potassium permanganate. Furthermore, this compound forms in freshly prepared solutions of the reagent as a consequence of the reaction of permanganate ion with organic matter and dust in the water used to prepare the solution. Removal of manganese dioxide by filtration before standardization markedly enhances the stability of permanganate solutions. Before filtration, the solution is allowed to stand for about 24 h or is heated for a brief period to hasten oxidation of the organic species generally present in small amounts in distilled and deionized water. A filtering crucible must be used for filtering because permanganate ion reacts with paper to form additional manganese dioxide.

Standardized permanganate solutions should be stored in the dark. Filtration and restandardization are required if any solid is detected in the solution or on the walls of the storage bottle. In any event, restandardization every one or two weeks is a good precautionary measure.

Solutions containing excess standard permanganate should never be heated because they decompose by oxidizing water. This decomposition cannot be compensated for with a blank. On the other hand, it is possible to titrate hot, acidic solutions of reductants with permanganate without error provided the reagent is added slowly enough so that large excesses do not accumulate.

> Permanganate solutions are moderately stable provided they are freed of manganese dioxide and stored in a dark container.

Example 17-2

Describe how you would prepare 2.0 L of an approximately 0.010 M solution of $KMnO_4$ (fw = 158.04 g).

$$\text{wt } KMnO_4 \text{ needed} = 2.0 \, L \times 0.010 \, \frac{\text{mol } KMnO_4}{L} \times \frac{158.04 \text{ g } KMnO_4}{\text{mol } KMnO_4}$$

$$= 3.16 \text{ g}$$

Dissolve about 3.2 g of $KMnO_4$ in a little water. After solution is complete, add water to bring the volume to about 2.0 L. Heat the solution to boiling for a brief period, and let stand until cool. Filter through a glass filtering crucible and store in a clean dark bottle.

Cerium(IV) Solutions. The most widely used compounds for preparing solutions of cerium(IV) are listed in Table 17-4. Primary-standard-grade cerium ammonium nitrate is available commercially and can be used to prepare standard solutions of the cation directly by weight. More commonly, less expensive reagent-grade cerium(IV) ammonium nitrate or cerium(IV) hydroxide is used to give a solution that is subsequently standardized. In either case, the reagent is dissolved in a solution that is at least 0.1 M in sulfuric acid to prevent the precipitation of basic salts.

Sulfuric acid solutions of tetravalent cerium are remarkably stable and can be stored for months or heated at 100°C for prolonged periods without change in concentration.

Primary Standards
Several excellent primary standards are available for the standardization of solutions of potassium permanganate and tetravalent cerium.

Sodium Oxalate. Sodium oxalate is widely used for standardizing permanganate and cerium(IV) solutions. In acidic solutions, the oxalate ion is converted to the undissociated acid. Thus, its reaction with the permanganate ion can be depicted as

$$2 MnO_4^- + 5 H_2C_2O_4 + 6 H^+ \rightleftharpoons 2 Mn^{2+} + 10 CO_2(g) + 8 H_2O$$

The reaction between Ce^{4+} and $H_2C_2O_4$ is

$$2 Ce^{4+} + H_2C_2O_4 \rightleftharpoons 2 Ce^{3+} + 2 H^+ + 2 CO_2$$

Carbon dioxide is also the product with cerium(IV).

The reaction between permanganate ion and oxalic acid is complex and proceeds slowly even at elevated temperature unless manganese(II) is present as a catalyst. Thus, when the first few milliliters of standard permanganate are added to a hot solution of oxalic acid, several seconds elapse before the color of the permanganate ion disappears. As the concentration of manganese(II) builds up, however, the reaction proceeds more and more rapidly as a result of autocatalysis.

It has been found that when solutions of sodium oxalate are titrated at 60 to 90°C, the consumption of permanganate is from 0.1 to 0.4% less than theoretical, probably due to the air-oxidation of a fraction of the oxalic acid. This small error can be avoided by adding 90 to 95% of the required

Table 17-4
ANALYTICALLY USEFUL CERIUM(IV) COMPOUNDS

Name	Formula	Formula Weight
Cerium(IV) ammonium nitrate	$Ce(NO_3)_4 \cdot 2 NH_4NO_3$	548.2
Cerium(IV) ammonium sulfate	$Ce(SO_4)_2 \cdot 2(NH_4)_2SO_4 \cdot 2 H_2O$	632.6
Cerium(IV) hydroxide	$Ce(OH)_4$	208.1
Cerium(IV) hydrogen sulfate	$Ce(HSO_4)_4$	528.4

permanganate to a cool solution of the oxalate. After the added permanganate is completely used up, as indicated by the disappearance of color, the solution is heated to about 60°C and titrated to a pink color that persists for about 30 s. The disadvantage of this procedure is that it requires a knowledge of the approximate concentration of the permanganate solutions so that a proper initial volume of it can be added. For most purposes, the direct titration of the hot oxalic acid solution provides perfectly adequate data (usually 0.2 to 0.3% high).

Tetravalent cerium standardizations against sodium oxalate are usually performed at 50°C in a hydrochloric acid solution containing iodine monochloride as a catalyst.

Solutions of $KMnO_4$ and Ce^{4+} can also be standardized with electrolytic iron wire and potassium iodide.

Example 17-3

You wish to standardize the solution in Example 17-2 against primary-grade $Na_2C_2O_4$ (fw = 134.00 g). If you want to use between 30 and 45 mL of the reagent for the standardization, what range of weights of the primary standard should you take?

For a 30-mL titration:

$$\text{amount } KMnO_4 = 30 \text{ mL } KMnO_4 \times 0.010 \frac{\text{mmol } KMnO_4}{\text{mL } KMnO_4}$$

$$= 0.30 \text{ mmol } KMnO_4$$

$$\text{wt } Na_2C_2O_4 = 0.30 \text{ mmol } KMnO_4 \times \frac{5 \text{ mmol } Na_2C_2O_4}{2 \text{ mmol } KMnO_4}$$

$$\times 0.134 \frac{\text{g } Na_2C_2O_4}{\text{mmol } Na_2C_2O_4} = 0.101 \text{ g } Na_2C_2O_4$$

Proceeding in the same way, we find for a 45-mL titration:

$$\text{wt } Na_2C_2O_4 = 45 \times 0.010 \times \frac{5}{2} \times 0.134 = 0.151 \text{ g } Na_2C_2O_4$$

Thus, your samples should weigh between 0.10 and 0.15 g.

Example 17-4

Exactly 33.31 mL of the solution in Example 17-2 was required to titrate a 0.1278-g sample of primary-standard $Na_2C_2O_4$. What is the molarity of the $KMnO_4$ reagent?

$$\text{amount } Na_2C_2O_4 = 0.1278 \text{ g } Na_2C_2O_4 \times \frac{1 \text{ mmol } Na_2C_2O_4}{0.13400 \text{ g } Na_2C_2O_4}$$

$$= 0.95373 \text{ mmol } Na_2C_2O_4$$

$$C_{KMnO_4} = 0.95373 \text{ mmol } Na_2C_2O_4 \times \frac{2 \text{ mmol } KMnO_4}{5 \text{ mmol } Na_2C_2O_4}$$

$$\times \frac{1}{33.31 \text{ mL } KMnO_4} = 0.01145 \text{ M } KMnO_4$$

Table 17-5

SOME APPLICATIONS OF POTASSIUM PERMANGANATE AND CERIUM(IV) IN ACID SOLUTION

Substance Analyzed	Half-Reaction	Special Conditions
Sn	$Sn^{2+} \rightleftharpoons Sn^{4+} + 2\ e^-$	Prereduction with Zn
H_2O_2	$H_2O_2 \rightleftharpoons O_2(g) + 2\ H^+ + 2\ e^-$	
Fe	$Fe^{2+} \rightleftharpoons Fe^{3+} + e^-$	Prereduction with $SnCl_2$ or with Jones or Walden reductor
$Fe(CN)_6^{4-}$	$Fe(CN)_6^{4-} \rightleftharpoons Fe(CN)_6^{3-} + e^-$	
V	$VO^{2+} + 3\ H_2O \rightleftharpoons V(OH)_4^+ + 2\ H^+ + e^-$	Prereduction with Bi amalgam or SO_2
Mo	$Mo^{3+} + 4\ H_2O \rightleftharpoons MoO_4^{2-} + 8\ H^+ + 3\ e^-$	Prereduction with Jones reductor
W	$W^{3+} + 4\ H_2O \rightleftharpoons WO_4^{2-} + 8\ H^+ + 3\ e^-$	Prereduction with Zn or Cd
U	$U^{4+} + 2\ H_2O \rightleftharpoons UO_2^{2+} + 4\ H^+ + 2\ e^-$	Prereduction with Jones reductor
Ti	$Ti^{3+} + H_2O \rightleftharpoons TiO^{2+} + 2\ H^+ + e^-$	Prereduction with Jones reductor
$H_2C_2O_4$	$H_2C_2O_4 \rightleftharpoons 2\ CO_2 + 2\ H^+ + 2\ e^-$	
Mg, Ca, Zn, Co, Pb, Ag	$H_2C_2O_4 \rightleftharpoons 2\ CO_2 + 2\ H^+ + 2\ e^-$	Sparingly soluble metal oxalates filtered, washed, and dissolved in acid; liberated oxalic acid titrated
HNO_2	$HNO_2 + H_2O \rightleftharpoons NO_3^- + 3\ H^+ + 2\ e^-$	15-min reaction time; excess $KMnO_4$ back-titrated
K	$K_2NaCo(NO_2)_6 + 6\ H_2O \rightleftharpoons$ $Co^{2+} + 6\ NO_3^- + 12\ H^+ + 2\ K^+ + Na^+ + 11\ e^-$	Precipitated as $K_2NaCo(NO_2)_6$; filtered and dissolved in $KMnO_4$; excess $KMnO_4$ back-titrated
Na	$U^{4+} + 2\ H_2O \rightleftharpoons UO_2^{2+} + 4\ H^+ + 2\ e^-$	Precipitated as $NaZn(UO_2)_3(OAc)_9$; filtered, washed, dissolved; U determined as above

Applications of Potassium Permanganate and Cerium(IV) Solutions
Table 17-5 lists some of the many applications of permanganate and cerium(IV) solutions to the volumetric determination of inorganic species. Both reagents have also been applied to the determination of organic compounds with oxidizable functional groups.

Example 17-5

Aqueous solutions containing approximately 3% (w/w) H_2O_2 are sold in drugstores as a disinfectant. Propose a method for determining the peroxide content of one of these preparations using the standard solution described in Example 17-4 and between 35 and 45 mL of the reagent per titration. The reaction is

$$5\ H_2O_2 + 2\ MnO_4^- + 6\ H^+ \rightarrow 5\ O_2 + 2\ Mn^{2+} + 8\ H_2O$$

The number of millimoles of $KMnO_4$ in 35 to 45 mL of the reagent is between

$$\text{amount } KMnO_4 = 35\ \cancel{mL} \times 0.01145\ \frac{\text{mmol } KMnO_4}{\cancel{mL}\ KMnO_4}$$

$$= 0.401\ \text{mmol } KMnO_4$$

and

$$\text{amount KMnO}_4 = 45 \times 0.01145 = 0.515 \text{ mmol KMnO}_4$$

The number of millimoles of H_2O_2 consumed by these amounts of $KMnO_4$ is

$$\text{amount H}_2\text{O}_2 = 0.401 \text{ mmol KMnO}_4 \times \frac{5 \text{ mmol H}_2\text{O}_2}{2 \text{ mmol KMnO}_4}$$

$$= 1.00 \text{ mmol H}_2\text{O}_2$$

and

$$\text{amount H}_2\text{O}_2 = 0.515 \times \frac{5}{2} = 1.29 \text{ mmol H}_2\text{O}_2$$

Thus we need to take samples that contain from 1.0 to 1.3 mmol H_2O_2.

$$\text{wt sample} = 1.0 \text{ mmol H}_2\text{O}_2 \times 0.03401 \frac{\text{g H}_2\text{O}_2}{\text{mmol H}_2\text{O}_2} \times \frac{100 \text{ g sample}}{3 \text{ g H}_2\text{O}_2}$$

$$= 1.1 \text{ g sample}$$

to

$$\text{wt sample} = 1.3 \times 0.03401 \times \frac{100}{3} = 1.5 \text{ g sample}$$

Thus our samples should weigh between 1.1 and 1.5 g. They should be diluted to perhaps 75 to 100 mL with water and made slightly acidic with dilute H_2SO_4 before titration.

17C-2 Potassium Dichromate

In its analytical applications, dichromate ion is reduced to green chromium(III) ion:

$$Cr_2O_7^{2-} + 14 \text{ H}^+ + 6 \text{ e}^- \rightleftarrows 2 \text{ Cr}^{3+} + 7 \text{ H}_2\text{O} \qquad E^0 = 1.33 \text{ V}$$

Dichromate titrations are generally carried out in solutions that are about 1 M in hydrochloric or sulfuric acid. In these media, the formal potential for the half-reaction is 1.0 to 1.1 V.

Potassium dichromate solutions are indefinitely stable, can be boiled without decomposition, and do not react with hydrochloric acid. Moreover, primary-standard reagent is available commercially at a modest cost. The disadvantages of potassium dichromate compared with cerium(IV) and permanganate ion are its lower electrode potential and the slowness of its reaction with certain reducing agents.

The Preparation, Properties, and Uses of Dichromate Solutions

For most purposes, reagent-grade potassium dichromate is sufficiently pure to permit the direct preparation of standard solutions; the solid is simply dried at 150 to 200°C before being weighed.

The orange color of a dichromate solution is not intense enough for use in end-point detection. However, diphenylamine sulfonic acid (Table 16-2) is an excellent indicator for titrations with this reagent. The oxidized form of the indicator is red-violet, and its reduced form is essentially colorless; thus, the color change observed in a direct titration is from the green of chromium(III) to violet.

Applications of Potassium Dichromate Solutions

The principal use of dichromate is for the volumetric titration of iron(II) based upon the reaction

$$Cr_2O_7^{2-} + 6\ Fe^{2+} + 14\ H^+ \rightarrow 2\ Cr^{3+} + 6\ Fe^{3+} + 7\ H_2O$$

Often, this titration is performed in the presence of moderate concentrations of hydrochloric acid.

The reaction of dichromate with iron(II) has been widely used for the indirect determination of a variety of oxidizing agents. In these applications, a measured excess of a solution of iron(II) is added to an acidic solution of the analyte. The excess iron(II) is then back-titrated with standard potassium dichromate (Section 17B-1). Standardization of the iron(II) solution by titration with the dichromate is performed concurrently because solutions of iron(II) tend to be air-oxidized. This method has been applied to the determination of nitrate, chlorate, permanganate, and dichromate ions as well as organic peroxides and several other oxidizing agents.

Example 17-6

A 5.00-mL sample of brandy was diluted to 1.000 L in a volumetric flask. The ethanol (C_2H_5OH) in a 25.00-mL aliquot of the diluted solution was distilled into 50.00 mL of 0.02000 M $K_2Cr_2O_7$, where oxidation to acetic acid occurred with heating:

$$3\ C_2H_5OH + 2\ Cr_2O_7^{2-} + 16\ H^+ \rightarrow 4\ Cr^{3+} + 3\ CH_3COOH + 11\ H_2O$$

After cooling, 20.00 mL of 0.1253 M Fe^{2+} was pipetted into the flask. The excess Fe^{2+} was then titrated with 7.46 mL of the standard $K_2Cr_2O_7$ to a diphenylamine sulfonic acid end point. Calculate the weight/volume percent of C_2H_5OH (fw = 46.07 g) in the brandy.

$$\text{total volume of } K_2Cr_2O_7 = 50.00\text{ mL} + 7.46\text{ mL} = 57.46\text{ mL}$$

$$\text{total amount } K_2Cr_2O_7 = 57.46\ \cancel{\text{mL } K_2Cr_2O_7} \times 0.02000\ \frac{\text{mmol } K_2Cr_2O_7}{\cancel{\text{mL } K_2Cr_2O_7}}$$

$$= 1.1492\text{ mmol } K_2Cr_2O_7$$

$$\text{amount } Fe^{2+} = 20.00 \text{ mL } Fe^{2+} \times 0.1253 \frac{\text{mmol } Fe^{2+}}{\text{mL } Fe^{2+}}$$

$$= 2.5060 \text{ mmol } Fe^{2+}$$

amount $K_2Cr_2O_7$ consumed by Fe^{2+}

$$= 2.5060 \text{ mmol } Fe^{2+} \times \frac{1 \text{ mmol } K_2Cr_2O_7}{6 \text{ mmol } Fe^{2+}}$$

$$= 0.41767 \text{ mmol } K_2Cr_2O_7$$

amount $K_2Cr_2O_7$ consumed by C_2H_5OH

$$= (1.1492 - 0.41767) \text{ mmol } K_2C_6O_7$$

$$= 0.73153 \text{ mmol } K_2Cr_2O_7$$

$$\text{wt } C_2H_5OH = 0.73153 \text{ mmol } K_2Cr_2O_7 \times \frac{3 \text{ mmol } C_2H_5OH}{2 \text{ mmol } K_2Cr_2O_7}$$

$$\times 0.04607 \frac{\text{g } C_2H_5OH}{\text{mmol } C_2H_5OH}$$

$$= 0.050552 \text{ g } C_2H_5OH$$

$$\% \text{ } C_2H_5OH \text{ (w/v)} = \frac{0.050552 \text{ g } C_2H_5OH}{5.00 \text{ mL sample} \times 25.00 \text{ mL}/1000 \text{ mL}} \times 100\%$$

$$= 40.44\% = 40.4\% \text{ } C_2H_5OH \text{ (w/v)}$$

17C-3 Iodine

Solutions of iodine are weak oxidizing agents that are used for the determination of strong reductants. The most accurate description of the half-reaction for iodine in these applications is

$$I_3^- + 2 e^- \rightleftarrows 3 I^- \qquad E^0 = 0.536 \text{ V}$$

The application of standard iodine solutions are more limited than those of the other oxidants we have described because of the significantly lower electrode potential. Occasionally, however, this low potential is advantageous because it imparts a degree of selectivity that makes possible the determination of strong reducing agents in the presence of weak ones. An important advantage of iodine is the availability of a sensitive and reversible indicator for the titrations. Unfortunately, iodine solutions lack stability and must be restandardized regularly.

The Preparation and Properties of Iodine Solutions
Iodine is not very soluble in water (≈ 0.001 M). In order to prepare solutions having analytically useful concentrations of the element, iodine is ordinarily dissolved in moderately concentrated solutions of potassium iodide. In this medium, iodine is reasonably soluble as a consequence of the reaction

Solutions prepared by dissolving iodine in a concentrated solution of potassium iodide are properly called triiodide solutions. In practice, however, they are usually termed iodine solutions because this terminology accounts for the stoichiometric behavior ($I_2 + 2 e^- \rightarrow 2 I^-$).

$$I_2(s) + I^- \rightleftarrows I_3^- \qquad K = 7.1 \times 10^2$$

Iodine dissolves only slowly in solutions of potassium iodide, particularly when the iodide concentration is low. To ensure complete solution, the solid is always dissolved in a small volume of concentrated potassium iodide, care being taken to avoid dilution of the concentrated solution until the last trace of solid iodine has disappeared. Otherwise, the concentration of iodine in the diluted solution gradually increases with time. This problem can be avoided by filtering the solution through a sintered glass crucible before standardization.

Iodine solutions lack stability for several reasons, one being the volatility of the solute. Losses of iodine from an open vessel occur in a relatively short time even in the presence of an excess of iodide ion. In addition, iodine slowly attacks most organic materials. Consequently, cork or rubber stoppers are never used to close containers of the reagent, and precautions must be taken to protect standard solutions from contact with organic dusts and fumes.

Air-oxidation of iodide ion also causes changes in the molarity of an iodine solution:

$$4\,I^- + O_2(g) + 4\,H^+ \rightarrow 2\,I_2 + 2\,H_2O$$

In contrast to the other effects, this reaction causes the molarity of the iodine to increase. Air-oxidation is promoted by acids, heat, and light.

The Standardization and Applications of Iodine Solutions

Iodine solutions can be standardized against anhydrous sodium thiosulfate or barium thiosulfate monohydrate, both of which are available commercially. Often solutions of iodine are standardized against solutions of sodium thiosulfate that have been standardized against potassium iodate or potassium dichromate (Section 17B-2).

Table 17-6 summarizes methods that use iodine as an oxidizing agent.

17C-4 Potassium Bromate as a Source of Bromine

Primary-standard potassium bromate is available from commercial sources and can be used directly to prepare standard solutions that are

Table 17-6
SOME APPLICATIONS OF IODINE SOLUTIONS

Substance Analyzed	Half-Reaction
As	$H_3AsO_3 + H_2O \leftrightarrows H_3AsO_4 + 2\,H^+ + 2\,e^-$
Sb	$H_3SbO_3 + H_2O \leftrightarrows H_3SbO_4 + 2\,H^+ + 2\,e^-$
Sn	$Sn^{2+} \leftrightarrows Sn^{4+} + 2\,e^-$
H_2S	$H_2S \leftrightarrows S(s) + 2\,H^+ + 2\,e^-$
SO_2	$SO_3^{2-} + H_2O \leftrightarrows SO_4^{2-} + 2\,H^+ + 2\,e^-$
$S_2O_3^{2-}$	$2\,S_2O_3^{2-} \leftrightarrows S_4O_6^{2-} + 2\,e^-$
N_2H_4	$N_2H_4 \leftrightarrows N_2(g) + 4\,H^+ + 4\,e^-$
Ascorbic acid*	$C_6H_8O_6 \rightarrow C_6H_6O_6 + 2\,H^+ + 2\,e^-$

*For the structure of ascorbic acid, see Section 28H-3.

stable indefinitely. Direct titrations with potassium bromate are relatively few. Instead, the reagent is a convenient and widely used stable source of bromine.[4] In this application, an unmeasured excess of potassium bromide is added to an acidic solution of the analyte. Upon introduction of a measured volume of standard potassium bromate, a stoichiometric quantity of bromine is produced.

$$BrO_3^- + 5\ Br^- + 6\ H^+ \rightarrow 3\ Br_2 + 3\ H_2O$$

Standard   Excess
Solution

This indirect generation circumvents the problems associated with the use of standard bromine solutions, which lack stability.

The primary use of standard potassium bromate is for the determination of organic compounds that react with bromine. Few of these reactions are rapid enough to make direct titration feasible. Instead, a measured excess of standard bromate is added to the solution that contains the sample plus an excess of potassium bromide. After acidification, the mixture is allowed to stand in a glass-stoppered vessel until the bromine/analyte reaction is judged complete. To determine the excess bromine, an excess of potassium iodide is introduced so that the following reaction occurs:

$$2\ I^- + Br_2 \rightarrow I_2 + 2\ Br^-$$

The liberated iodine is then titrated with standard sodium thiosulfate (Equation 17-1).

Bromine is incorporated into an organic molecule either by substitution or by addition.

Substitution Reactions

Halogen substitution involves the replacement of hydrogen in an aromatic ring by a halogen. Substitution methods have been successfully applied to the determination of aromatic compounds that contain strong ortho-para-directing groups, particularly amines and phenols.

Example 17-7

A 0.2891-g sample of an antibiotic powder was dissolved in HCl and the solution diluted to 100.0 mL. A 20.00-mL aliquot was transferred to a flask, and followed by 25.00 mL of 0.01767 M $KBrO_3$. An excess of KBr was added to form Br_2, and the flask was stoppered. After 10 min, during which time the Br_2 brominated the sulfanilamide, an excess of KI was added. The liberated iodine was then titrated with 12.92 mL of 0.1215 M

---

[4]For a discussion of bromate solutions and their applications, see M.R.F. Ashworth, *Titrimetric Organic Analysis,* Part I, pp. 118–130. New York: Interscience, 1964.

sodium thiosulfate. The reactions are

$$BrO_3^- + 5\ Br^- + 6\ H^+ \rightarrow 3\ Br_2 + 3\ H_2O$$

$$Br_2 + 2\ I^- \rightarrow 2\ Br^- + I_2 \quad (\text{excess KI})$$
$$I_2 + 2\ S_2O_3^{2-} \rightarrow S_4O_6^{2-} + 2\ I^-$$

Calculate the % $NH_2C_6H_4SO_2NH_2$ (fw = 172.21 g) in the powder.
First we determine the amount of Br_2:

$$\text{amount } Br_2 = 25.00\ \text{mL KBrO}_3 \times 0.01767\ \frac{\text{mmol KBrO}_3}{\text{mL KBrO}_3}$$

$$\times \frac{3\ \text{mmol } Br_2}{\text{mmol KBrO}_3}$$

$$= 1.32525\ \text{mmol } Br_2$$

We next calculate how much Br_2 was in excess over that required to brominate the analyte:

$$\text{amount excess } Br_2 = \text{amount } I_2$$

$$= 12.92\ \text{mL Na}_2S_2O_3 \times 0.1215\ \frac{\text{mmol Na}_2S_2O_3}{\text{mL Na}_2S_2O_3}$$

$$\times \frac{1\ \text{mmol } I_2}{2\ \text{mmol Na}_2S_2O_3}$$

$$= 0.78489\ \text{mmol } Br_2$$

The amount of Br_2 consumed by the sample is

$$\text{amount } Br_2 = 1.32525 - 0.78489 = 0.54036\ \text{mmol } Br_2$$

$$\text{wt analyte} = 0.54036\ \text{mmol } Br_2 \times \frac{1\ \text{mmol analyte}}{2\ \text{mmol } Br_2}$$

$$\times 0.17221\ \frac{\text{g analyte}}{\text{mmol analyte}}$$

$$= 0.046528\ \text{g analyte}$$

$$\% \text{ analyte} = \frac{0.046528\ \text{g analyte}}{0.2891\ \text{g sample} \times 20.00\ \text{mL}/100\ \text{mL}} \times 100\%$$

$$= 80.47\%\ \text{sulfanilamide}$$

An important example of the use of a bromine substitution reaction is the determination of 8-hydroxyquinoline:

In contrast to most bromine substitutions, this reaction takes place rapidly enough in hydrochloric acid solution to make direct titration feasible. The titration of 8-hydroxyquinoline with bromine is particularly significant because the former is an excellent precipitating reagent for cations (Section 5D-3). For example, aluminum can be determined according to the sequence

$$Al^{3+} + 3\ HOC_9H_6N \xrightarrow{pH\ 4-9} Al(OC_9H_6N)_3(s) + 3\ H^+$$

$$Al(OC_9H_6N)_3(s) + 3\ H^+ \xrightarrow{hot\ 4\ M\ HCl} 3\ HOC_9H_6N + Al^{3+}$$

$$3\ HOC_9H_6N + 6\ Br_2 \rightarrow 3\ HOC_9H_4NBr_2 + 6\ HBr$$

The stoichiometric relationships in this case are

$$1\ mol\ Al^{3+} \equiv 3\ mol\ HOC_9H_6N \equiv 6\ mol\ Br_2 \equiv 2\ mol\ KBrO_3$$

Addition Reactions

Addition reactions involve the opening of an olefinic double bond. For example, 1 mol of ethylene reacts with 1 mol of bromine in the reaction

The literature contains numerous references to the use of bromine for the estimation of olefinic unsaturation in fats, oils, and petroleum products.

17D QUESTIONS AND PROBLEMS

*17-1. Why is a Walden reductor always used with solutions that contain appreciable concentrations of HCl?

17-2. Why is zinc amalgam rather than pure zinc used in a Jones reductor?

*17-3. Write a balanced equation for the reduction of UO_2^{2+} in a Walden reductor.

17-4. Write a balanced equation for the reduction of TiO^{2+} in a Jones reductor.

*17-5. Why are standard $KMnO_4$ solutions seldom used for the titration of solutions containing HCl?

17-6. Why are Ce^{4+} solutions never used for the titrations of reductants in basic solutions?

*17-7. Write a balanced equation showing why $KMnO_4$ end points fade.

17-8. Why are $KMnO_4$ solutions filtered before they are standardized?

*17-9. What is the primary use of standard solutions of $K_2Cr_2O_7$?

17-10. Why are iodine solutions prepared by dissolving I_2 in concentrated KI?

*17-11. A standard solution of I_2 increased in molarity with standing. Write a balanced equation that accounts for the increase.

17-12. When a solution of $Na_2S_2O_3$ is introduced into a solution of HCl, a cloudiness develops almost immediately. Write a balanced equation explaining this phenomenon.

*17-13. Why are standard solutions of reductants less often used for titrations than standard solutions of oxidants?

17-14. Write a balanced equation describing the titration of hydrazine (N_2H_4) with standard iodine.

*17-15. In the titration of I_2 solutions with $Na_2S_2O_3$, starch indicator is never added until just before chemical equivalence. Why?

17-16. Why are solutions of $KMnO_4$ and $Na_2S_2O_3$ generally stored in dark bottles?

*17-17. When a solution of $KMnO_4$ was left standing in a buret for 3 h, a brownish ring formed at the surface of the liquid. Write a balanced equation that explains this observation.

17-18. Suggest a way in which a solution of KIO_3 is used as a source of known quantities of I_2.

*17-19. Write balanced equations showing how $KBrO_3$ could be used for standardizing solutions of $Na_2S_2O_3$.

17-20. Write balanced equations showing how $K_2Cr_2O_7$ could be used for standardizing solutions of $Na_2S_2O_3$.

*17-21. Write balanced equations to describe
(a) the oxidation of Mn^{2+} to MnO_4^- by ammonium peroxydisulfate.
(b) the oxidation of Ce^{3+} to Ce^{4+} by sodium bismuthate.
(c) the oxidation of U^{4+} to UO_2^{2+} by H_2O_2.
(d) the reaction of $V(OH)_4^+$ in a silver reductor.
(e) the titration of H_2O_2 with $KMnO_4$.
(f) the reaction between KI and ClO_3^- in acidic solution.

17-22. Write balanced equations to describe
(a) the reduction of Fe^{3+} to Fe^{2+} by SO_2.
(b) the reaction of H_2MoO_4 in a Jones reductor.
(c) the oxidation of HNO_2 by a solution of MnO_4^-.
(d) the reaction of aniline ($C_6H_4NH_2$) with a mixture of $KBrO_3$ and KBr in acidic solution.
(e) the air-oxidation of $HAsO_3^{2-}$ to $HAsO_4^{2-}$.
(f) the reaction of KI with HNO_2 in acidic solution.

*17-23. Describe the preparation of 2.500 L of 0.03500 M $K_2Cr_2O_7$ from the pure salt.

17-24. Describe the preparation of 500 mL of approximately 0.025 M $KMnO_4$ to be used for the titration of a reducing agent in an acidic solution.

*17-25. Describe the preparation of 750.0 mL of 0.02500 M $KBrO_3$ to be used as a source of Br_2.

17-26. Describe the preparation of 2.000 L of 0.01700 M I_3^- from pure I_2 and KI.

*17-27. To standardize a solution of $Na_2S_2O_3$, 0.1518 g of $K_2Cr_2O_7$ was dissolved in dilute HCl. An excess of KI was added, following which the liberated I_2 was titrated with 46.13 mL of the reagent. Calculate the molarity of the $Na_2S_2O_3$.

17-28. To standardize a solution of $Na_2S_2O_3$, 0.1017 g of $KBrO_3$ was dissolved in dilute HCl. An excess of KI was added, following which the liberated I_2 was titrated with 39.75 mL of the reagent. Calculate the molarity of the $Na_2S_2O_3$.

*17-29. The Sb(III) in a 1.080-g sample of stibnite ore required a 41.67-mL titration with 0.03134 M I_2. Express the results of this analysis in terms of (a) percentage of Sb and (b) percentage of stibnite (Sb_2S_3).

17-30. A 0.2236-g sample of limestone was dissolved in dilute HCl. After $(NH_4)_2C_2O_4$ was introduced and the pH of the resulting solution adjusted to permit the quantitative precipitation of CaC_2O_4, the solid was isolated by filtration, washed free of excess $C_2O_4^{2-}$, and dissolved in dilute H_2SO_4. Titration of the liberated $H_2C_2O_4$ required 26.77 mL of 0.02356 M $KMnO_4$. Calculate the percentage of CaO in the sample.

*17-31. An 8.13-g sample of an ant-control preparation was decomposed with a mixture of H_2SO_4 and HNO_3. The As in the residue was reduced to the trivalent state with hydrazine. After removal of the excess reducing agent, the As(III) required a 23.77-mL titration with 0.02425 M I_2 in a faintly alkaline medium. Express the results of this analysis in terms of percentage of As_2O_3 in the original sample.

17-32. A 4.971-g sample containing the mineral tellurite was dissolved and then treated with 50.00 mL of 0.03114 M $K_2Cr_2O_7$:

$$3\ TeO_2 + Cr_2O_7^{2-} + 8\ H^+ \rightarrow$$
$$3\ H_2TeO_4 + 2\ Cr^{3+} + H_2O$$

Upon completion of the reaction, the excess $Cr_2O_7^{2-}$ required a 10.05-mL back-titration with 0.1135 M Fe^{2+}. Calculate the percentage of TeO_2 in the sample.

*17-33. A 25.0-mL aliquot of a solution containing Tl(I) ion was treated with K_2CrO_4. The Tl_2CrO_4 was filtered, washed free of excess precipitating agent, and dissolved in dilute H_2SO_4. The $Cr_2O_7^{2-}$ produced was titrated with 40.60 mL of 0.1004 M Fe^{2+} solution. What was the weight of Tl in the sample? The reactions are

$$2\ Tl^+ + CrO_4^{2-} \rightarrow Tl_2CrO_4(s)$$
$$2\ Tl_2CrO_4(s) + 2\ H^+ \rightarrow 4\ Tl^+ + Cr_2O_7^{2-} + H_2O$$
$$Cr_2O_7^{2-} + 6\ Fe^{2+} + 14\ H^+ \rightarrow$$
$$6\ Fe^{3+} + 2\ Cr^{3+} + 7\ H_2O$$

17-34. A 0.306-g sample of an impure aluminum salt was dissolved in dilute acid, treated with an excess of $(NH_4)_2C_2O_4$, and slowly made alkaline through the addition of aqueous NH_3. The precipitated $Al_2(C_2O_4)_3$ was filtered, washed, dissolved in dilute acid, and titrated with 36.08 mL of 0.02413 M $KMnO_4$. Calculate the percentage of Al in the sample.

*17-35. A 1.065-g sample of stainless steel was dissolved in HCl (this treatment converts the Cr present to Cr^{3+}) and diluted to 500.0 mL in a volumetric flask. A 50.00-mL aliquot was passed through a Walden reductor and then titrated with 13.72 mL of 0.01920 M $KMnO_4$. A 100.0-mL aliquot was passed through a Jones reductor into 50 mL of 0.10 M Fe^{3+}. Titration of the resulting solution required 36.43 mL of the $KMnO_4$ solution. Calculate the percentages of Fe and Cr in the alloy.

17-36. A 2.559-g sample containing both Fe and V was dissolved under conditions that converted the elements to Fe(III) and V(V). The solution was diluted to 500.0 mL, and a 50.00-mL aliquot was passed through a Walden reductor and titrated with 17.74 mL of 0.1000 M Ce^{4+}. A second 50.00-mL aliquot was passed through a Jones reductor and required 44.67 mL of the same Ce^{4+} solution to reach an end point. Calculate the percentage of Fe_2O_3 and V_2O_5 in the sample.

*17-37. A sensitive method for I^- in the presence of Cl^- and Br^- entails oxidation of the I^- to IO_3^- with Br_2. The Br_2 is then removed by boiling or by reduction with formate ion. The IO_3^- is determined by addition of excess I^- and titration of the resulting I_2. A 1.204-g sample of mixed halides was dissolved and analyzed by the foregoing procedure; the titration required 20.66 mL of 0.05551 M thiosulfate. Calculate the percentage of KI in the sample.

17-38. The $KClO_3$ in a 0.1342-g sample of an explosive was determined by reaction with 50.00 mL of 0.09601 M Fe^{2+}:

$$ClO_3^- + 6\ Fe^{2+} + 6\ H^+ \rightarrow Cl^- + 3\ H_2O + 6\ Fe^{3+}$$

When the reaction was complete, the excess Fe^{2+} was back-titrated with 12.99 mL of 0.08362 M Ce^{4+}. Calculate the percentage of $KClO_3$ in the sample.

*17-39. The ethyl mercaptan concentration in a mixture was determined by shaking a 1.657-g sample of the mixture with 50.0 mL of 0.01194 M I_2 in a tightly stoppered flask:

$$2\ C_2H_5SH + I_2 \rightarrow C_2H_5SSC_2H_5 + 2\ I^- + 2\ H^+$$

The excess I_2 was back-titrated with 16.77 mL of 0.01325 M $Na_2S_2O_3$. Calculate the percentage of C_2H_5SH (fw = 62.13 g).

17-40. The CO concentration in a 3.21-L sample of air was obtained by passage over iodine pentoxide at 150°C:

$$I_2O_5 + 5\ CO(g) \rightleftarrows 5\ CO_2(g) + I_2(g)$$

The I_2 distilled at this temperature was collected in a solution of KI. The resulting I_3^- was titrated with 7.76 mL of 0.00221 M $Na_2S_2O_3$. Calculate the parts per million of CO in the gas, assuming an air density of 1.20×10^{-3} g/mL.

Chapter 18

Potentiometric Methods

In Chapter 15, we showed that the potential of an electrode relative to the standard hydrogen electrode is determined by the concentration of one or more of the species in the solution in which the electrode is immersed. The present chapter deals with the measurement of electrode potentials and how these data are used to determine the concentration of analytes.[1] Analytical methods based upon potential measurements are termed *potentiometric methods*.

18A GENERAL PRINCIPLES

In Feature 15-3, we showed that absolute values for individual half-cell potentials cannot be determined in the laboratory. That is, only *cell* potentials can be obtained experimentally. Figure 18-1 shows a typical cell for a potentiometric analysis. This cell can be depicted as

$$\underbrace{\text{reference electrode}}_{E_{\text{ref}}}|\underbrace{\text{salt bridge}}_{E_{\text{j}}}|\underbrace{\text{analyte solution}|\text{indicator electrode}}_{E_{\text{ind}}}$$

The *reference electrode* in this diagram is a half-cell with an accurately known electrode potential E_{ref} that is independent of the analyte concen-

[1]For further reading on potentiometric methods, see E. P. Serjeant, *Potentiometry and Potentiometric Titrations*. New York: Wiley, 1984.

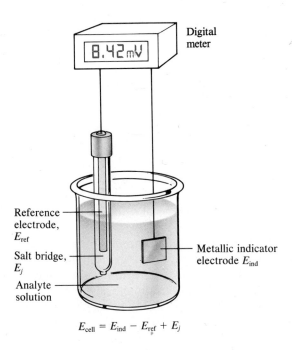

Digital meter

8.42 mV

Reference electrode, E_{ref}

Salt bridge, E_j

Analyte solution

Metallic indicator electrode E_{ind}

$$E_{cell} = E_{ind} - E_{ref} + E_j$$

Figure 18-1
A cell for potentiometric analysis.

tration or of any other ions in the solution under study. By convention, the reference electrode is always treated as the anode in potentiometric measurements.

The *indicator electrode,* which is immersed in the analyte solution, develops a potential E_{ind} that depends upon analyte activity. Most indicator electrodes used in potentiometry are highly selective in their responses.

The third component of a potentiometric cell is a salt bridge, which prevents the components of the analyte solution from mixing with those of the reference electrode. As noted in Chapter 15, a potential E_j develops across the two liquid junctions making up the salt bridge.

The potential of the cell we have just considered is given by the equation

$$E_{cell} = E_{ind} - E_{ref} + E_j \qquad (18\text{-}1)$$

The first term on the right contains the information we seek about the analyte concentration. A potentiometric analysis, then, involves measuring a cell potential, correcting this potential for the reference and junction potentials, and computing the analyte concentration from the indicator electrode potential.

In the sections that follow, we shall consider the sources of the three potentials that appear on the right side of Equation 18-1.

Reference electrodes are *always* treated as anodes in this text.

18B   REFERENCE ELECTRODES

The ideal reference electrode has a potential that is accurately known, constant, and completely insensitive to the composition of the analyte solution. In addition, a reference electrode should be rugged and easy to assemble and should maintain a constant potential while passing small currents.

18B-1 Calomel Electrodes

A calomel electrode can be represented schematically as

$$Hg \mid Hg_2Cl_2(sat'd), KCl(x~M) \mid$$

A standard hydrogen electrode is seldom used as a reference electrode for day-to-day potentiometric measurements because it is somewhat inconvenient and is also a fire hazard.

where x represents the molar concentration of potassium chloride in the solution. Three concentrations of potassium chloride are common: 0.1 M, 1 M, and saturated (about 4.6 M). The saturated calomel electrode (SCE) is the most widely used because it is so easily prepared. Its main disadvantage is its somewhat large temperature coefficient, which is important only in those rare circumstances where substantial temperature changes occur during a measurement. The *electrode* potential of the saturated calomel electrode is 0.2444 V at 25°C.

The "saturated" in a saturated calomel electrode refers to the KCl concentration. All calomel electrodes are saturated with Hg_2Cl_2 (calomel).

The electrode reaction in calomel half-cells is

$$Hg_2Cl_2(s) + 2~e^- \rightleftharpoons 2~Hg(l) + 2~Cl^-$$

The saturated calomel electrode shown in Figure 18-2 is a typical commercial electrode. It consists of an outer tube that is 5 to 15 cm in length and 0.5 to 1.0 cm in diameter. A mercury/mercury(I) chloride paste in saturated potassium chloride is contained in an inner tube and connected to the saturated potassium chloride solution in the outer tube through a small opening. Contact with the analyte solution is made through a fritted disk, a porous fiber, or a piece of porous Vycor ("thirsty glass") sealed in the end of the outer tube.

Figure 18-3 shows a saturated calomel electrode you can construct from materials available in your laboratory. A salt bridge (Section 15A-2) provides electrical contact with the analyte solution.

A salt bridge is readily constructed by filling a U-tube with a conducting gel prepared by heating about 5 g of agar in 100 mL of water containing about 35 g of potassium chloride. When the liquid cools, it sets up into a gel that is a good electrical conductor.

18B-2 Silver/Silver Chloride Electrodes

A system analogous to the saturated calomel electrode consists of a silver electrode immersed in a solution that is saturated in both potassium chloride and silver chloride:

$$Ag \mid AgCl(sat'd), KCl(sat'd) \mid$$

The half-reaction is

$$AgCl(s) + e^- \rightleftharpoons Ag(s) + Cl^-$$

At 25°C, the potential of the saturated calomel electrode versus the standard hydrogen electrode is 0.244 V; for the saturated silver/silver chloride electrode, this potential is 0.199 V.

The potential of this electrode is 0.199 V at 25°C.

Silver/silver chloride electrodes of various sizes and shapes are on the market. A simple and easily constructed electrode of this type is shown in Figure 18-4.

18C LIQUID-JUNCTION POTENTIALS

A liquid-junction potential develops across the boundary between two electrolyte solutions having different compositions. Figure 18-5 shows a very simple liquid junction consisting of a 1 M hydrochloric acid solution in contact with a solution that is 0.01 M in that acid. An inert porous barrier, such as a fritted glass plate, prevents the two solutions from mixing. Both hydrogen ions and chloride ions tend to diffuse across this boundary from the more concentrated to the more dilute solution. The driving force for each ion is proportional to the concentration difference between the two solutions. In the present example, hydrogen ions are substantially more mobile than chloride ions. Thus, hydrogen ions diffuse more rapidly than chloride ions, and, as shown in Figure 18-5, a separation of charge results. The more dilute side of the boundary becomes positively charged because of the more rapid diffusion of hydrogen ions. The concentrated side therefore acquires a negative charge from the excess of slower-moving chloride ions. The charge developed tends to counteract the differences in diffusion rates of the two ions so that a condition of equilibrium is attained rapidly. The potential difference resulting from this charge separation may be several hundredths of a volt.

The magnitude of the liquid-junction potential can be minimized by placing a salt bridge between the two solutions. The salt bridge is most effective if the mobilities of the negative and positive ions in the bridge are nearly equal to each other and if their concentrations are large. A saturated solution of potassium chloride is good from both standpoints. The net junction potential with such a bridge is typically a few millivolts.

All cells used for potentiometric analyses contain a salt bridge that connects the reference electrode to the analyte solution. As we shall see,

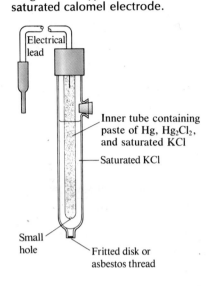

Figure 18-2
Diagram of a typical commercial saturated calomel electrode.

Electrical lead

Inner tube containing paste of Hg, Hg_2Cl_2, and saturated KCl

Saturated KCl

Small hole

Fritted disk or asbestos thread

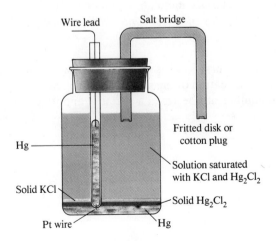

Wire lead

Salt bridge

Hg

Fritted disk or cotton plug

Solution saturated with KCl and Hg_2Cl_2

Solid KCl

Solid Hg_2Cl_2

Pt wire

Hg

Half-reaction:
$$Hg_2Cl_2(s) + 2 e^- \rightleftharpoons 2 Hg + 2 Cl^-$$

Figure 18-3
Diagram of an easily constructed saturated calomel electrode.

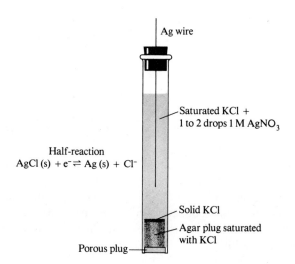

Ag wire

Saturated KCl +
1 to 2 drops 1 M $AgNO_3$

Half-reaction
$AgCl(s) + e^- \rightleftharpoons Ag(s) + Cl^-$

Solid KCl

Agar plug saturated
with KCl

Porous plug

Figure 18-4
Diagram of a silver/silver chloride electrode.

uncertainties in the magnitude of the junction potential across this bridge place a fundamental limit on the accuracy of potentiometric methods of analysis.

18D   INDICATOR ELECTRODES

An ideal indicator electrode responds rapidly and reproducibly to changes in the concentration of an analyte ion (or group of ions). Although no indicator electrode is absolutely specific in its response, a few are now available that are remarkably selective. There are two types of indicator electrodes: metallic and membrane.

18D-1 Metallic Indicator Electrodes

It is convenient to classify metallic indicator electrodes as *electrodes of the first kind, electrodes of the second kind,* and inert *redox electrodes.*

Electrodes of the First Kind
An electrode of the first kind is a piece of pure metal that is in direct equilibrium with the cation of the metal. A single reaction is involved. For example, the equilibrium between a metal X and its cation X^{n+} is

$$X^{n+} + ne^- \rightarrow X(s)$$

for which

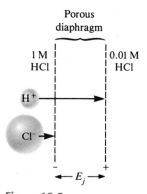

Porous
diaphragm

1 M
HCl

0.01 M
HCl

H^+

Cl^-

$\leftarrow E_j \rightarrow$

Figure 18-5
Schematic representation of a liquid junction, showing the source of the junction potential E_j. The length of the arrows corresponds to the relative mobility of the two ions.

$$E_{ind} = E^0_{X^{n+}} - \frac{0.0592}{n} \log \frac{1}{a_{X^{n+}}} = E^0_{X^{n+}} + \frac{0.0592}{n} \log a_{X^{n+}} \quad (18\text{-}2)$$

where E_{ind} is the electrode potential of the metal electrode and $a_{X^{n+}}$ is the activity of the ion (or approximately its molar concentration, $[X^{n+}]$).

We often express the electrode potential of the indicator electrode in terms of the p-function of the cation. Thus, substituting the definition of pX into Equation 18-2 gives

$$E_{ind} = E_{X^{n+}}^0 - \frac{0.0592}{n} pX \qquad (18\text{-}3)$$

Equation 18-3 accurately describes the behavior of a number of common metals that are used as indicator electrodes of the first kind. In contrast, certain harder metals—notably iron, chromium, tungsten, cobalt, and nickel—do not provide reproducible potentials. Moreover, plots of the electrode potential of a metal of this latter kind as a function of pX often yield slopes that differ significantly from the theoretical ($-0.0592/n$). The nonideal behavior of this type of electrode can be attributed to strains and deformations in the crystal structure of the metal or to the presence of an oxide film on the surface.

Electrodes of the Second Kind

Metals not only serve as indicator electrodes for their own cations but also respond to the concentration of anions that form sparingly soluble precipitates or stable complexes with such cations. The potential of a silver electrode, for example, correlates reproducibly with the concentration of chloride ion in a solution saturated with silver chloride. Here, the electrode reaction can be written as

$$AgCl(s) + e^- \rightleftarrows Ag(s) + Cl^- \qquad E_{AgCl}^0 = 0.222 \text{ V}$$

The Nernst expression for this process is

$$E_{ind} = E_{AgCl}^0 - 0.0592 \log [Cl^-] = E_{AgCl}^0 + 0.0592 \, pCl \qquad (18\text{-}4)$$

Equation 18-4 shows that the potential of a silver electrode is proportional to pCl, the negative logarithm of the chloride ion concentration. Thus, in a solution saturated with silver chloride, a silver electrode can serve as an indicator electrode of the second kind for chloride ion. Note that the sign of the logarithmic term for an electrode of this type is opposite that for an electrode of the first kind (see Equation 18-3).

Inert Metallic Electrodes for Redox Systems

As noted in Chapter 15, an inert metal—such as platinum, gold, palladium, or carbon—responds to the potential of redox systems with which it is in contact. For example, the potential of a platinum electrode immersed in a solution containing cerium(III) and cerium(IV) is

$$E_{ind} = E_{Ce(IV)}^0 - 0.0592 \log \frac{[Ce^{3+}]}{[Ce^{4+}]}$$

A platinum electrode is a convenient indicator electrode for titrations involving standard cerium(IV) solutions.

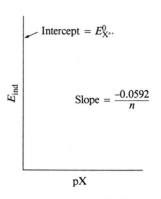

A plot of Equation 18-3 for an electrode of the first kind.

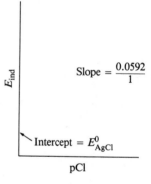

A plot of Equation 18-4 for an electrode of the second kind for Cl⁻.

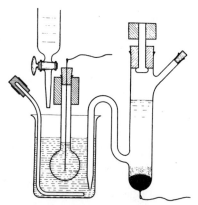

The first practical glass electrode of Haber and Klemensiewicz, *Z. Phys. Chem.*, **1909,** *65,* 385.

Metal electrodes function by the transfer of electrons across the interface between electrode and solution. Membrane electrodes function by the transfer of ions from one side of the membrane to the other.

18D-2 Membrane Electrodes[2]

For many years, the most convenient method for determining pH has involved measurement of the potential that develops across a thin glass membrane that separates two solutions with different hydrogen ion concentrations. The phenomenon upon which the measurement is based was first reported in 1906 and by now has been extensively studied by many investigators. As a result, the sensitivity and selectivity of glass membranes toward hydrogen ions are reasonably well understood. Furthermore, this understanding has led to the development of other types of membranes that respond selectively to more than two dozen other ions.

Membrane electrodes are sometimes called *p-ion electrodes* because the data obtained from them are usually presented as p-functions, such as pH, pCa, or pNO_3. The membranes used in constructing these electrodes are classified as *crystalline* and *noncrystalline*. Noncrystalline membranes can be further subdivided into *glass, liquid,* and *immobilized liquid.* In this section, we consider all of these p-ion membranes. In addition, a gas-sensing probe based on a noncrystalline membrane is described.

It is important to note at the outset of this discussion that membrane electrodes are *fundamentally different* from metal electrodes both in design and in principle. We shall use the glass electrode for pH measurements to illustrate these differences.

18D-3 The Glass Electrode for pH Measurements

Figure 18-6 shows a typical *cell* for measuring pH. The cell consists of a glass indicator electrode and a saturated calomel reference electrode, both immersed in the solution whose pH is to be determined. The indicator electrode consists of a thin, pH-sensitive glass membrane sealed onto one end of a heavy-walled glass or plastic tube. A small volume of dilute hydrochloric acid saturated with silver chloride is contained in the tube (in some electrodes this solution is a buffer containing chloride ion). A silver wire in this solution forms a silver/silver chloride reference electrode, which is connected to one of the terminals of a potential-measuring device. The calomel electrode is connected to the other terminal.

Figure 18-6 and the schematic representation of this cell in Figure 18-7 show that the system contains *two* reference electrodes: (1) the *external* calomel electrode and (2) the *internal* silver/silver chloride electrode. Although the internal reference electrode is part of the glass electrode, *it is not the pH-sensing element. Instead, it is the thin glass membrane at the tip of the electrode that responds to pH.*

The Composition of Glass Membranes

Much systematic investigation has been devoted to how glass composition affects the sensitivity of membranes to protons and other cations, and

[2]Some suggested sources for additional information on this topic are *Ion-Selective Electrodes in Analytical Chemistry,* H. Freiser, Ed. New York: Plenum Press, 1978, 1980; J. Vesely, D. Weiss, and K. Stulik, *Analysis with Ion-Selective Electrodes.* New York: Wiley, 1979; *Ion-Selective Methodology,* A. K. Covington, Ed. Boca Raton, FL: CRC Press, 1979; R. P. Buck, in *Comprehensive Treatise of Electrochemistry,* J. D. M. Bockris, B. C. Conway, H. E. Yeager, Eds., Vol. 8, Chapter 3. New York: Plenum Press, 1984.

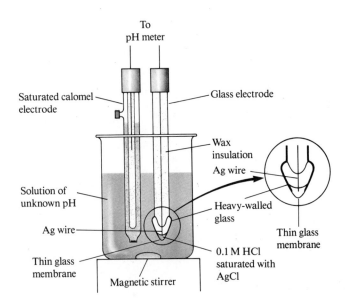

Figure 18-6
Typical electrode system for measuring pH.

a number of formulations are now used for the manufacture of electrodes. Corning 015 glass, which has been widely used for membranes, consists of approximately 22% Na_2O, 6% CaO, and 72% SiO_2. This membrane shows excellent specificity toward hydrogen ions up to a pH of about 9. At higher pH values, however, the glass becomes somewhat responsive to sodium as well as to other singly charged cations. Other glass formulations are now in use in which sodium and calcium ions are replaced to various degree by lithium and barium ions. These membranes have superior selectivity and lifetime.

The Structure of Membrane Glasses

As shown in Figure 18-8, a silicate glass used for membranes consists of an infinite three-dimensional network of SiO_4^{4-} groups in which each silicon is bonded to four oxygens and each oxygen is shared by two silicons. Within the interstices of this structure are sufficient cations to balance the negative charge of the silicate groups. Singly charged cations, such as sodium and lithium, are mobile in the lattice and are responsible for electrical conduction within the membrane.

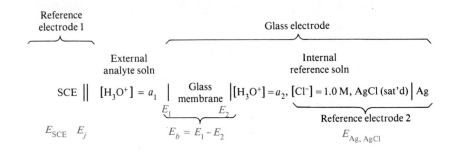

Figure 18-7
Diagram of a glass/calomel cell for measurement of pH.

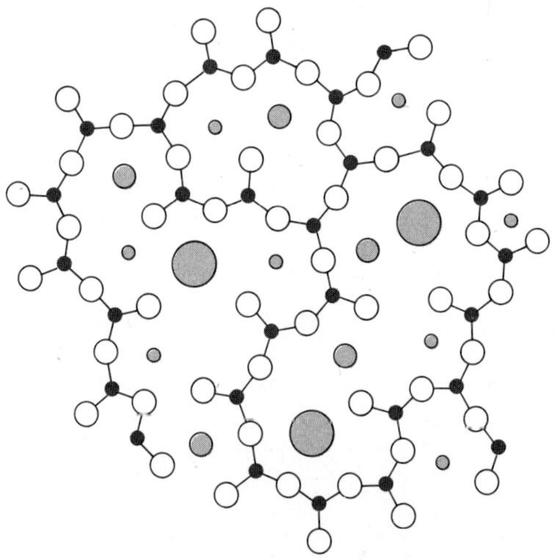

Figure 18-8
Cross-sectional view of a silicate glass structure. In addition to the three Si-O bonds shown, each silicon is bonded to an additional oxygen atom, either above or below the plane of the paper. (Adapted with permission from G. A. Perley, *Anal. Chem.*, **1949,** *21,* 395. Copyright 1949 American Chemical Society.)

● Silicon   ○ Oxygen   ○ ◯ Cations

Glasses that absorb water are said to be hygroscopic.

The Hygroscopicity of Glass Membranes

The surface of a glass membrane must be hydrated before it will function as a pH electrode. The amount of water involved is approximately 50 mg per cubic centimeter of glass. Nonhygroscopic glasses show no pH function. Even hygroscopic glasses lose their pH sensitivity after dehydration by storage over a desiccant. The effect is reversible, however, and response can be restored by soaking the membrane in water.

The hydration of a pH-sensitive glass membrane involves an ion-exchange reaction between singly charged cations in the glass lattice and protons from the solution in which the electrode is immersed. The process involves univalent cations exclusively because di- and trivalent cations are too strongly held within the silicate structure to exchange with ions in the solution. Typically, then, the ion-exchange reaction can be written as

$$H^+ + Na^+Gl^- \rightleftarrows Na^+ + H^+Gl^- \qquad (18\text{-}5)$$

Soln   Glass   Soln   Glass

This process is shown in the schematic of Figure 18-9. Oxygen atoms attached to only one silicon atom are negatively charged Gl^- sites. The equilibrium constant for this process is so large that the surface of a hydrated glass membrane ordinarily consists entirely of silicic acid (H^+Gl^-). An exception to this situation exists in highly alkaline media, where the hydrogen ion concentration is extremely small and the sodium ion concentration large; here, a significant fraction of the negatively charged SiO_4^{4-} sites are occupied by sodium ions.

Electrical Conduction Across Membranes

To serve as an indicator for cations, a glass membrane must conduct electricity. Conduction within the hydrated glass membrane involves the

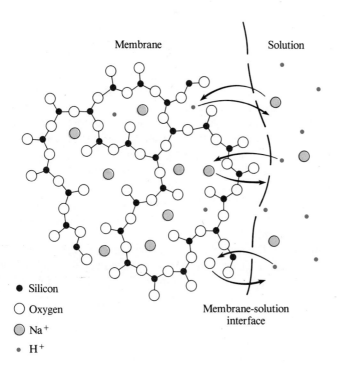

Membrane   Solution

- Silicon
○ Oxygen
◯ Na$^+$
· H$^+$

Membrane-solution
interface

Figure 18-9
Ion exchange at the membrane/
solution interface. Protons exchange with sodium ions in the
silicate structure of the glass
membrane.

movement of sodium and hydrogen ions. Sodium ions are the charge
carriers in the dry interior of the membrane, and the protons are mobile in
the gel layer. Conduction across the solution/gel interfaces occurs by the
reactions

$$H^+ \; + \; Gl^- \; \rightleftarrows \; H^+Gl^- \qquad\qquad (18\text{-}6)$$

Soln$_1$ Glass$_1$ Glass$_1$

$$H^+Gl^- \rightleftarrows H^+ \; + \; Gl^- \qquad\qquad (18\text{-}7)$$

Glass$_2$ Soln$_2$ Glass$_2$

where subscript 1 refers to the interface between glass and analyte solution and subscript 2 refers to the interface between internal solution and
glass. The positions of these two equilibria are determined by the hydrogen ion concentrations in the solutions on the two sides of the membrane.
Where these positions differ from each other, the surface at which the
greater dissociation has occurred is negative with respect to the other
surface. A boundary potential E_b thus develops across the membrane.
The magnitude of the boundary potential depends upon the ratio of the
hydrogen ion concentrations of the two solutions. It is this potential difference that serves as the analytical parameter in a potentiometric pH
measurement.

Membrane electrodes function by the
exchange of ions at the surfaces of the
membrane.

The membrane of a typical glass electrode (with a thickness of 0.03 to 0.1
mm) has an electrical resistance of 50
to 500 megaohms.

Membrane Potentials

The lower part of Figure 18-7 shows four potentials that develop in a cell when pH is being determined with a glass electrode. Two of these, $E_{Ag/AgCl}$ and E_{SCE}, are reference electrode potentials. There is a third potential across the salt bridge that separates the calomel electrode from the analyte solution. This interface and its associated *junction potential, E_j,* are found in all cells used for the potentiometric measurement of ion concentration. The fourth, and most important, potential shown in Figure 18-7 is the *boundary potential, E_b, which varies with the pH of the analyte solution.* The two reference electrodes simply provide electrical contacts with the solutions so that changes in the boundary potential can be measured.

Figure 18-7 reveals that the potential of a glass electrode has two components: the fixed potential of a silver/silver chloride electrode $E_{Ag/AgCl}$ and the pH-dependent boundary potential E_b.

Not shown in Figure 18-7 is a fifth potential, called the *asymmetry potential,* which is found in most membrane electrodes and which changes slowly with time. The sources of the asymmetry potential are obscure.

The Boundary Potential

As shown in Figure 18-7, the boundary potential consists of two potentials, E_1 and E_2, each of which is associated with one of the two gel/solution interfaces. The boundary potential is simply the difference between these potentials:

$$E_b = E_1 - E_2 \tag{18-8}$$

The significance of these potentials is shown in the potential profiles of Figure 18-10. The profiles are plotted across the membrane from the analyte solution on the left through the glass membrane to the internal reference solution on the right. The potential E_1 is determined by the ratio of the hydrogen ion activity a_1 in the analyte solution to the hydrogen ion activity in the gel layer and can be considered a measure of the driving force for the reaction shown in Equation 18-6. Similarly, E_2 is related to the ratio of the hydrogen ion activities in the internal reference solution and in the corresponding gel layer and is related to the driving force for the reaction shown in Equation 18-7.

The relationship between the boundary potential and the two hydrogen ion activities is

$$E_b = E_1 - E_2 = 0.0592 \log \frac{a_1}{a_2} \tag{18-9}$$

As shown in Figure 18-10a, if the hydrogen ion activity of the analyte is ten times the activity in the reference solution,

$$E_b = 0.0592 \log \frac{10a_2}{a_2} = 0.0592 \text{ V}$$

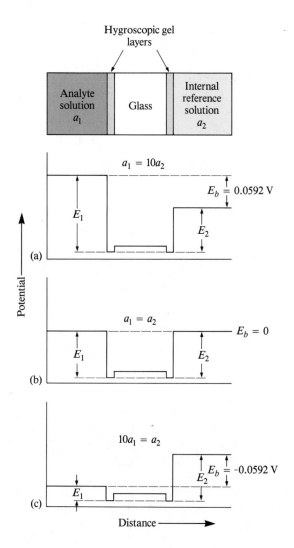

Figure 18-10
Potential profile across a glass membrane from the analyte solution to the internal reference solution. The reference electrode potentials are not shown.

If the two activities are equal, then $E_b = 0$, as illustrated in Figure 18-10b. Finally, Figure 18-10c shows that when $10a_1 = a_2$, $E_b = -0.0592$ V.

Thus, the boundary potential depends only upon the hydrogen ion activities of the solutions on either side of the membrane. For a glass pH electrode, the hydrogen ion activity of the internal solution a_2 is held constant, so that Equation 18-9 simplifies to

$$E_b = L' + 0.0592 \log a_1 = L' - 0.0592 \text{ pH} \qquad (18\text{-}10)$$

where

$$L' = -0.0592 \log a_2$$

The boundary potential is then a measure of the hydrogen ion activity of the external solution.

The Potential of the Glass Electrode

As noted earlier, the potential of a glass indicator electrode has three components: (1) the boundary potential, given by Equation 18-10, (2) the potential of the internal Ag/AgCl reference electrode, and (3) a small asymmetry potential, E_{asy}. In equation form,

$$E_{ind} = E_b + E_{Ag/AgCl} + E_{asy}$$

Substitution of Equation 18-10 for E_b gives

$$E_{ind} = L' + 0.0592 \log a_1 + E_{Ag/AgCl} + E_{asy}$$

or

$$E_{ind} = L + 0.0592 \log a_1 = L - 0.0592 \text{ pH} \qquad (18\text{-}11)$$

where L is a combination of the three constant terms. Compare Equations 18-11 and 18-3. Although these two equations are similar in form, remember that the sources of the electrode potential they describe *are totally different.*

The Alkaline Error

In basic solutions, glass electrodes respond to the concentration of both hydrogen ion and alkali metal ions. The magnitude of this *alkaline error* for four glass membranes is shown in Figure 18-11 (curves C to F). These curves refer to solutions in which the sodium ion concentration was held constant at 1 M while the pH was varied. Note that the error is negative (that is, the measured pH values are lower than the true values), which suggests that the electrode is responding to sodium ions as well as to protons. This observation is confirmed by data obtained for solutions

Figure 18-11
Acid and alkaline error for selected glass electrodes at 25°C. (From R. G. Bates, *Determination of pH*, 2nd ed., p. 365. New York: Wiley, 1973. With permission.)

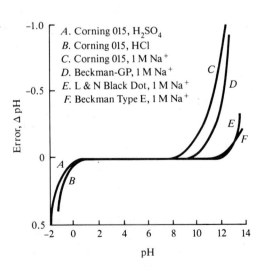

A. Corning 015, H_2SO_4
B. Corning 015, HCl
C. Corning 015, 1 M Na^+
D. Beckman-GP, 1 M Na^+
E. L & N Black Dot, 1 M Na^+
F. Beckman Type E, 1 M Na^+

containing different sodium ion concentrations. Thus at pH 12, the electrode with a Corning 015 membrane (curve C in Figure 18-11) registered a pH of 11.3 when immersed in a solution with a sodium ion concentration of 1 M but 11.7 in a solution that was 0.1 M in this ion. All singly charged cations induce an alkaline error whose magnitude depends upon both the cation in question and the composition of the glass membrane.

The alkaline error can be satisfactorily explained by assuming an exchange equilibrium between the hydrogen ions on the glass surface and the cations in solution. This process is simply the reverse of that shown in Equation 18-5:

$$H^+Gl^- + B^+ \rightleftarrows B^+Gl^- + H^+$$

$$\text{Glass} \qquad \text{Soln} \qquad \text{Glass} \qquad \text{Soln}$$

where B^+ represents some singly charged cation, such as sodium ion. In this case, the activity of the sodium ions relative to that of the hydrogen ions becomes so large that the electrode responds to both species.

Selectivity Coefficients

The effect of an alkali metal ion on the potential across a membrane can be accounted for by inserting an additional term in Equation 18-10 to give

$$E_b = L' + 0.0592 \log (a_1 + k_{H,B} b_1) \qquad (18\text{-}12)$$

In Equation 18-12, b_1 represents the activity of some singly charged cation such as Na^+ or K^+.

where $k_{H,B}$ is the *selectivity coefficient* for the electrode. Equation 18-12 applies not only to glass indicator electrodes for hydrogen ion but also to all other types of membrane electrodes. Selectivity coefficients range from zero (no interference) to values greater than unity. Thus, if an electrode for ion A responds 20 times more strongly to ion B than to ion A, $k_{A,B}$ has a value of 20. If the response of the electrode to ion C is 0.001 of its response to A (a much more desirable situation), $k_{A,C}$ is 0.001.

The selectivity coefficient is a measure of the response of an ion-selective electrode to other ions.

The product $k_{H,B} b_1$ for a glass pH electrode is ordinarily small relative to a_1 provided the pH is less than 9; under these conditions, Equation 18-12 simplifies to Equation 18-10. At high pH values and at high concentrations of a singly charged ion, however, the second term in Equation 18-12 assumes a more important role in determining E_b, and an alkaline error is encountered. For electrodes specifically designed for work in highly alkaline media (curve E in Figure 18-11), the magnitude of $k_{H,B} b_1$ is appreciably smaller than for ordinary glass electrodes.

The Acid Error

As shown in Figure 18-11, the typical glass electrode exhibits a *positive* error when the pH is less than about 0.5; pH readings tend to be too high in this region. The magnitude of the error depends upon a variety of factors and is generally not very reproducible. The causes of the acid error are not well understood.

18D-4 Glass Electrodes for Cations Other Than Protons

The alkaline error in early glass electrodes led to investigations concerning the effect of glass composition upon the magnitude of this error. One consequence has been the development of glasses for which the alkaline error is negligible below about pH 12. Other studies have discovered glass compositions that permit the determination of cations other than hydrogen. This latter application requires that the hydrogen ion activity a_1 in Equation 18-12 be negligible relative to $k_{H,B}b_1$; under these circumstances, the potential is independent of pH and is a function of pB instead. Incorporation of Al_2O_3 or B_2O_3 in the glass has the desired effect. Glass electrodes that permit the direct potentiometric measurement of such singly charged species as Na^+, K^+, NH_4^+, Rb^+, Cs^+, Li^+, and Ag^+ have been developed. Some of these glasses are reasonably selective toward particular singly charged cations. Glass electrodes for Na^+, Li^+, NH_4^+, and total concentration of univalent cations are now available from commercial sources.

18D-5 Liquid-Membrane Electrodes

The potential of a liquid-membrane electrode develops across the interface between the solution containing the analyte and a liquid-ion exchanger that selectively bonds with the analyte ion. These electrodes have been developed for the direct potentiometric measurement of numerous polyvalent cations as well as certain anions.

Figure 18-12 is a schematic of a liquid-membrane electrode for calcium. It consists of a conducting membrane that selectively bonds calcium ions, an internal solution containing a fixed concentration of calcium chloride, and a silver electrode coated with silver chloride to form an internal reference electrode. The active membrane ingredient is a calcium dialkyl phosphate ion exchanger that is nearly insoluble in water. In the electrode shown in Figure 18-12, the ion exchanger is dissolved in an immiscible

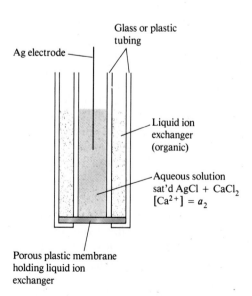

Ag electrode

Glass or plastic tubing

Liquid ion exchanger (organic)

Aqueous solution sat'd AgCl + $CaCl_2$ $[Ca^{2+}] = a_2$

Porous plastic membrane holding liquid ion exchanger

Figure 18-12
Diagram of a liquid-membrane electrode for Ca^{2+}.

organic liquid that is forced by gravity into the pores of a hydrophobic porous disk. This disk then serves as the membrane that separates the internal solution from the analyte solution. In a more recent design, the ion exchanger is immobilized in a tough polyvinyl chloride gel cemented to the end of a tube that holds the internal solution and reference electrode. In either design, there is set up at each membrane interface a dissociation equilibrium that is analogous to Equation 18-7:

$$[(RO)_2POO]_2Ca \rightleftarrows 2\ (RO)_2POO^- + Ca^{2+}$$

| Organic | Organic | Aqueous |

> Hydrophobic means water-hating. The hydrophobic disk is porous to organic liquids but repels water.

where R is a high-molecular-weight aliphatic group. As with the glass electrode, a potential develops across the membrane when the extent of dissociation at one surface differs from that at the other surface. This potential is a result of differences in the calcium ion activity of the internal and external solutions. The relationship between the membrane potential and the calcium ion activities is given by an equation that is similar to Equation 18-9:

$$E_b = E_1 - E_2 = \frac{0.0592}{2} \log \frac{a_1}{a_2} \qquad (18\text{-}13)$$

where a_1 and a_2 are the activities of calcium ion in the external and internal solutions, respectively. Since the calcium ion activity of the internal solution is constant.

> Ion-selective microelectrodes can be used to measure ion activities within a living organism.

$$E_b = N + \frac{0.0592}{2} \log a_1 = N - \frac{0.0592}{2} \text{pCa} \qquad (18\text{-}14)$$

where N is a constant (compare Equations 18-14 and 18-10). Note that, because calcium is divalent, a 2 appears in the denominator of the coefficient of the logarithmic term.

Figure 18-13 compares the structural features of a glass-membrane electrode and a commercially available liquid-membrane electrode for calcium ion. The sensitivity of the latter to calcium ion is reported to be 50 times greater than to magnesium ion and 1000 times greater than to sodium or potassium ions. Calcium ion activities as low as 5×10^{-7} M can be measured. Electrode performance is independent of pH in the range between 5.5 and 11. At lower pH levels, hydrogen ions undoubtedly replace some of the calcium ions on the exchanger; the electrode then becomes sensitive to pH as well as to pCa.

The calcium ion liquid-membrane electrode is a valuable tool for physiological investigations because this ion plays important roles in such processes as nerve conduction, bone formation, muscle contraction, cardiac expansion and contraction, renal tubular function, and perhaps hypertension. At least some of these processes are more influenced by calcium ion activity than by calcium ion concentration; activity, of course, is the parameter measured by the membrane electrode.

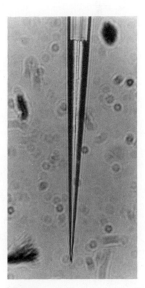

Photograph of a potassium liquid-ion exchanger microelectrode with 125 μm of ion exchanger inside the tip. The magnification of the original photo was 400×. From J. L. Walker, *Anal. Chem.*, **1971**, *43(3)N*, 91A. Reproduced by permission of the American Chemical Society.

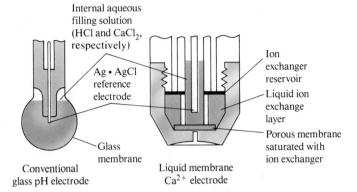

Figure 18-13

Comparison of a liquid-membrane calcium ion electrode with a glass electrode. (Courtesy of Orion Research, Boston, MA.)

Table 18-1

CHARACTERISTICS OF LIQUID-MEMBRANE ELECTRODES*

Analyte Ion	Concentration Range, M	Major Interferences
Ca^{2+}	10^0 to 5×10^{-7}	Pb^{2+}, Fe^{2+}, Ni^{2+}, Hg^{2+}, Sr^{2+}
Cl^-	10^0 to 5×10^{-6}	I^-, OH^-, SO_4^{2-}
NO_3^-	10^0 to 7×10^{-6}	ClO_4^-, I^-, ClO_3^-, CN^-, Br^-
ClO_4^-	10^0 to 7×10^{-6}	I^-, ClO_3^-, CN^-, Br^-
K^+	10^0 to 1×10^{-6}	Cs^+, NH_4^+, Tl^+
Water hardness (Ca^{2+} + Mg^{2+})	10^0 to 6×10^{-6}	Cu^{2+}, Zn^{2+}, Ni^{2+}, Sr^{2+}, Fe^{2+}, Ba^{2+}

*From *Orion Guide to Ion Analysis*. Boston, MA: Orion Research, 1983. With permission.

A liquid-membrane electrode specific for potassium ion is also of great value to physiologists because the transport of neural signals appears to involve movement of this ion across nerve membranes. Investigation of this process requires an electrode that can detect small concentrations of potassium ion in media with much larger concentrations of sodium ion. Several liquid-membrane electrodes show promise in meeting this requirement. One is based upon the antibiotic valinomycin, a cyclic ether that has a strong affinity for potassium ion. Of equal importance is the observation that a liquid membrane consisting of valinomycin in diphenyl ether is about 10^4 times as responsive to potassium ion as to sodium ion.[3]

Table 18-1 lists some liquid-membrane electrodes available from commercial sources. The anion-sensitive electrodes shown make use of a solution containing an anion-exchange resin in an organic solvent. Liquid-membrane electrodes in which the exchange liquid is held in a polyvinyl chloride gel have been developed for Ca^{2+}, K^+, NO_3^-, and BF_4^-. These electrodes look like crystalline electrodes, which are considered in the following section.

[3]M. S. Frant and J. W. Ross Jr., *Science*, **1970**, *167*, 987.

Feature 18-1

AN EASILY CONSTRUCTED LIQUID-MEMBRANE
ION-SELECTIVE ELECTRODE

You can make a liquid-membrane ion-selective electrode with glassware and chemicals available in most laboratories.[4] All you need are a pH meter, a pair of reference electrodes, a fritted-glass filter crucible or tube, trimethylchlorosilane, and a liquid ion exchanger.

First, cut the filter crucible (or alternatively, a fritted tube) as shown in Figure 18-14, page 344. Carefully clean and dry the crucible, and then draw a small amount of trimethylchlorosilane into the frit. This coating makes the glass in the frit hydrophobic. Rinse the frit with water, dry, and apply a commercial liquid ion exchanger to it. After a minute, remove the excess exchanger. Add a few milliliters of a 10^{-2} M solution of the ion of interest to the crucible, insert a reference electrode into the solution, and voilà, you have a very nice ion-selective electrode. The exact details of washing, drying, and preparing the electrode are provided in the original article.

Connect the ion-selective electrode and the second reference electrode to the pH meter as shown in Figure 18-14. Prepare a series of standard solutions of the ion of interest, measure the cell potential for each concentration, plot a working curve of E_{cell} versus log C, and perform a least-squares analysis on the data. Compare the slope of the line with the theoretical slope of $(0.0592 \text{ V})/n$. Measure the potential for an unknown solution of the ion and calculate the concentration from the least-squares parameters.

18D-6 Crystalline-Membrane Electrodes

Considerable work has been devoted to the development of solid membranes that are selective toward anions in the same way that some glasses respond to cations. We have seen that anionic sites on a glass surface account for the selectivity of a membrane toward certain cations. By analogy, a membrane with cationic sites might be expected to respond selectively toward anions.

Membranes prepared from cast pellets of silver halides have been used successfully in electrodes for the selective determination of chloride, bromide, and iodide ions. In addition, an electrode based upon a polycrystalline Ag_2S membrane is offered by one manufacturer for the determination of sulfide ion. In both types of membranes, silver ions are sufficiently mobile to conduct electricity through the solid medium. Mixtures of PbS, CdS, and CuS with Ag_2S provide membranes that are selective for Pb^{2+}, Cd^{2+}, and Cu^{2+}, respectively. Silver ion must be present in these membranes to conduct electricity because divalent ions are immobile in crystals. The potential that develops across crystalline solid-state electrodes is described by a relationship similar to Equation 18-14.

---

[4]T. K. Christopoulos and E. P. Diamandis, *J. Chem. Educ.*, **1988**, *65*, 648.

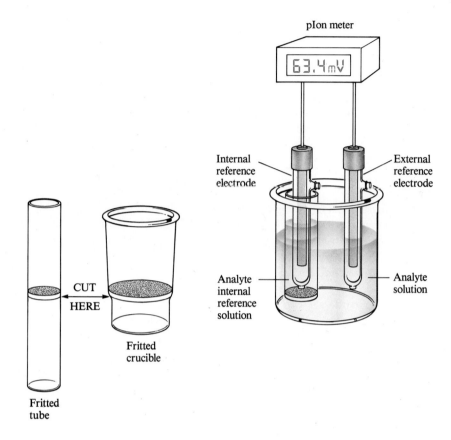

Figure 18-14

A homemade liquid-membrane electrode.

A crystalline electrode for fluoride ion is available from commercial sources. The membrane consists of a slice of a single crystal of lanthanum fluoride that has been doped with europium(II) fluoride to improve its conductivity. The membrane, supported between a reference solution and the solution to be measured, shows a theoretical response to changes in fluoride ion activity in the range from 10^0 to 10^{-6} M. The electrode is

Table 18-2

CHARACTERISTICS OF SOLID–STATE CRYSTALLINE ELECTRODES*

Analyte Ion	Concentration Range, M	Major Interferences
Br^-	10^0 to 5×10^{-6}	CN^-, I^-, S^{2-}
Cd^{2+}	10^{-1} to 1×10^{-7}	Fe^{2+}, Pb^{2+}, Hg^{2+}, Ag^+, Cu^{2+}
Cl^-	10^0 to 5×10^{-5}	CN^-, I^-, Br^-, S^{2-}
Cu^{2+}	10^{-1} to 1×10^{-8}	Hg^{2+}, Ag^+, Cd^{2+}
CN^-	10^{-2} to 1×10^{-6}	S^{2-}
F^-	Sat'd to 1×10^{-6}	OH^-
I^-	10^0 to 5×10^{-8}	
Pb^{2+}	10^{-1} to 1×10^{-6}	Hg^{2+}, Ag^+, Cu^{2+}
Ag^+/S^{2-}	Ag^+: 10^0 to 1×10^{-7}	Hg^{2+}
	S^{2-}: 10^0 to 1×10^{-7}	
SCN^-	10^0 to 5×10^{-6}	I^-, Br^-, CN^-, S^{2-}

*From *Orion Guide to Ion Analysis*. Cambridge, MA: Orion Research, 1983. With permission.

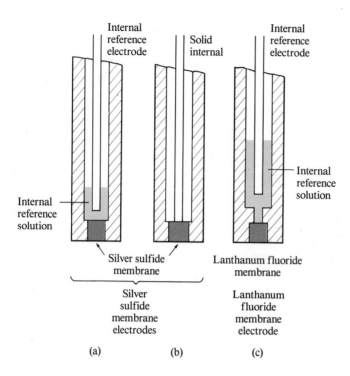

Figure 18-15

Types of crystalline ion-selective electrodes. (a) Solid-membrane electrode sensitive to Ag^+ and S^{2-}; (b) all solid-state solid-membrane electrode sensitive to Ag^+ and S^{2-}; (c) membrane configuration ion-sensing electrode for fluoride ion.

selective for fluoride ion over other common anions by several orders of magnitude; only hydroxide ion appears to offer serious interference.

Some solid-state electrodes available from commercial sources are listed in Table 18-2. Several types of crystalline electrodes are shown in Figure 18-15.

18D-7 Gas-Sensing Probes

Figure 18-16 illustrates the essential features of a potentiometric gas-sensing probe, which consists of a tube containing a reference electrode, a specific-ion electrode, and an electrolyte solution. A thin, replaceable,

A gas-sensing probe is a galvanic *cell* whose potential is related to the concentration of a gas in a solution.

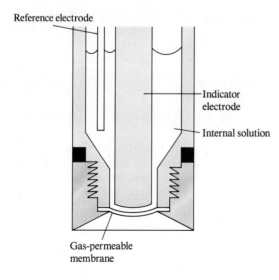

Figure 18-16

Diagram of a gas-sensing probe.

gas-permeable membrane attached to one end of the tube serves as a barrier between the internal and analyte solutions. As can be seen from Figure 18-16, this device is a complete electrochemical cell and is more properly referred to as a probe rather than an electrode.

Membrane Composition

A *microporous membrane* is fabricated from a hydrophobic polymer. As the name implies, the membrane is highly porous (the average pore size is less than 1 μm) and allows the free passage of gases; at the same time, the water-repellent polymer prevents water and solute ions from entering the pores. The thickness of the membrane is about 0.1 mm.

The Mechanism of Response

Using carbon dioxide as an example, we can represent the transfer of gas to the internal solution by the following set of equations:

$$CO_2(aq) \rightleftarrows CO_2(g)$$

| Analyte | Membrane |
| Solution | Pores |

$$CO_2(g) \rightleftarrows CO_2(aq)$$

| Membrane | Internal |
| Pores | Solution |

$$CO_2(aq) + 2\,H_2O \rightleftarrows HCO_3^- + H_3O^+$$

| Internal | Internal |
| Solution | Solution |

The last equilibrium causes the pH of the internal surface film to change. This change is then detected by the internal glass/calomel electrode system. A description of the overall process is obtained by adding the equations for the three individual equilibria to give

$$CO_2(aq) + 2\,H_2O \rightleftarrows H_3O^+ + HCO_3^-$$

| Analyte | Internal |
| Solution | Solution |

It can be shown that the potential of the internal cell is given by

Challenge: Derive Equation 18-15.

$$E_{cell} = L + 0.0592 \log [CO_2(aq)]_{ext} \qquad (18\text{-}15)$$

where L is a constant. Thus, the potential between the glass electrode and the reference electrode in the internal solution is determined by the CO_2 concentration in the external solution. *Note that no electrode comes in direct contact with the analyte solution.* Therefore, these devices are gas-sensing *cells*, or *probes*, rather than gas-sensing electrodes. Nevertheless, they continue to be called electrodes in some literature and many advertising brochures.

The only species that interfere are other dissolved gases that permeate

Although sold as gas-sensing electrodes, these devices are complete electrochemical cells that contain *two* electrodes and should be called gas-sensing probes.

the membrane and then affect the pH of the internal solution. The selectivity of gas probes depends upon the selectivity of the internal indicator electrode. Gas-sensing probes for CO_2, NO_2, H_2S, SO_2, HF, HCN, and NH_3 are now available from commercial sources.

18E   INSTRUMENTS FOR MEASURING CELL POTENTIALS

Most cells containing an ion electrode have very high electrical resistance (as much as 10^8 ohms or more). In order to measure potentials of such high-resistance circuits accurately, it is necessary that the voltmeter have an electrical resistance that is several orders of magnitude greater than the resistance of the cell being measured. If the meter resistance is too low, current is drawn from the cell, which has the effect of lowering its output potential, thus creating a negative error. This effect is shown in Figure 18-17, which is a plot of the relative error in potential reading as a function of the ratio of the resistance of a meter to the resistance of a cell. When the meter and the cell have the same resistance, a relative error of -50% results. When this ratio of meter resistance to cell resistance is 10, the error is about -9%. When it is 1000, the error is less than 0.1% relative.

Figure 18-17

Relative error in a cell-potential measurement as a function of the ratio of the electrical resistance of the meter R_M to the resistance of the cell R_{cell}.

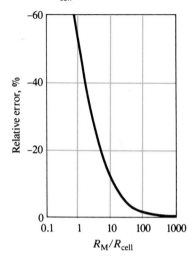

Feature 18-2
OPERATIONAL-AMPLIFIER VOLTAGE MEASUREMENTS

One of the most important developments in chemical instrumentation over the last several years has been the advent of compact, inexpensive, versatile integrated-circuit amplifiers (op amps).[5] These devices allow us to make potential measurements on high-resistance cells, such as those that contain a glass electrode, without drawing appreciable current. Even a small current (10^{-7} to 10^{-10} A) in a glass electrode results in a large error in the measured voltage. One of the most important uses for operational amplifiers is to isolate voltage sources from their measurement circuits. The basic *voltage follower,* which allows this type of measurement is shown in Figure 18-18a. This circuit has two important characteristics: the output voltage E_{out} is equal to the input voltage E_{in}, and the input current I_{in} is essentially zero (10^{-9} to 10^{-11} A).

A practical application of this circuit is in measuring cell potentials. The cell is connected to the op-amp input as shown in Figure 18-18b, and the output of the op amp is connected to a digital voltmeter to measure the voltage. Modern op amps are nearly ideal voltage-measurement devices and are incorporated in most ion meters and pH meters to monitor high-resistance indicator electrodes with little error.

---

[5]For a detailed description of op-amp circuits, see H. V. Malmstadt, C. G. Enke, and S. R. Crouch, *Electronics and Instrumentation for Scientists,* Chapter 5. Menlo Park, CA: Benjamin-Cummings, 1981.

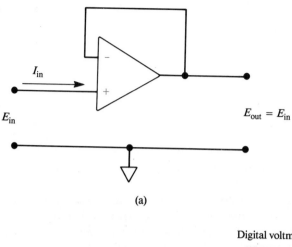

I_{in}

E_{in}

$E_{out} = E_{in}$

(a)

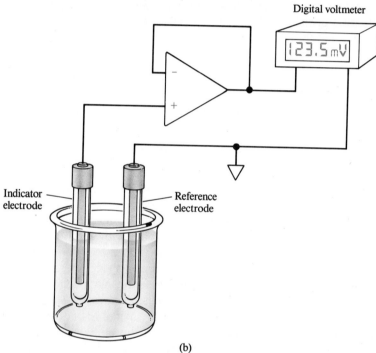

Digital voltmeter

123.5 mV

Indicator electrode

Reference electrode

(b)

Figure 18-18

(a) A voltage-follower operational amplifier. (b) Typical arrangement for potentiometric measurements with a membrane electrode.

Numerous high-resistance, direct-reading meters with internal resistances of 10^{11} to 10^{12} ohms are now on the market. These meters are commonly called *pH meters* but could more properly be referred to as *pIon meters* or *ion meters* since they are frequently used for the measurement of concentrations of other ions as well. Feature 18-2 describes a typical design of an ion meter.

The readout of ion meters is either digital or analog (in the latter, a needle sweeps a range on a scale from 0 to 14 pH units). Some meters are capable of precision on the order of 0.001 to 0.005 pH unit. Seldom is it possible to measure pH with a comparable degree of *accuracy*. Inaccuracies of ±0.02 to ±0.03 pH unit are typical.

18F   DIRECT POTENTIOMETRIC MEASUREMENTS

Direct potentiometric measurements provide a rapid and convenient method for determining the activity of a variety of cations and anions. The technique requires only the measurement of the potential of an indicator electrode when immersed in (1) the unknown and (2) a solution containing a known concentration of the analyte. If the response of the electrode is specific for the analyte, as it often is, no preliminary separation steps are required.

Direct potentiometric measurements are also readily adapted to applications requiring continuous and automatic recording of analytical data.

18F-1   The Sign Convention and Equations for Direct Potentiometry

The sign convention for potentiometry is consistent with the convention described in Chapter 15 for standard electrode potentials.[6] In this convention, the indicator electrode is *always* treated as the *cathode* and the reference electrode as the *anode*.[7] For direct potentiometric measurements, the potential of a cell can then be expressed in terms of the potentials developed by the indicator electrode, the reference electrode, and a junction potential:

$$E_{cell} = E_{ind} - E_{ref} + E_j \qquad (18\text{-}16)$$

In Section 18D, which deals with indicator electrodes, we describe the response of various types of indicator electrodes to analyte activities. For the cation X^{n+} at 25°C, the electrode response takes the general *Nernstian* form

$$E_{ind} = L - \frac{0.0592}{n}\,pX = L + \frac{0.0592}{n}\log a_X \qquad (18\text{-}17)$$

where L is a constant and a_X is the activity of the cation. For metallic indicator electrodes, L is ordinarily the standard electrode potential; for membrane electrodes, L is the summation of several constants, including the time-dependent asymmetry potential of uncertain magnitude.

Substitution of Equation 18-17 into Equation 18-16 yields, with rearrangement,

$$pX = -\log a_X = -\frac{E_{cell} - (E_j - E_{ref} + L)}{0.0592/n}$$

<hr/>

[6]According to Bates, the convention being described here has been endorsed by standardizing groups in the United States and Great Britain as well as IUPAC. See R. G. Bates, in *Treatise on Analytical Chemistry,* 2nd ed., I. M. Kolthoff and P. J. Elving, Eds., Part I, Vol. 1, pp. 831–832. New York: Wiley, 1978.

[7]In effect, the sign convention for electrode potentials described in Section 15C-3 also designates the indicator electrode as the cathode by stipulating that half-reactions always be written as reductions; the standard hydrogen electrode, which is the reference electrode in this case, is then the anode.

The constant terms in parentheses can be combined to give a new constant K:

$$\text{pX} = -\log a_X = -\frac{E_{cell} - K}{0.0592/n} \qquad (18\text{-}18)$$

For an anion A^{n-}, the sign of Equation 18-18 is reversed:

$$\text{pA} = \frac{E_{cell} - K}{0.0592/n} \qquad (18\text{-}19)$$

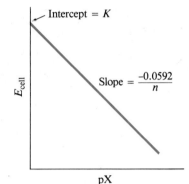

A plot of Equation 18-20 for cationic electrodes.

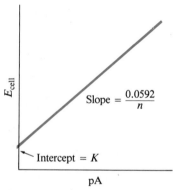

A plot of Equation 18-21 for anionic electrodes.

All direct potentiometric methods are based upon Equation 18-18 or 18-19. The difference in sign in the two equations has a subtle but important consequence in the way that ion-selective electrodes are connected to pH meters and pIon meters. When the two equations are solved for E_{cell}, we find that for cations

$$E_{cell} = K - \frac{0.0592}{n}\text{pX} \qquad (18\text{-}20)$$

and for anions

$$E_{cell} = K + \frac{0.0592}{n}\text{pA} \qquad (18\text{-}21)$$

Equation 18-20 shows that an increase in pX results in a *decrease* in E_{cell} with a cation-selective electrode. Thus, when a high-resistance voltmeter is connected to the cell in the usual way, with the indicator electrode attached to the positive terminal, the meter reading decreases as pX increases. To eliminate this problem, instrument manufacturers generally reverse the leads so that cation-sensitive electrodes are connected to the *negative* terminal of the voltage-measuring device. Meter readings then increase with increases in pX. Anion-selective electrodes, on the other hand, are connected to the *positive* terminal of the meter so that increases in pA also yield larger readings.

18F-2 The Electrode-Calibration Method

As we saw in Section 18D, the constant K in Equations 18-18 and 18-19 is made up of several constants, at least one of which, the junction potential, cannot be computed from theory or measured directly. So before these equations can be used for the determination of pX or pA, K must be evaluated *experimentally* with a standard solution of the analyte.

In the electrode-calibration method, K in Equations 18-18 and 18-19 is determined by measuring E_{cell} for one or more standard solutions of known pX or pA. The assumption is then made that K is unchanged when the standard is replaced by the analyte solution. The calibration is ordinarily performed at the time pX or pA for the unknown is determined. With membrane electrodes, recalibration may be required if measurements extend over several hours because of changes in the asymmetry potential.

The electrode-calibration method offers the advantages of simplicity, speed, and applicability to the continuous monitoring of pX or pA. It suffers, however, from a somewhat limited accuracy because of uncertainties in junction potentials.

Inherent Error in the Electrode-Calibration Procedure

A serious disadvantage of the electrode-calibration method is the inherent error that results from the assumption that K in Equations 18-18 and 18-19 remains constant after calibration. This assumption can seldom, if ever, be exactly true because the electrolyte composition of the unknown almost inevitably differs from that of the solution employed for calibration. The junction-potential term contained in K varies slightly as a consequence, even when a salt bridge is used. This error is frequently on the order of 1 mV or more. Unfortunately, because of the nature of the potential/activity relationship, such an uncertainty has an amplified effect on the inherent accuracy of the analysis. The relative error in analyte concentration can be estimated from the following equation:[8]

$$\% \text{ relative error} = \frac{\Delta a_1}{a_1} \times 100\% = 3.89 \times 10^3 n \Delta K\% \approx 4000 n \Delta K\%$$

The quantity $\Delta a_1/a_1$ is the relative error in a_1 associated with an absolute uncertainty ΔK in K. If, for example, ΔK is ± 0.001 V, a relative error in activity of about $\pm 4n\%$ can be expected. *It is important to appreciate that this error is characteristic of all measurements involving cells that contain a salt bridge and cannot be eliminated by even the most careful measurements of cell potentials or the most sensitive and precise measuring devices.*

Activity Versus Concentration

Electrode response is related to analyte activity rather than analyte concentration. We are usually interested in concentration, however, and the determination of this quantity from a potentiometric measurement requires activity-coefficient data. Activity coefficients are not often available because the ionic strength of the solution is either unknown or else is so large that the Debye-Hückel equation is not applicable.

The difference between activity and concentration is illustrated by Figure 18-19, in which the response of a calcium ion electrode is plotted

---

[8]For a derivation of this equation, see D. A. Skoog, D. M. West, and F. J. Holler, *Fundamentals of Analytical Chemistry,* 5th ed., p. 380. Philadelphia: Saunders College Publishing, 1988.

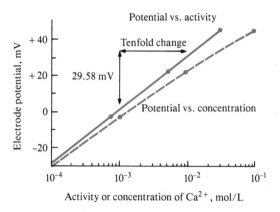

Figure 18-19

Response of a liquid-membrane electrode to variations in the concentration and activity of calcium ion. (Courtesy of Orion Research, Boston, MA.)

Many chemical reactions of physiological importance depend upon the activity of metal ions rather than their concentration.

TISAB (total ionic strength adjustment buffer) is used to control the ionic strength and the pH of samples and standards in ion-selective electrode measurements for fluoride.

against a logarithmic function of calcium chloride *concentration*. The nonlinearity is due to the increase in ionic strength—and the consequent decrease in calcium ion activity—with increasing electrolyte concentration. The upper curve is obtained when these concentrations are converted to activities. This straight line has the theoretical slope of 0.0296 (0.0592/2).

Activity coefficients for singly charged species are less affected by changes in ionic strength than are the coefficients for ions with multiple charges. Thus, the effect shown in Figure 18-19 is less pronounced for electrodes that respond to H^+, Na^+, and other univalent ions.

In potentiometric pH measurements, the pH of the standard buffer used for calibration is generally based on the activity of hydrogen ions. Thus, the results are also on an activity scale. If the unknown sample has a high ionic strength, the hydrogen ion *concentration* will differ appreciably from the activity measured.

An obvious way to convert potentiometric measurements from activity to concentration is to make use of an empirical calibration curve, such as the lower plot in Figure 18-19. For this approach to be successful, it is necessary to make the ionic composition of the standards essentially the same as that of the analyte solution. Matching the ionic strength of standards to that of samples is often difficult, particularly for samples that are chemically complex.

Where electrolyte concentrations are not too great, it is often useful to swamp both samples and standards with a measured excess of an inert electrolyte. The added effect of the electrolyte from the sample matrix becomes negligible under these circumstances, and the empirical calibration curve yields results in terms of concentration. This approach has been used, for example, in the potentiometric determination of fluoride ion in drinking water. Both samples and standards are diluted with a solution that contains sodium chloride, an acetate buffer, and a citrate buffer; the diluent is sufficiently concentrated so that the samples and standards have essentially identical ionic strengths. This method provides a rapid means for measuring fluoride concentrations in the part-per-million range with an accuracy of about 5% relative.

18F-3 The Standard-Addition Method

The standard-addition method involves determining the potential of the electrode system before and after a measured volume of a standard has been added to a known volume of the analyte solution. Often an excess of an electrolyte is incorporated into the analyte solution at the outset to prevent any major shift in ionic strength that might accompany the addition of standard. It is also necessary to assume that the junction potential remains constant during the two measurements.

Example 18-1

A cell consisting of a saturated calomel electrode and a lead ion electrode developed a potential of -0.4706 V when immersed in 50.00 mL of a sample. A 5.00-mL addition of standard 0.02000 M lead solution caused the potential to shift to -0.4490 V. Calculate the molar concentration of lead in the sample.

We shall assume that the activity of Pb^{2+} is approximately equal to $[Pb^{2+}]$ and apply Equation 18-18. Thus,

$$pPb = -\log[Pb^{2+}] = -\frac{E'_{cell} - K}{0.0592/2}$$

where E'_{cell} is the initial measured potential $(-0.4706$ V$)$.

After the standard solution is added, the potential becomes E''_{cell} $(-0.4490$ V$)$ and

$$-\log \frac{50.00 \times [Pb^{2+}] + 5.00 \times 0.0200}{50.00 + 5.00} = -\frac{E''_{cell} - K}{0.0592/2}$$

$$-\log (0.9091[Pb^{2+}] + 1.818 \times 10^{-3}) = -\frac{E''_{cell} - K}{0.0592/2}$$

Subtracting this equation from the first leads to

$$-\log \frac{[Pb^{2+}]}{0.9091[Pb^{2+}] + 1.818 \times 10^{-3}} = \frac{2(E''_{cell} - E'_{cell})}{0.0592}$$

$$= \frac{2[-0.4490 - (-0.4706)]}{0.0592} = 0.7297$$

Taking the antilog of both sides of this equation gives

$$\frac{[Pb^{2+}]}{0.9091[Pb^{2+}] + 1.818 \times 10^{-3}} = 0.1863$$

$$[Pb^{2+}] = 4.08 \times 10^{-4} \text{ M}$$

18F-4 Potentiometric pH Measurements with a Glass Electrode[9]

The glass electrode is unquestionably the most important indicator electrode for hydrogen ion. It is convenient to use and subject to few of the interferences that affect other pH-sensing electrodes.

The glass/calomel electrode system is a remarkably versatile tool for measuring pH under many conditions. It can be used without interference in solutions containing strong oxidants, strong reductants, proteins, and gases; the pH of viscous or even semisolid fluids can be determined. Electrodes for special applications are available. Included among these are small electrodes for pH measurements in one drop (or less) of solution, in a tooth cavity, or in the sweat on the skin; microelectrodes that permit the measurement of pH inside a living cell; rugged electrodes for insertion in a flowing liquid stream to provide a continuous monitoring of pH; and small electrodes that can be swallowed to indicate the acidity of the stomach contents (the calomel electrode is kept in the mouth).

Errors That Affect pH Measurements with the Glass Electrode

The ubiquity of the pH meter and the general applicability of the glass electrode tend to lull the chemist into the attitude that any measurement obtained with such equipment is surely correct. The reader must be alert to the fact that there are distinct limitations to the electrode, some of which were discussed in earlier sections:

1. *The alkaline error.* The ordinary glass electrode becomes somewhat sensitive to alkali metal ions and gives low readings at pH values greater than 9.
2. *The acid error.* Values registered by the glass electrode tend to be somewhat high when the pH is less than about 0.5.
3. *Dehydration.* Dehydration may cause erratic electrode performance.
4. *Error in unbuffered neutral solutions.* Because equilibrium between the bulk of the solution and the layer of solution at the surface of a membrane is achieved only slowly in poorly buffered, approximately neutral solutions, time must be allowed for equilibrium between the two to be established. Before being used to determine the pH of such solutions, the glass electrode should be thoroughly rinsed with water. Then both electrodes should be immersed in successive portions of the unknown until a constant pH reading is obtained.
5. *Variation in junction potential.* A fundamental source of uncertainty for which a correction cannot be applied is the junction-potential variation resulting from differences in the composition of the standard and the unknown solution.
6. *Error in the pH of the standard buffer.* Any inaccuracies in the preparation of the buffer used for calibration or any changes in its composition during storage cause an error in subsequent pH measurements. The action of bacteria on organic buffer components is a common cause for deterioration.

The most common instrumental technique in science is the measurement of pH.

---

[9]For a detailed discussion of potentiometric pH measurements, see R. G. Bates, *Determination of pH*, 2nd ed. New York: Wiley, 1973.

The Operational Definition of pH

The utility of pH as a measure of the acidity or alkalinity of aqueous media, the wide availability of commercial glass electrodes, and the relatively recent proliferation of inexpensive solid-state pH meters have made the potentiometric measurement of pH perhaps the most common analytical technique in all of science. It is thus extremely important that pH be defined in a manner that is easily duplicated at various times and in various laboratories throughout the world. To meet this requirement, it is necessary to define pH in operational terms—that is, by the way the measurement is made. Only then will the pH measured by one worker be the same as that measured by another.

The operational definition of pH endorsed by the National Institute of Standards and Technology (NIST), similar organizations in other countries, and the IUPAC is based upon the direct calibration of the meter with carefully prescribed standard buffers followed by potentiometric determination of the pH of unknown solutions.

Consider, for example, the glass/calomel system in Figures 18-6 and 18-7. When these electrodes are immersed in a standard buffer, Equation 18-18 applies and we can write

$$pH_S = -\frac{E_S - K}{0.0592}$$

where E_S is the cell potential when the electrodes are immersed in the standard buffer. Similarly, if the cell potential is E_U when the electrodes are immersed in a solution of unknown pH, we have

$$pH_U = -\frac{E_U - K}{0.0592}$$

By subtracting the first equation from the second and solving for pH_U, we find

$$pH_U = pH_S - \frac{(E_U - E_S)}{0.0592} \qquad (18\text{-}22)$$

> By definition, pH is what you measure with a glass electrode and a pH meter.

Equation 18-22 has been adopted throughout the world as the *operational definition of pH*.

> An operational definition of a quantity defines the quantity in terms of how it is measured.

Workers at the National Institute of Standards and Technology and elsewhere have used cells without liquid junctions to study primary-standard buffers extensively. Some of the properties of these buffers are presented and discussed in detail elsewhere.[10] For general use, the buffers can be prepared from relatively inexpensive laboratory reagents. For careful work, certified buffers can be purchased from the NIST.

---

[10]R. G. Bates, *Determination of pH,* 2nd ed., Chapter 4. New York: Wiley, 1973.

18G   POTENTIOMETRIC TITRATIONS

A *potentiometric titration* involves measuring the potential of a suitable indicator electrode as a function of titrant volume. The information provided by a potentiometric titration is not the same as that obtained from a direct potentiometric measurement. For example, the direct measurement of 0.100 M solutions of hydrochloric and acetic acids would yield two substantially different hydrogen ion concentrations because the latter is only partially dissociated. In contrast, the potentiometric titration of equal volumes of the two acids would require the same amount of standard base because both solutes have the same number of titratable protons.

Potentiometric titrations provide data that are more reliable than data from titrations that use chemical indicators and are particularly useful with colored or turbid solutions and for detecting the presence of unsuspected species. They suffer from the disadvantage of being more time-consuming than those involving indicators; on the other hand, they are readily automated.

Figure 18-20 illustrates a typical apparatus for performing a manual potentiometric titration. Its use involves measuring and recording the cell potential (in units of millivolts or pH, as appropriate) after each addition of reagent. The titrant is added in large increments at the outset and in smaller and smaller increments as the end point is approached (as indicated by larger changes in response per unit volume).

Automatic titrators for carrying out potentiometric titrations are available from several manufacturers.

18G-1 End-Point Detection

Several methods can be used to determine the end point of a potentiometric titration. The most straightforward involves a direct plot of potential as a function of reagent volume, as in Figure 18-21a; the midpoint in the

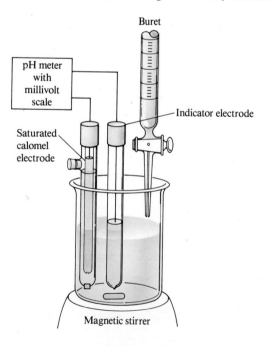

Buret

pH meter with millivolt scale

Indicator electrode

Saturated calomel electrode

Magnetic stirrer

Figure 18-20

Apparatus for a potentiometric titration.

steeply rising portion of the curve is estimated visually and taken as the end point. A second approach to end-point detection is to plot the change in potential per unit volume of titrant (that is, $\Delta E/\Delta V$) as a function of the average volume V. As shown in Figure 18-21b, the curve obtained has a maximum that corresponds to the end point.

Figure 18-21c shows that the second derivative for the titration data changes sign at the end point. This change is used as the analytical signal in some automatic titrators.

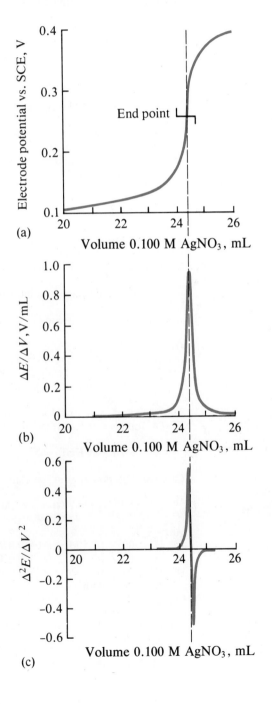

(a)

(b)

(c)

Figure 18-21

Titration of 2.433 mmol of chloride ion with 0.1000 M silver nitrate. (a) Titration curve. (b) First-derivative curve. (c) Second-derivative curve.

18H   QUESTIONS AND PROBLEMS

*18-1. Differentiate between an electrode of the first kind and an electrode of the second kind.

18-2. What occurs when a newly manufactured glass electrode is immersed in water?

18-3. What is the source of
   *(a) the boundary potential in a membrane electrode?
   (b) a junction potential in a glass/calomel electrode system?
   *(c) the potential of a crystalline membrane electrode used to determine the concentration of F^-?

*18-4. What is the alkaline error in pH measurement with a glass electrode?

18-5. List the advantages and disadvantages of a potentiometric titration relative to a titration with chemical indicators.

*18-6. A solution of ethylamine is titrated with HCl using a glass/calomel electrode system. Show how K_b can be obtained from the pH at the point of half-neutralization.

*18-7. (a) Calculate the standard potential for the reaction

$$CuSCN(s) + e^- \rightleftharpoons Cu(s) + SCN^-$$

   (b) Sketch a cell having a copper indicator electrode as a cathode and a saturated calomel electrode as an anode that could be used for the determination of SCN^-.
   (c) Derive an equation relating the measured potential of the cell in (b) to pSCN. (Assume the junction potential is zero.)
   (d) Calculate the pSCN of a thiocyanate-containing solution that is saturated with CuSCN and employed in conjunction with a copper electrode in the cell sketched in (b) if the resulting potential is -0.076 V.

18-8. (a) Calculate the standard potential for the reaction

$$Ag_2S(s) + 2\ e^- \rightleftharpoons 2\ Ag(s) + S^{2-}$$

   (b) Sketch a cell having a silver indicator electrode as the cathode and a standard calomel electrode as an anode that could be used for determining S^{2-}.
   (c) Derive an equation relating the measured potential of the cell in (b) to pS. (Assume the junction potential is zero.)
   (d) Calculate the pS of a solution that is saturated with Ag_2S and then employed with the cell sketched in (b) if the resulting potential is 0.538 V.

*18-9. Give a schematic representation of each of the following cells. Derive an equation relating cell potential to p-function. Assume the junction potential is negligible, treat the indicator electrode as the cath-

ode, and specify any necessary concentrations as 1.00×10^{-4} M.
   (a) A cell with a mercury indicator electrode for the determination of pCl.
   (b) A cell with a silver indicator electrode for the determination of pCO_3.
   (c) A cell with a platinum electrode for the determination of pSn(IV).

18-10. Give a schematic representation of each of the following cells. Derive an equation relating cell potential and p function. Assume the junction potential is negligible, treat the indicator electrode as the cathode, and specify any necessary concentrations as 1.00×10^{-4} M.
   (a) A cell with a lead electrode for the determination of $pCrO_4$.
   (b) A cell with a silver indicator electrode for the determination of $pAsO_4$.
   (c) A cell with a platinum electrode for the determination of pT1(III).

*18-11. The cell

$$SCE\|Ag_2CrO_4(sat'd),CrO_4^{2-}(x\ M)|Ag$$

is employed for the determination of $pCrO_4$. Calculate $pCrO_4$ when the cell potential is 0.402 V.

18-12. Calculate the potential of the cell

$$SCE\|aqueous\ solution|Hg$$

when the aqueous solution is
   (a) 7.40×10^{-3} M Hg^{2+}.
   (b) 7.40×10^{-3} M Hg_2^{2+}.
   (c) $Hg_2SO_4(sat'd),SO_4^{2-}(0.0250$ M).
   (d) $Hg^{2+}(2.00 \times 10^{-3}$ M),$OAc^-(0.100$ M)
       $[Hg^{2+} + 2\ OAc^- \rightleftharpoons Hg(OAc)_2(aq)$
       $K_f = 2.7 \times 10^8]$.

*18-13. The standard potential for the reduction of the EDTA complex of Hg(II) is

$$HgY^{2-} + 2\ e^- \rightleftharpoons Hg(l) + Y^{4-} \qquad E^0 = 0.210\ V$$

Calculate the potential of the cell

$$SCE\|HgY^{4-}(2.00 \times 10^{-4}\ M),Z|Hg$$

when Z is
   (a) $H^+(1.00 \times 10^{-4}$ M),EDTA(0.0200 M).
   (b) $H^+(1.00 \times 10^{-8}$ M),EDTA(0.0200 M).
   (c) $H^+(1.00 \times 10^{-10}$ M),$CaCl_2(0.0200$ M), EDTA(0.0200 M).

18-14. Calculate the potential of the cell described in Problem 18-13 when Z is
   (a) $H^+(1.00 \times 10^{-6}$ M),EDTA(0.0100 M).
   (b) $H^+(1.00 \times 10^{-10}$ M),EDTA(0.0100 M).
   (c) $H^+(1.00 \times 10^{-7}$ M), $Zn(NO_3)_2(0.0100$ M), EDTA(0.0100 M).

*18-15. The cell

$$SCE\|H^+(a = x)|glass\ electrode$$

has a potential of 0.2094 V when the solution in the right-hand compartment is a buffer of pH 4.006. The following potentials are obtained when the buffer is replaced with unknowns: **(a)** −0.3011 V and **(b)** +0.1163 V. Calculate the pH and the hydrogen ion activity of each unknown. **(c)** Assuming an uncertainty of ±0.002 V in the junction potential, what is the range of hydrogen ion activities within which the true value might be expected to lie?

18-16. The cell

$$SCE\|MgA_2(a_{Mg^{2+}} = 9.62 \times 10^{-3})|membrane$$
$$electrode\ for\ Mg^{2+}$$

has a potential of 0.367 V.
(a) When the solution of known magnesium activity is replaced with an unknown solution, the potential is −0.544 V. What is the pMg of this unknown solution?
(b) Assuming an uncertainty of ±0.002 V in the junction potential, what is the range of Mg^{2+} activities within which the true value might be expected to lie?

*18-17. The cell

$$SCE\|CdA_2(sat'd),A^-(0.0250\ M)|Cd$$

has a potential of −0.721 V. Calculate the solubility product of CdA_2, neglecting the junction potential.

18-18. The cell

$$SCE\|HA(0.250\ M),NaA(0.180\ M)|H_2(1.00\ atm),Pt$$

has a potential of −0.797 V. Calculate the dissociation constant of HA, neglecting the junction potential.

18-19. A 40.00-mL aliquot of 0.05000 M HNO_2 is diluted to 75.00 mL and titrated with 0.08000 M Ce^{4+}. The pH of the solution is maintained at 1.00 throughout the titration, and the formal potential of the cerium system is 1.44 V.
*(a) Calculate the potential of the indicator electrode with respect to a saturated calomel reference electrode after the addition of 5.00, 10.00, 15.00, 25.00, 40.00, 49.00, 50.00, 51.00, 55.00, and 60.00 mL of cerium(IV).
(b) Draw a titration curve for these data.

18-20. Calculate the potential of a silver cathode versus the standard calomel electrode after the addition of 5.00, 15.00, 25.00, 30.00, 35.00, 39.00, 40.00, 41.00, 45.00, and 50.00 mL of 0.1000 M $AgNO_3$ to 50.00 mL of 0.0800 M KSeCN. Construct a titration curve from these data.
(K_{sp} for AgSeCN $= 4.20 \times 10^{-16}$.)

*18-21. Quinhydrone is an equimolar mixture of quinone (Q) and hydroquinone (H_2Q). These two compounds react reversibly at a platinum electrode:

$$Q + 2\ H^+ + 2\ e^- \rightleftarrows H_2Q \qquad E^0 = 0.699\ V$$

The pH of a solution can be determined by saturating it with quinhydrone and making it a part of the cell

$$SCE\|quinhydrone(sat'd),H^+(x\ M)|Pt$$

If such a cell has a potential of 0.313 V, what is the pH of the solution, assuming the junction potential is zero?

Chapter 19

Electrogravimetric and Coulometric Methods

In this chapter we describe two related electroanalytical methods: *electrogravimetry* and *coulometry*. In both methods, an electrolysis is carried out long enough to ensure that the analyte is completely oxidized or reduced to a single product of known composition. In electrogravimetric methods, the product is weighed as a deposit on one of the electrodes (the *working electrode*). In coulometric procedures, the quantity of electrons needed to complete the electrolysis is measured.[1]

Electrogravimetry and coulometry are moderately sensitive, rapid, and among the most accurate and precise techniques available to the chemist. Like gravimetry, these methods require no preliminary calibration against standards because the functional relationship between the quantity measured and the analyte concentration can be derived from theory and atomic-weight data.

Electrogravimetry and coulometry differ from potentiometry in that they require a significant current throughout the analytical process. In contrast, potentiometric measurements are performed under conditions of essentially zero current. When there is a current in an electrochemical cell, the cell potential is no longer simply the difference between the electrode potentials of the cathode and the anode. Additional phenomena

Electrogravimetry and coulometry have relative errors of a few parts per thousand.

---

[1]For further information concerning the methods in this chapter, see J. A. Plambeck, *Electroanalytical Chemistry.* New York: Wiley, 1982; *Laboratory Techniques in Electroanalytical Chemistry,* P. T. Kissinger and W. R. Heineman, Eds. New York: Marcel Dekker, 1984.

require application of potentials greater than theoretical for an electrolytic cell and development of potentials smaller than theoretical in a galvanic cell. Before proceeding, we must examine these phenomena in detail.

19A   THE EFFECT OF CURRENT ON CELL POTENTIALS

Consider the electrolytic cell shown in Figure 19-1. A voltage $E_{applied}$ is applied to the cell in such a way that E_{cell} and $E_{applied}$ have opposing polarities. A current meter is placed in the circuit to indicate the magnitude and direction of any current in the cell. There will be a current I in the circuit in the indicated direction only when $E_{applied} > E_{cell}$. When the two potentials are equal, that is when $E_{applied} = E_{cell}$, $I = 0$.

When there is a current, the potential of the cell is less than the thermodynamic potential because one or more of three phenomena are operating: *IR drop, concentration polarization,* and *kinetic polarization.*

The potential of a galvanic cell also changes when a current is present. Here, *IR* drop, concentration polarization, and/or kinetic polarization make the output potential smaller than the equilibrium potential.

19A-1   Ohmic Potential: *IR* Drop

Electrochemical cells, like metallic conductors, resist the flow of charge. In both types of conduction, Ohm's law describes the effect of this resistance. The product of the resistance R of a cell in ohms (Ω) and the current I in amperes (A) is called the *ohmic potential* or the *IR drop* of the cell. In Figure 19-2 we have used a resistor R to represent the cell resistance in the circuit. If we begin with $E_{applied} = E_{cell}$ ($I = 0$) and gradually increase $E_{applied}$, a small current appears in the circuit. The current

By definition, an electric current is a flow of charge. In metallic conductors the charge is electrons. In electrolytes, the current involves the flow of positive ions in one direction and negative ions in the other. A convention for indicating the direction of currents is needed because charges of opposite sign move in opposite directions under the influence of a voltage. You should appreciate that a positive charge moving in one direction in a circuit is equivalent to a negative charge moving in the other. For simplicity and algebraic consistency, scientists assume that *all charge carriers are positive.* Thus, the current I in Figure 19-2 is given a positive sign and treated as if it were a flow of positive charge from the positive electrode of the battery to the negative electrode even though the actual current is made up of negative electrons.

If there is to be a current in an electrolytic cell, it is necessary that $E_{applied} > E_{cell}$.

When a galvanic cell produces a current, $E_{output} < E_{cell}$.

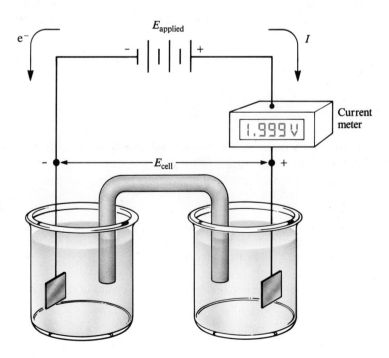

Figure 19-1

An electrolytic cell with a meter to measure the current during electrolysis.

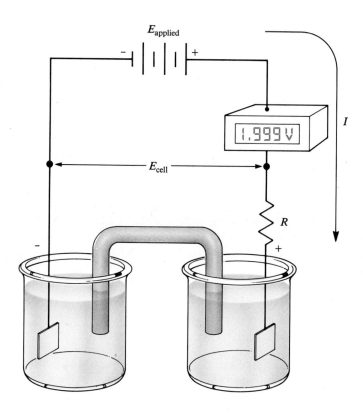

Figure 19-2
An electrolytic cell with a resistor
R to represent the cell resistance.

through the cell resistance R results in a potential drop of $-IR$ volts. In other words, the applied voltage must be greater than the theoretical cell potential by $-IR$ volts. Similarly, when a current of I amperes is produced by a galvanic cell with a resistance of R ohms, the output potential of the cell is decreased by $-IR$ volts. Thus, in the presence of a current, Equation 15-10 for a cell potential must be modified by the addition of the term $-IR$:

$$E_{cell} = E_{cathode} - E_{anode} - IR \qquad (19\text{-}1)$$

Ohm's law: $E = IR$ or $I = \left(\dfrac{1}{R}\right) E$

where $E_{cathode}$ and E_{anode} are electrode potentials computed with the Nernst equation discussed in Section 15C-4.

Example 19-1

Consider a cell consisting of a copper electrode in contact with 1.00 M Cu^{2+}, a cadmium electrode in contact with 1.00 M Cd^{2+}, and a connecting salt bridge. The cell has a resistance of 4.00 Ω.

(a) Calculate the potential needed to develop a current of 0.0200 A in the electrolytic cell

$$Cu\,|\,Cu^{2+}(1.00\text{ M})\,\|\,Cd^{2+}(1.00\text{ M})\,|\,Cd$$

Since both cation concentrations are 1.00 M, the electrode potentials and the standard electrode potentials are numerically equal. Substituting in Equation 19-1 gives

$$E_{cell} = E_{Cd}^0 - E_{Cu}^0 - IR$$
$$= -0.403 \text{ V} - 0.337 \text{ V} - 0.0200 \text{ A} \times 4.00 \text{ }\Omega$$
$$= -0.740 \text{ V} - 0.080 \text{ V} = -0.820 \text{ V}$$

Thus a potential 0.08 V more negative than the theoretical value is needed to deposit cadmium at the rate required to maintain a current of 0.0200 A.

(b) Calculate the cell potential when there is a current of 0.0200 A in the galvanic cell

$$Cd|Cd^{2+}(1.00 \text{ M})\|Cu^{2+}(1.00 \text{ M})|Cu$$

Here,

$$E_{cell} = E_{Cu}^0 - E_{Cd}^0 - IR$$
$$= 0.337 \text{ V} - (-0.403 \text{ V}) - 0.0200 \text{ A} \times 4.00 \text{ }\Omega$$
$$= 0.740 \text{ V} - 0.080 \text{ V} = 0.660 \text{ V}$$

The production of a current in this cell causes the output potential to be markedly less than the theoretical value because of the *IR* drop in the cell.

André Marie Ampère (1775–1836), French mathematician and physicist, was the first to apply mathematics to the study of electrical current. Consistent with Benjamin Franklin's definitions of positive and negative charge, Ampère defined a positive current to be the direction of flow of positive charge. Although we now know that negative electrons carry current in metals, Ampère's definition has survived to the present. The unit of current, the ampere, is named in his honor.

19A-2 Polarization Effects

Equation 19-1 can be rearranged to give

$$I = -\frac{1}{R} E_{cell} + \frac{1}{R} (E_{cathode} - E_{anode})$$

For small currents and brief periods of time, $E_{cathode}$ and E_{anode} remain relatively constant during an electrolysis. The cell behavior can then be approximated by the relationship

$$I = -\frac{1}{R} E_{cell} + k \qquad (19\text{-}2)$$

where k is a constant.

According to Equation 19-2, a plot of current in an electrolytic cell as a function of applied potential should be a straight line with a slope equal to the negative reciprocal of the resistance. As shown in Figure 19-3a, the plot is indeed linear with small currents. As the applied voltage increases, the current deviates significantly from linearity. Figure 19-3b shows that galvanic cells behave in an analogous way. Just as before, there is a linear

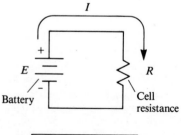

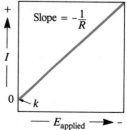

Illustration of Equation 19-2.

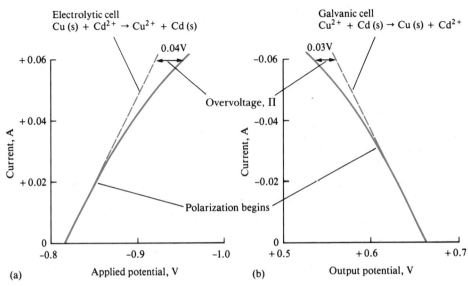

Figure 19-3
Current/voltage curves for (a) an electrolytic and (b) a galvanic cell. In both cells, the overvoltage at 0.06 A is symbolized by Π. For cell (a), $\Pi = -0.040$ V; for cell (b), $\Pi = -0.03$ V.

> Overvoltage is the potential difference between the theoretical cell potential and the actual cell potential at a given level of current.

Factors that influence polarization: (1) electrode size, shape, and composition; (2) composition of the electrolyte solution; (3) temperature and stirring rate; (4) current level; (5) physical state of species involved in the cell reaction.

relationship where small currents are involved, but significant departures from the straight line occur as the current becomes greater.

Cells that exhibit nonlinear behavior are said to be *polarized*, and the degree of polarization is given by an *overvoltage*, or *overpotential*, which is symbolized by Π in the figure. Note that polarization requires the application of a potential greater than the theoretical value to give a current of the expected magnitude. Thus, the overpotential required to achieve a current of 0.06 A in the electrolytic cell in Figure 19-3a is about -0.04 V. For the galvanic cell in Figure 19-3b, the polarization resulting from a 0.06 A current causes the output cell potential to decrease by about 0.03 V—that is, the overvoltage is -0.03 V. Note that in each case the overvoltage is negative. Thus, for a cell affected by overvoltage, Equation 19-1 becomes

$$E_{cell} = E_{cathode} - E_{anode} - IR - \Pi \qquad (19\text{-}3)$$

Polarization is an electrode phenomenon that may affect either or both of the electrodes in a cell. The degree of polarization of an electrode varies widely. In some instances it approaches zero, but in others it can be so nearly complete that the current in the cell becomes independent of potential. Polarization phenomena are conveniently divided into two categories: *concentration polarization* and *kinetic polarization*.

Concentration Polarization

Electron transfer between a reactive species in a solution and an electrode can take place only from a thin film of solution located immediately adjacent to the electrode surface; this film is only a few angstroms thick and contains a limited number of reactive ions or molecules. In order for there to be a steady current in a cell, this film must be continuously replenished with reactant from the bulk of the solution. That is, as reactant ions or molecules are consumed by the electrochemical reaction, more must be transported into the surface film at a rate that is sufficient to maintain the current. For example, for a current of 0.01 A in the cell described in Example 19-1a, it is necessary to transport cadmium ions to the cathode surface at a rate of about 5×10^{-8} mol/s, which is equal to 3×10^{16} cadmium ions per second. (Similarly, copper ions must be removed from the surface film of the anode at this same rate.)

Concentration polarization occurs when reactant species do not arrive at the cathode surface or product species do not leave the anode surface fast enough to maintain the desired current. When this happens, the current is limited to values less than those predicted by Equation 19-2.

Reactants are transported to an electrode surface by three mechanisms: (1) *diffusion*, (2) *migration*, and (3) *convection*. Products are removed from an electrode surface in the same ways. We will focus on mass-transport processes to the cathode, but our discussion applies equally to anodes.

> Reactants are transported to an electrode by (1) diffusion, (2) migration, and (3) convection.

Diffusion. When there is a concentration difference between two regions of a solution, ions or molecules move from the more concentrated region to the more dilute. This process is called *diffusion* and ultimately leads to a disappearance of the concentration difference. The rate of diffusion is directly proportional to the concentration difference. For example, when cadmium ions are deposited at a cathode by a current as illustrated in Figure 19-4, the concentration of Cd^{2+} at the electrode surface $[Cd^{2+}]_0$ is very small. The difference between $[Cd^{2+}]_0$ at the surface and $[Cd^{2+}]$ in the bulk solution creates a concentration gradient that causes cadmium ions to diffuse toward the surface film. The rate of diffusion is given by

> Diffusion is a process in which ions or molecules move from a more concentrated part of a solution to a more dilute one.

$$\text{rate of diffusion to cathode surface} = k'([Cd^{2+}] - [Cd^{2+}]_0) \quad (19\text{-}4)$$

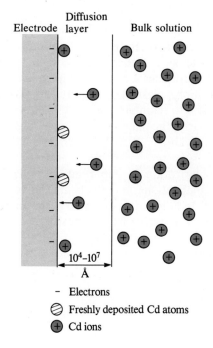

Diffusion layer

Electrode

Bulk solution

10^4–10^7
Å

- Electrons
⊘ Freshly deposited Cd atoms
⊕ Cd ions

Figure 19-4

Concentration changes at the surface of a cathode. As Cd^{2+} ions are reduced to Cd atoms at the electrode surface, the concentration of Cd^{2+} at the surface becomes very small. Ions then diffuse from the bulk of the solution to the surface as a result of the concentration gradient.

Figure 19-5

Migration is the movement of ions through a solution as a result of electrostatic attraction between the ions and the electrodes.

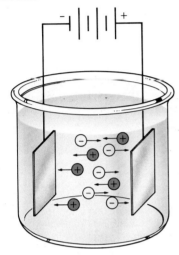

Convection is the mechanical transport of ions or molecules through a solution as a result of stirring, vibration, or temperature and/or density gradients.

Experimental variables that influence degree of concentration polarization: (1) reactant concentration, (2) electrolyte concentration, (3) mechanical agitation, (4) electrode size.

The current in a kinetically polarized cell is governed by the rate of electron transfer rather than the rate of mass transfer.

where $[Cd^{2+}]$ is the reactant concentration in the bulk of the solution, $[Cd^{2+}]_0$ is its equilibrium concentration at the surface of the cathode, and k' is a proportionality constant. The value of $[Cd^{2+}]_0$ at any instant is fixed by the *potential of the electrode* and can be calculated from the Nernst equation. In this example, the surface cadmium ion concentration is given by

$$E_{cathode} = E^0_{Cd^{2+}} - \frac{0.0592}{2} \log \frac{1}{[Cd^{2+}]_0}$$

where $E_{cathode}$ is the potential applied to the cathode. As the applied potential becomes more and more negative, $[Cd^{2+}]_0$ becomes smaller and smaller. The result is that the rate of diffusion and the current become correspondingly larger.

Migration. The process by which ions move under the influence of an electric field is called *migration*. This process, shown schematically in Figure 19-5, is the primary cause of mass transfer in the bulk of the solution in a cell. The rate at which ions migrate to or away from an electrode surface generally increases as the electrode potential increases. This charge movement constitutes a current, which also increases with potential.

Convection. Reactants can also be transferred to or from an electrode by mechanical means. Thus, forced *convection,* such as stirring or agitation, will tend to decrease the thickness of the diffuse layer at an electrode surface and thus decrease concentration polarization. Natural convection resulting from temperature or density differences also contributes to the transport of species to and from an electrode.

The Importance of Concentration Polarization. As noted earlier, concentration polarization sets in when the effects of diffusion, migration, and convection are insufficient to transport a reactant to or from an electrode surface at a rate that produces a current of the magnitude given by Equation 19-2. Concentration polarization requires applied potentials that are larger than theoretical to maintain a given current in an electrolytic cell (Figure 19-3a). Similarly, the phenomenon causes a galvanic cell potential to be smaller than the value predicted from theory and the *IR* drop (Figure 19-3b).

Concentration polarization is important in several electroanalytical methods. In some applications, its effects are deleterious, and steps are taken to eliminate it. In others, it is essential to the analytical method, and every effort is made to promote its occurrence.

Kinetic Polarization

In kinetic polarization, the magnitude of the current is limited by the rate of one or both electrode reactions—that is, the rate of electron transfer between reactants and electrodes. In order to offset kinetic polarization, an additional potential, or overvoltage, is required to overcome the energy barrier to the half-reaction.

Kinetic polarization is most pronounced for electrode processes that involve gaseous products and is often negligible for reactions that involve the deposition or solution of a metal. In common with *IR* drop, overvoltage effects cause the potential of a galvanic cell to be smaller than theoretically predicted and to require potentials greater than theoretically assumed in order to operate an electrolytic cell at a desired current.[2]

The overvoltages associated with the formation of hydrogen and oxygen are often 1 V or more and are of considerable importance because these molecules are frequently produced by electrochemical reactions. Of particular interest is the high overvoltage of hydrogen on such metals as copper, zinc, and mercury. These metals and several others can therefore be deposited without interference from hydrogen evolution. In theory, it is not possible to deposit zinc from a neutral aqueous solution because hydrogen forms at a potential that is considerably less than that required for zinc deposition. In fact, the metal can be deposited on a copper electrode with no significant hydrogen formation because the rate at which the gas forms on both zinc and copper is negligible, as shown by the high hydrogen overvoltage associated with these metals.

19B   ELECTROGRAVIMETRIC METHODS OF ANALYSIS

Electrolytic precipitation has been used for over a century for the gravimetric determination of metals. In most applications, the metal is deposited on a weighed platinum cathode, and the increase in weight is determined. Important exceptions include the anodic deposition of lead as lead dioxide on platinum and of chloride as silver chloride on silver.

There are two types of electrogravimetric methods. In one, the potential of the working electrode is not controlled and the applied cell potential is held at a more or less constant level that provides a large enough current to complete the electrolysis in a reasonable length of time. The second type is the *potentiostatic method*. This procedure is also called the *controlled-cathode-potential* or the *controlled-anode-potential method*, depending upon whether the working electrode is a cathode or an anode.

19B-1   Electrogravimetry Without Potential Control of the Working Electrode

Electrolytic procedures in which no effort is made to control the potential of the working electrode make use of simple and inexpensive equipment and require little operator attention. In this procedure, the potential applied across the cell is maintained at a more or less constant level throughout the electrolysis.

Instrumentation
As shown in Figure 19-6, the apparatus for an analytical electrodeposition without cathode-potential control consists of a suitable cell and a direct-

> A working electrode is the electrode at which the analytical reaction occurs.

> A potentiostatic method is an electrolytic procedure in which the potential of the working electrode is maintained at a constant level versus a reference electrode, such as an SCE.

---

[2]Overvoltage data for various gaseous species on different electrode surfaces have been compiled by J. A. Page, in *Handbook of Analytical Chemistry*, L. Meites, Ed., p. 5-184. New York: McGraw-Hill, 1963.

Figure 19-6

Apparatus for the electrodeposition of metals without cathode-potential control.

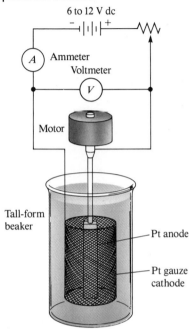

6 to 12 V dc

Ammeter

Voltmeter

Motor

Tall-form beaker

Pt anode

Pt gauze cathode

A depolarizer is a chemical that is easily oxidized (or reduced) and stabilizes the potential of a working electrode by minimizing concentration polarization.

current power supply. The dc power supply usually consists of an ac rectifier, although a 6-V storage battery can also be used. The voltage applied to the cell is controlled by a rheostat. An ammeter and a voltmeter indicate the approximate current and applied voltage. When an analytical electrolysis is performed with this apparatus, the applied voltage is adjusted with the rheostat to give a current of several tenths of an ampere. The voltage is then maintained at about the initial level until the deposition is judged complete.

Electrolysis Cells
Figure 19-6 shows a typical cell for the deposition of a metal on a solid electrode. Ordinarily, the working electrode is a platinum gauze cylinder 2 or 3 cm in diameter and perhaps 6 cm in length. Often, as shown, the anode takes the form of a solid platinum stirring paddle.

Applications
In practice, electrolysis at a constant cell potential is limited to the separation of an easily reduced cation from cations that are more difficult to reduce than hydrogen ion or some other easily reduced species, such as nitrate ion. The reason for this limitation is illustrated in Figure 19-7, which shows the changes in current, *IR* drop, and cathode potential during an electrolysis in the cell in Figure 19-6. The analyte here is copper(II) in a solution containing an excess of acid. Initially, the rheostat is adjusted so that the potential applied to the cell is about −2.5 V, which, as shown in Figure 19-7a, leads to a current of about 1.5A. The electrolytic deposition of copper is then completed at this applied potential.

As shown in Figure 19-7b, the *IR* drop decreases continually as the reaction proceeds. The reason for this decrease is primarily concentration polarization at the cathode, which limits the rate at which copper ions are brought to the electrode surface and thus limits the current. From Equation 19-1, it is apparent that the decrease in *IR* must be offset by an increase in the cathode potential since the cell potential is constant.

Ultimately, the decrease in current and increase in cathode potential are slowed at point *B* by the reduction of hydrogen ions. Because the solution contains a large excess of acid, the current is now no longer limited by concentration polarization, and codeposition of copper and hydrogen goes on simultaneously until the remainder of the copper ions are deposited. Under these conditions, the cathode is said to be *depolarized* by hydrogen ions.

Consider now the fate of a metal ion such as lead(II), which begins to deposit at point *A* on the cathode-potential curve. Clearly this ion would codeposit well before copper deposition was complete, and an interference would result. In contrast, a metal ion such as cobalt(II), which reacts at a cathode potential corresponding to point *C* on the curve, would not interfere because depolarization by hydrogen formation prevents the cathode from reaching this potential.

Codeposition of hydrogen during electrolysis often leads to the formation of analyte deposits that do not adhere to the electrode, a situation that is unsatisfactory for analytical purposes. For this reason, cathode depolarizers that are reduced at less negative cathode potentials than

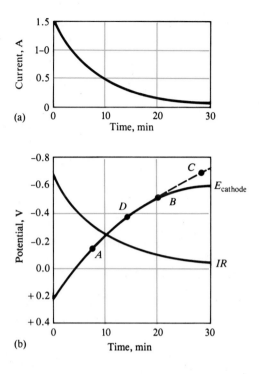

Figure 19-7
(a) Current, *IR* (b) drop, and cathode-potential change during the electrolytic deposition of copper at a constant applied cell potential.

hydrogen ion are introduced. Nitrate ion functions in this manner, its reaction beginning at point *C* in Figure 19-7b. The cathodic reduction of nitrate ion is described by the equation.

$$NO_3^- + 10\ H^+ + 8\ e^- \rightleftarrows NH_4^+ + 3\ H_2O \qquad (19\text{-}5)$$

Electrolytic methods performed without electrode-potential control, while limited by their lack of selectivity, do have several applications of practical importance. Table 19-1 lists the common elements that can be determined by this procedure.

Table 19-1
TYPICAL APPLICATIONS OF ELECTROGRAVIMETRIC
METHODS WITHOUT POTENTIAL CONTROL

Analyte	Weighed as	Cathode	Anode	Conditions
Ag^+	Ag	Pt	Pt	Alkaline CN^- solution
Br^-	AgBr (on anode)	Pt	Ag	
Cd^{2+}	Cd	Cu on Pt	Pt	Alkaline CN^- solution
Cu^{2+}	Cu	Pt	Pt	H_2SO_4/HNO_3 solution
Mn^{2+}	MnO_2 (on anode)	Pt	Pt dish	$HCOOH/HCOONa$ solution
Ni^{2+}	Ni	Cu on Pt	Pt	Ammoniacal solution
Pb^{2+}	PbO_2 (on anode)	Pt	Pt	Strong HNO_3 solution
Zn^{2+}	Zn	Cu on Pt	Pt	Acidic citrate solution

19B-2 Constant-Cathode-Potential Gravimetry

In the discussion that follows, the working electrode is assumed to be a cathode, at which an analyte is deposited as a metal. The remarks are readily extended to an anodic working electrode, however, or to products that do not form as metallic deposits.

Instrumentation

In order to separate species with electrode potentials that differ by only a few tenths of a volt, it is necessary to use a more elaborate technique than the one just described. These more elaborate techniques are required because concentration polarization at the cathode, if unchecked, causes the potential of that electrode to become so negative that codeposition of the other species begins before the analyte is completely deposited (Figure 19-7b). A large negative drift in the cathode potential can be avoided by employing a three-electrode system, such as that shown in Figure 19-8.

The controlled-potential apparatus shown in Figure 19-8 is made up of two independent electrical circuits that share a common electrode, the *working electrode* at which the analyte is deposited. The *electrolysis circuit* consists of a dc source, a potential divider (*ACB*) that permits continuous variation in the potential applied across the working electrode, a *counter electrode*, and a current meter. The *reference*, or *control*, *circuit* is made up of a reference electrode (often an SCE), a high-resistance digital voltmeter, and the working electrode. The electrical resistance of the reference circuit is so large that the electrolysis circuit supplies essentially all the current for the deposition.

The purpose of the reference circuit is to monitor continuously the potential between the working electrode and the reference electrode. When this potential reaches a level at which codeposition of an interfering species is about to begin, the potential across the working and counter

A controlled potential apparatus such as that shown in Figure 19-8 is called a potentiostat. Most potentiostats are automated.

A counter electrode has no effect on the reaction at the working electrode. It simply serves to feed electrons to the working cathode.

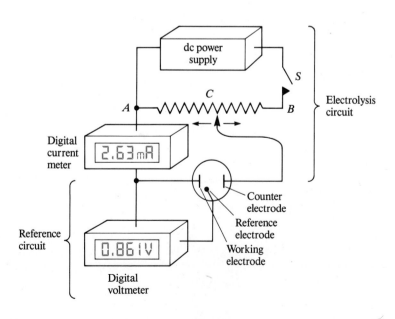

Figure 19-8

Apparatus for controlled-potential electrolysis. Contact *C* is adjusted as necessary to maintain the working electrode (cathode in this example) at a constant potential. The current in the reference-electrode circuit is essentially zero at all times.

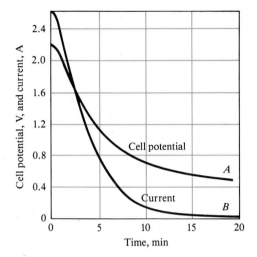

Figure 19-9
Changes in cell potential (*A*) and current (*B*) during a controlled-cathode-potential deposition of copper. The cathode is maintained at −0.36 V versus SCE throughout the experiment. (Data from J. J. Lingane, *Anal. Chem. Acta,* **1948,** *2,* 590. With permission.)

electrodes is decreased by moving contact *C* to the left. Since the potential of the counter electrode remains constant during this change, the cathode potential becomes smaller, thus preventing interference from codeposition.

The current and applied, or cell, voltage changes that occur in a typical constant-cathode-potential electrolysis are depicted in Figure 19-9. Note that the applied cell potential has to be decreased continuously throughout the electrolysis. As a consequence, the analyst would need to monitor the apparatus throughout the analysis. To avoid such waste of operator time, controlled-cathode-potential electrolyses are generally performed with automated instruments called *potentiostats,* which maintain a constant cathode potential electronically.

A potentiostat is an instrument that maintains the working electrode potential at a constant value.

Applications
The controlled-cathode-potential method is a potent tool for separating and determining metallic species having standard potentials that differ by only a few tenths of a volt. As shown by the example that follows, the feasibility of and conditions for accomplishing a given separation are readily derived from standard electrode potential data.

Example 19-2

Is a quantitative separation of Cu^{2+} and Pb^{2+} by electrolytic deposition theoretically feasible? If so, what range of cathode potentials (versus SCE) can be used? Assume that the sample solution is initially 0.1000 M in each ion and that quantitative removal of an ion is realized when only 1 part in 10,000 remains undeposited.

In Appendix 1, we find

$$Cu^{2+} + 2\ e^- \rightleftarrows Cu(s) \qquad E^0 = 0.337\ V$$
$$Pb^{2+} + 2\ e^- \rightleftarrows Pb(s) \qquad E^0 = -0.126\ V$$

It is apparent that copper will begin to deposit before lead. Let us first calculate the cathode potential required to reduce the Cu^{2+} concentration to 10^{-4} of its original concentration (that is, to 1.00×10^{-5} M). Substituting into the Nernst equation, we obtain

$$E = 0.337 - \frac{0.0592}{2} \log \frac{1}{1.00 \times 10^{-5}} = 0.189 \text{ V}$$

Similarly, we can derive the cathode potential at which lead begins to deposit:

$$E = -0.126 - \frac{0.0592}{2} \log \frac{1}{0.100} = -0.156 \text{ V}$$

Therefore, if the cathode potential is maintained between 0.189 and -0.156 V (versus SHE), a quantitative separation should in theory occur. To convert these potentials to potentials relative to a saturated calomel electrode, we treat the reference electrode as the anode and write

$$E_{cell} = E_{cathode} - E_{SCE} = 0.189 - 0.244 = -0.055 \text{ V}$$

and

$$E_{cell} = -0.156 - 0.244 = -0.400 \text{ V}$$

Therefore the cathode potential should be kept between -0.055 and -0.400 V versus the SCE.

An example illustrating the power of the controlled-cathode-potential method involves determining copper, bismuth, lead, cadmium, zinc, and tin in mixtures by successive deposition of the metals on a platinum cathode of known weight that is weighed after each deposition. The first three elements are deposited from a nearly neutral solution containing tartrate ion to complex the tin(IV) and prevent its deposition. First copper is reduced quantitatively by maintaining the cathode potential at -0.2 V with respect to a saturated calomel electrode. After being weighed, the copper-plated cathode is returned to the solution, and bismuth is removed at a potential of -0.4 V. After the electrode is again weighed lead is deposited quantitatively by increasing the cathode potential to -0.6 V. When lead deposition is complete, the solution is made strongly ammoniacal, and cadmium and zinc are deposited successively at -1.2 and -1.5 V. Finally, the solution is acidified in order to decompose the tin/tartrate complex by the formation of undissociated tartaric acid. Tin is then deposited at a cathode potential of -0.65 V. A fresh cathode must be used here because the zinc redissolves under these conditions.

A procedure such as this is particularly attractive for use with a potentiostat because little operator time is required for the complete analysis.

Table 19-2 lists some other separations performed by the controlled-cathode-potential method.

Table 19-2

SOME APPLICATIONS OF CONTROLLED-
CATHODE-POTENTIAL ELECTROLYSIS

Element Determined	Other Elements That Can Be Present
Ag	Cu and heavy metals
Cu	Bi, Sb, Pb, Sn, Ni, Cd, Zn
Bi	Cu, Pb, Zn, Sb, Cd, Sn
Sb	Pb, Sn
Sn	Cd, Zn, Mn, Fe
Pb	Cd, Sn, Ni, Zn, Mn, Al, Fe
Cd	Zn
Ni	Zn, Al, Fe

19C COULOMETRIC METHODS OF ANALYSIS

Coulometric methods are performed by measuring the quantity of electrical charge (electrons) required to convert an analyte quantitatively to a different oxidation state. Coulometric and gravimetric methods share the common advantage that the proportionality constant between the quantity measured and the analyte weight is derived from accurately known physical constants, thus eliminating the need for calibration standards. In contrast to gravimetric methods, coulometric procedures are usually rapid and do not require that the product of the electrochemical reaction be a weighable solid. Coulometric methods are as accurate as conventional gravimetric and volumetric procedures and in addition are readily automated.[3]

19C-1 The Quantity of Electrical Charge

Units for the quantity of charge include the coulomb (C) and the faraday (F). *The coulomb is the quantity of electrical charge transported by a constant current of one ampere in one second.* Thus, the number of coulombs (Q) resulting from a constant current of I amperes operated for t seconds is

1 coulomb = 1 ampere · second

$$Q = It \qquad (19\text{-}6)$$

For a variable current i, the number of coulombs is given by the integral

$$Q = \int_0^t i \, dt \qquad (19\text{-}7)$$

---

[3]For additional information about coulometric methods, see E. Bishop, in *Comprehensive Analytical Chemistry,* C. L. Wilson and D. W. Wilson, Eds., Vol. 11D. New York: Elsevier Scientific, 1975; J. A. Plambeck, *Electroanalytical Chemistry,* Chapter 12. New York: Wiley, 1982; D. J. Curran, in *Laboratory Technique in Electroanalytical Chemistry,* P. T. Kissinger and W. R. Heinemann, Eds., pp. 539–568. New York: Marcel Dekker, 1984.

A faraday of charge is equivalent to one mole of electrons, or 6.022×10^{23} electrons.

The equivalent weight of a substance involved in an oxidation/reduction reaction is the formula weight of the substance divided by the number of electrons it donates or accepts.

The faraday is the quantity of charge that produces one equivalent of chemical change at an electrode. Since the equivalent in an oxidation/reduction reaction is that amount of a substance that donates or accepts one mole of electrons, the faraday is equivalent to 6.022×10^{23} electrons. The faraday also equals 96,485 C. As shown in Example 19-3, we can use these definitions to calculate the weight of a chemical species formed at an electrode by a current of known magnitude.

Example 19-3

A constant current of 0.800 A is used to deposit copper at the cathode and oxygen at the anode of an electrolytic cell. Calculate the number of grams of each product formed in 15.2 min, assuming no other redox reaction.

The equivalent weights of copper and oxygen are determined from consideration of the two half-reactions

$$Cu^{2+} + 2\ e^- \rightarrow Cu(s)$$
$$2\ H_2O \rightarrow 4\ e^- + O_2(g) + 4\ H^+$$

Thus 1 mol of copper is equivalent to 2 mol of electrons, and 1 mol of oxygen represents 4 mol of electrons. Therefore the equivalent weight of copper is fw Cu/2 and the equivalent weight of oxygen is fw O_2/4.

Substituting into Equation 19-6 yields

$$Q = 0.800\ \text{A} \times 15.2\ \text{min} \times 60\ \text{s/min} = 729.6\ \text{A} \cdot \text{s} = 729.6\ \text{C}$$

To convert this quantity of electrical charge to moles of electrons, we write

$$\frac{729.6\ \cancel{C}}{96,485\ \cancel{C}/F} = 7.56 \times 10^{-3}\ F \equiv 7.562 \times 10^{-3}\ \text{mol of electrons}$$

From the definition of the faraday, 7.562×10^{-3} equivalent of copper is deposited on the cathode; a similar quantity of oxygen is evolved at the anode. Therefore,

$$\text{wt Cu} = 7.562 \times 10^{-3}\ \cancel{\text{mol}\ e^-} \times \frac{1\ \cancel{\text{mol Cu}}}{2\ \cancel{\text{mol}\ e^-}} \times \frac{63.54\ \text{g Cu}}{1\ \cancel{\text{mol Cu}}} = 0.240\ \text{g Cu}$$

$$\text{wt } O_2 = 7.562 \times 10^{-3}\ \cancel{\text{mol}\ e^-} \times \frac{1\ \cancel{\text{mol } O_2}}{4\ \cancel{\text{mol}\ e^-}} \times \frac{32.00\ \text{g } O_2}{1\ \cancel{\text{mol } O_2}} = 0.0605\ \text{g } O_2$$

Michael Faraday (1791–1867) was one of the foremost chemists and physicists of his time. Among his most important discoveries were Faraday's laws of electrolysis. Faraday, a simple man who lacked mathematical sophistication, was a superb experimentalist and an inspiring teacher and lecturer. The quantity of charge equal to a mole of electrons is named in his honor.

19C-2 Types of Coulometric Methods

Amperostatic coulometry is also called coulometric titrimetry.

Two types of methods have been developed that are based on the quantity of charge: *potentiostatic coulometry* and *amperostatic coulometry*. Potentiostatic methods are performed in much the same way as controlled-potential gravimetric methods, with the potential of the working electrode relative to a reference electrode being maintained constant throughout the electrolysis. Here however, the electrolysis current is recorded as a function of time to give a curve similar to curve *B* in Figure 19-9. The analysis is then completed by integrating the current/time curve to obtain the number of coulombs and thus the number of equivalents of analyte.

Coulometric titrations are similar to other titrimetric methods in that analyses are based on measuring the combining capacity of the analyte with a standard reagent. In the coulometric procedure, the reagent is electrons and the standard solution is a constant current of known magnitude. Electrons are added to the analyte (in this case via the direct current) or to some species that immediately reacts with the analyte until an end point is reached. At that point, the electrolysis is discontinued. The amount of analyte is determined from the magnitude of the current and the time required to complete the titration. The magnitude of the current in amperes is analogous to the molarity of a standard solution, and the time measurement is analogous to the volume measurement in conventional titrimetry.

Electrons are the reagent in a coulometric titration.

19C-3 Current-Efficiency Requirements

A fundamental requirement for all coulometric methods is 100% current efficiency; that is, each faraday of electricity must bring about one equivalent of chemical change in the analyte. Note that 100% current efficiency can be achieved without the analyte participating directly in electron transfer at an electrode. For example, chloride ions are readily determined by either of the two coulometric methods by the generating silver ions at a silver anode. These ions then react with the analyte to form a deposit of silver chloride. The quantity of electricity required to complete the silver chloride formation serves as the analytical parameter. In this instance, 100% current efficiency is realized because the number of moles of electrons is exactly equal to the number of moles of chloride ion in the sample despite the fact that these ions do not react directly at the electrode.

19C-4 Controlled-Potential Coulometry

In controlled-potential coulometry, the potential of the working electrode is maintained constant at a level that only the analyte is responsible for the conduction of charge across the electrode/solution interface. The number of coulombs of electricity required to convert the analyte to its reaction product is then determined by recording and integrating the current/time curve derived from the electrolysis or by means of a chemical coulometer.

Instrumentation
The instrumentation for potentiostatic coulometry consists of an electrolysis cell, a potentiostat, and a device for determining the number of coulombs of electricity consumed by the analyte.

Cells. Figure 19-10 illustrates two types of cells used for potentiostatic coulometry. The first consists of a platinum-gauze working electrode, a

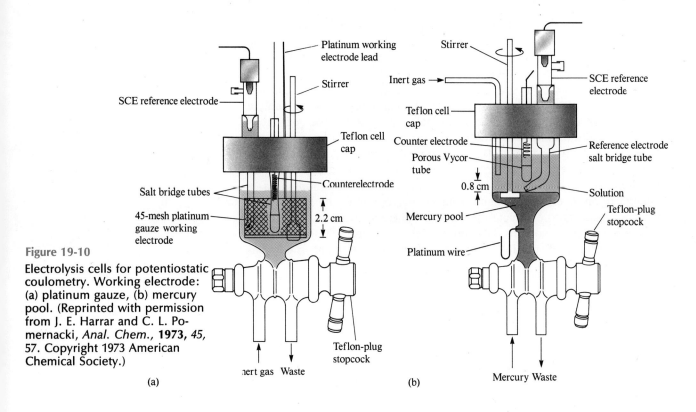

Figure 19-10

Electrolysis cells for potentiostatic coulometry. Working electrode: (a) platinum gauze, (b) mercury pool. (Reprinted with permission from J. E. Harrar and C. L. Pomernacki, *Anal. Chem.*, **1973**, *45*, 57. Copyright 1973 American Chemical Society.)

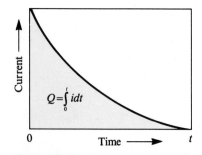

Figure 19-11

For a current that varies with time, the quantity of charge Q in a time t is the shaded area under the curve.

platinum-wire counter electrode, and a saturated calomel reference electrode. The counter electrode is separated from the analyte solution by a salt bridge that usually contains the same electrolyte as the solution being analyzed. This bridge prevents the reaction products formed at the counter electrode from diffusing into the analyte solution and interfering. For example, hydrogen gas is a common product at a cathodic counter electrode. Unless physically isolated from the analyte solution by the bridge, this species will react directly with many analytes determined by oxidation at the working anode.

The second type of cell, shown in Figure 19-10b, is a mercury-pool type. A mercury cathode is particularly useful for separating easily reduced elements as a preliminary step in an analysis. In addition, however, it has found considerable use in the coulometric determination of several metallic cations that form metals that are soluble in mercury. In these applications, little or no hydrogen evolution occurs even at high applied potentials because of the large overvoltage effects. A coulometric cell such as that shown in Figure 19-10b is also useful for the coulometric determination of certain types of organic compounds.

Potentiostats and Coulometers. For controlled-potential coulometry, a potentiostat similar to that shown in Figure 19-8 is required. Generally, however, the potentiostat is automated and is equipped with a recorder that provides a plot of current as a function of time as shown in Figure 19-11. The quantity of analyte is then determined from the area under the current-versus-time curve. This area is usually evaluated with an electronic integrator.

Applications

Controlled-potential coulometric methods have been applied to the determination of some 55 elements in inorganic compounds.[4] Mercury appears to be favored as the cathode, and methods for the deposition of two dozen or more metals at this electrode have been described. The method has found widespread use in the nuclear-energy field for the relatively interference-free determination of uranium and plutonium.

The controlled-potential coulometric procedure also offers possibilities for the electrolytic determination (and synthesis) of organic compounds. For example, trichloroacetic acid and picric acid are quantitatively reduced at a mercury cathode whose potential is suitably controlled:

$$Cl_3CCOO^- + H^+ + 2\ e^- \rightarrow Cl_2HCCOO^- + Cl^-$$

Coulometric measurements permit the determination of these compounds with a relative error of a few tenths of a percent.

19C-5 Coulometric Titrations[5]

Coulometric titrations are carried out with a constant-current source called an *amperostat,* which senses decreases in cell current and responds by increasing the potential applied to the cell until the current is restored to its original level. Because of the effects of concentration polarization, 100% current efficiency with respect to the analyte can be maintained in only one way: There must be a large excess of an auxiliary reagent that is oxidized or reduced at the electrode to give a product that reacts with the analyte. As an example, consider the coulometric titration of iron(II) at a platinum anode. At the beginning of the titration, the primary anodic reaction is

$$Fe^{2+} \rightarrow Fe^{3+} + e^-$$

Constant-current generators are called amperostats or galvanostats.

Auxiliary reagents are essential in coulometric titrations.

---

[4]For a summary of the applications, see J. E. Harrar, in *Electroanalytical Chemistry,* A. J. Bard, Ed., Vol. 8. New York: Marcel Dekker, 1975; E. Bishop, in *Comprehensive Analytical Chemistry,* C. L. Wilson and D. W. Wilson, Eds., Vol. 11D, Chapter XV. New York: Elsevier, 1975.

[5]For further details on this technique, see D. J. Curran, in *Laboratory Techniques in Electroanalytical Chemistry,* P. T. Kissinger and W. R. Heineman, Eds., Chapter 20. New York: Marcel Dekker, 1984.

As the concentration of iron(II) decreases, however, the requirement of a constant current results in an increase in the applied cell potential. Because of concentration polarization, this increase in potential causes the anode potential to increase to the point where the decomposition of water becomes a competing process:

$$2 \ H_2O \rightarrow O_2(g) + 4 \ H^+ + 4 \ e^-$$

The quantity of electricity required to complete the oxidation of iron(II) then exceeds that demanded by theory, and the current efficiency is less than 100%. The lowered current efficiency is avoided, however, by introducing at the outset an unmeasured quantity of cerium(III), which is oxidized at a lower potential than is water:

$$Ce^{3+} \rightarrow \ Ce^{4+} + e^-$$

With stirring, the cerium(IV) produced is rapidly transported from the surface of the electrode to the bulk of the solution, where it oxidizes an equivalent amount of iron(II):

$$Ce^{4+} + Fe^{2+} \rightarrow Ce^{3+} + Fe^{3+}$$

The net effect is an electrochemical oxidation of iron(II) with 100% current efficiency, even though only a fraction of that species is directly oxidized at the electrode surface.

End Points in Coulometric Titrations

Coulometric titrations, like their volumetric counterparts, require a means for determining when the reaction between analyte and reagent is complete. Generally, the end points described in the chapters on volumetric methods are applicable to coulometric titrations as well. Thus, for the titration of iron(II) just described, an oxidation/reduction indicator, such as 1,10-phenanthroline, can be used; as an alternative, the end point can be established potentiometrically. Similarly, an adsorption indicator or a potentiometric end point can be employed in the coulometric titration of chloride ion by the silver ions generated at a silver anode.

Instrumentation

As shown in Figure 19-12, the equipment required for a coulometric titration includes a source of constant current that has a range of one to several hundred milliamperes, a titration vessel, a switch, an electric timer, and a device for monitoring current. Movement of the switch to position 1 simultaneously starts the timer and initiates a current in the titration cell. When the switch is moved to position 2, the electrolysis and the timing are discontinued. With the switch in this position, however, electricity continues to be drawn from the source and passes through a dummy resistor R_D that has about the same electrical resistance as the cell. This arrangement ensures continuous operation of the source, which aids in maintaining the current at a constant level.

The constant-current source for a coulometric titration is often an am-

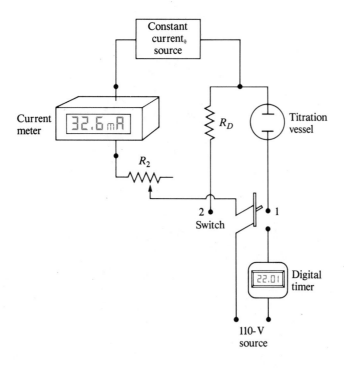

Figure 19-12
Diagram of a coulometric titration apparatus.

perostat, an electronic device capable of maintaining a current of 200 mA or more that is constant to a few hundredths percent of the current. Amperostats are available from several instrument manufacturers. A much less expensive source that produces reasonably constant currents can be constructed from several heavy-duty batteries connected in series to give an output potential of 100 to 300 V. This output is applied to the titration cell, which is in series with a resistor whose resistance is large relative to the cell resistance. Small changes in the cell conductance then have a negligible effect on the current in the resistor and that in the cell.

An ordinary motor-driven electric clock is unsatisfactory for the measurement of the electrolysis time because the rotor of such a device tends to coast when stopped and lag when started. Modern digital electronic timers eliminate this problem.

Cells. Figure 19-13 shows a typical coulometric titration cell consisting of a generator electrode at which the reagent is produced and a counter electrode to complete the circuit. The generator electrode—ordinarily a platinum rectangle, a coil of wire, or a gauze cylinder—has a relatively large surface area to minimize polarization effects. In most instances, the counter electrode is isolated from the reaction medium by a sintered disk or some other porous medium in order to prevent interference by any reaction products from this electrode.

A Comparison of Coulometric and Conventional Titrations

The various components of the titrator in Figure 19-12 have their counterparts in the reagents and apparatus required for a volumetric titration.

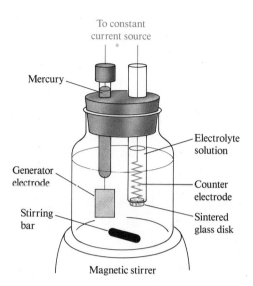

Figure 19-13
A typical coulometric titration cell.

The constant-current source of known magnitude serves the same function as the standard solution in a volumetric method. The electronic timer and switch correspond to the buret and stopcock, respectively. Electricity is passed through the cell for relatively long periods of time at the outset of a coulometric titration, but the time intervals are made smaller and smaller as chemical equivalence is approached. Note that these steps are analogous to the way a buret is operated in a conventional titration.

A coulometric titration offers several significant advantages over a conventional volumetric procedure. Principal among these is the elimination of the problems associated with the preparation, standardization, and storage of standard solutions. This advantage is particularly significant with labile reagents such as chlorine, bromine, and titanium(III) ion, which are sufficiently unstable in aqueous solution to seriously limit their use as volumetric reagents. In contrast, their utilization in a coulometric determination is straightforward because they are consumed as soon as they are generated.

Coulometric methods also excel when small amounts of analyte have to be titrated because tiny quantities of reagent are generated with ease and accuracy through the proper choice of current. In contrast, the use of very dilute solutions and the accurate measurement of small volumes are inconvenient at best.

A further advantage of the coulometric procedure is that a single constant-current source provides reagents for precipitation, complex formation, neutralization, or oxidation/reduction titrations. Finally, coulometric titrations are more readily automated since current control is more readily accomplished than is control of liquid flow.

The current/time measurements required for a coulometric titration are inherently as accurate as or more accurate than the comparable volume/molarity measurements of a conventional volumetric method, particularly where small quantities of reagent are involved. When the accuracy of a titration is limited by end-point sensitivity, the two titration methods have comparable accuracies.

Coulometric methods are as accurate and precise as comparable volumetric methods.

Applications

Coulometric titrations have been developed for all types of volumetric reactions.[6] Selected applications are described in this section.

Neutralization Titrations. Hydroxide ion can be generated at the surface of a platinum cathode immersed in a solution containing the analyte acid:

$$2\,H_2O + 2\,e^- \rightarrow 2\,OH^- + H_2(g)$$

The platinum anode must be isolated by some sort of diaphragm to eliminate potential interference from the hydrogen ions simultaneously produced by the anodic oxidation of water. As a convenient alternative, a silver wire can be substituted for the platinum anode, provided chloride or bromide ions are added to the analyte solution. The anode reaction then becomes

$$Ag(s) + Br^- \rightarrow AgBr(s) + e^-$$

Silver bromide does not interfere with the neutralization reaction.

Coulometric titrations of acids are much less susceptible to the carbonate error encountered in volumetric methods (Section 13A-3). The only measure required to avoid this type of error is to remove the carbon dioxide from the solvent by boiling it or by bubbling an inert gas, such as nitrogen, through the solution for a brief period.

Hydrogen ions generated at the surface of a platinum anode can be used for the coulometric titration of strong as well as weak bases:

$$2\,H_2O \rightarrow O_2 + 4\,H^+ + 4\,e^-$$

Here, the cathode must be isolated from the analyte solution to prevent interference from hydroxide ion.

Precipitation and Complex-Formation Reactions. Coulometric titrations with EDTA are carried out by reduction of the ammine mercury(II) EDTA chelate at a mercury cathode:

$$HgNH_3Y^{2-} + NH_4^+ + 2\,e^- \rightarrow Hg(l) + 2\,NH_3 + HY^{3-} \quad (19\text{-}8)$$

Because the mercury chelate is more stable than the corresponding complexes of such cations as calcium, zinc, lead, and copper, complexation of these ions occurs only after the ligand has been freed by the electrode process.

As shown in Table 19-3, several precipitating reagents can be generated coulometrically. The most widely used of these is silver ion, which is generated at a silver anode.

---

[6]For additional information, see E. Bishop, in *Comprehensive Analytical Chemistry*, C. L. Wilson and D. W. Wilson, Eds., Vol. 11D, Chapters XVIII to XXIV. New York: Elsevier, 1975; J. T. Stock, *Anal. Chem.*, **1984**, *56*, 1R, and **1980**, *52*, 1R.

Table 19-3

SUMMARY OF COULOMETRIC TITRATIONS INVOLVING NEUTRALIZATION, PRECIPITATION, AND COMPLEX-FORMATION REACTIONS

Species Determined	Generator Electrode Reaction	Secondary Analytical Reaction
Acids	$2 H_2O + 2 e^- \rightleftharpoons 2 OH^- + H_2$	$OH^- + H^+ \rightleftharpoons H_2O$
Bases	$H_2O \rightleftharpoons 2 H^+ + \frac{1}{2} O_2 + 2 e^-$	$H^+ + OH^- \rightleftharpoons H_2O$
Cl^-, Br^-, I^-	$Ag \rightleftharpoons Ag^+ + e^-$	$Ag^+ + Cl^- \rightleftharpoons AgCl(s)$, etc.
Mercaptans	$Ag \rightleftharpoons Ag^+ + e^-$	$Ag^+ + RSH \rightleftharpoons AgSR(s) + H^+$
Cl^-, Br^-, I^-	$2 Hg \rightleftharpoons Hg_2^{2+} + 2 e^-$	$Hg_2^{2+} + 2 Cl^- \rightleftharpoons Hg_2Cl_2(s)$, etc.
Zn^{2+}	$Fe(CN)_6^{3-} + e^- \rightleftharpoons Fe(CN)_6^{4-}$	$3 Zn^{2+} + 2 K^+ + 2 Fe(CN)_6^{4-} \rightleftharpoons$ $K_2Zn_3[Fe(CN)_6]_2(s)$
Ca^{2+}, Cu^{2+}, Zn^{2+}, Pb^{2+}	See Equation 19-8.	$HY^{3-} + Ca^{2+} \rightleftharpoons CaY^{2-} + H^+$, etc.

Oxidation/Reduction Titrations. Table 19-4 reveals that a variety of redox reagents can be generated coulometrically. Of particular interest is bromine, the coulometric generation of which forms the basis for a large number of methods. Of interest as well are reagents not ordinarily encountered in conventional volumetric analysis owing to the instability of their solutions; silver(II), manganese(III), and the chloride complex of copper(I) are examples.

Automatic Coulometric Titrators

A number of instrument manufacturers offer automatic coulometric titrators, most of which employ a potentiometric end point. Some of these instruments are multipurpose and can be used for the determination of a variety of species. Others are designed for a single type of analysis. Examples of the latter are chloride titrators, in which silver ion is generated coulometrically; sulfur dioxide monitors, where anodically generated bromine oxidizes the analyte to sulfate ions; and carbon dioxide monitors, in which the gas, absorbed in monoethanolamine, is titrated with coulometrically generated base.

Table 19-4

SUMMARY OF COULOMETRIC TITRATIONS INVOLVING OXIDATION/REDUCTION REACTIONS

Reagent	Generator Electrode Reaction	Substance Determined
Br_2	$2 Br^- \rightleftharpoons Br_2 + 2 e^-$	As(III), Sb(III), U(IV), Tl(I), I^-, SCN^-, NH_3, N_2H_4, NH_2OH, phenol, aniline, mustard gas, mercaptans, 8-hydroxyquinoline, olefins
Cl_2	$2 Cl^- \rightleftharpoons Cl_2 + 2 e^-$	As(III), I^-, styrene, fatty acids
I_2	$2 I^- \rightleftharpoons I_2 + 2 e^-$	As(III), Sb(III), $S_2O_3^{2-}$, H_2S, ascorbic acid
Ce^{4+}	$Ce^{3+} \rightleftharpoons Ce^{4+} + e^-$	Fe(II), Ti(III), U(IV), As(III), I^-, $Fe(CN)_6^{4-}$
Mn^{3+}	$Mn^{2+} \rightleftharpoons Mn^{3+} + e^-$	$H_2C_2O_4$, Fe(II), As(III)
Ag^{2+}	$Ag^+ \rightleftharpoons Ag^{2+} + e^-$	Ce(III), V(IV), $H_2C_2O_4$, As(III)
Fe^{2+}	$Fe^{3+} + e^- \rightleftharpoons Fe^{2+}$	Cr(VI), Mn(VII), V(V), Ce(IV)
Ti^{3+}	$TiO^{2+} + 2 H^+ + e^- \rightleftharpoons Ti^{3+} + H_2O$	Fe(III), V(V), Ce(IV), U(VI)
$CuCl_3^{2-}$	$Cu^{2+} + 3 Cl^- + e^- \rightleftharpoons CuCl_3^{2-}$	V(V), Cr(VI), IO_3^-
U^{4+}	$UO_2^{2+} + 4 H^+ + 2 e^- \rightleftharpoons U^{4+} + 2 H_2O$	Cr(VI), Ce(IV)

19-1. Define

*(a) coulometric titration.

(b) kinetic polarization.

*(c) concentration polarization.

(d) potentiostat.

*(e) faraday.

(f) overvoltage.

*(g) controlled-cathode-potential electrolysis.

(h) ohmic potential.

*(i) cathode depolarizer.

(j) coulomb.

*(k) working electrode.

(l) diffusion.

*19-2. Differentiate between amperostatic coulometry and potentiostatic coulometry.

19-3. Differentiate between kinetic polarization and concentration polarization.

*19-4. Describe three phenomena that cause ions to migrate to an electrode surface from the bulk of a solution.

19-5. Describe variables that tend to decrease concentration polarization.

*19-6. Under what circumstances is kinetic polarization likely to occur?

19-7. Why is an auxiliary reagent always required in a coulometric titration?

*19-8. How is a constant-cathode-potential electrolysis performed?

19-9. Compare a coulometric and a volumetric titration.

*19-10. Nickel is to be deposited from a solution that is 0.200 M in Ni^{2+} and buffered to pH 2.00. Oxygen is evolved at a partial pressure of 1.00 atm at a platinum anode. The cell has a resistance of 3.15 Ω; the temperature is 25°C. Calculate

(a) the thermodynamic potential needed to initiate the deposition of nickel.

(b) the *IR* drop for a current of 1.10 A.

(c) the applied potential when $[Ni^{2+}] = 0.00020$ (assuming all other variables are unchanged).

19-11. Calculate the minimum difference in standard electrode potentials needed to lower the concentration of the metal M_1 to 1.00×10^{-4} M in a solution that is 0.200 M in the less reducible metal M_2, where

*(a) M_2 is univalent and M_1 is divalent.

(b) M_1 and M_2 are both divalent.

*(c) M_2 is trivalent and M_1 is univalent.

(d) M_2 is divalent and M_1 is univalent.

(e) M_2 is divalent and M_1 is trivalent.

*19-12. A solution is 0.150 M in Co^{2+} and 0.0750 M in Cd^{2+}. Calculate

(a) the Co^{2+} concentration in the solution as the first cadmium starts to deposit.

(b) the cathode potential needed to lower the Co^{2+} concentration to 1×10^{-5} M.

19-13. A solution is 0.0500 M in BiO^+ and 0.0400 M in Co^{2+} and has a pH of 2.50.

(a) What is the concentration of the more readily reduced cation at the onset of deposition of the less reducible one?

(b) What is the potential of the cathode when the concentration of the more easily reduced species is 1.00×10^{-6} M?

19-14. Electrogravimetric analysis involving control of the cathode potential is proposed as a means of separating Bi^{3+} and Sn^{2+} in a solution that is 0.200 M in each ion and buffered to pH 1.50.

(a) Calculate the theoretical cathode potential at the start of deposition of the more readily reduced ion.

(b) Calculate the residual concentration of the more readily reduced species at the onset of deposition of the less readily reduced species.

(c) Propose a range (versus SCE), if such exists, within which the cathode potential should be maintained; consider a residual concentration less than 10^{-6} M as constituting quantitative removal.

*19-15. Halide ions can be deposited on a silver anode via the reaction

$$Ag(s) + X^- \rightarrow AgX(s) + e^-$$

(a) If 1.00×10^{-5} M is used as the criterion for quantitative removal, is it theoretically feasible to separate Br^- from I^- through control of the anode potential in a solution that is initially 0.250 M in each ion?

(b) Is a separation of Cl^- and I^- theoretically feasible in a solution that is initially 0.250 M in each ion?

(c) If a separation is feasible in either (a) or (b), what range of anode potentials (versus SCE) should be used?

*19-16. Calculate the time needed for a constant current of 0.961 A to deposit 0.500 g of Co(II) as

(a) elemental cobalt on the surface of a cathode.

(b) Co_3O_4 on an anode.

19-17. Calculate the time needed for a constant current of 1.20 A to deposit 0.500 g of

(a) Tl(III) as the element on a cathode.

(b) Tl(I) as Tl_2O_3 on an anode.

(c) Tl(I) as the element on a cathode.

*19-18. An excess of $HgNH_3Y^{2-}$ was introduced into 25.00 mL of well water. Express the water hardness in terms of ppm $CaCO_3$ if the EDTA needed for a titration was generated at a mercury cathode (Equation 19-8) in 2.02 min by a constant current of 31.6 mA.

19-19. A 0.1516-g sample of a purified organic acid was neutralized by the hydroxide ion produced in 5 min and 24 s by a constant current of 0.401 A. Calculate the equivalent weight of the acid.

*19-20. The nitrobenzene in 210 mg of an organic mixture was reduced to phenylhydroxylamine at a constant potential of -0.96 V (versus SCE) applied to a mercury cathode:

$$C_6H_5NO_2 + 4\ H^+ + 4\ e^- \rightarrow C_6H_5NHOH + H_2O$$

The sample was dissolved in 100 mL of methanol; after electrolysis for 30 min, the reaction was judged complete. An electronic coulometer in series with the cell indicated that the reduction required 26.74 C. Calculate the percentage of $C_6H_5NO_2$ in the sample.

19-21. Electrolytically generated I_2 was used to determine the amount of H_2S in 100.0 mL of brackish water. Following addition of excess KI, a titration required a constant current of 36.32 mA for 10.12 min. The reaction was

$$H_2S + I_2 \rightarrow S(s) + 2\ H^+ + 2\ I^-$$

Express the results of the analysis in terms of ppm H_2S.

*19-22. At a potential of -1.0 V (versus SCE), CCl_4 in methanol is reduced to $CHCl_3$ at a mercury cathode:

$$2\ CCl_4 + 2\ H^+ + 2\ e^- + 2\ Hg(l) \rightarrow$$
$$2\ CHCl_3 + Hg_2Cl_2(s)$$

At -1.80 V, the $CHCl_3$ further reacts to give CH_4:

$$2\ CHCl_3 + 6\ H^+ + 6\ e^- + 6\ Hg(l) \rightarrow$$
$$2\ CH_4(g) + 3\ Hg_2Cl_2(s)$$

A 0.750-g sample containing CCl_4, $CHCl_3$, and inert organic species was dissolved in methanol and electrolyzed at -1.0 V until the current approached zero. A coulometer indicated that 11.63 C was required to complete the reaction. The potential of the cathode was adjusted to -1.8 V. Completion of the titration at this potential required an additional 68.6 C. Calculate the percentage of CCl_4 and $CHCl_3$ in the mixture.

19-23. A 0.1309-g sample containing only $CHCl_3$ and CH_2Cl_2 was dissolved in methanol and electrolyzed in a cell containing a mercury cathode; the potential of the cathode was held constant at -1.80 V (versus SCE). Both compounds were reduced to CH_4 (see Problem 19-22 for the reaction). Calculate the percentage of $CHCl_3$ and CH_2Cl_2 if 306.7 C was required to complete the reduction.

*19-24. The phenol content of water downstream from a coking furnace was determined by coulometric analysis. A 100-mL sample was rendered slightly acidic, and an excess of KBr was introduced. To produce Br_2 for the reaction

$$C_6H_5OH + 3\ Br_2 \rightarrow Br_3C_6H_2OH(s) + 3\ HBr$$

a steady current of 0.0313 A for 7 min and 33 s was required. Express the results of this analysis in terms of parts of C_6H_5OH per million parts of water. (Assume that the density of water is 1.00 g/mL.)

19-25. The CN^- concentration of 10.0 mL of a plating solution was determined by titration with electrogenerated hydrogen ion to a methyl orange end point. A color change occurred after 3 min and 22 s with a current of 43.4 mA. Calculate the number of grams of NaCN per liter of solution.

*19-26. Traces of $C_6H_5NH_2$ can be determined by reaction with an excess of electrolytically generated Br_2:

The polarity of the working electrode is then reversed, and the excess Br_2 is determined by a coulometric titration involving the generation of Cu(I):

$$Br_2 + 2\ Cu^+ \rightarrow 2\ Br^- + 2\ Cu^{2+}$$

Suitable quantities of KBr and $CuSO_4$ were added to a 25.0-mL sample containing aniline. Calculate the number of micrograms of $C_6H_5NH_2$ in the sample from the data:

Working Electrode Functioning as	Generation Time with a Constant Current of 1.51 mA, min
anode	3.76
cathode	0.270

19-27. Quinone can be reduced to hydroquinone with an excess of electrolytically generated Sn(II):

The polarity of the working electrode is then reversed, and the excess Sn(II) is oxidized with Br_2 generated in a coulometric titration:

$$Sn^{2+} + Br_2 \rightleftarrows Sn^{4+} + 2\ Br^-$$

Appropriate quantities of $SnCl_4$ and KBr were added to a 50.0-mL sample. Calculate the weight of $C_6H_4O_2$ in the sample from the data:

Working Electrode Functioning as	Generation Time with a Constant Current of 1.062 mA, min
cathode	8.34
anode	0.691

CHAPTER 20

Molecular Ultraviolet/Visible Absorption Spectroscopy

Spectroscopic methods of analysis are based upon measurement of the electromagnetic radiation produced or absorbed by an analyte.[1] *Emission methods* make use of radiation given off when an analyte is excited by thermal, electrical, or radiant energy. *Absorption methods,* in contrast, are based upon the decrease in power (*attenuation*) of a beam of electromagnetic radiation as a consequence of its interaction with the analyte.

Spectroscopic methods are also classified according to region of the electromagnetic spectrum. These regions include X-ray, ultraviolet, visible, infrared, microwave, and radio-frequency. Chapters 20 and 21 deal with the absorption of ultraviolet and visible radiation by molecules (*molecular spectroscopy*). Chapter 22 treats fluorescence emission, another type of molecular spectroscopy. Chapter 23 is concerned with absorption and emission by atoms and elementary ions (*atomic spectroscopy*).

20A PROPERTIES OF ELECTROMAGNETIC RADIATION

Electromagnetic radiation is a form of energy that is transmitted through space at enormous velocities. Many of the properties of electromagnetic radiation are conveniently described in terms of sinusoidal waves that

---

[1]For further study, see E. J. Meehan, in *Treatise on Analytical Chemistry,* 2nd ed., Part I, Vol. 7, Chapters 1–3, P. J. Elving, E. J. Meehan, and I. M. Kolthoff, Eds. New York: Wiley, 1981; J. D. Ingle Jr. and S. R. Crouch, *Analytical Spectroscopy.* Englewood Cliffs, NJ: Prentice Hall, 1988; J. E. Crooks, *The Spectrum in Chemistry.* New York: Academic Press, 1978.

have wavelength, frequency, velocity, and amplitude. In contrast to other wave phenomena, such as sound, electromagnetic radiation requires no supporting medium for its transmission; thus, it readily passes through a vacuum.

The wave model fails to account for phenomena associated with the absorption and emission of radiant energy. For these processes, electromagnetic radiation must be treated as a stream of discrete particles, or wave packets, of energy called *photons* or *quanta*. The energy of a photon is proportional to the frequency of the radiation. These dual views of radiation as particles and as waves are not mutually exclusive but, rather, complementary. Indeed, the duality applies to the behavior of streams of electrons and other elementary particles such as protons and is completely rationalized by wave mechanics.

20A-1 Wave Properties of Electromagnetic Radiation

For many purposes, electromagnetic radiation is conveniently pictured as an electric field that undergoes sinusoidal oscillations in space. Figure 20-1 is a two-dimensional representation of a beam of monochromatic (that is, single-wavelength) radiation. The electric field is represented as a vector whose length is proportional to the field strength. The abscissa in this plot is either time as the radiation passes a fixed point in space or distance at a fixed time. Note that the direction in which the field oscillates is perpendicular to the direction in which the radiation is being propagated.

Wave Parameters

In Figure 20-1, the *amplitude A* of the sinusoidal wave is defined as the length of the electrical vector at the maximum in the wave. The time required for the passage of successive maxima (or minima) through a fixed point in space is called the *period p* of the radiation. The *frequency ν* is the number of oscillations of the field per second and is equal to $1/p$.

It is important to realize that *frequency* is determined by the radiation source and remains *constant* regardless of the medium traversed by the

> A wave in which the direction of displacement is perpendicular to the direction of propagation, such as the wave in Figure 20-1, is called a transverse wave.

> The unit of frequency is the hertz Hz, which corresponds to one oscillation per second.

The frequency of a beam of electromagnetic radiation never changes.

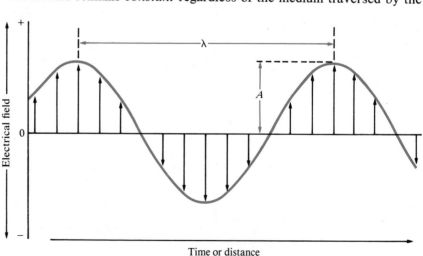

Figure 20-1

Representation of a beam of monochromatic radiation of wavelength λ and amplitude A. The arrows represent the electrical vector of the radiation.

radiation. In contrast, the *velocity* of propagation v_i, the rate at which the wavefront moves through a medium, is *dependent* upon both the medium and the frequency; the subscript "i" is employed to indicate this frequency dependence.

Another parameter of interest is the *wavelength* λ_i, which is the linear distance between successive maxima or minima of a wave. Multiplication of the frequency in waves per second by the wavelength in centimeters gives the velocity of propagation in centimeters per second:

$$v_i = \nu\lambda_i \qquad (20\text{-}1)$$

The Velocity of Radiation

In a vacuum, the velocity at which radiation is propagated becomes independent of wavelength and is at its maximum. This velocity, which is given the symbol c, has been determined to be 2.99792×10^{10} cm/s. The velocity of radiation in air differs only slightly from c (it is about 0.03% less). Thus, for a vacuum or for air, the velocity of light is conveniently rounded to

$$c = \nu\lambda = 3.00 \times 10^{10} \text{ cm/s} \qquad (20\text{-}2)$$

The rate of radiation propagation is less than c in a medium containing matter because the electromagnetic field of the radiation interacts with the electrons in the atoms or molecules of the medium and is slowed as a consequence. Since the radiant frequency is invariant and fixed by the source, the *wavelength must decrease* as radiation passes from a vacuum to a medium containing matter (Equation 20-1). This effect is illustrated in Figure 20-2 for a beam of visible radiation. Note that the wavelength shortens nearly 200 nm, or more than 30%, as the radiation passes from air into glass; a reverse change occurs when the radiation again enters air.

The *wavenumber* $\bar{\nu}$ is yet another way of describing electromagnetic

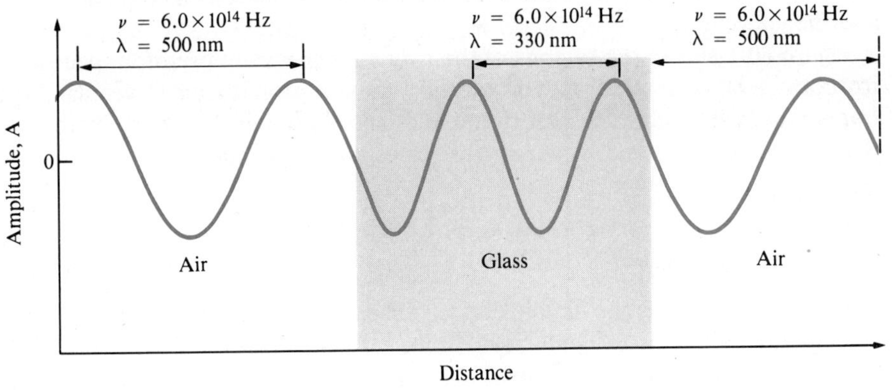

Wavelength Units for Various Spectral Regions

Region	Unit	Definition
X-ray	Angstrom, Å	10^{-10} m
Ultraviolet/ visible	Nanometer, nm	10^{-9} m
Infrared	Micrometer, μm	10^{-6} m

To three significant figures, Equation 20-2 is equally applicable in air and in vacuum.

In contrast to frequency, radiation velocity and wavelength both become smaller as the radiation passes from a vacuum or from air to a denser medium.

The refractive index η of a medium is a measure of its interaction with radiation and is defined by $\eta = c/v_i$. For example, the refractive index of water at room temperature is 1.33, which means that radiation passes through water at a rate of $c/1.33$, or 2.26×10^{10} cm/s.

$\nu = 6.0 \times 10^{14}$ Hz
$\lambda = 500$ nm

$\nu = 6.0 \times 10^{14}$ Hz
$\lambda = 330$ nm

$\nu = 6.0 \times 10^{14}$ Hz
$\lambda = 500$ nm

Amplitude, A

Air Glass Air

Distance

Figure 20-2

Change in wavelength as radiation passes from air into a dense glass and back to air.

The wavenumber $\bar{\nu}$ in cm^{-1} is generally used to describe infrared radiation. The most useful part of the infrared spectrum for the detection and determination of organic species is from 2.5 to 15 μm in wavelength, which corresponds to a wavenumber range of 4000 to 667 cm^{-1}.

A photon is a particle of electromagnetic radiation having zero mass and an energy of $h\nu$.

Both frequency and wavenumber are proportional to the energy of a photon.

We will on occasion speak of "a mole of photons," meaning 6.02×10^{23} packets of radiation.

The visible region of the spectrum extends from about 380 nm to 780 nm.

radiation. It is defined as the number of waves per centimeter and is equal to $1/\lambda$. By definition, $\bar{\nu}$ has units of cm^{-1}.

The *power P* is the energy of a beam of radiation that reaches a given area per second; the *intensity* is the power per unit solid angle. Both quantities are related to the amplitude of the radiation (Figure 20-1). Although it is not strictly correct to do so, power and intensity are frequently used interchangeably.

20A-2 Particulate Properties of Electromagnetic Radiation

Some interactions with matter require that we treat electromagnetic radiation as *photons*. The energy of a photon depends upon the frequency of the radiation and is given by

$$E = h\nu \qquad (20\text{-}3)$$

where h is the Planck constant (6.63×10^{-34} J s). We can also relate the energy of radiation to wavelength and wavenumber:

$$E = \frac{hc}{\lambda} = hc\bar{\nu} \qquad (20\text{-}4)$$

Note that the wavenumber, in common with frequency, is directly proportional to energy.

20A-3 The Electromagnetic Spectrum

The electromagnetic spectrum covers an enormous range of wavelengths (or energies). An X-ray photon (($\lambda \approx 10^{-10}$ m), for example, is approximately 10,000 times as energetic as one that is emitted by an incandescent tungsten wire ($\lambda \approx 10^{-6}$ m) and 10^{11} times as energetic as a photon in the radio-frequency range ($\lambda \approx 10^5$ m).

The major divisions of the electromagnetic spectrum are depicted in the color plate located inside the front cover of this book. Note that both the frequency and the wavelength scales are logarithmic; note also that the region to which the human eye is perceptive (the *visible spectrum*) is but a minute part of the entire spectrum. Such diverse radiations as gamma rays and radio waves differ from visible light only in the matter of frequency and, hence, energy.

Figure 20-3 shows the regions of the magnetic spectrum used for spectroscopic analyses. Also shown are the types of atomic and molecular transitions responsible for absorption and emission in each region. Note that the wavelength and wavenumber scales are logarithmic.

20B INTRODUCTION TO MOLECULAR ABSORPTION METHODS

Every molecular analyte is capable of absorbing certain characteristic wavelengths of electromagnetic radiation. In this process, the energy of the radiation is temporarily transferred to the molecule, and the intensity

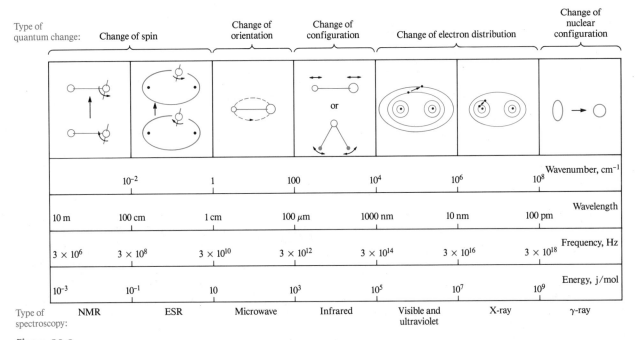

Type of quantum change:	Change of spin		Change of orientation	Change of configuration	Change of electron distribution		Change of nuclear configuration

Wavenumber, cm⁻¹

| 10^{-2} | 1 | 100 | 10^4 | 10^6 | 10^8 |

Wavelength

| 10 m | 100 cm | 1 cm | 100 μm | 1000 nm | 10 nm | 100 pm |

Frequency, Hz

| 3×10^6 | 3×10^8 | 3×10^{10} | 3×10^{12} | 3×10^{14} | 3×10^{16} | 3×10^{18} |

Energy, j/mol

| 10^{-3} | 10^{-1} | 10 | 10^3 | 10^5 | 10^7 | 10^9 |

| Type of spectroscopy: | NMR | ESR | Microwave | Infrared | Visible and ultraviolet | X-ray | γ-ray |

Figure 20-3

The regions of the electromagnetic spectrum. (From C. N. Banwell, *Fundamentals of Molecular Spectroscopy*, 3rd ed., p. 7. New York: McGraw-Hill, 1983.)

Attenuate means to weaken.

of the radiation is decreased as a consequence. When absorption occurs, the radiation is said to be *attenuated*. When visible radiation is absorbed by a solution of an analyte, the solution becomes colored, and the intensity of the color gives a measure of the analyte concentration. *Colorimetric methods* in which the color of an analyte solution is compared visually with the color of one or more standard solutions have been used since early in the nineteenth century for the determination of metal ions. Methods of this kind still find use for rough analysis of water, soils, and other commercial materials.

Today analyses based upon molecular absorption are generally carried out with photoelectric instruments called *spectrophotometers* and are therefore called *spectrophotometric methods*. This and the following chapter deal with molecular spectrophotometry.

In colorimetry, the color produced in a solution by the analyte is compared visually to the color of one or more standards.

For example, you can purchase colorimetric water test kits for determining a dozen or more species, including chlorine, oxygen, fluorine, and iron as well as total hardness. The relative errors associated with such measurements is perhaps 10 to 20%, which is entirely adequate in many cases.

Spectrophotometric methods are based upon the absorption of ultraviolet, visible, and infrared radiation by the analyte. Measurements are made photoelectrically.

20B-1 Terms Employed in Absorption Spectroscopy

Table 20-1 lists the common terms and symbols used in absorption spectroscopy. This nomenclature is recommended by the American Society for Testing Materials as well as the American Chemical Society. Column 3 contains alternative symbols encountered in the older literature. Because a standard nomenclature is highly desirable in order to avoid ambiguities, the reader is urged to learn and use the recommended terms and symbols and avoid those in column 3.

In Chapters 20 through 23, we follow the recommendation of the American Chemical Society and use lower case c to represent concentration. In earlier chapters we symbolized molar concentration by capital C.

Table 20-1

IMPORTANT TERMS AND SYMBOLS IN ABSORPTION MEASUREMENT

Term and Symbol*	Definition	Alternative Name and Symbol
Radiant power P, P_0	Energy of radiation (in ergs) impinging on a 1-cm^2 area of a detector per second	Radiation intensity I, I_0
Absorbance A	$\log \dfrac{P_0}{P}$	Optical density D; extinction E
Transmittance T	$\dfrac{P}{P_0}$	Transmission T
Path length of radiation† b	—	l, d
Absorptivity† a	$\dfrac{A}{bc}$	Extinction coefficient k
Molar absorptivity‡ ε	$\dfrac{A}{bc}$	Molar extinction coefficient

*Terminology recommended by the American Chemical Society (*Anal. Chem.*, **1952**, *24*, 1349; **1976**, *48*, 2298).

†c may be expressed in g/L or in other specified concentration units; b may be expressed in cm or in other units of length.

‡c is expressed in mol/L; b is expressed in cm.

Transmittance

Figure 20-4 depicts the attenuation of a beam of parallel radiation before and after it has passed through a layer of solution with a thickness of b cm and a concentration c of an absorbing species. As a consequence of interactions between the photons and absorbing species in the solution, the power of the beam is attenuated from P_0 to P. The *transmittance T* of the solution is defined as the fraction of incident radiation transmitted by the solution:

$$T = P/P_0 \tag{20-5}$$

Transmittance is often expressed as a percentage.

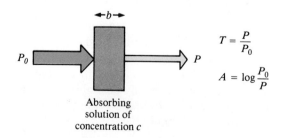

Figure 20-4

Attenuation of a beam of radiation by an absorbing solution.

Absorbance

The absorbance of a solution is defined by the equation

$$A = -\log_{10} T = \log \frac{P_0}{P}$$ (20-6)

Note that, in contrast to transmittance, the absorbance of a solution increases as the attenuation of the beam becomes greater. If the readout meter of a spectrophotometer is calibrated in both transmittance and absorbance, Equation 20-6 requires that the absorbance scale be logarithmic. Figure 20-5 shows a typical meter scale.

Experimental Transmittances and Absorbances

Transmittance and absorbance, as defined by Equations 20-5 and 20-6, cannot be measured in the laboratory because the solution to be studied must be held in some sort of container. Interaction between the radiation and the container walls is inevitable, with losses in power occurring at each interface as a result of reflection and possibly absorption (Figure 20-6). Reflection losses are substantial. For example, it can be shown that about 8.5% of a beam of yellow light is lost by reflection in passing vertically through a glass-walled container holding an aqueous solution.[2] In addition to reflective losses, scattering by large molecules or inhomogeneities in the solvent may cause a decrease in the power of the beam as it traverses the solution.

To compensate for these effects, the power of a beam that has been transmitted through the analyte solution is compared with the power of a similar beam after passing through an identical cell containing only solvent. An experimental absorbance that closely approximates the true absorbance for the solution is thus obtained; that is,

$$A = \log \frac{P_0}{P} \approx \log \frac{P_{\text{solvent}}}{P_{\text{solution}}}$$ (20-7)

The terms P_0 and P, when used henceforth, refer to the power of radiation after it has passed through cells containing the solvent and the analyte, respectively.

20B-2 Beer's Law

According to Beer's law, absorbance is linearly related to the concentration c of the absorbing species and to the path length b of the radiation in the absorbing medium. That is,

$$A = \log(P_0/P) = abc$$ (20-8)

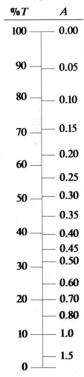

Figure 20-5
A spectrophotometer readout. The readout scales on some spectrophotometers are linear in percent transmittance. As shown below, the absorbance scales must then be logarithmic.

%T	A
100	0.00
90	0.05
80	0.10
70	0.15
60	0.20
	0.25
50	0.30
	0.35
40	0.40
	0.45
30	0.50
	0.60
20	0.70
	0.80
10	1.0
	1.5
0	

Beer's law is sometimes paraphrased:
The deeper the glass
The darker the brew
The smaller the amount
of light that gets through.

[2]See D. A. Skoog, *Principles of Instrumental Analysis*, 3rd ed., p. 101. Philadelphia: Saunders College Publishing, 1985.

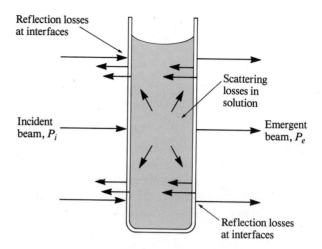

Figure 20-6
Reflection and scattering losses.

where a is a proportionality constant called the *absorptivity*. Because absorbance is a unitless quantity, the absorptivity must have units to make the right side of this equation dimensionless.

When the concentration in Equation 20-8 is expressed in moles per liter and b is in centimeters, the proportionality constant is called the *molar absorptivity* and is given the special symbol ε. Thus,

$$A = \varepsilon bc \qquad \text{(20-9)}$$

where ε has the units of L cm^{-1} mol^{-1}.

Example 20-1

A 7.50×10^{-5} M solution of potassium permanganate has a percent transmittance of 36.4% when measured in a 1.05-cm cell at a wavelength of 525 nm. Calculate (a) the absorbance of this solution; (b) the molar absorptivity of $KMnO_4$.

(a) $A = -\log T = -\log 0.364 = -(-0.4389) = 0.439$

(b) Rearranging Equation 20-9,

$\varepsilon = A/bc = 0.439/(1.05 \text{ cm} \times 7.50 \times 10^{-5} \text{ mol/L})$
$= 5.57 \times 10^3 \text{ cm}^{-1} \text{ mol}^{-1} \text{ L}$

20B-3 The Application of Beer's Law to Mixtures

Beer's law also applies to solutions containing more than one kind of absorbing substance. Provided no interaction occurs between the various species, the total absorbance for a multicomponent system is

$$A_{\text{total}} = A_1 + A_2 + \cdots + A_n = \varepsilon_1 b c_1 + \varepsilon_2 b c_2 + \cdots + \varepsilon_n b c_n$$

Absorbances are additive.

(20-10)

where the subscripts refer to absorbing components 1, 2, . . . , n.

20B-4 Limitations to the Applicability of Beer's Law

The linear relationship between absorbance and path length at a fixed concentration is a generalization for which there are few exceptions. On the other hand, deviations from the direct proportionality between absorbance and concentration (when b is constant) are frequently observed. Some of these deviations are fundamental and represent real limitations to the law. Others occur as a consequence of the manner in which the absorbance measurements are made (*instrumental deviations*) or as a result of chemical changes associated with concentration changes (*chemical deviations*).

Real Limitations

Beer's law is successful in describing the absorption behavior of dilute solutions only and in this sense is a *limiting law*. At high concentrations (usually > 0.01 M), the average distances between particles of the absorbing species are diminished to the point where each particle affects the charge distribution of its neighbors. This interaction can alter each particle's ability to absorb a given wavelength of radiation. Because the extent of interaction depends upon concentration, the occurrence of this phenomenon causes deviations from the linear relationship between absorbance and concentration.

Real limitations to Beer's law are encountered only in relatively concentrated solutions of the analyte or in concentrated electrolyte solutions.

A similar effect is sometimes encountered in dilute solutions of absorbers that contain high concentrations of other species, particularly electrolytes. The close proximity of ions to the absorber alters the molar absorptivity of the latter by electrostatic interactions, which leads to departures from Beer's law.

Chemical Deviations

Apparent deviations from Beer's law appear when the analyte undergoes association, dissociation, or reaction with the solvent to give products with absorbing characteristics that differ from those of the analyte. As shown in Example 20-2, the extent of such departures can be predicted from the molar absorptivities of the absorbing species and the equilibrium constants for the reactions involved.

Chemical deviations from Beer's law are encountered when the absorbing species participates in a concentration-dependent equilibrium such as a dissociation or association reaction.

Example 20-2

Consider the dissociation equilibrium for the acid/base indicator HIn:

$$HIn + H_2O \rightleftarrows H_3O^+ + In^-$$

for which

$$K_{HIn} = \frac{[H_3O^+][In^-]}{[HIn]} = 1.42 \times 10^{-5}$$

At 430 and 570 nm, the individual absorbances of HIn and In^- are linear with respect to concentration, as demonstrated by measurements in 0.10 M HCl (where, for all practical purposes, the indicator exists solely as HIn) and in 0.10 M NaOH (where In^- is the predominant species). Molar absorptivities are

	ε_{430}	ε_{570}
HIn (HCl soln)	6.30×10^2	7.12×10^3
In^- (NaOH soln)	2.06×10^4	9.60×10^2

Calculate the absorbance of a 2.00×10^{-5} M solution of the indicator at 430 and 570 nm in a 1.00-cm cell.

The molar concentrations of HIn and In^- in an unbuffered solution can be calculated through the use of Equation 7-16. Here, however, we are interested in $[In^-]$ rather than $[H_3O^+]$; that is,

$$[In^-] = [H_3O^+]$$

$$[HIn] = c_{HIn} - [In^-]$$

Substitution of these quantities into the expression for K_{HIn} and rearrangement yield the quadratic equation

Absorbance Data at Other Concentrations of HIn c_{HIn}

$M \times 10^5$	A_{430}	A_{570}
2.00	0.236	0.073
4.00	0.381	0.175
8.00	0.596	0.401
12.00	0.771	0.640
16.00	0.922	0.887

$$[In^-]^2 + K_{HIn}[In^-] - K_{HIn}c_{HIn} = 0$$

Thus, when c_{HIn} is 2.00×10^{-5} M,

$$[In^-]^2 + 1.42 \times 10^{-5}[In^-] - 1.42 \times 10^{-5} \times 2.00 \times 10^{-5} = 0$$

$$[In^-] = \frac{-1.42 \times 10^{-5} + \sqrt{(1.42 \times 10^{-5})^2 - 4(-1.42 \times 10^{-5} \times 2.00 \times 10^{-5})}}{2}$$

$$= 1.12 \times 10^{-5}$$

$$[HIn] = 2.00 \times 10^{-5} - 1.12 \times 10^{-5} = 8.8 \times 10^{-6}$$

When the measurements are made in 1.00-cm cells,

$$A_{430} = (2.06 \times 10^4)(1.00)(1.12 \times 10^{-5}) + (6.30 \times 10^2)(1.00)(8.8 \times 10^{-6})$$
$$= 0.236$$

$$A_{570} = (9.60 \times 10^2)(1.00)(1.12 \times 10^{-5}) + (7.12 \times 10^3)(1.00)(8.8 \times 10^{-6})$$
$$= 0.073$$

The data for the plot in Figure 20-7 were computed by the method shown in Example 20-2 and illustrate chemical deviations from Beer's

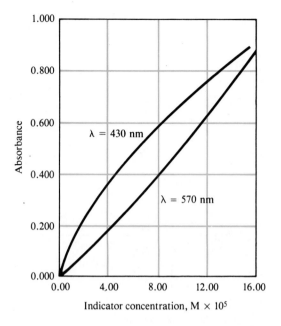

Figure 20-7
Chemical deviations from Beer's law for unbuffered solutions of the indicator HIn. For data, see Example 20-2.

law. Note that the direction of curvature for one plot is opposite that for the other.

Instrumental Deviations with Polychromatic Radiation

Beer's law is also a limiting law in the sense that it applies only when absorbance is measured with monochromatic radiation. Truly monochromatic sources, such as lasers, are not practical for routine analytical measurements, however. Instead, a polychromatic continuous source is employed in conjunction with a grating or a filter that isolates a more or less symmetric band of wavelengths around the desired one (page 411). The following derivation illustrates how such a source may lead to deviations from Beer's law.

Consider a beam made up of just two wavelengths λ' and λ'', and assume that Beer's law applies strictly to each. With this assumption, we can write for radiation at λ'

> Deviations from Beer's law often occur when polychromatic radiation is used to measure absorbance.

$$A' = \log \frac{P_0'}{P'} = \varepsilon' bc$$

$$\frac{P_0'}{P'} = 10^{\varepsilon' bc} \quad \text{and} \quad P' = P_0' 10^{-\varepsilon' bc}$$

Similarly, for λ''

$$\frac{P_0''}{P''} = 10^{\varepsilon'' bc} \quad \text{and} \quad P'' = P_0'' 10^{-\varepsilon'' bc}$$

When absorbance is measured with radiation composed of both wavelengths, the power of the beam emerging from the solution is given by

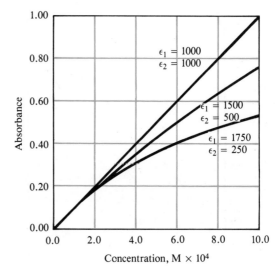

Figure 20-8
Deviations from Beer's law with polychromatic light. The absorber has the indicated molar absorptivities at the two wavelengths λ' and λ''.

$P' + P''$ and that of the beam emerging from the solvent by $P_0' + P_0''$. Therefore, the measured absorbance is

$$A_m = \log \frac{P_0' + P_0''}{P' + P''}$$

which can be rewritten as

$$A_m = \log \frac{P_0' + P_0''}{P_0'10^{-\varepsilon'bc} + P_0''10^{-\varepsilon''bc}}$$
$$= \log(P_0' + P_0'') - \log(P_0'10^{-\varepsilon'bc} + P_0''10^{-\varepsilon''bc})$$

Now, when $\varepsilon' = \varepsilon''$, this equation simplifies to

$$A_m = \varepsilon bc$$

and Beer's law is followed. As shown in Figure 20-8, however, the relationship between A_m and concentration is no longer linear when the molar absorptivities differ. Moreover, departures from linearity become greater

Figure 20-9
The effect of polychromatic radiation upon Beer's law. Band A shows little deviation because ε does not change greatly throughout the band. Band B shows marked deviation because ε undergoes significant changes in this region.

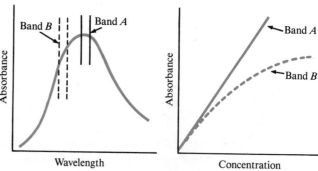

as the difference between ε' and ε'' increases. When this treatment is expanded to include additional wavelengths, the effect remains the same.

It is an experimental fact that deviations from Beer's law resulting from the use of a polychromatic beam are not appreciable, provided the radiation used does not encompass a spectral region in which the absorber exhibits large changes in absorbance as a function of wavelength. This observation is illustrated in Figure 20-9.

20B-5 Absorption Spectra

An *absorption spectrum* is a plot of the absorption characteristics of an analyte as a function of wavelength or wavenumber (or occasionally frequency). As shown in Figure 20-10 the vertical axis of such plots may be percent transmittance, absorbance, or log absorbance. (Occasionally ε, a, log ε, and log a serve as the vertical axis.) A plot with log A as ordinate leads to loss of spectral detail but is convenient for comparing solutions of different concentrations since the curves are displaced equally along the vertical axis.

Figure 20-10

Methods for plotting spectral data. The numbers for the curves indicate ppm of $KMnO_4$ in the solution; $b = 2.00$ cm. (From M. G. Mellon, *Analytical Absorption Spectroscopy*, pp. 104–106. New York: Wiley, 1950. With permission.)

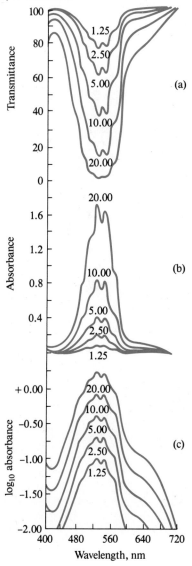

Feature 20-1
WHY IS A RED SOLUTION RED?

A solution such as $FeSCN^{2+}$ is red not because the complex adds red radiation to the solvent. Instead, this complex absorbs the green component from the incoming white radiation but transmits the red component unaltered. Thus in a colorimetric analysis of iron based upon its thiocyanate complex, the maximum change in absorbance with concentration occurs with green radiation; the absorbance change with red radiation is negligible. In general, then, the radiation used for a colorimetric analysis should be the complementary color of the analyte solution. The table below shows this relationship for various parts of the visible spectrum.

THE VISIBLE SPECTRUM

Wavelength Region, nm	Color	Complementary Color
400–435	Violet	Yellow-green
435–480	Blue	Yellow
480–490	Green-blue	Orange
490–500	Blue-green	Red
500–560	Green	Purple
560–580	Yellow-green	Violet
580–595	Yellow	Blue
595–650	Orange	Green-blue
650–750	Red	Blue-green

	## 20C THEORY OF MOLECULAR ABSORPTION
The lowest energy state of a molecule is called its ground state.	According to quantum theory, every molecular species has a unique set of energy states, the lowest of which is the *ground* state. At room temperature, most molecules are in their ground state. When a photon of radiation passes near a molecule, absorption becomes probable if (and only if) the energy of the photon matches *exactly* the energy difference between the ground state and one of the higher energy states of the molecule. Under these circumstances, the energy of the photon is transferred to the molecule, converting it to the higher energy state, which is termed an *excited state*.

Excitation of a species M to its excited state M* can be depicted by the equation

$$M + h\nu \rightarrow M^*$$

<table>
<tr>
<td>Excitation is a process in which a chemical species absorbs thermal, electrical, or radiant energy and is promoted to a higher energy state.</td>
<td></td>
</tr>
</table>

After a brief period (10^{-8} to 10^{-9} s), the excited species *relaxes* to its original, or ground, state, most commonly by transferring its excess energy to other atoms or molecules in the medium. This process, which causes a small rise in the temperature of the surroundings, is described by the equation

$$M^* \rightarrow M + heat$$

Relaxation is a process in which an excited species gives up its excess energy and returns to a lower energy state.	Relaxation may also occur by *photochemical decomposition* of M* to form new species or by the *fluorescent* or *phosphorescent* reemission of radiation.

It is important to note that the lifetime of M* is so very short that its concentration at any instant is ordinarily negligible. Furthermore, the amount of thermal energy released during relaxation is usually so small as to be undetectable. Thus, absorption measurements have the advantage of creating minimal disturbance of the system under study.

20C-1 Types of Molecular Transitions

Molecules undergo three types of quantized transitions when excited by ultraviolet, visible, and infrared radiation. For ultraviolet and visible radiation, excitation involves promoting an electron residing in a low-energy molecular or atomic orbital to a higher-energy orbital. In order for this transition to occur, the energy $h\nu$ of the photon must be exactly the same as the energy difference between the two orbital energies. The transition of an electron between two orbitals is termed an *electronic transition* and the absorption process is called *electronic absorption*.

In addition to electronic transitions, molecules exhibit two other types of radiation-induced transitions: *vibrational transitions* and *rotational transitions*. Vibrational transitions come about because molecules have a multitude of quantized energy levels (or *vibrational states*) associated with the bonds that hold the molecule together.

An idea of the nature of vibrational states can be pictured by treating a bond as a flexible spring with atoms attached at both ends. In Figure 20-11a, two types of stretching vibration are shown. With each vibration,

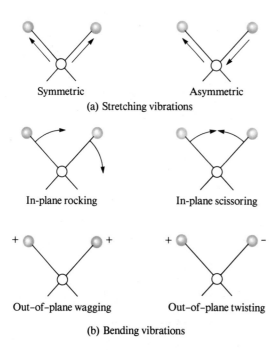

Symmetric   Asymmetric

(a) Stretching vibrations

In-plane rocking   In-plane scissoring

Out–of–plane wagging   Out–of–plane twisting

(b) Bending vibrations

Figure 20-11
Types of molecular vibrations.
The plus sign indicates motion
from the page toward the reader;
the minus sign indicates motion
away from the reader.

atoms first approach and then move away from one another. The potential energy of the system at any instant depends upon the extent to which the spring-like bond is stretched or compressed. For an ordinary spring, the energy of the system varies continually and reaches a maximum when the spring is fully stretched or fully compressed. In contrast, the energy of a spring system having atomic dimensions can assume only certain discrete energies called vibrational energy levels.

Figure 20-11b shows other types of molecular vibrations. The energies associated with these vibrational states usually differ from one another and from the energies associated with stretching vibrations.

In addition to quantized vibrational states, a molecule has a host of quantized rotational states that are associated with the rotational motion of a molecule around its center of gravity.

The overall energy E associated with an orbital of a molecule is then given by

$$E = E_{electronic} + E_{vibrational} + E_{rotational} \qquad (20\text{-}11)$$

where $E_{electronic}$ is the energy associated with the electrons in the various outer orbitals of the molecule and $E_{vibrational}$ is the energy of the molecule as a whole due to interatomic vibrations. The last term accounts for the energy associated with rotation of the molecule about its center of gravity.

20C-2 Energy-Level Diagrams for Molecules

Figure 20-12 is a partial energy-level diagram that depicts some of the processes that occur when a polyatomic species absorbs infrared, visible, and ultraviolet radiation. The energies E_1 and E_2 of two of the several

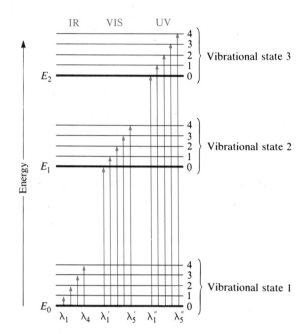

Figure 20-12

Energy-level diagram showing some of the energy changes that occur during absorption.

electronically excited states of a molecule are shown relative to the energy of the molecule's ground state E_0. In addition, the relative energies of a few of the many vibrational states associated with each electronic state are indicated by the lighter horizontal lines labeled 1, 2, 3, and 4 (the lowest vibrational levels are labeled 0). Note that energy differences between vibrational states are significantly smaller than the differences between energy levels of the electronic states (typically, an order of magnitude smaller).

Although they are not shown, a set of rotational energy states is superimposed on each of the vibrational states shown in Figure 20-12. The energy differences between these states are smaller than those between vibrational states by about an order of magnitude.

$$\Delta E_{electronic} \approx 10\, E_{vibrational}$$
$$\approx 100\, E_{rotational}$$

Infrared radiation is not sufficiently energetic to cause electronic transitions.

Infrared Absorption

Infrared radiation generally is not sufficiently energetic to cause electronic transitions but can induce transitions in the vibrational and rotational states associated with *the ground electronic state* of a molecule. Four of these transitions are depicted in the lower left part of Figure 20-12. For absorption to occur, the analyte must be irradiated with wavelengths λ_1 through λ_4. The absorbed frequencies correspond exactly to the energies indicated by the lengths of the four arrows.

Absorption of Ultraviolet and Visible Radiation

The center arrows in Figure 20-12 suggest that the molecule under consideration absorbs visible radiation of five wavelengths (λ_1' through λ_5'), thereby promoting electrons to the five vibrational levels of the excited electronic level E_1. Ultraviolet photons, which are more energetic than visible photons, are required to produce the absorption indicated by the five arrows on the right (λ_1'' through λ_5'').

As suggested by Figure 20-12, molecular absorption in the ultraviolet and visible regions consists of absorption *bands* made up of closely spaced lines. (A real molecule has many more vibrational energy levels than shown here; thus the typical absorption band consists of a multitude of lines.) In a solution, the absorbing species are surrounded by solvent, and the band nature of molecular absorption often becomes blurred because collisions tend to spread the energies of the quantum states, thus giving smooth and continuous absorption peaks, as shown in Figure 20-13c.

20C-3 Molecular Species That Absorb Ultraviolet and Visible Radiation

Absorption measurements in the visible and ultraviolet regions of the spectrum provide qualitative and quantitative information about organic, inorganic, and biochemical molecules.

Absorption by Organic Compounds

Absorption of radiation by organic molecules in the wavelength region between 180 and 780 nm results from interactions between photons and those electrons that either (1) participate directly in bond formation (and

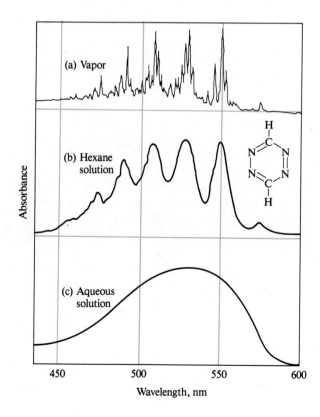

Figure 20-13

Typical ultraviolet absorption spectra. The compound is 1,2,4,5-tetrazine. Note the effects physical state and solvent have on the curves. (From S. F. Mason, *J. Chem. Soc.*, **1959**, 1265. With permission.)

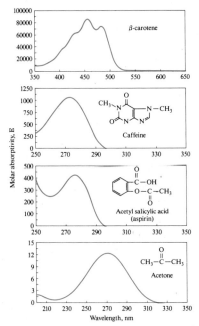

Spectra for typical organic compounds.

are thus associated with more than one atom) or (2) are localized about such atoms as oxygen, sulfur, nitrogen, and the halogens.

The wavelength at which an organic molecule absorbs depends upon how tightly its several electrons are bound. The shared electrons in such single bonds as carbon-carbon and carbon-hydrogen are so firmly held that absorption occurs only in a region of the ultraviolet spectrum ($\lambda <$ 180 nm) where the components of air also absorb (this region is known as the *vacuum ultraviolet*). Owing to the difficulty of making measurements in this region, such absorption is seldom used for analytical purposes.

Electrons involved in double and triple bonds of organic molecules are not as strongly held and are therefore more easily excited by radiation; thus species with unsaturated bonds generally exhibit useful absorption peaks. Unsaturated organic functional groups that absorb in the ultraviolet or visible region are known as *chromophores*. Table 20-2 lists common chromophores and the approximate wavelengths at which they absorb. The data for position and peak intensity can only serve as a rough guide for identification purposes, since both are influenced by solvent effects as well as other structural details of the molecule. Moreover, conjugation between two (or more) chromophores tends to shift peak maxima to longer wavelengths. Finally, vibrational effects broaden absorption peaks

A chromophore is an unsaturated organic functional group that absorbs ultraviolet or visible radiation.

Table 20-2
ABSORPTION CHARACTERISTICS OF SOME COMMON ORGANIC CHROMOPHORES

Chromophore	Example	Solvent	λ_{max}, nm	ε_{max}
Alkene	$C_6H_{13}CH{=}CH_2$	*n*-Heptane	177	13,000
Conjugated alkene	$CH_2{=}CHCH{=}CH_2$	*n*-Heptane	217	21,000
Alkyne	$C_5H_{11}C{\equiv}C{-}CH_3$	*n*-Heptane	178	10,000
			196	2000
			225	160
Carbonyl	CH_3CCH_3	*n*-Hexane	186	1000
			280	16
	CH_3CH	*n*-Hexane	180	Large
			293	12
Carboxyl	CH_3COH	Ethanol	204	41
Amido	CH_3CNH_2	Water	214	60
Azo	$CH_3N{=}NCH_3$	Ethanol	339	5
Nitro	CH_3NO_2	Isooctane	280	22
Nitroso	C_4H_9NO	Ethyl ether	300	100
			665	20
Nitrate	$C_2H_5ONO_2$	Dioxane	270	12
Aromatic	Benzene	*n*-Hexane	204	7900
			256	200

in the ultraviolet and visible regions, which often makes precise determination of a maximum difficult.

The unshared electrons in such elements as sulfur, bromine, and iodine are less strongly held than the shared electrons of a saturated bond. As a result, organic molecules incorporating these elements frequently exhibit useful peaks in the ultraviolet region.

Absorption by Inorganic Species

The spectra for most inorganic ions and molecules have broad absorption maxima and little fine structure. The ions of the lanthanide and actinide series represent an important exception to this statement. The electrons responsible for absorption by these elements ($4f$ and $5f$, respectively) are shielded from external influences by electrons that occupy orbitals with larger principal quantum numbers. As a result, the bands tend to be narrow and relatively unaffected by the species bonded by the outer electrons (Figure 20-14).

In general, the ions and complexes of elements in the first two transition series absorb broad bands of visible radiation in at least one of their oxidation states and are, as a consequence, colored (Figure 20-15). In the complexes of transition elements, absorption involves transitions between filled and unfilled d-orbitals, with energies that depend upon the ligands bonded to the metal ions. The energy differences between these d-orbitals (and thus the position of the corresponding absorption peak) depend upon the position of the element in the periodic table, its oxidation state, and the nature of the ligand bonded to it.

Charge-transfer absorption is of particular importance from the analytical standpoint because molar absorptivities of charge-transfer peaks are usually large ($\varepsilon_{\max} > 10,000$). Thus, methods based upon this type of absorption are highly sensitive. Many organic and inorganic complexes exhibit charge-transfer absorption and are known as *charge-transfer complexes*. Common examples include the thiocyanate and phenolic complexes of iron(III), the iodide complex of molecular iodine, the iron(II)

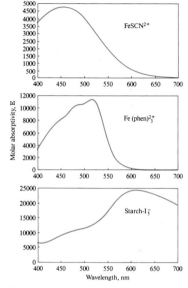

Charge-transfer spectra.

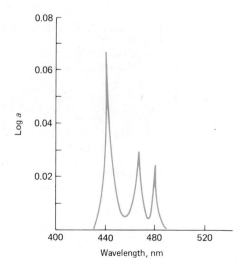

Figure 20-14
The absorption spectrum of a praseodymium chloride solution; a = absorptivity in L cm^{-1} g^{-1}.

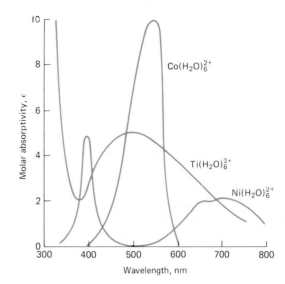

Figure 20-15
Absorption spectra of some transition-metal ions.

complex of orthophenanthroline, and the ferro/ferricyanide complex responsible for the color Prussian blue.

Charge-transfer complexes generally contain both an electron-donor group and an electron-acceptor group. Absorption of radiation involves transition of an electron from the donor group to an orbital that is largely associated with the acceptor group. The excited state is thus the product of an internal oxidation/reduction process.[3]

---

[3]For further information concerning charge transfer, see C.N.R. Rao, *Ultra-Violet Spectroscopy, Chemical Applications,* 3rd ed., Chapter 11. London: Butterworth and Co., 1975.

20D   QUESTIONS AND PROBLEMS

20-1. What kind of transitions are responsible for the
 *(a) absorption of ultraviolet radiation?
 (b) absorption of infrared radiation?

20-2. Define
 *(a) ground state.
 (b) excited electronic state.
 *(c) photon.
 (d) chromophore.
 *(e) percent transmittance.
 (f) absorbance.
 *(g) molar absorptivity.
 (h) absorptivity.
 *(i) wavenumber.
 (j) relaxation.

20-3. Calculate the frequency in hertz of
 *(a) an X-ray beam with a wavelength of 2.65 Å.
 (b) an emission line for copper at 211.0 nm.
 *(c) the line at 694.3 nm produced by a ruby laser.
 (d) the output of a CO_2 gas laser at 10.6 μm.
 *(e) an infrared absorption peak at 19.6 μm.
 (f) a microwave beam at 1.86 cm.

20-4. Calculate the wavelength in centimeters and the wavenumber of
 *(a) an airport tower transmitting at 118.6 MHz.
 (b) a VOR (radio navigation aid) transmitting at 114.10 kHz.
 *(c) an NMR signal at 105 MHz.
 (d) an infrared absorption peak having a wavenumber of 1210 cm^{-1}.

*20-5. A typical simple infrared spectrophotometer covers a wavelength range from 3 to 15 μm. Express its range (a) in wavenumbers and (b) in hertz.

20-6. A sophisticated ultraviolet/visible/near-IR instrument has a wavelength range of 185 to 3000 nm. What are its wavenumber and frequency ranges?

*20-7. What color radiation should be used in the colorimetric determination of Cu^{2+} based upon adding ammonia to give the deep blue ammine complex?

20-8. Express the following absorbances in terms of percent transmittance:

*(a) 0.064.   (d) 0.209.
*(b) 0.765.   (e) 0.437.
*(c) 0.318.   (f) 0.413.

20-9. Convert the following transmittance data to absorbances:

*(a) 19.4%.   (d) 4.51%.
*(b) 86.3%.   (e) 10.0%.
*(c) 27.2%.   (f) 79.8%.

20-12. Use the following data to evaluate the missing quantities. Wherever necessary assume that the molecular weight of the analyte is 250.

20-10. Calculate the percent transmittance of solutions having twice the absorbance of the solutions in Problem 20-8.

20-11. Calculate the absorbances of solutions having half the percent transmittance of those in Problem 20-9.

	A	% T	ε	b, cm	c, M	c, ppm	a, $cm^{-1}\ ppm^{-1}$
*(a)	0.416			1.40	1.25×10^{-4}		
(b)		45.5		2.10	8.15×10^{-3}		
*(c)	1.424			0.996			0.137
(d)		19.6	5.42×10^3		2.50×10^{-4}		
*(e)			3.46×10^3	2.50		3.33	
(f)			1.214×10^4	1.25	7.77×10^{-4}		
*(g)		48.3		0.250		6.72	
(h)	0.842		7.73×10^3	2.00			
*(i)		76.3		1.10			0.0631
(j)		6.54	9.82×10^2		8.64×10^{-3}		

*20-13. A solution containing 4.48 ppm $KMnO_4$ has a transmittance of 0.309 in a 1.00-cm cell at 520 nm. Calculate the molar absorptivity of $KMnO_4$.

20-14. A solution containing 3.75 mg/100 mL of the organic compound AB (fw = 220 g) has a transmittance of 39.6% in a 1.50-cm cell at 480 nm. Calculate the molar absorptivity of AB.

20-15. A solution containing the complex formed between Bi(III) and thiourea has a molar absorptivity of 9.32×10^3 L cm^{-1} mol^{-1} at 470 nm.

(a) What is the absorbance of a 6.24×10^{-5} M solution of the complex at 470 nm in a 1.00-cm cell?

(b) What is the percent transmittance of the solution described in (a)?

(c) What is the molar concentration of the complex in a solution that has the absorbance described in (a) when measured at 470 nm in a 5.00-cm cell?

*20-16. At 580 nm, which is the wavelength of its maximum absorption, the complex $FeSCN^{2+}$ has a molar absorptivity of 7.00×10^3 L cm^{-1} mol^{-1}. Calculate

(a) the absorbance of a 2.50×10^{-5} M solution of the complex at 580 nm in a 1.00-cm cell.

(b) the absorbance of a solution in which the concentration of the complex is twice that in (a).

(c) the transmittance of the solutions described in (a) and (b).

(d) the absorbance of a solution that has half the transmittance of that described in (a).

*20-17. A 2.50-mL aliquot of a solution that contains 3.8 ppm iron(III) is treated with an appropriate excess of KSCN and diluted to 50.0 mL. What is the absorbance of the resulting solution at 580 nm in a 2.50-cm cell? See Problem 20-16 for absorptivity data.

20-18. Zinc(II) and the ligand L form a product that absorbs strongly at 600 nm. As long as the molar concentration of L exceeds that of zinc(II) by a factor of 5, the absorbance is dependent only on the cation concentration. Neither zinc(II) nor L absorbs at 600 nm. A solution that is 1.60×10^{-4} M in zinc(II) and 1.00×10^{-3} M in L has an absorbance of 0.464 in a 1.00-cm cell at 600 nm. Calculate

(a) the percent transmittance of this solution.

(b) the percent transmittance of this solution in a 2.50-cm cell.

(c) the molar absorptivity of the complex.

*20-19. The equilibrium constant for the conjugate acid/base pair

$$HIn + H_2O \rightleftarrows H_3O^+ + In^-$$

is 8.00×10^{-5}. From the additional information

| Species | Absorption Maximum, nm | Molar Absorptivity | |
		430 nm	600 nm
HIn	430	8.04×10^3	1.23×10^3
In$^-$	600	0.775×10^3	6.96×10^3

(a) calculate the absorbance at 430 nm and 600 nm for the following indicator concentrations: 3.00×10^{-4} M, 2.00×10^{-4} M, 1.00×10^{-4} M, 0.500×10^{-4} M, and 0.250×10^{-4} M.
(b) plot absorbance as a function of indicator concentration.

20-20. The equilibrium constant for the reaction

$$2\,CrO_4^{2-} + 2\,H^+ \rightleftarrows Cr_2O_7^{2-} + H_2O$$

is 4.2×10^{14}. The molar absorptivities for the two principal species in a solution of $K_2Cr_2O_7$ are

λ, nm	$\varepsilon_1(CrO_4^{2-})$	$\varepsilon_2(Cr_2O_7^{2-})$
345	1.84×10^3	10.7×10^2
370	4.81×10^3	7.28×10^2
400	1.88×10^3	1.89×10^2

Four solutions were prepared by dissolving 4.00×10^{-4}, 3.00×10^{-4}, 2.00×10^{-4}, and 1.00×10^{-4} mole of $K_2Cr_2O_7$ in water and diluting to 1.00 L with a pH 5.60 buffer. Derive theoretical absorbance values (1.00-cm cells) for each solution and plot the data for (a) 345 nm, (b) 370 nm, (c) 400 nm.

Chapter 21

Applications of Molecular-Absorption Spectroscopy

In this chapter, we first describe instruments that currently find use for ultraviolet and visible molecular-absorption spectroscopy. We then turn to their applications for determining inorganic, organic, and biochemical species.

21A INSTRUMENTS FOR ULTRAVIOLET AND VISIBLE ABSORPTION MEASUREMENTS

Two types of instruments are used to perform quantitative absorption analyses: *spectrophotometers* and *photometers*. Spectrophotometers employ a grating or a prism to provide a narrow band of radiation for measurements. Photometers, in contrast, use an absorption filter or an interference filter for this purpose. Spectrophotometers offer the considerable advantage that the wavelength used can be varied continuously, thus making it possible to obtain entire absorption spectra.

In the past, photometers were often favored for quantitative work because of their simplicity, ruggedness, and low cost. Today, however, the cost and dependability of spectrophotometers rival those of photometers, and few photometers are manufactured for general absorption analysis. Here and in the chapters that follow, therefore, we shall focus our attention on spectrophotometric measurements, although from time to time we

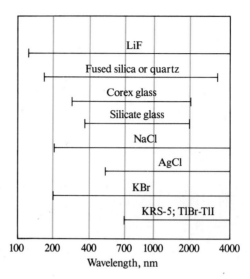

Figure 21-1
Transmittance range for various optical materials.

shall mention certain specialized applications where filter photometers are important.

Modern spectrophotometers vary widely in appearance, performance characteristics, and sophistication, with some costing $100 or less and others costing more than 100 times this figure. We shall confine our discussion largely to simple, nonrecording instruments, such as you are likely to encounter in your undergraduate laboratory.

21A-1 Optical Materials

The cells, windows, lenses, and wavelength-dispersing elements in a spectrophotometer or photometer must be transparent in the wavelength region being used for analyses. Figure 21-1 shows the wavelength ranges for several optical materials that find use in the ultraviolet, visible, and infrared regions of the spectrum. For the visible region, ordinary silicate glass is completely adequate and has the considerable advantage of low cost. Below about 380 nm, glass begins to absorb, and fused silica or quartz must be used instead.

21A-2 Instrument Components

All instruments for absorption measurements with ultraviolet, visible, and infrared radiation are made up of the five components shown in Figure 21-2:

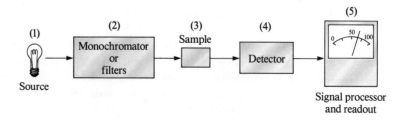

Figure 21-2
Components of instruments for absorption spectroscopy.

1. A stable source of radiant energy.
2. A monochromator (or sometimes a filter) to isolate a narrow wavelength region from the source.
3. Transparent containers to hold the sample and blank in the beam of radiation from the monochromator.
4. A radiation detector, or *transducer,* that converts the radiant energy to a measurable electrical signal.
5. A signal processor and readout device that amplifies the electrical signal and displays the magnitude of the amplified signal.

Radiation Sources

A radiation source for molecular-absorption measurements must generate a beam with sufficient power in the wavelength region of interest to permit ready detection and measurement. In addition, the source should provide *continuous radiation* that is made up of all wavelengths within the spectral region to be used. (See Figure 21-3b for an example of continuous radiation.) Finally, the source must provide a stable output for the period needed to measure P_0 and P. Only under these conditions are reproducible absorbance measurements obtained. Some instruments are designed to measure P_0 and P simultaneously, or nearly so, thus making the stability requirement less stringent.

Source of Visible Radiation. An ordinary tungsten filament lamp provides a continuous spectrum from 320 to 2500 nm (Figure 21-3). The intensity of a tungsten bulb varies as the fourth power of voltage. For this reason, an electronically stabilized voltage supply is required unless P and P_0 are measured simultaneously.

Sources of Ultraviolet Radiation. Deuterium (and also hydrogen) lamps provide continuous radiation in the range from 160 to 380 nm and are the most common sources used in ultraviolet spectrophotometry. A deuterium lamp consists of a cylindrical tube with a quartz window from which the radiation exits (Figure 21-4). The tube contains deuterium at a low pressure. Excitation is carried out by applying about 40 V between a heated oxide-coated electrode and a metal electrode. A regulated power supply is required to produce radiation of constant intensity.

The emission spectrum for a typical deuterium lamp is shown in Figure 21-5.

Monochromators and Filters

Instruments for absorption measurements generally require a device that restricts the wavelength to be used to a narrow band that is absorbed by the analyte. A narrow band of radiation offers three advantages. First, Beer's law is much more likely to be obeyed (Section 20B-4). In addition, a greater selectivity is ensured since substances with absorption peaks in other wavelength regions are less likely to interfere. Finally, use of a narrow band of wavelengths causes the greatest change in absorbance per increment of change in concentration; sensitivity is thus increased.

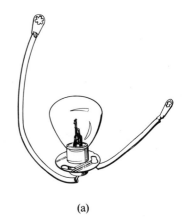

(a)

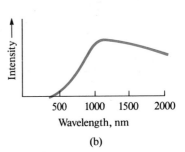

(b)

Figure 21-3
(a) A tungsten lamp. (b) Its emission spectrum.

A continuous source emits radiation of all wavelengths within a spectral region.

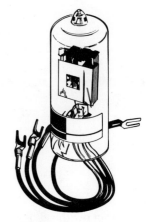

Figure 21-4
A deuterium lamp.

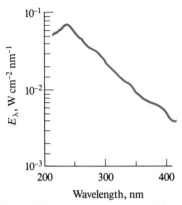

Figure 21-5
Output from a deuterium lamp.

Angular dispersion is a process by which radiation is reflected from a grating at different angles depending upon its wavelength.

Monochromators. Figure 21-6a shows the design of a typical grating monochromator. Radiation from a source enters the monochromator via a narrow rectangular opening or *slit*. The radiation is then collimated by a concave mirror, a device that produces a parallel beam that then strikes the surface of a *reflection grating*. For illustrative purposes, the radiation is shown as being made up of just two wavelengths λ_1 and λ_2 where λ_1 is longer than λ_2. The dashed line is the path of λ_1 after it is reflected from the grating. The solid line shows the path of λ_2. Note that the shorter wavelength radiation λ_2 is reflected off the grating at a sharper angle than is λ_1. That is, *angular dispersion* of the radiation takes place. Angular dispersion results from diffraction, which occurs at the reflective surface (see Feature 21-1). The two wavelengths are focused by another concave mirror onto the *focal plane* of the monochromator, where they appear as two images of the entrance slit, one for λ_1 and the other for λ_2 (Figure 21-6b). By rotating the grating, either λ_1 or λ_2 can be focused on the exit slit of the monochromator.

If a detector is located at the exit slit of the monochromator shown in Figure 21-6a and the grating is rotated so that one of the lines shown (say, λ_1) is *scanned* across the slit from $\lambda_1 - \Delta\lambda$ to $\lambda_1 + \Delta\lambda$, where $\Delta\lambda$ is a small wavelength difference, the output of the detector takes the Gaussian shape shown in Figure 21-7. The effective bandwidth of the monochromator, which is defined in the figure, depends upon the size and quality of the dispersing element, the slit widths, and the focal length of the monochromator. A high-quality monochromator will exhibit an effective bandwidth of a few tenths of a nanometer or less in the ultraviolet and visible regions. The effective bandwidth of a monochromator that is satisfactory for most quantitative applications is from about 1 to 20 nm.

Many monochromators are equipped with adjustable slits to permit some control over the bandwidth. A narrow slit decreases the effective

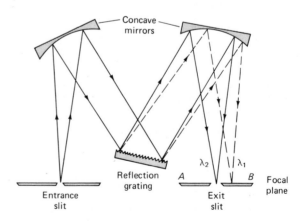

Figure 21-6
(a) View from overhead of a grating monochromator ($\lambda_1 > \lambda_2$). (b) End view of the focal plane.

Feature 21-1

DISPERSION BY REFLECTION GRATINGS

A reflection grating consists of a flat, reflecting aluminum surface that is a few centimeters in length and contains a large number of parallel and closely spaced grooves. The surface is covered with a coating of transparent silica to prevent corrosion of the metal. A magnified cross section of a few of the grooves is shown in Figure 21-8. For ultraviolet and visible radiation, a grating typically contains 1200 to 1400 grooves/mm. For infrared radiation, 10 to 200 grooves/mm is common.

In the grating shown in Figure 21-8, parallel beams of monochromatic radiation labeled 1 and 2 strike two of the broad faces at an incident angle i to the *grating normal*. Maximum constructive interference is shown as occurring at the reflected angle r. It is evident that beam 2 travels a greater distance than beam 1 and this distance is equal to $\overline{CD} - \overline{AB}$. For constructive interference to occur, this distance must equal $n\lambda$:

$$n\lambda = \overline{CD} - \overline{AB}$$

where n, a small whole number, is called the diffraction *order*. Note, however, that angle CAD is equal to angle i and that angle BDA is identical to angle r. Therefore, from trigonometry,

$$\overline{CD} = d \sin i$$

where d is the spacing between the reflecting surfaces. It is also seen that

$$\overline{AB} = -d \sin r$$

The minus sign by convention indicates that the angle of reflection r lies on the opposite side of the grating normal from the incident angle i (as in Figure 21-8); angle r is positive when it is on the same side as i.

Substitution of the last two expressions into the first gives the condition for constructive interference:

$$n\lambda = d(\sin i + \sin r)$$

This equation suggests that several values of λ exist for a given diffraction angle r. Thus, if a first-order line ($n = 1$) of 900 nm is found at r, second-order (450 nm) and third-order (300 nm) lines also appear at this angle. Ordinarily, the first-order line is the most intense; indeed, it is possible to design gratings that concentrate as much as 90% of the incident intensity in this order. The higher-order lines can generally be removed by filters. For example, glass, which absorbs radiation below 350 nm, eliminates the high-order spectra associated with first-order radiation in most of the visible region.

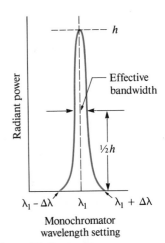

Figure 21-7

Output of an exit slit as monochromator is scanned from $\lambda_1 - \Delta\lambda$ to $\lambda_1 + \Delta\lambda$.

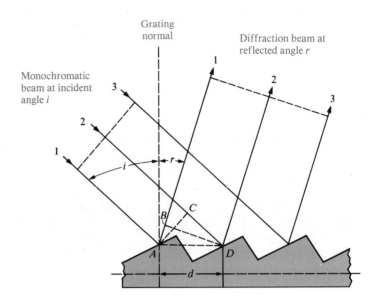

Figure 21-8
Diffraction at a reflection grating.

bandwidth but also diminishes the power of the emergent beam. Thus, the minimum practical bandwidth may be limited by the sensitivity of the detector. For qualitative analysis, narrow slits and minimum effective bandwidths are required if a spectrum has narrow peaks. For quantitative work, on the other hand, wider slits permit operation of the detector system at lower amplification, which in turn provides greater reproducibility of response.

Radiation Filters. Two types of filters are used in photometers. *Absorption filters* are colored glass plates that remove substantial portions of the visible spectrum by molecular absorption. Figure 21-9 shows the transmission characteristics of two absorption filters.

Interference filters provide narrower bands of radiation (as low as 10 nm) and greater transmission of the desired wavelength than do absorption filters. As their name implies, these filters remove most of the radiation in a beam by destructive interference. The transmission characteristics of three interference filters are shown in Figure 21-10.

Sample Containers

The *cells,* or *cuvettes,* that hold samples are fabricated from a material that is transparent to radiation in the spectral region of interest. The best cells have windows that are exactly perpendicular to the light path—an arrangement that minimizes reflection losses. Most instruments are provided with a pair of cells that have been carefully matched with respect to light path and transmission characteristics in order to make possible an accurate comparison of the radiant power transmitted through the sample and the solvent. The most common cell path for work in the ultraviolet and visible regions is 1 cm. Other path lengths ranging from less than 0.1 cm to 20 cm are encountered. Transparent spacers for decreasing the path length of 1-cm cells to 0.1 cm are also available.

For reasons of economy, cylindrical cells are sometimes used for mea-

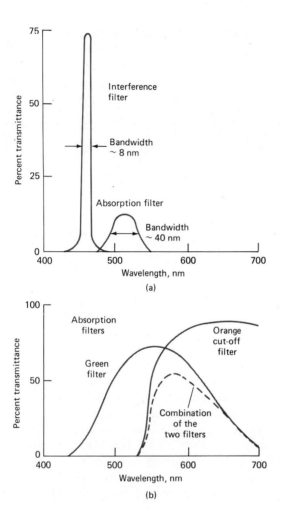

Figure 21-9

Transmission characteristics of two glass absorption filters.

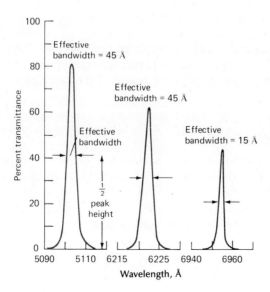

Figure 21-10

Transmission characteristics of three interference filters.

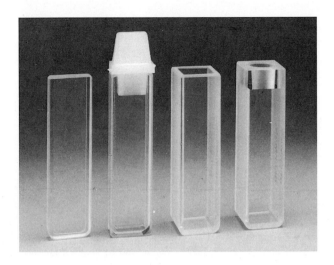

Figure 21-11
Typical commercially available cells.

surements in the ultraviolet and visible regions. Care must be taken to duplicate the position of such cells with respect to the light path; otherwise, variations in path length and in reflection losses will introduce appreciable error.

Figure 21-11 shows some typical commercially available cells.

The quality of absorbance data is critically dependent upon the care and maintenance of matched cells. Scratches, fingerprints, grease, or other deposits will alter the transmission characteristics of a cell markedly. Thorough cleaning before and after use is imperative. Care must be taken to avoid touching the windows during use. Matched cells should be calibrated against each other regularly with an absorbing solution. They should never be dried by heating in an oven or over a flame because such treatment may cause physical damage or a change in path length.

Avoid touching or scratching the windows of cuvettes.

Radiation Detectors

Radiation detectors, or transducers, convert electromagnetic energy to an electrical signal. Radiation detectors should respond rapidly to a broad wavelength range, be sensitive to low levels of radiation, produce an electrical signal that is readily amplified, and have a relatively low noise level. Finally, it is highly desirable that the signal produced by the detector be directly proportional to the power of the radiation it receives; that is,

$$G = kP$$

"Noise" refers to small, random, and unwanted fluctuations in the signal source, detector, amplifier, or readout device. The term has its origins in radio technology, where such fluctuations appear as audible static. Common causes for noise include vibration, pickup from 60-Hz lines, temperature variations, and frequency or voltage fluctuations in the power supply.

where G is the response of the detector in units of current, resistance, or potential and P is the radiant power received by the detector. The constant k measures the sensitivity of the detector in terms of electrical response per unit of radiant power. When this expression is substituted into the definition of absorbance, we obtain

$$A = \log \frac{P_0}{P} = \log \frac{G_0/k}{G/k} = \log \frac{G_0}{G} \qquad (21\text{-}1)$$

Many detectors exhibit a small constant response, known as a *dark current,* even in the absence of radiation. Such instruments are ordinarily equipped with a compensating circuit to permit application of a counter-signal to decrease the dark current to zero just before a measurement is made.

> A dark current is a current produced by a photoelectric detector in the absence of light.

Phototubes. Figure 21-12 shows a typical phototube with its electrical circuitry. The tube consists of a semicylindrical cathode and a wire anode sealed inside an evacuated, transparent envelope. The concave surface of the cathode supports a layer of material that emits electrons upon being irradiated. This coating determines the spectral response of the phototube. Photoemissive materials include the alkali metals and the alkali-metal oxides, alone or combined with other oxides. Application of a potential between the electrodes causes the emitted *photoelectrons* to migrate to the wire anode, producing a current that is proportional to the number of photons. This current is readily amplified and measured.

> Photoelectrons are electrons that are ejected from a photosensitive surface as a result of absorption of electromagnetic radiation.

Photomultiplier Tubes. Figure 21-13 is a schematic of a photomultiplier tube, a device that offers a significant advantage over a phototube when radiation intensities are low. The composition of the cathode surface of a photomultiplier tube is similar to that of a phototube, electrons being emitted upon exposure to radiation. The tube also contains additional electrodes, called *dynodes* (these are labeled 1 to 9 in Figure 21-13). Dynode 1 is maintained at a potential 90 V more positive than the cathode, and photoelectrons are accelerated toward it as a consequence. Each photoelectron that strikes the dynode causes the emission of several additional electrons; the emitted electrons are, in turn, accelerated toward dynode 2, which is 90 V more positive than dynode 1. Again, several electrons are emitted for each electron that strikes dynode 2. By the time this process has been repeated nine times, 10^6 to 10^7 electrons have been generated for each photon that originally struck the cathode; this cascade is finally collected at the anode. The resulting current is then amplified and measured.

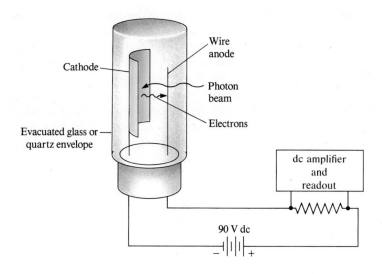

Figure 21-12

A phototube and accessory circuit. The photocurrent induced by the radiation causes a potential drop in the resistor; this drop is then amplified to drive a meter or recorder.

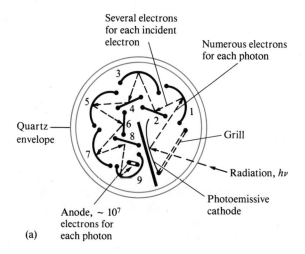

(a)

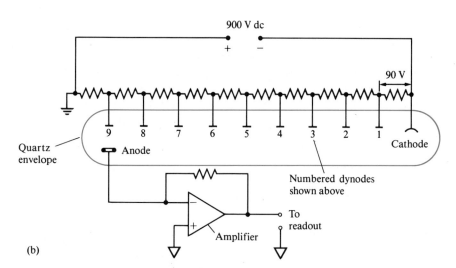

(b)

Figure 21-13

Diagram of a photomultiplier tube. (a) Cross section of the tube. (b) Electrical circuit.

Use of a photomultiplier tube is limited to measurement of low-power radiation because intense light will cause irreversible changes in its performance. For this reason, the device is housed in a light-tight compartment and care is taken to eliminate the possibility of its being exposed even momentarily to daylight or other strong light.

Silicon Photodiodes. A silicon diode is a semiconductor device originally developed by the electronics industry as a current rectifier. Its electrical resistance in one direction is 10^6 to 10^8 times that in the other. Thus, it can be used to convert ac electricity to dc.

A silicon diode is manufactured by forming a *pn* junction on a silicon crystal using modern semiconductor technology. The junction consists of an *n* region, which contains mobile electrons that can conduct electricity, and a *p* region, which contains positive holes that are also free to move and conduct.

Figure 21-14a is a diagram of a silicon diode. The *pn* junction is shown

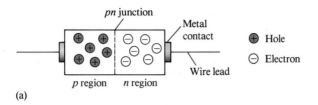

(a)

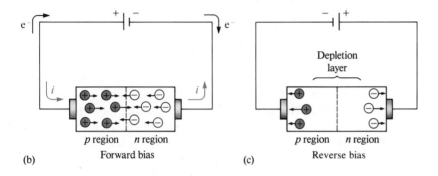

(b)   Forward bias

(c)   Reverse bias

Figure 21-14
(a) Schematic of a silicon diode. (b) Flow of electricity under forward bias. (c) Formation of depletion layer, which prevents flow of electricity under reverse bias.

as a dashed line through the middle of the crystal. Electrical wires are attached to both ends of the device. Figure 21-14b shows the junction in its conduction mode, wherein the positive terminal of a dc source is connected to the *p* region and the negative terminal to the *n* region (the diode is said to be *forward-biased* under these conditions). The excess electrons in the *n* region and the positive holes in the *p* region move toward the junction, where they combine and annihilate each other. The negative terminal of the source injects new electrons into the *n* region, which can continue the conduction process. The positive terminal extracts electrons from the *p* region, thus creating new holes that are free to migrate toward the *pn* junction.

Figure 21-14c illustrates the behavior of a silicon diode under *reverse biasing*. Here, the holes and electrons are drawn away from the junction, leaving a nonconductive *depletion layer*. The conductance under reverse bias is only about 10^{-6} to 10^{-8} times that under forward biasing.

A reverse-biased silicon diode can serve as a radiation detector because ultraviolet and visible photons are sufficiently energetic to create additional electrons and holes when they strike the depletion layer of a *pn* junction. The resulting increase in conductivity is readily measured and is directly proportional to radiant power. A silicon photodiode detector is more sensitive than a simple vacuum phototube but less sensitive than a photomultiplier tube.

21A-3 Spectrophotometer Designs

The components discussed in the previous sections have been combined in various ways to produce dozens of instruments for absorption measurements. The design of these instruments runs the gamut from very simple to very sophisticated; not surprisingly, substantial differences also exist

> In electronics, a bias is a dc voltage that is inserted in series with a circuit element.

> A silicon photodiode is a reverse-biased silicon diode that is used for measuring radiant power.

in their cost. No single instrument is best for all purposes. Selection must be determined by the type of work for which the instrument is intended.

Single-Beam Instruments

Figure 21-15 shows a simple and inexpensive spectrophotometer, the Spectronic 20. The original version of this instrument first appeared on the market in the mid-1950s, and the modified version shown in the figure is still being manufactured and widely sold. Undoubtedly, more of these instruments are currently in use throughout the world than any other single spectrophotometer model.

The Spectronic 20 employs a tungsten-filament light source operated by a stabilized power supply that provides radiation of constant intensity for sufficient time to provide good reproducibility for absorbance readings. The radiation from the source passes through a fixed slit to the surface of a diffraction grating. A portion of the diffracted radiation then passes through an exit slit to the sample or reference cell and thence to a photo-tube. The amplified electrical signal from the detector powers a meter with a $5\frac{1}{2}$-in scale that is linear in percent transmittance and logarithmic in absorbance.

The Spectronic 20 is equipped with an occluder, which is a vane that

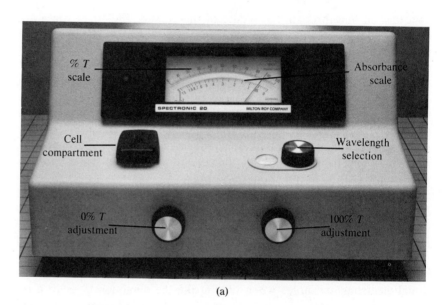

Figure 21-15

(a) The Spectronic 20 spectropho-tometer. (b) Its optical diagram. (Courtesy of Milton Roy Company, Analytical Products Division, Rochester, NY.)

(a)

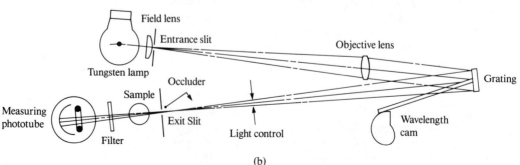

(b)

Feature 21-2

A PROBE-TYPE PHOTOMETER

Figure 21-16 shows the optics of an interesting, commercially available dipping-type filter photometer, which uses optical fibers to transmit light from a source to a layer of solution lying between a mirror and the glass seal at the end of the fiber. The reflected radiation from the mirror passes to a photodiode via a second glass fiber. The photometer uses an amplifier with an electronic chopper that is synchronized with the light source; as a result, the photometer responds only to radiation from the source. Six interference filters can be interchanged by means of a knob located on the instrument panel. Light path lengths of from 1 to 10 mm are available.

Transmittance is measured by first dipping the probe into the solvent and setting the meter to read 100% transmittance. The probe is then dipped into the analyte solution and the percent transmittance read off the meter. This instrument is particularly convenient for photometric titrations (Section 21B-3).

automatically falls between the beam and the detector whenever the cuvette is removed from its holder. The light-control device, located between the grating and the occluder, is a V-shaped aperture that is moved in and out of the beam in order to control the intensity of the beam falling on the photocell.

To obtain a percent transmittance or an absorbance reading, the pointer of the meter is first zeroed with the sample compartment empty so that the occluder blocks off the beam and no radiation reaches the detector. This process is called the *0% T calibration*. A cell containing pure solvent is then inserted into the cell holder, and the pointer is brought to the 100% *T* mark by means of the light-control knob that adjusts the position of the light-control aperture. This adjustment is called the *100% T calibration*. Finally, the sample is placed in the cell compartment, and the percent transmittance or the absorbance is read directly off the meter scale.

> This calibration procedure should be performed before each absorbance measurement.

The spectral range of the Spectronic 20 is 340 to 625 nm (an accessory phototube extends the range to 950 nm). Other specifications for the instrument include an effective bandwidth of 20 nm and a wavelength accuracy of ±2.5 nm.

Single-beam instruments, such as the Spectronic 20, are well suited for quantitative absorption measurements at a single wavelength. Here, simplicity of instrumentation, low cost, and ease of maintenance offer distinct advantages.

Several instrument manufacturers offer single-beam instruments for both ultraviolet and visible measurements. The lower wavelength extremes for these instruments range from 190 to 210 nm and the upper from 800 to 1000 nm. All are equipped with interchangeable tungsten and deuterium or hydrogen lamps. Most employ photomultiplier tubes for detectors and gratings for dispersing radiation. Many are equipped with digital readout devices; others employ large meters. Prices for these instruments range from $2000 to $8000.

Double-Beam Instruments

In double-beam spectrophotometers, radiation from the source is split so that approximately half traverses the cell containing the sample while the other half passes through a second cell containing the blank. Every effort is made to ensure that the path lengths through the two cells are identical. The power of the two beams is then compared electronically and converted to transmittance or absorbance of the solution. Because this comparison is made simultaneously or nearly simultaneously, a double-beam instrument compensates for all but the most transitory electrical fluctuations as well as for irregular performance of the source and the detector. Moreover, this design is readily adapted to automatic recording of spectra.

Figure 21-17 shows the design of one type of double-beam spectrophotometer. Radiation from the monochromator is mechanically split with a rotating *chopper,* which alternately directs the beam through the solvent blank and the sample. The detector thus receives an interrupted signal, which it converts to an ac electrical signal that is readily amplified. A variable *attenuator* is used to match the power of the light passing through the solvent blank with that which passes through the sample. The attenua-

Feature 21-3
DIODE-ARRAY SPECTROMETERS

Diode-array spectrometers (Figure 21-18) are a product of modern integrated circuit wizardry that makes it possible to record an entire ultraviolet or visible spectrum in a second or less. The heart of these instruments is an array of several hundred silicon diode detectors and associated electronic circuitry fabricated side by side on a single silicon chip. Typically, a chip is 1 to 6 cm in length, and the widths of the individual diodes are 0.015 to 0.050 mm. The chip also contains a capacitor and an electronic switch for each diode.

A computer-driven shift register sequentially closes each switch momentarily, which causes each capacitor to be charged to −5 V. Radiation impinging upon the diode surface causes partial discharge of its capacitor. This lost charge is replaced during the next switching cycle. The resulting charging currents, which are proportional to radiant power, are amplified, digitized, and stored in computer memory. With one or two of these diode arrays placed along the length of the focal plane of a grating monochromator, all wavelengths can be monitored simultaneously and data for an entire spectrum collected and stored in a second or less.

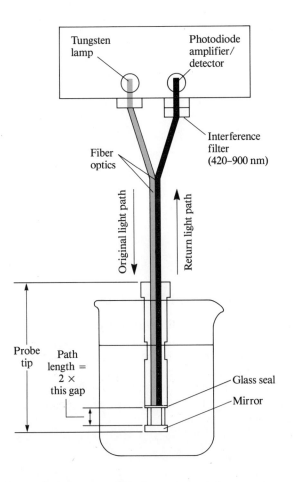

Figure 21-16
A probe-type photometer. (Courtesy of Brinkman Instrument Company, Division of Sybron Corporation, Westbury, NY 11590.)

tor, in turn, is calibrated to register the transmittance (or the absorbance) when the power of the two beams is equal and the output of the detector is a direct current. Since P_0 and P are measured nearly simultaneously, only the most rapid fluctuations in source output have any effect on the measurement.

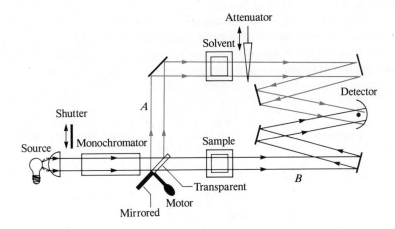

Figure 21-17
Schematic of alternating light paths through a double-beam spectrophotometer: (*A*) through the solvent blank; (*B*) through the sample.

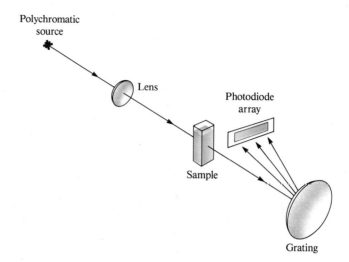

Figure 21-18

Diagram of a spectrophotometer based upon a grating and a photodiode-array detector.

In the health sciences alone, over 95% of all analyses are performed by spectrophotometry. This represents over 3,000,000 daily determinations in the United States alone.

21B   QUANTITATIVE ABSORPTION ANALYSIS

Absorption spectroscopy in the ultraviolet and visible regions is one of the most widely used tools for quantitative analysis.[1] Important characteristics of this rapid and convenient method include wide applicability to organic, inorganic, and biochemical systems; good sensitivity (10^{-4} to 10^{-7} M); moderate to high selectivity; and reasonable accuracy and precision (relative errors in the 1 to 3% range and, with special techniques, as low as a few tenths of a percent).

21B-1   Scope

The applications of absorption analysis are not only numerous but also touch upon every area in which quantitative information is sought. The reader can obtain a notion of the scope of spectrophotometry by consulting the series of review articles published biennially in *Analytical Chemistry*[2] as well as monographs on the subject.[3]

Applications to Absorbing Species

Table 20-2 lists many common organic chromophoric groups. Spectrophotometric determination of organic compounds containing one or more of these groups is thus potentially feasible; many such applications can be found in the literature.

---

[1] For a wealth of detailed, practical information on spectrophotometric practices, see *Techniques in Visible and Ultraviolet Spectrometry*, Vol. I, *Standards in Absorption Spectroscopy*, C. Burgess and A. Knowles, Eds. London: Chapman and Hall, 1981; and J. R. Edisbury, *Practical Hints on Absorption Spectrometry*. New York: Plenum Press, 1968.

[2] L. G. Hargis and J. A. Howell, *Anal. Chem.*, **1980**, *52*, 306R; **1982**, *54*, 171R; **1984**, *56*, 225R; **1986**, *58*, 108R; **1988**, *60*, 131R.

[3] See, for example, E. B. Sandell and H. Onishi, *Photometric Determination of Traces of Metals*, 4th ed. New York: Wiley, 1978; *Colorimetric Determination of Nonmetals*, 2nd ed., D. F. Boltz, Ed. New York: Interscience, 1978; F. D. Snell, *Photometric and Fluorometric Methods of Analysis*. New York: Wiley, 1978.

A number of inorganic species also absorb. We have noted that many transition-metal ions are colored in solution and can thus be determined spectrophotometrically. In addition, a number of other species show characteristic absorption peaks, including nitrite, nitrate, and chromate ions, the oxides of nitrogen, the elemental halogens, and ozone.

Applications to Nonabsorbing Species

Many nonabsorbing analytes can be determined photometrically by causing them to react with chromophoric reagents to produce products that absorb strongly in the ultraviolet or visible region. The successful application of these reagents usually requires that their reaction with the analyte be forced to near-completion.

Typical inorganic reagents include: thiocyanate ion for iron, cobalt, and molybdenum; the anion of hydrogen peroxide for titanium, vanadium, and chromium; and iodide ion for bismuth, palladium, and tellurium. Of even greater importance are organic chelating reagents that form stable colored complexes with cations. Common examples are 1,10-phenanthroline for the determination of iron, dimethylglyoxime for nickel, diethyldithiocarbamate for copper, and diphenylthiocarbazone for lead. Figure 21-19 shows the color-forming reaction for three of these reagents.

Other reagents that react with organic functional groups to produce colors useful for quantitative analysis are available. For example, the red color of the 1:1 complexes between low-molecular-weight aliphatic alcohols and cerium(IV) can be used for the quantitative estimation of such alcohols.

21B-2 Procedural Details

A first step in any photometric or spectrophotometric analysis is the development of conditions that yield a reproducible relationship (preferably linear) between absorbance and analyte concentration.

Wavelength Selection

In order to realize maximum sensitivity, spectrophotometric absorbance measurements are ordinarily made at a wavelength corresponding to an absorption peak because the change in absorbance per unit of concentration is greatest at this point. In addition, the absorption curve is often flat at a maximum, which leads to good adherence to Beer's law (Figure 20-9) and less uncertainty from failure to reproduce precisely the wavelength setting of the instrument.

Variables That Influence Absorbance

Common variables that influence the absorption spectrum of a substance include nature of the solvent, pH of the solution, temperature, high electrolyte concentrations, and presence of interfering substances. The effects of these variables must be known and conditions for the analysis chosen such that the absorbance is not materially affected by small, uncontrolled variations in their magnitudes.

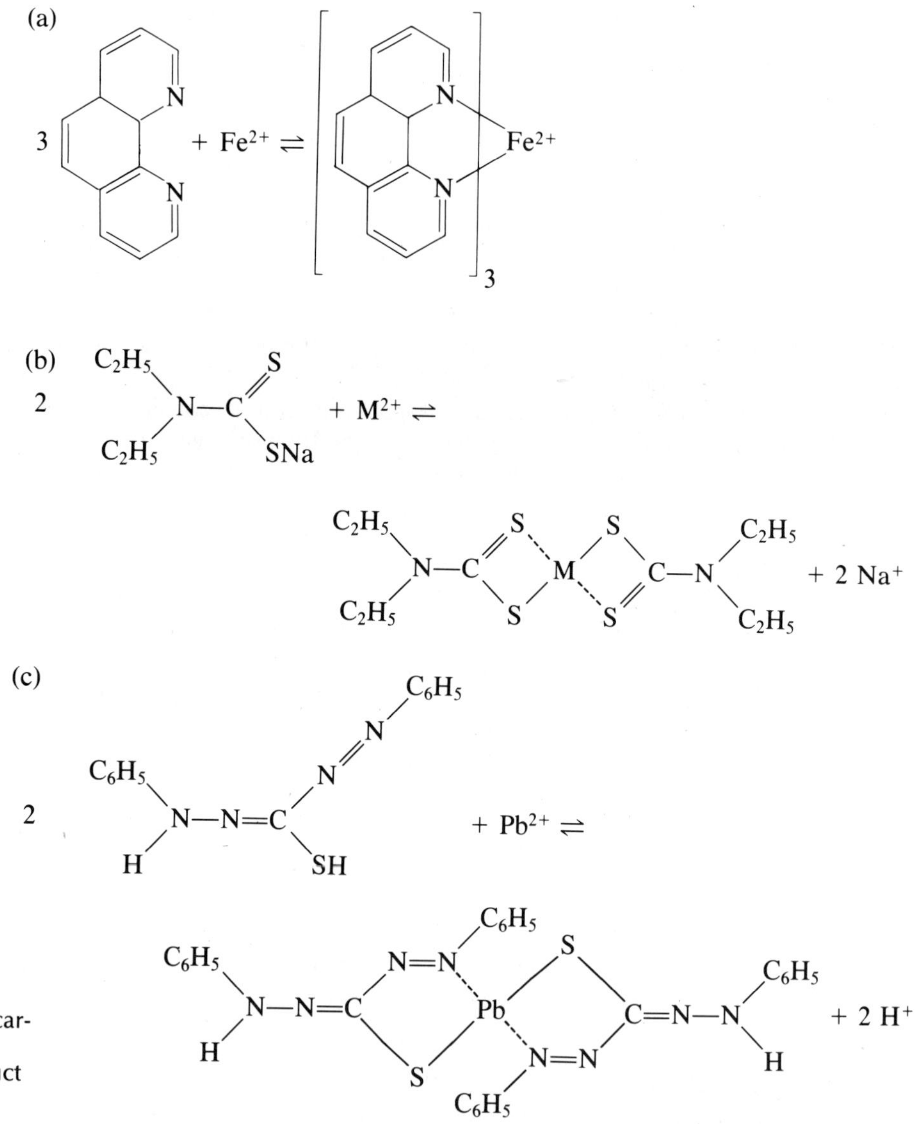

(a)

(b)

(c)

Figure 21-19
Typical chelating reagents for absorption analysis: (a) 1,10-phenanthroline; (b) diethyl-dithiocarbamate; (c) diphenylcar-bazone. (See Section 5D-3 for dimethylglyoxime and its adduct with nickel.)

Cleaning and Handling of Cells

Accurate spectrophotometric analysis requires the use of good-quality, matched cells. These should be regularly calibrated against one another to detect differences that can arise from scratches, etching, and wear. Equally important is the use of proper cell cleaning and drying techniques.

Determination of the Relationship Between Absorbance and Concentration

The calibration standards for a photometric or a spectrophotometric analysis should approximate as closely as possible the overall composition of the samples and should encompass a reasonable range of analyte concentrations. Seldom, if ever, is it safe to assume adherence to Beer's law and use only a single standard to determine molar absorptivity; it is even less prudent to base the results of an analysis on a literature value for molar absorptivity.

The difficulties that attend production of standards with an overall composition closely resembling that of the sample can be formidable, if not insurmountable. Under such circumstances, the *standard-addition approach* may prove useful. Here, a known amount of analyte is introduced to a second aliquot of the sample. Provided Beer's law is obeyed (and this must be confirmed experimentally), the difference in absorbance is used to calculate the analyte concentration of the sample.

Example 21-1

A 2.00-mL urine specimen was treated with reagents to generate color with phosphate, following which the sample was diluted to 100 mL. Photometric measurement for the phosphate in a 25.0-mL aliquot yielded an absorbance of 0.428. Addition of 1.00 mL of a solution containing 0.0500 mg of phosphate to a second 25.0-mL aliquot resulted in an absorbance of 0.517. Use these data to calculate the concentration in milligrams of phosphate in each milliliter of the specimen.

The absorbance of the second measurement must be corrected for dilution. Thus,

$$\text{corrected absorbance} = 0.517 \times \frac{26.0 \text{ mL}}{25.0 \text{ mL}} = 0.538$$

$$\text{absorbance caused by 0.0500 mg phosphate} = 0.538 - 0.428 = 0.110$$

$$\text{weight phosphate in } \frac{25.0 \text{ mL}}{100 \text{ mL}} \text{ of specimen} = \frac{0.428}{0.110} \times 0.050 = 0.195 \text{ mg}$$

Finally, then,

$$\text{mg phosphate/mL of specimen} = \frac{100}{25.0} \times 0.195 \times \frac{1}{2.00} = 0.390$$

Analysis of Mixtures

The total absorbance of a solution at any given wavelength is equal to the sum of the absorbances of the individual components in the solution (Equation 20-10). This relationship makes it possible in principle to determine the concentrations of the individual components of a mixture even if total overlap in their spectra exists. For example, Figure 21-20 shows the spectrum of a solution containing a mixture of species M and species N as well as spectra for the individual components. Clearly, there is no wavelength at which the absorbance is due to just one of these components. To analyze the mixture, molar absorptivities for M and N are first determined at wavelengths λ_1 and λ_2; this step is carried out with enough standards to be sure that Beer's law is obeyed over an absorbance range that encom-

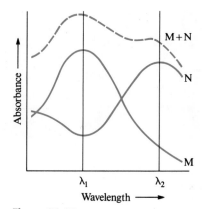

Figure 21-20
Absorption spectrum of a two-component mixture (M + N), with spectra of the individual components.

passes the absorbance of the sample. Note that the wavelengths selected are ones at which the two spectra differ significantly. Thus, at λ_1, the molar absorptivity of component M is much larger than that of component N. The reverse is true for λ_2. To complete the analysis, the absorbance of the mixture is determined at the same two wavelengths. Example 21-2 demonstrates how the composition of the mixture is derived from data of this kind.

Example 21-2

Palladium(II) and gold(III) can be determined simultaneously through reaction with methiomeprazine ($C_{19}H_{24}N_2S_2$). The absorption maximum for the Pd complex occurs at 480 nm, while that for the Au complex is at 635 nm. Molar absorptivity data at these wavelengths are

	Molar Absorptivity, ε	
	480 nm	635 nm
Pd complex	3.55×10^3	5.64×10^2
Au complex	2.96×10^3	1.45×10^4

A 25.0-mL sample was treated with an excess of methiomeprazine and subsequently diluted to 50.0 mL. Calculate the molar concentrations of Pd(II), c_{Pd}, and Au(III), c_{Au}, in the sample if the diluted solution had an absorbance of 0.533 at 480 nm and 0.590 at 635 nm when measured in a 1.00-cm cell.

At 480 nm,

$$0.533 = (3.55 \times 10^3)(1.00)c_{Pd} + (2.96 \times 10^3)(1.00)c_{Au}$$

or

$$c_{Pd} = \frac{0.533 - 2.96 \times 10^3\, c_{Au}}{3.55 \times 10^3}$$

At 635 nm,

$$0.590 = (5.64 \times 10^2)(1.00)c_{Pd} + (1.45 \times 10^4)(1.00)c_{Au}$$

Substitution for c_{Pd} in this expression gives

$$0.590 = \frac{5.64 \times 10^2(0.533 - 2.96 \times 10^3 c_{Au})}{3.55 \times 10^3} + 1.45 \times 10^4\, c_{Au}$$

$$= 0.0847 - 4.70 \times 10^2\, c_{Au} + 1.45 \times 10^4\, c_{Au}$$

$$c_{Au} = (0.590 - 0.0847)/(1.403 \times 10^4) = 3.60 \times 10^{-5}\ M$$

and

$$c_{Pd} = \frac{0.533 - (2.96 \times 10^3)(3.60 \times 10^{-5})}{3.55 \times 10^3} = 1.20 \times 10^{-4}\ M$$

Since the analysis involved a twofold dilution, the concentrations of Pd(II) and Au(III) in the original sample are 7.20×10^{-5} and 2.40×10^{-4} M, respectively.

Mixtures containing more than two absorbing species can be analyzed, in principle at least, if one additional absorbance measurement is made for each added component. The uncertainties in the resulting data become greater, however, as the number of measurements increases. Some of the newer computerized spectrophotometers are capable of minimizing these uncertainties by overdetermining the system; that is, these instruments use many more data points than unknowns and effectively match the entire spectrum of the unknown as closely as possible by deriving synthetic spectra for various concentrations of the components. The derived spectra are then compared with that of the analyte until a close match is found. The spectrum for a standard solution of each component is required, of course.

Computer Application 21-1
SOLVING EXAMPLE 21-2 WITH EUREKA: THE SOLVER

The pair of simultaneous equations in Example 21-2 are readily solved with a Eureka program similar to that described in Computer Application 5-1. The printed report of this calculation has the following appearance:

```
**************************************************************
Eureka: The Solver, Version 1.0
Friday October 6, 1989, 5:56 pm.
Name of input file: C:\EU\SIMULT.EKA
**************************************************************

;Simultaneous Determination of Two Components
;
; At 480 nm
;
0.533 = 3.55e3 * 1.00 * cPd + 2.96e3 * 1.00 * cAu
;
; At 635 nm
;
0.590 = 5.64e2 * 1.00 * cPd + 1.45e4 * 1.00 * cAu
;
**************************************************************

Solution

 Variables    Values

   cAu      = .000036017826

   cPd      = .00012010908




**************************************************************
```

Use Eureka to solve Problems 21-14, 21-16, and 21-22.

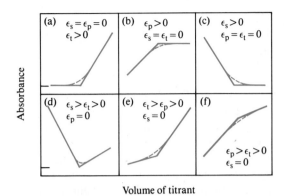

Figure 21-21
Typical photometric titration curves. Molar absorptivities of the substance titrated, product, and titrant are ε_s, ε_p, and ε_t.

21B-3 Spectrophotometric and Photometric Titrations

Spectrophotometric and photometric measurements are useful for locating the equivalence points of titrations.[4] This application of absorption measurements requires either that one or more of the reactants or products absorb radiation or that an absorbing indicator be present.

Titration Curves

A photometric titration curve is a plot of absorbance (corrected for volume change) as a function of titrant volume. If conditions are chosen properly, the curve consists of two straight-line regions with different slopes, one occurring at the outset of the titration and the other located well beyond the equivalence-point region; the end point is taken as the intersection of extrapolated linear portions of the two lines.

Figure 21-21 shows typical photometric titration curves. Figure 21-21a is the curve for the titration of a nonabsorbing species with an absorbing titrant that is decolorized by the reaction. An example is the titration of thiosulfate ion with triiodide ion. The titration curve for the formation of an absorbing product from colorless reactants is shown in Figure 21-21b. An example is the titration of iodide ion with a standard solution of iodate ion to form triiodide. The remaining figures illustrate the curves obtained with various combinations of absorbing analytes, titrants, and products.

In order to obtain titration curves with linear portions that can be extrapolated, the absorbing system(s) must obey Beer's law. Furthermore, absorbances must be corrected for volume changes by multiplying the observed absorbance by $(V + v)/V$, where V is the original volume of the solution and v is the volume of added titrant.

Instrumentation

Spectrophotometric and photometric titrations are ordinarily performed with an instrument that has been modified so that the titration vessel is held in the light path.[5] After the instrument is set to a suitable wavelength,

---

[4]For further information, see J. B. Headridge, *Photometric Titrations*. New York: Pergamon Press, 1961.

[5]Titration flasks and cells for use in the Spectronic 20 are available from the Kontes Manufacturing Corp., Vineland, NJ 08360.

the 0% T adjustment is made in the usual way. With radiation passing through the analyte solution to the detector, the instrument is then adjusted to a convenient absorbance reading by varying the source intensity or the detector sensitivity. Ordinarily, no attempt is made to measure the true absorbance since relative values are perfectly adequate for end-point detection. Titration data are then collected without alteration of the instrument settings. The power of the radiation source and the response of the detector must remain constant during a photometric titration.

The probe-type photometer shown in Figure 21-16 is particularly well suited for performing titrations of this type.

Applications of Spectrophotometric and Photometric Titrations

Spectrophotometric and photometric titrations often provide more accurate results than a direct photometric determination because the data from several measurements are pooled in determining the end point. Furthermore, the presence of other absorbing species may not interfere since only a change in absorbance is being measured.

One advantage of a photometric end point is that the experimental data are taken well away from the equivalence-point region. Consequently, the equilibrium constants of the reactions need not be as favorable as those required for a titration that depends upon observations near the equivalence point (for example, potentiometric or indicator end points). For the same reason, solutions with relatively low analyte concentrations may be titrated.

The photometric end point has been applied to all types of reactions.[6] For example, most standard oxidizing agents have characteristic absorption spectra and thus produce photometrically detectable end points. Although standard acids or bases do not absorb, the introduction of acid/base indicators permits photometric neutralization titrations. The photometric end point has also been used to great advantage in titrations with EDTA and other complexing agents. For example, Figure 21-22 illustrates the successive titration of bismuth(III) and copper(II). At

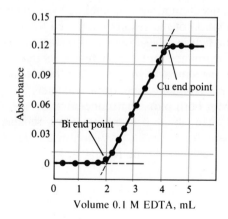

Volume 0.1 M EDTA, mL

Figure 21-22
Photometric titration curve at 745 nm for 100 mL of a solution that was 2.0×10^{-3} M in Bi^{3+} and Cu^{2+}. The titrant was 0.1 M EDTA. (A. L. Underwood, *Anal. Chem.,* **1954,** *26,* 1322. With permission of the American Chemical Society.)

---

[6]See, for example, the review by A. L. Underwood in *Advances in Analytical Chemistry and Instrumentation,* C. N. Reilley, Ed., Vol. 3, pp. 31–104. New York: Interscience, 1964.

Figure 21-23
Experimental curves relating relative concentration uncertainties to absorbance for two spectrophotometers. Data obtained with (a) a Spectronic 20, a low-cost instrument, and (b) a Cary 118, a research-quality instrument. (From W. E. Harris and B. Kratochvil, *An Introduction to Chemical Analysis*, p. 384. Philadelphia: Saunders College Publishing, 1981. With permission.)

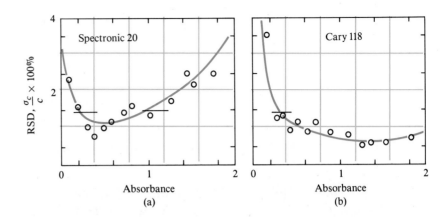

745 nm, the cations, the reagent, and the bismuth complex formed in the first part of the titration do not absorb but the copper complex does. Thus, the solution exhibits no absorbance until essentially all the bismuth has been titrated. With the first formation of the copper complex, an increase in absorbance occurs. The increase continues until the copper equivalence point is reached. Further reagent additions cause no further absorbance change. Two well defined end points result.

21C   ERRORS IN SPECTROPHOTOMETRIC ANALYSES[7]

In most of the analytical methods we have considered so far, the relative error in the result can be directly related to the relative instrumental error associated with the measurement. Thus, in a volumetric analysis, a relative error of 0.03 mL in a 10-mL titration will lead to a relative uncertainty in the concentration of the analyte of 0.03 mL/10 mL, or 3 ppt. This straightforward relationship between instrumental error and concentration error does not exist in spectrophotometric methods, as shown by the experimental data plotted in Figure 21-23. To obtain these plots, a series of standard solutions of an absorbing analyte were analyzed and the relative standard deviation of the results computed for various analyte concentrations. In the figures, these relative standard deviations are plotted as a function of absorbance for two spectrophotometers. One was a Spectronic 20 like the one shown in Figure 21-15; the second was a Cary 118, an expensive, research-quality spectrophotometer.

It is evident from both plots that concentration measurements at absorbances lower than about 0.1 are not very reliable and should be avoided. The reason for the poor precision in this region can be understood by writing Beer's law in the form.

$$A = abc = \log \frac{P_0}{P} = \log P_0 - \log P$$

---

[7]For a detailed discussion of errors associated with spectrophotometric measurements in the visible and ultraviolet regions, see L. D. Rothman, S. R. Crouch, and J. D. Ingle Jr., *Anal. Chem.*, **1972**, *44*, 1375.

Note that concentration is directly proportional to the *difference* between two measured quantities, $\log P_0$ and $\log P$. At low concentrations, $\log P$ is nearly as large as $\log P_0$, and so A is a small difference between two large numbers. Therefore, the relative uncertainties in A and c are large.

Note in Figure 21-23a that large errors are also encountered with the Spectronic 20 when the measured absorbance is above about 1.2. Here, the power of the beam is so very low after it has passed through the analyte solution that it cannot be measured precisely. As shown in Figure 21-23b, this type of error is much less severe with a highly sensitive research-quality instrument.

21D QUESTIONS AND PROBLEMS

*21-1. Define the term "effective bandwidth of a spectrophotometer."

21-2. Describe how a spectrophotometer and a photometer differ from each other.

*21-3. Why do quantitative and qualitative analyses often require different monochromator slit widths?

21-4. Describe the differences between the following and list any particular advantages possessed by one over the other:
  *(a) filters and monochromators as wavelength selectors.
  (b) phototubes and photomultiplier tubes.
  *(c) spectrophotometers and photometers.
  (d) single-beam and double-beam instruments for absorbance measurements.
  *(e) conventional and diode-array spectrometers.

21-5. Define
  *(a) chromophore.
  (b) standard-addition method.
  *(c) noise in a spectrophotometric measurement.
  (d) photometric titration.
  *(e) charge-transfer absorption.
  (f) dark current.

*21-6. Iron(III) reacts with SCN^- to form the red complex $FeSCN^{2+}$. Sketch a photometric titration curve for Fe(III) with SCN^- ion when a photometer with a green filter is used to collect data. Why is a green filter used?

21-7. Sketch a photometric titration curve for the titration of Sn^{2+} with MnO_4^-. What color radiation should be used for this titration? Explain.

*21-8. A portable photometer with a linear response to radiation registered a current of 73.6 μA with a blank solution in the light path. Replacement of the blank with an absorbing solution yielded a response of 24.9 μA. Calculate
  (a) the percent transmittance of the sample solution.
  (b) the absorbance of the sample solution.
  (c) the transmittance to be expected for a solution in which the concentration of the absorber is one third of the original sample solution.

(d) the transmittance to be expected for a solution that has twice the concentration of the sample solution.

21-9. A photometer with a linear response to radiation gave a potential reading of 685 mV with a blank in the light path and 179 mV when the blank was replaced by an absorbing solution. Calculate
  (a) the percent transmittance and absorbance of the absorbing solution.
  (b) the expected transmittance if the concentration of absorber is one half that of the original solution.
  (c) the transmittance to be expected if the light path through the original solution is doubled.

*21-10. A 4.97-g petroleum specimen was decomposed by wet-ashing and subsequently diluted to 500 mL in a volumetric flask. Cobalt was determined by treating 25.00-mL aliquots of this diluted solution as follows:

Reagent Volume, mL			
3.00 ppm Co(II)	Ligand	H₂O	Absorbance
0.00	20.00	5.00	0.398
5.00	20.00	0.00	0.510

Assume that the Co(II)/ligand chelate obeys Beer's law, and calculate the percentage of cobalt in the original sample.

21-11. A two-tablet sample of vitamin/mineral supplement weighing 6.08 g was wet-ashed to eliminate organic matter and then diluted to 1.00 L. Two 10.00-mL aliquots were then analyzed. Calculate the average weight of iron in each tablet, based upon the following information:

Reagent Volume, mL			
1.00 ppm Fe(III)	Ligand	H₂O	Absorbance
0.00	25.00	15.00	0.492
15.00	25.00	0.00	0.571

*21-12. Ethylenediaminetetraacetic acid abstracts bismuth(III) from its thiourea complex:

$$Bi(tu)_6^{3+} + H_2Y^{2-} \rightarrow BiY^- + 6\ tu + 2\ H^+$$

where tu is the thiourea molecule, $(NH_2)_2CS$. Predict the shape of a photometric titration curve based on this process, given that the Bi(III)/thiourea complex is the only species in the system that absorbs at 465 nm, the wavelength selected for the analysis.

21-13. The accompanying data (1.00-cm cells) were obtained for the spectrophotometric titration of 10.00 mL of Pd(II) with 2.44×10^{-4} M Nitroso R (O. W. Rollins and M. M. Oldham, *Anal. Chem.*, **1971**, *43*, 262).

Volume of Nitroso R, mL	A_{500}
0	0
1.00	0.147
2.00	0.271
3.00	0.375
4.00	0.371
5.00	0.347
6.00	0.325
7.00	0.306
8.00	0.289

Calculate the concentration of the Pd(II) solution, given that the ligand-to-cation ratio in the colored product is 2:1.

21-14. A. J. Mukhedkar and N. V. Deshpande (*Anal. Chem.*, **1963**, *35*, 47) report on a simultaneous determination for cobalt and nickel based upon absorption by the two 8-quinolinol complexes. Molar absorptivities are $\varepsilon_{Co} = 3529$ and $\varepsilon_{Ni} = 3228$ at 365 nm and $\varepsilon_{Co} = 428.9$ and $\varepsilon_{Ni} = 0$ at 700 nm. Calculate the concentration of nickel and cobalt in each of the following solutions (1.00-cm cells):

Solution	A_{365}	A_{700}
*1	0.724	0.0710
2	0.614	0.0744
3	0.693	0.0460

*21-15. Solutions of P and Q individually obey Beer's law over a large concentration range. Spectral data for these species in 1.00-cm cells are

	Absorbance	
λ, nm	8.55×10^{-5} M P	2.37×10^{-4} M Q
400	0.078	0.550
420	0.087	0.592
440	0.096	0.599
460	0.102	0.590
480	0.106	0.564
500	0.110	0.515
520	0.113	0.433
540	0.116	0.343
560	0.126	0.255
580	0.170	0.170
600	0.264	0.100
620	0.326	0.055
640	0.359	0.030
660	0.373	0.030
680	0.370	0.035
700	0.346	0.063

(a) Plot an absorption spectrum for a solution that is 8.55×10^{-5} M in P and 2.37×10^{-4} M in Q.
(b) Calculate the absorbance (1.00-cm cells) at 440 nm of a solution that is 4.00×10^{-5} M in P and 3.60×10^{-4} M in Q.
(c) Calculate the absorbance (1.00-cm cells) at 620 nm for a solution that is 1.61×10^{-4} M in P and 7.35×10^{-4} M in Q.

21-16. Use the data in the previous problem to calculate the molar concentration of P and Q in each of the following solutions:

	A_{440}	A_{620}
*(a)	0.357	0.803
(b)	0.830	0.448
*(c)	0.248	0.333
(d)	0.910	0.338
*(e)	0.480	0.825
(f)	0.194	0.315

21-17. The indicator HIn has an acid dissociation constant of 4.80×10^{-6} at ordinary temperatures. The accompanying absorbance data are for 8.00×10^{-5} M solutions of the indicator measured in 1.00-cm cells in strongly acidic and strongly alkaline media.

	Absorbance	
λ, nm	pH 1.00	pH 13.00
420	0.535	0.050
445	0.657	0.068
450	0.658	0.076
455	0.656	0.085
470	0.614	0.116
510	0.353	0.223
550	0.119	0.324
570	0.068	0.352
585	0.044	0.360
595	0.032	0.361
610	0.019	0.355
650	0.014	0.284

Estimate the wavelength at which absorption by the indicator becomes independent of pH (that is, the isosbestic point).

21-18. Calculate the absorbance (1.00-cm cells) at 450 nm of a solution in which the total molar concentration of the indicator described in Problem 21-17 is 8.00×10^{-5} and the pH is *(a) 4.92, (b) 5.46, *(c) 5.93, (d) 6.16.

21-19. What is the absorbance at 595 nm (1.00-cm cells) of a solution that is 1.25×10^{-4} M in the indicator of Problem 21-17 and has a pH of *(a) 5.30, (b) 5.70, *(c) 6.10?

21-20. Several buffer solutions were made 1.00×10^{-4} M in the indicator of Problem 21-17. Absorbance data (1.00-cm cells) are

Solution	A_{450}	A_{595}
*A	0.344	0.310
B	0.508	0.212
*C	0.653	0.136
D	0.220	0.380

Calculate the pH of each solution.

21-21. Construct an absorption spectrum for an 8.00×10^{-5} M solution of the indicator of Problem 21-17 when measurements are made with 1.00-cm cells and $[HIn]/[In^-]$ = *(a) 3, (b) 1, (c) 0.333.

21-22. Molar absorptivity data for the cobalt and nickel complexes with 2,3-quinoxalinedithiol are ε_{Co} = 36,400 and ε_{Ni} = 5520 at 510 nm and ε_{Co} = 1240 and ε_{Ni} = 17,500 at 656 nm. A 0.425-g sample was dissolved and diluted to 50.0 mL. A 25.0-mL aliquot was treated to eliminate interferences; after addition of 2,3-quinoxalinedithiol, the volume was adjusted to 50.0 mL. This solution had an absorbance of 0.446 at 510 nm and 0.326 at 656 nm in a 1.00-cm cell. Calculate the parts per million of cobalt and nickel in the sample.

*21-23. The logarithm of the molar absorptivity for acetone in ethanol is 2.75 at 366 nm. Calculate the range of acetone concentrations that can be used if the percent transmittance is to be greater than 10% and less than 90% with a 1.50-cm cell.

21-24. The logarithm of the molar absorptivity of phenol in aqueous solution is 3.812 at 211 nm. Calculate the range of phenol concentrations that can be used if the absorbance is to be greater than 0.100 and less than 2.000 with a 1.25-cm cell.

*21-25. A standard solution was put through appropriate dilutions to give the concentrations of iron shown below. The iron(II)/1,10-phenanthroline complex was then developed in 25.0-mL aliquots of these solutions, following which each was diluted to 50.0 mL. The following absorbances (1.00-cm cells) were recorded at 510 nm:

Fe(II) Concentration in Original Solution, ppm	A_{510}
4.00	0.160
10.0	0.390
16.0	0.630
24.0	0.950
32.0	1.260
40.0	1.580

(a) Sketch a calibration curve from these data.
(b) Use the method of least squares to derive an equation relating absorbance and the concentration of iron(II).
(c) Calculate the standard deviation about regression.
(d) Calculate the standard deviation of the slope.

21-26. The method developed in Problem 21-25 was used for the routine determination of iron in 25.0-mL aliquots of ground water. Express the concentration (as ppm Fe) in samples that yielded the following absorbance data (1.00-cm cell). Calculate the relative standard deviation of the result. Repeat the calculation assuming the absorbance data are the means of three measurements.

*(a) 0.143 *(c) 0.068 *(e) 1.512
(b) 0.675 (d) 1.009 (f) 0.546

Chapter 22

Molecular-Fluorescence Spectroscopy

Fluorescence is an analytically important emission process in which atoms or molecules are excited by the absorption of a beam of electromagnetic radiation. The excited species then relax to the ground state, giving up their excess energy as photons. In this chapter we consider applications of molecular fluorescence.

Fluorescence emission is over in 10^{-5} s or less. In contrast, phosphorescence may go on for several minutes or even hours. Fluorescence is much more widely used for analyses than phosphorescence.

22A THEORY OF MOLECULAR FLUORESCENCE

Figure 22-1 is the partial energy diagram for a hypothetical molecular species. Three electronic energy states are shown, E_0, E_1, and E_2, where E_0 is the ground state and E_1 and E_2 are excited electronic states. Each of the electronic states is shown as having four excited vibrational states.

Figure 22-1a shows that when molecules of this species are exposed to a band of visible radiation, λ'_1 through λ'_5, all of the five vibrational states of the lower energy electronic state E_1 are momentarily populated. Similarly, when the molecules are irradiated with a band of ultraviolet radiation made up of shorter wavelengths λ''_1 through λ''_5, the five vibrational levels of the higher energy electronic state E_2 become briefly populated.

22A-1 Relaxation Processes

As noted in Section 20C, the lifetime of an excited species is brief because there are several ways an excited atom or molecule can give up its excess

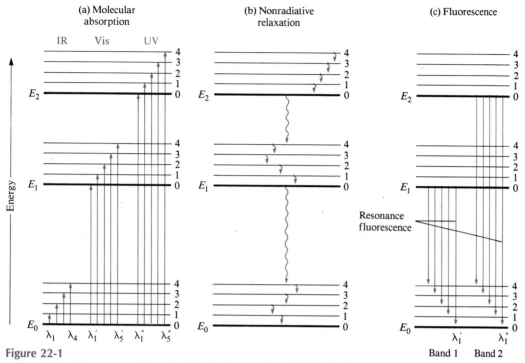

Figure 22-1

Band 1   Band 2

Energy-level diagram showing some of the energy changes that occur during (a) absorption, (b) nonradiative relaxation, and (c) fluorescence by a molecular species.

energy and relax to its ground state. Two of the most important of these mechanisms, nonradiative relaxation and fluorescent relaxation, are illustrated in Figure 22-1b and c.

Two types of nonradiative relaxation are shown in Figure 22-1b. *Vibrational deactivation,* or *relaxation,* depicted by the short wavy arrows between vibrational energy levels, takes place during collisions between excited molecules and molecules of the solvent. During the collisions, the excess vibrational energy is transferred to solvent molecules in a series of steps, as indicated in the figure. The gain in vibrational energy of the solvent is reflected in a slight increase in the temperature of the medium. Vibrational relaxation is such an efficient process that the average lifetime of an excited *vibrational* state is only about 10^{-15} s.

Nonradiative relaxation between the lowest vibrational level of an excited electronic state and the upper vibrational level of another electronic state can also occur. This type of relaxation, which is sometimes called *internal conversion,* is depicted by the two longer wavy arrows in Figure 22-1b. Internal conversion is much less efficient than vibrational relaxation, and so the average lifetime of an electronic excited state is between 10^{-6} and 10^{-9} s. The mechanisms by which this type of relaxation occurs are not fully understood, but the net effect is again a tiny rise in the temperature of the medium.

Figure 22-1c depicts another relaxation process: fluorescence. Note that bands of radiation are produced when molecules fluoresce because the electronically excited molecules can relax to any of the several vibra-

Vibrational relaxation takes 10^{-15} s or less.

Internal conversion takes from 10^{-6} to 10^{-9} s.

tional states of the ground electronic state. Like molecular-absorption bands, molecular-fluorescence bands are made up of a multitude of closely spaced lines that are often difficult to resolve.

Resonance Lines and the Stokes Shift

Note that the lines that terminate the two fluorescence bands on the short-wavelength, or high-energy, side (λ_1' and λ_1'') are identical in energy to the two lines labeled λ_1' and λ_1'' in the absorption diagram (Figure 22-1a). These lines are termed *resonance lines* because the fluorescence and absorption wavelengths are identical. Note also that molecular-fluorescence bands are made up largely of lines that are longer in wavelength, or lower in energy, than the band of absorbed radiation responsible for their excitation. This shift to longer wavelengths is sometimes called the *Stokes shift*.

To develop a better understanding of Stokes shifts, let us consider what occurs when the molecule under consideration is irradiated by a single wavelength λ_5''. As shown in Figure 22-1a, absorption of this radiation promotes an electron into vibrational level 4 of the second excited electronic state E_2. In 10^{-15} s or less, vibrational relaxation to the zero vibrational level of E_2 occurs (Figure 22-1b). At this point, further relaxation can follow either the nonradiative route depicted in Figure 22-1b or the radiative route shown in Figure 21-1c. If the radiative route is followed, relaxation to any of the several vibrational levels of the ground state takes place, giving a band (band 2) of emitted wavelengths, as shown. All of these lines are lower in energy, or longer in wavelength, than the excitation line λ_5''.

Let us now turn to those molecules in excited state E_2 that undergo internal conversion to electronic state E_1. As before, further relaxation can take a nonradiative or a radiative route to the ground state. In the latter case, band 1 of fluorescence is produced. Note that the Stokes shift is from ultraviolet radiation to visible. Note also that band 1 can be produced not only by the mechanism just described but also by the absorption of visible radiation of wavelengths λ_1' through λ_5' (Figure 22-2a).

Relationship Between Excitation Spectra and Fluorescence Spectra

Because the energy differences between vibrational states is about the same for both ground and excited states, the absorption spectrum, or, excitation spectrum, and the fluorescence spectrum for a compound often appear as approximate mirror images of one another with overlap occurring at the resonance line. This effect is demonstrated by the spectra in Figure 22-2.

22A-2 Fluorescent Species

As shown in Figure 22-1, fluorescence is one of several mechanisms by which a molecule returns to the ground state after it has been excited by absorption of radiation. Thus all absorbing molecules have the potential to fluoresce. Most do not, however, because their structure provides radiationless pathways by which relaxation can occur faster than fluorescent emission.

The wavelength of resonance fluorescence is identical to the wavelength of radiation that caused the fluorescence.

The wavelength of Stokes-shifted fluorescence is longer than that of the radiation that caused the fluorescence.

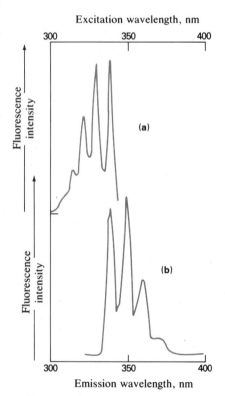

Figure 22-2
Fluorescence spectra for 1 ppm anthracene in alcohol: (a) excitation spectrum; (b) emission spectrum.

The *quantum yield* of molecular fluorescence is simply the ratio of the number of molecules that fluoresce to the total number of excited molecules (or the ratio of photons emitted to photons absorbed). Highly fluorescent molecules, such as fluorescein, have quantum efficiencies that approach unity under some conditions. Nonfluorescent species have efficiencies that are essentially zero.

Quantum yield $= \dfrac{r_f}{r_f + r_r}$, where r_f is the rate of fluorescent relaxation and r_r is the rate of radiationless relaxation.

Fluorescence and Structure

Compounds containing aromatic rings give the most intense and most useful molecular fluorescent emission. While certain aliphatic and alicyclic carbonyl compounds as well as highly conjugated double-bonded structures also fluoresce, their numbers are small in comparison with the number of fluorescent compounds that incorporate aromatic systems.

Most unsubstituted aromatic hydrocarbons fluoresce in solution, with the quantum efficiency increasing with the number of rings and their degree of condensation. The simplest heterocyclics, such as pyridine, furan, thiophene, and pyrrole, do not exhibit molecular fluorescence, but fused-ring structures containing these rings often do.

Substitution on an aromatic ring causes shifts in the wavelength of absorption maxima and corresponding changes in the fluorescence peaks. In addition, substitution frequently affects the fluorescence efficiency. These effects are demonstrated by the data in Table 22-1.

Many aromatic compounds fluoresce.

The Effect of Structural Rigidity

It is found experimentally that fluorescence is particularly favored in rigid molecules. For example, under similar conditions of measurement, the quantum efficiency of fluorene is nearly 1.0, whereas that of biphenyl is about 0.2:

fluorene   biphenyl

The difference in behavior appears to be largely a result of the increased rigidity furnished by the bridging methylene group in fluorene. This rigidity lowers the rate of nonradiative relaxation to the point where relaxation by fluorescence has time to occur. Many similar examples can be cited. In addition, enhanced emission frequently results when fluorescing dyes are adsorbed on a solid surface; here again, the added rigidity provided by the solid may account for the observed effect.

The influence of rigidity has also been invoked to account for the increase in fluorescence of certain organic chelating agents when they are

Table 22-1

EFFECT OF SUBSTITUTION ON THE FLUORESCENCE OF BENZENE DERIVATIVES*

Compound	Relative Intensity of Fluorescence
Benzene	10
Toluene	17
Propylbenzene	17
Fluorobenzene	10
Chlorobenzene	7
Bromobenzene	5
Iodobenzene	0
Phenol	18
Phenolate ion	10
Anisole	20
Aniline	20
Anilinium ion	0
Benzoic acid	3
Benzonitrile	20
Nitrobenzene	0

*In ethanol solution. Taken from W. West, *Chemical Applications of Spectroscopy* (*Techniques of Organic Chemistry*, Vol. IX, p. 730). New York: Interscience, 1956. Reprinted by permission of John Wiley & Sons.

complexed with a metal ion. For example, the fluorescence intensity of 8-hydroxyquinoline is much less than that of the zinc complex:

Temperature and Solvent Effects

In most molecules, the quantum efficiency of fluorescence decreases with increasing temperature because the increased frequency of collision at elevated temperatures improves the probability of collisional relaxation. A decrease in solvent viscosity leads to the same result.

Rigid molecules and complexes tend to fluoresce.

22B   THE EFFECT OF CONCENTRATION ON FLUORESCENCE INTENSITY

The power of fluorescent radiation F is proportional to the radiant power of the excitation beam absorbed by the system:

$$F = K'(P_0 - P) \tag{22-1}$$

where P_0 is the power of the beam incident on the solution and P is its power after it traverses a length b of the medium. The constant K' depends upon the quantum efficiency of the fluorescence. In order to relate F to the concentration c of the fluorescing particle, we write Beer's law in the form

$$\frac{P}{P_0} = 10^{-\varepsilon bc} \tag{22-2}$$

where ε is the molar absorptivity of the fluorescing species and εbc is the absorbance A. By substituting Equation 22-2 into Equation 22-1, we obtain

$$F = K'P_0(1 - 10^{-\varepsilon bc}) \tag{22-3}$$

Expansion of the exponential term in Equation 22-3 leads to

$$F = K'P_0 \left[2.3\varepsilon bc - \frac{(-2.3\varepsilon bc)^2}{2!} - \frac{(-2.3\varepsilon bc)^3}{3!} - \cdots \right] \tag{22-4}$$

Provided $\varepsilon bc = A < 0.05$, all the subsequent terms in the brackets are small with respect to the first, and so we can write

$$F = 2.3K'\varepsilon bcP_0 \tag{22-5}$$

or, at constant P_0,

$$F = Kc \qquad (22\text{-}6)$$

where K is a new constant that is equal to 2.3 $K'\varepsilon b P_0$. Thus, a plot of the fluorescence power of a solution versus the concentration of the emitting species is linear at low concentrations, as shown in Figure 22-3.

When c becomes great enough that the absorbance is larger than about 0.05 (or the transmittance is smaller than about 90%), linearity is lost and F lies below an extrapolation of the straight-line plot. This effect is a result of *self-quenching,* a phenomenon in which analyte molecules absorb the fluorescence produced by other analyte molecules. Indeed, at very high concentrations, F reaches a maximum and then begins to decrease with increasing concentration.

22C   FLUORESCENCE INSTRUMENTS

Figure 22-4 shows a typical configuration for the components of *fluorometers* and *spectrofluorometers*. These components are identical to the ones described in Section 21A-2 for ultraviolet and visible spectroscopy. A fluorometer, like a photometer, employs filters for wavelength selection. Most spectrofluorometers, in contrast, employ a filter to limit the excitation radiation and a grating monochromator to disperse the fluorescence from the sample. A few spectrofluorometers have two monochromator systems, one for the excitation radiation and one for the fluorescence.

As shown in Figure 22-4, fluorescence instruments are usually double-beam in design in order to compensate for fluctuations in source power. The beam to the sample first passes through a primary filter or a primary monochromator, which transmits radiation that causes fluorescence but

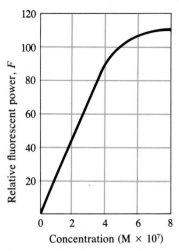

Figure 22-3

Calibration curve for the spectro-fluorometric determination of tryptophan in soluble proteins from the lens of a mammalian eye.

Spectrofluorometers with two mono-chromators are used to obtain fluorescence emission and fluorescence excitation spectra. To obtain emission spectra, the excitation monochromator is set to an absorption peak and the spectrum of the emitted radiation is obtained with the other monochromator. An excitation spectrum is derived by setting the emission monochromator at a fluorescence peak and scanning the sample with the excitation monochromator.

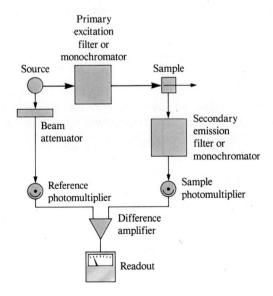

Figure 22-4

Components of a fluorometer or a spectrofluorometer.

excludes or limits radiation that corresponds to the fluorescence wavelengths. Fluorescence radiation is propagated from the sample in all directions but is most conveniently observed at right angles to the excitation beam; at other angles, increased scattering from the solution and the cell walls may cause large errors in the intensity measurement. The emitted radiation reaches a photoelectric detector after passing through the secondary filter or monochromator, which isolates a fluorescence peak for measurement.

The reference beam passes through an attenuator to decrease its power to approximately that of the fluorescence radiation (the power reduction is usually by a factor of 100 or more). The signals from the reference and sample phototubes are then processed by a difference amplifier whose output is displayed on a meter or recorder. Many fluorescence instruments are of the null type, this state being achieved by optical or electrical attenuators.

The sophistication, performance characteristics, and cost of fluorometers and spectrofluorometers differ as widely as do those of the corresponding instruments for absorption measurements. In many regards, filter-type instruments are better suited for quantitative analytical work than are the more elaborate instruments based on monochromators. Generally, fluorometers are more sensitive than spectrofluorometers because filters have a higher radiation throughput than do monochromators. In addition, source and detector can be positioned closer to the sample in the simpler instrument, a factor that enhances sensitivity.

22D   APPLICATIONS OF FLUORESCENCE METHODS

Fluorescence methods are 10 to 1000 times more sensitive than absorption methods.

Fluorescence methods are generally one to three orders of magnitude more sensitive than methods based upon absorption because the sensitivity of the former can be enhanced either by increasing the power of the excitation beam (Equation 22-5) or amplifying the detector signal. Neither of these options improves the sensitivity of methods based upon absorption, however, because the concentration-related parameter in this case is a ratio:

$$c = k \log(P_0/P)$$

where $k = 1/ab$ (Equation 20-8). Increasing P_0 increases P proportionately and thus has no effect on sensitivity. Similarly, increasing the amplification of the detector signal affects the two measured quantities in an identical way and thus provides no improvement.

22D-1   Methods for Inorganic Species

Inorganic fluorometric methods are of two types. Direct methods are based upon the reaction of the analyte with a chelating agent to form a complex that fluoresces. In contrast, indirect methods depend upon the decrease, or *quenching,* of fluorescence of a reagent as a result of its reaction with the analyte. Quenching is used primarily for the determination of anions.

Table 22-2

SELECTED FLUOROMETRIC METHODS FOR INORGANIC SPECIES*

Ion	Reagent	Wavelength, nm		Sensitivity, $\mu g/mL$	Interference
		Absorption	Fluorescence		
Al^{3+}	Alizarin garnet R	470	500	0.007	Be, Co, Cr, Cu, F$^-$, NO$_3^-$, Ni, PO$_4^{3-}$, Th, Zr
F^-	Al complex of Alizarin garnet R (quenching)	470	500	0.001	Be, Co, Cr, Cu, Fe, Ni, PO$_4^{3-}$, Th, Zr
$B_4O_7^{2-}$	Benzoin	370	450	0.04	Be, Sb
Cd^{2+}	2-(o-Hydroxyphenyl)-benzoxazole	365	Blue	2	NH$_3$
Li^+	8-Hydroxyquinoline	370	580	0.2	Mg
Sn^{4+}	Flavanol	400	470	0.1	F$^-$, PO$_4^{3-}$, Zr
Zn^{2+}	Benzoin	—	Green	10	B, Be, Sb, colored ions

*From *Handbook of Analytical Chemistry*, L. Meites, Ed., pp. 6-178 to 6-181. New York: McGraw-Hill, 1963. With permission.

The most successful fluorometric reagents for the determination of cations are aromatic compounds with two or more donor functional groups that permit chelate formation with the metal ion. A typical example is 8-hydroxyquinoline, the structure of which is given in Section 5D-3. A few other fluorometric reagents and their applications are listed in Table 22-2. With most of these reagents, the cation is extracted into a solution of the reagent in an immiscible organic solvent, such as chloroform. The fluorescence of the organic solvent is then measured. For a more complete summary of fluorescent chelating reagents, see the handbook by Meites.[1]

Nonradiative relaxation of transition-metal chelates is so efficient that fluorescence of these species is seldom encountered. It is noteworthy that most transition metals absorb in the ultraviolet or visible region, whereas nontransition-metal ions do not. For this reason, fluorometry often complements spectrophotometry as a method for the determination of cations.

22D-2 Methods for Organic and Biochemical Species

The number of applications of fluorometric methods to organic problems is impressive. Weissler and White have summarized the most important of these in several tables.[2] More than 100 entries are found under the heading *Organic and General Biochemical Substances*, including such diverse compounds as adenine, anthranilic acid, aromatic polycyclic hydrocarbons, cysteine, guanidine, indole, naphthols, certain nerve gases,

[1]L. Meites, *Handbook of Analytical Chemistry*, pp. 6-178 to 6-181. New York: McGraw-Hill, 1963.

[2]A. Weissler and C. E. White, in *Handbook of Analytical Chemistry*, L. Meites, Ed., pp. 6-182 to 6-196. New York: McGraw-Hill, 1963.

proteins, salicylic acid, skatole, tryptophan, uric acid, and warfarin. Some 50 medicinal agents that can be determined fluorometrically are listed. Included among these are adrenaline, alkylmorphine, chloroquin, digitalis principles, lysergic acid diethylamide (LSD), penicillin, phenobarbital, procaine, and reserpine. Methods for the analysis of ten steroids and an equal number of enzymes and coenzymes are also listed in these tables. Some of the plant products listed are chlorophyll, ergot alkaloids, rauwolfia serpentian alkaloids, flavonoids, and rotenone.

Without question, the most important application of fluorometry is in the analysis of food products, pharmaceuticals, clinical samples, and natural products. The sensitivity and selectivity of the method make it a particularly valuable tool in these fields.

22E　QUESTIONS AND PROBLEMS

22-1. Define the following terms:
　　*(a) resonance fluorescence.
　　(b) vibrational relaxation.
　　*(c) internal conversion.
　　(d) quantum yield.
　　*(e) Stokes shift.
　　(f) self-quenching.
*22-2. Why is spectrofluorometry potentially more sensitive than spectrophotometry?
22-3. Which compound below is expected to have a greater fluorescence quantum yield? Explain.

phenolphthalein

fluorescein

*22-4. Why do some absorbing compounds fluoresce and others not?
22-5. Describe the characteristics of organic compounds that fluoresce.
*22-6. Explain why molecular fluorescence often occurs at a wavelength that is longer than the wavelength of the exciting radiation.

22-7. Describe the components of a fluorometer.
*22-8. Why are most fluorescence instruments double-beam in design?
22-9. Why are fluorometers often more useful than spectrofluorometers for quantitative analysis?
*22-10. The reduced form of nicotinamide adenine dinucleotide (NADH) is an important and highly fluorescent coenzyme. It has an absorption maximum at 340 nm and an emission maximum at 465 nm. Standard solutions of NADH gave the following fluorescence intensities:

Concn NADH, μmol/L	Relative Intensity
0.100	2.24
0.200	4.52
0.300	6.63
0.400	9.01
0.500	10.94
0.600	13.71
0.700	15.49
0.800	17.91

(a) Construct a calibration curve for NADH.
(b) Derive a least-squares equation for the plot in part (a).
(c) Calculate the standard deviation of the slope and about regression for the curve.
(d) An unknown exhibits a relative fluorescence of 12.16. Calculate the concentration of NADH.
(e) Calculate the relative standard deviation for the result in part (d).
(f) Calculate the relative standard deviation for the result in part (d) if the reading of 12.16 was the mean of three measurements.

22-11. The following volumes of a solution containing 1.10 ppm of Zn^{2+} were pipetted into five separatory funnels, each of which contained 5.00 mL of an unknown zinc solution: 0.00, 1.00, 4.00, 7.00, and 11.00 mL. Each solution was then extracted with three 5-mL aliquots of CCl_4 containing an excess of 8-hydroxyquinoline. The extracts were then diluted to 25.0 mL and their fluorescence measured. The results were

mL of Std Zn^{2+}	Fluorometer Reading
0.000	6.12
4.00	11.16
8.00	15.68
12.00	20.64

(a) Plot the data.
(b) Derive by least squares an equation for the plot.
(c) Calculate the standard deviation of the slope and about regression.
(d) Calculate the concentration of zinc in the sample.
(e) Calculate a standard deviation for the result in part (d).

*22-12. Fluoride ion quenches the fluorescence of the Al-acid Alizarin Garnet R complex, and the reduction in fluorescence intensity provides a measure of aluminum concentration. To four 10.0-mL aliquots of a water sample were added 0.00, 1.00, 2.00, and 3.00 mL of a standard NaF solution containing 10.0 ppb F^-. Exactly 5.00-mL portions of a solution containing an excess of the aluminum complex were added to each, and the solutions were then diluted to 50.0 mL. The fluorescent intensity of the four solutions plus a blank were:

mL Sample	mL of Std F^-	Meter Reading
5.00	0.00	68.2
5.00	1.00	55.3
5.00	2.00	41.3
5.00	3.00	28.8

(a) Plot the data.
(b) By least squares, derive an equation relating the decrease in fluorescence to the volume of standard reagent.
(c) Calculate the standard deviation of the slope and about regression.
(d) Calculate the ppb F^- in the sample.

CHAPTER 23

ATOMIC SPECTROSCOPY BASED UPON FLAMES

Atomic spectroscopy is used for the qualitative and quantitative determination of perhaps 70 elements. Sensitivities of atomic methods lie typically in the parts-per-million to parts-per-billion range. In addition, these methods are rapid, convenient, and unusually selective.

Spectroscopic determination of atomic species can be performed only on a gaseous medium in which the individual atoms (or sometimes, elementary ions) are well separated from one another. Consequently, the first step in all atomic spectroscopic procedures is *atomization,* a process in which a sample is volatilized and decomposed to produce an atomic gas. The efficiency and reproducibility of the atomization step in large measure determine the method's sensitivity, precision, and accuracy; that is, atomization is by far the most critical step in atomic spectroscopy.

As shown in Table 23-1, several methods are used to atomize samples for atomic spectroscopic studies. The most widely used of these is flame atomization, which is discussed in this chapter. Note that flame-atomized samples produce atomic absorption, emission, and fluorescence spectra. We shall limit ourselves largely to consideration of flame methods based upon absorption and emission spectra.

23A SOURCES OF ATOMIC SPECTRA

Atomic emission, absorption, and fluorescence spectra are much simpler than the corresponding molecular spectra because vibrational and rota-

> Throughout this chapter, the word "atomic" will refer not only to neutral atoms but also to elementary ions, such as K^+, Ba^+, and Al^+.

> Atomization is a process in which a sample is converted to gaseous atoms.

Table 23-1

CLASSIFICATION OF ATOMIC SPECTRAL METHODS

Atomization Method	Typical Atomization Temperature, °C	Basis for Method	Common Name and Abbreviation of Method
Flame	1700–3150	Absorption	Atomic absorption spectroscopy, AAS
		Emission	Atomic emission spectroscopy, AES
		Fluorescence	Atomic fluorescence spectroscopy, AFS
Electrothermal	1200–3000	Absorption	Electrothermal atomic absorption spectroscopy
		Fluorescence	Electrothermal atomic fluorescence spectroscopy
Inductively coupled argon plasma	6000–8000	Emission	Inductively coupled plasma spectroscopy, ICP
		Fluorescence	Inductively coupled plasma fluorescence spectroscopy
Direct-current argon plasma	6000–10,000	Emission	DC plasma spectroscopy, DCP
Electric arc	4000–5000	Emission	Arc-source emission spectroscopy
Electric spark	40,000(?)	Emission	Spark-source emission spectroscopy

tional states and transitions cannot exist in the absence of chemical bonds. Thus atomic emission, absorption, and fluorescence spectra are made up of a limited number of narrow peaks—so narrow, in fact, that they are often referred to as *lines*.

23A-1 Emission Spectra

Figure 23-1 is a partial energy-level diagram for atomic sodium, showing the source of three of its most prominent emission lines. These lines are generated by heating gaseous sodium to 2000 to 3000°C in a flame. The heat promotes the single outer electrons of the atoms from their ground-state 3s orbitals to 3p, 4p, or 5p excited-state orbitals. After a microsecond or less, the excited atoms relax to the ground state, giving up their energy as photons of visible or ultraviolet radiation. As shown, the wavelengths of the emitted radiation are 590, 330, and 285 nm.

23A-2 Absorption Spectra

Figure 23-2a shows three of several absorption peaks for sodium vapor. The source of these peaks is indicated in the partial energy-level diagram shown in Figure 23-2b. Here, absorption of radiation of 285, 330, and 590 nm excites the single outer electron of sodium from its ground-state 3s energy level to the excited 3p, 4p, and 5p orbitals, respectively. After a few microseconds, the excited atoms relax to their ground state by transferring their excess energy to other atoms or molecules in the medium. Alternatively, relaxation may take the form of fluorescence.

23A-3 Fluorescence Spectra

Figure 23-3 reveals that atoms can emit both resonance and Stokes-shifted fluorescence radiation just as can molecular species. For example,

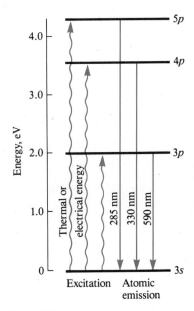

Figure 23-1

Source of three emission lines for sodium.

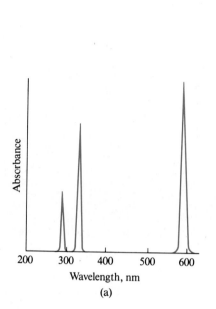

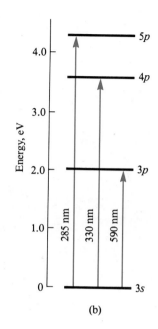

Figure 23-2

(a) Partial absorption specturm for sodium vapor. (b) Electronic transitions responsible for the lines in (a).

when sodium atoms absorb radiation of 285 nm and are promoted to the 5p state, relaxation can take several pathways. One of these is the emission of resonance fluorescence at 285 nm, as shown at the right in Figure 23-3a. Other pathways are shown in Figure 23-3b. One involves radiationless relaxation to the 4p state followed by fluorescence at 330 nm. Alternatively, relaxation to the 3p state may be followed by fluorescence at 590 nm. Thus two Stokes-shifted lines are possible.

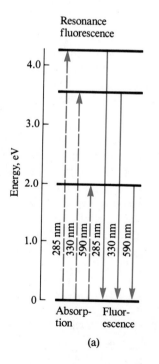

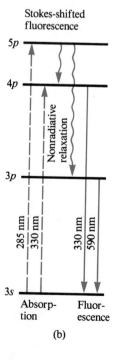

Figure 23-3

Two types of atomic fluorescence shown as solid lines: (a) resonance and (b) Stokes-shifted.

The absorption, emission, and fluorescence spectra for sodium are relatively simple and consist of perhaps 40 peaks. For elements that have several outer electrons that can be excited, absorption spectra may be much more complex and consist of hundreds or even thousands of peaks.

Note that the wavelengths of the absorption, emission, and fluorescence peaks for sodium are all identical.

23B   FLAME ATOMIZATION

In flame atomization, a solution of the analyte (usually aqueous) is converted to a mist, or *nebulized,* and carried into the flame by a flow of gaseous oxidant or fuel. Emission and absorption spectra are generated in the resulting hot, gaseous medium.

Nebulize means to reduce to a fine spray.

23B-1   Flame Atomizers

As shown in Figure 23-4, two types of burners are used in flame spectroscopy: *turbulent-flow burners* and *laminar-flow burners.* Figure 23-4a is a

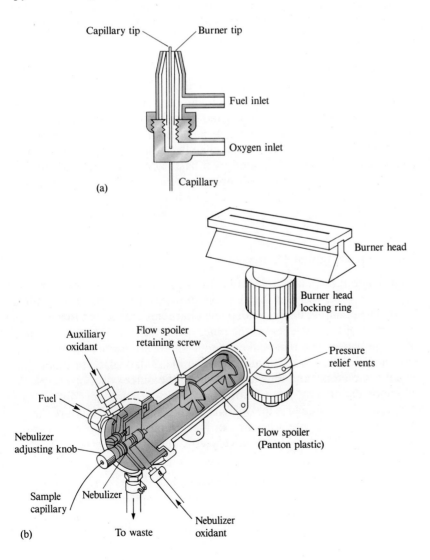

Figure 23-4

Diagrams of burners for atomic flame spectroscopy. (a) Turbulent-flow, or total-consumption, burner. (Courtesy of Beckman Instruments, Fullerton, CA.) (b) Laminar-flow, or premix, burner. (Courtesy of Perkin-Elmer Corporation, Norwalk, CT.)

diagram of a commercially available turbulent-flow, or *total-consumption,* burner. Note that the nebulizer and burner are combined in a single unit. The sample is drawn up the capillary and nebulized by the Venturi action caused by the flow of gases around the capillary tip. Typical sample flow rates are 1 to 3 mL/min.

Turbulent-flow burners offer the advantage of introducing a relatively large and representative sample into the flame. Disadvantages include a short path length through the flame and problems with clogging of the tip. In addition, these burners are noisy from both electronic and auditory standpoints. Turbulent-flow burners find little use in present-day absorption instruments.

Figure 23-4b is a diagram of a typical commercial laminar-flow, or *premix,* burner. The sample is nebulized by the flow of oxidant past a capillary tip. The resulting aerosol is then mixed with fuel and flows past a series of baffles that remove all but the finest droplets. As a result of the baffles, much of the sample collects in the bottom of the mixing chamber, where it drains into a waste container. The aerosol, oxidant, and fuel are then fed into a slotted burner that provides a flame that is usually 5 to 10 cm in length.

Laminar-flow burners provide a relatively quiet flame and a significantly longer path length. These properties tend to enhance sensitivity and reproducibility. Furthermore, clogging is seldom a problem. Disadvantages include a lower rate of sample introduction (which may offset the longer-path-length advantage) and the possibility of selective evaporation of mixed solvents in the mixing chamber, which can lead to analytical uncertainties. Finally, the mixing chamber contains a potentially explosive mixture that can be ignited by a flashback. Note that the burner in Figure 23-4b is equipped with pressure-relief vents for this reason. In addition, the burner head is sometimes held in place by stainless steel cables.

23B-2 Properties of Flames

When a nebulized sample is carried into a flame, the solvent evaporates in the *base region,* which is located just above the tip of the burner (Figure 23-5). The resulting finely divided solid particles are carried to a region in the center of the flame called the *inner cone.* Here, in this hottest part of the flame, gaseous atoms and elementary ions are formed from the solid particles. Excitation of atomic emission also takes place in this region. Finally, the atoms and ions are carried to the outer edge, or *outer cone,* where oxidation may occur before the atomization products disperse into the atmosphere. Because the velocity of the fuel/oxidant mixture through the flame is high, only a fraction of the sample undergoes all these processes; indeed, a flame is not a very efficient atomizer.

Temperatures in Various Types of Flames
Several combinations of fuel and oxidant are employed in flame spectroscopy. Low-temperature flames (1750 to 1850°C), which are obtained with propane or natural gas as the fuel and air as the oxidant, have sufficient

Modern atomic absorption instruments use laminar flow burners almost exclusively.

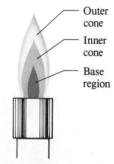

Outer cone
Inner cone
Base region

Figure 23-5
Regions in a flame.

energy to provide satisfactory spectra for the alkali metals but are cool enough to prevent significant ionization of analyte atoms and accompanying loss of sensitivity. Furthermore, few other elements are excited in this type of flame, and so spectra are simple and analyte lines readily isolated even with inexpensive absorption filters. Figure 23-6 shows the temperatures in various parts of a natural gas/air flame.

Air/acetylene flames, which have temperatures in the 2000 to 2400°C range, are useful for many atomic-absorption methods. This mixture is not satisfactory, however, with elements such as aluminum, silicon, the alkaline earths, and vanadium, which form refractory oxides that are incompletely atomized at these temperatures.

To obtain emission spectra for most of the elements, acetylene is employed as fuel with oxygen or nitrous oxide as oxidant; these mixtures produce flames having temperatures of 2950 to 3050°C.

Air/hydrogen (2100°C) and oxygen/hydrogen (2700°C) flames are useful for observing lines in the shorter-wavelength ultraviolet because both flames are transparent in this region. Acetylene and other hydrocarbon flames are not transparent here because of the presence of absorbing carbon-containing molecules.

The Effects of Flame Temperature

Both emission and absorption spectra are affected in a complex way by variations in flame temperature. For both, higher temperatures increase the total atom population of the flame and thus the sensitivity. With certain elements, however, this increase in atom population is more than offset by the loss of atoms by ionization.

Flame temperature also determines the relative number of excited and unexcited atoms in a flame. In an air/acetylene flame, for example, the

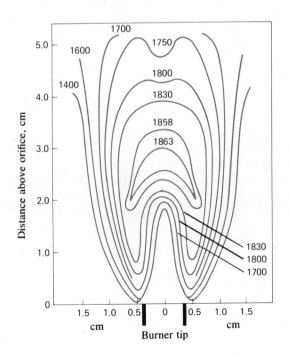

Figure 23-6
Temperature profiles (in °C) for a natural gas/air flame. (From: B. Lewis and G. van Elbe, *J. Chem. Phys.*, **1943**, *11*, 94. With permission.)

Atomic emission methods require closer control of flame temperature than do atomic absorption methods.

ratio of excited to unexcited magnesium atoms can be computed to be about 10^{-8}, whereas in an oxygen/acetylene flame, which is about 700°C hotter, this ratio is about 10^{-6}. Control of temperature is thus of prime importance in flame-emission methods. For example, with a 2500°C flame, a temperature increase of 10°C causes the number of sodium atoms in the excited $3p$ state to increase by about 3%. In contrast, the corresponding *decrease* in the much larger number of ground-state atoms is only about 0.002%. Therefore, emission methods, based as they are on the population of *excited atoms*, require much closer control of flame temperature than do absorption procedures, in which the analytical signal depends upon the number of *unexcited atoms*.

The number of unexcited atoms in a typical flame exceeds the number of excited ones by a factor of 10^3 to 10^{10} or more. This fact suggests that absorption methods should be significantly more sensitive than emission methods. In fact, however, several other variables also influence sensitivity, and the two methods tend to complement each other in this regard. Table 23-2 illustrates this point.

Absorption and Emission Spectra in Flames

Both atomic and molecular emission and absorption spectra are found when a sample is atomized in a flame. A typical flame-emission spectrum is shown in Figure 23-7. Atomic emissions in this spectrum are made up of narrow peaks, such as that for sodium at about 330 nm, potassium at approximately 404 nm, and calcium at 423 nm. Also present are broad emission bands that result from excitation of such molecular species as MgOH, MgO, CaOH, and OH. Here, vibrational transitions superimposed upon electronic transitions produce closely spaced lines that are not completely resolved by the spectrometer.

Atomic absorption spectra are seldom recorded because to do so would require a sophisticated monochromator that produces radiation of unusually narrow bandwidths. Such spectra have much the same appearance as Figure 23-7, with both atomic and molecular absorption peaks being present. The vertical axis of the absorption spectrum is absorbance rather than relative power.

The width of atomic emission peaks is about 10^{-4} Å.

Table 23-2

COMPARISON OF DETECTION LIMITS FOR VARIOUS ELEMENTS BY FLAME-ABSORPTION AND FLAME-EMISSION METHODS

Flame Emission More Sensitive	Sensitivity About the Same	Flame Absorption More Sensitive
Al, Ba, Ca, Eu, Ga, Ho, In, K, La, Li, Lu, Na, Nd, Pr, Rb, Re, Ru, Sm, Sr, Tb, Tl, Tm, W, Yb	Cr, Cu, Dy, Er Gd, Ge, Mn, Mo, Nb, Pd, Rh, Sc, Ta, Ti, V, Y, Zr	Ag, As, Au, B, Be, Bi, Cd, Co, Fe, Hg, Ir, Mg, Ni, Pb, Pt, Sb, Se, Si, Sn, Te, Zn

Adapted with permission from E. E. Pickett and S. R. Koirtyohann, *Anal. Chem.*, **1969**, *41* (14), 42A. Copyright 1969 American Chemical Society.

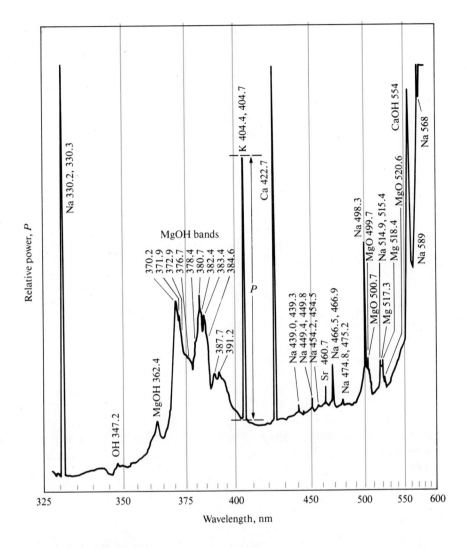

Figure 23-7
Emission spectrum of a brine obtained with an oxyhydrogen flame. (R. Hermann and C. T. J. Alkemade, *Chemical Analysis by Flame Photometry,* 2nd ed., p. 484. New York: Interscience, 1963. With permission.)

Ionization in Flames

All elements ionize to some degree in a flame, which leads to a mixture of atoms, ions, and electrons in the hot medium. For example, when a sample containing barium is atomized, the equilibrium

$$Ba \rightleftarrows Ba^+ + e^-$$

is established in the inner cone of the flame. The position of this equilibrium depends on the temperature of the flame, on the total concentration of barium, and on the concentration of the electrons produced from the ionization of *all elements* in the sample. At the temperatures of the hottest flames (> 3000 K), nearly half of the barium is present in ionic form. The emission and absorption spectra of Ba and Ba^+ are, however, totally different from one another. Thus, in a high-temperature flame, two spectra for barium are generated, one for the atom and one for its ion. This is another reason why control of the flame temperatures is important in flame spectroscopy.

Ionization of an atomic species in a flame is an equilibrium process that can be treated by the law of mass action.

The spectrum of an atom is entirely different from that of its ions.

23C   FLAME ABSORPTION SPECTROSCOPY

Flame atomic absorption spectroscopy is currently the most widely used of all the atomic methods listed in Table 23-1 because of its simplicity, effectiveness, and relatively low cost. The general use of this technique by chemists for elemental analysis began in the early 1950s and grew explosively after that. The reason that atomic-absorption methods were not widely used until that time was directly related to problems created by the very narrow widths of atomic absorption lines.

23C-1   Atomic Absorption Line Widths

The natural width of an atomic absorption or an atomic emission line is on the order of 10^{-5} nm. Two effects, however, cause line widths to be broadened by a factor of 100 (or more). As will become apparent, line broadening is an important consideration in the design of atomic absorption instruments.

Doppler Broadening

Doppler broadening results from the rapid motion of atoms as they emit or absorb radiation. Atoms moving toward the detector emit wavelengths that are slightly shorter than the wavelengths emitted by atoms moving at right angles to the detector (Figure 23-8). This difference is a manifestation of the well known Doppler shift; the effect is reversed for atoms moving away from the detector. The net effect is an increase in the width of the emission line. For precisely the same reason, the Doppler effect also causes broadening of absorption lines. This type of broadening becomes more pronounced as the flame temperature increases because of the consequent increased rate of motion of the atoms.

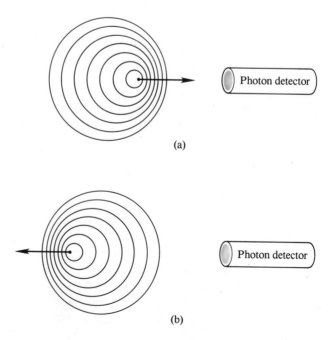

(a)

Figure 23-8

Cause of Doppler broadening. (a) When an atom moves toward a photon detector and emits radiation, the detector sees wave crests more often and detects radiation of higher frequency. (b) When an atom moves away from a photon detector and emits radiation, the detector sees crests less frequently and thus detects radiation of lower frequency.

(b)

Pressure Broadening

Pressure broadening results from collisions between atoms that result in slight variations in their ground-state energies and thus slight energy differences between their ground and excited states. Like Doppler broadening, pressure broadening becomes greater with increases in temperature. Therefore, broader absorption and emission peaks are always encountered at elevated temperatures.

The Effect of Narrow Line Widths on Absorbance Measurements

Since transition energies for atomic absorption lines are unique for each element, analytical methods based upon atomic absorption are highly selective. The narrow lines do, however, create a problem in quantitative analysis that is not encountered in molecular absorption.

No ordinary monochromator is capable of yielding a band of radiation as narrow as the peak width of an atomic absorption line (0.002 to 0.005 nm). As a result, the use of radiation that has been isolated from a continuous source by a monochromator inevitably causes instrumental departures from Beer's law (see the discussion of instrument deviations from Beer's law in Section 20B-4). In addition, since the fraction of radiation absorbed from such a beam is small, the detector receives a signal that is only slightly attenuated (that is, $P \rightarrow P_0$), and consequently the sensitivity of the measurement is reduced.

> The widths of atomic absorption peaks are much less than the effective bandwidths of most monochromators.

The problem created by narrow absorption peaks was surmounted in the mid-1950s by the use of radiation from a source that emits not only a *line of the same wavelength* as the one selected for absorption measurements but also one that is *narrower*. For example, a mercury-vapor lamp is selected as the external radiation source for the determination of mercury. Gaseous mercury atoms electrically excited in such a lamp return to the ground state by *emitting* radiation with wavelengths that are identical to the wavelengths *absorbed* by the analyte mercury atoms in the flame. Since the lamp is operated at a temperature lower than that of the flame, the Doppler broadening and pressure broadening of the mercury emission lines from the lamp are less than the corresponding broadening of the analyte absorption peaks in the hot flame that holds the sample. The effective bandwidths of the lines emitted by the lamp are therefore significantly less than the corresponding bandwidths of the absorption peaks for the analyte in the flame.

Figure 23-9 illustrates the strategy generally used in measuring absorbances in atomic-absorption methods. Figure 23-9a shows four narrow *emission* lines from a typical atomic lamp source. Also shown is how one of these lines is isolated by a filter or monochromator. Figure 23-9b shows the flame *absorption spectrum* for the analyte between the wavelengths λ_1 and λ_2; note that the width of the absorption peak in the flame is significantly greater than the width of the emission line from the lamp. As shown in Figure 23-9c, the intensity of the incident beam P_0 has been decreased to P by passage through the sample. Since the bandwidth of the emission line from the lamp is significantly less than the bandwidth of the absorption peak in the flame, log P_0/P is likely to be linearly related to concentration.

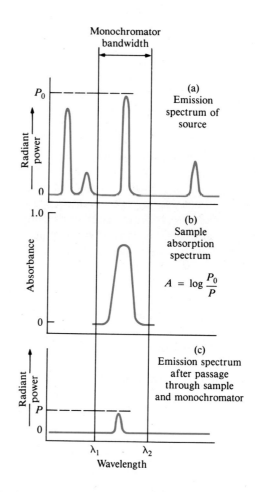

Figure 23-9

Atomic absorption of a resonance line.

23C-2 Line Sources

Two types of lamps are used in atomic-absorption instruments: *hollow-cathode lamps* and *electrodeless discharge lamps*.

Hollow-Cathode Lamps

The most useful radiation source for atomic absorption spectroscopy is the *hollow-cathode lamp,* shown schematically in Figure 23-10. It consists of a tungsten anode and a cylindrical cathode sealed in a glass tube containing an inert gas, such as argon, at a pressure of 1 to 5 torr. The cathode

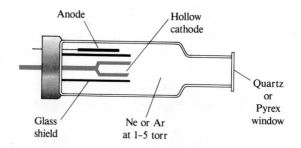

Figure 23-10

Diagram of a hollow-cathode lamp.

is either fabricated from the analyte metal or else serves as a support for a coating of that metal.

The application of a potential of about 300 V across the electrodes causes ionization of the argon and generation of a current of 10 to 30 mA as the argon cations and electrons migrate to the two electrodes. If the potential is sufficiently large, the argon cations strike the cathode with sufficient energy to dislodge some of the metal atoms and thereby produce an atomic cloud; this process is called *sputtering*. A fraction of the sputtered metal atoms is in an excited state and emit their characteristic wavelengths as they return to the ground state. It is important to recall that the atoms producing emission lines in the lamp are at a significantly lower temperature than the analyte atoms in the flame. Thus the emission lines from the lamp are broadened less than the absorption peaks in the flame.

Sputtering is a process in which atoms or ions are ejected from a surface by a beam of charged particles.

The sputtered metal atoms in a lamp eventually diffuse back to the cathode surface (or to the walls of the lamp) and are deposited.

Hollow-cathode lamps for about 40 elements are available from commercial sources. Some are fitted with a cathode containing more than one element; such lamps provide spectral lines for the determination of several species. The development of the hollow-cathode lamp is widely regarded as the single most important event in the evolution of atomic absorption spectroscopy.

Hollow cathode lamps made atomic absorption spectroscopy practical.

Electrodeless Discharge Lamps

Electrodeless discharge lamps are useful sources of atomic line spectra and provide radiant intensities that are usually one to two orders of magnitude greater than their hollow-cathode counterparts. A typical lamp is constructed from a sealed quartz tube containing an inert gas, such as argon, at a pressure of a few torr and a small quantity of the analyte metal (or its salt). The lamp contains no electrode but instead is energized by an intense field of radio-frequency or microwave radiation. The argon ionizes in this field, and the ions are accelerated by the high-frequency component of the field until they gain sufficient energy to excite (by collision) the atoms of the metal whose spectrum is sought.

Electrodeless discharge lamps are available commercially for several elements. Their performance does not appear to be as reliable as that of the hollow-cathode lamp.

Source Modulation

In an atomic-absorption measurement, it is necessary to discriminate between radiation from the source lamp and radiation from the flame. Much of the latter is eliminated by the monochromator, which is always located between the flame and the detector. The thermal excitation of a fraction of the analyte atoms in the flame, however, produces radiation of the wavelength at which the monochromator is set. Since such radiation is not removed, it acts as a potential source of interference.

The effect of analyte emission is overcome by *modulating* the output from the hollow-cathode lamp so that its intensity fluctuates at a constant frequency. The detector thus receives an alternating signal from the hollow-cathode lamp and a continuous signal from the flame and converts

Modulation is defined as changing some property of signal such as frequency, amplitude, or wavelength. In AAS, the frequency of the source is modulated from dc to ac.

these signals to the corresponding types of electric current. A relatively simple electronic system then eliminates the unmodulated dc signal produced by the flame and passes the ac signal from the source to an amplifier and finally to the readout device.

Modulation is most often accomplished by interposing a motor-driven circular chopper between the source and the flame (Figure 23-11). Segments of the metal chopper have been removed so that radiation passes through the device half of the time and is reflected the other half. Rotation of the chopper at a constant speed causes the beam reaching the flame to vary periodically from zero intensity to some maximum intensity and then back to zero.

As an alternative, the power supply for the source can be designed for intermittent (or ac) operation.

> Beam choppers are widely used in spectroscopy to produce pulsed beams of radiation.

23C-3 Instruments

An atomic-absorption instrument contains the same basic components as an instrument designed for molecular-absorption measurements: a source, a sample container (here, a flame reservoir), a wavelength selector, and a detector/readout system. Both single- and double-beam instruments are offered by numerous manufacturers. The range of sophistication and the cost (upward from a few thousand dollars) are both substantial.

Photometers

As a minimum, an instrument for atomic absorption spectroscopy must be capable of providing a sufficiently narrow bandwidth to isolate the line chosen for a measurement from other lines that may interfere with or diminish the sensitivity of the method. A photometer equipped with a hollow-cathode source and filters is satisfactory for measuring concentrations of the alkali metals, which have only a few widely spaced resonance lines in the visible region. A more versatile photometer is sold with readily interchangeable interference filters and lamps. Each element has a separate filter and lamp. Satisfactory results for the analysis of 22 metals are claimed.

Spectrophotometers

Most atomic-absorption measurements are made with instruments equipped with an ultraviolet/visible grating monochromator. Figure 23-11 is a schematic of a typical double-beam instrument. Radiation from the

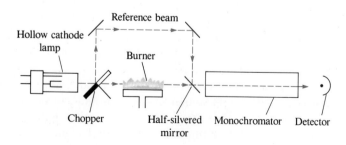

Figure 23-11

Optical paths in a double-beam atomic absorption spectrophotometer.

hollow-cathode lamp is chopped and mechanically split into two beams, one of which passes through the flame and the other around the flame. A half-silvered mirror returns both beams to a single path, by which they pass alternately through the monochromator and to the detector. A signal processor then separates the ac signal generated by the chopped light source from the dc signal produced by the flame. The logarithm of the ratio between the reference and sample components of the ac signal is then computed and sent to the readout device for display as absorbance.

23C-4 Interferences

Two types of interference are encountered in atomic-absorption methods. *Spectral interferences* occur when particulate matter from the atomization scatters the incident radiation from the source or when the absorption wavelength of an interfering species is so close to the analyte wavelength that overlap of absorption peaks occurs. *Chemical interferences* result from various chemical processes that occur during atomization and alter the absorption characteristics of the analyte.

Spectral Interferences

Interference due to overlapping lines is rare because the emission lines of hollow-cathode sources are so very narrow. Spectral interferences do arise, however, from the presence of either *molecular* combustion products that exhibit broad-band absorption or particulate products that scatter radiation. Both diminish the power of the transmitted beam and lead to positive analytical errors. Where the source of these products is the fuel/oxidant mixture alone, corrections are readily made with a blank aspirated into the flame.

> Spectral interferences are encountered in AAS due to scattering by potential products in the flame and absorption by molecular species in the flame.

A much more troublesome problem is encountered when the source of absorption or scattering originates in the sample matrix. In this type of interference, the power of the transmitted beam P is reduced by the matrix components, but the incident beam power P_0 is not; a positive error in absorbance and thus concentration results.

Fortunately, spectral interferences by matrix products are not widely encountered with flame atomization and usually can be eliminated by variations in such analytical parameters as temperature and fuel-to-oxidant ratio. Alternatively, if the source of interference is known, an excess of the interfering substance can be added to both sample and standards; provided the excess is large with respect to the concentration from the sample matrix, the contribution of the latter will become insignificant. The added substance is sometimes called a *radiation buffer*.

> A radiation buffer is a substance that is added in large excess to both samples and standards in atomic spectroscopy to swamp the effect of matrix species and thus prevent interference.

Chemical Interferences

Anions often interfere in flame methods by forming compounds of low volatility with the analyte, thus decreasing the rate at which it is atomized. Low results are the consequence. An example is the decrease in calcium absorbance observed with increasing concentrations of sulfate or phosphate ions, which form nonvolatile compounds with calcium ion.

Interferences due to the formation of species of low volatility can often be eliminated or moderated by use of higher temperatures. Alternatively,

> Chemical interferences in AAS arise from reactions between the analyte and the interference that reduce the analyte concentration in the flame.

Releasing agents are cations that react selectively with anions and thus prevent their interfering in the determination of a cationic analyte.

Protective agents are reagents that form stable volatile complexes with an analyte and thus prevent interference by anions that form nonvolatile compounds with it.

releasing agents, which are cations that react preferentially with the interference and prevent its interaction with the analyte, can be introduced. For example, the addition of excess strontium or lanthanum ion minimizes interference by phosphate in the determination of calcium. Here, the strontium or lanthanum replaces the analyte in the nonvolatile compound formed with the interfering species.

Protective agents prevent interference by preferentially forming stable but volatile species with the analyte. Three common reagents for this purpose are EDTA, 8-hydroxyquinoline, and APDC (the ammonium salt of 1-pyrrolidinecarbodithioic acid). For example, the presence of EDTA has been shown to eliminate interference by silicon, phosphate, and sulfate in the determination of calcium.

23C-5 Ionization Effects

The ionization of atoms and molecules is usually inconsequential in combustion mixtures that involve air as the oxidant. In high-temperature flames, however, where oxygen or nitrous oxide serves as the oxidant, ionization becomes appreciable, and a significant concentration of free electrons exists as a consequence of the equilibrium

$$M \rightleftarrows M^+ + e^-$$

where M represents a neutral atom or molecule and M^+ is its ion. Ordinarily, the spectrum of M^+ is quite different from that of M, so that ionization of the analyte leads to low results. It is important to appreciate that treating the ionization process as an equilibrium—with free electrons as one of the products—implies that the degree of ionization of an analyte atom is strongly influenced by the presence of other ionizable metals in the flame. Thus, if the medium contains not only species M but species B as well, and if B ionizes according to the equation

$$B \rightleftarrows B^+ + e^-$$

An ionization suppressor is an easily ionized species that represses ionization of an analyte by providing a high concentration of electrons in a flame.

then the degree of ionization of M is decreased by the mass-action effect of the electrons formed from B. The errors caused by analyte ionization can frequently be eliminated by addition of an *ionization suppressor,* which provides a relatively high concentration of electrons to the flame; suppression of analyte ionization results. Potassium salts are frequently used as ionization suppressors because of the low ionization energy of this element.

23C-6 Quantitative Flame-Absorption Analyses

Atomic absorption spectroscopy provides a sensitive means of determining more than 60 elements. The method is well suited for routine measurements by relatively unskilled operators.

Region of Flame for Quantitative Measurements
Figure 23-12 shows the absorbance of three elements as a function of distance above the burner tip. For magnesium and silver, the initial rise in

absorbance is a consequence of the longer exposure to the heat, which leads to a greater concentration of atoms in the radiation path. The absorbance for magnesium, however, reaches a maximum near the center of the flame and then falls off as oxidation to magnesium oxide takes place. The effect is not seen with silver because this element is much more resistant to oxidation. For chromium, which forms very stable oxides, maximum absorbance lies immediately above the burner tip. For this element, oxide formation begins as soon as chromium atoms are formed.

It is clear from Figure 23-12 that the part of a flame to be used in an analysis must vary from element to element and furthermore that the position of the flame with respect to the source must be reproduced closely during calibration and analysis. Generally, the flame position is adjusted to yield a maximum absorbance reading.

> The analyte absorbance varies from one part of a flame to another.

Calibration
Quantitative atomic-absorption methods are usually based on calibration curves, which in principle are linear. Departures from linearity occur, however, and analyses should *never* be based on the measurement of a single standard with the assumption that Beer's law is being followed. In addition, the production of an atomic vapor involves a sufficient number of uncontrollable variables to warrant measuring the absorbance of at least one standard solution each time an analysis is performed. Any deviation of the standard from its original calibration value can then be applied as a correction to the analytical results.

> In AAS, it is common practice to run at least one standard every time an analysis is performed.

Standard-Addition Method
The standard-addition method is extensively used in atomic absorption spectroscopy. In this procedure, two or more aliquots of the sample are transferred to volumetric flasks. One is diluted to volume directly, and a known amount of analyte is introduced into the other before dilution to the same volume. The absorbance of each is measured (several different standard additions are recommended if the method is unfamiliar). If a linear relationship exists between absorbance and concentration (and this must be verified experimentally), the following relationships apply:

$$A_x = \frac{kV_x c_x}{V_T}$$

$$A_T = \frac{kV_x c_x}{V_T} + \frac{kV_s c_s}{V_T} \qquad (23\text{-}1)$$

where V_x and c_x are the volume and concentration of the analyte solution, V_s and c_s are the volume and concentration of the standard, V_T is the total volume, k is a proportionality constant, and A_x and A_T are the absorbances of the sample alone and the sample plus standard, respectively. These two equations can be combined to give

$$c_x = \frac{A_x}{A_T - A_x} \times \frac{c_s V_s}{V_x}$$

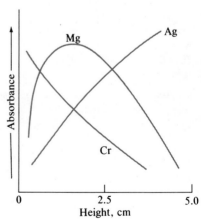

Figure 23-12

Flame-absorbance profile for three elements.

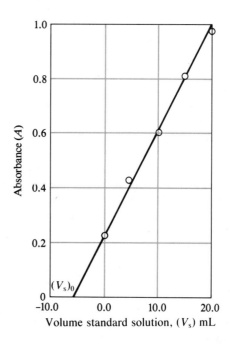

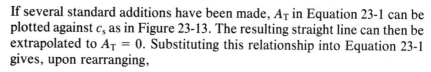

Figure 23-13

A plot of Equation 23-1 for a standard-addition data.

If several standard additions have been made, A_T in Equation 23-1 can be plotted against c_s as in Figure 23-13. The resulting straight line can then be extrapolated to $A_T = 0$. Substituting this relationship into Equation 23-1 gives, upon rearranging,

$$c_x = \frac{-c_s(V_s)_0}{V_x}$$

where $(V_s)_0$ is the extrapolated volume of standard. Use of the standard-addition method tends to compensate for variations caused by spectral and chemical interferences in the analyte solution.

Sensitivity and Accuracy

Table 23-3 shows detection limits for a number of common elements by flame atomic absorption and also by flame-emission methods.

Under usual conditions, the relative error of flame-absorption analysis is of the order of 1 to 2%. With special precautions, this figure can be lowered to a few tenths of one percent.

23D   FLAME EMISSION SPECTROSCOPY

Atomic emission spectroscopy employing a flame (also called flame emission spectroscopy or flame photometry) has found widespread application in elemental analysis. Its most important uses are in the determination of sodium, potassium, lithium, and calcium, particularly in biological fluids and tissues. Because of its convenience, speed, and relative freedom from interferences, flame emission spectroscopy has become the method of choice for these elements, which are otherwise difficult to determine. The

Gustav Robert Kirchhoff (1824–1877) was a German physicist who, along with his colleague, chemist Robert Wilhelm Bunsen (1811–1899), discovered spectroscopic analysis. Bunsen's burner was used to atomize samples of the elements, and Kirchhoff's prism spectroscope was used to analyze the light given off by the incandescent samples. Using this technique, they were able to identify several new elements and demonstrate the presence of a number of elements in the sun by analyzing sunlight. This discovery is depicted on the Vatican stamp shown above. Shown on the stamp are the sun's corona, one of Kirchhoff's spectroscopes, and the absorption spectrum of hydrogen.

method has also been applied, with various degrees of success, to the determination of perhaps half the elements in the periodic table.

In addition to its quantitative applications, flame emission spectroscopy is also useful for qualitative analysis. Complete spectra are readily recorded; identification of the elements present is then based upon the peak wavelengths, which are unique for each element. In this respect, flame emission has a clear advantage over flame absorption, which does not provide complete absorption spectra because of the discontinuous nature of the radiation sources that must be used.

23D-1 Instruments

Instruments for flame-emission work are similar in design to flame-absorption instruments except that in the former the flame acts as the radiation source; a hollow-cathode lamp and chopper are therefore unnecessary. Much of the early work in atomic-emission analyses was accomplished with turbulent-flow burners. Laminar-flow burners, however, are becoming more and more widely used.

Photometers

Simple filter photometers often suffice for routine determinations of the alkali and alkaline earth metals. A low-temperature flame is employed to prevent excitation of most other metals. As a consequence, the spectra

One of Kirchhoff's early prism spectro-scopes. From his paper in Abhandl. Berlin Akad., 1862, 227.

Table 23-3

DETECTION LIMITS OF FLAME SPECTROSCOPIC METHODS FOR SELECTED ELEMENTS*

Element	Absorption§	Emission§
Al	30	5
As	100	0.0005
Ca	1	0.1
Cd	1	800
Cr	3	4
Cu	2	10
Fe	5	30
Hg	500	0.0004
Mg	0.1	5
Mn	2	5
Mo	30	100
Na	2	0.1
Ni	5	20
Pb	10	100
Sn	20	300
V	20	10
Zn	2	0.0005

*From V. A. Fassel and R. N. Kniseley, *Anal. Chem.,* **1974,** *46,* 1111A. With permission of the American Chemical Society.

§All values in nanograms/milliliter = 10^{-3} μg/mL = 10^{-3} ppm.

Atomic-emission and atomic-absorption instruments are similar except no lamp source is required for emission measurements.

are simple, and interference filters can be used to isolate the desired emission line.

Several instrument manufacturers supply flame photometers designed specifically for the analysis of sodium, potassium, and lithium in blood serum and other biological samples. In these instruments, the radiation from the flame is split into three beams of approximately equal power. Each beam then passes into a separate photometric system consisting of an interference filter (which transmits an emission line of one of the elements while absorbing those of the other two), a phototube, and an amplifier. The outputs can be measured separately if desired. Ordinarily, however, lithium serves as an *internal standard* for the analysis. For this purpose, a fixed amount of lithium is introduced into each standard and sample. The output ratio between the sodium transducer and the lithium transducer and that between the potassium transducer and the lithium transducer then serve as analytical parameters. This system provides improved accuracy because the intensities of the three lines are affected in the same way by most analytical variables, such as flame temperature, fuel flow rate, and background radiation. Clearly, lithium must be absent from the sample.

> An internal standard is a pure substance that is introduced in known amount into each standard and sample of the analyte. The ratio of the analyte signal to the internal standard signal is then used to determine the analyte concentration.

Automated Flame Photometers

Fully automated photometers are now widely used in clinical laboratories for the determination of sodium and potassium. In these instruments, samples are withdrawn sequentially from a sample turntable, dialyzed to remove protein and particulates, diluted with a lithium internal standard, and aspirated into a flame. Calibration is performed automatically after every few samples.

Spectrophotometers

For nonroutine analysis, a recording ultraviolet/visible spectrophotometer with a bandwidth of perhaps 0.05 nm is desirable. Such an instrument

Figure 23-14

A portion of the emission spectrum for potassium in a hydrogen/oxygen flame. The potassium emission is superimposed on the background emission from the flame. For clarity, the spectrum for the blank has been displaced downward.

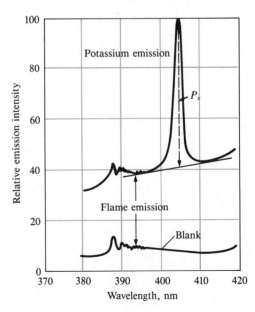

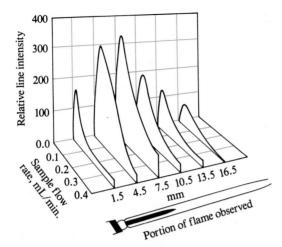

Figure 23-15
Flame profile for calcium line in a cyanogen/oxygen flame for different sample flow rates. (Reprinted with permission from K. Fuwa, R. E. Thiers, B. L. Vallee, and M. R. Baker, *Anal. Chem.*, **1959**, *31*, 2041. Copyright 1959 American Chemical Society.)

can provide complete emission spectra that are useful for the identification of the elements present in a sample. Figure 23-7 is an example of a typical flame emission spectrum excited by an oxyhydrogen flame. The sample was a brine, and spectral lines and bands for several elements are identified. Figure 23-14 illustrates how background corrections are made with a recorded spectrum.

23D-2 Interferences

The interferences encountered in flame emission spectroscopy have the same sources as those encountered in atomic-absorption methods (Section 23C-4); the severity of any given interference often differs for the two procedures, however.

23D-3 Analytical Techniques

Figure 23-15 demonstrates that in flame emission spectroscopy, as in flame absorption procedures, care is required in positioning the flame if maximum sensitivity and precision are to be realized.

The analytical techniques for flame emission spectroscopy are similar to those described for atomic absorption spectroscopy. Both calibration curves and the standard-addition method are employed. In addition, internal standards can be used to compensate for flame variables.

23E QUESTIONS AND PROBLEMS

*23-1. Describe the basic differences between atomic emission and atomic absorption spectroscopy.

23-2. Define *(a) atomization. (b) pressure broadening. *(c) Doppler broadening. (d) turbulent-flow nebulizer. *(e) laminar-flow nebulizer. (f) hollow-cathode lamp. *(g) sputtering. (h) ionization suppressor. *(i) spectral interference. (j) chemical interference. *(k) radiation buffer. (l) releasing agent. *(m) protective agent.

*23-3. Why is atomic emission more sensitive to flame instability than atomic absorption?

23-4. Why is source modulation needed for atomic absorption spectroscopy?

*23-5. In a hydrogen/oxygen flame, an atomic absorption peak for iron decreased in the presence of large concentrations of sulfate ion.
  (a) Suggest an explanation for this observation.
  (b) Suggest three possible methods for overcoming

the potential interference of sulfate in a quantitative determination of iron.

23-6. Why are the lines from a hollow-cathode lamp generally narrower than the lines emitted by atoms in a flame?

*23-7. In the concentration range from 500 to 2000 ppm of U, a linear relationship is observed between absorbance at 351.5 nm and concentration. At lower concentrations, the relationship is nonlinear unless about 2000 ppm of an alkali metal salt is introduced. Explain.

23-8. What is the purpose of an internal standard in flame-emission methods?

*23-9. A 5.00-mL sample of blood was treated with trichloroacetic acid to precipitate proteins. After centrifugation, the resulting solution was brought to pH 3 and extracted with two 5-mL portions of methyl isobutyl ketone containing the organic lead-complexing agent APCD. The extract was aspirated directly into an air/acetylene flame and yielded an absorbance of 0.502 at 283.3 nm. Five-milliliter aliquots of standard solutions containing 0.400 and 0.600 ppm of lead were treated in the same way and yielded absorbances of 0.396 and 0.599. Calculate the parts per million of lead in the sample assuming that Beer's law is followed.

23-10. The sodium in a series of cement samples was determined by flame emission spectroscopy. The flame photometer was calibrated with a series of standards containing 0, 20.0, 40.0, 60.0, and 80.0 μg Na_2O per milliliter. The instrument readings for these solutions were 3.1, 21.5, 40.9, 57.1, and 77.3.
(a) Plot the data.
(b) Derive a least-squares line for the data.
(c) Calculate standard deviations for the slope and about regression for the line in (b).
(d) The following data were obtained for replicate 1.00-g samples of cement dissolved in HCl and diluted to 100.0 mL after neutralization:

Replicate	Blank	Emission Reading		
		Sample A	Sample B	Sample C
1	5.1	28.6	40.7	73.1
2	4.8	28.2	41.2	72.1
3	4.9	28.9	40.2	spilled

Calculate the % Na_2O in each sample. What are the absolute and relative standard deviations for the average of each determination?

*23-11. The chromium in an aqueous sample was determined by pipetting 10.0 mL of the unknown into each of five 50.0-mL volumetric flasks. Various volumes of a standard containing 12.2 ppm Cr were added to the flasks, and the solutions were then diluted to volume.

Unknown, mL	Standard, mL	Absorbance
10.0	0.0	0.201
10.0	10.0	0.292
10.0	20.0	0.378
10.0	30.0	0.467
10.0	40.0	0.554

(a) Plot absorbance as a function of volume of standard V_s.
(b) Derive an expression relating absorbance to the concentrations of standard and unknown (c_s and c_x) and the volumes of the standard and unknown (V_s and V_x) as well as the volume to which the solutions were diluted (V_T).
(c) Derive expressions for the slope and intercept of the straight line obtained in (a) in terms of the variables listed in (b).
(d) Show that the concentration of the analyte is given by the relationship $c_x = ac_s/bV_x$, where a and b are the slope and intercept, respectively, of the straight line in (a).
(e) Determine values for a and b by the method of least squares.
(f) Calculate the standard deviation for the slope and about regression in (e).
(g) Calculate ppm Cr in the sample using the relationship given in (d).
(h) The relative variance of a result obtained in this way is approximately equal to the sum of the relative variances of the slope and the intercept. Calculate the absolute standard deviation for the result obtained in (g).

Chapter 24

Analytical Separations

An interference in a chemical analysis is a species in the sample matrix that either produces a signal that is indistinguishable from that of the analyte or, alternatively, attenuates the analyte signal. Few if any analytical signals are so specific as to be free of interference. As a consequence, most analytical methods require one or more preliminary steps to eliminate the effects of interferences.

Two general methods are available for dealing with interferences. The first makes use of a *masking agent* to immobilize or chemically bind the interfering species in a form in which it no longer contributes to or attenuates the signal from the analyte.[1] Clearly, a masking agent must not affect the behavior of the analyte significantly. In earlier chapters, we encountered several masking agents. An example is the use of fluoride ion to prevent iron(III) from interfering in the iodometric determination of copper(II). Here, masking results from the strong tendency of fluoride ions to complex iron(III) but not copper(II). The consequence is a decrease in the electrode potential of the iron(III) system to the point where only copper(II) ions from the sample oxidize iodide to iodine.

The second approach in dealing with an interference involves separating the analyte and the interference into two phases. The classical way of performing this type of separation was based upon precipitating the analyte selectively with an appropriate chemical reagent, such as hydrogen sulfide (Section 9C) or any of several organic precipitating agents (Section 5D-3). Another method of removing the analyte as a separate phase in-

> A masking agent is a reagent that chemically binds an interference and prevents it from causing errors in an analysis.

> Chemical species are generally separated by converting them to different phases that can then be mechanically isolated.

---

[1]For a monograph on masking agents, see D. D. Perrin, *Masking and Demasking Reactions*. New York: Wiley-Interscience, 1970.

volves electrolysis at controlled electrode potential, as described in Section 19B-2.

This chapter is concerned with two other methods of isolating an analyte from interferences: *extraction* and *ion exchange*. In the two chapters that follow, we describe various types of chromatographic separations. These methods are also based upon distributing the analyte and interference between two phases.

24A   SEPARATION BY EXTRACTION

The extent to which solutes, both inorganic and organic, distribute themselves between two immiscible liquids differs enormously, and these differences have been used for decades to accomplish separations of chemical species. This section considers applications of the distribution phenomenon to analytical separations.

24A-1   Theory

The partition of a solute between two immiscible phases is an equilibrium phenomenon governed by the *distribution law*. If the solute species A is allowed to distribute itself between water and an organic phase, the resulting equilibrium may be written as

$$A_{aq} \rightleftharpoons A_{org}$$

where the subscripts refer to the aqueous and organic phases. Ideally, the ratio of activities for A in the two phases will be constant and independent of the total quantity of A; that is, at any given temperature,

$$K_d = \frac{[A_{org}]}{[A_{aq}]} \qquad (24\text{-}1)$$

The equilibrium constant K_d is known as the *partition*, or *distribution*, *coefficient*. As with other equilibria, molar concentrations can be substituted for activities without serious error. Generally, the numerical value for K_d is approximately the ratio of the solubilities of A in the two solvents.

Partition coefficients are useful because they provide guidance as to the most efficient way to perform an extractive separation. Consider, for example, a simple system that is described by Equation 24-1.[2] Suppose further that a_0 mmol of the solute A in V_{aq} mL of aqueous solution is extracted with V_{org} mL of an immiscible organic solvent. At equilibrium, a_1 mmol of A remains in the aqueous layer and $(a_0 - a_1)$ mmol has been

---

[2]This treatment can be modified to take account of other equilibria; see H. A. Laitinen and W. E. Harris, *Chemical Analysis,* 2nd ed., pp. 443–453. New York: McGraw-Hill, 1975.

transferred to the organic layer. The concentrations of A in the two layers are then

$$[A_{aq}] = \frac{a_1}{V_{aq}}$$

and

$$[A_{org}] = \frac{(a_0 - a_1)}{V_{org}}$$

Substitution of these quantities into Equation 24-1 and rearrangement give

$$a_1 = \left(\frac{V_{aq}}{V_{org}K_d + V_{aq}}\right) a_0 \qquad (24\text{-}2)$$

Similarly, the number of millimoles a_2 remaining after a second extraction with the same volume of solvent is

$$a_2 = \left(\frac{V_{aq}}{V_{org}K_d + V_{aq}}\right) a_1$$

Substitution of Equation 24-2 for a_1 in this expression gives

$$a_2 = \left(\frac{V_{aq}}{V_{org}K_d + V_{aq}}\right)^2 a_0$$

By the same argument, the number of millimoles a_n remaining after n extractions is given by the expression

$$a_n = \left(\frac{V_{aq}}{V_{org}K_d + V_{aq}}\right)^n a_0 \qquad (24\text{-}3)$$

Finally, Equation 24-3 can be written in terms of the initial and final concentrations of a in the aqueous layer by substituting the relationships

$$a_n = [A_{aq}]_n V_{aq} \quad \text{and} \quad a_0 = [A_{aq}]_0 V_{aq}$$

Thus

$$[A_{aq}]_n = \left(\frac{V_{aq}}{V_{org}K_d + V_{aq}}\right)^n [A_{aq}]_0 \qquad (24\text{-}4)$$

Example 24-1 demonstrates that several small volumes provide a more efficient extraction than does a single large volume.

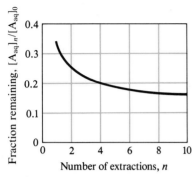

Figure 24-1

Plot of Equation 24-4, with $K_d = 2$ and $V_{aq} = 100$ mL. The total volume of organic solvent is also 100 mL; thus, $V_{org} = 100/n$.

It is always better to use several small portions of solvent to extract a sample than to extract with one large portion.

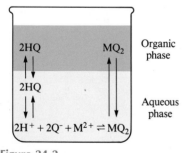

Figure 24-2

Equilibria in the extraction of an aqueous cation M^{2+} into an immiscible organic solvent containing 8-hydroxyquinoline.

Example 24-1

The distribution coefficient for iodine between CCl_4 and H_2O is 85. Calculate the concentration of I_2 remaining in the aqueous layer after extraction of 50.0 mL of 1.00×10^{-3} M I_2 with the following quantities of CCl_4: (a) 50.0 mL; (b) two 25.0-mL portions; (c) five 10.0-mL portions.

Substitution into Equation 24-4 gives

(a) $[I_{2\ aq}] = \left(\dfrac{50.0}{50.0 \times 85 + 50.0}\right)^1 \times 1.00 \times 10^{-3} = 1.16 \times 10^{-5}$ M

(b) $[I_{2\ aq}] = \left(\dfrac{50.0}{25.0 \times 85 + 50.0}\right)^2 \times 1.00 \times 10^{-3} = 5.28 \times 10^{-7}$ M

(c) $[I_{2\ aq}] = \left(\dfrac{50.0}{10.0 \times 85 + 50.0}\right)^5 \times 1.00 \times 10^{-3} = 5.29 \times 10^{-10}$ M

Figure 24-1 shows that the improved efficiency of multiple extractions falls off rapidly as a total fixed volume is subdivided into smaller and smaller portions. Clearly, little is to be gained by dividing the extracting solvent into more than five or six portions.

24A-2 Applications

An extraction is frequently more attractive than a precipitation for separating inorganic species. The processes of equilibration and separation of phases in a separatory funnel are less tedious and time-consuming than conventional precipitation, filtration, and washing. Moreover, difficulties associated with coprecipitation are avoided. Finally, and in contrast to the precipitation process, extraction procedures are ideally suited for the isolation of trace quantities of analytes.

The Extractive Separation of Metal Ions as Chelates

Many organic chelating agents are weak acids that react with metal ions to give uncharged complexes that are highly soluble in such organic solvents as ethers, hydrocarbons, ketones, and chlorinated species (including chloroform and carbon tetrachloride). The metal chelates, on the other hand, are usually nearly insoluble in water. Similarly, the chelating agents themselves are often quite soluble in organic solvents but of limited solubility in water.

Figure 24-2 shows the equilibria that develop when an aqueous solution of a divalent cation, such as zinc(II), is extracted with an organic solution containing a large excess of 8-hydroxyquinoline (see Section 5D-3 for the structure and reactions of this chelating agent). Four equilibria are shown. The first, which is not very favorable, involves distribution of the 8-hydroxyquinoline, HQ, between the organic and aqueous layers. The second is the acid dissociation of the HQ in water to give H^+ and Q^- ions. The third equilibrium is the complex-formation reaction giving MQ_2 (were it not for the fourth equilibrium, MQ_2 would precipitate out of the solution). Finally there is a second distribution equilibrium involving the complex between water and the organic phase.

The overall equilibrium is the sum of these four reactions:

$$2\ HQ(org) + M^{2+}(aq) \rightleftarrows MQ_2(org) + 2\ H^+(aq)$$

The equilibrium constant for this reaction is

$$K' = \frac{[MQ_2(org)][H^+(aq)]^2}{[HQ(org)]^2[M^{2+}(aq)]}$$

Ordinarily, HQ in the organic layer is in large excess with respect to M^{2+} in the aqueous phase, so that $[HQ(org)]$ remains essentially constant during the extraction. Therefore, the equilibrium-constant expression can be simplified to

$$K'[HQ(org)]^2 = K = \frac{[MQ_2(org)][H^+(aq)]^2}{[M^{2+}(aq)]}$$

or

$$\frac{[MQ_2(org)]}{[M^{2+}(aq)]} = \frac{K}{[H^+(aq)]^2}$$

Thus, we see that the ratio of the concentrations of the cation in the two layers is inversely proportional to the square of the hydrogen ion concentration of the aqueous layer. Equilibrium constants K vary widely from metal ion to metal ion, and these differences make it possible to selectively separate one cation from another by buffering the aqueous solution at a level where one is extracted nearly completely into the organic phase and the second remains largely in the aqueous phase.

Several useful extractive separations with 8-hydroxyquinoline have been developed. Furthermore, numerous chelating agents that behave in a similar way are described in the literature. As a consequence, pH-controlled extractions provide a powerful method for separating metallic ions.

The Extraction of Metal Chlorides and Nitrates

A number of inorganic species can be separated by extraction with suitable solvents. For example, a single ether extraction of a 6 M hydrochloric acid solution will cause better than 50% of several ions to be transferred to the organic medium; included among these are iron(III), antimony(V), titanium(III), gold(III), molybdenum(VI), and tin(IV). Other ions, such as aluminum(III) and the divalent cations of cobalt, lead, manganese, and nickel, are not extracted.

Uranium(VI) can be separated from such elements as lead and thorium by ether extraction of a solution that is 1.5 M in nitric acid and saturated with ammonium nitrate. Bismuth and iron(III) are also extracted to some extent from this medium.

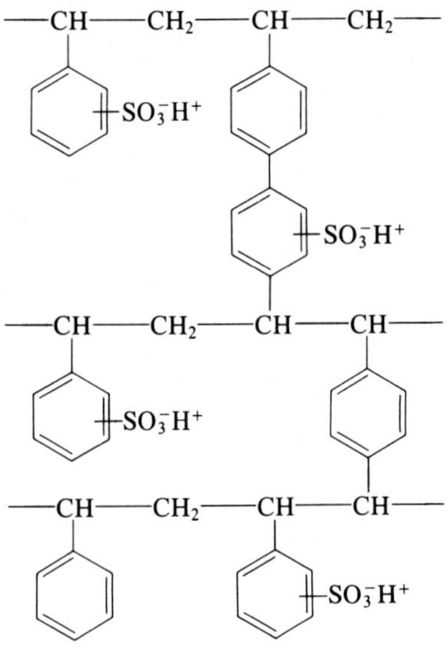

Figure 24-3

Structure of a cross-linked poly-styrene ion-exchange resin. Similar resins are used in which the $-SO_3^-H^+$ group is replaced by $-COO^-H^+$, $-NH_3^+OH^-$, and $-N(CH_3)_3^+OH^-$ groups.

24B   SEPARATION BY ION EXCHANGE

Ion exchange is a process by which ions held on an essentially insoluble solid are exchanged for ions in an aqueous solution that is brought in contact with the solid. The ion-exchange properties of clays and zeolites have been recognized and studied for more than a century. Synthetic ion-exchange resins were first produced in 1935 and have since found widespread application in water softening, water deionization, solution purification, and ion separation.

24B-1   Ion-Exchange Resins

Synthetic ion-exchange resins are high-molecular-weight polymers that contain large numbers of an ionic functional group per molecule. Cation-exchange resins have acidic groups, while anion-exchange resins have basic groups. Strong-acid exchangers have sulfonic acid groups ($-SO_3^-H^+$) attached to the polymeric matrix and have wider application than weak-acid type exchangers, which owe their action to carboxylic acid ($-COOH$) groups. Similarly, strong-base anion exchangers contain quaternary amine $[-N(CH_3)_3^+OH^-]$ groups, whereas weak-base exchangers contain secondary or tertiary amines.

Cation exchange is illustrated by the equilibrium

$$x\ RSO_3^-H^+ + M^{x+} \rightleftarrows (RSO_3^-)_xM^{x+} + x\ H^+$$

Solid   Soln   Solid   Soln

where M^{x+} represents a cation and R represents *that part of a resin molecule that contains one sulfonic acid group*. An ion exchanger of this type is shown in Figure 24-3.

The analogous equilibrium involving a strong-base anion exchanger and an anion A^{x-} is

$$x\ RN(CH_3)_3^+OH^- + A^{x-} \rightleftarrows [RN(CH_3)_3^+]_xA^{x-} + x\ OH^-$$

24B-2   Ion-Exchange Equilibria

Ion-exchange equilibria can be treated by the law of mass action. For example, when a dilute solution containing M^{2+} ions is passed through a column packed with a sulfonic acid resin, the following equilibrium is established:

$$M_{aq}^{2+} + 2\ H_{res}^+ \rightleftarrows M_{res}^{2+} + 2\ H_{aq}^+$$

Soln   Solid   Solid   Soln

for which

$$K = \frac{[M_{res}^{2+}][H_{aq}^+]^2}{[M_{aq}^{2+}][H_{res}^+]^2} \tag{24-5}$$

As usual, the bracketed terms are molar concentrations (strictly, activities) of the species in the two phases. Note that $[M_{res}^{2+}]$ and $[H_{res}^{+}]$ are molar concentrations of the two ions *in the solid phase*. In contrast to most solids, however, these concentrations can vary from zero to some maximum value when all the negative sites on the resin are occupied by one species only.

Ion-exchange separations are ordinarily performed under conditions in which one ion predominates in *both* phases. Thus, in the removal of M^{2+} ions from a dilute and somewhat acidic solution, the M^{2+} ion concentration will be much smaller than that of hydrogen ion in both the aqueous and resin phases; that is,

$$[M_{res}^{2+}] \ll [H_{res}^{+}]$$

and

$$[M_{aq}^{2+}] \ll [H_{aq}^{+}]$$

As a consequence, the hydrogen ion concentration is essentially constant in both phases, and Equation 24-5 can be rearranged to

$$\frac{[M_{res}^{2+}]}{[M_{aq}^{2+}]} = K \frac{[H_{res}^{+}]^2}{[H_{aq}^{+}]^2} = K_D \qquad (24\text{-}6)$$

where K_D is a distribution constant analogous to the constant that governs an extraction equilibrium (Equation 24-1). Note that K_D in Equation 24-6 represents the affinity of the resin for M^{2+} ion relative to another ion (here, H^+). In general, where K_D for an ion is large, a strong tendency for the stationary phase to retain that ion exists; where K_D is small, the opposite is true. Selection of a common reference ion (such as H^+) permits a comparison of distribution ratios for various ions on a given type of resin. Such experiments reveal that polyvalent ions are much more strongly retained than singly charged species. Within a given charge group, differences that exist among values for K_D appear to be related to the size of the hydrated ion, as well as other properties. Thus, for a typical sulfonated cation-exchange resin, values of K_D for univalent ions decrease in the order $Ag^+ > Cs^+ > Rb^+ > K^+ > NH_4^+ > Na^+ > H^+ > Li^+$. For divalent cations, the order is $Ba^{2+} > Pb^{2+} > Sr^{2+} > Ca^{2+} > Ni^{2+} > Cd^{2+} > Cu^{2+} > Co^{2+} > Zn^{2+} > Mg^{2+} > UO_2^{2+}$.

24B-3 Applications

Ion-exchange resins are used to eliminate ions that would otherwise interfere with an analysis. For example, iron(III) and aluminum(III), as well as many other cations, tend to coprecipitate with barium sulfate during the determination of sulfate ion. Passage of a solution containing sulfate through a cation-exchange resin results in the retention of all these cations and the release of an equivalent number of hydrogen ions. Sulfate ions pass freely through the column and can be precipitated as barium sulfate from the effluent.

Another valuable application of ion-exchange resins involves the concentration of ions from a very dilute solution. Thus, traces of metallic elements in large volumes of natural waters can be collected on a cation-exchange column and subsequently liberated from the resin by treatment with acid; the result is a considerably more concentrated solution for analysis.

The total salt content of a sample can be determined by titrating the hydrogen ion released as an aliquot of sample passes through a cation exchanger in the acidic form. Similarly, a standard hydrochloric acid solution can be prepared by diluting to known volume the effluent resulting from treatment of a cation-exchange resin with a known weight of sodium chloride. Substitution of an anion-exchange resin in its hydroxide form will permit the preparation of a standard base solution.

As shown in Section 26B-4, ion-exchange resins are particularly useful for the chromatographic separation of both inorganic and organic ionic species.

24C   QUESTIONS AND PROBLEMS

*24-1. What is a masking agent and how does it function?

24-2. How do strong-acid synthetic ion-exchange resins differ in structure from their weak-acid counterparts?

*24-3. The distribution coefficient for X between chloroform and water is 9.6. Calculate the concentration of X remaining in the aqueous phase after extractions of 50.0 mL of 0.150 M X with the following quantities of chloroform:
(a) one 40.0-mL portion.
(b) two 20.0-mL portions.
(c) four 10.0-mL portions.
(d) eight 5.00-mL portions.

24-4. The distribution coefficient for Z between n-hexane and water is 6.25. Calculate the percent of Z remaining in 25.0 mL of water that was originally 0.0600 M in Z after extraction with the following volumes of n-hexane:
(a) one 25.0-mL portion.
(b) two 12.5-mL portions.
(c) five 5.00-mL portions.
(d) ten 2.50-mL portions.

*24-5. What volume of $CHCl_3$ is required to decrease the concentration of X in the aqueous phase of Problem 24-3 to 1.00×10^{-4} M if 25.0 mL of 0.0500 M X is extracted with
(a) 25.0-mL portions of $CHCl_3$?
(b) 10.0-mL portions of $CHCl_3$?
(c) 2.0-mL portions of $CHCl_3$?

24-6. What volume of n-hexane is required to decrease the concentration of Z in the aqueous phase of Problem 24-4 to 1.00×10^{-5} M if 40.0 mL of 0.0200

M Z is extracted with
(a) 50.0-mL portions of n-hexane?
(b) 25.0-mL portions?
(c) 10.0-mL portions?

*24-7. What is the minimum distribution coefficient that permits removal of 99% of a solute from 50.0 mL of water with
(a) two 25.0-mL extractions with benzene?
(b) five 10.0-mL extractions with benzene?

24-8. If 30.0 mL 0.0500 M aqueous Q is to be extracted with four 10.0-mL portions of an immiscible organic solvent, what is the minimum distribution coefficient that allows transfer of all but the following percentages of the solute to the organic layer:
*(a) 1.00×10^{-4}?
(b) 1.00×10^{-2}?
(c) 1.00×10^{-3}?

*24-9. A 0.150 M aqueous solution of the weak organic acid HA was prepared from the pure compound, and three 50.0-mL aliquots were transferred to 100-mL volumetric flasks. Solution 1 was diluted to 100 mL with 1.0 M $HClO_4$, solution 2 was diluted to the mark with 1.0 M NaOH, and solution 3 was diluted to the mark with water. A 25.0-mL aliquot of each was extracted with 25.0 mL of n-hexane. The extract from solution 2 contained no detectable trace of A-containing species, indicating that A^- is not soluble in the organic solvent. The extract from solution 1 contained no ClO_4^- or $HClO_4$ but was found to be 0.0454 M in HA (by extraction with standard NaOH and back-titration with standard HCl). The extract from solution 3 was found to be 0.0225 M in

HA. Assume that HA does not associate or dissociate in the organic solvent, and calculate

(a) the distribution ratio for HA between the two solvents.

(b) the concentration of the *species* HA and A^- in aqueous solution 3 after extraction.

(c) the dissociation constant of HA in water.

24-10. To determine the equilibrium constant for the reaction

$$I_2 + 2\ SCN^- \rightleftharpoons I(SCN)_2^- + I^-$$

25.0 mL of a 0.0100 M aqueous solution of I_2 was extracted with 10.0 mL of CCl_4. After extraction, spectrophotometric measurements revealed that the I_2 concentration *of the aqueous layer* was 1.12×10^{-4} M. An aqueous solution that was 0.0100 M in I_2 and 0.100 M in KSCN was then prepared. After extraction of 25.0 mL of this solution with 10.0 mL of CCl_4, the concentration of I_2 *in the CCl_4 layer* was found from spectrophotometric measurement to be 1.02×10^{-3} M.

(a) What is the distribution coefficient for I_2 between CCl_4 and H_2O?

(b) What is the formation constant for $I(SCN)_2^-$?

*24-11. The total cation content of natural water is often determined by exchanging the cations for hydrogen ions on a strong-acid ion-exchange resin. A 25.0-mL sample of a natural water was diluted to 100 mL with distilled water, and 2.0 g of a cation-exchange resin was added. After stirring, the mixture was filtered, and the solid remaining on the filter paper was washed with three 15.0-mL portions of water. The filtrate and washings required 15.3 mL of 0.0202 M NaOH to give a bromocresol green end point.

(a) Calculate the number of milliequivalents of cation present in exactly 1 L of sample. (Here, the equivalent weight of a cation is its formula weight divided by its charge.)

(b) Report the results in terms of milligrams of $CaCO_3$ per liter.

24-12. An organic acid was isolated and purified by recrystallization of its barium salt. To determine the equivalent weight of the acid, a 0.393-g sample of the salt was dissolved in about 100 mL of water. The solution was passed through a strong-acid ion-exchange resin, and the column was then washed with water; the eluate and washings were titrated with 18.1 mL of 0.1006 M NaOH to a phenolphthalein end point.

(a) Calculate the equivalent weight of the organic acid.

(b) A potentiometric titration curve of the solution resulting when a second sample was treated in the same way revealed two end points, one at pH 5 and the other at pH 9. What is the molecular weight of the acid?

*24-13. Describe the preparation of exactly 2 L of 0.1500 M HCl from primary-standard-grade NaCl using a cation-exchange resin.

24-14. An aqueous solution containing $MgCl_2$ and HCl was analyzed by first titrating a 25.00-mL aliquot to a bromocresol green end point with 18.96 mL of 0.02762 M NaOH. A 10.00-mL aliquot was then diluted to 50.00 mL with distilled water and passed through a strong-acid ion-exchange resin. The eluate and washings required 36.54 mL of the NaOH solution to reach the same end point. Calculate the molar concentrations of HCl and $MgCl_2$ in the sample.

Chapter 25

An Introduction to Chromatographic Methods

Chromatography was invented by the Russian botanist Mikhail Tswett shortly after the turn of the century. He used the technique to separate various plant pigments, such as chlorophylls and xanthophylls, by passing solutions of them through glass columns packed with finely divided calcium carbonate. The separated species appeared as colored bands on the column, which accounts for the name he chose for the method (Greek *chroma*, meaning "color," and *graphein*, meaning "to write").

Chromatography is a technique in which the components of a mixture are separated based upon the rates at which they are carried through a stationary phase by a gaseous or liquid mobile phase.

Chromatography is widely used for the separation, identification, and determination of the chemical components in complex mixtures. No other separation method is as powerful and as generally applicable as is chromatography.[1]

25A A GENERAL DESCRIPTION OF CHROMATOGRAPHY

The word "chromatography" is difficult to define rigorously because the term has been applied to such a variety of systems and techniques. All of these methods, however, have in common the use of a *stationary phase* and a *mobile phase*. Components of a mixture are carried through the stationary phase by the flow of a gaseous or a liquid mobile phase, separations being based on differences in migration rates among the sample components.

---

[1]General references on chromatography include *Chromatography: Fundamentals and Applications of Chromatography and Electrophotometric Methods, Part A: Fundamentals, Part B: Applications*, E. Heftmann, Ed. New York: Elsevier, 1983; P. Sewell and B. Clarke, *Chromatographic Separations*. New York: Wiley, 1988; *Chromatographic Theory and Basic Principles*, J. A. Jonsson, Ed. New York: Marcel Dekker, 1987; R. M. Smith, *Gas and Liquid Chromatography in Analytical Chemistry*. New York: Wiley, 1988.

25A-1 Classification of Chromatographic Methods

Chromatographic methods are of two types. In *column chromatography,* the stationary phase is held in a narrow tube, and the mobile phase is forced through the tube under pressure or by gravity. In *planar chromatography,* the stationary phase is supported on a flat plate or in the pores of a paper. Here the mobile phase moves through the stationary phase by capillary action or under the influence of gravity. We will consider column chromatography only.

As shown in the first column of Table 25-1, chromatographic methods fall into three categories based upon the nature of the mobile phase. The three types of phases include liquids, gases, and supercritical fluids. The second column of the table reveals that there are five types of liquid chromatography and three types of gas chromatography that differ in the nature of the stationary phase and the types of equilibria between phases.

Planar and column chromatography are based upon the same types of equilibria.

Liquid chromatography can be performed in columns and on planar surfaces, but gas chromatography and supercritical fluid chromatography are restricted to column procedures.

25A-2 Elution Chromatography

Figure 25-1 shows schematically how two components A and B are resolved on a column by *elution chromatography.* Elution involves washing a solute through a column by additions of fresh solvent. A single portion of the sample dissolved in the mobile phase is introduced at the head of the column (at time t_0 in Figure 25-1), whereupon components A and B distribute themselves between the two phases. Introduction of additional mobile phase (the *eluent*) forces the dissolved portion of the sample down the column, where further partition between the mobile phase and fresh portions of the stationary phase occurs (time t_1).

Elution is a process in which solutes are washed through a stationary phase by the movement of a mobile phase.

Table 25-1
CLASSIFICATION OF COLUMN CHROMATOGRAPHIC METHODS

General Classification	Specific Method	Stationary Phase	Type of Equilibrium
Liquid chromatography (LC) (mobile phase: liquid)	Liquid-liquid, or partition	Liquid adsorbed on a solid	Partition between immiscible liquids
	Liquid-bonded phase	Organic species bonded to a solid surface	Partition between liquid and bonded surface
	Liquid-solid, or adsorption	Solid	Adsorption
	Ion exchange	Ion-exchange resin	Ion exchange
	Size exclusion	Liquid in interstices of a polymeric solid	Partition/sieving
Gas chromatography (GC) (mobile phase: gas)	Gas-liquid	Liquid adsorbed on a solid	Partition between gas and liquid
	Gas-bonded phase	Organic species bonded to a solid surface	Partition between liquid and bonded surface
	Gas-solid	Solid	Adsorption
Supercritical-fluid chromatography (SFC) (mobile phase: supercritical fluid)		Organic species bonded to solid surface	Partition between supercritical fluid and bonded surface

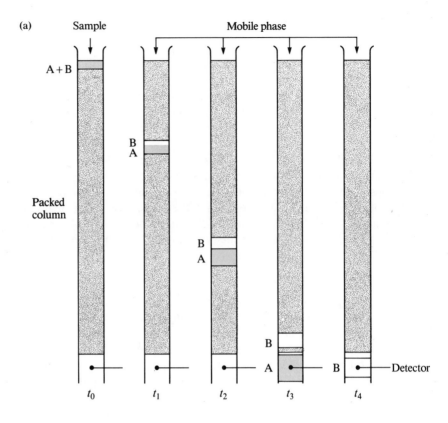

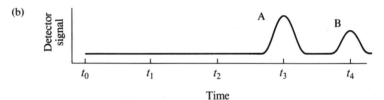

Figure 25-1

(a) Diagram showing the separation of a mixture of components A and B by column elution chromatography. (b) The output of the signal detector at the various stages of elution shown in (a).

An eluent is a solvent used to carry the components of a mixture through a stationary phase.

Further additions of solvent carry solute molecules down the column in a continuous series of transfers between the two phases. Because solute movement can occur only in the mobile phase, the average *rate* at which a solute migrates *depends upon the fraction of time it spends in that phase.* This fraction is small for solutes that are strongly retained by the stationary phase (component B in Figure 25-1, for example) and large where retention in the mobile phase is more likely (component A). Ideally, the resulting differences in rates cause the components in a mixture to separate into *bands,* or *zones,* along the length of the column (see Figure 25-2). Isolation of the separated species is then accomplished by passing a sufficient quantity of mobile phase through the column to cause the individual bands to pass out the end (to be *eluted* from the column), where they can be collected (times t_3 and t_4 in Figure 25-1).

Chromatograms

If a detector that responds to solute concentration is placed at the end of the column and its signal is plotted as a function of time (or of volume of

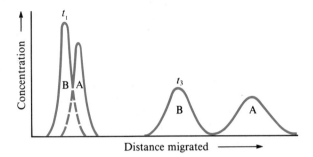

Figure 25-2
Concentration profiles of solute bands A and B at two different times in their migration down the column in Figure 25-1. The times t_1 and t_3 are indicated in Figure 25-1.

added mobile phase), a series of symmetric peaks is obtained, as shown in the lower part of Figure 25-1. Such a plot, called a *chromatogram,* is useful for both qualitative and quantitative analysis. The positions of the peaks on the time axis can be used to identify the components of the sample; the areas under the peaks provide a quantitative measure of the amount of each species.

> A chromatogram is a plot of some function of solute concentration versus elution time or elution volume.

The Effects of Relative Migration Rates and Band Broadening on Resolution

Figure 25-2 shows concentration profiles for the bands containing solutes A and B on the column in Figure 25-1 at time t_1 and at a later time t_3.[2] Because B is more strongly retained by the stationary phase than is A, B lags during the migration. Clearly, the distance between the two increases as they move down the column. At the same time, however, broadening of both bands takes place, which lowers the efficiency of the column as a separating device. While band broadening is inevitable, conditions can often be found where it occurs more slowly than band separation. Thus, as shown in Figure 25-2, a clean resolution of species is possible provided the column is sufficiently long.

Several chemical and physical variables influence the rates of band separation and band broadening. As a consequence, improved separations can often be realized by the control of variables that either increase the rate of band separation or decrease the rate of band spreading. These alternatives are illustrated in Figure 25-3.

The variables that influence the relative rates at which solutes migrate through a stationary phase are described in the next section. Following this discussion, we shall turn to those factors that play a part in zone broadening.

25B   MIGRATION RATES OF SOLUTES

The effectiveness of a chromatographic column in separating two solutes depends in part upon the relative rates at which the two species are

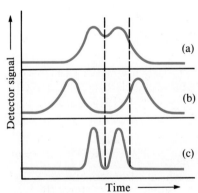

Figure 25-3
Two-component chromatographs illustrating two methods of improving separation: (a) original chromatogram with overlapping peaks; improvement brought about by (b) an increase in band separation and (c) a decrease in bandwidth.

---

[2]Note that the relative positions of the bands for A and B in the concentration profile in Figure 25-2 are reversed from their positions in the lower part of Figure 25-1. The difference is that the abscissa is distance along the column in Figure 25-2 but time in Figure 25-1. Thus, in Figure 25-1, the *front* of a peak lies to the left and the *tail* to the right; in Figure 25-2, the reverse is true.

eluted. These rates are in turn determined by the partition ratios of the solutes between the two phases.

25B-1 Partition Ratios in Chromatography

All chromatographic separations are based upon differences in the extent to which solutes are partitioned between the mobile and the stationary phase. For solute species A, the equilibrium involved is described by the equation

$$A_{mobile} \rightleftarrows A_{stationary}$$

The equilibrium constant K for this reaction is called a *partition ratio,* or *partition coefficient,* and is defined as

$$K = \frac{C_S}{C_M} \tag{25-1}$$

where C_S is the molar analytical concentration of the solute in the stationary phase and C_M is its analytical concentration in the mobile phase. Ideally, the partition ratio is constant over a wide range of solute concentrations; that is, C_S is directly proportional to C_M.

25B-2 Retention Time

Figure 25-4 is a typical chromatogram for a sample containing a single analyte. The time it takes after sample injection for the analyte peak to reach the detector is called the *retention time* and is given the symbol t_R. The small peak on the left is for a species that is *not* retained by the column. Often the sample or the mobile phase will contain an unretained species. When they do not, such a species may be added to aid in peak identification. The time t_M for the unretained species to reach the detector is sometimes called the *dead time*. The rate of migration of the unretained species is the same as the average rate of motion of the mobile phase molecules.

> Retention time is the time between sample injection and the appearance of a solute peak at the detector of a chromatographic column.

> The dead time is the time it takes for an unretained species to pass through a column.

Figure 25-4

A typical chromatogram for a two-component mixture. The small peak on the left represents a solute that is not retained on the column and so reaches the detector almost immediately after elution is started. Thus its retention time t_M is approximately equal to the time required for a molecule of the mobile phase to pass through the column.

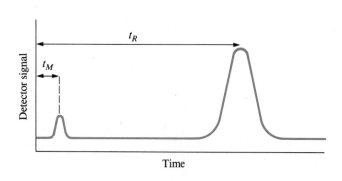

The average linear rate of solute migration $\bar{v}$ is

$$\bar{v} = \frac{L}{t_R} \qquad (25\text{-}2)$$

where L is the length of the column packing. Similarly, the average linear rate of movement u of the molecules of the mobile phase is

$$u = \frac{L}{t_M} \qquad (25\text{-}3)$$

where t_M is the time required for a molecule of the mobile phase to pass through the column.

25B-3 The Relationship Between Retention Time and Partition Ratio

In order to relate the retention time of a solute to its partition ratio, we express its migration rate as a fraction of the velocity of the mobile phase:

$$\bar{v} = u \times \text{fraction of time solute spends in mobile phase}$$

This fraction, however, equals the average number of moles of solute in the mobile phase at any instant divided by the total number of moles of solute in the column:

$$\bar{v} = u \times \frac{\text{moles of solute in mobile phase}}{\text{total moles of solute}}$$

The total number of moles of solute in the mobile phase is equal to the molar concentration C_M of the solute in that phase multiplied by its volume V_M. Similarly, the number of moles of solute in the stationary phase is given by the product of the concentration C_S of the solute in the stationary phase and its volume V_S. Therefore,

$$\bar{v} = u \times \frac{C_M V_M}{C_M V_M + C_S V_S} = u \times \frac{1}{1 + C_S V_S / C_M V_M}$$

Substitution of Equation 25-1 into this equation gives an expression for the rate of solute migration as a function of its partition ratio and as a function of the volumes of the stationary and mobile phases:

$$\bar{v} = u \times \frac{1}{1 + K V_S / V_M} \qquad (25\text{-}4)$$

The two volumes can be estimated from the method by which the column is prepared.

25B-4 The Rate of Solute Migration: The Capacity Factor

The *capacity factor* is an important parameter that is widely used to describe the migration rates of solutes on columns. For a solute A, the capacity factor k'_A is defined as

$$k'_A = \frac{K_A V_S}{V_M} \tag{25-5}$$

where K_A is the partition ratio for the species A. Substitution of Equation 25-5 into 25-4 yields

$$\bar{v} = u \times \frac{1}{1 + k'_A} \tag{25-6}$$

In order to show how k'_A can be derived from a chromatogram, we substitute Equations 25-2 and 25-3 into Equation 25-6:

$$\frac{L}{t_R} = \frac{L}{t_M} \times \frac{1}{1 + k'_A} \tag{25-7}$$

This equation rearranges to

$$k'_A = \frac{t_R - t_M}{t_M} \tag{25-8}$$

As shown in Figure 25-4, t_R and t_M are readily obtained from a chromatogram. When the capacity factor for a solute is much less than unity, elution occurs so rapidly that accurate determination of the retention times is difficult. When the capacity factor is larger than perhaps 20 to 30, elution times become inordinately long. Ideally, separations are performed under conditions in which the capacity factors for the solutes in a mixture lie in the range between 1 and 5.

The capacity factors in gas chromatography can be varied by changing the temperature and the column packing. In liquid chromatography, capacity factors can often be manipulated to give better separations by varying the composition of the mobile phase and the stationary phase.

Ideally, the capacity factor for analytes in a sample is between 1 and 5.

25B-5 Differential Migration Rates: The Selectivity Factor

The *selectivity factor* α of a column for the two species A and B is defined as

$$\alpha = \frac{K_B}{K_A} \tag{25-9}$$

where K_B is the partition ratio for the more strongly retained species B

The selectivity factor for two analytes in a column provides a measure of how well the column will separate the two.

and K_A is the partition ratio for the less strongly held, or more rapidly eluted, species A. By this definition, α *is always greater than unity*.

Substitution of Equation 25-5 and the analogous equation for solute B into Equation 25-9 provides, after rearrangement, a relationship between the selectivity factor for two solutes and their capacity factors:

$$\alpha = \frac{k'_B}{k'_A} \tag{25-10}$$

where k'_B and k'_A are the capacity factors for B and A, respectively. Substitution of Equation 25-8 for the two solutes in Equation 25-10 gives an expression that permits the determination of α from an experimental chromatogram:

$$\alpha = \frac{(t_R)_B - t_M}{(t_R)_A - t_M} \tag{25-11}$$

In Section 25D-1 we show how to use the selectivity factor to compute the resolving power of a column.

25C   THE EFFICIENCY OF CHROMATOGRAPHIC COLUMNS

The efficiency of a chromatographic column refers to the amount of band broadening that occurs when a compound passes through the column. Before defining column efficiency in more quantitative terms, let us examine the reasons that bands become broader as they move down a column.

25C-1 The Rate Theory of Chromatography

The *rate theory* of chromatography describes the shapes and breadths of elution peaks in quantitative terms based on a random-walk mechanism for the migration of molecules through a column. A detailed discussion of the rate theory is beyond the scope of this text. We can, however, give a qualitative picture of why bands broaden and what variables improve column efficiency.

If you examine the chromatograms shown in this and the next chapter, you will see that the elution peaks look very much like the Gaussian error curves that you encountered in Chapters 3 and 4. As shown in Section 3D, normal error curves are rationalized by assuming that the uncertainty associated with any single measurement is the summation of a much larger number of small, individually undetectable and random uncertainties, each of which has an equal probability of being positive or negative. In a similar way, the typical Gaussian shape of a chromatographic band can be attributed to the additive combination of the random motions of the myriad molecules in the band as it moves down the column.

It is instructive to consider a single solute molecule as it undergoes many thousands of transfers between the stationary and mobile phases during elution. Residence time in either phase is highly irregular. Transfer from one phase to the other requires energy, and the molecule must acquire this energy from its surroundings. Thus, the residence time in a given phase may be transitory for some molecules and relatively long for others. Recall that movement down the column can occur *only while the molecule is in the mobile phase.* As a consequence, certain molecules travel rapidly by virtue of their accidental inclusion in the mobile phase for a majority of the time, whereas others lag because they happen to be incorporated in the stationary phase for a greater-than-average period. The result of these random individual processes is a symmetric spread of velocities around the mean value, which represents the behavior of the average analyte molecule.

Diffusion also contributes to band broadening. Recall from Section 19A-2 that molecules tend to move from a more concentrated region of a solution to a more dilute part, the rate of diffusion being proportional to the concentration difference. In the center of a chromatographic band the concentration of a species is high, while at the two edges the concentration approaches zero. Therefore, molecules tend to migrate to either edge of the band. Band broadening is the result. Note that half of the diffusion will be in the direction of flow and the other half in the opposite direction.

Another cause of band broadening is shown in Figure 25-5, which shows that individual molecules in the mobile phase follow paths of different lengths as they traverse the column. Thus, their arrival time at the detector differs, and bands are broadened as a consequence.

The breadth of a band increases as the band moves down the column because more time is allowed for spreading to occur as a result of these various mechanisms. Thus, zone breadth is directly related to residence time in the column and inversely related to the velocity of the mobile phase.

25C-2 A Quantitative Definition of Column Efficiency

Two related terms are widely used as quantitative measures of chromatographic column efficiency: (1) *plate height H* and (2) *number of theoretical plates N*. The two are related by the equation

$$N = L/H \qquad (25\text{-}12)$$

The plate height H is also known as the height equivalent of a theoretical plate (HETP).

where L is the length (usually in centimeters) of the column packing. Feature 25-1 describes how these measures of column efficiency got their names.

The efficiency of a column is great when H is small and N is large.

The efficiency of chromatographic columns increases as the number of plates becomes greater and as the plate height becomes smaller. Enormous differences in efficiencies are encountered in columns as a result of differences in column type and in mobile and stationary phases. Efficiencies in terms of plate numbers can vary from a few hundred to several hundred thousand; plate heights ranging from a few tenths to one thousandth of a centimeter or smaller are not uncommon.

The Definition of Plate Height
In Section 4A-2, we pointed out that the breadth of a Gaussian curve is best described by the standard deviation σ and the variance σ^2. Because

chromatographic bands are also Gaussian and because the efficiency of a column is reflected in the breadth of chromatographic peaks, the variance per unit length of column is used by chromatographers as a measure of column efficiency. That is, the plate height H is given by

$$H = \frac{\sigma^2}{L} \qquad (25\text{-}13)$$

This definition of column efficiency is illustrated in Figure 25-6a, which shows a column having a packing L cm in length. Above this schematic is a plot showing the distribution of molecules along the length of the column at the moment the analyte peak reaches the end of the packing (that is, at the retention time t_R). The curve is Gaussian, and the locations of $L + 1\sigma$ and $L - 1\sigma$ are indicated as broken vertical lines. Note that L carries units of centimeters and σ^2 units of centimeters squared; thus H represents a linear distance in centimeters (Equation 25-13). In fact, the plate height can be thought of as the length of column that contains a fraction of the analyte that lies between L and $L - \sigma$. Because the area under a normal error curve bounded by $\pm\sigma$ is about 68% of the total area (page 48), the plate height, as defined, contains 34% of the analyte.

The Experimental Evaluation of H and N

Figure 25-7 is a typical chromatogram with time as the abscissa. The variance of the solute peak, which can be obtained by a simple graphical procedure, has units of seconds squared and is usually designated as τ^2 to distinguish it from σ^2, which has units of centimeters squared. The two standard deviations τ and σ are related by

$$\tau = \frac{\sigma}{L/t_R} \qquad (25\text{-}14)$$

where L/t_R is the average linear velocity of the solute in centimeters per second.

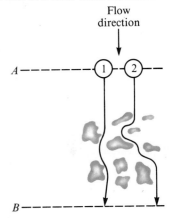

Figure 25-5

Typical pathways of two molecules during elution. Note that the distance traveled by molecule 2 is greater than that traveled by molecule 1. Thus, molecule 2 will arrive at B later than molecule 1.

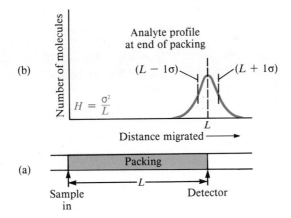

Figure 25-6

Definition of plate height $H = \sigma^2/L$.

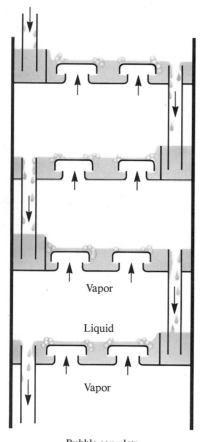

Vapor

Liquid

Vapor

Bubble cap plate
column

Plates in a fractionating column.

Feature 25-1

WHERE DO THE TERMS "PLATE" AND "PLATE HEIGHT" COME FROM?

The 1952 Nobel Prize in Chemistry was awarded to two Englishmen, A. J. P. Martin and R. L. M. Synge, for their work in the development of modern chromatography. In their theoretical studies, they adapted a model that was first developed in the early 1920s to describe separations on fractional distillation columns. Fractionating columns, which were first used in the petroleum industry for separating closely related hydrocarbons, consist of numerous interconnected bubble-cap plates (see margin) at which vapor-liquid equilibria are established when a column is operated under reflux conditions.

Martin and Synge treated a chromatographic column as if it were made up of a long series of bubble-cap-like plates, within which equilibrium conditions always prevail. This plate model successfully accounts for the Gaussian shape of chromatographic peaks as well as for factors that influence differences in solute-migration rates. It cannot account for zone broadening, however, because of its basic assumption that equilibrium conditions prevail throughout a column during elution. This assumption can never be valid in the dynamic state that exists in a chromatographic column, where phases are moving past one another at such a pace that sufficient time is not available for equilibration.

Because the plate model is such a poor representation of a chromatographic column, we urge you (1) to avoid attaching any real or imaginary significance to the terms "plate" and "plate height" and (2) to view these terms as designators of column efficiency that are retained for historic reasons only. They have no physical significance. Unfortunately, both terms are so well entrenched in the chromatographic literature that their replacement by more appropriate designations is unlikely.

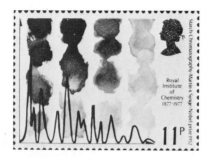

Stamp in honor of biochemists Archer J. P. Martin (1910–) and Richard L. M. Synge (1914–) who were awarded the 1952 Nobel Prize in chemistry for their contributions to the development of modern chromatography.

Figure 25-7 illustrates a simple means for approximating τ and σ from an experimental chromatogram. Tangents at the inflection points on the two sides of the chromatographic peak are extended to form a triangle with the baseline of the chromatogram. The area of this triangle can be shown to be approximately 96% of the total area under the peak. In Section 4A-2 it was shown that about 96% of the area under a Gaussian peak is included within plus or minus two standard deviations ($\pm 2\sigma$) of its maximum. Thus, the intercepts shown in Figure 25-7 occur at approximately $\pm 2\tau$ from the maximum, and $W = 4\tau$, where W is the magnitude of the base of the triangle. Substituting this relationship into Equation 25-14 and rearranging yield

$$\sigma = \frac{LW}{4t_R}$$

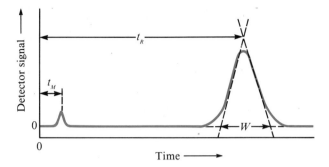

Figure 25-7
Determination of the standard deviation τ from a chromatographic peak: $W = 4\tau$.

Substitution of this equation for σ into Equation 25-13 gives

$$H = \frac{LW^2}{16t_R^2} \qquad (25\text{-}15)$$

To obtain N, we substitute into Equation 25-12 and rearrange to get

$$N = 16\left(\frac{t_R}{W}\right)^2 \qquad (25\text{-}16)$$

Thus, N can be calculated from two time measurements, t_R and W; to obtain H, the length of the column packing L must also be known.

The number of theoretical plates N and the plate height H are widely used in the literature and by instrument manufacturers as measures of column performance.

25C-3 Variables That Affect Column Efficiency

Band broadening, and thus loss of column efficiency, are the consequences of the finite rates at which several mass-transfer processes occur as a solute migrates down a column. Some of the variables that affect these rates are controllable and can be exploited to improve separations.

The Effect of Mobile-Phase Flow Rate
The extent of band broadening depends upon the length of time the mobile phase is in contact with the stationary phase. Therefore, as shown in Figure 25-8, column efficiency depends upon the flow rate of the mobile phase. Note that for both liquid and gas-liquid chromatography, minimum plate heights (maximum efficiencies) occur at relatively low flow rates. Maximum efficiencies for liquid chromatography occur at rates that are well below those for gas chromatography. In fact, the minima for liquid chromatography are often so low that they are not observed under ordinary operating conditions.

Generally, liquid chromatograms are obtained at lower flow rates than gas chromatograms. Furthermore, as shown in Figure 25-8, plate heights for liquid chromatographic columns are an order of magnitude or more smaller than those for gas chromatographic columns. Offsetting this ad-

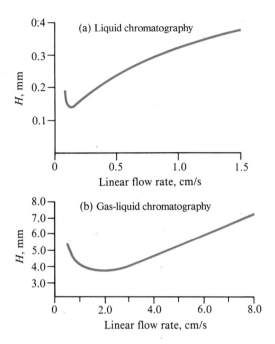

Figure 25-8

Effect of mobile-phase flow rate on plate height for (a) liquid chromatography and (b) gas chromatography.

vantage, however, is the difference in lengths of the two types of columns. Gas chromatographic columns can be 50 m or more in length. Liquid columns, on the other hand, can be no longer than 25 to 50 cm because of the high pressure drops along their lengths. As a consequence, the number of plates in a gas chromatographic column may exceed that in a liquid chromatographic column by a factor of several hundred.

Other Variables

It has been found that column efficiency can be increased by decreasing the particle size of column packings, by employing thinner layers of the immobilized film (where the stationary phase is an adsorbed liquid), and by lowering the mobile-phase viscosity. Increases in temperature also decrease band broadening under most circumstances. Figure 25-9 illustrates how particle size affects plate heights.

Figure 25-9

Effect of particle size on plate height. The numbers to the right are particle diameters. (From J. Boheman and J. H. Purnell, in *Gas Chromatography 1958*, D. H. Desty, Ed. New York: Academic Press, 1958. With permission of Butterworths, Stoneham, MA.)

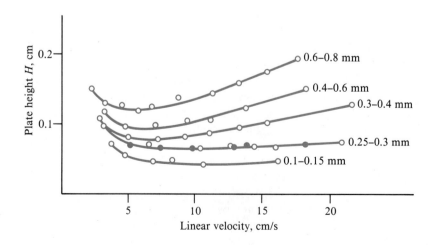

25D COLUMN RESOLUTION

The *resolution* R_s of a column provides a quantitative measure of the column's ability to separate two analytes. The significance of this term is illustrated in Figure 25-10, which consists of chromatograms for species A and B on three columns with different resolving powers. The resolution of each column is defined as

$$R_s = \frac{2\,\Delta Z}{W_A + W_B} = \frac{2[(t_R)_B - (t_R)_A]}{W_A + W_B} \qquad (25\text{-}17)$$

where all the terms on the right side are as defined in the figure.

It is evident from Figure 25-10 that a resolution of 1.5 gives an essentially complete separation of A and B, whereas a resolution of 0.75 does not. At a resolution of 1.0, zone A contains about 4% B and zone B contains about 4% A. At a resolution of 1.5, the overlap is about 0.3%. The resolution for a given stationary phase can be improved by lengthening the column, thus increasing the number of plates. An adverse consequence of the added plates, however, is an increase in the time required for the resolution.

25D-1 The Effect of Capacity and Selectivity Factors on Resolution

A useful equation is readily derived that relates the resolution of a column to the number of plates it contains as well as to the capacity and selectiv-

Michael Tswett (1872–1919), A Russian botanist, discovered the basic principles of column chromatography. He separated plant pigments by eluting a mixture of the pigments on a column of powdered alumina. The various pigments separated into colored bands on the column; hence the name chromatography.

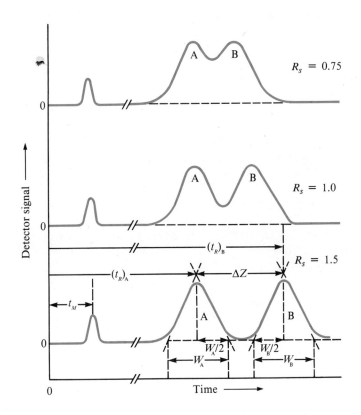

Figure 25-10
Separation at three resolutions: $R_s = 2\Delta Z/(W_A + W_B)$.

ity factors of a pair of solutes on the column. Thus, it is readily shown[3] that for the two solutes A and B in Figure 25-10, the resolution is given by the equation

$$R_s = \frac{\sqrt{N}}{4}\left(\frac{\alpha - 1}{\alpha}\right)\left(\frac{k_B'}{1 + k_B'}\right) \tag{25-18}$$

where k_B' is the capacity factor of the slower-moving species and α is the selectivity factor. This equation can be rearranged to give the number of plates needed to realize a given resolution:

$$N = 16R_s^2\left(\frac{\alpha}{\alpha - 1}\right)^2\left(\frac{1 + k_B'}{k_B'}\right)^2 \tag{25-19}$$

25D-2 The Effect of Resolution on Retention Time

As mentioned earlier, the goal in chromatography is the highest possible resolution in the shortest possible elapsed time. Unfortunately, these goals tend to be incompatible, and a compromise between the two is usually necessary. The time $(t_R)_B$ required to elute the two species in Figure 25-10 with a resolution of R_s is

$$(t_R)_B = \frac{16R_s^2H}{u}\left(\frac{\alpha}{\alpha - 1}\right)^2\frac{(1 + k_B')^3}{(k_B')^2} \tag{25-20}$$

where u is the linear rate of movement of the mobile phase.

Example 25-1

Substances A and B have retention times of 16.40 and 17.63 min, respectively, on a 30.0-cm column. An unretained species passes through the column in 1.30 min. The peak widths (at base) for A and B are 1.11 and 1.21 min, respectively. Calculate the (a) column resolution, (b) average number of plates in the column, (c) plate height, (d) length of column required to achieve a resolution of 1.5, and (e) time required to elute substance B on the longer column.

(a) Employing Equation 25-17, we find

$$R_s = 2(17.63 - 16.40)/(1.11 + 1.21) = 1.06$$

(b) Equation 25-16 permits computation of N:

$$N = 16\left(\frac{16.40}{1.11}\right)^2 = 3493 \quad \text{and} \quad N = 16\left(\frac{17.63}{1.21}\right)^2 = 3397$$

$$N_{av} = (3493 + 3397)/2 = 3445 = 3.4 \times 10^3$$

---

[3]See D. A. Skoog, *Principles of Instrumental Analysis*, 3rd ed., pp. 741–743. Philadelphia: Saunders College Publishing, 1985.

(c) $H = L/N = 30.0/3445 = 8.7 \times 10^{-3}$ cm

(d) Since k' and α do not change greatly with increasing N and L, substituting N_1 and N_2 into Equation 25-19 and dividing one of the resulting equations by the other yield

$$\frac{(R_s)_1}{(R_s)_2} = \frac{\sqrt{N_1}}{\sqrt{N_2}}$$

where the subscripts 1 and 2 refer to the original and longer columns, respectively. Substituting the appropriate values for N_1, $(R_s)_1$, and $(R_s)_2$ gives

$$\frac{1.06}{1.5} = \frac{\sqrt{3445}}{\sqrt{N_2}}$$

$$N_2 = 3445 \left(\frac{1.5}{1.06}\right)^2 = 6.9 \times 10^3$$

But

$$L = NH = 6.9 \times 10^3 \times 8.7 \times 10^{-3} = 60 \text{ cm}$$

(e) Substituting $(R_s)_1$ and $(R_s)_2$ into Equation 25-20 and dividing yields

$$\frac{(t_R)_1}{(t_R)_2} = \frac{(R_s)_1^2}{(R_s)_2^2} = \frac{17.63}{(t_R)_2} = \frac{(1.06)^2}{(1.5)^2}$$

$$(t_R)_2 = 35 \text{ min}$$

Thus, to obtain the improved resolution, the separation time must be doubled.

25E APPLICATIONS OF CHROMATOGRAPHY

Chromatography has grown to be a remarkably powerful and versatile tool for separating closely related chemical species. In addition, it can be employed for the qualitative identification and quantitative determination of separated species.

25E-1 Qualitative Analysis

Chromatography is widely used to confirm the presence or absence of components in mixtures that contain a limited number of possible species whose identities are known. For example, 30 or more amino acids in a protein hydrolysate can be detected with a reasonable degree of certainty by means of a chromatogram.

On the other hand, because a chromatogram provides but a single piece of information about each species in a mixture (the retention time), the application of the technique to the qualitative analysis of complex sam-

Here are some hyphenated techniques: Gas chromatography–mass spectrometry (GC-MS). High performance liquid chromatography–mass spectrometry (HPLC-MS). Gas chromatography–infrared spectrometry (GC-IR)

ples of unknown composition is limited. For this reason, it has become common practice to couple chromatographic columns directly with ultraviolet, infrared, and mass spectrometers. The resulting *hyphenated instruments* are powerful tools for identifying the components of complex mixtures.

It is important to note that while a chromatogram may not lead to positive identification of the species in a sample, it often provides sure evidence of the *absence* of species. Thus, failure of a sample to produce a peak at the same retention time as a standard processed under identical conditions is strong evidence that the compound in question is absent (or present at a concentration below the detection limit of the procedure).

25E-2 Quantitative Analysis

Chromatography owes its enormous growth in part to its speed, simplicity, relatively low cost, and wide applicability as a separating tool. It is doubtful, however, that its use would have become so widespread had it not been for the fact that it can also provide quantitative information about separated species.

Quantitative chromatography is based upon a comparison of either the height or the area of the analyte peak with that of one or more standards. If conditions are properly controlled, both of these parameters vary linearly with concentration.

Analyses Based on Peak Height

The height of a chromatographic peak is obtained by connecting the baselines on the two sides of the peak by a straight line and measuring the perpendicular distance from this line to the peak. This measurement can ordinarily be made with reasonably high precision and yields accurate results, provided variations in column conditions do not alter peak width during the period required to obtain chromatograms for sample and standards. The variables that must be controlled closely are column temperature, eluent flow rate, and rate of sample injection. In addition, care must be taken to avoid overloading the column. The effect of sample-injection rate is particularly critical for the early peaks of a chromatogram. Relative errors of 5 to 10% due to this cause are not unusual with syringe injection.

Analyses Based on Peak Area

Peak area is independent of broadening effects caused by the variables mentioned in the previous paragraph. From this standpoint, therefore, area is a more satisfactory analytical parameter than peak height. On the other hand, peak heights are more easily measured and, for narrow peaks, more accurately determined.

Many modern chromatographic instruments are equipped with electronic integrators that provide precise measurements of relative peak areas. If such equipment is not available, a manual estimate must be made. A simple method that works well for symmetric peaks of reasonable widths is to multiply peak height by the width at one-half peak height. Alternatively, the recorded peaks for sample and standards can be cut out and weighed, and the weights then compared.

Calibration with Standards

The most straightforward method for quantitative chromatographic analyses involves the preparation of a series of standard solutions that approximate the composition of the unknown. Chromatograms for the standards are then obtained, and peak heights or areas are plotted as a function of concentration. A plot of the data should yield a straight line passing through the origin; analyses are based upon this plot. Frequent restandardization is necessary for highest accuracy.

The most important source of error in analyses by the method based on calibration standards is usually the uncertainty in the volume of sample; occasionally, the rate of injection is also a factor. Samples are ordinarily small ($\approx 1\ \mu L$), and the uncertainties associated with injecting a reproducible volume of this size with a microsyringe can amount to several percent relative. The situation is even worse in gas-liquid chromatography, where the sample must be injected into a heated sample port. In this circumstance, evaporation from the needle tip may lead to large variations in the volume injected.

Errors in sample volume can be reduced to perhaps 1 to 2% relative by means of a rotary sample valve such as that shown in Figure 25-11. The sample loop ACB in (a) is filled with liquid; rotation of the valve by 45 degrees then introduces a reproducible volume of sample (the volume originally contained in ACB) into the mobile-phase stream.

The Internal-Standard Method

The highest precision for quantitative chromatography is obtained by using internal standards because the uncertainties introduced by sample injection are avoided. In this procedure, a carefully measured quantity of an internal standard is introduced into each standard and sample, and the ratio of analyte peak area (or height) to internal-standard peak area (or height) is the analytical parameter. For this method to be successful, it is necessary that the internal-standard peak be well separated from the peaks of all other components in the sample; it must appear close to the analyte peak, however. With a suitable internal standard, precisions of 0.5 to 1% relative are reported.

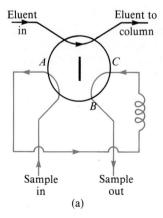

(a)

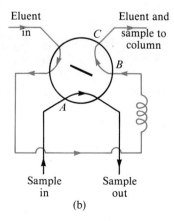

(b)

Figure 25-11

A rotary sample valve: (a) valve position for filling sample loop ACB; (b) Valve position for introducing sample into column.

25F   QUESTIONS AND PROBLEMS

25-1. Define
*(a) elution.
(b) mobile phase.
*(c) stationary phase.
(d) partition ratio.
*(e) retention time.
(f) capacity factor.
*(g) selectivity factor.
(h) plate height.

25-2. List the variables that lead to zone broadening.

*25-3. What is the difference between gas-liquid and liquid-liquid chromatography?

25-4. What is the difference between liquid-liquid and liquid-solid chromatography?

*25-5. Describe a method for determining the number of plates in a column.

25-6. Name two general methods for improving the resolution of two substances on a chromatographic column.

*25-7. The following data are for a liquid chromatographic column:

Length of packing	24.7 cm
Flow rate	0.313 mL/min
V_M	1.37 mL
V_S	0.164 mL

A chromatogram of a mixture of species A, B, C, and D provided the following data:

	Retention Time, min	Width of Peak Base (W), min
Nonretained	3.1	—
A	5.4	0.41
B	13.3	1.07
C	14.1	1.16
D	21.6	1.72

Calculate
(a) the number of plates from each peak.
(b) the mean and the standard deviation for N.
(c) the plate height for the column.

*25-8. From the data in Problem 25-7, calculate for A, B, C, and D
(a) the capacity factor.
(b) the partition coefficient.

*25-9. From the data in Problem 25-7, calculate for species B and C
(a) the resolution.
(b) the selectivity factor.
(c) the length of column necessary to separate the two species with a resolution of 1.5.
(d) the time required to separate the two species on the column in part (c).

*25-10. From the data in Problem 25-7, calculate for species C and D
(a) the resolution.
(b) the length of column necessary to separate the two species with a resolution of 1.5.

25-11. The following data were obtained by gas-liquid chromatography on a 40-cm packed column:

Compound	t_R, min	$W_{1/2}$, min
Air	1.9	—
Methylcyclohexane	10.0	0.76
Methylcyclohexene	10.9	0.82
Toluene	13.4	1.06

Calculate
(a) an average number of plates from the data.
(b) the standard deviation for the average in (a).
(c) an average plate height for the column.
(d) the capacity factor for each of the three species.

25-12. Referring to Problem 25-11, calculate the resolution for
(a) methylcyclohexene and methylcyclohexane.
(b) methylcyclohexene and toluene.
(c) methylcyclohexane and toluene.

25-13. If a resolution of 1.5 is desired in separating methylcyclohexane and methylcyclohexene in Problem 25-11,
(a) how many plates are required?
(b) how long must the column be if the same packing is employed?
(c) what is the retention time for methylcyclohexene on the column in Problem 25-11?

*25-14. If V_S and V_M for the column in Problem 25-11 are 19.6 and 62.6 mL, respectively, and a nonretained air peak appears after 1.9 min, calculate the
(a) capacity factor for each compound.
(b) partition coefficient for each compound.
(c) selectivity factor for methylcyclohexane and methylcyclohexene.

*25-15. From distribution studies, species M and N are known to have water/hexane partition coefficients of 6.01 and 6.20 ($K = [M]_{H_2O}/[M]_{hex}$). The two species are to be separated by elution with hexane in a column packed with silica gel containing adsorbed water. The ratio V_S/V_M for the packing is 0.422.
(a) Calculate the capacity factor for each solute.
(b) Calculate the selectivity factor.
(c) How many plates are needed to provide a resolution of 1.5?
(d) How long should a column be if the plate height of the packing is 2.2×10^{-3} cm?
(e) If a flow rate of 7.10 cm/min is employed, how long will it take to elute the two species?

25-16. Repeat the calculations in Problem 25-15 assuming $K_M = 5.81$ and $K_N = 6.20$.

CHAPTER 26

Applications of Chromatography

In this chapter, we discuss how analytical separations are carried out by gas-liquid chromatography and by high-performance liquid chromatography.

26A GAS-LIQUID CHROMATOGRAPHY

In *gas-liquid chromatography* (GLC), the components of a vaporized sample are fractionated as a consequence of being partitioned between a mobile *gaseous* phase and a liquid-stationary phase held in a column.[1]

26A-1 Apparatus

The basic components of a typical instrument for performing gas chromatography are shown in Figure 26-1 and are described briefly in this section.

Carrier-Gas Supply
The gaseous mobile phase must be chemically inert. Helium is the most common mobile phase, although argon, nitrogen, and hydrogen are also

In all types of chromatography, liquid stationary phases are immobilized on solid surfaces by adsorption or by chemical bonding.

Helium is the most common mobile phase in gas-liquid chromatography.

[1]For detailed treatment of GLC, see J. Willet, *Gas Chromatography*. New York: Wiley, 1987; *Modern Practice of Gas Chromatography*, 2nd ed., R. L. Grob, Ed. New York: Wiley, 1985; J. A. Perry, *Introduction to Analytical Gas Chromatography*. New York: Marcel Dekker, 1981.

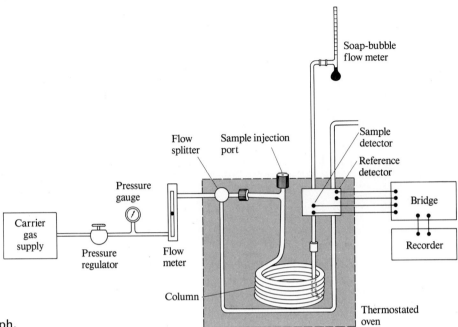

Figure 26-1

Diagram of a gas chromatograph.

A soap-bubble flow meter. (Courtesy of Chrompack Inc., Raritan, NJ.)

used. These gases are available in pressurized tanks. Pressure regulators, gauges, and flow meters are required to control the flow rate of the gas.

Pressures at the column inlet usually range from 10 to 50 psi (lb/in^2 above room pressure) and provide flow rates of 25 to 50 mL/min. Flow rates are generally measured by a simple soap-bubble meter, such as the one shown at the end of the column in Figure 26-1 and in the margin. A soap film is formed in the path of the gas when the rubber bulb containing a solution of soap is squeezed; the time required for this film to move between two graduations on the buret is measured and converted to a flow rate.

Sample-Injection System

Column efficiency requires that the sample be of a suitable size and that it be introduced as a "plug" of vapor; slow injection or oversized samples cause band spreading and poor resolution. A microsyringe is used to inject liquid samples through a rubber or silicone diaphragm, or septum, into a heated sample port located at the head of the column (the sample port is ordinarily about 50°C above the boiling point of the least volatile component of the sample). For ordinary packed analytical columns, sample sizes range from a few tenths of a microliter to 20 μL. Capillary columns require samples that are smaller by a factor of 100 or more. When these columns are used, a sample splitter is often needed to deliver only a small known fraction (1 : 100 to 1 : 500) of the injected sample, with the remainder going to waste.

Packed Columns

Two types of columns are encountered in gas-liquid chromatography: *packed* and *open-tubular* (also called *capillary*). Packed columns, which are discussed in this section, can accommodate larger samples than open-tubular columns and are generally more convenient to use. Capillary columns which are described in the next section, are of considerable importance because of their unparalleled resolution.

Modern packed columns are fabricated from glass or metal tubing; they are typically 2 to 3 m long and have inside diameters of 2 to 4 mm. The tubes are ordinarily formed as coils with diameters of roughly 15 cm to permit convenient thermostating in an oven.

The packing, or support, for a column holds the liquid stationary phase in place, so that the surface area exposed to the mobile phase is as large as possible. The ideal solid packing consists of small, uniform, spherical particles with good mechanical strength and with a specific surface of at least 1 m²/g. In addition, the material should be inert at elevated temperatures and uniformly wetted by the liquid phase. No substance that meets all these criteria perfectly is yet available.

The earliest, and still the most widely used, packings for gas chromatography are prepared from naturally occurring diatomaceous earth, which consists of the skeletons of thousands of species of single-celled plants that inhabited ancient lakes and seas. These support materials are often treated chemically with dimethylchlorosilane, which gives a surface layer of methyl groups. This treatment reduces the tendency of the packing to adsorb polar molecules.

The particle size of packings for gas chromatography typically range from 60 to 80 mesh (250 to 170 μm) or 80 to 100 mesh (170 to 149 μm). The use of smaller particles is not practical because the pressure drop across the column becomes prohibitively high.

Open Tubular Columns

Open tubular, or capillary, columns were first described in the 1950s, when it became apparent from theoretical considerations that such columns should provide separations that were unprecedented in terms of speed and number of theoretical plates.[2] At that time, several investigators demonstrated that columns with 300,000 plates or more were practical. Despite such spectacular results, the use of capillary columns was delayed until recently because of problems associated with their use. These problems have now become manageable, and a number of instrument vendors offer open tubular equipment for routine use.

Capillary columns, which are generally constructed of glass or fused silica, typically have inside diameters of 0.25 to 0.50 mm and lengths of 25 to 50 m. Their inner surfaces are coated with a thin layer of the stationary phase, which may be any of the liquids described in Section 26A-2.

The manufacture of fused-silica columns is based on techniques developed for the production of optical fibers. Silica capillaries, which have much thinner walls than their glass or metal counterparts, have outside diameters of about 0.3 mm. The tubes are given added strength by an outside polyimide coating. The resulting columns are quite flexible and

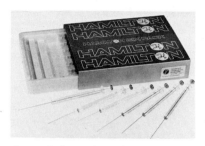

A set of microsyringes for sample injection. (Courtesy of Chrompack, Inc., Raritan, NJ.)

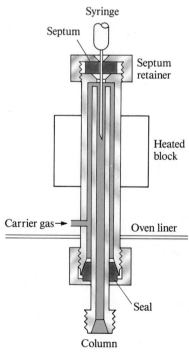

A heated sample port. (From H. H. Willard, L. L. Merritt, J. A. Dean, and F. A. Settle, *Instrumental Methods of Analysis,* 7th ed., p. 542. Belmont, CA: Wadsworth, 1988. With permission.)

---

[2]For a detailed description of open tubular columns, see M. L. Lee, F. J. Yang, and K. D. Bartle, *Open Tubular Column Gas Chromatography.* New York: Wiley, 1984.

In 1987, a column manufacturer reported drawing a fused-silica column 1300 m in length with over 2 million theoretical plates.

strong, and can be bent into coils with diameters of a few inches. An important advantage of fused-silica columns is their minimal tendency to adsorb analyte molecules.

Column Thermostating

Reproducible retention times require control of the column temperature to within a few tenths of a degree. For this reason, the coiled column is ordinarily housed in a thermostated oven. The optimum temperature depends upon the boiling points of the sample components. A temperature roughly equal to or slightly above the average boiling point of a sample results in a reasonable elution period (2 to 30 min). For samples with a broad boiling range, it may be necessary to employ temperature programming, whereby the column temperature is increased either continuously or in steps as the separation proceeds.

In general, optimum resolution is associated with minimal temperature. Lower temperatures, however, result in longer elution times and hence slower analyses. Figure 26-2a and b illustrates this principle. Often improved separation can be realized by temperature programming in which the column temperature is raised linearly or in steps as elution is carried out. The value of temperature programming is illustrated in Figure 26-2c.

Temperature programming is a technique in which the temperature of a gas-chromatographic column is increased continuously or in steps during elution.

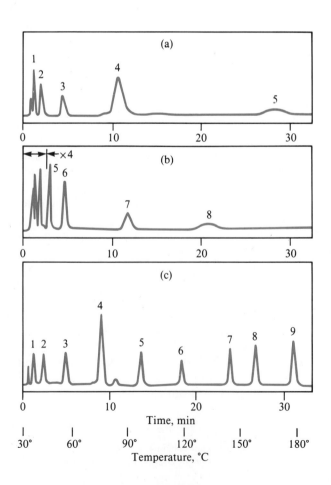

Figure 26-2

Effect of temperature on gas chromatograms. (a) Isothermal at 45°C; (b) isothermal at 145°C; (c) programmed from 30 to 180°C. (From W. E. Harris and H. W. Habgood, *Programmed Temperature Gas Chromatography*, p. 10. New York: Wiley, 1966. Reprinted with permission.)

Detectors

Detection devices for gas-liquid chromatography must respond rapidly to minute concentrations of solutes as they exit the column. The solute concentration in the carrier gas at any instant is no more than a few parts per thousand and often is smaller by one or two orders of magnitude. Moreover, the time during which a peak passes the detector is typically 1 s (or less), which requires that the device be capable of exhibiting its full response during this brief period.

Other desirable properties for a detector include linear response, stability, and uniform response for a wide variety of chemical species or, alternatively, a predictable and selective response toward one or more classes of solutes. No single detector fulfills all of these requirements. Two of the most widely used detectors are discussed in the sections that follow.

The Thermal-Conductivity Detector. The *thermal-conductivity detector,* or *katharometer,* which was one of the earliest detectors for gas chromatography, still finds wide application. This device consists of an electrically heated source whose temperature at constant electric power depends upon the thermal conductivity of the surrounding gas. The heated element may be a fine platinum, gold, or tungsten wire or, alternatively, a small thermistor. The electrical resistance of this element depends on the thermal conductivity of the gas. Twin detectors are ordinarily used, one being located ahead of the sample injection chamber and the other immediately beyond the column; alternatively, the gas stream is split as in Figure 26-1. The detectors are incorporated in two arms of a simple bridge circuit such that the thermal conductivity of the carrier is canceled. In addition, the effects of variations in temperature, pressure, and electric power are minimized.

A 25-m fused-silica capillary column. (Courtesy of Chrompack, Inc., Raritan, NJ.)

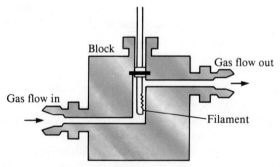

A typical thermal conductivity detector. (Courtesy of Varian Instrument Division, Palo Alto, CA.)

The thermal conductivities of helium and hydrogen are roughly six to ten times greater than those of most organic compounds. Thus, even small amounts of organic species cause relatively large decreases in the thermal conductivity of the column effluent, which results in a marked rise in detector temperature. Detection by thermal conductivity is less satisfactory with carrier gases whose conductivities more closely resemble those of most samples.

The advantages of the thermal-conductivity detector include its simplicity, its large linear dynamic range (about five orders of magnitude), its general response to both organic and inorganic species, and its nondestructive character, which permits collection of solutes after detection. Its chief limitation is its relatively low sensitivity ($\approx 10^{-8}$ g/mL of carrier). Other detectors exceed this sensitivity by factors of 10^4 to 10^7.

The Flame-Ionization Detector. Flame-ionization detectors are perhaps the most widely used of all detectors for gas chromatography. These devices are based upon the fact that most organic compounds, when pyrolyzed in a hot flame, produce ionic intermediates that conduct electricity through the flame. Hydrogen is used as the carrier gas with this detector, and the eluent is mixed with oxygen and combusted in a burner equipped with a pair of electrodes. Detection involves monitoring the conductivity of the combustion products. The ionization detector exhibits a high sensitivity ($\approx 10^{-13}$ g/mL), a large linear response, and low noise. It is also rugged and easy to use. A disadvantage is that this detector destroys the sample.

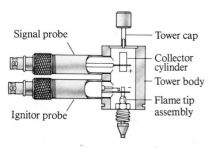

A flame-ionization detector. (Courtesy of Varian Instrument Division, Palo Alto, CA.)

The polarities of common organic functional groups, in increasing order, are as follows: aliphatic hydrocarbons < olefins < aromatic hydrocarbons < halides < sulfides < ethers < nitro compounds < esters $\approx$ aldehydes $\approx$ ketones < alcohols $\approx$ amines < sulfones < sulfoxides < amides < carboxylic acids < water.

26A-2 Liquid Phases for Gas-Liquid Chromatography

Several hundred liquids have been used as the stationary phase in gas-liquid chromatography. The successful separation of closely related compounds is often critically dependent upon the proper choice of stationary phase. The retention time for a solute depends upon its partition coefficient, which in turn is related to the nature of the stationary phase. Clearly, to be useful in gas-liquid chromatography, the immobilized liquid must generate different partition coefficients for different solutes; additionally, however, these coefficients must be neither extremely large nor extremely small. The time required for elution becomes prohibitive for species with large coefficients, while compounds with low coefficients pass so rapidly through the column that little or no separation occurs.

To have a reasonable residence time in the column, a species must show at least some degree of compatibility (solubility) with the stationary phase. Thus, the polarities of the two substances should be at least somewhat alike. For example, a stationary phase such as squalene (a high-molecular-weight, nonpolar, saturated hydrocarbon) might be chosen for separation of members of a nonpolar homologous series, such as hydrocarbons or ethers. On the other hand, a more polar stationary phase, such as polyethylene glycol, would be more effective for separating alcohols or amines. For aromatic hydrocarbons, benzyldiphenyl is a more appropriate stationary phase.

Among analytes of similar polarity, the elution order usually follows the order of boiling points; where these differ sufficiently, clean separations are feasible. Solutes with nearly identical boiling points but different polarities frequently require a stationary phase that will selectively retain one or more of the components by dipole interaction or by adduct formation. Another important interaction that often enhances selectivity is hydrogen bonding. For this effect to operate, the solute must be a proton donor and the stationary phase must contain a proton-acceptor group (such as oxygen, fluorine, or nitrogen), or the converse. Table 26-1 lists a few of the most widely used stationary phases.

Table 26-1

SOME COMMON STATIONARY PHASES FOR GAS-LIQUID CHROMATOGRAPHY

Trade Name	Chemical Composition	Maximum Temperature, °C	Polarity*	Type of Separation
Squalene	$C_{30}H_{62}$	150	NP	Hydrocarbons
OV-1	Polymethyl siloxane	350	NP	General purpose nonpolar
DC 710	Polymethylphenyl siloxane	300	NP	Aromatics
QF-1	Polytrifluoropropylmethyl siloxane	250	P	Amino acids, steroids, nitrogen compounds
XE-30	Polycyanomethyl siloxane	275	P	Alkaloids, halogenated compounds
Carbowax 20M	Polyethylene glycol	250	P	Alcohols, esters, essential oils
DEG adipate	Diethylene glycol adipate	200	SP	Fatty acids, esters
	Dinonyl phthalate	150	SP	Ketones, ethers, sulfur compounds

*NP = nonpolar; SP = semipolar; P = polar.

26A-3 Applications of Gas-Liquid Chromatography

Gas-liquid chromatography is applicable to species that are appreciably volatile and thermally stable at temperatures up to a few hundred degrees Celsius. An enormous number of compounds of commercial and scientific interest possess these qualities. Consequently, gas chromatography has been widely applied to the separation and determination of the components in a variety of sample types. Figure 26-3 shows chromatograms for a few such applications. The chromatogram in Figure 26-3f was obtained with an open tubular column and illustrates the power of this technique. Here, methane and nine heptane isomers were separated in just under one minute.

26B HIGH-PERFORMANCE LIQUID CHROMATOGRAPHY

Early columns for liquid chromatography were glass tubes with diameters of perhaps 10 to 50 mm that held 50- to 500-cm lengths of solid particles of the stationary phase. To ensure reasonable flow rates, the particle size of the solid was kept larger than 150 to 200 μm; even then, flow rates were, at best, a few tenths of a milliliter per minute. Attempts to speed up this classic procedure by application of vacuum or pressure were not effective because increases in flow rates were accompanied by increases in plate heights and accompanying decreases in column efficiency.

Early in the development of liquid chromatography, it was realized that large decreases in plate heights could be expected to accompany decreases in the particle size of packings. It was not until the late 1960s, however, that the technology for producing and using packings with particle diameters as small as 5 to 10 μm was developed. This technology required sophisticated instruments that contrasted markedly with the simple devices that preceded them. The name *high-performance liquid chro-*

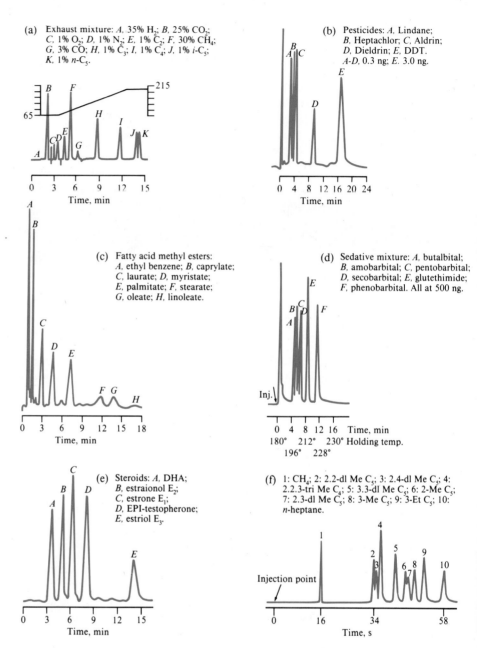

(a) Exhaust mixture: *A*, 35% H$_2$; *B*, 25% CO$_2$; *C*, 1% O$_2$; *D*, 1% N$_2$; *E*, 1% C$_2$; *F*, 30% CH$_4$; *G*, 3% CO; *H*, 1% C$_3$; *I*, 1% C$_4$; *J*, 1% *i*-C$_5$; *K*, 1% *n*-C$_5$.

(b) Pesticides: *A*, Lindane; *B*, Heptachlor; *C*, Aldrin; *D*, Dieldrin; *E*, DDT. *A-D*, 0.3 ng; *E*, 3.0 ng.

(c) Fatty acid methyl esters: *A*, ethyl benzene; *B*, caprylate; *C*, laurate; *D*, myristate; *E*, palmitate; *F*, stearate; *G*, oleate; *H*, linoleate.

(d) Sedative mixture: *A*, butalbital; *B*, amobarbital; *C*, pentobarbital; *D*, secobarbital; *E*, glutethimide; *F*, phenobarbital. All at 500 ng.

(e) Steroids: *A*, DHA; *B*, estraionol E$_2$; *C*, estrone E$_1$; *D*, EPI-testopherone; *E*, estriol E$_3$.

(f) 1: CH$_4$; 2: 2.2-dl Me C$_5$; 3: 2.4-dl Me C$_5$; 4: 2.2.3-tri Me C$_4$; 5: 3.3-dl Me C$_5$; 6: 2-Me C$_5$; 7: 2.3-dl Me C$_5$; 8: 3-Me C$_5$; 9: 3-Et C$_5$; 10: *n*-heptane.

Figure 26-3

Some examples of gas chromatographic separations.

matography (HPLC) is often employed to distinguish these newer procedures from their predecessors, which still find considerable use for preparative purposes.[3]

[3]References for HPLC include L. R. Snyder and J. J. Kirkland, *Introduction to High-Performance Liquid Chromatography,* 2nd ed. New York: Chapman and Hall, 1982; P. Brown and R. A. Hartwick, *High-Performance Liquid Chromatography.* New York: Wiley, 1988; S. Lindsey, *High-Performance Liquid Chromatography.* New York: Wiley, 1987; V. Meyer, *Practical High-Performance Liquid Chromatography.* New York: Wiley, 1988.

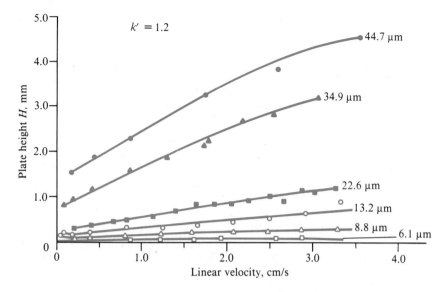

Figure 26-4
Effect of packing particle size and flow rate on plate height in liquid chromatography. Column dimensions: 30 cm × 2.4 mm. Solute: N,N-diethyl-*n*-aminoazobenzene. Mobile phase: mixture of hexane, methylene chloride, and isopropyl alcohol. (From R. E. Majors, *J. Chromatogr. Sci.*, **1973,** *11,* 92. With permission.)

Figure 26-4 illustrates the improved column efficiency that can be obtained with smaller-diameter particles. It is of interest to note that in none of these plots is the minimum shown in Figure 25-8a reached. The reason for this difference is that diffusion in liquids is much slower than in gases; consequently, band broadening because of diffusion in the mobile phase is observed only at extremely low flow rates.

26B-1 Apparatus

Pumping pressures of several hundred atmospheres are required to achieve reasonable flow rates with packings in the 3 to 10 μm size range, sizes common in modern liquid chromatography. As a consequence of these high pressures, the equipment for high-performance liquid chromatography tends to be considerably more elaborate and expensive than that encountered in other types of chromatography. Figure 26-5 (page 504) is a diagram showing the important components of a typical high-performance chromatograph.

Mobile-Phase Reservoirs and Solvent-Treatment Systems

A modern HPLC apparatus is equipped with one or more glass or stainless steel reservoirs, each of which contains 500 mL or more of a solvent. Provisions are often included to remove dissolved gases and dust from the liquids. The former produce bubbles in the column and thereby cause band spreading; in addition, both bubbles and dust interfere with detector performance. Degassers may consist of a vacuum pumping system, a distillation system, a device for heating and stirring, or, as shown in Figure 26-5, a system for *sparging,* in which the dissolved gases are swept out of solution by fine bubbles of an inert gas that is not soluble in the mobile phase.

An elution with a single solvent of constant composition is termed *isocratic.* In *gradient elution,* two (and sometimes more) solvent systems that differ significantly in polarity are employed. The ratio of the two solvents is varied in a programmed way, sometimes continuously and

> Sparging is a process in which dissolved gases are swept out of a solvent by bubbles of an inert, insoluble gas.

> An isocratic elution in HPLC is one in which the composition of the solvent remains constant.

A gradient elution in HPLC is one in which the composition of the solvent is changed either continuously or in a series of steps.

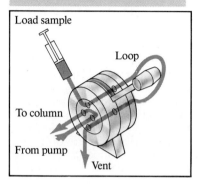

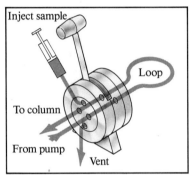

A sampling loop for liquid chromatography. (Courtesy of Beckman Instruments, Fullerton, CA.)

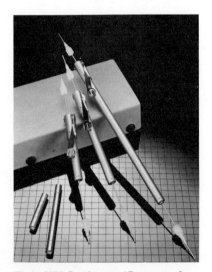

Three HPLC columns. (Courtesy of Chrompak, Inc., Raritan, NJ.)

sometimes in a series of steps. Gradient elution frequently improves separation efficiency, just as temperature programming helps in gas chromatography. Modern high-performance liquid-chromatography instruments are often equipped with proportionating valves that introduce liquids from two or more reservoirs at rates that vary continuously (Figure 26-5).

Pumping Systems

The requirements for the pumps used in liquid chromatography are severe and include (1) the generation of pressure up to 6000 psi, (2) pulse-free output, (3) flow rates ranging from 0.1 to 10 mL/min, (4) flow reproducibilities of 0.5% relative or better, and (5) resistance to corrosion by a variety of solvents.

It should be noted that the high pressures generated by liquid-chromatography pumps do not constitute an explosion hazard because liquids are not very compressible. Thus, rupture of a component results only in solvent leakage. To be sure, such leakage may constitute a fire hazard.

Sample-Injection Systems

Although syringe injection through an elastomeric septum is often used in liquid chromatography, this procedure is not very reproducible and is limited to pressures less than about 1500 psi. In *stop-flow* injection, the solvent flow is stopped momentarily, a fitting at the column head is removed, and the sample is injected directly onto the head of the packing by means of a syringe.

Although syringe injection finds considerable use owing to its simplicity, the most widely used method of sample introduction in liquid chromatography is based upon sampling loops like the one shown in the margin. These devices are often an integral part of modern liquid-chromatography equipment and have interchangeable loops that provide a choice of sample sizes ranging from 5 to 500 μL. The reproducibility of injections with a typical sampling loop is a few tenths of a percent relative.

Columns for High-Performance Liquid Chromatography

Liquid-chromatography columns are usually constructed from stainless steel tubing, although heavy-walled glass tubing is sometimes employed for lower-pressure applications (< 600 psi). Most columns range in length from 10 to 30 cm and have inside diameters of 4 to 10 mm. Column packings typically have particle sizes of 5 or 10 μm. Columns of this type often contain 40,000 to 60,000 plates/m. Recently, high-performance microcolumns with inside diameters of 1 to 4.6 mm and lengths of 3 to 7.5 cm have become available. These columns, which are packed with 3- or 5-μm particles, contain as many as 100,000 plates/m and have the advantage of speed and minimal solvent consumption.

The most common packing for liquid chromatography is prepared from silica particles, which are synthesized by agglomerating submicrometer silica particles under conditions that lead to larger particles with highly uniform diameters. The resulting particles are often coated with a thin organic film, which is chemically or physically bonded to the surface. Other packing materials include alumina particles, porous polymer particles, and ion-exchange resins.

Detectors

No highly sensitive, universal detector system, such as those for gas chromatography, is available for high-performance liquid chromatography. Thus, the system used will depend upon the nature of the sample. Table 26-2 lists some of the common detectors and their properties.

The most widely used detectors for liquid chromatography are based upon absorption of ultraviolet or visible radiation. Both photometers and spectrophotometers, specifically designed for use with chromatographic columns, are available from commercial sources. The former often make use of the 254- and 280-nm lines from a mercury source because many organic functional groups absorb in this region. Deuterium- or tungsten-filament sources with interference filters also provide a simple means of detecting absorbing species. Some modern instruments are equipped with filter wheels with several interference filters that can be rapidly switched into place. Spectrophotometric detectors are considerably more versatile than photometers and are also widely used in high-performance instruments.

Another detector that has found considerable application is based upon the changes in solvent refractive index that are caused by analyte molecules. In contrast to most of the other detectors listed in Table 26-2, the refractive-index indicator is general rather than selective and responds to the presence of all solutes.

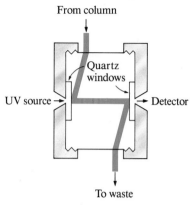

A UV detector for HPLC.

26B-2 High-Performance Partition Chromatography

Partition chromatography has become the most widely used of all liquid chromatographic procedures. This technique can be subdivided into *liquid-liquid* and *liquid–bonded-phase* chromatography. The difference between the two lies in the method by which the stationary phase is held on the support particles of the packing. With liquid-liquid, retention is by physical adsorption; with bonded-phase, covalent bonds are involved.

In liquid-liquid partition chromatography, the stationary phase is a solvent that is held in place by adsorption on the surface of packing particles.

Table 26-2

CHARACTERISTICS OF DETECTORS FOR HIGH-PERFORMANCE LIQUID CHROMATOGRAPHY*

Basis for Detection	Type†	Maximum Sensitivity‡	Flow-Rate Sensitive?	Temperature Sensitivity	Applicable to Gradient Elution?	Available Commercially?
UV absorption	S	2×10^{-10}	No	Low	Yes	Yes
IR absorption	S	10^{-6}	No	Low	Yes	Yes
Fluorometry	S	10^{-11}	No	Low	Yes	Yes
Refractive index	G	1×10^{-7}	No	$\pm 10^{-4} \, g \cdot mL^{-1}/°C$	No	Yes
Electrical conductivity	S	10^{-8}	Yes	2%/°C	No	Yes
Mass spectrometry	G	10^{-10}	No	None	Yes	Yes
Electrochemical	S	10^{-12}	Yes	1.5%/°C	No	Yes
Radiochemical	S	—	No	None	Yes	No

*Most data from L. R. Snyder and J. J. Kirkland, *Introduction to Modern Liquid Chromatography,* 2nd ed., p. 162. New York: Wiley-Interscience, 1979. With permission.

†G = general; S = selective.

‡Sensitivity for a favorable sample in grams per milliliter.

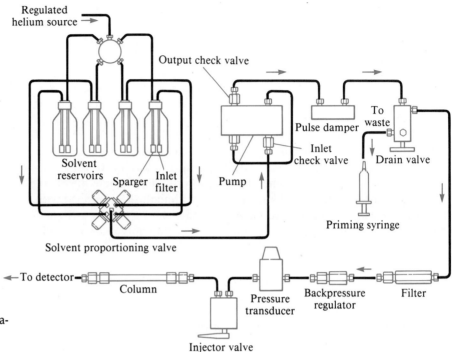

Figure 26-5
Schematic of an apparatus for high-performance liquid chromatography. (Courtesy of Perkin-Elmer, Norwalk, CT.)

In liquid–bonded-phase partition chromatography, the stationary phase is an organic species that is attached to the surface of the packing particles by

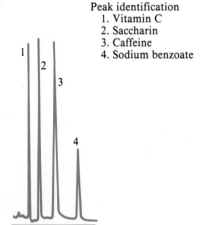

Peak identification
1. Vitamin C
2. Saccharin
3. Caffeine
4. Sodium benzoate

Time, min

Liquid–bonded-phase chromatogram of soft-drink additives. (Courtesy of Du-Pont Biotechnology Systems, Wilmington, DE.)

Early partition chromatography was exclusively liquid-liquid; now, however, bonded-phase methods predominate, with liquid-liquid separations being relegated to certain special applications.

Bonded-Phase Packings
Most bonded-phase packings are prepared by reaction of an organo-chlorosilane with the —OH groups formed on the surface of silica particles by hydrolysis in hot, dilute hydrochloric acid. The product is an organosiloxane. The reaction for one such SiOH site on the surface of a particle can be written as

$$-\overset{|}{\underset{|}{Si}}-O-H \;+\; Cl-\overset{\overset{\displaystyle CH_3}{|}}{\underset{\underset{\displaystyle CH_3}{|}}{Si}}-R \;\rightarrow\; -\overset{|}{\underset{|}{Si}}-O-\overset{\overset{\displaystyle CH_3}{|}}{\underset{\underset{\displaystyle CH_3}{|}}{Si}}-R \;+\; HCl$$

where R is often a straight-chain octyl or octyldecyl group. Other organic functional groups that have been bonded to silica surfaces include aliphatic amines, ethers, and nitriles, as well as aromatic hydrocarbons. Thus, a variety of polarities for the bonded stationary phase are available.

Bonded-phase packings have the advantage of markedly greater stability compared with physically held stationary phases. With the latter, periodic recoating of the solid surfaces is required because the stationary phase is gradually dissolved away in the mobile phase. Furthermore, gradient elution is not practical with liquid-liquid packings, again because of losses resulting from solubility to the mobile phase. The main disadvantage of bonded-phase packings is their somewhat limited sample capacity.

Normal- and Reversed-Phase Packings

Two types of partition chromatography are distinguishable, based upon the relative polarities of the mobile and stationary phases. Early work in liquid chromatography was based upon highly polar stationary phases such as triethylene glycol or water; a relatively nonpolar solvent such as hexane or *i*-propyl ether then served as the mobile phase. For historic reasons, this type of chromatography is now called *normal-phase chromatography*. In *reversed-phase chromatography*, the stationary phase is nonpolar, often a hydrocarbon, and the mobile phase is a relatively polar solvent (such as water, methanol, or acetonitrile).[4] In normal-phase chromatography, the *least* polar component is eluted first; *increasing* the polarity of the mobile phase then *decreases* the elution time. In the reversed-phase method, the *most* polar component elutes first, and *increasing* the mobile-phase polarity *increases* the elution time.

It has been estimated that more than three-quarters of all HPLC separations are currently performed with reversed-phase, bonded, octyl or octyldecyl siloxane packings. With such preparations, the long-chain hydrocarbon groups are aligned parallel to one another and perpendicular to the surface of the particle, giving a brushlike, nonpolar, hydrocarbon surface. The mobile phase used with these packings is often an aqueous solution containing various concentrations of such solvents as methanol, acetonitrile, and tetrahydrofuran.

Choice of Mobile and Stationary Phases

Successful partition chromatography requires a proper balance of intermolecular forces among the three participants in the separation process—the solute, the mobile phase, and the stationary phase. These intermolecular forces are described qualitatively in terms of the relative polarity possessed by each of the three reactants (see the margin note on page 498 for relative polarities of organic functional groups).

As a rule, most chromatographic separations are achieved by matching the polarity of the analyte to that of the stationary phase; a mobile phase of considerably different polarity is then used. This procedure is generally more successful than one in which the polarities of the analyte and the mobile phase are matched but are different from that of the stationary phase. This is because in the latter method, the stationary phase often cannot compete successfully for the sample components; retention times then become too short for practical application. At the other extreme is the situation where the polarities of the analyte and stationary phase are too much alike; here, retention times become inordinately long.

Applications

The two chromatograms in the margin illustrate typical applications of bonded-phase partition chromatography. Table 26-3 further illustrates the variety of samples to which the technique is applicable.

---

[4]For a detailed discussion of reversed-phase HPLC, see A. M. Krstulovic and P. R. Brown, *Reversed-Phase High-Performance Liquid Chromatography*. New York: Wiley, 1982.

In normal-phase partition chromatography, the stationary phase is polar and the mobile phase nonpolar. In reversed-phase partition chromatography, the polarity of these phases is reversed.

In normal-phase chromatography, the least polar analyte is eluted first. In reversed-phase chromatography, the least polar analyte is eluted last.

Peak identification
1. Methyl parathion
2. Ciodrin
3. Parathion
4. Dyfonate
5. Diazinon
6. EPN
7. Ronnel
8. Trithion

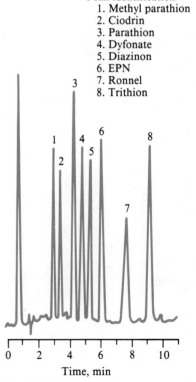

Liquid–bonded-phase chromatogram of organophosphate additives. (Courtesy of DuPont Biotechnology Systems, Wilmington, DE.)

Table 26-3

TYPICAL APPLICATIONS OF HIGH-PERFORMANCE PARTITION CHROMATOGRAPHY

Field	Typical Mixtures
Pharmaceuticals	Antibiotics, sedatives, steroids, analgesics
Biochemical	Amino acids, proteins, carbohydrates, lipids
Food products	Artificial sweeteners, antioxidants, aflatoxins, additives
Industrial chemicals	Condensed aromatics, surfactants, propellants, dyes
Pollutants	Pesticides, herbicides, phenols, PCBs
Forensic chemistry	Drugs, poisons, blood alcohol, narcotics
Clinical medicine	Bile acids, drug metabolites, urine extracts, estrogens

26B-3 High-Performance Adsorption Chromatography

All of the pioneering work in chromatography was based upon adsorption of analyte species on a solid surface. Here, the stationary phase is the surface of a finely divided polar solid. With such a packing, the analyte competes with the mobile phase for sites on the surface of the packing, and retention is the result of adsorption forces.

In adsorption chromatography, analyte species are adsorbed onto the surface of a polar packing.

Stationary and Mobile Phases

Finely divided silica and alumina are the only stationary phases that find extensive use in adsorption chromatography. Silica is preferred for most (but not all) applications because of its higher sample capacity and its wider range of useful forms. The adsorption characteristics of the two substances parallel one another. For both, retention times become longer as the polarity of the analyte increases.

In adsorption chromatography, the only variable that affects the partition coefficient of analytes is the composition of the mobile phase (in contrast to partition chromatography, where the polarity of the stationary phase can also be varied). Fortunately, enormous variations in retention and thus resolution accompany variations in the solvent system, and only rarely is a suitable mobile phase unavailable.

Applications of Adsorption Chromatography

Currently, liquid-solid HPLC is used extensively for the separation of relatively nonpolar, water-insoluble organic compounds with molecular weights that are less than about 5000. A particular strength of adsorption chromatography not shared by other methods is its ability to resolve isomeric mixtures such as meta- and para-substituted benzene derivatives.

26B-4 High-Performance Ion-Exchange Chromatography

We have already considered some of the applications of ion-exchange resins to analytical separations. In addition, these materials are useful as stationary phases for liquid chromatography, where they are used to sepa-

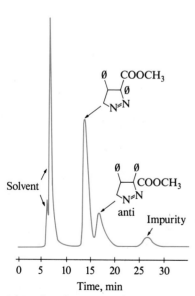

Adsorption-chromatography separation of pyrazoline isomers.

rate charged species.[5] In most cases, conductivity measurements are used for detecting eluents.

Two types of ion chromatography are currently in use: *suppressor-based* and *single-column*. They differ in the method used to prevent the conductivity of the eluting electrolyte from interfering with the measurement of analyte conductivities.

Ion Chromatography Based on Suppressors

Conductivity detectors have many of the properties of the ideal detector. They can be highly sensitive, they are universal for charged species, and as a general rule, they respond in a predictable way to concentration changes. Furthermore, such detectors are simple to operate, inexpensive to construct and maintain, easy to miniaturize, and ordinarily give prolonged, trouble-free service. The only limitation to the use of conductivity detectors, a limitation that delayed their general application to ion chromatography until the mid-1970s, arises from the high electrolyte concentrations required to elute most analyte ions in a reasonable time. As a consequence, the conductivity from the mobile-phase components tends to swamp that from the analyte ions, thus greatly reducing the detector sensitivity.

In 1975, the problem created by the high conductance of eluents was solved by the introduction of an *eluent-suppressor column* immediately following the ion-exchange column.[6] The suppressor column is packed with a second ion-exchange resin that effectively converts eluent ions to a molecular species of limited ionization without affecting the conductivity due to analyte ions. For example, when cations are being separated and determined, hydrochloric acid is chosen as the eluting reagent, and the suppressor column is an anion-exchange resin in the hydroxide form. The product of the reaction between the eluent and the suppressor is water:

$$H^+(aq) + Cl^-(aq) + resin^+OH^-(s) \rightarrow resin^+Cl^-(s) + H_2O$$

The analyte cations are of course not retained by this second column.

For anion separations, the suppressor packing is the acidic form of a cation-exchange resin and sodium bicarbonate or carbonate is the eluting agent. The reaction in the suppressor is

$$Na^+(aq) + HCO_3^-(aq) + resin^-H^+(s) \rightarrow resin^-Na^+(s) + H_2CO_3(aq)$$

The largely undissociated carbonic acid does not contribute significantly to the conductivity.

---

[5]References: F. C. Smith Jr. and R. C. Chang, *The Practice of Ion Chromatography*. New York: Wiley, 1982; R. E. Smith, *Ion Chromatography Applications*. Boca Raton, FL: CRC Press, 1987; *Ion Chromatography*, J. T. Tarter, Ed. New York: Marcel Dekker, 1987.

[6]H. Small, T. S. Stevens, and W. C. Bauman, *Anal. Chem.*, **1975,** *47,* 1801.

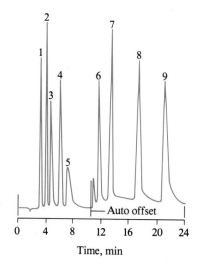

	ppm
1. Li^+	0.5
2. Na^+	2
3. NH_4^+	3
4. K^+	3
5. Morpholine	30
6. Cyclohexylamine	10
7. Mg^{2+}	1
8. Ca^{2+}	2
9. Sr^{2+}	10

Ion chromatogram of a mixture of cations. (Courtesy of Dionex, Sunnyvale, CA.)

1. SiO_3^{2-}	2 ppm
2. F^-	0.4 ppm
3. Formate	1 ppm
4. Cl^-	2 ppm
5. NO_2^-	2 ppm
6. Br^-	2 ppm
7. NO_3^-	4 ppm

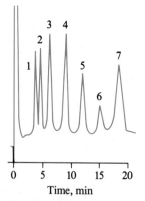

Ion chromatogram of a mixture of anions. (Courtesy of Dionex, Sunnyvale, CA.)

Single-Column Ion Chromatography

Recently, equipment has become available commercially for ion chromatography in which no suppressor column is used. This new approach depends upon the small differences in conductivity between the eluted sample ions and the prevailing eluent ions. Low-capacity exchangers that permit elution with dilute eluent solutions are used to amplify these differences. Furthermore, eluents of low equivalent conductance are chosen.[7]

Single-column ion chromatography offers the advantage of not requiring special equipment for suppression. It is, however, a somewhat less sensitive method for determining anions than suppressor-column methods.

Applications

The chromatograms in the margin illustrate two applications of ion chromatography based upon a suppressor column and conductometric detection. In each, the ions were present in the parts-per-million range. The method is particularly important for anion analysis because no other rapid and convenient method for handling mixtures of this type now exists.

26B-5 High-Performance Size-Exclusion Chromatography

Size-exclusion, or gel, chromatography is the newest liquid-chromatography procedure. It is a powerful technique that is particularly applicable to high-molecular-weight species.[8]

Packings

Packings for size-exclusion chromatography consist of small (≈ 10 μm) silica or polymer particles containing a network of uniform pores into which solute and solvent molecules can diffuse. While in the pores, molecules are effectively trapped and removed from the mobile phase. The average residence time of analyte molecules depends upon their effective size. Molecules that are significantly larger than the average pore size of the packing are excluded and thus suffer no retention; that is, they travel through the column at the rate of the mobile phase. Molecules that are appreciably smaller than the pores can penetrate throughout the pore maze and are thus entrapped for the greatest time; they are last to be eluted. Between these two extremes are intermediate-size molecules whose average penetration into the packing pores depends upon their diameters. The fractionation that occurs within this group is directly related to molecular size and, to some extent, molecular shape.

Note that size-exclusion separations differ from the other chromatographic procedures in that no chemical or physical interactions between analytes and the stationary phase are involved. Indeed, every effort is made to avoid such interactions because they lead to impaired column efficiencies.

[7]See R. M. Becker, *Anal. Chem.*, **1980,** *52,* 1510; J. R. Benson, *Amer. Lab.*, **1985,** (6), 30; and T. Jupille, *Amer. Lab.*, **1986,** (5), 114.

[8]See W. W. Yao, J. J. Kirkland, and D. D. Bly, *Modern Size Exclusion Liquid Chromatography.* New York: Wiley, 1979.

Numerous size-exclusion packings are on the market. Some are hydrophilic for use with aqueous mobile phases; others are hydrophobic and are used with nonpolar organic solvents. Chromatography based on the hydrophilic packings is sometimes called *gel filtration*, while techniques based on hydrophobic packings are termed *gel permeation*. With both types of packings, a wide variety of pore diameters are available. Ordinarily, a given packing will accommodate a 2- to 2.5-decade range of molecular weight. The average molecular weight suitable for a given packing may be as small as a few hundred or as large as several million.

Applications

The two chromatograms in the margin illustrate typical applications of size-exclusion chromatography. In the first, a hydrophilic packing was used to exclude molecular weights greater than 1000. The second chromatogram was obtained with a hydrophobic packing in which the eluent was tetrahydrofuran and the sample was a commercial epoxy resin in which each monomer unit had a molecular weight of 280 (n = number of monomer units).

Another important application of size-exclusion chromatography is in the rapid determination of the molecular weight or the molecular-weight distribution of large polymers or natural products. Here, the elution volumes of the sample are compared with elution volumes for a series of standard compounds that have similar chemical characteristics.

26C COMPARISON OF HIGH-PERFORMANCE LIQUID CHROMATOGRAPHY WITH GAS-LIQUID CHROMATOGRAPHY

Table 26-4 provides a comparison between high-performance liquid chromatography and gas-liquid chromatography. When either is applicable, gas-liquid chromatography offers the advantage of speed and simplicity of

Gel filtration is size-exclusion chromatography with a hydrophilic packing. It is used for separating polar species.

Gel permeation is size-exclusion chromatography with a hydrophobic packing. It is used to separate nonpolar species.

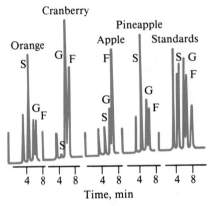

Gel-filtration chromatogram for glucose (G), fructose (F), and sucrose (S) in canned juices. (Courtesy of DuPont Biotechnology Systems, Wilmington, DE.)

Table 26-4
COMPARISON OF HIGH-PERFORMANCE LIQUID CHROMATOGRAPHY AND GAS-LIQUID CHROMATOGRAPHY

Characteristics of both methods
Efficient, highly selective, widely applicable
Only small sample required
May be nondestructive of sample
Readily adapted to quantitative analysis
Advantages of HPLC
Can accommodate nonvolatile and thermally unstable samples
Generally applicable to inorganic ions
Advantages of GLC
Simple and inexpensive equipment
Rapid
Unparalleled resolution (with capillary columns)
Easily interfaced with mass spectroscopy

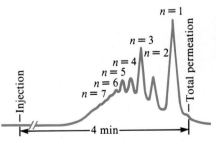

Gel-permeation separation of components in an epoxy resin. (Courtesy of Du Pont Biotechnology Systems, Wilmington, DE.)

equipment. On the other hand, high-performance liquid chromatography is applicable to nonvolatile substances (including inorganic ions) and thermally unstable materials, whereas gas-liquid chromatography is not. Often, the two methods are complementary.

26D QUESTIONS AND PROBLEMS

*26-1. What are the effects of slow injection of a sample onto a gas-chromatographic column?

26-2. How do planar and column chromatography differ?

*26-3. What is meant by temperature programming in gas chromatography?

26-4. What is a chromatogram?

*26-5. What is the fundamental difference between N and H as measures of column separation efficiencies?

26-6. What variables must be controlled if satisfactory quantitative data are to be obtained from chromatograms?

*26-7. Describe the physical differences between open tubular and packed columns. What are the advantages and disadvantages of each?

26-8. Define the following terms used in HPLC:
  *(a) sparging   *(e) stop-flow injection
   (b) gradient elution   (f) sampling loops
  *(c) isocratic elution   *(g) bonded-phase packings
   (d) reversed-phase   (h) gel filtration
       packing

26-9. Indicate the order in which the following compounds would be eluted from an HPLC column with a reversed-phase packing:
  *(a) benzene, diethyl ether, n-hexane
   (b) acetone, dichloroethane, acetamide

26-10. Indicate the order of elution of the following compounds from a normal-phase HPLC column:
  *(a) ethyl acetate, acetic acid, dimethylamine
  *(b) propylene, hexane, benzene, dichlorobenzene

*26-11. Describe the fundamental difference between adsorption and partition chromatography.

26-12. Describe the fundamental difference between ion-exchange and size-exclusion chromatography.

*26-13. Describe the difference between gel-filtration and gel-permeation chromatography.

26-14. What types of species can be separated by HPLC but not by GLC?

*26-15. One method for quantitative determination of the concentration of constituents in a sample analyzed by gas chromatography is the area-normalization method. Here, complete elution of all sample constituents is necessary. The area under each peak is then measured and corrected for differences in detector response to the different eluates. This correction involves multiplication of the area by an empirically determined correction factor. The concentration of the analyte is found from the ratio of its corrected area to the total corrected area under all peaks. For a chromatogram containing three peaks, the relative areas were found to be 16.4, 45.2, and 30.2, in the order of increasing retention time. Calculate the percentage of each compound if the relative detector responses were 0.60, 0.78, and 0.88, respectively.

*26-16. Peak areas and relative detector responses are to be used to determine the concentration of the five species in a sample. The area-normalization method described in Problem 26-15 is to be used. The relative areas for the five gas-chromatographic peaks are given below. Also shown are the relative responses of the detector. Calculate the percentage of each component in the mixture.

Compound	Relative Peak Area	Relative Detector Response
A	32.5	0.70
B	20.7	0.72
C	60.1	0.75
D	30.2	0.73
E	18.3	0.78

Chapter 27

Chemicals, Apparatus, and Unit Operations

This chapter is concerned with the practical aspects of the unit operations encountered in an analytical laboratory as well as with the apparatus and chemicals used in these operations.

27A THE SELECTION AND HANDLING OF REAGENTS AND OTHER CHEMICALS

The purity of reagents has an important bearing upon the accuracy attained in any analysis. It is therefore essential that the quality of a reagent be consistent with the use for which it is intended.

27A-1 The Classification of Chemicals

Reagent Grade
Reagent-grade chemicals conform to the minimum standards set forth by the Reagent Chemical Committee of the American Chemical Society[1] and are used wherever possible in analytical work. Some suppliers label their products with the maximum limits of impurity allowed by the ACS specifications; others print actual assay for the various impurities.

---

[1]Committee on Analytical Reagents, *Reagent Chemicals*, 7th ed. Washington, D.C.: American Chemical Society, 1986.

The National Institute of Standards and Technology (NIST) is the current name of what was formerly the National Bureau of Standards (NBS). The 1988–1989 catalog of reference materials still bears the NBS label.

Primary-Standard Grade

The qualities required of a *primary standard*—in addition to extraordinary purity—are set forth in Section 6A-3. Primary-standard reagents have been carefully analyzed by the supplier, and the assay is printed on the container label. The National Institute of Standards and Technology is an excellent source for primary standards. This agency also provides *reference standards*, which are complex substances that have been exhaustively analyzed.[2]

Special-Purpose Reagent Chemicals

Chemicals that have been prepared for a specific application are also available. Included among these are solvents for spectrophotometry and high-performance liquid chromatography. Information pertinent to the intended use is supplied with these reagents. Data provided with a spectrophotometric solvent, for example, might include its absorbance at selected wavelengths and its ultraviolet cutoff wavelength.

27A-2 Rules for Handling Reagents and Solutions

High quality in a chemical analysis requires reagents and solutions of established purity. A freshly opened bottle of a reagent-grade chemical can ordinarily be used with confidence; whether this same confidence is justified when the bottle is half empty depends entirely on the way it was handled after being opened. The following rules should be observed to prevent the accidental contamination of reagents and solutions.

1. Select the best grade of chemical available for analytical work. Whenever possible, pick the smallest bottle that will supply the desired quantity.
2. Replace the top of every container *immediately* after removal of the reagent; do not rely on someone else to do this.
3. Hold the stoppers of reagent bottles between your fingers; never set a stopper down on any work surface.
4. *Unless specifically directed otherwise, never return any excess reagent to a bottle.* The money saved by returning excesses is seldom worth the risk of contaminating the entire bottle.
5. Unless directed otherwise, never insert spatulas, spoons, or knives into a bottle that contains a solid chemical. Instead, shake the capped bottle vigorously or tap it gently against a wooden table to break up any encrustation; then pour out the desired quantity. These measures are occasionally ineffective, and in such cases a clean porcelain spoon should be used to scoop out the chemical.
6. Keep the reagent shelf and the laboratory balance clean and neat. Clean up any spillages immediately, even though someone else is waiting to use the same chemical or reagent.
7. Observe local regulations concerning the disposal of surplus reagents and solutions.

---

[2]U.S. Department of Commerce, *NBS Standard Reference Materials Catalog, 1988-89, NBS Special Publication 260*. Washington, D.C. 20234: Government Printing Office, 1988.

27B THE CLEANING AND MARKING OF LABORATORY WARE

A chemical analysis is ordinarily performed in duplicate or triplicate. Thus, each vessel that holds a sample must be marked so that its contents can be positively identified. Flasks, beakers, and some crucibles have small etched areas on which semipermanent markings can be made with a pencil.

Special marking inks are available for porcelain surfaces. The marking is baked permanently into the glaze by heating at a high temperature. A saturated solution of iron(III) chloride, while not as satisfactory as the commercial preparations, can also be used for marking.

Every beaker, flask, or crucible that will contain the sample must be thoroughly cleaned before being used. The apparatus should be washed with a hot detergent solution and then rinsed—initially with copious amounts of tap water and finally with several small portions of distilled water.[3] Properly cleaned glassware will be coated with a uniform and unbroken film of water. *It is seldom necessary to dry the interior surface of glassware before use;* drying is ordinarily a waste of time at best and a potential source of contamination at worst.

An organic solvent, such as benzene or acetone, may be effective in removing grease films. Chemical suppliers also market preparations for the elimination of such films.

27C THE EVAPORATION OF LIQUIDS

It is frequently necessary to decrease the volume of a solution that contains a nonvolatile solute. Figure 27-1 illustrates how this operation is performed. The ribbed cover glass permits vapors to escape and protects the remaining solution from accidental contamination. Less satisfactory is the use of glass hooks to provide space between the rim of the beaker and a conventional cover glass.

Evaporation is frequently difficult to control, owing to the tendency of some solutions to overheat locally. The bumping that results can be sufficiently vigorous to cause partial loss of the solution. Careful and gentle heating will minimize the danger of such loss. Where their use is permissible, glass beads will also minimize bumping.

Some unwanted species can be eliminated during evaporation. For example, chloride and nitrate can be removed from a solution by adding sulfuric acid and evaporating until copious white fumes of sulfur trioxide are given off (this operation must be performed in a hood). Urea is effective in removing nitrate ion and nitrogen oxides from acidic solutions. The elimination of ammonium chloride is best accomplished by adding concentrated nitric acid and evaporating the solution to a small volume. Heating causes rapid oxidation of ammonium ion; the solution is then evaporated to dryness.

Organic constituents can frequently be eliminated from a solution by adding sulfuric acid and heating to the appearance of sulfur trioxide fumes

> Bumping is sudden, often violent boiling that tends to spatter solution out of its container.

Figure 27-1

Arrangement for the evaporation of a liquid.

---

[3]References to distilled water in this chapter and the one that follows apply equally to deionized water.

Wet-ashing is the oxidation of the organic constituents of a sample with oxidizing reagents such as nitric acid, sulfuric acid, hydrogen peroxide, aqueous bromine, or a combination of these reagents.

(hood); this process is known as *wet-ashing*. Nitric acid can be added toward the end of heating to hasten oxidation of the last traces of organic matter.

27D   THE MEASUREMENT OF MASS

In most analyses, an *analytical balance* must be used to obtain highly accurate weights. Less accurate *laboratory balances* are also employed in the analytical laboratory for weighings where the demands for reliability are not critical.

27D-1 The Distinction Between Mass and Weight

Mass is an invariant measure of the amount of matter in an object. Weight is the force of gravitational attraction between that object and the earth.

You should understand the difference between mass and weight. *Mass* is an invariant measure of the amount of matter in an object. *Weight* is the force of attraction that exists between an object and its surroundings, principally the earth. Because gravitational attraction varies with geographic location, the weight of an object depends upon where you weigh it. For example, a crucible weighs less in Denver than in Atlantic City (both cities are at approximately the same latitude) because the attractive force between crucible and earth is smaller at the higher altitude. Similarly, the crucible weighs more in Seattle than in Panama (both cities are at sea level) because the earth is somewhat flattened at the poles, and the force of attraction increases measurably with latitude. The mass of the crucible, however, remains constant regardless of where you measure it.

Weight and mass are related by the familiar expression

$$W = Mg$$

where W is the weight of an object, M is its mass, and g is the acceleration due to gravity.

A chemical analysis is always based upon mass so that the results will not depend upon locality. A balance is used to compare the weight of an object with the weight of one or more standard masses. Because g affects both unknown and known in the same way, an equality in their weights also indicates an equality in mass.

The distinction between mass and weight is often lost in common usage, and the process of comparing masses is ordinarily called *weighing*. In addition, the object of known mass as well as the results of weighing are called *weights*. You should always remember however, that analytical data are based upon mass rather than weight.

27D-2 Types of Analytical Balances

An analytical balance has a maximum capacity that ranges from 1 g to several kilograms and a precision at maximum capacity of at least one part in 10^5.

By definition, an *analytical balance* is a weighing instrument with a maximum capacity that ranges from 1 g to a few kilograms with a precision of at least 1 part in 10^5 at maximum capacity. The precision and accuracy of many modern analytical balances exceed one part in 10^6 at full capacity.

The most commonly encountered analytical balances (*macrobalances*) have a maximum capacity ranging between 160 and 200 g; measurements can be made with a standard deviation of ±0.1 mg. *Semimicrobalances* have a maximum loading of 10 to 30 g with a precision of ±0.01 mg. A typical *microanalytical balance* has a capacity of 1 to 3 g and a precision of ±0.001 mg.

The analytical balance has undergone a dramatic evolution over the past several decades. The traditional analytical balance had two pans attached to either end of a lightweight beam that pivoted about a knife edge located in the center of the beam. The object to be weighed was placed on one pan; sufficient standard weights were then added to the other pan to restore the beam to its original position. Weighing with such an *equal-arm balance* was tedious and time-consuming.

The first *single-pan analytical balance* appeared on the market in 1946. The speed and convenience of weighing with this balance were vastly superior to what could be realized with the traditional equal-arm balance. Consequently, this balance rapidly replaced the latter in most laboratories. The single-pan balance is still extensively used. The design and operation of a single-pan balance are discussed in Section 27D-3.

The single-pan balance is currently being replaced by the *electronic analytical balance,* which has neither a beam nor a knife edge; this type of balance is discussed briefly in Section 27D-4.

27D-3 The Single-Pan Mechanical Analytical Balance

Components
Although they differ considerably in appearance and performance characteristics, all mechanical balances—equal-arm as well as single-pan—have several common components. Figure 27-2 is a diagram of a typical single-

> The most common type of analytical balance, a macrobalance, has a maximum load of 160 to 200 g and a precision of 0.1 mg.

> A semimicrobalance has a maximum load of 10 to 30 g and a precision of 0.01 mg.

> A microbalance has a maximum load of 1 to 3 g and a precision of 0.001 mg, or 1 μg.

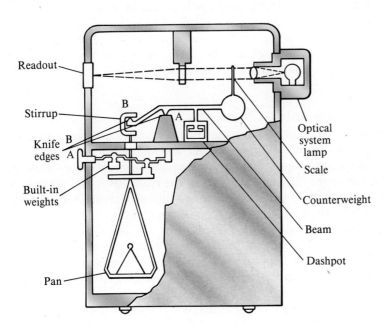

Figure 27-2

Modern single-pan analytical balance. (From R. M. Schoonover, *Anal. Chem.,* **1982,** *54,* 973A. Published 1982 American Chemical Society.)

pan balance. Fundamental to this instrument is a lightweight *beam* that is supported on a planar surface by a prism-shaped knife edge (A). Attached to the left end of the beam is a pan to hold the object to be weighed and a full set of weights held in place by hangers. These weights can be lifted from the beam one at a time by a mechanical arrangement controlled by a set of knobs on the exterior of the balance case. The right end of the beam holds a counterweight of such a size as to just balance the pan and weights on the left end.

A second knife edge (B) is located near the left end of the beam and serves to support a second planar surface, which is located in the inner side of a *stirrup* that couples the pan to the beam. The two knife edges and their planar surfaces are fabricated from extraordinarily hard materials (agate or synthetic sapphire) and form two bearings that permit the beam and pan to move with a minimum of friction. The performance of a mechanical balance is critically dependent upon the perfection of these two bearings.

Single-pan balances are also equipped with a *beam arrest* and a *pan arrest*. The beam arrest is a mechanical device that (1) raises the beam so that the central knife edge no longer touches its bearing surface and (2) simultaneously frees the stirrup from contact with the outer knife edge. The purpose of both arrest mechanisms is to prevent damage to the bearings while objects are being placed upon or removed from the pan. When engaged, the pan arrest supports most of the weight of the pan and its contents and thus prevents oscillation. Both arrests are controlled by a single lever mounted on the outside of the balance case and should be engaged whenever the balance is not in use.

An *air damper* (also known as a *dashpot*) is mounted near the end of the beam opposite the pan. This device consists of a piston that moves within a concentric cylinder attached to the balance case. Air in the cylinder undergoes expansion and contraction as the beam is set in motion; the beam rapidly comes to rest as a result of this opposition to motion.

Protection from air currents is needed to permit discrimination between small differences in weight (<1 mg). An analytical balance is thus always enclosed in a case equipped with doors to permit the introduction or removal of objects.

Weighing with a Single-Pan Balance

The beam of a properly adjusted balance assumes an essentially horizontal position with no object on the pan and all the weights in place. When the pan and beam arrests are disengaged, the beam is free to rotate around the knife edge. Placing an object on the pan causes the left end of the beam to move downward. Weights are then removed systematically one by one from the beam until the imbalance is less than 100 mg. The angle of deflection of the beam with respect to its original horizontal position is directly proportional to the milligrams of additional weight that must be removed to restore the beam to its original horizontal position. The optical system shown in the upper part of Figure 27-2 measures this angle of deflection and converts it to milligrams. A *reticle*, which is a small transparent screen mounted on the beam, is scribed with a scale that reads 0 to 100 mg. A beam of light passes through the scale to an enlarging lens, which in turn forcuses a small part of the enlarged scale onto a frosted glass plate located on the front of the balance. A vernier makes it possible to read this scale to the nearest 0.1 mg.

Instructions for the Use of a Single-Pan Balance

1. Zero the empty balance. Rotate the arrest control to the fully released position. Adjust the vernier so that the scale reading is zero.
2. Arrest the balance again, and place the object to be weighed on the pan. Turn the arrest knob to the partially released position. Rotate the knob controlling the heaviest likely weight for the object until the illuminated scale changes or the instruction "remove weight" appears; then turn the weight knob back one stop. Repeat this procedure with all the other knobs, working systematically through the lighter weights.
3. Turn the arrest control to its fully released position and allow time for the balance to come to equilibrium. The weight of the object is the sum of the weights indicated on the dials in addition to that displayed on the illuminated scale. Use the vernier to establish the weight to the nearest 0.1 mg.

Note. Directions for operation differ somewhat with make and model. Consult the instructor for any modifications to this procedure that may be required.

Summary of Rules for the Use of a Single-Pan Balance

The single-pan mechanical analytical balance is a delicate instrument. In using it you should adhere to the following rules.

To Minimize Wear or to Avoid Damage.

1. Be certain that you have engaged the arresting mechanisms for the beam whenever the loading is being changed and whenever the balance is not in use.
2. Center the load on the pan insofar as possible.
3. Protect the balance from corrosion. Objects to be placed on the pan should be limited to nonreactive metals, nonreactive plastics, and vitreous materials.
4. Observe special precautions (Section 27E-6) for the weighing of liquids.
5. Consult with the instructor if the balance appears to need adjustment.
6. Keep the balance and its case scrupulously clean. A camel's-hair brush is useful for the removal of spilled material or dust.

To Obtain Reliable Weighing Data.

7. Always allow an object that has been heated to return to room temperature before weighing it.
8. Use tongs or finger pads when handling weighing containers that have been heated to dryness. Dried objects can pick up enough moisture from your fingers to cause a detectable change in weight.

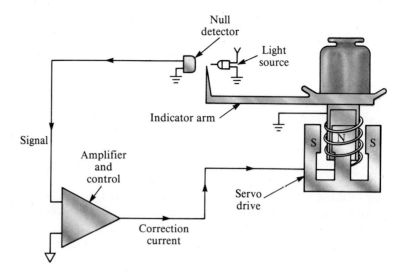

Figure 27-3
Electronic analytical balance.
(From R. M. Schoonover, *Anal. Chem.*, **1982**, *54*, 974A. Published 1982 American Chemical Society.)

A servo system is a device in which a small electric signal causes a mechanical system to return to a null position.

27D-4 The Electronic Analytical Balance[4]

Figure 27-3 is a diagram of an electronic analytical balance. The pan rides above a hollow metal cylinder that is surrounded by a coil and fits over the inner pole of a cylindrical permanent magnet. An electric current in the coil creates a magnetic field that supports, or levitates, the cylinder, the pan and indicator arm, and whatever load is on the pan. The current is adjusted so that the level of the indicator arm is in the null position when the pan is empty. Placing an object on the pan causes the pan and indicator arm to move downward, which increases the amount of light striking the photocell of the null detector. The increased current from the photocell is amplified and fed into the coil, creating a larger magnetic field, which returns the pan to its original null position. A device such as this, in which a small electric current causes a mechanical system to maintain a null position, is called a *servo system*. The current required to keep the pan and object in the null position is directly proportional to the weight of the object and is readily measured, digitized, and displayed. Calibration of an electronic balance involves the use of a standard mass and adjustment of the current so that the mass of the standard is exhibited on the display.

Figure 27-4 shows the configurations for two electronic analytical balances. In each, the pan is tethered to a system of constraints known collectively as a *cell*. The cell incorporates several *flexures* that permit limited movement of the pan and prevent torsional forces (resulting from off-center loading) from disturbing the alignment of the balance mechanism. At null, the beam is parallel to the gravitational horizon and each flexure pivot is in a relaxed position.

Figure 27-4a shows an electronic balance with the pan located below the cell. Higher precision is achieved with this arrangement than with the top-loading design shown in Figure 27-4b. Even so, top-loading electronic

[4]For a more detailed discussion, see R. M. Schoonover, *Anal. Chem.*, **1982**, *54*, 973A; K. M. Lang, *Amer. Lab.*, **1983**, *15* (3), 72.

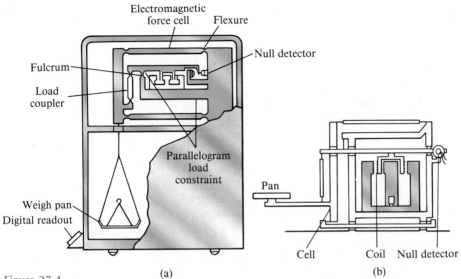

Figure 27-4
Electronic analytical balances. (a) Classical configuration with pan beneath the cell. (From R. M. Schoonover, *Anal. Chem.,* **1982,** *54,* 976A. Published 1982 American Chemical Society.) (b) A top-loading design. Note that the mechanism is enclosed in a windowed case. (Reprinted with permission from *Amer. Lab.,* **1983,** *15* (3), 72. Copyright 1983 by International Scientific Communications, Inc.)

balances have a precision that equals or exceeds that of the best mechanical balances and additionally provide unencumbered access to the pan.

Electronic balances generally feature an automatic *taring control* that causes the display to read zero with a container (such as a boat or weighing bottle) on the pan. Most balances permit taring to 100% of capacity.

Some electronic balances have dual capacities and dual precisions. These features permit the capacity to be decreased from that of a macrobalance to that of a semimicrobalance (30 g) with a concomitant gain in precision to 0.01 mg. Thus, the chemist has effectively two balances in one.

A modern electronic analytical balance provides unprecedented speed and ease of use. For example, one model is controlled by touching a single bar at various positions along its length. One position on the bar turns the instrument on or off, another automatically calibrates the balance against a standard mass, and a third zeros the display, either with or without an object on the pan. Reliable weighing data are obtainable with little or no instruction.

27D-5 Sources of Error in Weighing

Correction for Buoyancy[5]
A *buoyancy error* will affect data if the density of the object being weighed differs significantly from that of the standard weights. This error has its origin in the fact that the buoyant force exerted by the medium (air)

> A tare is the mass of an empty sample container. Taring is the process of setting a balance to read zero in the presence of the tare.

> A buoyancy error is the weighing error that develops when the density of the object weighed is significantly different from that of the standard weights.

_____
[5]For further information, see R. Battino and A. G. Williamson, *J. Chem. Educ.,* **1984,** *64,* 51.

on the object being weighed is different from the buoyant force exerted on the weights. Correction for buoyancy is accomplished with the equation

$$W_1 = W_2 + W_2 \left(\frac{d_{air}}{d_{obj}} - \frac{d_{air}}{d_{wts}} \right) \qquad (27\text{-}1)$$

where W_1 is the corrected mass of the object, W_2 is the mass of the standard weights, d_{obj} is the density of the object, d_{wts} is the density of the weights, and d_{air} is the density of the air displaced by them; d_{air} has a value of 0.0012 g/cm^3.

The consequences of Equation 27-1 are shown in Figure 27-5, in which the relative error due to buoyancy is plotted against the density of objects weighed in air against stainless steel weights. Note that this error is less than 0.1% for objects that have a density of 2 g/cm^3 or greater. It is thus seldom necessary to apply a correction to the weight of most solids. The same cannot be said for low-density solids, liquids, or gases, however; for these, the effects of buoyancy are significant, and a correction must be applied.

The weights used in single-pan balances have densities that range from 7.8 to 8.4 g/cm^3, depending upon the manufacturer. Use of 8 g/cm^3 is adequate for most purposes. If a greater accuracy is required, the specifications for the balance to be used should be consulted for the necessary density data.

Example 27-1

A bottle weighed 7.6500 g empty and 9.9700 g after introduction of an organic liquid with a density of 0.92 g/cm^3. The balance was equipped with stainless steel weights ($d = 8.0$ g/cm^3). Correct the weight of the sample for the effects of buoyancy.

The apparent weight of the liquid is $9.9700 - 7.6500 = 2.3200$ g. The same buoyant force acts on the container during both weighings; thus, we need to consider only the force that acts on the 2.3200 g of liquid. Substi-

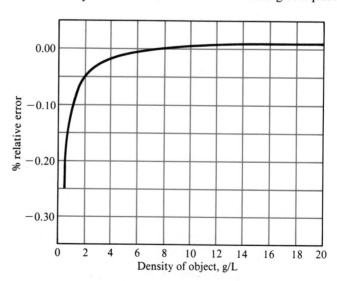

Figure 27-5

Effect of buoyancy on weighing data (stainless steel weights). Plot of relative error as a function of the density of the object weighed.

tution of 0.0012 g/cm³ for d_{air}, 0.92 g/cm³ for d_{obj}, and 8.0 g/cm³ for d_{wts} in Equation 27-1 gives

$$W_1 = 2.3200 + 2.3200 \left(\frac{0.0012}{0.92} - \frac{0.0012}{8.0} \right) = 2.3227 \text{ g}$$

Temperature Effects

Attempts to weigh an object whose temperature is different from that of its surroundings will result in a significant error. Failure to allow sufficient time for a heated object to return to room temperature is the commonest source of this problem. Errors due to a difference in temperature have two sources. First, convection currents within the balance case exert a buoyant effect on the pan and object. Second, warm air trapped in a closed container weighs less than the same volume at a lower temperature. Both effects cause the apparent weight of the object to be low. This error can amount to as much as 10 or 15 mg for a typical porcelain filtering crucible or a weighing bottle (Figure 27-6).

Heated objects must always be cooled to room temperature before being weighed.

Other Sources of Error

A porcelain or glass object will occasionally acquire a static charge that is sufficient to cause a balance to perform erratically; this problem is particularly serious when the relative humidity is low. Spontaneous discharge frequently occurs after a short period. A low-level source of radioactivity (such as a photographer's brush) in the balance case will provide sufficient ions to relieve the charge. Alternatively, the object can be wiped with a faintly damp chamois.

The optical scale of a single-pan balance should be checked regularly for accuracy, particularly under loading conditions that require the full scale range. A standard 100-mg weight is used for this check.

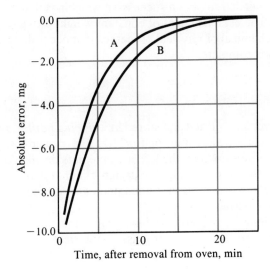

Figure 27-6
Effect of temperature on weighing data. Absolute error in weight as a function of time after the object was removed from a 110°C drying oven. A: porcelain filtering crucible. B: weighing bottle containing about 7.5 g of KCl.

27D-6 Auxiliary Balances

Balances less precise than analytical balances find extensive use in the analytical laboratory. They offer the advantages of speed, ruggedness, large capacity, and convenience; they should be used whenever high sensitivity is not required.

Top-loading auxiliary balances are particularly convenient. A sensitive top-loading balance will accommodate 150 to 200 g with a precision of about 1 mg—an order of magnitude less than an analytical balance. Some balances of this type tolerate loads as great as 25,000 g with a precision of ±0.05 g. Most are equipped with a taring device that brings the balance reading to zero with an empty container on the pan. Some are fully automatic, require no manual dialing or weight handling, and provide a digital readout of the weight. Modern top-loading balances are electronic.

A triple-beam balance with a sensitivity less than that of a typical top-loading auxiliary balance is also useful. This is a single-pan balance with three decades of weights that slide along individual calibrated scales. The precision of a triple-beam balance may be one or two orders of magnitude less than that of a top-loading instrument but is adequate for many weighing operations. This type of balance offers the advantages of simplicity, durability, and low cost.

27E THE EQUIPMENT AND MANIPULATIONS ASSOCIATED WITH WEIGHING

The weight of many solids changes with humidity, owing to their tendency to absorb weighable amounts of moisture. This effect is especially pronounced when a large surface area is exposed, as with a reagent chemical or a sample that has been ground to a fine powder. The first step in a typical analysis, then, involves drying the sample so that the results will not be affected by the humidity of the surrounding atmosphere.

> Weighing to constant weight is a process in which a solid is cycled through heating, cooling, and weighing steps until its weight becomes constant to within 0.2 to 0.3 mg.

A sample, a precipitate, or a container is brought to *constant weight* by a cycle that involves heating (ordinarily for one hour or more) at an appropriate temperature, cooling, and weighing. This cycle is repeated as many times as needed to obtain successive weighings that agree to within 0.2 to 0.3 mg. The establishment of constant weight provides some assurance that the chemical or physical processes that occur during the heating (or ignition) are complete.

27E-1 Weighing Bottles

Solids are conveniently dried and stored in *weighing bottles*, two common varieties of which are shown in Figure 27-7. The ground-glass portion of the cap-style bottle shown on the left is on the outside and does not come into contact with the contents; this design eliminates the possibility of some of the sample becoming entrained upon and subsequently lost from the ground-glass surface.

Plastic weighing bottles are available; ruggedness is the principal advantage of these bottles over their glass counterparts.

Figure 27-7
Typical weighing bottles.

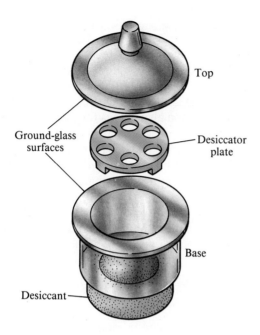

Top

Ground-glass
surfaces

Desiccator
plate

Base

Desiccant

Figure 27-8
Components of a typical desiccator.

27E-2 Desiccators and Desiccants

Oven drying is the most common way of removing moisture from solids. This approach is not appropriate for substances that decompose or for those from which water is not removed at the temperature of the oven.

Dried materials are stored in a *desiccator* while they cool in order to minimize the uptake of moisture. Figure 27-8 shows the components of a typical desiccator. The base section contains a chemical drying agent, such as anhydrous calcium chloride, calcium sulfate (Drierite), anhydrous magnesium perchlorate (Anhydrone or Dehydrite), or phosphorus pentoxide. The ground-glass surfaces are lightly coated with grease.

When you remove or replace the lid of a desiccator, use a sliding motion to minimize the likelihood of disturbing the sample. An airtight seal is achieved by slight rotation and downward pressure upon the positioned lid.

When you place a heated object in a desiccator, the increase in pressure as the enclosed air is warmed may be sufficient to break the seal between lid and base. Conversely, if the seal is not broken, the cooling of the heated object can cause development of a partial vacuum. Both of these conditions can cause the contents of the desiccator to be physically lost or contaminated. Although it defeats the purpose of the desiccator somewhat, you should allow cooling to occur before the lid is seated. It is also helpful to break the seal once or twice during cooling to relieve any excessive vacuum that develops. Finally, you should lock the lid in place with your thumbs while moving the desiccator from one place to another.

Very hygroscopic materials should be stored in containers equipped with snug covers, such as weighing bottles; the covers remain in place while in the desiccator. Most other solids can be safely stored uncovered.

A desiccator is a device for drying substances or objects.

Figure 27-9
Arrangement for the drying of samples.

27E-3 The Manipulation of Weighing Bottles

Heating at 105 to 110°C is sufficient to remove the moisture from the surface of most solids. Figure 27-9 depicts the arrangement recommended for drying a sample. The weighing bottle is contained in a labeled beaker with a ribbed cover glass. This arrangement protects the sample from accidental contamination and also allows for the free access of air. Crucibles containing a precipitate that can be freed of moisture by simple drying can be treated similarly. The beaker containing the weighing bottle or crucible to be dried must be carefully marked to permit identification.

You should avoid handling a dried object with your fingers because detectable amounts of water or oil from your skin may be transferred to the object. This problem is avoided by using tongs, chamois finger cots, clean cotton gloves, or strips of paper to handle dried objects for weighing. Figure 27-10 shows how a weighing bottle is manipulated with strips of paper.

27E-4 Weighing by Difference

Weighing by difference is a simple method for determining a series of sample weights. First the bottle and its contents are weighed. One sample is then transferred from the bottle to a container; gentle tapping of the bottle with its top and slight rotation of the bottle provide control over the amount of sample removed. Following transfer, the bottle and its residual contents are weighed. The weight of the sample is the difference between the two weighings. It is essential that all the solid removed from the weighing bottle be transferred without loss to the container.

27E-5 Weighing Hygroscopic Solids

Hygroscopic substances rapidly absorb moisture from the atmosphere and therefore require special handling. You need a weighing bottle for each sample to be weighed. Place the approximate amount of sample needed in the individual bottles and heat for an appropriate time. When heating is complete, quickly cap the bottles and cool in a desiccator. Weigh one of the bottles after opening it momentarily to relieve any vacuum. Quickly empty the contents of the bottle into its receiving vessel, cap immediately, and weigh the bottle again (along with any solid that did not get transferred). Repeat for each sample and determine the sample weights by difference.

27E-6 Weighing Liquids

Figure 27-10
Method for quantitative transfer of a solid sample. Note the use of paper strips to avoid contact between glass and skin.

The weight of a liquid is always obtained by difference. Liquids that are noncorrosive and relatively nonvolatile can be transferred to previously weighed containers with snugly fitting covers (such as weighing bottles); the weight of the container plus residual liquid is subtracted from the total weight.

A volatile or corrosive liquid should be sealed in a weighed glass ampoule in the following manner. The empty ampoule is heated, and the neck is then immersed in the sample; as cooling occurs, the liquid is drawn into the bulb. The ampoule is then inverted and the neck sealed off with a small flame. The ampoule and its contents, along with any glass removed during sealing, are cooled to room temperature and weighed. The ampoule is then transferred to an appropriate container and broken. A volume correction for the glass of the ampoule may be needed if the receiving vessel is a volumetric flask.

27F THE EQUIPMENT AND MANIPULATIONS FOR FILTRATION AND IGNITION

27F-1 Apparatus

Simple Crucibles

Simple crucibles serve only as containers. Those made of porcelain, aluminum oxide, silica, and platinum maintain constant weight—within the limits of experimental error—and are used principally to convert a precipitate to a suitable weighing form. The solid is first collected on filter paper. The filter and contents are then transferred to a weighed crucible, and the paper is ignited.

Simple crucibles of nickel, iron, silver, and gold are used as containers for the high-temperature fusion of samples that are not soluble in aqueous reagents. Attack by both the atmosphere and the contents may cause these crucibles to suffer weight changes. Moreover, such attack will contaminate the sample with species derived from the crucible. In such cases, the chemist selects the crucible whose products will offer the least interference in subsequent steps of the analysis.

Filtering Crucibles

Filtering crucibles serve not only as containers but also as filters. A vacuum is used to hasten the filtration; a tight seal between crucible and filtering flask is accomplished with any of several types of rubber adaptors (Figure 27-11; a complete filtration train is shown in Figure 27-16). Collection of a precipitate with a filtering crucible is frequently less time-consuming than with paper.

Sintered-glass (also called *fritted-glass*) crucibles are manufactured in fine, medium, and coarse porosities (marked f, m, and c). The upper temperature limit for a sintered-glass crucible is ordinarily about 200°C. Filtering crucibles made entirely of quartz can tolerate substantially higher temperatures without damage. The same is true for crucibles with unglazed porcelain or aluminum oxide frits. The latter are not as costly as quartz.

A *Gooch crucible* has a perforated bottom that supports a fibrous mat. Asbestos was at one time the filtering medium of choice for a Gooch crucible; current regulations concerning this material have virtually eliminated its use. Small circles of glass matting have now replaced asbestos; they are used in pairs to protect against disintegration during the filtration. Glass mats can tolerate temperatures in excess of 500°C and are substantially less hygroscopic than asbestos.

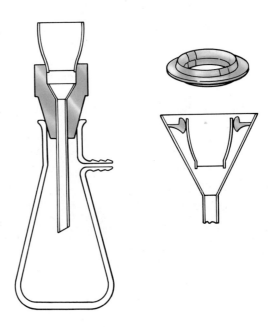

Figure 27-11
Adaptors for filtering crucibles.

Filter Paper

Paper is an important filtering medium. Ashless paper is manufactured from cellulose fibers that have been treated with hydrochloric and hydrofluoric acids to remove metallic impurities and silica; ammonia is then used to neutralize the acids. The residual ammonium salts in many filter papers may be sufficient to affect the analysis for nitrogen by the Kjeldahl method.

All papers tend to pick up moisture from the atmosphere, and ashless paper is no exception. It is thus necessary to destroy the paper by ignition if the precipitate collected on it is to be weighed. Typically, 9- or 11-cm circles of ashless paper leave a residue that weighs less than 0.1 mg, an amount that is ordinarily negligible. Ashless paper can be obtained in several porosities.

Gelatinous precipitates, such as hydrous iron(III) oxide, clog the pores of any filtering medium. A coarse-porosity ashless paper is most effective for the filtration of such solids, but even here clogging occurs. This problem can be minimized by mixing a dispersion of ashless filter paper with the precipitate prior to filtration. Filter-paper pulp is available in tablet form from chemical suppliers; if necessary, the pulp can be made in the laboratory by treating a piece of ashless paper with concentrated hydrochloric acid and washing the disintegrated mass free of acid.

Table 27-1 summarizes the characteristics of common filtering media. None satisfies all requirements.

Heating Equipment

Many precipitates can be weighed directly after being brought to constant weight in a low-temperature drying oven. Such an oven is electrically heated and capable of maintaining a constant temperature to within 1°C

Table 27-1

COMPARISON OF FILTERING MEDIA FOR GRAVIMETRIC ANALYSES

| Characteristic | Paper | Gooch Crucible | | Glass Crucible | Porcelain Crucible | Aluminum Oxide Crucible |
		Asbestos Mat*	Glass Mat			
Speed of filtration	Slow	Rapid	Rapid	Rapid	Rapid	Rapid
Convenience and ease of preparation	Troublesome, inconvenient	Troublesome, inconvenient	Convenient	Convenient	Convenient	Convenient
Maximum ignition temperature, °C	None	1200	>500	200–500	1100	1450
Chemical reactivity	Carbon has reducing properties	Inert	Inert	Inert	Inert	Inert
Porosity	Many available	Difficult to control	Several available	Several available	Several available	Several available
Convenience with gelatinous precipitates	Satisfactory	Unsuitable; filter tends to clog	Unsuitable; filter tends to clog	Unsuitable; filter tends to clog	Unsuitable; filter tends to clog	Unsuitable; filter tends to clog
Cost	Low	Low	Low	High	High	High

*The use of asbestos, a confirmed carcinogen, is subject to stringent control in many areas.

(or better). The maximum attainable temperature ranges from 140 to 260°C, depending upon make and model; for many precipitates, 110°C is a satisfactory drying temperature. The efficiency of a drying oven is greatly increased by the forced circulation of air. The passage of predried air through an oven designed to operate under a partial vacuum represents an additional improvement.

Microwave laboratory ovens are currently appearing on the market. Where applicable, these greatly shorten drying cycles. For example, slurry samples that require 12 to 16 h for drying in a conventional oven are reported to be dried within 5 to 6 min in a microwave oven.[6]

An ordinary heat lamp can be used to dry a precipitate that has been collected on ashless paper and to char the paper as well. The process is conveniently completed by ignition at an elevated temperature in a muffle furnace.

Burners are convenient sources of intense heat. The maximum attainable temperature depends upon the design of the burner and the combustion properties of the fuel. Of the three common laboratory burners, the Meker provides the highest temperatures, followed by the Tirrill and Bunsen types.

A heavy-duty electric furnace (*muffle furnace*) is capable of maintaining controlled temperatures of 1100°C or higher. Long-handled tongs and heat-resistant gloves are needed for protection when transferring objects to or from such a furnace.

[6]D. G. Kuehn, R. L. Brandvig, D. C. Lundean, and R. H. Jefferson, *Amer. Lab.*, **1986**, *18* (7), 31. See also *Anal. Chem.*, **1986**, *58*, 1424A; E. S. Beary, *Anal. Chem.*, **1988**, *60*, 742.

27F-2 The Manipulations Associated with Filtration and Ignition

The Preparation of Crucibles

A crucible used to convert a precipitate to a form suitable for weighing must maintain—within the limits of experimental error—a constant weight throughout the drying and/or ignition. The crucible is first cleaned thoroughly (filtering crucibles are conveniently cleaned by backwashing on a filtration train, see Figure 27-16) and subjected to the same regimen of heating and cooling as that required for the precipitate. This process is repeated until constant weight (page 522) has been achieved, that is, until consecutive weighings differ by 0.3 mg or less.

The Filtration and Washing of Precipitates

The steps involved in filtering an analytical precipitate are *decantation, washing,* and *transfer.* In decantation, as much supernatant liquid as possible is passed through the filter while the precipitated solid is kept essentially undisturbed in the beaker where it was formed. This procedure speeds the overall filtration rate by delaying the time at which the pores of the filtering medium become clogged with precipitate. A stirring rod is used to direct the flow of decantate (Figure 27-12). When as much of the liquid as possible has been decanted, the drop of liquid at the end of the pouring spout is collected with the stirring rod and returned to the beaker.

> Backwashing a filtering crucible is done by turning the crucible upside down in the adaptor (Figure 27-11) and drawing water through the inverted crucible.

> Decantation is the process of pouring a liquid gently so as to not disturb a solid in the bottom of the container.

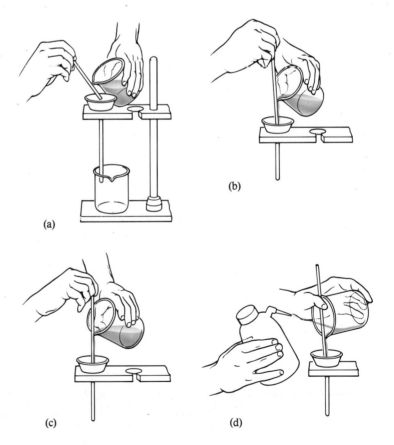

Figure 27-12

Steps in filtering: (a and b) washing by decantation; (c and d) transfer of the precipitate.

Wash liquid is next added to the beaker and thoroughly mixed with the precipitate. The solid is allowed to settle, following which this liquid is also decanted through the filter. Several such washings may be required, depending upon the precipitate. Most washings should be carried out *before* the solid is transferred; this results in a more thoroughly washed precipitate and a more rapid filtration.

The transfer process is illustrated in Figure 27-12c and d. The bulk of the precipitate is moved from beaker to filter by suitably directed streams of wash liquid. As in decantation and washing, a stirring rod provides direction for the flow of material to the filtering medium.

The last traces of precipitate that cling to the inside of the beaker are dislodged with a *rubber policeman,* which is a small section of rubber tubing that has been crimped on one end. The open end of the tubing is fitted onto the end of a stirring rod, and the policeman is wetted with wash liquid before use. Any solid collected with it is combined with the main portion on the filter.

Small pieces of ashless paper can be used to wipe the last traces of hydrous oxide precipitates from the wall of the beaker; these papers are ignited along with the paper that holds the bulk of the precipitate.

Many precipitates have an exasperating tendency to *creep,* which means that they spread over a wetted surface against the force of gravity. Filters are never filled to more than three quarters of capacity, owing to the possibility that some of the precipitate could be lost as the result of creeping. The addition of a small amount of nonionic detergent, such as Triton X-100, to the supernatant liquid or to the wash liquid can be helpful in minimizing creeping.

> Creeping is a process in which a solid moves up the side of a wetted container or filter paper.

A gelatinous precipitate must be completely washed before it is allowed to dry. These precipitates shrink and develop cracks as they dry. Further additions of wash liquid simply pass through these cracks and accomplish little or no washing.

27F-3 Directions for the Filtration and Ignition of Precipitates

The Preparation of a Filter Paper

Figure 27-13 shows the sequence followed in folding a filter paper and seating it in a 60-degree funnel. The paper is folded exactly in half (a), firmly creased, and folded again (b). A triangular piece from one of the corners is torn off parallel to the second fold (c). The paper is then opened so that the untorn quarter forms a cone (d). The cone is fitted into the funnel, and the second fold is creased (e). Seating is completed by dampening the cone with water from a wash bottle and *gently* patting it with a finger. There will be no leakage of air between the funnel and a properly seated cone; in addition, the stem of the funnel will be filled with an unbroken column of liquid.

The Transfer of Paper and Precipitate to a Crucible

After filtration and washing have been completed, the filter and its contents must be transferred from the funnel to a crucible that has been brought to constant weight. Ashless paper has very low wet strength and must be handled with care during the transfer. The danger of tearing is

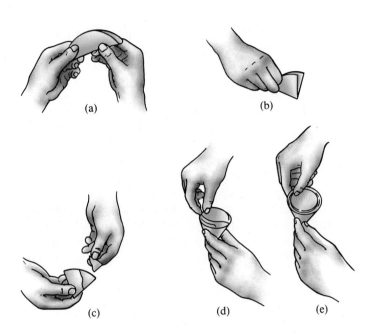

Figure 27-13
Folding and seating a filter paper.

lessened considerably if you allow the paper to dry somewhat before it is removed from the funnel.

Figure 27-14 illustrates the transfer process. The triple-thick portion of the filter paper is pulled to the opposite side of the funnel to flatten the cone along its upper edge (a); next the collapsed cone is carefully lifted out of the funnel and its corners are folded inward (b); the top edge is then folded over (c). Finally, the paper and its contents are eased into the crucible (d) so that the bulk of the precipitate is near the bottom.

The Ashing of a Filter Paper

If a heat lamp is to be used for ashing, the crucible is placed on a clean, nonreactive surface, such as a wire screen covered with aluminum foil. The lamp is then positioned about 1 cm above the rim of the crucible and turned on. Charring takes place without further attention. The process is

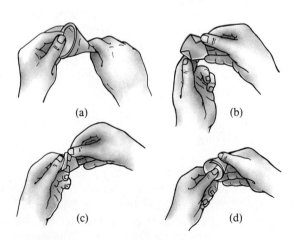

Figure 27-14
Transferring a filter paper and precipitate to a crucible.

considerably accelerated if the paper is moistened with no more than one drop of concentrated ammonium nitrate solution. Elimination of the residual carbon is accomplished with a burner, as described in the next paragraph.

Considerably more attention must be paid if you use a burner to ash a filter paper. The burner produces much higher temperatures than a heat lamp. The possibility thus exists for the mechanical loss of precipitate if moisture is expelled too rapidly in the initial stages of heating or if the paper bursts into flame. Also, partial reduction of some precipitates can occur through reaction with the hot carbon of the charring paper; such reduction is a serious problem if reoxidation following ashing is inconvenient. These difficulties can be minimized by positioning the crucible as illustrated in Figure 27-15; the tilted position allows for the ready access of air. A clean crucible cover should be available to extinguish any flame that might appear.

Heating is begun with a small flame. The temperature is gradually increased as moisture is evolved and the paper begins to char. The intensity of heating that can be tolerated can be gauged by the amount of smoke given off. Thin wisps are normal. A significant increase in the amount of smoke indicates that the paper is about to flash and that heating should be temporarily discontinued. Any flame that does appear in the crucible should be immediately extinguished with a crucible cover. (The cover may become discolored, owing to the condensation of carbonaceous products; these products must ultimately be removed from the cover by ignition to confirm the absence of entrained particles of precipitate.) When no further smoking can be detected, heating is increased to eliminate the residual carbon. Strong heating, as necessary, can then be undertaken.

This sequence ordinarily precedes the final ignition of a precipitate in a muffle furnace, where a reducing atmosphere is equally undesirable.

The Use of Filtering Crucibles

A vacuum filtration train (Figure 27-16) is used where a filtering crucible can be used instead of paper. The trap isolates the filter flask from the source of vacuum.

27F-4 Rules for the Manipulation of Heated Objects

Careful adherence to the following rules will minimize the possibility of accidental loss of a precipitate.

1. Practice unfamiliar manipulations before putting them to use.
2. *Never* place a heated object on the benchtop; instead, place it on a wire gauze or a heat-resistant ceramic plate.
3. Allow a crucible that has been subjected to the full flame of a burner or to a muffle furnace to cool momentarily (on a wire gauze or ceramic plate) before transferring it to the desiccator.
4. Keep the tongs and forceps used to handle heated objects scrupulously clean. In particular, do not allow the tips to touch the benchtop.

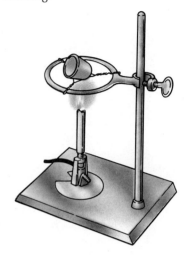

Figure 27-15
Ignition of a precipitate. Proper crucible position for preliminary charring.

To vacuum

Trap

Figure 27-16
Train for vacuum filtration.

27G   THE MEASUREMENT OF VOLUME

The precise measurement of volume is as important to many analytical methods as is the precise measurement of mass.

27G-1   Units of Volume

The unit of volume is the *liter* (L), defined as one thousand cubic centimeters. The *milliliter* (mL) is one one-thousandth of a liter and is used where the liter represents an inconveniently large volume unit.

27G-2   The Effect of Temperature on Volume Measurements

The volume occupied by a given mass of liquid varies with temperature, as does the container that holds the liquid during measurement. Most volumetric measuring devices are made of glass, however, which fortunately has a small coefficient of expansion. Consequently, variations in the volume of a glass container with temperature need not be considered in ordinary analytical work.

The coefficient of expansion for dilute aqueous solutions (approximately 0.025%/°C) is such that a 5°C change has a measurable effect upon the reliability of ordinary volumetric measurements.

Example 27-2

A 40.00-mL sample is taken from an aqueous solution at 5°C; what volume does it occupy at 20°C?

$$V_{20°} = V_{5°} + 0.00025(20 - 5)(40.00) = 40.00 + 0.15 = 40.15 \text{ mL}$$

Volumetric measurements must be referred to some standard temperature; this reference point is ordinarily 20°C. The ambient temperature of most laboratories is sufficiently close to 20°C to eliminate the need for temperature corrections in volume measurements for aqueous solutions. In contrast, the coefficient of expansion for organic liquids may require corrections for temperature differences of 1°C or less.

27G-3   Apparatus for the Precise Measurement of Volume

Volume is measured reliably with a *pipet*, a *buret*, and a *volumetric flask*.

Volumetric equipment is marked by the manufacturer to indicate not only the manner of calibration (usually TD for "to deliver" or TC for "to contain") but also the temperature at which the calibration strictly applies. Pipets and burets are ordinarily calibrated to deliver specified volumes, whereas volumetric flasks are calibrated on a to-contain basis.

Pipets

Pipets permit the transfer of accurately known volumes from one container to another. Common types are shown in Figure 27-17; information

A hand-held, battery-operated motorized pipet. (Courtesy of Rainin Instrument Co., Inc., Woburn, MA.)

concerning their use is given in Table 27-2. A *volumetric*, or *transfer*, pipet (Figure 27-17a) delivers a single, fixed volume between 0.5 and 200 mL. Many such pipets are color-coded by volume for convenience in identification and sorting. *Measuring* pipets (Figure 27-17b and c) are calibrated in convenient units to permit delivery of any volume up to a maximum capacity ranging from 0.1 to 25 mL.

Volumetric and measuring pipets are filled to a calibration mark at the outset; the manner in which the transfer is completed depends upon the particular type of pipet. Because an attraction exists between most liquids and glass, a small amount of liquid tends to remain in the tip after the pipet is emptied. This residual liquid is never blown out of a volumetric pipet or from some measuring pipets; it is blown out of other types of pipets (Table 27-2).

Hand-held Eppendorf micropipets (Figure 27-17d) deliver adjustable microliter volumes of liquid. With these pipets, a known and adjustable volume of air is displaced from the plastic disposable tip by depressing the pushbutton on the top of the pipet to a first stop. This button operates a spring-loaded piston that forces air out of the pipet. The volume of displaced air can be varied by a locking digital micrometer adjustment located on the front of the device. The plastic tip is then inserted into the liquid and the pressure on the button released, causing liquid to be drawn into the tip. The tip is then placed against the wall of the receiving vessel, and the pushbutton is again depressed to the first stop. After one second, the pushbutton is depressed further to a second stop, which completely empties the tip. The range of volumes and precision of typical pipets of this type are shown in the margin.

Numerous *automatic* pipets are available for situations that call for the repeated delivery of a particular volume. In addition, a motorized, computer-controlled microliter pipet is now available. This device is programmed to function as a pipet, a dispenser of multiple volumes, a buret, and a means for diluting samples. The volume desired is entered on a keyboard and is displayed on a panel. A motor-driven piston dispenses the liquid. Maximum volumes range from 10 to 2500 μL.

Burets

Burets, like measuring pipets, enable the analyst to deliver any volume up to a maximum capacity. The precision attainable with a buret is substantially greater than with a pipet.

A buret consists of a calibrated tube to hold titrant plus a valve arrangement by which the flow of titrant is controlled. This valve is the principal source of difference among burets. The simplest valve consists of a close-fitting glass bead inside a short length of rubber tubing that connects the buret and its tip; only when the tubing is deformed does liquid flow past the bead.

A buret equipped with a glass stopcock for a valve relies upon a lubricant between the ground-glass surfaces of stopcock and barrel for a liquid-tight seal. Some solutions, notably bases, cause a glass stopcock to freeze upon long contact; therefore thorough cleaning is needed after each use. Valves made of Teflon are commonly encountered; these are unaffected by most common reagents and require no lubricant.

TOLERANCES, CLASS A TRANSFER PIPETS

Capacity, mL	Tolerances, mL
0.5	±0.006
1	±0.006
2	±0.006
5	±0.01
10	±0.02
20	±0.03
25	±0.03
50	±0.05
100	±0.08

RANGE AND PRECISION OF TYPICAL EPPENDORF MICROPIPETS

Volume range, μL	Standard Deviation, μL
1–20	<0.04 @ 2 μL
	<0.06 @ 20 μL
10–100	<0.10 @ 15 μL
	<0.15 @ 100 μL
20–200	<0.15 @ 25 μL
	<0.30 @ 200 μL
100–1000	<0.6 @ 250 μL
	<1.3 @ 1000 μL
500–5000	<3 @ 1.0 mL
	<8 @ 5.0 mL

TOLERANCES, CLASS A BURETS

Volume, mL	Tolerances, mL
5	±0.01
10	±0.02
25	±0.03
50	±0.05
100	±0.20

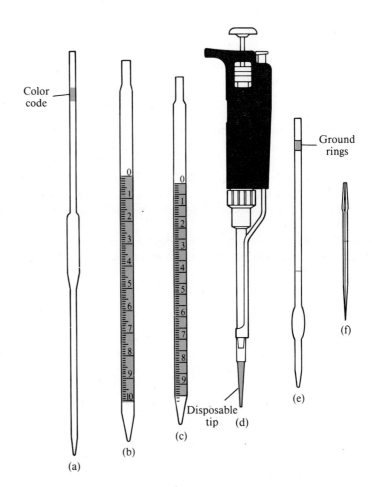

Figure 27-17
Typical pipets: (a) volumetric,
(b) Mohr, (c) serological,
(d) Eppendorf micropipet,
(e) Ostwald-Folin, (f) lambda.

Table 27-2
CHARACTERISTICS OF PIPETS

Name	Type of Calibration*	Function	Available Capacity, mL	Type of Drainage
Volumetric	TD	Delivery of fixed volume	1–200	Free
Mohr	TD	Delivery of variable volume	1–25	To lower calibration line
Serological	TD	Delivery of variable volume	0.1–10	Blow out last drop†
Serological	TD	Delivery of variable volume	0.1–10	To lower calibration line
Ostwald-Folin	TD	Delivery of fixed volume	0.5–10	Blow out last drop†
Lambda	TC	Containment of fixed volume	0.001–2	Wash out with suitable solvent
Lambda	TD	Delivery of fixed volume	0.001–2	Blow out last drop†
Eppendorf	TD	Delivery of variable or fixed volume	0.001–1	Tip emptied by air displacement

*TD–to deliver; TC–to contain.

†A frosted ring near the top of recently manufactured pipets indicates that the last drop is to be blown out.

Volumetric Flasks

Volumetric flasks are manufactured with capacities ranging from 5 mL to 5 L and are usually calibrated to contain a specified volume when filled to a line etched on the neck. They are used for the preparation of standard solutions and for the dilution of samples to a fixed volume prior to taking aliquots with a pipet. Some are also calibrated on a to-deliver basis; these are readily distinguished by the two reference lines on the neck. If delivery of the stated volume is desired rather than containment, the flask is filled to the upper line.

27G-4 General Considerations Concerning the Use of Volumetric Equipment

Volume markings are blazed upon clean volumetric equipment by the manufacturer. An equal degree of cleanliness is needed in the laboratory if these markings are to have their stated meanings. Only clean glass surfaces support a uniform film of liquid. Dirt or oil causes breaks in this film; the existence of breaks is a certain indication of an unclean surface.

Cleaning

A brief soaking in a warm detergent solution is usually sufficient to remove the grease and dirt responsible for water breaks. Prolonged soaking should be avoided because a rough area or ring is likely to develop at a detergent/air interface. This ring cannot be removed and causes a film break that destroys the usefulness of the equipment.

After being cleaned, the apparatus must be thoroughly rinsed with tap water and then with three or four portions of distilled water. It is seldom necessary to dry volumetric ware.

Avoiding Parallax

The top surface of a liquid confined in a narrow tube exhibits a marked curvature, or *meniscus*. It is common practice to use the bottom of the meniscus as the point of reference in calibrating and using volumetric equipment. This minimum can be established more exactly by holding an opaque card or piece of paper behind the graduations (Figure 27-18).

In reading volumes, the eye must be at the level of the liquid surface in

TOLERANCES, CLASS A VOLUMETRIC FLASKS	
Capacity, mL	Tolerances, mL
5	±0.02
10	±0.02
25	±0.03
50	±0.05
100	±0.08
250	±0.12
500	±0.20
1000	±0.30
2000	±0.50

A meniscus is the curved surface of a liquid at its interface with the atmosphere.

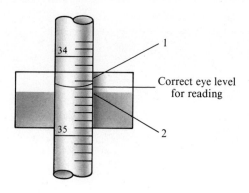

Figure 27-18
Method for reading a buret. The eye should be level with the meniscus. The reading shown is 34.39 mL. If viewed from position 1, the reading appears smaller than 34.39 mL; from position 2, it appears larger.

Parallax is the apparent displacement of a liquid level or of a pointer as an observer changes position. It occurs when the observer's line of vision is not perpendicular to the surface of the calibrated scale being read.

order to avoid an error due to *parallax*, a condition that causes the volume to appear smaller than its actual value if the meniscus is viewed from above and larger if the meniscus is viewed from below (Figure 27-18).

27G-5 Directions for the Use of a Pipet

The following directions pertain specifically to volumetric pipets but can be modified for the use of other types as well.

Liquid is drawn into a pipet through the application of a slight vacuum. *The mouth should never be used for suction because of the possibility of accidentally ingesting the liquid being pipetted.* Instead, a rubber suction bulb or a rubber tube connected to a vacuum source should be used (Figure 27-19a).

Cleaning
Use a rubber bulb to draw detergent solution to a level 2 to 3 cm above the calibration mark of the pipet. Drain this solution and then rinse the pipet

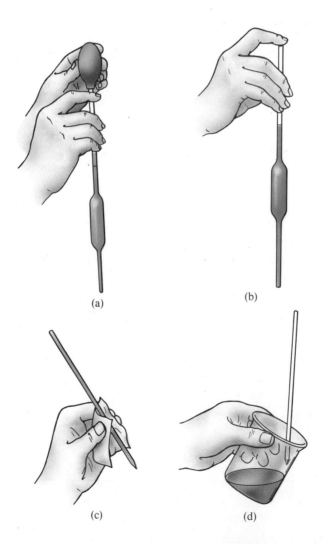

(a)

(b)

(c)

(d)

Figure 27-19
Steps in dispensing an aliquot.

with several portions of tap water. Inspect for film breaks; repeat this detergent/rinse water portion of the cleaning cycle if necessary. Finally, fill the pipet with distilled water to perhaps one third of its capacity and carefully rotate it so that the entire interior surface is wetted. Repeat this rinsing step at least twice.

Measurement of an Aliquot

Use a rubber bulb to draw a small volume of the liquid to be sampled into the pipet and thoroughly wet the entire interior surface. Discard this liquid, and repeat with *at least* two additional portions. Then carefully fill the pipet to a level somewhat above the graduation mark (Figure 27-19a). Quickly replace the bulb with a *forefinger* to arrest the outflow of liquid (Figure 27-19b). Make certain there are no bubbles in the bulk of the liquid or foam at the surface. Tilt the pipet slightly from the vertical and wipe the exterior free of adhering liquid (Figure 27-19c). Touch the tip of the pipet to the wall of a glass vessel (*not* the container into which the aliquot is to be transferred), and slowly allow the liquid level to drop by partially releasing your forefinger (Note 1). Halt further flow as the bottom of the meniscus coincides exactly with the graduation mark. Then place the pipet tip well within the clean receiving vessel for the aliquot, and allow the liquid to drain. When free flow ceases, rest the tip against the inner wall of the receiver for a full 10 s (Figure 27-19d). Finally, withdraw the pipet with a rotating motion to remove any liquid adhering to the tip. *The small volume remaining inside the tip of a volumetric pipet should not be blown or rinsed into the receiving vessel* (Note 2).

Notes

1. The liquid can best be held at a constant level if the forefinger is *faintly* moist. Too much moisture makes control impossible.
2. Rinse the pipet thoroughly after use.

27G-6 Directions for the Use of a Buret

Before it is placed in service, a buret must be scrupulously clean; in addition, its valve must be liquid-tight.

Cleaning

Thoroughly clean the tube of the buret with detergent and a long brush. Rinse thoroughly with tap water and then with distilled water. Inspect for water breaks. Repeat the treatment if necessary.

Lubrication of a Glass Stopcock

Carefully remove all old grease from a glass stopcock and its barrel with a paper towel, and dry both parts completely. Lightly grease the stopcock, taking care to avoid the area adjacent to the hole. Insert the stopcock into the barrel and rotate it vigorously with slight inward pressure. A proper amount of lubricant has been used when (1) the area of contact between stopcock and barrel appears nearly transparent, (2) the seal is liquid-tight, and (3) no grease has worked its way into the tip.

Notes

1. Grease films that are unaffected by cleaning solution may yield to such organic solvents as acetone or benzene. Thorough washing with detergent should follow such treatment. The use of silicone lubricants is not recommended; contamination by such preparations is difficult—if not impossible—to remove.
2. So long as the flow of liquid is not impeded, fouling of a buret tip with stopcock grease is not a serious matter. Removal is best accomplished with organic solvents. A stoppage during a titration can be freed by *gentle* warming of the tip with a lighted match.
3. Before a buret is returned to service after reassembly, it is advisable to test for leakage. Simply fill the buret with water and establish that the volume reading does not change with time.

Filling

Make certain the stopcock is closed. Add 5 to 10 mL of the titrant, and carefully rotate the buret to wet the interior completely. Allow the liquid to drain through the tip. *Repeat this procedure at least two more times.* Then fill the buret well above the zero mark. Free the tip of air bubbles by rapidly rotating the stopcock and permitting small quantities of the titrant to pass. Finally, lower the level of the liquid just to or somewhat below the zero mark. Allow for drainage ($\approx$1 min), and then record the initial volume reading, estimating to the nearest 0.01 mL.

Titration

Figure 27-20 illustrates the preferred method for manipulating a stopcock; when the hand is held as shown, any tendency for lateral movement by the stopcock will be in the direction of firmer seating. Be sure the tip of the buret is well within the titration vessel (ordinarily a flask). Introduce the titrant in increments of about 1 mL. Swirl (or stir) constantly to ensure thorough mixing. Decrease the size of the increments as the titration progresses; add titrant dropwise in the immediate vicinity of the end point (Note 2). When it is judged that only a few more drops are needed, rinse the walls of the container (Note 3). Allow for drainage (at least 30 s) at the completion of the titration. Then record the final volume, again to the nearest 0.01 mL.

Notes

1. When unfamiliar with a particular titration, many chemists prepare an extra sample. No care is lavished on its titration since its functions are to reveal the nature of the end point and to provide a rough estimate of titrant requirements. This deliberate sacrifice of one sample frequently results in an overall saving of time.
2. Increments smaller than one drop can be taken by allowing a small volume of titrant to form on the tip of the buret and then touching the tip to the wall of the flask. This partial drop is then combined with the bulk of the liquid as in Note 3.

Figure 27-20
Recommended method for manipulation of a buret stopcock.

3. Instead of being rinsed toward the end of a titration, the flask can be tilted and rotated so that the bulk of the liquid picks up any drops that adhere to the inner surface.

27G-7 Directions for the Use of a Volumetric Flask

Before being put into use, volumetric flasks should be washed with detergent and thoroughly rinsed. Only rarely do they need to be dried. If required, however, drying is best accomplished by clamping the flask in an inverted position. Insertion of a glass tube connected to a vacuum line hastens drying.

Direct Weighing into a Volumetric Flask

The direct preparation of a standard solution requires the introduction of a known weight of solute to a volumetric flask. Use a powder funnel to minimize the possibility of losing solid during the transfer. Rinse the funnel thoroughly; collect the washings in the flask.

The foregoing procedure is inappropriate if heating is needed to dissolve the solute. Instead, weigh the solid into a beaker or flask, add solvent, heat to dissolve the solute, and allow the solution to cool to room temperature. Transfer this solution quantitatively to the volumetric flask, as described in the next section.

The Quantitative Transfer of Liquid to a Volumetric Flask

Insert a funnel into the neck of the volumetric flask; use a stirring rod to direct the flow of liquid from the beaker into the funnel. Tip off the last drop of liquid on the spout of the beaker with the stirring rod. Rinse both the stirring rod and the interior of the beaker with distilled water and transfer the washings to the volumetric flask, as before. Repeat the rinsing process *at least* two more times.

Dilution to the Mark

After the solute has been transferred, fill the flask about half-full and swirl the contents to hasten solution. Add more solvent and again mix well. Bring the liquid level almost to the mark, and allow time for drainage (≈ 1 min); then use a medicine dropper to make such final additions of solvent as are necessary (Note). Firmly stopper the flask, and invert it repeatedly to ensure thorough mixing. Transfer the contents to a storage bottle that either is dry or has been thoroughly rinsed with several small portions of the solution from the flask.

Note. If, as sometimes happens, the liquid level accidentally exceeds the calibration mark, the solution can be saved by correcting for the excess volume. Use a gummed label to mark the actual location of the meniscus. After the flask has been emptied, carefully refill to the manufacturer's etched mark with water. Use a buret to determine the additional volume needed to bring the meniscus to the gummed-label mark. This volume must be added to the nominal volume of the flask when calculating the concentration of the solution.

27H   THE CALIBRATION OF VOLUMETRIC WARE

Volumetric glassware is calibrated by measuring the mass of a liquid (usually distilled water) of known density and temperature that is contained in (or delivered by) the volumetric ware. In carrying out a calibration, a buoyancy correction must be made (Section 27D-5) since the density of water is quite different from that of the weights.

The calculations associated with calibration, while not difficult, are somewhat involved. The raw weighing data are first corrected for buoyancy with Equation 27-1. Next, the volume of the apparatus at the temperature of calibration (T) is obtained by dividing the density of the liquid at that temperature into the corrected weight. Finally, this volume is corrected to the standard temperature of 20°C as in Example 27-2.

Table 27-3 is provided to ease the computational burden of calibration. Corrections for buoyancy with respect to stainless steel or brass weights (the density difference between the two is small enough to be neglected) and for the volume change of water and of glass containers have been incorporated into these data. Multiplication by the appropriate factor from Table 27-3 converts the mass of water at temperature T to (1) the corresponding volume at that temperature or (2) the volume at 20°C.

Table 27-3

VOLUME OCCUPIED BY 1.0000 g OF WATER WEIGHED IN AIR AGAINST STAINLESS STEEL WEIGHTS

Temperature, T °C	Volume, mL	
	At T	Corrected to 20°C
10	1.0013	1.0016
11	1.0014	1.0016
12	1.0015	1.0017
13	1.0016	1.0018
14	1.0018	1.0019
15	1.0019	1.0020
16	1.0021	1.0022
17	1.0022	1.0023
18	1.0024	1.0025
19	1.0026	1.0026
20	1.0028	1.0028
21	1.0030	1.0030
22	1.0033	1.0032
23	1.0035	1.0034
24	1.0037	1.0036
25	1.0040	1.0037
26	1.0043	1.0041
27	1.0045	1.0043
28	1.0048	1.0046
29	1.0051	1.0048
30	1.0054	1.0052

Corrections for buoyancy (stainless steel weights) and change in container volume have been applied.

Example 27-3

A 25-mL pipet delivers 24.976 g of water weighed against stainless steel weights at 25°C. Use the data in Table 27-3 to calculate the volume delivered by this pipet at 25 and 15°C.

At 25°C: $V = 24.976 \text{ g} \times 1.0040 \text{ mL/g} = 25.08 \text{ mL}$

At 15°C: $V = 24.976 \text{ g} \times 1.0019 \text{ mL/g} = 25.02 \text{ mL}$

27H-1 General Directions for Calibration Work

All volumetric ware should be painstakingly freed of water breaks before being calibrated. Burets and pipets need not be dry; volumetric flasks should be thoroughly drained and dried at room temperature. The water used for calibration should be in thermal equilibrium with its surroundings. This condition is best established by drawing the water well in advance, noting its temperature at frequent intervals, and waiting until no further changes occur.

Although an analytical balance can be used for calibration, weighings to the nearest milligram are perfectly satisfactory for all but the very smallest volumes. Thus a top-loading balance is more convenient. Weighing bottles or small, well-stoppered conical flasks can serve as receivers for the calibration liquid.

The Calibration of a Volumetric Pipet

Determine the empty weight of the stoppered receiver to the nearest milligram. Transfer a portion of temperature-equilibrated water to the receiver with the pipet, weigh the receiver and its contents (again, to the nearest milligram), and calculate the weight of water delivered from the difference in these weights. Calculate the volume delivered with the aid of Table 27-3. Repeat the calibration several times; calculate the mean volume delivered and its standard deviation.

The Calibration of a Buret

Fill the buret with temperature-equilibrated water and make sure that no air bubbles are trapped in the tip. Allow about 1 min for drainage; then lower the liquid level to bring the bottom of the meniscus to the 0.00-mL mark. Touch the tip to the wall of a beaker to remove any adhering drop. Wait 10 min and recheck the volume; if the stopcock is tight, there should be no perceptible change. During this interval, weigh (to the nearest milligram) a 125-mL conical flask fitted with a rubber stopper.

Once tightness of the stopcock has been established, slowly transfer (at about 10 mL/min) approximately 10 mL of water to the flask. Touch the tip to the wall of the flask. Wait 1 min, record the volume that was apparently delivered, and refill the buret. Weigh the flask and its contents to the nearest milligram; the difference between this weight and the initial value gives the mass of water delivered. Use Table 27-3 to convert this mass to the true volume. Subtract the apparent volume from the true volume. This difference is the correction that should be applied to the apparent volume to give the true volume. Repeat the calibration until agreement within ±0.02 mL is achieved.

Starting again from the zero mark, repeat the calibration, this time delivering about 20 mL to the receiver. Test the buret at 10-mL intervals over its entire volume. Prepare a plot of the correction to be applied as a function of volume delivered. The correction associated with any interval can be determined from this plot.

The Calibration of a Volumetric Flask

Weigh the clean, dry flask to the nearest milligram. Then fill to the mark with temperature-equilibrated water and reweigh. Calculate the volume contained with the aid of Table 27-3.

The Calibration of a Volumetric Flask Relative to a Pipet

The calibration of a volumetric flask relative to a pipet provides an excellent method for partitioning a sample into aliquots. These directions pertain to a 50-mL pipet and a 500-mL volumetric flask; other combinations are equally convenient.

Carefully transfer ten 50-mL aliquots from the pipet to a dry 500-mL volumetric flask. Mark the location of the meniscus with a gummed label. Cover with a label varnish to ensure permanence. Anytime this volumetric flask is used later, dilution to the label permits the same pipet to deliver precisely a one-tenth aliquot of the solution in the flask. Note that recalibration is necessary if another pipet is to be used.

An aliquot is a measured fraction of the volume of a liquid sample.

27I   THE LABORATORY NOTEBOOK

A laboratory notebook is needed to record measurements and observations concerning an analysis. The book should be permanently bound with consecutively numbered pages (if necessary, the pages should be hand-numbered before any entries are made). Most notebooks have more than ample room; there is no need to crowd entries.

The first few pages should be saved for a table of contents that is updated as entries are made.

27I-1   Rules for the Maintenance of a Laboratory Notebook

1. *Record all data and observations directly into the notebook in ink.* Neatness is desirable, but you should not realize neatness by transcribing data from a sheet of paper to the notebook or from one notebook to another. The risk of misplacing—or incorrectly transcribing––crucial data and thereby ruining an experiment is unacceptable.
2. Supply each entry or series of entries with a heading or label. A series of weighing data for a set of empty crucibles should carry the heading "empty crucible weights" (or something similar), for example, and the weight of each crucible should be identified by the same number or letter used to label the crucible.
3. Date each page of the notebook as it is used.
4. *Never* attempt to erase or obliterate an incorrect entry. Instead, cross

An entry in a laboratory notebook should never be erased but should be crossed out instead.

Gravimetric Determination of Chloride OB

The Chloride in a soluble sample was precipitated as AgCl and weighed as such

Sample weights	1	2	3
Wt. Bottle plus sample, g	27.6115	27.2185	26.8105
− less sample, g	27.2185	26.8105	26.4517
wt. sample, g	0.3930	0.4080	0.3588
Crucible weights, empty	~~20.7925~~	~~22.8311~~	~~21.2488~~
	20.7926	22.8311	~~21.2482~~
			21.2483
Crucible weights, with AgCl, g	~~21.4294~~	~~23.4920~~	~~21.8324~~
	~~21.4297~~	~~23.4914~~	21.8323
	21.4296	23.4915	
Weight of AgCl, g	0.6370	0.6604	0.5840
Percent Cl⁻	40.10	40.04	40.27
Average percent Cl⁻		40.12	
Relative standard deviation	3.0 parts per thousand		
Date Started	1-9-88		
Date Completed	1-16-88		

Figure 27-21
Laboratory notebook data page.

it out with a single horizontal line and locate the correct entry as nearby as possible. Do not write over incorrect numbers; with time, it may become impossible to distinguish the correct entry from the incorrect one.

5. Never remove a page from the notebook. Draw diagonal lines across any page that is to be disregarded. Provide a brief rationale for disregarding the page.

27I-2 Format

The instructor should be consulted concerning the format to be used in keeping the laboratory notebook.[7] One convention involves using each page consecutively for the recording of data and observations as they occur. The completed analysis is then summarized on the next available page spread (that is, left and right facing pages). As shown in Figure

[7]See also H. M. Kanare, *Writing the Laboratory Notebook*. Washington, D.C. 20036: The American Chemical Society, 1985.

27-21, the first of these two facing pages should contain the following entries:

1. The title of the experiment ("The Gravimetric Determination of Chloride").
2. A brief statement of the principles upon which the analysis is based.
3. A complete summary of the weighing, volumetric, and/or instrument-response data needed to calculate the results.
4. A report of the best value for the set and a statement of its precision.

The second page should contain the following items:

1. Equations for the principal reactions in the analysis.
2. An equation showing how the results were calculated.
3. A summary of observations that appear to bear upon the validity of a particular result or the analysis as a whole. *Any such entry must have been originally recorded in the notebook at the time the observation was made.*

27J   SAFETY IN THE LABORATORY

Work in a chemical laboratory necessarily involves a degree of risk; accidents can and do happen. Strict adherence to the following rules will go far toward preventing (or minimizing the effect of) accidents.

1. At the outset, learn the location of the nearest eye fountain, fire blanket, shower, and fire extinguisher. Learn the proper use of each, and do not hesitate to use this equipment should the need arise.
2. *WEAR EYE PROTECTION AT ALL TIMES.* The potential for serious and perhaps permanent eye injury makes it mandatory that adequate eye protection be worn at all times by students, instructors, and visitors. Eye protection should be donned before entering the laboratory and should be used continuously until it is time to leave. Serious eye injuries have occurred to people performing such innocuous tasks as computing or writing in a laboratory notebook; such incidents are usually the result of someone else's losing control of an experiment. Regular prescription glasses are not adequate substitutes for eye protection approved by the Office of Safety and Health Administration (OSHA). Contact lenses should *never* be worn in the laboratory because laboratory fumes may react with them and have a harmful effect on the eyes.
3. Most of the chemicals in a laboratory are toxic; some are very toxic, and some—such as concentrated solutions of acids and bases—are highly corrosive. Avoid contact between these liquids and your skin. In the event of such contact, *immediately* flood the affected area with copious quantities of water. If a corrosive solution is spilled on clothing, remove the garment immediately. Time is of the essence; modesty cannot be a matter of concern.
4. *NEVER* perform an unauthorized experiment. Such activity is grounds for disqualification at many institutions.

5. Never work alone in the laboratory; be certain that someone is always within earshot.
6. Never bring food or beverages into the laboratory. Do not drink from laboratory glassware. Do not smoke in the laboratory.
7. Always use a bulb to draw liquids into a pipet; *NEVER* use your mouth to provide suction.
8. Wear adequate foot covering (no sandals). Confine long hair with a net or cap. A laboratory coat or apron will provide some protection and may be required.
9. Be extremely tentative in touching objects that have been heated; hot glass looks just like cold glass.
10. Always fire-polish the ends of freshly cut glass tubing. *NEVER* attempt to force glass tubing through the hole of a stopper. Instead, make sure that both tubing and hole are wet with soapy water. Protect hands with several layers of towel while inserting glass into a stopper.
11. Use fume hoods whenever toxic or noxious gases are likely to be evolved. Be cautious in testing for odors; use your hand to waft vapors above containers toward your nose.
12. Notify the instructor in the event of an injury.
13. Dispose of solutions and chemicals as instructed. It is illegal to flush solutions containing heavy metal ions or organic liquids down the drain in many localities; alternative arrangements are required for the disposal of such liquids.

CHAPTER 28

Selected Methods of Analysis

This chapter contains detailed directions for performing a variety of chemical analyses. The methods have been chosen to introduce you to a variety of analytical techniques that are widely used by chemists. For most of these analyses, the composition of the samples is known to the instructor. Thus, you will be able to judge how well you are mastering these techniques.

Your chances of success in the laboratory will greatly improve if you take time before beginning any analysis to read carefully and *understand* each step in the method and to develop a plan for how and when you will perform each step. For greatest efficiency, such study and planning should take place *before you enter the laboratory*.

The discussion in this section is aimed at helping you develop efficient work habits in the laboratory and also at providing you with some general information about an analytical chemistry laboratory.

Background Information

Before undertaking an analysis, you should understand the significance of each step in the procedure in order to avoid the pitfalls and potential sources of error that exist in all analytical methods. Information about these steps can usually be found in preliminary discussion sections, in earlier chapters that are referred to in the discussion section, and in the "Notes" that follow many of the procedures. If, after reading these materials, you still don't understand the reason for one or more of the steps in the method, consult your instructor before you begin laboratory work.

The Accuracy of Measurements

In looking over an analytical procedure, you should identify the measurements that must be made with maximum precision, and thus with maximum care, and those that can be carried out rapidly with little concern for precision. Generally, measurements that appear in the equation used to compute the results must be performed with maximum precision. The remaining measurements can *and should* be made less carefully to conserve time. The words "about" and "approximately" are frequently used to indicate that a measurement does not have to be done carefully. You should not waste time and effort to measure, let us say, a volume to ±0.02 mL when an uncertainty of ±0.5 mL or even ±5 mL will have no discernible effect on the results.

In some procedures, a statement such as "weigh three 0.5-g samples to the nearest 0.1 mg . . ." is encountered. Here, samples of perhaps 0.4 to 0.6 g are acceptable, but their weights must be known to the nearest 0.1 mg. The number of significant figures in the specification of a volume or a weight is also a guide as to the care you should take in making a measurement. For example, the statement "add 10.00 mL of a solution to the beaker" indicates that you should measure the volume carefully with a buret or a pipet with the aim of limiting the uncertainty to perhaps ±0.02 mL. In contrast, if the directions read "add 10 mL," the measurement can be made with a graduated cylinder.

Time Utilization

You should carefully study the time requirements of the several unit operations involved in an analysis *before you begin work*. Such study will reveal operations that require considerable elapsed, or clock, time but little or no operator time—for example, when a sample is dried in an oven, cooled in a desiccator, or evaporated on a hot plate. The experienced chemist plans to use such periods of waiting to perform other operations or perhaps to begin a new analysis. Some people find it worthwhile to prepare a written time schedule for each laboratory period to avoid times when no work can be done.

Time planning is also needed to identify places where an analysis can be interrupted for overnight or longer, as well as those operations that must be completed without a break.

Reagents

Directions for the preparation of reagents accompany many of the procedures described in this chapter. Before preparing such reagents, be sure to check to see if they are already prepared and available for general use.

If a reagent is known to pose a hazard, you should plan *in advance of the laboratory period* the steps to be taken to minimize injury or damage.

Acquaint yourself with the rules that apply in your laboratory for the disposal of waste liquids and solids. These rules vary from one part of the country to another and even among laboratories in the same locale. Your instructor will inform you of the rules that apply.

Water

Some laboratories use deionizers to purify water; others use stills for this purpose. The terms "distilled water" and "deionized water" are used interchangeably in the directions that follow. Either type is satisfactory for analytical work.

You should use tap water only for preliminary cleaning of glassware. Then rinse the cleaned glassware with at least three small portions of distilled or deionized water.

28A   GRAVIMETRIC METHODS OF ANALYSIS

General aspects, calculations, and typical applications of gravimetric analysis are discussed in Chapter 5.

28A-1 The Gravimetric Determination of Chloride in a Soluble Sample

Discussion
The chloride content of a soluble salt can be determined by precipitation of the anion as silver chloride:

$$Ag^+ + Cl^- \rightarrow AgCl(s)$$

The precipitate is collected in a weighed filtering crucible and washed. Its weight is then determined after it has been dried to constant weight at 110°C.

The solution containing the sample is kept somewhat acidic during the precipitation to eliminate possible interference from anions of weak acids (such as CO_3^{2-}) that form sparingly soluble silver salts in a neutral environment. A moderate excess of silver ion is needed to diminish the solubility of silver chloride, but a large excess is avoided to minimize coprecipitation of silver nitrate.

Silver chloride forms first as a colloid and is subsequently coagulated with heat. Nitric acid and the small excess of silver nitrate promote coagulation by providing a moderately high electrolyte concentration. Nitric acid in the wash solution maintains the electrolyte concentration and eliminates the possibility of peptization during the washing step; the acid subsequently decomposes to give volatile products when the precipitate is dried. See Section 5A-2 for additional information concerning the properties and treatment of colloidal precipitates.

In common with other silver halides, finely divided silver chloride undergoes photodecomposition:

$$2 AgCl(s) \rightarrow 2 Ag(s) + Cl_2(g)$$

The elemental silver produced in this reaction is responsible for the violet color that develops in the precipitate. In principle, this reaction leads to low results for chloride ion. In practice, however, its effect is negligible provided you avoid direct and prolonged exposure of the precipitate to sunlight.

If photodecomposition of silver chloride occurs before filtration, the additional reaction

$$3 \, Cl_2(aq) + 3 \, H_2O + 5 \, Ag^+ \rightarrow 5 \, AgCl(s) + ClO_3^- + 6 \, H^+$$

tends to cause high results.

Some photodecomposition of silver chloride is inevitable the way the analysis is ordinarily performed. It is worthwhile to minimize exposure of the solid to intense sources of light as far as possible.

Because silver nitrate is expensive, any unused reagent should be collected in a storage container; similarly, precipitated silver chloride should be retained after the analysis is complete.[1]

Procedure

Clean three sintered-glass or porcelain filtering crucibles by allowing about 5 mL of concentrated HNO_3 to stand in each for about 5 min. Use a vacuum (Figure 27-16) to draw the acid through the crucible. Rinse each crucible with three portions of tap water, and then discontinue the vacuum. Next, add about 5 mL of 6 M NH_3 and wait about 5 min before drawing it through the filter. Finally, rinse each crucible with six to eight portions of distilled water. Provide each crucible with an identifying mark. Bring the crucibles to constant weight by heating at 110°C while the other steps in the analysis are being carried out. The first drying should be for at least 1 h; subsequent heating periods can be somewhat shorter (30 to 40 min).

Transfer the unknown to a weighing bottle and dry at 110°C (Figure 27-9) for 1 to 2 h; allow the bottle and contents to cool to room temperature in a desiccator. Weigh (to the nearest 0.1 mg) individual samples by difference into 400-mL beakers (Note 1). Dissolve each sample in about 100 mL of distilled water to which 2 to 3 mL of 6 M HNO_3 has been added.

Slowly, and with good stirring, add 0.2 M $AgNO_3$ to each of the cold sample solutions until AgCl is observed to coagulate (Notes 2 and 3); then introduce an additional 3 to 5 mL of $AgNO_3$. Heat almost to boiling, and digest the solids for about 10 min. Add a few drops of $AgNO_3$ to confirm that precipitation is complete. If more precipitate forms, add about 3 mL of $AgNO_3$, digest, and again test for completeness of precipitation. Pour any unused $AgNO_3$ into a waste container (NOT into the original reagent bottle). Cover each beaker, and store in a dark place for at least 2 h and preferably until the next laboratory period.

Read the instructions for filtration in Section 27F-2. Decant the supernatant liquids through weighed filtering crucibles. Wash the precipitates several times (while they are still in the beaker) with a solution consisting of 2 to 5 mL of 6 M HNO_3 per liter of distilled water; decant these washings through the filters. Quantitatively transfer the AgCl from the beakers to the individual crucibles with fine streams of wash solution; use

Digest means to heat without stirring a precipitate and the solution from which it is formed.

---

[1]Silver can be recovered from silver chloride and from surplus reagent by reduction with ascorbic acid; see J. W. Hill and L. Bellows, *J. Chem. Educ.,* **1986,** *63* (4), 357; see also J. P. Rawat and S. Iqbal M. Kamoonpuri, *ibid.,* **1986,** *63* (6), 537, for recovery (as $AgNO_3$) based on ion exchange. See also D. D. Perrin, W. L. F. Armarego, and D. R. Perrin, *Chemistry International,* **1987,** *9* (1), 3, concerning a potential hazard in the recovery of silver nitrate.

rubber policemen to dislodge any particles that adhere to the walls of the beakers. Continue washing until the filtrates are essentially free of Ag^+ ion (Note 4).

Dry the precipitates at 110°C for at least 1 h. Store the crucibles in a desiccator while they cool. Determine the weight of the crucibles and their contents. Repeat the cycle of heating, cooling, and weighing until consecutive weighings agree to within 0.2 mg. Calculate the percentage of Cl^- in the sample.

Upon completion of the analysis, remove the precipitates by gently tapping the crucibles over a piece of glazed paper. Transfer the collected AgCl to a container for silver wastes. Remove the last traces of AgCl by filling the crucibles with 6 M NH_3 and allowing them to stand.

Notes

1. Consult with the instructor concerning an appropriate sample size.
2. Determine the approximate amount of $AgNO_3$ needed by calculating the volume that would be required if the unknown were pure NaCl.
3. Use a separate stirring rod for each sample and leave it in its beaker throughout the determination.
4. To test the washings for Ag^+, collect a small volume in a test tube and add a few drops of HCl. Washing is judged complete when little or no turbidity develops.

28A-2 The Gravimetric Determination of Nickel in Steel

Discussion

The nickel in a steel sample can be precipitated from a slightly alkaline medium with an alcoholic solution of dimethylglyoxime (Section 5D-3). Interference from iron(III) is eliminated by masking with tartaric acid. The product is freed of moisture by drying at 110°C.

Tartrate ion forms a stable complex with Fe(III) and thus prevents formation of $Fe_2O_3 \cdot x\ H_2O$, which would otherwise precipitate as the solution is made basic.

The bulky character of nickel dimethylglyoxime limits the weight of nickel that can be accommodated conveniently and thus the sample weight. Care must also be taken to control the excess of alcoholic dimethylglyoxime used. If too much is added, the alcohol concentration becomes sufficient to dissolve appreciable amounts of the nickel dimethylglyoxime, a condition that leads to low results. If the alcohol concentration becomes too low, however, some of the reagent may precipitate and cause a positive error.

Preparation of Solutions

a. *Dimethylglyoxime, 1% (w/v)*. Dissolve 10 g of dimethylglyoxime in 1 L of ethanol. (Sufficient for about 50 precipitations.)
b. *Tartaric acid, 15% (w/v)*. Dissolve 225 g of tartaric acid in sufficient water to give 1500 mL of solution. Filter before use if the solution is not clear. (Sufficient for about 50 precipitations.)

Procedure

Clean and mark three medium-porosity sintered-glass crucibles (Note 1); bring them to constant weight by drying at 110°C for at least 1 h.

Weigh (to the nearest 0.1 mg) samples containing between 30 and 35 mg of nickel into individual 400-mL beakers (Note 2). Dissolve each sample in about 50 mL of 6 M HCl with gentle warming (HOOD). Carefully add approximately 15 mL of 6 M HNO_3, and boil gently to expel any oxides of nitrogen that may have been produced. Dilute to about 200 mL and heat to boiling. Introduce about 30 mL of 15% tartaric acid and sufficient concentrated NH_3 to produce a faint odor of NH_3 in the vapors over the solutions (Note 3); then add another 1 to 2 mL of NH_3. If the solutions are not clear at this stage, proceed as directed in Note 4. Make the solutions acidic with HCl (no odor of NH_3), heat to 60 to 80°C, and add about 20 mL of the 1% dimethylglyoxime solution. With good stirring, add 6 M NH_3 until a slight excess exists (faint odor of NH_3) plus an additional 1 to 2 mL. Digest the precipitates for 30 to 60 min, cool for at least 1 h, and filter through the weighed crucibles.

Wash the solids with water until the washings are free of Cl^- (Note 5). Bring the crucibles and their contents to constant weight at 110°C. Report the percentage of nickel in the sample. The dried precipitate has the composition $Ni(C_4H_7O_2N_2)_2$ (fw = 288.93).

Notes

1. Medium-porosity porcelain filtering crucibles or Gooch crucibles with glass pads can be substituted for sintered-glass crucibles in this determination.
2. Use a separate stirring rod for each sample and leave it in the beaker throughout.
3. The existence or absence of excess NH_3 is readily established by odor; use a waving motion with your hand to waft the vapors toward your nose.
4. If $Fe_2O_3 \cdot x\,H_2O$ forms upon addition of NH_3, acidify the solution with HCl, introduce additional tartaric acid, and neutralize again. Alternatively, remove the solid by filtration. Thorough washing with a hot NH_3/NH_4Cl solution is required; the washings are combined with the solution containing the bulk of the sample.
5. Test the washings for Cl^- by collecting a small portion in a test tube, acidifying with HNO_3, and adding a drop or two of 0.1 M $AgNO_3$. Washing is judged complete when little or no turbidity develops.

28B PRECIPITATION TITRATIONS

As noted in Chapter 10, most precipitation titrations make use of a standard silver nitrate solution as titrant. Directions follow for the volumetric titration of chloride ion using an adsorption indicator.

28B-1 Preparation of a Standard Silver Nitrate Solution

Procedure

Use a top-loading balance to transfer the approximate weight of $AgNO_3$ to a weighing bottle (Note 1). Dry at 110°C for about 1 h but not much longer (Note 2), and then cool to room temperature in a desiccator. Weigh the bottle and contents (to the nearest 0.1 mg). Using a powder funnel, transfer the bulk of the $AgNO_3$ to a volumetric flask. Cap the weighing bottle and reweigh it and any solid that remains. Rinse the powder funnel thoroughly. Dissolve the $AgNO_3$, dilute to the mark with water, and mix well (Note 3). Calculate the molar concentration of this solution.

Notes

1. Consult with the instructor concerning the volume and concentration of $AgNO_3$ to be prepared. The weight of $AgNO_3$ to be taken is as follows:

Silver Ion Concentration, M	Approximate Weight (g) of $AgNO_3$ Needed to Prepare		
	1000 mL	500 mL	250 mL
0.10	16.9	8.5	4.2
0.05	8.5	4.2	2.1
0.02	3.4	1.8	1.0

2. Prolonged heating causes partial decomposition of $AgNO_3$. Some discoloration may occur, even after only 1 h at 110°C; the effect of this decomposition on the purity of the reagent is ordinarily imperceptible.
3. Silver nitrate solutions should be stored in a dark place when not in use.

28B-2 The Determination of Chloride by Titration with an Adsorption Indicator

Discussion

In this titration, the anionic adsorption indicator dichlorofluorescein is used to locate the end point. With the first excess of titrant, the indicator becomes incorporated in the counter-ion layer surrounding the silver chloride and imparts color to the solid (Section 10B-3). In order to obtain a satisfactory color change, it is desirable to maintain the particles of silver chloride in the colloidal state. Dextrin is added to the solution to stabilize the colloid and prevent its coagulation.

Preparation of Solutions

Dichlorofluorescein indicator (sufficient for several hundred titrations). Dissolve 0.2 g of dichlorofluorescein in a solution prepared by mixing 75 mL of ethanol and 25 mL of water.

Procedure

Dry the unknown at 110°C for about 1 h; allow it to return to room temperature in a desiccator. Weigh individual samples (to the nearest 0.1 mg) into individual conical flasks, and dissolve them in appropriate volumes of distilled water (Note 1). To each, add about 0.1 g of dextrin and 5 drops of indicator. Titrate (Note 2) with $AgNO_3$ to the first permanent pink color of silver dichlorofluoresceinate. Report the percentage of Cl^- in the unknown.

Notes

1. Use 0.25-g samples for 0.1 M $AgNO_3$ and about half that amount for 0.05 M reagent. Dissolve the former in about 200 mL of distilled water and the latter in about 100 mL. If 0.02 M $AgNO_3$ is to be used, weigh a 0.4-g sample into a 500-mL volumetric flask, and take 50-mL aliquots for titration.

2. Colloidal AgCl is sensitive to photodecomposition, particularly in the presence of the indicator; attempts to perform the titration in direct sunlight will fail. If photodecomposition appears to be a problem, establish the approximate end point with a rough preliminary titration, and use this information to estimate the volumes of $AgNO_3$ needed for the other samples. For each subsequent sample, add the indicator and dextrin only after most of the $AgNO_3$ has been added, and then complete the titration without delay.

28C NEUTRALIZATION TITRATIONS

Neutralization titrations are performed with standard solutions of strong acids or bases. Although a single solution (of either acid or base) is sufficient for the titration of a given type of analyte, it is convenient to have standard solutions of both acid and base available in the event a back-titration is needed to locate end points more exactly. The concentration of one solution is established by titration against a primary standard; the concentration of the other is then determined from the acid/base ratio (that is, the volume of acid needed to neutralize 1.000 mL of the base).

28C-1 The Effect of Atmospheric Carbon Dioxide on Neutralization Titrations

Water in equilibrium with the atmosphere is about 1×10^{-5} M in carbonic acid as a consequence of the equilibrium

$$CO_2(g) + H_2O \rightleftarrows H_2CO_3(aq)$$

At this concentration level, the amount of 0.1 M base consumed by the carbonic acid in a typical titration is negligible. With more dilute reagents (<0.05 M), however, the water used as a solvent for the analyte and in the preparation of reagents must be freed of carbonic acid by boiling for a brief period.

Water that has been purified by distillation rather than by deionization is often supersaturated with carbon dioxide and may thus contain sufficient acid to affect the results of an analysis.[2] The instructions that follow are based upon the assumption that the amount of carbon dioxide in the water supply can be neglected without causing serious error. For further discussion on the effects of carbon dioxide in neutralization titrations, see Section 13A-3.

28C-2 Preparation of Indicator Solutions for Neutralization Titrations

Discussion
The theory of acid/base indicators is discussed in Section 11A-1. An indicator exists for virtually any pH range between 1 and 13.[3] Directions follow for the preparation of indicator solutions suitable for most neutralization titrations.

Procedure

Stock solutions ordinarily contain between 0.5 and 1.0 g of indicator per liter. (One liter of indicator is sufficient for hundreds of titrations.)

a. *Methyl orange, methyl red, and sulphonthaleins.* Dissolve the sodium salt directly in distilled water. (The common sulphonthaleins include bromocresol green, bromothymol blue, bromophenol blue, thymol blue, cresol red, and phenol red.)
b. *Phenolphthalein, thymolphthalein.* Dissolve the solid indicator in a solution consisting of 800 mL of ethanol and 200 mL of distilled or deionized water.

28C-3 Preparation of Dilute Hydrochloric Acid Solutions

Discussion
The preparation and standardization of acids are considered in Sections 13A-1 and 13A-2.

Procedure

For a 0.1 M solution, add about 8 mL of concentrated HCl to about 1 L of distilled water (Note). Mix thoroughly, and store in a glass-stoppered bottle.

---

[2]Water that is to be used for neutralization titrations can be tested by adding 5 drops of phenolphthalein to a 500-mL portion. Less than 0.2 to 0.3 mL of 0.1 M OH⁻ should suffice to produce the first faint pink color of the indicator. If a larger volume is needed, the water should be boiled and cooled before it is used to prepare standard solutions or to dissolve samples.

[3]See, for example, J. Beukenkemp and W. Rieman III, in *Treatise on Analytical Chemistry*, I. M. Kolthoff and P. J. Elving, Eds., Part I, Vol. 11, pp. 6987–7001. New York: Wiley, 1974.

Note

It is advisable to eliminate CO_2 from the water by a preliminary boiling if very dilute solutions (<0.05 M) are being prepared.

28C-4 Preparation of Carbonate-Free Sodium Hydroxide

Discussion

See Sections 13A-3 and 13A-4 for information concerning the preparation and standardization of bases.

Procedure

If so directed by the instructor, prepare a bottle for protected storage (Figure 13-1; also Note 1). Transfer 1 L of distilled water to the storage bottle (Note 2). Decant 4 to 5 mL of 50% NaOH into a small container (Note 3), add it to the water, and *mix thoroughly*. USE EXTREME CARE IN HANDLING 50% NaOH, which is highly corrosive. If the reagent comes into contact with skin, IMMEDIATELY flush the area with *copious* amounts of water.

Protect the solution from unnecessary contact with the atmosphere.

Notes

1. A solution of base that will be used up within two weeks can be stored in a tightly capped polyethylene bottle. After each removal of base, squeeze the bottle while tightening the cap to minimize the air space above the reagent. The bottle will become brittle after extensive use as a container for bases.
2. It is advisable to eliminate CO_2 from the water by a preliminary boiling if very dilute solutions (<0.05 M) are being prepared.
3. Be certain that any solid Na_2CO_3 in the 50% NaOH has settled to the bottom of the container and that the decanted liquid is absolutely clear. If necessary, filter the base through a glass mat in a Gooch crucible; collect the clear filtrate in a test tube inserted in the filter flask.

28C-5 The Determination of the Acid/Base Ratio

Discussion

If both acid and base solutions have been prepared, it is useful to determine their volumetric combining ratio. Knowing this ratio and the concentration of one solution permits calculation of the molarity of the other.

Procedure

Instructions for placing a buret into service are given in Sections 27G-4 and 27G-6; consult these instructions if necessary.

Fill one 50-mL buret with the standard acid solution and a second with the standard base solution. Place a test tube or small beaker over the top of the buret that holds the NaOH solution to minimize contact between the solution and the atmosphere.

Record the initial volumes of acid and base in the burets to the nearest 0.01 mL. Deliver 35 to 40 mL of the acid into a 250-mL conical flask. Touch the tip of the buret to the inside wall of the flask, and rinse down with a little distilled water. Add two drops of phenolphthalein (Note 1) and then sufficient base to render the solution a definite pink. Introduce acid dropwise to discharge the color, and again rinse down the walls of the flask. Carefully add base until the solution again acquires a faint pink hue that persists for at least 30 s (Notes 2 and 3). Record the final buret volumes (again, to nearest 0.01 mL). Repeat the titration. Calculate the acid/base volume ratio. The ratios for duplicate titrations should agree to within 1 to 2 ppt. Perform additional titrations, if necessary, to achieve this order of precision.

Notes

1. The volume ratio can also be determined with an indicator that has an acidic transition range, such as bromocresol green. If the NaOH is contaminated with carbonate, the ratio obtained with this indicator will differ significantly from the value obtained with phenolphthalein. In general, the acid/base ratio should be evaluated with the indicator that is to be used in subsequent titrations.
2. Fractional drops can be formed on the buret tip, touched to the wall of the flask, and rinsed down with a small amount of water.
3. The phenolphthalein end point fades as CO_2 is absorbed from the atmosphere.

28C-6 Standardization of Hydrochloric Acid Against Sodium Carbonate

Discussion
See Section 13A-2.

Procedure

Dry a quantity of primary-standard Na_2CO_3 for about 2 h at 110°C (Figure 27-9), and cool in a desiccator. Weigh individual 0.20- to 0.25-g samples (to the nearest 0.1 mg) into 250-mL conical flasks, and dissolve each in about 50 mL of distilled water. Introduce 3 drops of bromocresol green, and titrate with HCl until the solution just begins to change from blue to green. Boil the solution for 2 to 3 min, cool to room temperature (Note 1), and complete the titration (Note 2).

Determine an indicator correction by titrating approximately 100 mL of 0.05 M NaCl and 3 drops of indicator. Boil briefly, cool, and complete the titration. Subtract any volume needed for the blank from the titration volumes. Calculate the concentration of the HCl solution.

Notes

1. The indicator should change from green to blue as CO_2 is evolved during heating. If no color change occurs, an excess of acid was added originally. This excess can be back-titrated with base, provided the acid/base ratio is known; otherwise, the sample must be discarded.

2. It is permissible to back-titrate with base to establish the end point with greater certainty.

28C-7 Standardization of Sodium Hydroxide Against Potassium Hydrogen Phthalate

Discussion
See Section 13A-4.

Procedure

Dry a quantity of primary-standard potassium hydrogen phthalate (KHP) for about 2 h at 110°C (Figure 27-9), and cool in a desiccator. Weigh individual 0.7- to 0.8-g samples (to the nearest 0.1 mg) into 250-mL conical flasks, and dissolve each in 50 to 75 mL of distilled water. Add 2 drops of phenolphthalein; titrate with base until the pink color of the indicator persists for 30 s (Note). Calculate the concentration of the NaOH solution.

Note
It is permissible to back-titrate with acid to establish the end point more precisely. Record the volume used in the back-titration. Use the acid/base ratio to calculate the net volume of base used in the standardization.

28C-8 The Determination of Potassium Hydrogen Phthalate in an Impure Sample

Discussion
The unknown is a mixture of KHP and a neutral salt. This analysis is conveniently performed concurrently with the standardization of the base.

Procedure

Consult with the instructor concerning an appropriate sample size. Then follow the directions in Section 28C-7.

28C-9 The Determination of the Acid Content of Vinegars and Wines

Discussion
The total acid content of a vinegar or a wine is readily determined by titration with a standard base. It is customary to report the acid content of vinegar in terms of acetic acid, the principal acidic constituent, even though other acids are present. Similarly, the acid content of a wine is expressed as percent tartaric acid, even though there are other acids in

the sample. Most vinegars contain about 5% acid (w/v) expressed as acetic acid; wines ordinarily contain somewhat under 1% acid (w/v) expressed as tartaric acid.

Procedure

a. *If the unknown is a vinegar* (Note 1), pipet 25.00 mL into a 250-mL volumetric flask and dilute to the mark with distilled water. Mix thoroughly, and pipet 50.00-mL aliquots into 250-mL conical flasks. Add about 50 mL of water and 2 drops of phenolphthalein (Note 2) to each, and titrate with standard 0.1 M NaOH to the first permanent ($\approx$30 s) pink color. Report the acidity of the vinegar as percent (w/v) CH_3COOH (fw = 60.053).

b. *If the unknown is a wine,* pipet 50.00-mL aliquots into 250-mL conical flasks, add about 50 mL of distilled water and 2 drops of phenolphthalein to each (Note 2), and titrate to the first permanent ($\approx$30 s) pink color. Express the acidity of the sample as percent (w/v) tartaric acid $C_2H_4O_2(COOH)_2$ (fw = 150.09) (Note 3).

Notes

1. The acidity of bottled vinegar tends to decrease on exposure to air. It is recommended that unknowns be stored in individual vials with snug covers.
2. The amount of indicator used should be increased as necessary to make the color change visible in colored samples.
3. Tartaric acid has two acidic hydrogens, both of which are titrated at a phenolphthalein end point.

28C-10 The Determination of Sodium Carbonate in an Impure Sample

Discussion

The titration of sodium carbonate is discussed in Section 13A-2 in connection with its use as a primary standard; the same considerations apply for the determination of carbonate in an unknown that has no interfering contaminants.

Procedure

Dry the unknown at 110°C for 2 h, and then cool in a desiccator. Consult with your instructor for an appropriate sample size. Then follow the instructions in Section 28C-6.

Report the percentage of Na_2CO_3 in the sample.

28C-11 The Determination of Amine Nitrogen by the Kjeldahl Method

Discussion

These directions are suitable for the determination of protein in materials such as blood meal, wheat flour, pasta products, dry cereals, and pet foods. A simple modification permits the analysis of unknowns that contain more highly oxidized forms of nitrogen.[4] The chemistry of the Kjeldahl method is described in Section 13B-1.

Procedure

Preparation of Samples

Consult with the instructor on sample size. *If the unknown is powdered* (such as blood meal), weigh samples onto individual 9-cm filter papers (Note 1). Fold the paper around the sample and drop each into a separate Kjeldahl flask (the paper keeps the samples from clinging to the neck of the flask). *If the unknown is not powdered* (such as breakfast cereals or pasta), the samples can be weighed directly into the Kjeldahl flasks.

Add 25 mL of concentrated H_2SO_4, 10 g of powdered K_2SO_4, and the catalyst (Note 2) to each flask.

Digestion

Clamp the flasks in a slanted position in a hood or vented digestion rack. Heat carefully to boiling. Discontinue heating briefly if foaming becomes excessive; never allow the foam to reach the neck of the flask. Once foaming ceases and the acid is boiling vigorously, the samples can be left unattended; prepare the distillation apparatus during this time. Continue digestion until the solution becomes colorless or faintly yellow; up to 3 h may be needed for some materials. If necessary, *cautiously* replace the acid lost by evaporation.

When digestion is complete, discontinue heating, and allow the flasks to cool to room temperature; swirl the flasks if the contents show signs of solidifying. Cautiously add 250 mL of water to each flask and again allow the solution to cool to room temperature. If mercury has been used as the catalyst, introduce 25 mL of 4% (w/v) Na_2S solution (Note 3).

Distillation of Ammonia

Arrange a distillation apparatus similar to that shown in Figure 13-2b. Pipet 50.00 mL of standard 0.1 M HCl into the receiver flask (Note 4). Clamp the flask so that the tip of the adapter extends below the surface of the standard acid. Circulate water through the condenser jacket.

Hold the Kjeldahl flask at an angle and gently introduce about 60 mL of 50% (w/v) NaOH solution, taking care to minimize mixing with the solution in the flask. *The concentrated caustic solution is highly corrosive and should be handled with great care* (Note 5). Add several pieces of granulated zinc (Note 6) and a small piece of litmus paper. *Immediately* connect the Kjeldahl flask to the spray trap. Cautiously mix the contents by gentle swirling. The litmus paper should be blue after mixing is complete, indicating that the solution is basic.

Bring the solution to a boil, and distill at a steady rate until one half to one third of the original volume remains. Control the rate of heating to prevent the liquid in the receiver flask from being drawn back into the Kjeldahl flask. After distillation is judged complete, lower the receiver

---

[4]See *Official Methods of Analysis,* 14th ed., p. 16. Washington, D.C.: Association of Official Analytical Chemists, 1984.

flask to bring the adapter well clear of the liquid. Discontinue heating, disconnect the apparatus, and rinse the inside of the condenser with small portions of distilled water, collecting the washings in the receiver flask. Add 2 drops of bromocresol green to the receiver flask, and titrate the residual HCl with standard 0.1 M NaOH to the color change of the indicator.

Report the percentage of nitrogen and the percentage of protein (Note 7) in the unknown.

Notes

1. If filter paper is used to hold the sample, carry a similar piece through the analysis as a blank. Acid-washed filter paper is frequently contaminated with measurable amounts of ammonium ion and should be avoided if possible.
2. Any of the following catalyze the digestion: a drop of mercury, 0.5 g of HgO, a crystal of $CuSO_4$, 0.1 g of selenium, 0.2 g of $CuSeO_3$. The catalyst can be omitted, if desired.
3. Mercury(II) ions must be precipitated as the sulfide to prevent retention of some of the ammonia as an ammine complex.
4. A modification of this procedure uses about 50 mL of 4% boric acid solution in lieu of the standard HCl in the receiver flask (page 229). After distillation is complete, the ammonium borate produced is titrated with standard 0.1 M HCl, with 2 to 3 drops of bromocresol green as indicator.
5. If any sodium hydroxide solution comes into contact with the skin, the affected area should be washed IMMEDIATELY with copious amounts of water.
6. Granulated zinc (10 to 20 mesh) is added to minimize bumping during the distillation; it reacts slowly with the base to give small bubbles of hydrogen that prevent superheating of the liquid.
7. The percentage of protein in the unknown is calculated by multiplying the % N by an appropriate factor: 5.70 for cereals, 6.25 for meats, and 6.38 for dairy products.

28D   COMPLEX-FORMATION TITRATIONS WITH EDTA

See Chapter 14 for a discussion of the analytical uses of EDTA as a chelating reagent. Directions follow for a direct titration of magnesium and a determination of the hardness of a natural water.

28D-1 Preparation of Solutions

A pH-10 buffer and an indicator solution are needed for these titrations.

a. *Buffer solution, pH 10.* (Sufficient for about 50 titrations). Dilute 57 mL of concentrated NH_3 and 7 g of NH_4Cl in sufficient distilled water to give 100 mL of solution.
b. *Eriochrome Black T indicator.* (Sufficient for about 100 titrations.) Dissolve 100 mg of the solid in a solution containing 15 mL of ethanolamine and 5 mL of absolute ethanol. This solution should be freshly prepared every two weeks; refrigeration slows its deterioration.

28D-2 Preparation of Standard 0.01 M EDTA Solution

Discussion
See Section 14B-1 for a description of the properties of reagent-grade $Na_2H_2Y \cdot 2H_2O$ and its use in the direct preparation of standard EDTA solutions.

Procedure

Dry about 4 g of the purified dihydrate $Na_2H_2Y \cdot 2H_2O$ (Note 1) at 80°C to remove superficial moisture. Cool to room temperature in a desiccator. Weigh (to the nearest milligram) about 3.8 g into a 1-L volumetric flask (Note 2). Use a powder funnel to ensure quantitative transfer; rinse the funnel well with water before removing it from the flask. Add 600 to 800 mL of water (Note 3) and swirl periodically. Dissolution may take 15 min or longer. When all the solid has dissolved, dilute to the mark with water and mix well (Note 4). In calculating the molarity of the solution, correct the weight of the salt for the 0.3% moisture it ordinarily retains after drying at 80°C.

Notes
1. Directions for the purification of the disodium salt are described by W. J. Blaedel and H. T. Knight, *Anal. Chem.*, **1954,** *26* (4), 741.
2. The solution can be prepared from the anhydrous disodium salt, if desired. The weight taken should be about 3.6 g.
3. Water used in the preparation of standard EDTA solutions must be totally free of polyvalent cations. If any doubt exists concerning its quality, pass the water through a cation-exchange resin before use.
4. As an alternative, an EDTA solution that is approximately 0.01 M can be prepared and standardized by direct titration against a Mg^{2+} solution of known concentration (using the directions in Section 28D-3).

28D-3 The Determination of Magnesium by Direct Titration

Discussion
See Section 14B-7.

Procedure

Submit a clean 500-mL volumetric flask to receive the unknown, dilute to the mark with water, and mix thoroughly. Transfer 50.00-mL aliquots to 250-mL conical flasks, add 1 to 2 mL of pH-10 buffer and 3 to 4 drops of Eriochrome Black T indicator to each. Titrate with 0.01 M EDTA until the color changes from red to pure blue (Notes 1 and 2).

Express the results as parts per million of Mg^{2+} in the sample.

Notes
1. The color change tends to be slow in the vicinity of the end point. Care must be taken to avoid overtitration.
2. Other alkaline earths, if present, are titrated along with the Mg^{2+}; removal of Ca^{2+} and Ba^{2+} can be accomplished with $(NH_4)_2CO_3$. Most

polyvalent cations are also titrated. Precipitation as hydroxides or the use of a masking reagent may be needed to eliminate this source of interference.

28D-4 The Determination of Hardness in Water

Discussion
See Section 14B-9.

Procedure

Acidify 100.0-mL aliquots of the sample with a few drops of HCl, and boil gently for a few minutes to eliminate CO_2. Cool, add 3 to 4 drops of methyl red, and neutralize with 0.1 M NaOH. Introduce 2 mL of pH-10 buffer, 3 to 4 drops of Eriochrome Black T indicator, and titrate with standard 0.01 M Na_2H_2Y to a color change from red to pure blue (Note).

   Report the results in terms of milligrams of $CaCO_3$ per liter of water.

Note
The color change is sluggish if Mg^{2+} is absent. In this event, add 1 to 2 mL of 0.1 M MgY^{2-} before starting the titration. This reagent is prepared by adding 2.645 g of $MgSO_4 \cdot 7H_2O$ to 3.722 g of $Na_2H_2Y \cdot 2H_2O$ in 50 mL of distilled water. The solution is rendered faintly alkaline to phenolphthalein and diluted to 100 mL. A small portion mixed with pH-10 buffer and a few drops of Eriochrome Black T indicator should have a dull violet color. A single drop of 0.01 M EDTA solution should cause a color change to blue, while an equal volume of 0.01 M Mg^{2+} should cause a change to red. If necessary, adjust the composition with EDTA or with Mg^{2+} until these criteria are met.

28E TITRATIONS WITH POTASSIUM PERMANGANATE

The properties and uses of potassium permanganate are described in Section 17C-1. Directions follow for the determination of iron in an ore and calcium in a limestone.

28E-1 Preparation of 0.02 M Potassium Permanganate

Discussion
See page 313 for a discussion of the precautions needed in the preparation and storage of permanganate solutions.

Procedure

Dissolve about 3.2 g of $KMnO_4$ in 1 L of distilled water. Keep the solution at a gentle boil for about 1 h. Cover and let stand overnight. Remove MnO_2 by filtration (Note 1) through a fine-porosity filtering crucible (Note

2) or through a Gooch crucible fitted with glass mats. Transfer the solution to a clean glass-stoppered bottle; store in the dark when not in use.

Notes

1. The heating and filtering can be omitted if the permanganate solution is standardized and used on the same day.
2. Remove the MnO_2 that collects on the fritted plate with 1 M H_2SO_4 containing a few milliliters of 3% H_2O_2, followed by a rinse with copious quantities of water.

28E-2 Standardization of Potassium Permanganate Solutions

Discussion

See Section 17C-1 for a discussion of primary standards for permanganate solutions. Directions follow for standardization with sodium oxalate.

Procedure

Dry about 1.5 g of primary-standard $Na_2C_2O_4$ at 110°C for at least 1 h. Cool in a desiccator; weigh (to the nearest 0.1 mg) individual 0.2- to 0.3-g samples into 400-mL beakers. Dissolve each in about 250 mL of 1 M H_2SO_4. Heat each solution to 80 to 90°C, and titrate with $KMnO_4$ while stirring with a thermometer. The pink color imparted by one addition should be permitted to disappear before further titrant is introduced (Notes 1 and 2). Reheat if the temperature drops below 60°C. Take the first persistent ($\approx$30 s) pink color as the end point (Notes 3 and 4). Determine a blank by titrating an equal volume of the 1 M H_2SO_4.

Correct the titration data for the blank, and calculate the concentration of the permanganate solution (Note 5).

Notes

1. Promptly wash any $KMnO_4$ that spatters onto the walls of the beaker into the bulk of the liquid with a stream of water.
2. Finely divided MnO_2 will form along with Mn^{2+} if the $KMnO_4$ is added too rapidly and will cause the solution to acquire a faint brown discoloration. Precipitate formation is not a serious problem so long as sufficient oxalate remains to reduce the MnO_2 to Mn^{2+}; the titration is simply discontinued until the brown color disappears. The solution must be free of MnO_2 at the end point.
3. The surface of the permanganate solution rather than the bottom of the meniscus can be used to measure titrant volumes. Alternatively, backlighting with a flashlight or a match permits reading of the meniscus in the conventional manner.
4. A permanganate solution should not be allowed to stand in a buret any longer than necessary because partial decomposition to MnO_2 may occur. Freshly formed MnO_2 can be removed from a glass surface with 1 M H_2SO_4 containing a small amount of 3% H_2O_2.
5. As noted on page 314, this procedure yields molarities that are a few tenths of a percent low. For more accurate results, introduce from a buret sufficient permanganate to react with 90 to 95% of the oxalate

(about 40 mL of 0.02 M $KMnO_4$ for a 0.3-g sample). Let the solution stand until the permanganate color disappears. Then warm to about 60°C and complete the titration, taking the first permanent pink ($\approx$30 s) as the end point (Notes 3 and 4). Determine a blank by titrating an equal volume of the 1 M H_2SO_4.

28E-3 The Determination of Calcium in a Limestone

Discussion

In common with a number of other cations, calcium is conveniently determined by precipitation with oxalate ion. The solid calcium oxalate is filtered, washed free of excess precipitating reagent, and dissolved in dilute acid. The oxalic acid liberated in this step is then titrated with standard permanganate or some other oxidizing reagent. This method is applicable to samples that contain magnesium and the alkali metals. Most other cations must be absent since they either precipitate or coprecipitate as oxalates and cause positive errors in the analysis.

Factors Affecting the Composition of Calcium Oxalate Precipitates. It is essential that the mole ratio between calcium and oxalate be exactly unity in the precipitate and thus in solution at the time of titration. A number of precautions are needed to ensure this condition. For example, the calcium oxalate formed in a neutral or ammoniacal solution is likely to be contaminated with calcium hydroxide or a basic calcium oxalate, either of which will cause low results. The formation of these compounds is prevented by adding the oxalate to an acidic solution of the sample and slowly forming the precipitate by the dropwise addition of ammonia. The coarsely crystalline calcium oxalate that is produced under these conditions is readily filtered. Losses resulting from the solubility of calcium oxalate are negligible above pH 4, provided washing is limited to freeing the precipitate of excess oxalate.

Coprecipitation of sodium oxalate becomes a source of positive error in the determination of calcium whenever the concentration of sodium in the sample exceeds that of calcium. The error from this source can be eliminated by reprecipitation.

Magnesium, if present in high concentration, may also be a source of contamination. An excess of oxalate ion helps prevent this interference through the formation of soluble oxalate complexes of magnesium. Prompt filtration of the calcium oxalate can also help prevent interference because of the pronounced tendency of magnesium oxalate to form supersaturated solutions from which precipitate formation occurs only after an hour or more. These measures do not suffice for samples that contain more magnesium than calcium. Here, reprecipitation of the calcium oxalate becomes necessary.

The Composition of Limestones. Limestones are composed principally of calcium carbonate; dolomitic limestones contain large amounts of magnesium carbonate as well. Calcium and magnesium silicates are also present

Reprecipitation is a method for reducing coprecipitation errors by dissolving the initial precipitate and then reforming the precipitate.

in smaller amounts, along with the carbonates and silicates of iron, aluminum, manganese, titanium, sodium, and other metals.

Hydrochloric acid is an effective solvent for most limestones. Only silica, which does not interfere with the analysis, remains undissolved. Some limestones are more readily decomposed after they have been ignited; a few yield only to a carbonate fusion.

The method that follows is remarkably effective for determining calcium in most limestones. Iron and aluminum in amounts equivalent to that of calcium do not interfere. Small amounts of manganese and titanium can also be tolerated.

Procedure

Sample Preparation

Dry the unknown for 1 to 2 h at 110°C, and cool in a desiccator. If the material is readily decomposed in acid, weigh 0.25- to 0.30-g samples (to the nearest 0.1 mg) into 250-mL beakers. Add 10 mL of water to each sample and cover with a watch glass. Add 10 mL of concentrated HCl dropwise, taking care to avoid losses due to spattering as the acid is introduced.

Precipitation of Calcium Oxalate

Add 5 drops of saturated bromine water to oxidize any iron in the samples and boil gently (HOOD) for 5 min to remove the excess Br_2. Dilute each sample solution to about 50 mL, heat to boiling, and add 100 mL of hot 6% (w/v) $(NH_4)_2C_2O_4$ solution. Add 3 to 4 drops of methyl red, and precipitate CaC_2O_4 by slowly adding 6 M NH_3. As the indicator starts to change color, add the NH_3 at a rate of one drop every 3 to 4 s. Continue until the solutions turn to the intermediate yellow-orange color of the indicator (pH 4.5 to 5.5). Allow the solutions to stand for no more than 30 min (Note) and filter; medium-porosity filtering crucibles or Gooch crucibles with glass mats are satisfactory. Wash the precipitates with several 10-mL portions of cold water. Rinse the outside of the crucibles to remove residual $(NH_4)_2C_2O_4$ and return them to the beakers in which the CaC_2O_4 was formed.

Titration

Add 100 mL of water and 50 mL of 3 M H_2SO_4 to each of the beakers containing the precipitated calcium oxalate and the crucible. Heat to 80 to 90°C, and titrate with 0.02 M permanganate. The temperature should be above 60°C throughout the titration; reheat if necessary.

Report the percentage of CaO in the unknown.

Note

The period of standing can be longer if the unknown contains no Mg^{2+}.

28E-4 The Determination of Iron in an Ore

Discussion

The common ores of iron are hematite (Fe_2O_3), magnetite (Fe_3O_4), and limonite ($2Fe_2O_3 \cdot 3H_2O$). Steps in the analysis of these ores are (1) dissolution of the sample, (2) reduction of iron to the divalent state, and (3) titration of iron(II) with a standard oxidant.

The Decomposition of Iron Ores. Iron ores often decompose completely in hot concentrated hydrochloric acid. The rate of attack by this reagent is increased by the presence of a small amount of tin(II) chloride. The tendency of iron(II) and iron(III) to form chloro complexes accounts for the effectiveness of hydrochloric acid over nitric or sulfuric acid as a solvent for iron ores.

Many iron ores contain silicates that may not be entirely decomposed by treatment with hydrochloric acid. Incomplete silicate decomposition is indicated by a dark residue that remains after prolonged treatment with the acid. A white residue of hydrated silica, which does not interfere in any way, is indicative of complete decomposition.

The Prereduction of Iron. Because part or all of the iron is in the trivalent state after decomposition of the sample, reduction to iron(II) must precede titration with the oxidant. Any of the methods described in Section 17A-1 can be used. Perhaps the most satisfactory prereductant for iron is tin(II) chloride:

$$2 \ Fe^{3+} + Sn^{2+} \rightarrow 2 \ Fe^{2+} + Sn^{4+}$$

The only other common species reduced by this reagent are the high oxidation states of arsenic, copper, mercury, molybdenum, tungsten, and vanadium.

The excess reducing agent is eliminated by the addition of mercury(II) chloride:

$$Sn^{2+} + 2 \ HgCl_2(aq) \rightarrow Hg_2Cl_2(s) + Sn^{4+} + 2 \ Cl^-$$

The slightly soluble mercury(I) chloride does not reduce permanganate, nor does the excess mercury(II) chloride reoxidize iron(II). Care must be taken, however, to prevent the occurrence of the alternative reaction

$$Sn^{2+} + HgCl_2(aq) \rightarrow Hg(l) + Sn^{4+} + 2 \ Cl^-$$

Elemental mercury reacts with permanganate and causes the results of the analysis to be high. The formation of mercury, which is favored by an appreciable excess of tin(II), is prevented by careful control of this excess and by the rapid addition of excess mercury(II) chloride. A proper reduction is indicated by the appearance of a small amount of a silky white precipitate after the addition of mercury(II). Formation of a gray precipitate at this juncture indicates the presence of metallic mercury; the total absence of a precipitate indicates that an insufficient amount of tin(II) chloride was used. In either event, the sample must be discarded.

The Titration of Iron(II). The reaction of iron(II) with permanganate is smooth and rapid. The presence of iron(II) in the reaction mixture, however, *induces* oxidation of chloride ion by permanganate, a reaction that does not ordinarily proceed rapidly enough to cause serious error. High results are obtained if this parasitic reaction is not controlled. Its effects can be eliminated through removal of the hydrochloric acid by evaporation with sulfuric acid or by introduction of *Zimmermann-Reinhardt reagent,* which contains manganese(II) in a fairly concentrated mixture of sulfuric and phosphoric acids.

The oxidation of chloride ion during a titration is believed to involve a direct reaction between Cl⁻ and the manganese(III) ions that form as an intermediate in the reduction of permanganate ion by iron(II). The manganese(II) in the Zimmermann-Reinhardt reagent is believed to inhibit the formation of chlorine by decreasing the potential of the manganese(III)/manganese(II) couple. Phosphate ion is believed to exert a similar effect by forming stable manganese(III) complexes. Moreover, phosphate ions react with iron(III) to form nearly colorless complexes so that the yellow color of the iron(II)/chloro complexes does not interfere with the end point.[5]

Preparation of Solutions

The following solutions suffice for about 100 titrations.

a. *Tin(II) chloride, 0.25 M.* Dissolve 60 g of iron-free $SnCl_2 \cdot 2H_2O$ in 100 mL of concentrated HCl; warm if necessary. After the solid has dissolved, dilute to 1 L with distilled water and store in a well stoppered bottle. Add a few pieces of mossy tin to help preserve the solution.

b. *Mercury(II) chloride, 5% (w/v).* Dissolve 50 g of $HgCl_2$ in 1 L of distilled water.

c. *Zimmermann-Reinhardt reagent.* Dissolve 300 g of $MnSO_4 \cdot 4H_2O$ in 1 L of water. Cautiously add 400 mL of concentrated H_2SO_4 and 400 mL of 85% H_3PO_4, and dilute to 3 L.

Procedure

Sample Preparation

Dry the ore at 110°C for at least 3 h, and then allow it to cool to room temperature in a desiccator. Consult with the instructor for a sample size that will require from 25 to 40 mL of standard 0.02 M $KMnO_4$. Weigh samples into 500-mL conical flasks. To each, add 10 mL of concentrated HCl and about 3 mL of 0.25 M $SnCl_2$ (Note 1). Cover each flask with a small watch glass or Tuttle flask cover. Heat the flasks in a hood at just below boiling until the samples are decomposed and the undissolved solid—if any—is pure white (Note 2). Use another 1 to 2 mL of $SnCl_2$ to eliminate any yellow color that may develop as the solutions are heated.

---

[5]The mechanism by which Zimmermann-Reinhardt reagent acts has been the subject of much study. For a discussion of this work, see H. A. Laitinen, *Chemical Analysis,* pp. 369–372. New York: McGraw-Hill, 1960.

Heat a blank consisting of 10 mL of HCl and 3 mL of $SnCl_2$ for the same amount of time.

After the ore has been decomposed, remove the excess Sn(II) by the dropwise addition of 0.02 M $KMnO_4$ until the solutions become faintly yellow. Dilute to about 15 mL. Add sufficient $KMnO_4$ solution to impart a faint pink color to the blank; then decolorize with one drop of the $SnCl_2$ solution.

Take samples and blank individually through subsequent steps to minimize air-oxidation of iron(II).

Reduction of Iron

Heat the sample solution nearly to boiling, and make dropwise additions of 0.25 M $SnCl_2$ until the yellow color just disappears; then add 2 more drops (Note 3). Cool to room temperature, and *rapidly* add 10 mL of 5% $HgCl_2$ solution. A small amount of silky white Hg_2Cl_2 should precipitate (Note 4). The blank should be treated with the $HgCl_2$ solution.

Titration

Following addition of the $HgCl_2$, wait 2 to 3 min. Then add 25 mL of Zimmermann-Reinhardt reagent and 300 mL of water. Titrate *immediately* with standard 0.02 M $KMnO_4$ to the first faint pink that persists for 15 to 20 s. Do not add the $KMnO_4$ rapidly at any time. Correct the titrant volume for the blank.

Report the percentage of Fe_2O_3 in the sample.

Notes

1. The $SnCl_2$ hastens decomposition of the ore by reducing iron(III) oxides to iron(II). Insufficient $SnCl_2$ is indicated by the appearance of yellow iron(III)/chloride complexes.

2. If dark particles persist after the sample has been heated with acid for several hours, filter the solution through ashless paper, wash the residue with 5 to 10 mL of 6 M HCl, and retain the filtrate and washings. Ignite the paper and its contents in a small platinum crucible. Mix 0.5 to 0.7 g of Na_2CO_3 with the residue and heat until a clear melt is obtained. Cool, add 5 mL of water, and then cautiously add a few milliliters of 6 M HCl. Warm the crucible until the melt has dissolved, and combine the contents with the original filtrate. Evaporate the solution to 15 mL and continue the analysis.

3. The solution may not be entirely colorless but instead may acquire a faint yellow-green hue. Further additions of $SnCl_2$ will not alter this color. If too much $SnCl_2$ is added, it can be removed by adding 0.2 M $KMnO_4$ and repeating the reduction.

4. The absence of precipitate indicates that insufficient $SnCl_2$ was used and that the reduction of iron(III) was incomplete. A gray residue indicates the presence of elemental mercury, which reacts with $KMnO_4$. The sample must be discarded in either event.

5. These directions can be used to standardize a permanganate solution against primary standard iron. Weigh (to the nearest 0.1 mg) 0.2-g lengths of electrolytic iron wire into 250-mL conical flasks and dissolve in about 10 mL of concentrated HCl. Dilute each sample to about 75 mL. Then take each individually through the reduction and titration steps.

28F TITRATIONS WITH IODINE

The oxidizing properties of iodine, the composition and stability of triiodide solutions, and the applications of this reagent in volumetric analysis
are discussed in Section 17C-3.

28F-1 The Preparation of an Iodine Solution

The titrant concentration in iodine titrations is usually 0.05 M, and starch
is ordinarily used as the indicator.

Iodimetric methods make use of
standard iodine solutions as oxidizing reagents. Iodometric methods
make use of standard sodium
thiosulfate solutions to titrate the
iodine produced in the reaction of
an oxidizing analyte with an excess
of potassium iodide.

Preparation of Solutions

a. *Iodine, approximately 0.05 M.* Weigh about 40 g of KI into a 100-mL
   beaker. Add 12.7 g of I_2 and 10 mL of water. Stir for several minutes
   (Note 1). Introduce an additional 20 mL of water, and stir again for
   several minutes. Carefully decant the bulk of the liquid into a storage
   bottle containing 1 L of distilled water. It is essential that any undissolved iodine remain in the beaker (Note 2).
b. *Starch indicator.* (Sufficient for about 100 titrations.) Rub 1 g of soluble starch and 15 mL of water into a paste. Dilute to about 500 mL with
   boiling water, and heat until the mixture is clear. Cool; store in a
   tightly stoppered bottle (Note 3). For most titrations, 3 to 5 mL of the
   indicator is used.

Notes
1. Iodine dissolves slowly in the KI solution. Thorough stirring is needed
   to hasten the process.
2. Any solid I_2 inadvertently transferred to the storage bottle will cause
   the concentration of the solution to increase gradually. Filtration
   through a sintered-glass crucible eliminates this potential source of
   difficulty.
3. The indicator is readily attacked by airborne organisms and should be
   freshly prepared every few days.

28F-2 Standardization of Iodine Solutions

Discussion
Arsenic(III) oxide, long a favored primary standard for iodine solutions,
is now seldom used because of the elaborate federal regulations governing
the use of even small amounts of arsenic-containing compounds. Barium
thiosulfate monohydrate and anhydrous sodium thiosulfate have been
proposed as alternative standards.[6,7] Perhaps the most convenient method
for determining the concentration of an iodine solution is the titration of
aliquots with a sodium thiosulfate solution that has been standardized
against pure potassium iodate. Instructions for this method follow.

---

[6]W. M. McNevin and O. H. Kriege, *Anal. Chem.,* **1953,** 25 (5), 767.

[7]A. A. Woolf, *Anal. Chem.,* **1982,** 54 (12), 2134.

Preparation of Solutions

a. *Sodium thiosulfate, 0.1 M.* Follow the directions in Sections 28G-1 and 28G-2 for the preparation and standardization of this solution.
b. *Starch indicator.* See Section 28F-1.

Procedure

Transfer 25.00-mL aliquots of the iodine solution to 250-mL conical flasks, and dilute to about 50 mL. Introduce approximately 1 mL of 3 M H_2SO_4, and titrate immediately with standard sodium thiosulfate until the solution becomes a faint straw yellow. Add about 5 mL of starch indicator, and complete the titration, taking as the end point the change in color from blue to colorless (Note).

Note

The blue color of the starch/iodine complex may reappear after the titration has been completed, owing to the air-oxidation of iodide ion.

28F-3 The Determination of Antimony in Stibnite

Discussion

The analysis of stibnite, a common antimony ore, is a typical application of iodimetry and is based upon the oxidation of Sb(III) to Sb(V):

$$SbO_3^{3-} + I_2 + H_2O \rightleftharpoons SbO_4^{3-} + 2\,I^- + 2\,H^+$$

The position of this equilibrium is strongly dependent upon the hydrogen ion concentration. In order to force the reaction to the right, it is common practice to carry out the titration in the presence of an excess of sodium hydrogen carbonate, which consumes the hydrogen ions as they form.

Stibnite is an antimony sulfide ore containing silica and other contaminants. Provided the material is free of iron and arsenic, the analysis of stibnite for its antimony content is straightforward. Samples are decomposed in hot concentrated hydrochloric acid to eliminate sulfide as gaseous hydrogen sulfide. Care is needed to prevent loss of volatile antimony(III) chloride during this step. The addition of potassium chloride helps by favoring formation of nonvolatile chloro complexes such as $SbCl_4^-$ and $SbCl_6^{3-}$.

Sparingly soluble basic antimony salts, such as SbOCl, often form when the excess hydrochloric acid is neutralized; these react incompletely with iodine and cause low results. The difficulty is overcome by adding tartaric acid, which forms a soluble complex ($SbOC_4H_4O_6^-$) from which antimony is rapidly oxidized by the reagent.

Procedure

Dry the unknown at 110°C for 1 h, and allow it to cool in a desiccator. Weigh individual samples (Note 1) into 500-mL conical flasks. Introduce about 0.3 g of KCl and 10 mL of concentrated HCl to each flask. Heat the mixtures (HOOD) to just below boiling until only a white or slightly gray residue of SiO_2 remains.

Add 3 g of tartaric acid to each sample and heat for an additional 10 to 15 min. Then, with good swirling, add water (Note 2) from a pipet or buret until the volume is about 100 mL. If reddish Sb_2S_3 forms, discontinue dilution and heat further to eliminate H_2S; add more HCl if necessary.

Add 3 drops of phenolphthalein, and neutralize with 6 M NaOH to the first faint pink of the indicator. Discharge the color by the dropwise addition of 6 M HCl, and then add 1 mL in excess. Introduce 4 to 5 g of $NaHCO_3$, taking care to avoid losses of solution by spattering during the addition. Add 5 mL of starch indicator, rinse down the inside of the flask, and titrate with standard 0.05 M I_2 to the first blue color that persists for 30 s.

Report the percentage of Sb_2S_3 in the unknown.

Notes

1. Samples should contain between 1.5 and 2 mmol of antimony; consult with the instructor for an appropriate sample size. Weighings to the nearest milligram are adequate for samples larger than 1 g.

2. The slow addition of water, with efficient stirring, is essential to prevent the formation of SbOCl.

28G   TITRATIONS WITH SODIUM THIOSULFATE

Numerous analytical methods are based upon the reducing properties of iodide ion:

$$2\ I^- \rightarrow I_2 + 2\ e^-$$

Iodine, the reaction product, is ordinarily titrated with a standard sodium thiosulfate solution, with starch serving as the indicator:

$$I_2 + 2\ S_2O_3^{2-} \rightarrow 2\ I^- + S_4O_6^{2-}$$

A discussion of iodometric methods is found in Section 17B-2.

28G-1 Preparation of 0.1 M Sodium Thiosulfate

Procedure

Boil about 1 L of distilled water for 10 to 15 min. Allow the water to cool to room temperature; then add about 25 g of $Na_2S_2O_3 \cdot 5H_2O$ and 0.1 g of Na_2CO_3. Stir until the solid has dissolved. Transfer the solution to a clean glass or plastic bottle, and store in a dark place.

28G-2 Standardization of Sodium Thiosulfate Against Potassium Iodate

Discussion

Solutions of sodium thiosulfate are conveniently standardized by titration of the iodine produced when an unmeasured excess of potassium iodide is added to a known volume of an acidified standard potassium iodate solution. The reaction is

$$IO_3^- + 5\,I^- + 6\,H^+ \rightarrow 3\,I_2 + 3\,H_2O$$

Note that each formula weight of iodate results in the production of three formula weights of iodine. The procedure that follows is based upon this reaction.

Preparation of Solutions

a. *Potassium iodate, 0.0100 M.* Dry about 1.2 g of primary-standard KIO_3 at 110°C for at least 1 h and cool in a desiccator. Weigh (to the nearest 0.1 mg) about 1.1 g into a 500-mL volumetric flask; use a powder funnel to ensure quantitative transfer of the solid. Rinse the funnel well, dissolve the KIO_3 in about 200 mL of distilled water, dilute to the mark, and mix thoroughly.

b. *Starch indicator.* See Section 28F-1.

Procedure

Pipet 50.00-mL aliquots of standard iodate solution into 250-mL conical flasks. *Treat each sample individually from this point to minimize error resulting from the air-oxidation of iodide ion.* Introduce 2 g of iodate-free KI, and swirl the flask to hasten solution. Add 2 mL of 6 M HCl, and immediately titrate with thiosulfate until the solution becomes pale yellow. Introduce 5 mL of starch indicator, and titrate with constant stirring to the disappearance of the blue color. Calculate the molarity of the thiosulfate solution.

28G-3 Standardization of Sodium Thiosulfate Against Copper

Discussion

Thiosulfate solutions can also be standardized against pure copper wire or foil. This procedure is advantageous when the solution is to be used for the determination of copper because it cancels any determinate error in the method.

Copper(II) is reduced quantitatively to copper(I) by iodide ion:

$$2\,Cu^{2+} + 4\,I^- \rightarrow 2\,CuI(s) + I_2$$

The importance of CuI formation in forcing this reaction to completion can be seen from the following standard electrode potentials:

$$Cu^{2+} + e^- \rightleftarrows Cu^+ \qquad E^0 = 0.15 \text{ V}$$

$$I_2 + 2\,e^- \rightleftarrows 2\,I^- \qquad E^0 = 0.54 \text{ V}$$

$$Cu^{2+} + I^- + e^- \rightleftarrows CuI(s) \qquad E^0 = 0.86 \text{ V}$$

The first two potentials suggest that iodide should have no tendency to reduce copper(II); the formation of CuI, however, favors the reduction. The solution must contain at least 4% excess iodide to force the reaction to completion. Moreover, the pH must be below 4 to prevent the formation of basic copper species that react slowly and incompletely with iodide ion. The acidity of the solution cannot be greater than about 0.3 M, however, because of the tendency of iodide ion to undergo air-oxidation, a process catalyzed by copper salts.

Nitrogen oxides also catalyze the air-oxidation of iodide ion. A common source of these oxides is the nitric acid ordinarily used to dissolve metallic copper and other copper-containing solids. Urea is used to scavenge nitrogen oxides from solutions:

$$(NH_2)_2CO + 2\,HNO_2 \rightarrow 2\,N_2(g) + CO_2(g) + 3\,H_2O$$

The titration of iodine by thiosulfate tends to yield slightly low results owing to the adsorption of small but measurable quantities of iodine upon solid CuI. The adsorbed iodine is released only slowly, even when excess thiosulfate is present; transient and premature end points result. This difficulty is largely overcome by the addition of thiocyanate ion. The sparingly soluble copper(I) thiocyanate replaces part of the copper iodide at the surface of the solid:

$$CuI(s) + SCN^- \rightarrow CuSCN(s) + I^-$$

Accompanying this reaction is the release of the adsorbed iodine, which thus becomes available for titration. The addition of thiocyanate must be delayed until most of the iodine has been titrated to prevent interference from a slow reaction between the two species, possibly

$$2\,SCN^- + I_2 \rightarrow 2\,I^- + (SCN)_2$$

Preparation of Solutions

a. *Urea, 5% (w/v)*. Dissolve about 5 g of urea in sufficient water to give 100 mL of solution. Approximately 10 mL will be needed for each titration.

b. *Starch indicator*. See Section 28F-1.

Procedure

Use scissors to cut copper wire or foil into 0.20- to 0.25-g portions. Wipe the metal free of dust and grease with a filter paper; do not dry it. The pieces of copper should be handled with paper strips, cotton gloves, or tweezers to prevent contamination by contact with the skin.

Use a weighed watch glass or weighing bottle to obtain the weight of individual copper samples by difference (to the nearest 0.1 mg). Transfer each sample to a 250-mL conical flask. Add 5 mL of 6 M HNO_3, cover with a small watch glass, and warm gently (HOOD) until the metal has dissolved. Dilute with about 25 mL of distilled water, add 10 mL of 5% (w/v) urea, and boil briefly to eliminate nitrogen oxides. Rinse the watch glass, collecting the rinsings in the flask. Cool.

Add concentrated NH_3 dropwise and with thorough mixing to produce the intensely blue $Cu(NH_3)_4^{2+}$; the solution should smell faintly of ammonia (Note). Make dropwise additions of 3 M H_2SO_4 until the color of the complex just disappears, and then add 2.0 mL of 85% H_3PO_4. Cool to room temperature.

Treat each sample individually from this point on to minimize the air-oxidation of iodide ion. Add 4.0 g of KI to the sample, and titrate immediately with $Na_2S_2O_3$ until the solution becomes pale yellow. Add 5 mL of starch indicator, and continue the titration until the blue color becomes faint. Add 2 g of KSCN; swirl vigorously for 30 s. Complete the titration, using the disappearance of the blue starch/I_2 color as the end point.

Calculate the molarity of the $Na_2S_2O_3$ solution.

Note

Do not sniff vapors directly from the flask; instead, waft them toward your nose with a waving motion of your hand.

28G-4 The Determination of Copper in Brass

Discussion

The standardization procedure described in Section 28G-3 is readily adapted to the determination of copper in brass, an alloy that also contains appreciable amounts of tin, lead, and zinc (and perhaps minor amounts of nickel and iron). The method is relatively simple and applicable to brasses with less than 2% iron. A weighed sample is treated with nitric acid, which causes the tin to precipitate as a hydrated oxide of uncertain composition. Evaporation with sulfuric acid to the appearance of sulfur trioxide eliminates the excess nitrate, redissolves the tin compound, and possibly causes the formation of lead sulfate. The pH is adjusted through the addition of ammonia, followed by acidification with a measured amount of phosphoric acid. An excess of potassium iodide is added, and the liberated iodine is titrated with standard thiosulfate. See Section 28G-3 for additional discussion.

Procedure

If so directed, free the metal of oils by treatment with an organic solvent; briefly heat in an oven to drive off the solvent. Weigh (to the nearest 0.1 mg) 0.3-g samples into 250-mL conical flasks, and introduce 5 mL of 6 M HNO_3 into each; warm (HOOD) until solution is complete. Add 10 mL of concentrated H_2SO_4, and evaporate (HOOD) until copious white fumes of SO_3 are given off. Allow the mixture to cool. Cautiously add 30 mL of distilled water, boil for 1 to 2 min, and again cool.

Follow the instructions in the third and fourth paragraphs of the *Procedure* in Section 28G-3.

Report the percentage of copper in the sample.

28H   TITRATIONS WITH POTASSIUM BROMATE

Applications of standard bromate solutions to the determination of organic functional groups are described in Section 17C-4. Directions follow for the determination of ascorbic acid in vitamin C tablets.

28H-1 Preparation of Solutions

a. *Potassium bromate, 0.015 M.* Transfer about 1.5 g of reagent-grade potassium bromate to a weighing bottle, and dry at 110°C for at least 1 h. Cool in a desiccator. Weigh approximately 1.3 g (to the nearest 0.1 mg) into a 500-mL volumetric flask; use a powder funnel to ensure quantitative transfer of the solid. Rinse the funnel well, and dissolve the $KBrO_3$ in about 200 mL of distilled water. Dilute to the mark, and mix thoroughly. *Solid potassium bromate can cause a fire if it comes into contact with damp organic material (such as paper toweling in a waste container).* Consult with the instructor concerning the disposal of any excess.

b. *Sodium thiosulfate, 0.05 M.* Follow the directions in Section 28G-1; use about 12.5 g of $Na_2S_2O_3 \cdot 5H_2O$ per liter of solution.

c. *Starch indicator.* See Section 28F-1.

28H-2 Standardization of Sodium Thiosulfate Against Potassium Bromate

Discussion

Iodine is generated by the reaction between a known volume of standard potassium bromate and an unmeasured excess of potassium iodide:

$$BrO_3^- + 6\ I^- + 6\ H^+ \rightarrow Br^- + 3\ I_2 + 3\ H_2O$$

The iodine produced is titrated with the sodium thiosulfate solution.

Procedure

Pipet 25.00-mL aliquots of the $KBrO_3$ solution into 250-mL conical flasks and rinse the interior wall with distilled water. *Treat each sample individually beyond this point.* Introduce 2 to 3 g of KI and about 5 mL of 3 M H_2SO_4. Immediately titrate with $Na_2S_2O_3$ until the solution is pale yellow. Add 5 mL of starch indicator, and titrate to the disappearance of the blue color.

Calculate the concentration of the thiosulfate solution.

28H-3 The Determination of Ascorbic Acid in Vitamin C Tablets by Titration with Potassium Bromate

Discussion

Ascorbic acid, $C_6H_8O_6$, is cleanly oxidized to dehydroascorbic acid by bromine:

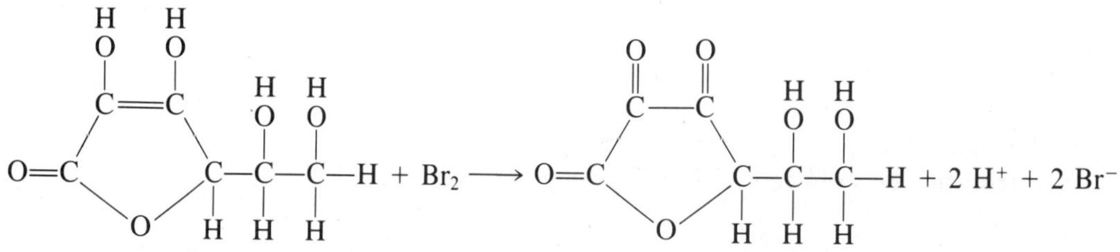

An unmeasured excess of potassium bromide is added to an acidified solution of the sample. The solution is titrated with standard potassium bromate to the first permanent appearance of excess bromine; this excess is then determined iodometrically with standard sodium thiosulfate. The entire titration must be performed without delay to prevent air-oxidation of the ascorbic acid.

Procedure

Weigh (to the nearest milligram) three to five vitamin C tablets (Note 1). Pulverize them thoroughly in a mortar, and transfer the powder to a dry weighing bottle. Weigh individual 0.40- to 0.50-g samples (to the nearest 0.1 mg) into dry 250-mL conical flasks. *Treat each sample individually beyond this point.* Dissolve the sample (Note 2) in 50 mL of 1.5 M H_2SO_4; then add about 5 g of KBr. Titrate immediately with standard $KBrO_3$ to the first faint yellow due to excess Br_2. Record the volume of $KBrO_3$ used. Add 3 g of KI and 5 mL of starch indicator; back-titrate (Note 3) with standard 0.05 M $Na_2S_2O_3$.

Calculate the average weight (in milligrams) of ascorbic acid (fw = 176.13) in each tablet.

Notes

1. This method is not applicable to chewable vitamin C tablets.
2. The binder in many vitamin C tablets remains in suspension throughout the analysis. If the binder is starch, the characteristic color of the complex with iodine appears upon the addition of KI.
3. The volume of thiosulfate needed for the back-titration seldom exceeds a few milliliters.

28I POTENTIOMETRIC METHODS

Potentiometric measurements provide a highly selective method for the quantitative determination of numerous cations and anions. A discussion of the principles and applications of potentiometric measurements is found in Chapter 18. Detailed instructions are given in this section on how to use potentiometric measurements to locate end points in volumetric titrations. In addition, a procedure for the direct potentiometric determination of fluoride ion in drinking water and in toothpaste is described.

28I-1 General Directions for Performing a Potentiometric Titration

The procedure that follows is applicable to the titrimetric methods described in this section. With the proper choice of indicator electrode, it can also be applied to most of the volumetric methods given in Sections 28B through 28H.

1. Dissolve the sample in 50 to 250 mL of water. Rinse a suitable pair of electrodes with distilled water, and immerse them in the sample solution. Provide magnetic (or mechanical) stirring. Position the buret so that reagent can be delivered without splashing.
2. Connect the electrodes to the meter, commence stirring, and record the initial buret volume and the initial potential (or pH).
3. Record the meter reading and buret volume after each addition of titrant. Introduce fairly large volumes (about 5 mL) at the outset. Withhold each succeeding addition until the meter reading remains constant within 1 to 2 mV (or 0.05 pH unit) for at least 30 s (Note). Judge the volume of reagent to be added by estimating a value of $\Delta E/\Delta V$ after each addition. In the immediate vicinity of the equivalence point, introduce the reagent in 0.1-mL increments. Continue the titration 2 to 3 mL beyond the equivalence point, increasing the volume increments as $\Delta E/\Delta V$ again becomes smaller.

Note. Stirring motors occasionally cause erratic meter readings; it may be advisable to turn off the motor while meter readings are made.

28I-2 The Potentiometric Titration of Chloride and Iodide in a Mixture

Discussion
The potentiometric titration of halide mixtures is discussed in Feature 10-1. The silver indicator electrode can be a commercial billet type or simply a polished wire. A calomel electrode can be used as reference, although diffusion of chloride ion from the salt bridge may cause the results of the titration to be measurably high. This source of error can be eliminated by placing the calomel electrode in a potassium nitrate solution that is in contact with the analyte solution by means of a KNO_3 salt bridge. Alternatively, the analyte solution can be made slightly acidic with several drops of nitric acid; a glass electrode can then serve as the reference electrode because the pH of the solution and thus its potential remain essentially constant throughout the titration.

The titration of I^-/Cl^- mixtures demonstrates how a potentiometric titration can have multiple end points. The potential of the silver electrode is proportional to pAg. Thus, a plot of E_{Ag} against titrant volume yields an experimental curve with the same shape as the theoretical curve shown in Figure 10-5 (the ordinate units will be different, of course).

Experimental curves for the titration of I^-/Cl^- mixtures do not show the sharp discontinuity that occurs at the first equivalence point of the theoretical curve (Figure 10-5). More important, the volume of silver nitrate needed to reach the I^- end point is generally somewhat greater than theoretical. This effect is the result of coprecipitation of the more soluble AgCl during formation of the less soluble AgI. An overconsumption of reagent thus occurs in the first part of the titration. The total volume closely approaches the correct amount.

Despite this coprecipitation error, the potentiometric method is useful for the analysis of halide mixtures. With approximately equal quantities of iodide and chloride, relative errors can be kept within 2%.

Preparation of Solutions

a. *Silver nitrate, 0.05 M.* Follow the instructions in Section 28B-1.
b. *Potassium nitrate salt bridge.* Bend an 8-mm glass tube into a U-shape with arms that are long enough to extend nearly to the bottom of two 100-mL beakers. Heat 50 mL of water to boiling, and stir in 1.8 g of powdered agar; continue to heat and stir until a uniform suspension is formed. Dissolve 12 g of KNO_3 in the hot suspension. Allow the mixture to cool somewhat. Clamp the U-tube with the openings facing up, and use a medicine dropper to fill it with the warm agar suspension. Cool the tube under a cold-water tap to form the gel. When the bridge is not in use, immerse the ends in 2.5 M KNO_3.

Procedure

Obtain the unknown in a clean 250-mL volumetric flask; dilute to the mark with water, and mix well.

Transfer 50.00 mL of the sample to a clean 100-mL beaker, and add a drop or two of concentrated HNO_3. Place about 25 mL of 2.5 M KNO_3 in a second 100-mL beaker, and make contact between the two solutions with the agar salt bridge. Immerse a silver electrode in the analyte solution and a calomel reference electrode in the second beaker. Titrate with $AgNO_3$ as described in Section 28I-1. Use small increments of titrant in the vicinity of the two end points.

Plot the data, and establish end points for the two analyte ions. Plot a theoretical titration curve, assuming the measured concentrations of the two constituents to be correct.

Report the number of milligrams of I^- and Cl^- in the sample or as otherwise instructed.

28I-3 The Potentiometric Determination of Solute Species in a Carbonate Mixture

Discussion

A glass/calomel electrode system can be used to locate end points in neutralization titrations and to estimate dissociation constants. As a preliminary step to the titrations, the electrode system is standardized against a buffer of known pH.

The unknown is issued as an aqueous solution prepared from one or perhaps two adjacent members of the following series: $NaHCO_3$, Na_2CO_3, and NaOH (see Section 13B-2). The object is to determine which of these components were used to prepare the unknown as well as the weight percent of each solute.

Most unknowns require a titration with either standard acid or standard base. A few may require separate titrations, one with acid and one with base. The initial pH of the unknown provides guidance concerning the appropriate titrant(s); a study of Figure 12-4 may be helpful in interpreting the data.

Preparation of Solutions

Standardized 0.1 M HCl and/or 0.1 M NaOH. Follow the directions in Sections 28C-3 through 28C-7.

Procedure

Obtain the unknown in a clean 250-mL volumetric flask. Dilute to the mark and mix well. Transfer a small amount of the diluted unknown to a beaker, and determine its pH. Titrate a 50.00-mL aliquot with standard acid or standard base (or perhaps both). Use the resulting titration curves to select indicator(s) suitable for end-point detection, and perform duplicate titrations with these.

Identify the solute species in the unknown, and report the weight/volume percent of each. Calculate the approximate dissociation constant that can be obtained for any carbonate-containing species from the titration data.

28I-4 The Direct Potentiometric Determination of Fluoride Ion

Discussion

The solid-state fluoride electrode (Section 18D-6) has found extensive use in the determination of fluoride in a variety of materials. Directions follow for the determination of this ion in drinking water and in toothpaste. A total-ionic-strength adjustment buffer (TISAB) is used to adjust all unknowns and standards to essentially the same ionic strength; with this reagent present, the *concentration* of fluoride, rather than its activity, is measured. The pH of the buffer is about 5, a level at which F^- is the predominant fluorine-containing species. The buffer also contains cyclohexylaminedinitrilotetraacetic acid, which forms stable chelates with iron(III) and aluminum(III), thus freeing fluoride ion from its complexes with these cations.

Review Section 18F-2 before undertaking these experiments.

Preparation of Solutions

a. *Total-ionic-strength adjustment buffer.* This solution is marketed commercially under the trade name TISAB.[8] Sufficient buffer for 15 to 20 determinations can be prepared by mixing (with stirring) 57 mL of glacial acetic acid, 58 g of NaCl, 4 g of cyclohexylaminedinitrilotetraacetic acid, and 500 mL of distilled water in a 1-L beaker. Cool the contents in a water or ice bath, and carefully add 6 M NaOH to a pH of 5.0 to 5.5. Dilute to 1 L with water, and store in a plastic bottle. This is sufficient for about 20 measurements.

b. *Standard fluoride solution, 100 ppm.* Dry a quantity of NaF at 110°C for 2 h. Cool in a desiccator; then weigh (to the nearest milligram) 0.22 g into a 1-L volumetric flask. (CAUTION! NaF IS HIGHLY TOXIC. *Immediately* wash any skin touched by this compound with copious quantities of water.) Dissolve in water, dilute to the mark, mix well, and store in a plastic bottle. Calculate the exact concentration of fluoride in parts per million. A standard F⁻ solution can be purchased from commercial sources.

Procedure

The apparatus for this experiment consists of a solid-state fluoride electrode, a saturated calomel electrode, and a pH meter. A sleeve-type calomel electrode is needed for the toothpaste determination because the measurement is made on a suspension that tends to clog the liquid junction. The sleeve must be loosened momentarily to renew the interface after each series of measurements.

Determination of Fluoride in Drinking Water

Transfer 50.00-mL portions of the water to 100-mL volumetric flasks, and dilute to the mark with TISAB solution.

Prepare a 5-ppm F⁻ solution by diluting 25.0 mL of the 100-ppm standard to 500 mL in a volumetric flask. Transfer 5.00-, 10.0-, 25.0-, and 50.0-mL aliquots of the 5-ppm solution to 100-mL volumetric flasks, add 50 mL of TISAB solution, and dilute to the mark. (These solutions correspond to 0.5, 1.0, 2.5, and 5.0 ppm F⁻ in the sample.)

After thorough rinsing and drying with paper tissue, immerse the electrodes in the 0.5-ppm standard. Stir mechanically for 3 min; then record the potential. Repeat with the remaining standards and samples.

Plot the measured potential against the log of the concentration of the standards. Use this plot to determine the concentration in parts per million of fluoride in the unknown.

Determination of Fluoride in Toothpaste[9]

Weigh (to the nearest milligram) 0.2 g of toothpaste into a 250-mL beaker. Add 50 mL of TISAB solution, and boil for 2 min with good mixing. Cool and then transfer the suspension quantitatively to a 100-mL volumetric flask, dilute to the mark with distilled water, and mix well. Follow the

---

[8]Orion Research, Boston, MA.

[9]From T. S. Light and C. C. Cappuccino, *J. Chem. Educ.,* **1975,** *52,* 247.

directions for the analysis of drinking water, beginning with the second paragraph.

Report the parts per million of F⁻ in the sample.

28J ELECTROGRAVIMETRIC METHODS

A convenient example of an electrogravimetric method is the simultaneous determination of copper and lead in a sample of brass. Additional information concerning electrogravimetric methods is found in Section 19B.

28J-1 The Electrogravimetric Determination of Copper and Lead in Brass

Discussion

This procedure is based upon the deposition of metallic copper on a cathode and of lead as PbO_2 on an anode. As a first step, the hydrous oxide of tin ($SnO_2 \cdot xH_2O$) that forms when the sample is treated with nitric acid must be removed by filtration. Lead dioxide is deposited quantitatively at the anode from a solution with a high nitrate ion concentration; copper is only partially deposited on the cathode under these conditions. It is therefore necessary to eliminate the excess nitrate after deposition of the PbO_2 is complete. Removal is accomplished through the addition of urea:

$$6\ NO_3^- + 6\ H^+ + 5\ (NH_2)_2CO \rightarrow 8\ N_2(g) + 5\ CO_2(g) + 13\ H_2O$$

Copper then deposits quantitatively from the solution after the nitrate ion concentration has been decreased.

Procedure

Preparation of Electrodes

Immerse the platinum electrodes in hot 6 M HNO_3 for about 5 min (Note 1). Wash them thoroughly with distilled water, rinse with several small portions of acetone or ethanol, and dry in an oven at 110°C for 2 to 3 min. Cool and weigh both anode and cathode to the nearest 0.1 mg.

Preparation of Samples

It is not necessary to dry the unknown. A rinse with acetone is recommended if there is evidence of oil on the surface of the metal. Weigh (to the nearest 0.1 mg) 1-g samples into 250-mL beakers. Cover each beaker with a watch glass. Cautiously add about 35 mL of 6 M HNO_3 (HOOD). Digest for at least 30 min; add more acid if necessary. Evaporate to about 5 mL but never to dryness (Note 2).

To each sample, add 5 mL of 3 M HNO_3, 25 mL of water, and one quarter of a tablet of filter-paper pulp; digest without boiling for about 45 min. Filter off the $SnO_2 \cdot xH_2O$, using a fine-porosity filter paper (Note 3); collect the filtrate in a tall-form electrolysis beaker. Use many small washes with hot 0.3 M HNO_3 to remove the last traces of copper; test for completeness with a few drops of NH_3. The final volume of filtrate and washings should be between 100 and 125 mL; either add water or evaporate to attain this volume.

Electrolysis

With the current switch off, attach the cathode to the negative terminal and the anode to the positive terminal of the electrolysis apparatus. Briefly turn on the stirring motor to be sure the electrodes do not touch. Cover the beaker with a split watch glass and commence the electrolysis. Maintain a current of 1.3 A for 35 min.

Rinse the cover glass and add 10 mL of 3 M H_2SO_4 followed by 5 g of urea to each beaker. Maintain a current of 2 A until the solution is colorless. To test for completeness of the electrolysis, remove one drop of the solution with a medicine dropper, and mix it with a few drops of NH_3 in a small test tube. If the mixture turns blue, rinse the contents of the tube back into the electrolysis vessel, and continue the electrolysis for an additional 10 min. Repeat the test until no blue $Cu(NH_3)_4^{2+}$ is produced.

When electrolysis is complete, discontinue stirring but leave the current on. Rinse the electrodes thoroughly with water as they emerge from the solution. After rinsing is complete, turn off the electrolysis apparatus (Note 4), disconnect the electrodes, and dip them in acetone. Dry the cathode for about 3 min and the anode for about 15 min at 110°C. Allow the electrodes to cool in air, and then weigh them.

Report the percentages of lead (Note 5) and copper in the brass.

Notes

1. Alternatively, grease and organic materials can be removed by heating platinum electrodes to redness in a flame. Electrode surfaces should not be touched with the fingers after cleaning because grease and oil cause nonadherent deposits that can flake off during washing and weighing.

2. Chloride ion must be totally excluded from this determination because it attacks the platinum anode during electrolysis. This reaction is not only destructive but also causes positive errors in the analysis by codepositing platinum with copper on the cathode.

3. If desired, the tin content can be determined by ignition of the $SnO_2 \cdot xH_2O$.

4. It is important to maintain a potential between the electrodes until they have been removed from the solution and washed. Some copper may redissolve if this precaution is not observed.

5. Experience has shown that a small amount of moisture is retained by the PbO_2 and that better results are obtained if 0.8643 is used for the stoichiometric factor (fw Pb/fw PbO_2) instead of 0.8662.

28K   METHODS BASED ON THE ABSORPTION OF RADIATION

Molecular absorption methods are discussed in Chapter 21. Directions follow for (1) the use of a calibration curve for the determination of iron in water, (2) the use of a standard-addition procedure for the determination of manganese in steel, and (3) a spectrophotometric determination of the pH of a buffer solution.

28K-1 The Cleaning and Handling of Cells

The accuracy of spectrophotometric measurements is critically dependent upon the availability of good-quality matched cells. These should be calibrated against one another at regular intervals to detect differences resulting from scratches, etching, and wear. Equally important is the proper cleaning of the exterior sides (the *windows*) just before the cells are inserted into a photometer or spectrophotometer. The preferred method is to wipe the windows with a lens paper soaked in methanol; the methanol is then allowed to evaporate, leaving the windows free of contaminants. It has been shown that this method is far superior to the usual procedure of wiping the windows with a dry lens paper, which tends to leave a residue of lint and a film on the window.[10]

28K-2 The Determination of Iron in a Natural Water

Discussion

The red-orange complex that forms between iron(II) and 1,10-phenanthroline (orthophenanthroline) is useful for determining iron in water supplies. The reagent is a weak base that reacts to form phenanthrolinium ion, PhenH$^+$, in acidic media. Complex formation with iron is thus best described by the equation

$$Fe^{2+} + 3\ PhenH^+ \rightleftarrows Fe(Phen)_3^{2+} + 3\ H^+$$

The formation constant for this equilibrium is 2.5×10^6 at 25°C. Iron(II) is quantitatively complexed in the pH range between 3 and 9. A pH of about 3.5 is ordinarily recommended to prevent precipitation of iron salts, such as phosphates.

An excess of a reducing reagent, such as hydroxylamine or hydroquinone, is needed to maintain iron in the +2 state. The complex, once formed, is very stable.

This determination can be performed with a spectrophotometer set at 508 nm or with a photometer equipped with a green filter.

Preparation of Solutions

a. *Standard iron solution, 0.01 mg/mL.* Weigh (to the nearest 0.2 mg) 0.0702 g of reagent-grade $Fe(NH_4)_2(SO_4)_2 \cdot 6H_2O$ into a 1-L volumetric flask. Dissolve in 50 mL of water that contains 1 to 2 mL of concentrated sulfuric acid; dilute to the mark, and mix well.

b. *Hydroxylamine hydrochloride.* (Sufficient for 80 to 90 measurements.) Dissolve 10 g of $H_2NOH \cdot HCl$ in about 100 mL of distilled water.

[10]For further information, see J. O. Erickson and T. Surles, *Amer. Lab.,* **1976,** *8* (6), 50.

c. *Orthophenanthroline solution.* (Sufficient for 80 to 90 measurements.) Dissolve 1.0 g of orthophenanthroline monohydrate in about 1 L of water. Warm slightly if necessary. Each milliliter is sufficient for no more than about 0.09 mg of Fe. Prepare no more reagent than needed; it darkens on standing and must then be discarded.

d. *Sodium acetate, 1.2 M.* (Sufficient for 80 to 90 measurements.) Dissolve 166 g of $NaOAc \cdot 3H_2O$ in 1 L of distilled water.

Procedure

Preparation of a Calibration Curve

Transfer 25.00 mL of the standard iron solution to a 100-mL volumetric flask and 25 mL of distilled water to a second 100-mL volumetric flask. Add 1 mL of hydroxylamine, 10 mL of sodium acetate, and 10 mL of orthophenanthroline to each flask. Allow the mixtures to stand for 5 min; dilute to the mark and mix well.

Clean a pair of matched cells for the instrument. Rinse each cell with at least three portions of the solution it is to contain. Determine the absorbance of the standard with respect to the blank.

Repeat the above procedure with at least three other volumes of the standard iron solution; attempt to encompass an absorbance range between 0.1 and 1.0. Plot a calibration curve.

Determination of Iron

Transfer 10.00 mL of the unknown to a 100-mL volumetric flask; treat in exactly the same way as the standards, measuring the absorbance with respect to the blank. Alter the volume of unknown taken to obtain absorbance measurements for replicate samples that are within the range of the calibration curve.

Report the parts per million of iron in the unknown.

28K-3 The Determination of Manganese in Steel

Discussion

Small quantities of manganese are readily determined photometrically by the oxidation of Mn(II) to the intensely colored permanganate ion. Potassium periodate is an effective oxidizing reagent for this purpose. The reaction is

$$5\ IO_4^- + 2\ Mn^{2+} + 3\ H_2O \rightarrow 5\ IO_3^- + 2\ MnO_4^- + 6\ H^+$$

Permanganate solutions that contain an excess of periodate are quite stable.

Interferences to the method are few. The presence of most colored ions can be compensated for with a blank. Cerium(III) and chromium(III) are exceptions; these yield oxidation products with periodate that absorb to some extent at the wavelength used for the measurement of permanganate.

The method given here is applicable to steels that do not contain large amounts of chromium. The sample is dissolved in nitric acid. Any carbon in the steel is oxidized with peroxodisulfate. Iron(III) is eliminated as a source of interference by complexation with phosphoric acid. The standard-addition method (page 585) is used to establish the relationship between absorbance and amount of manganese in the sample.

A spectrophotometer set at 525 nm or a photometer with a green filter can be used for the absorbance measurements.

Preparation of Solutions

Standard manganese(II) solution. (Sufficient for several hundred analyses.) Weigh 0.1 g (to the nearest 0.1 mg) of manganese into a 50-mL beaker, and dissolve in about 10 mL of 6 M HNO_3 (HOOD). Boil gently to eliminate oxides of nitrogen. Cool; then transfer the solution quantitatively to a 1-L volumetric flask. Dilute to the mark with water, and mix thoroughly. The manganese in 1 mL of the standard solution, after being converted to permanganate, causes a volume of 50 mL to increase in absorbance by about 0.09.

Procedure

The unknown does not require drying. If there is evidence of oil, rinse with acetone and dry briefly. Weigh (to the nearest 0.1 mg) duplicate samples (Note 1) into 150-mL beakers. Add about 50 mL of 6 M HNO_3 to each and boil gently (HOOD); heating for 5 to 10 min should suffice. Cautiously add about 1 g of ammonium peroxodisulfate, and boil gently for an additional 10 to 15 min. If the solutions are pink or have a deposit of MnO_2, add 1 mL of NH_4HSO_3 (or 0.1 g of $NaHSO_3$) and heat for 5 min. Cool; transfer quantitatively (Note 2) to 250.0-mL volumetric flasks. Dilute to the mark with water, and mix well. Use a 20.00-mL pipet to transfer three aliquots of each sample to individual beakers. Treat as follows:

Aliquot	Volume of 85% H_3PO_4, mL	Volume Standard Mn, mL	Weight of KIO_4, g
1	5	0.00	0.4
2	5	5.00 (Note 3)	0.4
3	5	0.00	0.0

Boil each solution gently for 5 min, cool, and transfer quantitatively to a 50-mL volumetric flask. Mix well. Measure the absorbance of aliquots 1 and 2 using aliquot 3 as the blank (Note 4).

Report the percentage of manganese in the unknown.

Notes

1. The sample size depends upon the manganese content of the unknown; consult with the instructor.
2. If there is turbidity, filter the solutions as they are transferred to the volumetric flasks.
3. The volume of the standard addition may be dictated by the absorbance of the sample. It is useful to obtain a rough estimate by generating

permanganate in about 20 mL of sample, diluting to about 50 mL, and measuring the absorbance.

4. A single blank can be used for all measurements, provided the samples weigh within 50 mg of one another.

28K-4 The Spectrophotometric Determination of pH

Discussion

The pH of an unknown buffer is determined by addition of an acid/base indicator and spectrophotometric measurement of the absorbance of the resulting solution. Because overlap exists between the spectra for the acid and base forms of the indicator, it is necessary to evaluate individual molar absorptivities for each form at two wavelengths. See page 425 for further discussion.

The relationship between the two forms of bromocresol green in an aqueous solution is described by the equilibrium

$$HIn + H_2O \rightleftharpoons H_3O^+ + In^-$$

for which

$$K_a = \frac{[H_3O^+][In^-]}{[HIn]} = 1.6 \times 10^{-5}$$

The spectrophotometric evaluation of $[In^-]$ and $[HIn]$ permits the calculation of $[H_3O^+]$.

Preparation of Solutions

a. *Bromocresol green, 1.0×10^{-4} M.* (Sufficient for about five determinations.) Dissolve 40.0 mg (to the nearest 0.1 mg) of the sodium salt of bromocresol green (fw = 720) in water, and dilute to 500 mL in a volumetric flask.

b. *HCl, 0.5 M.* Dilute about 4 mL of concentrated HCl to approximately 100 mL with water.

c. *NaOH, 0.4 M.* Dilute about 7 mL of 6 M NaOH to about 100 mL with water.

Procedure

Determination of Individual Absorption Spectra

Transfer 25.00-mL aliquots of the bromocresol green indicator solution to two 100-mL volumetric flasks. To one, add 25 mL of 0.5 M HCl; to the other, add 25 mL of 0.4 M NaOH. Dilute to the mark and mix well.

Obtain the absorption spectra for the acid and base forms of the indicator between 400 and 600 nm, using water as a blank. Record absorbance values at 10-nm intervals routinely and at closer intervals as needed to define maxima and minima. Evaluate molar absorptivities for HIn and In$^-$ at wavelengths corresponding to their absorption maxima.

Determination of the pH of an Unknown Buffer

Transfer 25.00 mL of the stock bromocresol green indicator to a 100-mL volumetric flask. Add 50.0 mL of the unknown buffer, dilute to the mark, and mix well. Measure the absorbance of the diluted solution at the wavelengths for which absorptivity data were calculated.

Report the pH of the buffer.

28L   ATOMIC SPECTROSCOPY

Several methods of analysis based upon atomic spectroscopy are discussed in Chapter 23. One such application is atomic absorption, which is demonstrated in the experiment that follows.

28L-1   The Determination of Lead in Brass by Atomic Absorption Spectroscopy

Discussion

Brasses and other copper-based alloys contain from 0 to about 10% lead as well as tin and zinc. Atomic absorption spectroscopy permits the quantitative estimation of these elements. The accuracy of this procedure is not as great as that obtainable with gravimetric or volumetric measurements, but the time needed to acquire the analytical information is considerably less.

A weighed sample is dissolved in a mixture of nitric and hydrochloric acids, the latter being needed to prevent the precipitation of tin as metastannic acid, $SnO_2 \cdot xH_2O$. After suitable dilution, the sample is aspirated into a flame, and the absorption of radiation from a hollow-cathode lamp is measured.

Preparation of Solutions

Standard lead solution, 100 mg/L. (This is sufficient for about 20 measurements.) Dry a quantity of reagent-grade $Pb(NO_3)_2$ for about 1 h at 110°C. Cool; weigh (to the nearest 0.1 mg) 0.17 g into a 1-L volumetric flask. Dissolve in a solution of 5 mL water and 1 to 3 mL of concentrated HNO_3. Dilute to the mark with distilled water, and mix well.

Procedure

Weigh duplicate samples of the unknown (Note 1) into 150-mL beakers. Cover with watch glasses, and then dissolve (HOOD) in a mixture consisting of about 4 mL of concentrated HNO_3 and 4 mL of concentrated HCl (Note 2). Boil gently to remove oxides of nitrogen. Cool; transfer the solutions quantitatively to 250-mL volumetric flasks, dilute to the mark with water, and mix well.

Use a buret to deliver 0-, 5-, 10-, 15-, and 20-mL portions of the standard lead solution to individual 50-mL volumetric flasks. Add 4 mL of concentrated HNO_3 and 4 mL of concentrated HCl to each, and dilute to the mark with water.

Transfer 10.00-mL aliquots of each sample to 50-mL volumetric flasks, add 4 mL of concentrated HCl and 4 mL of concentrated HNO_3, and dilute to the mark with water.

Set the monochromator at 283.3 nm, and measure the absorbance for the standards and samples at that wavelength. Take at least three—and preferably more—readings from each flask.

Plot the calibration data. Report the percentage of lead in the brass.

Notes

1. The weight of sample depends on the lead content of the brass and on the sensitivity of the instrument used for the absorption measurements. A sample containing 6 to 10 mg of lead is reasonable. Consult with the instructor.

2. Brasses that contain a large percentage of tin require additional HCl to prevent the formation of metastannic acid. The diluted samples may develop some turbidity on prolonged standing; a slight turbidity has no effect on the determination of lead.

28L-2 The Determination of Sodium, Potassium, and Calcium in Mineral Waters by Atomic Emission Spectroscopy

Discussion

A radiation buffer is a solution that contains an excess of the constituents of a sample matrix. The buffer is added to both sample and standards in an atomic emission analysis to prevent chemical interference by the matrix components.

A convenient method for the determination of alkali and alkaline earth metals in water and in blood serum is based upon the characteristic spectra these elements emit upon being aspirated into a natural gas/air flame. The following directions are suitable for the analysis of the three elements in water samples. Radiation buffers (Section 23C-4) are used to minimize the effect of each element upon the emission intensity of the others.

Preparation of Solutions

a. *Standard calcium solution, approximately 500 ppm.* Dry a quantity of $CaCO_3$ for about 1 h at 110°C. Cool in a desiccator; weigh (to the nearest milligram) 1.25 g into a 600-mL beaker. Add about 200 mL of distilled water, and about 10 mL of concentrated HCl. Cover the beaker with a watch glass, to avoid loss due to spattering. After reaction is complete, transfer the solution quantitatively to a 1-L volumetric flask, dilute to the mark, and mix well.

b. *Standard potassium solution, approximately 500 ppm.* Dry a quantity of KCl for about 1 h at 110°C. Cool; weigh (to the nearest milligram) about 0.95 g into a 1-L volumetric flask. Dissolve in distilled water, and dilute to the mark.

c. *Standard sodium solution, approximately 500 ppm.* Proceed as in step b, using 1.25 g (to the nearest milligram) of dried NaCl.

d. *Radiation buffer for the determination of calcium.* Prepare about 100 mL of a solution that has been saturated first with NaCl, then with KCl, and finally with $MgCl_2$.

e. *Radiation buffer for the determination of potassium.* Prepare about 100 mL of a solution that has been saturated first with NaCl, then with $CaCl_2$, and finally with $MgCl_2$.

f. *Radiation buffer for the determination of sodium.* Prepare about 100 mL of a solution that has been saturated first with $CaCl_2$, then with KCl, and finally with $MgCl_2$.

Procedure

Preparation of Working Curves

Add 5.00 mL of the appropriate radiation buffer to each of a series of 100-mL volumetric flasks. Add volumes of standard that will produce solutions that range from 0 to 10 ppm in the cation to be determined. Dilute to the mark with water, and mix well.

Measure the emission intensity for each solution, taking at least three readings for each. Aspirate distilled water between each set of measurements. Correct the average values for background luminosity, and prepare a working curve from the data.

Repeat for the other two cations.

Analysis of a Water Sample

Prepare duplicate aliquots of the unknown as directed for preparation of working curves. If necessary, use a standard to calibrate the response of the instrument to the working curve; then measure the emission intensity of the unknown. Correct the data for background. Determine the cation concentration in the unknown by comparison with the working curve.

28M   THE SEPARATION OF CATIONS BY ION EXCHANGE

The application of ion-exchange resins to the separation of ionic species of opposite charge is discussed in Section 24B. Directions follow for the ion-exchange separation of nickel(II) from zinc(II) based upon converting the zinc ions to negatively charged chloride complexes. After separation, each of the cations is determined by EDTA titration.

Discussion

The separation of the two cations is based upon differences in their tendency to form anionic complexes. Stable chlorozincate(II) complexes (such as $ZnCl_3^-$ and $ZnCl_4^{2-}$) are formed in 2 M hydrochloric acid and retained on an anion-exchange resin. In contrast, nickel(II) is not complexed appreciably in this medium and passes rapidly through such a column. After separation is complete, elution with water effectively decomposes the chloro complexes and permits removal of the zinc.

Both nickel and zinc are determined by titration with standard EDTA at pH 10. Eriochrome Black T is the indicator for the zinc titration. Bromopyrogallol red or murexide is used for the nickel titration.

Preparation of Solutions

a. *Standard EDTA.* See Section 28D-2.
b. *pH-10 buffer.* See Section 28D-1.
c. *Eriochrome Black T indicator.* See Section 28D-1.
d. *Bromopyrogallol red indicator.* (Sufficient for 100 titrations.) Dissolve 0.5 g of the solid indicator in 100 mL of 50% (v/v) ethanol.
e. *Murexide indicator.* The solid is approximately 0.2% indicator by weight in NaCl. Approximately 0.2 g is needed for each titration. The solid preparation is used because solutions of the indicator are quite unstable.

Preparation of Ion-Exchange Columns

A typical ion-exchange column is a cylinder 25 to 40 cm in length and 1 to 1.5 cm in diameter. A stopcock at the lower end permits adjustment of the flow of liquid through the column. A buret makes a convenient column. It is recommended that two columns be prepared to permit the simultaneous treatment of duplicate samples.

Insert a plug of glass wool to retain the resin particles. Then introduce sufficient strong-base anion-exchange resin (Note) to give a 10- to 15-cm column. Wash the column with about 50 mL of 6 M NH_3, followed by 100 mL of water and 100 mL of 2 M HCl. At the end of this cycle, the flow should be stopped so that the liquid level remains about 1 cm above the resin column. *At no time should the liquid level be allowed to drop below the top of the resin.*

Note

Amberlite® CG 400 or its equivalent can be used.

Procedure

Obtain the unknown, which should contain between 2 and 4 mmol of Ni^{2+} and Zn^{2+}, in a clean 100-mL volumetric flask. Add 16 mL of 12 M HCl, dilute to the mark with distilled water, and mix well. The resulting solution is approximately 2 M in acid. Transfer 10.00 mL of the diluted unknown onto the column. Place a 250-mL conical flask beneath the column, and slowly drain until the liquid level is barely above the resin. Rinse the interior of the column with several 2 to 3 mL portions of the 2 M HCl; lower the liquid level to just above the resin surface after each washing. Elute the nickel with about 50 mL of 2 M HCl at a flow rate of 2 to 3 mL/min.

Elute the Zn(II) by passing about 100 mL of water through the column, using the same flow rate; collect the liquid in a 500-mL conical flask.

Titration of Nickel

Evaporate the solution containing the nickel to dryness to eliminate excess HCl. Avoid overheating; the residual $NiCl_2$ must not be permitted to decompose to NiO. Dissolve the residue in 100 mL of distilled water, and add 10 to 20 mL of pH-10 buffer. Add 15 drops of bromopyrogallol red indicator or 0.2 g of murexide. Titrate to the color change (blue to purple for bromopyrogallol red, yellow to purple for murexide).

Calculate the weight in milligrams of nickel in the unknown.

Titration of Zinc

Add 10 to 20 mL of pH-10 buffer and 1 to 2 drops of Eriochrome Black T to the eluate. Titrate with standard EDTA solution to a color change from red to blue.

Calculate the weight in milligrams of zinc in the unknown.

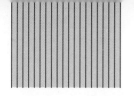

Appendix 1 Some Standard and Formal Electrode Potentials

Half-Reaction	E^0, V*	Formal Potential, V†
Aluminum		
$Al^{3+} + 3\,e^- \rightleftharpoons Al(s)$	-1.662	
Antimony		
$Sb_2O_5(s) + 6\,H^+ + 4\,e^- \rightleftharpoons 2\,SbO^+ + 3\,H_2O$	$+0.581$	
Arsenic		
$H_3AsO_4 + 2\,H^+ + 2\,e^- \rightleftharpoons H_3AsO_3 + H_2O$	$+0.559$	0.577 in 1 M HCl, HClO$_4$
Barium		
$Ba^{2+} + 2\,e^- \rightleftharpoons Ba(s)$	-2.906	
Bismuth		
$BiO^+ + 2\,H^+ + 3\,e^- \rightleftharpoons Bi(s) + H_2O$	$+0.320$	
$BiCl_4^- + 3\,e^- \rightleftharpoons Bi(s) + 4\,Cl^-$	$+0.16$	
Bromine		
$Br_2(l) + 2\,e^- \rightleftharpoons 2\,Br^-$	$+1.065$	1.05 in 4 M HCl
$Br_2(aq) + 2\,e^- \rightleftharpoons 2\,Br^-$	$+1.087\ddagger$	
$BrO_3^- + 6\,H^+ + 5\,e^- \rightleftharpoons \frac{1}{2}\,Br_2(l) + 3\,H_2O$	$+1.52$	
$BrO_3^- + 6\,H^+ + 6\,e^- \rightleftharpoons Br^- + 3\,H_2O$	$+1.44$	
Cadmium		
$Cd^{2+} + 2\,e^- \rightleftharpoons Cd(s)$	-0.403	
Calcium		
$Ca^{2+} + 2\,e^- \rightleftharpoons Ca(s)$	-2.866	
Carbon		
$C_6H_4O_2\ (\text{quinone}) + 2\,H^+ + 2\,e^- \rightleftharpoons C_6H_4(OH)_2$	$+0.699$	0.696 in 1 M HCl, HClO$_4$, H$_2$SO$_4$
$2\,CO_2(g) + 2\,H^+ + 2\,e^- \rightleftharpoons H_2C_2O_4$	-0.49	
Cerium		
$Ce^{4+} + e^- \rightleftharpoons Ce^{3+}$		+1.70 in 1 M HClO$_4$; +1.61 in 1 M HNO$_3$; +1.44 in 1 M H$_2$SO$_4$
Chlorine		
$Cl_2(g) + 2\,e^- \rightleftharpoons 2\,Cl^-$	$+1.359$	
$HClO + H^+ + e^- \rightleftharpoons \frac{1}{2}\,Cl_2(g) + H_2O$	$+1.63$	
$ClO_3^- + 6\,H^+ + 5\,e^- \rightleftharpoons \frac{1}{2}\,Cl_2(g) + 3\,H_2O$	$+1.47$	
Chromium		
$Cr^{3+} + e^- \rightleftharpoons Cr^{2+}$	-0.408	
$Cr^{3+} + 3\,e^- \rightleftharpoons Cr(s)$	-0.744	
$Cr_2O_7^{2-} + 14\,H^+ + 6\,e^- \rightleftharpoons 2\,Cr^{3+} + 7\,H_2O$	$+1.33$	
Cobalt		
$Co^{2+} + 2\,e^- \rightleftharpoons Co(s)$	-0.277	
$Co^{3+} + e^- \rightleftharpoons Co^{2+}$	$+1.808$	

Half-Reaction	E^0, V*	Formal Potential, V†
Copper		
$Cu^{2+} + 2\,e^- \rightleftharpoons Cu(s)$	+0.337	
$Cu^{2+} + e^- \rightleftharpoons Cu^+$	+0.153	
$Cu^+ + e^- \rightleftharpoons Cu(s)$	+0.521	
$Cu^{2+} + I^- + e^- \rightleftharpoons CuI(s)$	+0.86	
$CuI(s) + e^- \rightleftharpoons Cu(s) + I^-$	−0.185	
Fluorine		
$F_2(g) + 2\,H^+ + 2\,e^- \rightleftharpoons 2\,HF(aq)$	+3.06	
Hydrogen		
$2\,H^+ + 2\,e^- \rightleftharpoons H_2(g)$	0.000	−0.005 in 1 M HCl, HClO$_4$
Iodine		
$I_2(s) + 2\,e^- \rightleftharpoons 2\,I^-$	+0.5355	
$I_2(aq) + 2\,e^- \rightleftharpoons 2\,I^-$	+0.615‡	
$I_3^- + 2\,e^- \rightleftharpoons 3\,I^-$	+0.536	
$ICl_2^- + e^- \rightleftharpoons \frac{1}{2}\,I_2(s) + 2\,Cl^-$	+1.056	
$IO_3^- + 6\,H^+ + 5\,e^- \rightleftharpoons \frac{1}{2}\,I_2(s) + 3\,H_2O$	+1.196	
$IO_3^- + 6\,H^+ + 5\,e^- \rightleftharpoons \frac{1}{2}\,I_2(aq) + 3\,H_2O$	+1.178‡	
$IO_3^- + 2\,Cl^- + 6\,H^+ + 4\,e^- \rightleftharpoons ICl_2^- + 3\,H_2O$	+1.24	
$H_5IO_6 + H^+ + 2\,e^- \rightleftharpoons IO_3^- + 3\,H_2O$	+1.601	
Iron		
$Fe^{2+} + 2\,e^- \rightleftharpoons Fe(s)$	−0.440	
$Fe^{3+} + e^- \rightleftharpoons Fe^{2+}$	+0.771	0.700 in 1 M HCl; 0.732 in 1 M HClO$_4$; 0.68 in 1 M H$_2$SO$_4$
$Fe(CN)_6^{3-} + e^- \rightleftharpoons Fe(CN)_6^{4-}$	+0.36	0.71 in 1 M HCl; 0.72 in 1 M HClO$_4$, H$_2$SO$_4$
Lead		
$Pb^{2+} + 2\,e^- \rightleftharpoons Pb(s)$	−0.126	−0.14 in 1 M HClO$_4$; −0.29 in 1 M H$_2$SO$_4$
$PbO_2(s) + 4\,H^+ + 2\,e^- \rightleftharpoons Pb^{2+} + 2\,H_2O$	+1.455	
$PbSO_4(s) + 2\,e^- \rightleftharpoons Pb(s) + SO_4^{2-}$	−0.350	
Lithium		
$Li^+ + e^- \rightleftharpoons Li(s)$	−3.045	
Magnesium		
$Mg^{2+} + 2\,e^- \rightleftharpoons Mg(s)$	−2.363	
Manganese		
$Mn^{2+} + 2\,e^- \rightleftharpoons Mn(s)$	−1.180	
$Mn^{3+} + e^- \rightleftharpoons Mn^{2+}$		1.51 in 7.5 M H$_2$SO$_4$
$MnO_2(s) + 4\,H^+ + 2\,e^- \rightleftharpoons Mn^{2+} + 2\,H_2O$	+1.23	
$MnO_4^- + 8\,H^+ + 5\,e^- \rightleftharpoons Mn^{2+} + 4\,H_2O$	+1.51	
$MnO_4^- + 4\,H^+ + 3\,e^- \rightleftharpoons MnO_2(s) + 2\,H_2O$	+1.695	
$MnO_4^- + e^- \rightleftharpoons MnO_4^{2-}$	+0.564	
Mercury		
$Hg_2^{2+} + 2\,e^- \rightleftharpoons Hg(l)$	+0.788	0.274 in 1 M HCl; 0.776 in 1 M HClO$_4$; 0.674 in 1 M H$_2$SO$_4$
$2\,Hg^{2+} + 2\,e^- \rightleftharpoons Hg_2^{2+}$	+0.920	0.907 in 1 M HClO$_4$
$Hg^{2+} + 2\,e^- \rightleftharpoons 2\,Hg(l)$	+0.854	
$Hg_2Cl_2(s) + 2\,e^- \rightleftharpoons 2\,Hg(l) + 2\,Cl^-$	+0.268	0.244 in sat'd KCl; 0.282 in 1 M KCl; 0.334 in 0.1 M KCl
$Hg_2SO_4(s) + 2\,e^- \rightleftharpoons 2\,Hg(l) + SO_4^{2-}$	+0.615	

Half-Reaction	E^0, V*	Formal Potential, V†
Nickel		
$Ni^{2+} + 2\,e^- \rightleftharpoons Ni(s)$	-0.250	
Nitrogen		
$N_2(g) + 5\,H^+ + 4\,e^- \rightleftharpoons N_2H_5^+$	-0.23	
$HNO_2 + H^+ + e^- \rightleftharpoons NO(g) + H_2O$	$+1.00$	
$NO_3^- + 3\,H^+ + 2\,e^- \rightleftharpoons HNO_2 + H_2O$	$+0.94$	0.92 in 1 M HNO_3
Oxygen		
$H_2O_2 + 2\,H^+ + 2\,e^- \rightleftharpoons 2\,H_2O$	$+1.776$	
$HO_2^- + H_2O + 2\,e^- \rightleftharpoons 3\,OH^-$	$+0.88$	
$O_2(g) + 4\,H^+ + 4\,e^- \rightleftharpoons 2\,H_2O$	$+1.229$	
$O_2(g) + 2\,H^+ + 2\,e^- \rightleftharpoons H_2O_2$	$+0.682$	
$O_3(g) + 2\,H^+ + 2\,e^- \rightleftharpoons O_2(g) + H_2O$	$+2.07$	
Palladium		
$Pd^{2+} + 2\,e^- \rightleftharpoons Pd(s)$	$+0.987$	
Platinum		
$PtCl_4^{2-} + 2\,e^- \rightleftharpoons Pt(s) + 4\,Cl^-$	$+0.73$	
$PtCl_6^{2-} + 2\,e^- \rightleftharpoons PtCl_4^{2-} + 2\,Cl^-$	$+0.68$	
Potassium		
$K^+ + e^- \rightleftharpoons K(s)$	-2.925	
Selenium		
$H_2SeO_3 + 4\,H^+ + 2\,e^- \rightleftharpoons Se(s) + 3\,H_2O$	$+0.740$	
$SeO_4^{2-} + 4\,H^+ + 2\,e^- \rightleftharpoons H_2SeO_3 + H_2O$	$+1.15$	
Silver		
$Ag^+ + e^- \rightleftharpoons Ag(s)$	$+0.799$	0.228 in 1 M HCl; 0.792 in 1 M $HClO_4$; 0.77 in 1 M H_2SO_4
$AgBr(s) + e^- \rightleftharpoons Ag(s) + Br^-$	$+0.073$	
$AgCl(s) + e^- \rightleftharpoons Ag(s) + Cl^-$	$+0.222$	0.228 in 1 M KCl
$Ag(CN)_2^- + e^- \rightleftharpoons Ag(s) + 2\,CN^-$	-0.31	
$Ag_2CrO_4(s) + 2\,e^- \rightleftharpoons 2\,Ag(s) + CrO_4^{2-}$	$+0.446$	
$AgI(s) + e^- \rightleftharpoons Ag(s) + I^-$	-0.151	
$Ag(S_2O_3)_2^{3-} + e^- \rightleftharpoons Ag(s) + 2\,S_2O_3^{2-}$	$+0.017$	
Sodium		
$Na^+ + e^- \rightleftharpoons Na(s)$	-2.714	
Sulfur		
$S(s) + 2\,H^+ + 2\,e^- \rightleftharpoons H_2S(g)$	$+0.141$	
$H_2SO_3 + 4\,H^+ + 4\,e^- \rightleftharpoons S(s) + 3\,H_2O$	$+0.450$	
$SO_4^{2-} + 4\,H^+ + 2\,e^- \rightleftharpoons H_2SO_3 + H_2O$	$+0.172$	
$S_4O_6^{2-} + 2\,e^- \rightleftharpoons 2\,S_2O_3^{2-}$	$+0.08$	
$S_2O_8^{2-} + 2\,e^- \rightleftharpoons 2\,SO_4^{2-}$	$+2.01$	
Thallium		
$Tl^+ + e^- \rightleftharpoons Tl(s)$	-0.336	-0.551 in 1 M HCl; -0.33 in 1 M $HClO_4$, H_2SO_4
$Tl^{3+} + 2\,e^- \rightleftharpoons Tl^+$	$+1.25$	0.77 in 1 M HCl
Tin		
$Sn^{2+} + 2\,e^- \rightleftharpoons Sn(s)$	-0.136	-0.16 in 1 M $HClO_4$
$Sn^{4+} + 2\,e^- \rightleftharpoons Sn^{2+}$	$+0.154$	0.14 in 1 M HCl
Titanium		
$Ti^{3+} + e^- \rightleftharpoons Ti^{2+}$	-0.369	
$TiO^{2+} + 2\,H^+ + e^- \rightleftharpoons Ti^{3+} + H_2O$	$+0.099$	0.04 in 1 M H_2SO_4

Half-Reaction	E^0, V*	Formal Potential, V†
Uranium		
$UO_2^{2+} + 4\ H^+ + 2\ e^- \rightleftharpoons U^{4+} + 2\ H_2O$	+0.334	
Vanadium		
$V^{3+} + e^- \rightleftharpoons V^{2+}$	−0.256	−0.21 in 1 M $HClO_4$
$VO^{2+} + 2\ H^+ + e^- \rightleftharpoons V^{3+} + H_2O$	+0.359	
$V(OH)_4^+ + 2\ H^+ + e^- \rightleftharpoons VO^{2+} + 3\ H_2O$	+1.00	1.02 in 1 M HCl, $HClO_4$
Zinc		
$Zn^{2+} + 2\ e^- \rightleftharpoons Zn(s)$	−0.763	

*G. Milazzo, S. Caroli, and V. K. Sharma, *Tables of Standard Electrode Potentials*. London: Wiley, 1978. Reprinted by permission of John Wiley & Sons, Ltd.

†E. H. Swift and E. A. Butler, *Quantitative Measurements and Chemical Equilibria*. New York: Freeman, 1972.

‡These potentials are hypothetical because they correspond to solutions that are 1.00 M in Br_2 or I_2. The solubilities of these two compounds at 25°C are 0.18 M and 0.0020 M, respectively. In saturated solutions containing an excess of $Br_2(l)$ or $I_2(s)$, the standard potentials for the half-reaction $Br_2(l) + 2\ e^- \rightleftharpoons 2\ Br^-$ or $I_2(s) + 2\ e^- \rightleftharpoons 2\ I^-$ should be used. In contrast, at Br_2 and I_2 concentrations less than saturation, these hypothetical electrode potentials should be employed.

Appendix 2   Solubility-Product Constants

Substance	Formula	K_{sp}
Aluminum hydroxide	$Al(OH)_3$	2×10^{-32}
Barium carbonate	$BaCO_3$	5.1×10^{-9}
Barium chromate	$BaCrO_4$	1.2×10^{-10}
Barium iodate	$Ba(IO_3)_2$	1.57×10^{-9}
Barium manganate	$BaMnO_4$	2.5×10^{-10}
Barium oxalate	BaC_2O_4	2.3×10^{-8}
Barium sulfate	$BaSO_4$	1.3×10^{-10}
Bismuth oxide chloride	$BiOCl$	7×10^{-9}
Bismuth oxide hydroxide	$BiOOH$	4×10^{-10}
Cadmium carbonate	$CdCO_3$	2.5×10^{-14}
Cadmium hydroxide	$Cd(OH)_2$	5.9×10^{-15}
Cadmium oxalate	CdC_2O_4	9×10^{-8}
Cadmium sulfide	CdS	2×10^{-28}
Calcium carbonate	$CaCO_3$	4.8×10^{-9}
Calcium fluoride	CaF_2	4.9×10^{-11}
Calcium oxalate	CaC_2O_4	2.3×10^{-9}
Calcium sulfate	$CaSO_4$	2.6×10^{-5}
Copper(I) bromide	$CuBr$	5.2×10^{-9}
Copper(I) chloride	$CuCl$	1.2×10^{-6}
Copper(I) iodide	CuI	1.1×10^{-12}
Copper(I) thiocyanate	$CuSCN$	4.8×10^{-15}
Copper(II) hydroxide	$Cu(OH)_2$	1.6×10^{-19}
Copper(II) sulfide	CuS	6×10^{-36}
Iron(II) hydroxide	$Fe(OH)_2$	8×10^{-16}
Iron(II) sulfide	FeS	6×10^{-18}
Iron(III) hydroxide	$Fe(OH)_3$	4×10^{-38}
Lanthanum iodate	$La(IO_3)_3$	6.2×10^{-12}
Lead carbonate	$PbCO_3$	3.3×10^{-14}
Lead chloride	$PbCl_2$	1.6×10^{-5}
Lead chromate	$PbCrO_4$	1.8×10^{-14}
Lead hydroxide	$Pb(OH)_2$	2.5×10^{-16}
Lead iodide	PbI_2	7.1×10^{-9}
Lead oxalate	PbC_2O_4	4.8×10^{-10}
Lead sulfate	$PbSO_4$	1.6×10^{-8}
Lead sulfide	PbS	7×10^{-28}
Magnesium ammonium phosphate	$MgNH_4PO_4$	3×10^{-13}
Magnesium carbonate	$MgCO_3$	1×10^{-5}
Magnesium hydroxide	$Mg(OH)_2$	1.8×10^{-11}
Magnesium oxalate	MgC_2O_4	8.6×10^{-5}
Manganese(II) hydroxide	$Mn(OH)_2$	1.9×10^{-13}
Manganese(II) sulfide	MnS	3×10^{-13}
Mercury(I) bromide	Hg_2Br_2	5.8×10^{-23}

Substance	Formula	K_{sp}
Mercury(I) chloride	Hg_2Cl_2	1.3×10^{-18}
Mercury(I) iodide	Hg_2I_2	4.5×10^{-29}
Silver arsenate	Ag_3AsO_4	1×10^{-22}
Silver bromide	$AgBr$	5.2×10^{-13}
Silver carbonate	Ag_2CO_3	8.1×10^{-12}
Silver chloride	$AgCl$	1.82×10^{-10}
Silver chromate	Ag_2CrO_4	1.1×10^{-12}
Silver cyanide	$AgCN$	7.2×10^{-11}
Silver iodate	$AgIO_3$	3.0×10^{-8}
Silver iodide	AgI	8.3×10^{-17}
Silver oxalate	$Ag_2C_2O_4$	3.5×10^{-11}
Silver sulfide	Ag_2S	6×10^{-50}
Silver thiocyanate	$AgSCN$	1.1×10^{-12}
Strontium oxalate	SrC_2O_4	5.6×10^{-8}
Strontium sulfate	$SrSO_4$	3.2×10^{-7}
Thallium(I) chloride	$TlCl$	1.7×10^{-4}
Thallium(I) sulfide	Tl_2S	1×10^{-22}
Zinc hydroxide	$Zn(OH)_2$	1.2×10^{-17}
Zinc oxalate	ZnC_2O_4	7.5×10^{-9}
Zinc sulfide	ZnS	4.5×10^{-24}

From R. P. Frankenthal, in *Handbook of Analytical Chemistry,* L. Meites, Ed., pp. **1**-13–**1**-19. New York: McGraw-Hill, 1963. With permission.

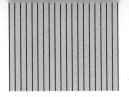

Appendix 3 Dissociation Constants for Acids

		Dissociation Constant at 25°C		
Acid	Formula	K_1	K_2	K_3
Acetic	CH_3COOH	1.75×10^{-5}		
Arsenic	H_3AsO_4	6.0×10^{-3}	1.05×10^{-7}	3.0×10^{-12}
Arsenous	H_3AsO_3	6.0×10^{-10}	3.0×10^{-14}	
Benzoic	C_6H_5COOH	6.14×10^{-5}		
Boric	H_3BO_3	5.83×10^{-10}		
1-Butanoic	$CH_3CH_2CH_2COOH$	1.51×10^{-5}		
Carbonic	H_2CO_3	4.45×10^{-7}	4.7×10^{-11}	
Chloroacetic	$ClCH_2COOH$	1.36×10^{-3}		
Citric	$HOOC(OH)C(CH_2COOH)_2$	7.45×10^{-4}	1.73×10^{-5}	4.02×10^{-7}
Ethylenediamine-tetraacetic	H_4Y	1.0×10^{-2}	2.1×10^{-3}	6.9×10^{-7}
			$K_4 = 5.5 \times 10^{-11}$	
Formic	$HCOOH$	1.77×10^{-4}		
Fumaric	$trans$-$HOOCCH:CHCOOH$	9.6×10^{-4}	4.1×10^{-5}	
Glycolic	$HOCH_2COOH$	1.48×10^{-4}		
Hydrazoic	HN_3	1.9×10^{-5}		
Hydrogen cyanide	HCN	2.1×10^{-9}		
Hydrogen fluoride	H_2F_2	7.2×10^{-4}		
Hydrogen peroxide	H_2O_2	2.7×10^{-12}		
Hydrogen sulfide	H_2S	5.7×10^{-8}	1.2×10^{-15}	
Hypochlorous	$HOCl$	3.0×10^{-8}		
Iodic	HIO_3	1.7×10^{-1}		
Lactic	$CH_3CHOHCOOH$	1.37×10^{-4}		
Maleic	cis-$HOOCCH:CHCOOH$	1.20×10^{-2}	5.96×10^{-7}	
Malic	$HOOCCHOHCH_2COOH$	4.0×10^{-4}	8.9×10^{-6}	
Malonic	$HOOCCH_2COOH$	1.40×10^{-3}	2.01×10^{-6}	
Mandelic	$C_6H_5CHOHCOOH$	3.88×10^{-4}		
Nitrous	HNO_2	5.1×10^{-4}		
Oxalic	$HOOCCOOH$	5.36×10^{-2}	5.42×10^{-5}	
Periodic	H_5IO_6	2.4×10^{-2}	5.0×10^{-9}	
Phenol	C_6H_5OH	1.00×10^{-10}		
Phosphoric	H_3PO_4	7.11×10^{-3}	6.34×10^{-8}	4.2×10^{-13}
Phosphorous	H_3PO_3	1.00×10^{-2}	2.6×10^{-7}	
o-Phthalic	$C_6H_4(COOH)_2$	1.12×10^{-3}	3.91×10^{-6}	
Picric	$(NO_2)_3C_6H_2OH$	5.1×10^{-1}		
Propanoic	CH_3CH_2COOH	1.34×10^{-5}		
Pyruvic	$CH_3COCOOH$	3.24×10^{-3}		
Salicylic	$C_6H_4(OH)COOH$	1.05×10^{-3}		
Sulfamic	H_2NSO_3H	1.03×10^{-1}		
Sulfuric	H_2SO_4	Strong	1.20×10^{-2}	
Sulfurous	H_2SO_3	1.72×10^{-2}	6.43×10^{-8}	
Succinic	$HOOCCH_2CH_2COOH$	6.21×10^{-5}	2.32×10^{-6}	
Tartaric	$HOOC(CHOH)_2COOH$	9.20×10^{-4}	4.31×10^{-5}	
Trichloroacetic	Cl_3CCOOH	1.29×10^{-1}		

From L. Meites, *Handbook of Analytical Chemistry*, p. 1-21. New York: McGraw-Hill, 1963. With permission.

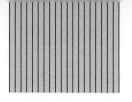

Appendix 4 DISSOCIATION CONSTANTS FOR BASES

Base	Formula	Dissociation Constant at 25°C
Ammonia	NH_3	1.76×10^{-5}
Aniline	$C_6H_5NH_2$	3.94×10^{-10}
1-Butylamine	$CH_3(CH_2)_2CH_2NH_2$	4.0×10^{-4}
Dimethylamine	$(CH_3)_2NH$	5.9×10^{-4}
Ethanolamine	$HOC_2H_4NH_2$	3.18×10^{-5}
Ethylamine	$CH_3CH_2NH_2$	4.28×10^{-4}
Ethylenediamine	$NH_2C_2H_4NH_2$	$K_1 = 8.5 \times 10^{-5}$
		$K_2 = 7.1 \times 10^{-8}$
Hydrazine	H_2NNH_2	1.3×10^{-6}
Hydroxylamine	$HONH_2$	1.07×10^{-8}
Methylamine	CH_3NH_2	4.8×10^{-4}
Piperidine	$C_5H_{11}N$	1.3×10^{-3}
Pyridine	C_5H_5N	1.7×10^{-9}
Trimethylamine	$(CH_3)_3N$	6.25×10^{-5}

From L. Meites, *Handbook of Analytical Chemistry*, p. 1-21. New York: McGraw-Hill, 1963. With permission.

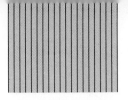

Appendix 5　Use of Exponential Numbers

Scientists frequently find it necessary (or convenient) to use exponential notation to express numerical data. A brief review of this notation follows.

EXPONENTIAL NOTATION

An exponent is used to describe the process of repeated multiplication or division. For example, 3^5 means

$$3 \times 3 \times 3 \times 3 \times 3 = 3^5 = 243$$

The 5 is the exponent of the number (or base) 3; thus, 3 raised to the fifth power is equal to 243.

A negative exponent represents repeated division. For example, 3^{-5} means

$$\frac{1}{3} \times \frac{1}{3} \times \frac{1}{3} \times \frac{1}{3} \times \frac{1}{3} = \frac{1}{3^5} = 3^{-5} = 0.00412$$

Note that changing the sign of the exponent yields the *reciprocal* of the number; that is,

$$3^{-5} = \frac{1}{3^5} = \frac{1}{243} = 0.00412$$

It is important to note that a number raised to the first power is the number itself, and any number raised to the zero power has a value of 1. For example,

$$4^1 = 4$$
$$4^0 = 1$$
$$67^0 = 1$$

Fractional Exponents

A fractional exponent symbolizes the process of extracting the root of a number. The fifth root of 243 is 3; this process is expressed exponentially as

$$(243)^{1/5} = 3$$

Other examples are

$$25^{1/2} = 5$$

$$25^{-1/2} = \frac{1}{25^{1/2}} = \frac{1}{5}$$

The Combination of Exponential Numbers in Multiplication and Division

Multiplication and division of exponential numbers having the same base are accomplished by adding and subtracting the exponents. For example,

$$3^3 \times 3^2 = (3 \times 3 \times 3)(3 \times 3) = 3^{(3+2)} = 3^5 = 243$$

$$3^4 \times 3^{-2} \times 3^0 = (3 \times 3 \times 3 \times 3)(\tfrac{1}{3} \times \tfrac{1}{3}) \times 1 = 3^{(4-2+0)} = 3^2 = 9$$

$$\frac{5^4}{5^2} = \frac{5 \times 5 \times 5 \times 5}{5 \times 5} = 5^{(4-2)} = 5^2 = 25$$

$$\frac{2^3}{2^{-1}} = \frac{(2 \times 2 \times 2)}{1/2} = 2^4 = 16$$

Note that in the last equation the exponent is given by the relationship

$$3 - (-1) = 3 + 1 = 4$$

Extraction of the Root of an Exponential Number

To obtain the root of an exponential number, the exponent is divided by the desired root. Thus,

$$(5^4)^{1/2} = (5 \times 5 \times 5 \times 5)^{1/2} = 5^{(4/2)} = 5^2 = 25$$

$$(10^{-8})^{1/4} = 10^{(-8/4)} = 10^{-4}$$

$$(10^9)^{1/2} = 10^{(9/2)} = 10^{4.5}$$

THE USE OF EXPONENTS IN SCIENTIFIC NOTATION

Scientists and engineers are frequently called upon to use very large or very small numbers for which ordinary decimal notation is either awkward or impossible. For example, to express Avogadro's number in decimal notation would require 21 zeros following the number 602. In scientific notation the number is written as a multiple of two numbers, the one number in decimal notation and the other expressed as a power of 10. Thus, Avogadro's number is written as 6.02×10^{23}. Other examples are

$$4.32 \times 10^3 = 4.32 \times 10 \times 10 \times 10 = 4320$$

$$4.32 \times 10^{-3} = 4.32 \times \frac{1}{10} \times \frac{1}{10} \times \frac{1}{10} = 0.00432$$

$$0.002002 = 2.002 \times \frac{1}{10} \times \frac{1}{10} \times \frac{1}{10} = 2.002 \times 10^{-3}$$

$$375 = 3.75 \times 10 \times 10 = 3.75 \times 10^2$$

It should be noted that the scientific notation for a number can be expressed in any of several equivalent forms. Thus,

$$4.32 \times 10^3 = 43.2 \times 10^2 = 432 \times 10^1 = 0.432 \times 10^4 = 0.0432 \times 10^5$$

The number in the exponent is equal to the number of places the decimal must be shifted to convert a number from scientific to purely decimal notation. The shift is to the right if the exponent is positive and to the left if it is negative. The process is reversed when decimal numbers are converted to scientific notation.

ARITHMETIC OPERATIONS WITH SCIENTIFIC NOTATION

The use of scientific notation is helpful in preventing decimal errors in arithmetic calculations. Some examples follow.

Multiplication

Here, the decimal parts of the numbers are multiplied and the exponents are added; thus,

$$420,000 \times 0.0300 = (4.20 \times 10^5)(3.00 \times 10^{-2})$$
$$= 12.60 \times 10^3 = 1.26 \times 10^4$$
$$0.0060 \times 0.000020 = 6.0 \times 10^{-3} \times 2.0 \times 10^{-5}$$
$$= 12 \times 10^{-8} = 1.2 \times 10^{-7}$$

Division

Here, the decimal parts of the numbers are divided; the exponent in the denominator is subtracted from that in the numerator. For example,

$$\frac{0.015}{5000} = \frac{15 \times 10^{-3}}{5 \times 10^3} = 3.0 \times 10^{-6}$$

Addition and Subtraction

Addition or subtraction in scientific notation requires that all numbers be expressed to a common power of 10. The decimal parts are then added or subtracted, as appropriate. Thus,

$$2.00 \times 10^{-11} + 4.00 \times 10^{-12} - 3.00 \times 10^{-10} =$$
$$2.00 \times 10^{-11} + 0.400 \times 10^{-11} - 30.0 \times 10^{-11} = -27.6 \times 10^{-11}$$
$$= -2.76 \times 10^{-10}$$

Raising to a Power a Number Written in Exponential Notation

Here, each part of the number is raised to the power separately. For example,

$$(2 \times 10^{-3})^4 = (2.0)^4 \times (10^{-3})^4 = 16 \times 10^{-(3 \times 4)}$$
$$= 16 \times 10^{-12} = 1.6 \times 10^{-11}$$

**Extraction of the Root of a Number Written
in Exponential Notation**

In this case, the number is written in such a way that the exponent of 10 is evenly divisible by the root. Thus,

$$(4.0 \times 10^{-5})^{1/3} = \sqrt[3]{40 \times 10^{-6}} = \sqrt[3]{40} \times \sqrt[3]{10^{-6}}$$
$$= 3.4 \times 10^{-2}$$

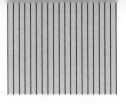

Appendix 6

Volumetric Calculations Using Normality and Equivalent Weight

The *normality* of a solution expresses the number of equivalents of solute contained in 1 L of solution or the number of milliequivalents in 1 mL. The equivalent and milliequivalent, like the mole and millimole, are units for describing the amount of a chemical species. The former are defined, however, in such a way that it is possible to state that, at the equivalence point in *any* titration,

$$\text{no. meq analyte present} = \text{no. meq standard reagent added} \quad \text{(A6-1)}$$

or

$$\text{no. eq analyte present} = \text{no. eq standard reagent added} \quad \text{(A6-2)}$$

As a consequence, stoichiometric ratios such as those described in Section 6C-3 need not be derived every time a volumetric calculation is performed. Instead, the stoichiometry is taken into account by the way equivalent or milliequivalent weight is defined.

A6-1 THE DEFINITION OF EQUIVALENT AND MILLIEQUIVALENT

In contrast to the mole, the amount of a substance contained in one equivalent can vary from reaction to reaction. Consequently, the weight of one equivalent of a compound can never be computed *without reference to a chemical reaction* in which that compound is, directly or indirectly, a participant. Similarly, the normality of a solution can never be specified *without knowledge about how the solution will be used.*

Equivalent Weights in Neutralization Reactions

One equivalent weight of a substance participating in a neutralization reaction is that amount of substance (molecule, ion, or paired ion such as NaOH) that either contributes or consumes 1 mol of hydrogen ions *in that reaction.*[1] A milliequivalent is simply 1/1000 of an equivalent.

---

[1] An alternative definition, proposed by the International Union of Pure and Applied Chemistry, is as follows: an equivalent is "that amount of a substance, which, in a specified reaction, releases or replaces that amount of hydrogen that is combined with 3 g of carbon-12 in methane $^{12}CH_4$" (see *Information Bulletin* No. 36, International Union of Pure and Applied Chemistry, August 1974). This definition applies to acids. For other types of reactions and reagents, the amount of hydrogen referred to may be replaced by the equivalent amount of hydroxide ions, electrons, or cations. The reaction to which the definition is applied must be specified.

The relationship between equivalent weight (eqw) and formula weight is straightforward for strong acids or bases and for other acids or bases that contain a single reactive hydrogen or hydroxide ion. For example, the equivalent weights of potassium hydroxide, hydrochloric acid, and acetic acid are equal to their formula weights because each has but a single reactive hydrogen ion or hydroxide ion. Barium hydroxide, which contains two identical hydroxide ions, reacts with two hydrogen ions in any acid/base reaction, and so its equivalent weight is one half its formula weight:

$$\text{eqw } Ba(OH)_2 = \frac{\text{fw } Ba(OH)_2}{2}$$

The situation becomes more complex for acids or bases that contain two or more reactive hydrogen or hydroxide ions with different tendencies to dissociate. With certain indicators, for example, only the first of the three protons in phosphoric acid is titrated:

$$H_3PO_4 + OH^- \rightarrow H_2PO_4^- + H_2O$$

With certain other indicators, a color change occurs only after two hydrogen ions have reacted:

$$H_3PO_4 + 2\ OH^- \rightarrow HPO_4^{2-} + 2\ H_2O$$

For a titration involving the first reaction, the equivalent weight of phosphoric acid is equal to the formula weight; for the second, the equivalent weight is one half the formula weight. (Because it is not practical to titrate the third proton, an equivalent weight that is one third the formula weight is not generally encountered for H_3PO_4.) If it is not known which of these reactions is involved, an unambiguous definition of the equivalent weight for phosphoric acid *cannot be made*.

Equivalent Weights in Oxidation/Reduction Reactions

The equivalent weight of a participant in an oxidation/reduction reaction is that weight which directly or indirectly produces or consumes 1 mol of electrons. The numerical value for the equivalent weight is conveniently established by dividing the formula weight of the substance of interest by the change in oxidation number associated with its reaction. As an example, consider the oxidation of oxalate ion by permanganate ion:

$$5\ C_2O_4^{2-} + 2\ MnO_4^- + 16\ H^+ \rightarrow 10\ CO_2 + 2\ Mn^{2+} + 8\ H_2O \quad \text{(A6-3)}$$

In this reaction, the change in oxidation number for manganese is 5 because the element passes from the $+7$ to the $+2$ state; the equivalent weights for MnO_4^- and Mn^{2+} are therefore one fifth their formula weights. Each carbon atom in the oxalate ion is oxidized from the $+3$ to the $+4$ state, leading to the production of two electrons by that species. Therefore, the equivalent weight of sodium oxalate is one half the formula weight. It is also possible to assign an equivalent weight to the carbon dioxide produced by the reaction. Since this molecule contains but a single carbon atom and since that carbon undergoes a change in oxidation number of 1, the formula weight and equivalent weight of the two are identical.

It is important to note that in evaluating the equivalent weight of a substance, *only* its change in oxidation number during the titration is considered. For example, suppose the manganese content of a sample containing Mn_2O_3 is to be determined by a titration based upon the reaction given in Equation A6-3. The fact that each manganese in the Mn_2O_3 has an oxidation number of +3 plays no part in determining equivalent weight. That is, we must assume that by suitable treatment, all the manganese is oxidized to the +7 state before the titration is begun. Each manganese from the Mn_2O_3 is then reduced from the +7 to the +2 state in the titration step. The equivalent weight is thus the formula weight of Mn_2O_3 divided by $2 \times 5 = 10$.

As in neutralization reactions, the equivalent weight for a given oxidizing or reducing agent is not invariant. Potassium permanganate, for example, reacts under some conditions to give MnO_2:

$$MnO_4^- + 3\,e^- + 2\,H_2O \rightarrow MnO_2(s) + 4\,OH^-$$

The change in the oxidation state of manganese in this reaction is from +7 to +4, and the equivalent weight of potassium permanganate is now equal to its formula weight divided by 3 (instead of 5 as in the earlier example).

Equivalent Weights in Precipitation and Complex-Formation Reactions

The equivalent weight of a participant in a precipitation or a complex-formation reaction is that weight which reacts with or provides one mole of the *reacting* cation if it is univalent, one half mole if it is divalent, one third mole if it is trivalent, and so on. It is important to note that the cation referred to in this definition is always *the cation directly involved in the analytical reaction* and not necessarily the cation contained in the compound whose equivalent weight is being defined.

Example A6-1

Define equivalent weights for $AlCl_3$ and $BiOCl$ if the two compounds are determined by a precipitation titration with $AgNO_3$:

$$Ag^+ + Cl^- \rightarrow AgCl(s)$$

In this instance, the equivalent weight is based upon the number of moles of *silver ions* consumed in the titration of each compound. Since 1 mol of Ag^+ reacts with 1 mol of Cl^- provided by one third mole of $AlCl_3$, we can write

$$\text{eqw } AlCl_3 = \frac{\text{fw } AlCl_3}{3}$$

Because each mole of BiOCl reacts with only 1 Ag^+ ion,

$$eqw\ BiOCl = \frac{fw\ BiOCl}{1}$$

Note that the fact that Bi^{3+} (or Al^{3+}) is trivalent has no bearing because the definition is based *upon the cation involved in the titration:* Ag^+.

A6-2 THE DEFINITION OF NORMALITY

The normality C_N of a solution expresses the number of milliequivalents of solute contained in 1 mL of solution or the number of equivalents contained in 1 L. Thus, a 0.20 N hydrochloric acid solution contains 0.20 meq of HCl in each milliliter of solution or 0.20 eq in each liter.

The normal concentration of a solution is defined by equations analogous to Equation 2-1. Thus, for a solution of the species A, the normality $C_{N(A)}$ is given by the equations

$$C_{N(A)} = \frac{no.\ meq\ A}{vol\ soln\ (mL)} \tag{A6-4}$$

$$C_{N(A)} = \frac{no.\ eq\ A}{vol\ soln\ (L)} \tag{A6-5}$$

A6-3 SOME USEFUL ALGEBRAIC RELATIONSHIPS

Two pairs of algebraic equations, analogous to Equations 6-1 and 6-2 as well as 6-3 and 6-4, apply when normal concentrations are being used:

$$amount\ A = no.\ meq\ A = \frac{wt\ A\ (g)}{meqw\ A\ (g/meq)} \tag{A6-6}$$

$$amount\ A = no.\ eq\ A = \frac{wt\ A\ (g)}{eqw\ A\ (g/eq)} \tag{A6-7}$$

$$amount\ A = no.\ meq\ A = vol\ (mL) \times C_{N(A)}\ (meq/mL) \tag{A6-8}$$

$$amount\ A = no.\ eq\ A = vol\ (L) \times C_{N(A)}\ (eq/L) \tag{A6-9}$$

A6-4 CALCULATION OF THE NORMALITY OF STANDARD SOLUTIONS

Example A6-2 shows how the normality of a standard solution is computed from preparatory data. Note the similarity between this example and Example 6-1 (Chapter 6).

Example A6-2

Describe the preparation of 5.000 L of 0.1000 N Na_2CO_3 from the primary-standard solid, assuming the solution is to be used for titrations in which the chemical reaction is

$$CO_3^{2-} + 2\ H^+ \rightarrow H_2O + CO_2$$

Applying Equation A6-9 gives

$$\text{amount } Na_2CO_3 = \text{vol soln (L)} \times C_{N(Na_2CO_3)} \text{ (eq/L)}$$
$$= 5.000\ L \times 0.1000\ eq/L = 0.5000\ eq\ Na_2CO_3$$

Rearranging Equation A6-7 gives

$$\text{wt } Na_2CO_3 = \text{no. eq } Na_2CO_3 \times \text{eqw } Na_2CO_3$$

But 2 eq of Na_2CO_3 are contained in each mole of the compound; therefore,

$$\text{wt } Na_2CO_3 = 0.5000\ eq\ Na_2CO_3 \times \frac{105.99\ g\ Na_2CO_3}{2\ eq\ Na_2CO_3} = 26.50\ g$$

Therefore, dissolve 26.50 g in water and dilute to 5.000 L.

It is worth noting that, when the carbonate ion reacts with two protons, the weight of sodium carbonate required to prepare a 0.10 N solution is just one half that required to prepare a 0.10 M solution.

A6-5   THE TREATMENT OF TITRATION DATA WHEN NORMALITY IS USED

Calculation of Normalities from Titration Data

Examples A6-3 and A6-4 illustrate how normality is computed from standardization data. Note that these examples are similar to Examples 6-5 and 6-6.

Example A6-3

Exactly 50.00 mL of an HCl solution required 29.71 mL of 0.03926 N $Ba(OH)_2$ to give an end point with bromocresol green indicator. Calculate the normality of the HCl.

Note that the molarity of $Ba(OH)_2$ is one half its normality. That is,

$$C_{Ba(OH)_2} = 0.03926\ \frac{meq}{mL} \times \frac{1\ mmol}{2\ meq} = 0.01963\ M$$

Because milliequivalents are the units employed, we write

$$\text{no. meq HCl} = \text{no. meq } Ba(OH)_2$$

The number of milliequivalents of standard is obtained by substituting into Equation A6-8:

$$\text{amount Ba(OH)}_2 = 29.71 \; \cancel{\text{mL Ba(OH)}_2} \times 0.03926 \; \frac{\text{meq Ba(OH)}_2}{\cancel{\text{mL Ba(OH)}_2}}$$

To obtain the number of milliequivalents of HCl, we write

$$\text{amount HCl} = (29.71 \times 0.03926) \; \cancel{\text{meq Ba(OH)}_2} \times \frac{1 \; \text{meq HCl}}{1 \; \cancel{\text{meq Ba(OH)}_2}}$$

Equating this result to Equation A6-8 yields

$$\text{amount HCl} = 50.00 \; \text{mL HCl} \times C_{\text{N(HCl)}} = (29.71 \times 0.03926 \times 1) \; \text{meq HCl}$$

$$C_{\text{N(HCl)}} = \frac{(29.71 \times 0.03926 \times 1) \; \text{meq HCl}}{50.00 \; \text{mL HCl}} = 0.02333 \; \text{N}$$

Example A6-4

A 0.2121-g sample of pure $Na_2C_2O_4$ (fw = 134.00) was titrated with 43.31 mL of $KMnO_4$. What was the normality of the $KMnO_4$ solution? The chemical reaction is

$$2 \; MnO_4^- + 5 \; C_2O_4^{2-} + 16 \; H^+ \rightarrow 2 \; Mn^{2+} + 10 \; CO_2 + 8 \; H_2O$$

By definition, at the equivalence point in the titration,

$$\text{no. meq } Na_2C_2O_4 = \text{no. meq } KMnO_4$$

Substituting Equations A6-8 and A6-6 into this relationship gives

$$\text{vol } KMnO_4 \times C_{\text{N(KMnO}_4)} = \frac{\text{wt } Na_2C_2O_4 \; (g)}{\text{meqw } Na_2C_2O_4 \; (g/\text{meq})}$$

$$43.31 \; \text{mL } KMnO_4 \times C_{\text{N(KMnO}_4)} = \frac{0.2121 \; \cancel{\text{g } Na_2C_2O_4}}{0.13400 \; \cancel{\text{g } Na_2C_2O_4}/2 \; \text{meq}}$$

$$C_{\text{N(KMnO}_4)} = \frac{0.2121 \; \cancel{\text{g } Na_2C_2O_4}}{43.31 \; \text{mL } KMnO_4 \times 0.1340 \; \cancel{\text{g } Na_2C_2O_4}/2 \; \text{meq}}$$

$$= 0.073093 \; \text{meq/mL } KMnO_4 = 0.07309 \; \text{N}$$

Note that the normality found here is five times the molarity computed in Example 6-6.

Calculation of the Quantity of Analyte from Titration Data

The examples that follow illustrate how analyte concentrations are computed when normalities are involved.

Example A6-5

A 0.8040-g sample of an iron ore was dissolved in acid. The iron was then reduced to Fe^{2+} and titrated with 47.22 mL of 0.1121 N (0.02242 M) $KMnO_4$ solution. Calculate the results of this analysis in terms of (a) % Fe (fw = 55.847) and (b) % Fe_3O_4 (fw = 231.54). The reaction of the analyte with the reagent is described by the equation

$$MnO_4^- + 5\ Fe^{2+} + 8\ H^+ \rightarrow Mn^{2+} + 5\ Fe^{3+} + 4\ H_2O$$

(a) At the equivalence point, we know that

$$\text{no. meq } KMnO_4 = \text{no. meq } Fe^{2+} = \text{no. meq } Fe_3O_4$$

Substituting Equations A6-8 and A6-6 leads to

$$\text{vol } KMnO_4\ (\text{mL}) \times C_{N(KMnO_4)}\ (\text{meq/mL}) = \frac{\text{wt } Fe^{2+}\ (\text{g})}{\text{meqw } Fe^{2+}\ (\text{g/meq})}$$

Substituting numerical data into this equation gives, after rearranging,

$$\text{wt } Fe^{2+} = 47.22\ \text{mL } KMnO_4 \times 0.1121\ \frac{\text{meq}}{\text{mL } KMnO_4} \times \frac{0.055847\ \text{g}}{1\ \text{meq}}$$

Note that the milliequivalent weight of the Fe^{2+} is equal to its milliformula weight. The percentage of iron is

$$\% Fe^{2+} = \frac{(47.22 \times 0.1121 \times 0.055847)\ \text{g } Fe^{2+}}{0.8040\ \text{g sample}} \times 100\% = 36.77\%$$

(b) Here,

$$\text{no. meq } KMnO_4 = \text{no. meq } Fe_3O_4$$

and

$$\text{vol } KMnO_4\ (\text{mL}) \times C_{N(KMnO_4)}\ (\text{meq/mL}) = \frac{\text{wt } Fe_3O_4\ (\text{g})}{\text{meqw } Fe_3O_4\ (\text{g/meq})}$$

Substituting numerical data and rearranging give

$$\text{wt } Fe_3O_4 = 47.22\ \text{mL} \times 0.1121\ \frac{\text{meq}}{\text{mL}} \times 0.23154\ \frac{\text{g } Fe_3O_4}{3\ \text{meq}}$$

Note that the milliequivalent weight of Fe_3O_4 is one third its formula weight because each Fe^{2+} undergoes a one-electron change and the compound is converted to 3 Fe^{2+} before titration. The percentage of Fe_3O_4 is then

$$\% Fe_3O_4 = \frac{(47.22 \times 0.1121 \times 0.23154/3)\ \text{g } Fe_3O_4}{0.8040\ \text{g sample}} \times 100\% = 50.81\%$$

Note that the answers to this example are identical to those in Example 6-7.

Example A6-6

A 0.4755-g sample containing $(NH_4)_2C_2O_4$ and inert compounds was dissolved in water and made alkaline with KOH. The liberated NH_3 was distilled into 50.00 mL of 0.1007 N (0.05035 M) H_2SO_4. The excess H_2SO_4 was back-titrated with 11.13 mL of 0.1214 N NaOH. Calculate the percentage of N (fw = 14.007) and of $(NH_4)_2C_2O_4$ (fw = 124.10) in the sample.

At the equivalence point, the number of milliequivalents of acid and base are equal. In this titration, however, two bases are involved: NaOH and NH_3. Thus,

$$\text{no. meq } H_2SO_4 = \text{no. meq } NH_3 + \text{no. meq NaOH}$$

After rearranging,

$$\text{no. meq } NH_3 = \text{no. meq N} = \text{no. meq } H_2SO_4 - \text{no. meq NaOH}$$

Substituting Equations A6-6 and A6-8 for the number of milliequivalents of N and H_2SO_4, respectively, yields

$$\frac{\text{wt N (g)}}{\text{meqw N (g/meq)}} = 50.00 \text{ mL } H_2SO_4 \times 0.1007 \frac{\text{meq}}{\text{mL } H_2SO_4}$$
$$- 11.13 \text{ mL NaOH} \times 0.1214 \frac{\text{meq}}{\text{mL NaOH}}$$

$$\text{wt N} = (50.00 \times 0.1007 - 11.13 \times 0.1214) \text{ meq} \times 0.014007 \text{ g N/meq}$$

$$\% \text{ N} = \frac{(50.00 \times 0.1007 - 11.13 \times 0.1214) \times 0.014007 \text{ g N}}{0.4755 \text{ g sample}} \times 100\%$$

$$= 10.85\%$$

The number of milliequivalents of $(NH_4)_2C_2O_4$ is equal to the number of milliequivalents of NH_3 and N, but the milliequivalent weight of the $(NH_4)_2C_2O_4$ is equal to one half its formula weight. Thus,

$$\text{wt } (NH_4)_2C_2O_4 = (50.00 \times 0.1007 - 11.13 \times 0.1214) \text{ meq} \times 0.12410 \text{ g/2 meq})$$

$$\% (NH_4)_2C_2O_4$$
$$= \frac{(50.00 \times 0.1007 - 11.13 \times 0.1214) \times 0.06205 \text{ g } (NH_4)_2C_2O_4}{0.4755 \text{ g sample}}$$
$$\times 100\% = 48.07\%$$

Note that the results obtained here are identical to those obtained in Example 6-9.

Answers to Asterisked Questions and Problems

Chapter 2

2-1. (a) The *millimole* is a unit of amount of an elementary species, such as an atom, an ion, a molecule, or an electron. One millimole contains 6.02×10^{20} particles of the species.

(b) The *milliformula weight* of a species is the weight in grams of one millimole. It is equal to fw $\times 10^{-3}$.

2-3. 1 L = 10^{-3} m³; M = mol/10^{-3} m³; Å = 10^{-10} m

2-4. (a) 1.5 MHz **(c)** 62.3 mmol **(e)** 96.494 kC

2-5. 5.09×10^{22} ions

2-7. (a) 9.82×10^{-2} mol **(c)** 3.82×10^{-2} mol
(b) 7.76×10^{-4} mol **(d)** 5.26×10^{-4} mol

2-9. (a) 5.52 mmol **(c)** 6.58×10^{-3} mmol
(b) 31.2 mmol **(d)** 966 mmol

2-11. (a) 4.20×10^4 mg **(c)** 1.52×10^6 mg
(b) 1.21×10^4 mg **(d)** 2.92×10^6 mg

2-13. (a) 1.33×10^3 mg **(b)** 519 mg

2-14. (a) 2.51 g **(b)** 2.88×10^{-3} g

2-15. (a) pNa = 0.618; pCl = 0.94; pOH = 0.90
(c) pH = -0.176; pZn = 0.92; pCl = -0.241
(e) pK = 4.249; pOH = 4.385; pFe(CN)$_6$ = 5.421

2-16. (a) 6.2×10^{-10} M **(e)** 4.8×10^{-8} M
(c) 3.5×10^{-1} M **(g)** 1.6 M

2-17. (a) pNa = pBr = 2.00; pH = pOH = 7.00
(c) pBa = 2.46; pOH = 2.16; pH = 11.84
(e) pCa = 2.28; pBa = 2.44; pCl = 1.75;
pH = pOH = 7.00

2-18. (a) $[H_3O^+] = 2.1 \times 10^{-9}$ M **(e)** 2.09 M
(c) $[Br^-]$ = 0.92 M **(g)** 0.99 M

2-19. (a) $[Na^+]$ = 0.0479 M; $[SO_4^{2-}]$ = 0.00287 M
(b) pNa = 1.320; pSO$_4$ = 2.543

2-21. (a) 0.01821 M **(d)** 0.5060% **(g)** 1.740
(b) 0.01821 M **(e)** 1.366 mmol **(h)** 1.263
(c) 0.05463 M **(f)** 712 ppm

2-23. (a) 0.3460 M **(b)** 1.038 M **(c)** 83.7 g/L

2-25. (a) Dilute 32.5 g of C_2H_5OH to 500 mL with water.
(b) Mix 32.5 g of C_2H_5OH with $(500 - 32.5) = 467.5$ g of H_2O.
(c) Dilute 32.5 mL of C_2H_5OH with water to 500 mL.

2-27. Dilute about 310 mL of the 85% reagent to approximately 750 mL.

2-29. (a) Dissolve 6.37 g of $AgNO_3$ in water and dilute to 500 mL.
(b) Dilute 52.5 mL of the 6.00 M reagent to 1.00 L.

(c) Dissolve 4.56 g of $K_4Fe(CN)_6$ in water and dilute to 600 mL.
(d) Dilute 144 mL of the 0.400 M $BaCl_2$ to 400 mL.
(e) Dilute about 25.1 mL of the concentrated reagent to 2.00 L.
(f) Dissolve 1.668 g of Na_2SO_4 in water and dilute to 9.00 L.

2-31. 3.35 g of $La(IO_3)_2$

2-33. (a) 4.65 g of CO_2 **(b)** 0.0286 M $HClO_4$

2-35. (a) 1.601 g of SO_2 **(b)** 0.0386 M $HClO_4$

2-37. 318.4 mL

Chapter 3

3-1. (a) *Accuracy* is the closeness of a result to its true or accepted value. *Precision* is the closeness of a result to other results measured in exactly the same way.

(c) The *mean* of a set of data is obtained by summing the results and dividing the sum by the number of results. The *median* of a set of data is the middle result when the data are arranged in order of size. For an even number of data, the median is obtained by averaging the middle two data.

(e) *Constant errors* are the same in magnitude regardless of the size of the sample. *Proportional errors* are proportional in size to the sample size.

3-2. (a) The *range* of a set of data is the numerical difference between the largest and the smallest value in the set.

(c) A *histogram* is a bar graph that is obtained by dividing a set of data into equal-size cells and plotting the percentage of measurements falling into each cell as a function of the measured quantity.

(e) The *standard error of a mean* σ_m is

$$\sigma_m = \frac{\sigma}{\sqrt{N}}$$

where σ is the standard deviation for a sample of data made up of N measurements.

3-3. Instrumental error, method error, and personal error

3-5. Constant determinate errors can be detected by varying sample size because the effect of the error becomes smaller as the sample size increases.

3-7.

	A	C	E
(a) mean	2.08 = 2.1	0.09180	69.53
(b) median	2.1	0.0885	69.635
(c) spread	0.9	0.0116	0.44
(d) std dev	0.35	0.0055	0.22
(e) CV	17%	6.0%	0.31%

3-8.

	A	C	E
(a) abs error	0.08	−0.0012	0.48
(b) rel error	40 ppt	−13 ppt	7.0 ppt

3-9. (a) $\dfrac{-0.3}{800} \times 100 = -0.04\%$

(c) $\dfrac{-0.3}{100} \times 100 = -0.3\%$

3-10. (a) 13 g

(c) 3 g

3-11.

	s	CV, %	Rounded Result
(a)	0.030	5.2	0.57(±0.03)
(b)	0.089	0.42	21.3(±0.1)
(c)	0.14×10^{-16}	2.0	$6.9(\pm 0.1) \times 10^{-16}$
(d)	1.4×10^{3}	0.77	$1.84(\pm 0.01) \times 10^{5}$
(e)	0.51×10^{-2}	8.5	$6.0(\pm 0.5) \times 10^{-2}$ or $7(\pm 1) \times 10^{-2}$
(f)	0.11×10^{-3}	1.3	$8.1(\pm 0.1) \times 10^{-3}$

3-13. (a) 0.238 (c) 23.7796 (e) 3.4×10^{-4} (g) 9.8

Chapter 4

4-1. For A, 95% CI = 2.08 ± 0.44 or 2.1 ± 0.4
For C, 95% CI = 0.0918 ± 0.0088 or 0.092 ± 0.009
For E, 95% CI = 69.53 ± 0.34 or 69.5 ± 0.3

4-2. For A, 95% CI = 2.08 ± 0.18 or 2.1 ± 0.2
For C, 95% CI = 0.0918 ± 0.0069 or 0.092 ± 0.007
For E, 95% CI = 69.53 ± 0.15 or 69.5 ± 0.2

4-4.

Sample	s, %K
1	0.095
2	0.12
3	0.11
4	0.10
5	0.10

$s_{\text{pooled}} = 0.11\%$ K$^+$

4-6. $s_{\text{pooled}} = 0.29\%$ heroin

4-8. (a) 80% CI = $18.5 \pm 1.29 \times 2.4/\sqrt{1}$
= 18.5 ± 3.1 μg/mL
95% CI = $18.5 \pm 1.96 \times 2.4/\sqrt{1}$
= 18.5 ± 4.7 μg/mL

(b) 80% CI = $18.5 \pm 1.29 \times 2.4/\sqrt{2}$
= 18.5 ± 2.2 μg/mL
95% CI = $18.5 \pm 1.96 \times 2.4/\sqrt{2}$
= 18.5 ± 3.3 μg/mL

(c) 80% CI = $18.5 \pm 1.29 \times 2.4/\sqrt{4}$
= 18.5 ± 1.5 μg/mL
95% CI = $18.5 \pm 1.96 \times 2.4/\sqrt{4}$
= 18.5 ± 2.4 μg/mL

4-10. 10 and 17 measurements

4-12. $\bar{x} = 3.22$, $s = 0.061$
(a) 95% CI = 3.22 ± 0.15 meq/L
(b) 95% CI = 3.22 ± 0.06 meq/L

4-14. (a) 12 measurements

4-15. (a) $Q_{\text{expt}} = 0.34/0.57 = 0.60$,
$Q_{\text{crit}} = 0.85$; therefore retain
(b) $Q_{\text{expt}} = 0.093/0.104 = 0.89$,
$Q_{\text{crit}} = 0.85$; therefore reject

4-17. (a) See broken line in the figure below.
(b) $R = 0.162 + 0.232C_x$
(c) See the solid line in the figure in part (a).
(d) $s_r = 0.28$; $s_b = 0.017$
(e) 15.1 mg SO$_4^{2-}$/mL; $s_c = 1.4$ mg SO$_4^{2-}$/mL; CV = 9.3%
(f) $s_c = 0.81$ mg SO$_4^{2-}$/mL; CV = 5.4%

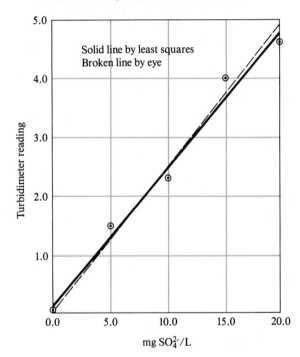

Chapter 5

5-1. (a) The individual particles of a *colloid* are smaller than about 10^{-5} mm in diameter, while those of a crystalline precipitate are larger. As a consequence, *crystalline precipitates* settle out of solution relatively rapidly, whereas colloidal particles do not unless they can be caused to agglomerate.

(c) *Precipitation* is the process by which a solid phase forms and is carried out of solution when the solubility product of a species is exceeded. *Coprecipitation* is the process in which a *normally soluble* species is carried out of solution during the formation of a precipitate.

(e) *Occlusion* is a type of coprecipitation in which an impurity is entrapped in a pocket formed by a rapidly growing crystal. *Mixed-crystal formation* is a type of coprecipitation in which a foreign ion is incorporated into a growing crystal in a lattice position that is ordinarily occupied by one of the ions of the precipitate. The contaminant has the same charge and is about the same size as the ion it replaces.

5-2. (a) *Digestion* is a process for improving the purity and filterability of a precipitate by heating the solid in contact with the solution from which it is formed (the mother liquor).

(c) In *reprecipitation,* a precipitate is filtered, washed, redissolved, and then reformed from the new solution. Because the concentration of contaminant is lower in this new solution than in the original, the second precipitate contains less coprecipitated impurity.

(e) The *counter-ion layer* is a layer of solution surrounding a colloidal particle that contains a sufficient excess of ions of opposite charge to just equal the number of primarily adsorbed ions on the particle.

(g) *Supersaturation* is a metastable state in which the concentration of solute molecules is larger than the equilibrium solubility of the solute.

5-3. A *chelating agent* is an organic compound that contains two electron-donor groups located in such a configuration that a five- or six-membered ring is formed when both groups complex a cation.

5-5. (a) plus   (b) adsorbed Ag^+ (c) NO_3^-

5-7. *Peptization* is the process in which a coagulated colloid returns to its original dispersed state as a consequence of a reduction in the electrolyte concentration of the solution in contact with the precipitate. Peptization during the washing of a coagulated colloid can be avoided by washing with an electrolyte solution rather than with pure water.

5-9. (a) wt SO_3 = wt $BaSO_4 \times \dfrac{\text{fw } SO_3}{\text{fw } BaSO_4}$

(c) wt In = wt $In_2O_3 \times \dfrac{2 \text{ fw In}}{\text{fw } In_2O_3}$

(e) wt CuO = wt $Cu_2(SCN)_2 \times \dfrac{2 \text{ fw CuO}}{\text{fw } CU_2(SCN)_2}$

(g) wt Pb_3O_4 = wt $PbO_2 \times \dfrac{\text{fw } Pb_3O_4}{3 \text{ fw } PbO_2}$

(i) wt $Na_2B_4O_7 \cdot 10 \ H_2O$ =

$$\text{wt } B_2O_3 \times \frac{\text{fw } Na_2B_4O_7 \cdot 10 \ H_2O}{2 \text{ fw } B_2O_3}$$

5-10. 95.36% Cl

5-12. 0.662 g of $Cu(IO_3)_2$

5-14. 0.124 g of AgI

5-16. 14.03% Mg

5-18. 18.99% C

5-20. 38.82% Hg_2Cl_2

5-22. 46.40% NH_3

5-24. minimum weight = 0.617 g;
weight of precipitate = 0.825 g

5-26. (a) 0.239 g of sample   (b) 0.494 g of AgCl
(c) 0.407 g of sample

5-28. % I^- = 27.05; % Cl^- = 4.72

5-30. 0.395 g of CO_2

5-32. (a) 0.369 g of $Ba(IO_3)_2$
(b) 0.0148 g of $BaCl_2 \cdot 2 \ H_2O$

Chapter 6

6-1. The *equivalence point* in a titration is that point at which sufficient titrant has been added so that stoichiometrically equivalent amounts of analyte and titrant are present. The *end point* in a titration is the point at which an observable physical change that signals the equivalence point occurs.

6-3. A *secondary standard* is a material or solution whose concentration has been established by careful analysis. Secondary standards are employed when a reagent is not available in primary-standard quality. For example, solid sodium hydroxide is hygroscopic and cannot be used to prepare a standard solution directly. A secondary-standard solution of the reagent is readily prepared, however, by standardizing a solution of sodium hydroxide against a primary-standard reagent such as potassium acid phthalate.

6-4. 3.03 M H_2SO_4

6-6. (a) Dilute 60.0 mL of 4.00 M H_2SO_4 to 3.00 L.
(b) Dilute 181 g of 13.0% solution to 3.00 L.
(c) Dilute 13.5 mL of the concentrated reagent to 3.00 L.

6-8. 0.09151 M

6-9. 0.1108 M

6-11. 0.2979 M $HClO_4$; 0.3258 M NaOH

6-12. 346 ppm S

6-14. 7.317% $(NH_2)_2CS$

6-16. 5.472% As_2O_3

6-17. (a) 9.36×10^{-3} M   (b) 1.9×10^{-5} M
(c) absolute error = -2.76×10^{-5} M;
relative error = -2.95 ppt

Chapter 7

7-1. (a) A *salt* is the product (other than water) of an acid/base reaction.

(c) The *hydronium ion* is the product of the reaction of a proton with one molecule of water. Its formula is H_3O^+.

(e) A *base* is a proton acceptor.

(g) The *Le Châtelier principle* states that the position of a chemical equilibrium is always shifted in such a direction as to relieve the effect of an applied stress.

(i) An *amphiprotic solvent* is one that undergoes self-ionization to form a pair of ionic species. For example,

$$2 \ H_2O \rightleftarrows H_3O^+ + OH^-$$

$$2 \ C_2H_5OH \rightleftarrows C_2H_5OH_2^+ + C_2H_5O^-$$

7-2. Water is a *leveling solvent* toward mineral acids because all of these acids are completely dissociated in that medium and are thus equal in strength. In ethanol, the degree of dissociation of the various mineral acids differs significantly. Thus, ethanol is termed a *differentiating* solvent because it differentiates among the strengths of the acids.

7-3. This conclusion is invalid because the units of K_{sp} for AgCl are mol^2/L^2, whereas the units for K_{sp} for Ag_2CO_3 are mol^3/L^3. Thus,

$$s_{AgCl} = (1.8 \times 10^{-10})^{1/2} = 1.3 \times 10^{-5}$$

$$s_{Ag_2CO_3} = (8.1 \times 10^{-12}/4)^{1/3} = 1.3 \times 10^{-4}$$

and Ag_2CO_3 is the more soluble.

7-4. (a) $K_b = \dfrac{[C_2H_5NH_3^+][OH^-]}{[C_2H_5NH_2]} = 4.28 \times 10^{-4}$

(c) $K_a = \dfrac{[C_5H_5N][H_3O^+]}{[C_5H_5NH^+]} = \dfrac{K_w}{K_b} = \dfrac{1.00 \times 10^{-14}}{1.7 \times 10^{-9}}$
$= 5.9 \times 10^{-6}$

(e) $\dfrac{[H_3O^+]^3[AsO_4^{3-}]}{[H_3AsO_4]} = K_1K_2K_3 = 6.0 \times 10^{-3} \times$
$1.05 \times 10^{-7} \times 3.0 \times 10^{-12} = 1.9 \times 10^{-21}$

7-5. $K_{sp} =$ (a) 4.0×10^{-16} (c) 2.0×10^{-2}
(e) 3.5×10^{-10}

7-6. (a) 0.056 g/100 mL
(b) 8.2×10^{-4} g/100 mL
(c) 0.014 g/100 mL

7-8. (a) 2.18×10^{-7} M   (b) 0.98 M

7-10. (a) 0.025 M   (c) 1.9×10^{-3} M
(b) 1.7×10^{-2} M   (d) 7.6×10^{-7} M

7-12.

	Quadratic Method		Approximate Method	
	$[H_3O^+]$	$[OH^-]$	$[H_3O^+]$	$[OH^-]$
(a)	2.4×10^{-5}	4.1×10^{-10}	2.4×10^{-5}	4.1×10^{-10}
(c)	1.05×10^{-12}	9.6×10^{-3}	1.02×10^{-12}	9.8×10^{-3}
(e)	5.0×10^{-11}	2.0×10^{-4}	5.0×10^{-11}	2.0×10^{-4}
(g)	3.05×10^{-4}	3.28×10^{-11}	3.06×10^{-4}	3.27×10^{-11}

7-13. $[H_3O^+] = [OH^-] = 3.38 \times 10^{-8}$

7-14. (a) 1.10×10^{-2} (quadratic);
1.17×10^{-2} (approximate)
(b) 1.17×10^{-8} (quadratic);
1.17×10^{-8} (approximate)
(e) 1.47×10^{-4} (quadratic);
1.59×10^{-4} (approximate)

7-15. (a) PbI_2 $(1.2 \times 10^{-3}) >$ TlI $(2.5 \times 10^{-4}) >$
BI_3 $(1.3 \times 10^{-5}) >$ AgI (9.1×10^{-9})
(b) PbI_2 $(7.1 \times 10^{-7}) >$ TlI $(6.5 \times 10^{-7}) >$
AgI $(8.3 \times 10^{-16}) >$ BiI_3 (8.1×10^{-16})
(c) PbI_2 $(1.3 \times 10^{-4}) >$ BiI_3 $(6.7 \times 10^{-7}) >$
TlI $(6.5 \times 10^{-7}) >$ AgI (8.3×10^{-16})

Chapter 8

8-1. *Thermodynamic equilibrium constants* are activity-based equilibrium constants that are largely independent of ionic strength.

8-3. Applying Equation 8-4 to the two equilibria gives

$$[H_3O^+][OH^-] = \frac{K_w}{f_{H_3O^+} \cdot f_{OH^-}} = K'_w$$

$$[Ba^{2+}][SO_4^{2-}] = \frac{K_{sp}}{f_{Ba^{2+}} \cdot f_{SO_4^{2-}}} = K'_{sp}$$

The activity coefficients for doubly charged Ba^{2+} and SO_4^{2-} decrease much more rapidly with increases in the concentration of NaCl than do those for singly charged H_3O^+ and OH^-. Thus K'_{sp} increases much more rapidly than does K'_w as the concentration of NaCl becomes larger.

8-5. (a) $\mu = 0.16$ (c) $\mu = 1.20$
8-6. (a) $f_\pm = 0.67$ (c) $f_\pm = 0.46$
8-7. (a) $f_{Fe^{3+}} = 0.20$ (c) $f_{Ce^{4+}} = 0.073$
8-8. (a) $f_{Fe^{3+}} = 0.20$ (c) $f_{Ce^{4+}} = 0.079$
8-9. (a) 1.7×10^{-12} (c) 4.7×10^{-11}
8-10. (b) 5.3×10^{-5}
8-11. (a) (1) 1.4×10^{-6} M   (2) 1.0×10^{-6} M
(b) (1) 2.0×10^{-3} M   (2) 1.2×10^{-3} M
(c) (1) 3.1×10^{-5} M   (2) 1.1×10^{-5} M
(d) (1) 1.4×10^{-5} M   (2) 2.0×10^{-6} M
8-13. (a) (1) 3.0×10^{-3} M   (2) 2.9×10^{-3} M
(b) (1) 3.6×10^{-3} M   (2) 2.9×10^{-3} M
(c) (1) 1.11×10^{-11} M   (2) 1.08×10^{-11} M
8-14. (a) (1) 5.4×10^{-6} M   (2) 5.4×10^{-6} M
(c) (1) 8.5×10^{-13} M   (2) 2.5×10^{-12} M
(e) (1) 1.1×10^{-11} M   (2) 3.1×10^{-12} M

Chapter 9

9-1. (a) $0.10 = [H_3PO_4] + [H_2PO_4^-] + [HPO_4^{2-}] + [PO_4^{3-}]$
(c) $0.100 = 0.0500 = [HNO_2] + [NO_2^-]$
$[Na^+] = C_{NaNO_2} = 0.0500$
(e) $0.100 = [Na^+] + [OH^-] - 2[Zn(OH)_4^{2-}]$
(g) $[Ca^{2+}] = \frac{1}{2}([F^-] + [HF])$
$[HF] - [OH^-]$

9-2. (a) $[H_3O^+] = [OH^-] + [H_2PO_4^-] +$
$2[HPO_4^{2-}] + 3[PO_4^{3-}]$
(c) $[Na^+] + [H_3O^+] = [OH^-] + [NO_2^-]$
(e) $2[Zn^{2+}] + [H_3O^+] + [Na^+] =$
$2[Zn(OH)_4^{2-}] + [OH^-]$
(g) $2[Ca^{2+}] + [H_3O^+] = [OH^-] + [F^-]$

9-3. (a) 5.2×10^{-3} M   (c) 3.6×10^{-4} M
9-4. (a) 1.5×10^{-4} M   (c) 7.4×10^{-5} M
9-5. (a) 2.5×10^{-13} M   (b) 2.3×10^{-6} M
9-7. (a) 3×10^{-8} M   (b) 3×10^{-11} M
9-9. (a) 0.7 M   (b) 2×10^{-3} M
9-11. (a) 8.7×10^{-2} M   (d) 9.0×10^{-4} M
(b) 8.7×10^{-3} M   (e) 1.2×10^{-3} M
(c) 1.0×10^{-3} M
9-13. 8.3×10^{-2} M
9-15. 1.8×10^{-6} M
9-17. (a) 8.3×10^{-11} M   (c) 1.3×10^4
(b) 1.6×10^{-11} M   (d) 1.3×10^4
9-19. (a) not feasible   (b) feasible   (c) feasible
(d) not feasible

Chapter 10

10-1. In the Volhard method for Cl^-, it is necessary to filter the AgCl from the solution before back-titration with KSCN because AgCl is more soluble than AgSCN. As a consequence, SCN^- present during the back-titration reacts with the precipitated AgCl and gives a fading end point:

$$AgCl(s) + SCN^- \rightleftarrows AgSCN(s) + Cl^-$$

The filtration step is unnecessary when Br^- is being determined because AgBr is less soluble than AgSCN and thus does not react with SCN^-.

10-3. Potassium is determined by precipitation with an excess of a standard solution of sodium tetraphenyl boron. An excess of standard $AgNO_3$ is then added to precipitate the excess tetraphenyl boron ion. The excess $AgNO_3$ is then titrated with a standard solution of SCN^-. The reactions are

$$K^+ + B(C_6H_5)_4^- \rightleftarrows KB(C_6H_5)_4(s)$$
(measured excess of $B(C_6H_5)_4^-$)

$$B(C_6H_5)_4^- + Ag^+ \rightleftarrows AgB(C_6H_5)_4(s)$$
(measured excess of $AgNO_3$)

The excess $AgNO_3$ is then determined by a Volhard titration with KSCN.

10-5. (a) 0.08368 M (c) 0.1558 M (e) 0.3130 M

10-7. 28.30%

10-9. 116.7 mg

10-10. 44.41%

10-12. 15.60 mg/tablet

10-14. One of the seven chlorine atoms is titrated.

10-19. 0.4348%

10-20. 10.60% Cl^-; 55.65% ClO_4^-

10-22. 26.8% K_2SO_4; 73.2% KCl

10-23.

(a)

Vol NH_4SCN, mL	$[Ag^+]$	$[SCN^-]$	pAg
30.00	9.09×10^{-3}	1.2×10^{-10}	2.04
40.00	3.85×10^{-3}	2.9×10^{-10}	2.42
49.00	3.38×10^{-4}	3.4×10^{-9}	3.47
50.00	1.05×10^{-6}	1.05×10^{-10}	5.98
51.00	3.3×10^{-9}	3.28×10^{-4}	8.48
60.00	3.7×10^{-10}	2.94×10^{-3}	9.43
70.00	2.1×10^{-10}	5.26×10^{-3}	9.68

(c)

Vol NaCl, mL	$[Ag^+]$	$[Cl^-]$	pAg
10.00	3.75×10^{-2}	4.85×10^{-9}	1.43
20.00	1.50×10^{-2}	1.21×10^{-8}	1.82
29.00	1.27×10^{-3}	1.43×10^{-7}	2.90
30.00	1.35×10^{-5}	1.35×10^{-5}	4.87
31.00	1.48×10^{-7}	1.23×10^{-3}	6.83
40.00	1.70×10^{-8}	1.07×10^{-2}	7.77
50.00	9.71×10^{-9}	1.88×10^{-2}	8.01

(e)

Vol Na_2SO_4, mL	$[Ba^{2+}]$	$[SO_4^{2-}]$	pBa
0.00	2.50×10^{-2}	0.0	1.60
10.00	1.00×10^{-2}	1.3×10^{-8}	2.00
19.00	8.48×10^{-4}	1.5×10^{-7}	3.07
20.00	1.1×10^{-5}	1.1×10^{-5}	4.94
21.00	1.6×10^{-7}	8.20×10^{-4}	6.80
30.00	1.8×10^{-8}	7.14×10^{-3}	7.74
40.00	1.0×10^{-8}	1.25×10^{-2}	7.98

10-24.

Vol $AgNO_3$, mL	$[Ag^+]$	pAg
5.00	1.63×10^{-11}	10.79
40.00	7.2×10^{-7}	6.14
45.00	2.6×10^{-3}	2.58

10-25.

Vol NaCl, mL	$[Hg_2^{2+}]$	pHg_2
10.00	3.33×10^{-2}	1.48
30.00	9.09×10^{-3}	2.04
40.00	6.9×10^{-7}	6.16
50.00	5.5×10^{-15}	14.26

Chapter 11

11-1. The initial pH of the NH_3 solution is less than that of the solution containing NaOH. With the first addition of titrant, the pH of the NH_3 solution decreases rapidly and then levels off and becomes nearly constant throughout the middle part of the titration. In contrast, additions of standard acid to the NaOH solution cause the pH of the NaOH solution to decrease gradually and nearly linearly until the equivalence point is approached. The equivalence-point pH for the NH_3 solution is well below 7, whereas that for the NaOH solution is exactly 7. Beyond the equivalence point, the two curves are identical.

11-3. Physiological fluids, such as blood, must remain more or less constant in pH in order for an animal to survive. Acidic and basic compounds are constantly being introduced into these fluids as a consequence of the intake of food and air. The buffers in the fluids maintain a constant pH despite these additions.

11-5. Mixture (a) has the greater buffer capacity because the concentration of buffering reagents is greater.

11-7. The *equivalence point* is the theoretical point in a titration at which the amount of reagent added is chemically equivalent to the amount of analyte. The *end point* is that point where an observable physical change related to the condition of chemical equivalence occurs.

11-9. The standard reagents in neutralization titrations are always strong acids or strong bases because the reactions with these types of reagents are more complete than are the reactions with their weaker counterparts. Sharper end points are the consequence of this difference.

11-11. 3.25

11-12. (a) NaOCl (c) hydrazine

11-13. (a) malic acid/sodium hydrogen malate
(c) NH_3/NH_4Cl

11-14. 7.47 at 0°C; 6.63 at 50°C; 6.16 at 100°C

11-16. −1.12

11-18. (a) 1.05 (b) 1.05 (c) 1.81 (d) 1.81 (e) 12.60

11-20. 6.62

11-22. (a) $[H_3O^+] = 0.0500$; pH = 1.30
(b) $a_{H_3O^+} = 0.0430$; pH = 1.37

11-24. (a) 1.94 (b) 2.44 (c) 3.52

11-26. (a) 5.12 (b) 5.62 (c) 6.62

11-28. (a) 10.26   (b) 9.76   (c) 8.76

11-30. (a) 2.41   (b) 8.35   (c) 12.35   (d) 3.85

11-32. (a) 3.85   (b) 4.06   (c) 2.64   (d) 2.08

11-34. (a) 0.00   (c) -1.00 (e) 0.500   (g) 0.000
(b) 1.00   (d) -0.509 (f) -0.003

11-35. (a) -5.00 (c) -0.097 (e) -3.273 (g) -0.017
(b) -0.079 (d) -1.124 (f) -0.176

11-36. (a) 5.00   (c) 0.079   (e) 3.37   (g) 0.018
(b) 0.097   (d) 1.032   (f) 0.176

11-38. 15.2 g

11-40. 196 mL

11-42. equivalence-point pH range = 6.5 to 9.8; cresol purple

11-44.

Vol, mL	pH	Vol, mL	pH
0.00	13.00	49.00	11.00
10.00	12.82	50.00	7.00
25.00	12.52	51.00	3.00
40.00	12.05	55.00	2.32
45.00	11.72	60.00	2.04

11-45.

Vol, mL	(a) pH	(c) pH
0.00	2.16	3.12
5.00	2.49	4.28
15.00	2.95	4.86
25.00	3.31	5.23
40.00	3.90	5.83
45.00	4.25	6.18
49.00	4.99	6.92
50.00	8.00	8.96
51.00	11.00	11.00
55.00	11.68	11.68
60.00	11.96	11.96

11-46.

Vol, mL	pH	Vol, mL	pH
0.00	11.12	49.00	7.56
5.00	10.20	50.00	5.27
15.00	9.61	51.00	3.00
25.00	9.25	55.00	2.32
40.00	8.64	60.00	2.04
45.00	8.29		

11-47.

Vol, mL	(a) pH	(c) pH
0.00	2.80	4.26
5.00	3.64	6.57
15.00	4.23	7.15
25.00	4.60	7.52
40.00	5.20	8.12
49.00	6.29	9.21
50.00	8.65	10.11
51.00	11.00	11.00
55.00	11.68	11.68
60.00	11.96	11.96

11-48. (a) $\alpha_0 = 0.152$; $\alpha_1 = 0.848$
(c) $\alpha_0 = 0.478$; $\alpha_1 = 0.522$

11-50.

	C_T	pH	[HA]	[A$^-$]	α_0	α_1
Lactic	0.120	3.61	0.768	0.432	0.640	0.360
Butanoic	0.162	5.00	0.0644	0.0097	0.398	0.602
Nitrous	0.179	3.45	0.074	0.105	0.413	0.587

Chapter 12

12-1. An amphiprotic substance is one that is capable of behaving as both an acid and a base.

12-2. (a) very slightly basic   (c) acidic   (e) basic
(g) acidic

12-4. equivalence-point pH = 9.2; therefore phenolphthalein

12-6. (a) equivalence-point pH $\approx$ 8.3; cresol purple
(c) equivalence-point pH $\approx$ 8.5; cresol purple
(e) equivalence-point pH $\approx$ 4.1; bromocresol green
(g) equivalence-point pH $\approx$ 9.9; phenolphthalein or thymolphthalein

12-7. (a) 4.23   (c) 2.07   (e) 11.35
(b) 1.16   (d) 12.78   (f) 8.52

12-9. (a) 9.92   (b) 2.95   (c) 4.56   (d) 8.39

12-11. (a) 1.54   (b) 1.99   (c) 12.07   (d) 12.01

12-13. (a) $[HSO_3^-]/[SO_3^{2-}] = 15.6$
(b) $[HCit^{2-}]/[Cit^{3-}] = 2.5$
(c) $[HM^-]/[M^{2-}] = 0.498$
(d) $[HT^-]/[T^{2-}] = 0.0232$

12-15. (a) 2.06   (b) 7.20   (c) 10.63   (d) 2.55   (e) 2.01

12-17. (a) 2.11   (b) 7.37

12-19. Mix 442 mL of 0.300 M Na_2CO_3 with 558 mL of 0.2000 M HCl.

12-22. 50.3 g

12-24.

mL Reagent	(a) pH	(c) pH
0.00	11.66	0.96
12.50	10.33	1.26
20.00	9.73	1.48
24.00	8.95	1.61
25.00	8.34	1.64
26.00	7.73	1.67
37.50	6.35	2.14
45.00	5.75	2.65
49.00	4.97	3.40
50.00	3.83	7.31
51.00	2.70	11.30
60.00	1.74	12.26

12-25.

Vol HClO$_4$, mL	pH	Vol HClO$_4$, mL	pH
0.00	13.00	35.00	8.11
10.00	12.70	44.00	6.84
20.00	12.15	45.00	4.74
24.00	11.43	46.00	2.68
25.00	10.42	50.00	2.00
26.00	9.39		

12-27. Let $D = [H_3O^+]^3 + K_1[H_3O^+]^2 + K_1K_2[H_3O^+] + K_1K_2K_3$. Then

$$\alpha_0 = [H_3O^+]^3/D$$
$$\alpha_1 = K_1[H_3O^+]^2/D$$
$$\alpha_2 = K_1K_2[H_3O^+]/D$$
$$\alpha_3 = K_1K_2K_3/D$$

12-28. (a) $\dfrac{[H_3AsO_4][HAsO_4^{2-}]}{[H_2AsO_4^-]^2} = \dfrac{K_2}{K_1}$

where

$$K_2 = \frac{[H_3O^+][HAsO_4^{2-}]}{[H_2AsO_4^-]}$$

$$K_1 = \frac{[H_3O^+][H_2AsO_4^-]}{[H_3AsO_4]}$$

12-29. $\dfrac{[H_3O^+][OAc^-]}{[HOAc]} = 1.75 \times 10^{-5}$

$\dfrac{[NH_4^+][OH^-]}{[NH_3]} = 1.76 \times 10^{-5}$

Multiplying the two equations together gives

$$\frac{[H_3O^+][OH^-][OAc^-][NH_4^+]}{[HOAc][NH_3]}$$
$$= 1.75 \times 10^{-5} \times 1.76 \times 10^{-5} = 3.08 \times 10^{-11}$$

$$\frac{[OAc^-][NH_4^+]}{[HOAc][NH_3]} = \frac{3.08 \times 10^{-11}}{[H_3O^+][OH^-]}$$
$$= \frac{3.08 \times 10^{-11}}{1.00 \times 10^{-14}} = 3.08 \times 10^3$$

Inverting this equation gives the desired constant. Thus,

$$\frac{[HOAc][NH_3]}{[NH_4^+][OAc^-]} = \frac{1}{3.08 \times 10^3} = 3.25 \times 10^{-4}$$

12-30.

	pH	α_0	α_1	α_2	α_3
(a)	2.00	0.899	0.101	3.94×10^{-5}	
	6.00	1.82×10^{-4}	0.204	0.796	
	10.00	2.28×10^{-12}	2.56×10^{-5}	1.000	
(c)	2.00	0.931	6.93×10^{-2}	1.20×10^{-4}	4.82×10^{-9}
	6.00	5.32×10^{-5}	3.96×10^{-2}	0.685	0.275
	10.00	1.93×10^{-16}	1.44×10^{-9}	2.49×10^{-4}	1.000
(e)	2.00	0.500	0.500	1.30×10^{-5}	
	6.00	7.94×10^{-5}	0.794	0.206	
	10.00	3.84×10^{-12}	3.84×10^{-4}	1.000	

Chapter 13

13-1. When carbon dioxide is dissolved in water, it is largely present as CO_2 and H_2CO_3. The H_2CO_3, upon heating, decomposes to give CO_2 (and H_2O). The CO_2 is not strongly bonded by water molecules, however, and thus is readily volatilized from aqueous media. Gaseous HCl molecules, on the other hand, are fully dissociated into H_3O^+ and Cl^- when dissolved in water; neither of these species is volatile.

13-3. Primary-standard Na_2CO_3 can be obtained by heating primary-standard-grade $NaHCO_3$ for about 1 h at 270 to 300°C. The reaction is

$$2\,NaHCO_3(s) \rightarrow Na_2CO_3(s) + H_2O(g) + CO_2(g)$$

13-5. The *carbonate error* arises in acid/base titrations whenever the pH range of the indicator used is different from that of the indicator used for reagent standardization. For example, if an indicator with an acidic range is used in a standardization of carbonate-containing base, each carbonate ion present reacts with two hydronium ions. If this solution is then used with an indicator with a basic range of color change, the carbonate present consumes but a single proton. Thus the molarity of the base has been effectively lowered by the change in indicator. This type of carbonate error is avoided by using carbonate-free solutions of base or by using the same indicator for standardization and analyses. When very dilute reagent solutions are being used, the equilibrium concentration of carbon dioxide in water may provide sufficient titratable protons to affect the results of acid/base titrations. This type of carbonate error is avoided by boiling and cooling the water to be used for preparing reagents and dissolving samples.

13-7. Treat an acidic solution of the sample with a good reducing agent, which converts the HNO_3 to NH_4^+. Then carefully make the solution basic as in the Kjeldahl method, and distill the NH_3 into a standard solution of acid. Back-titrate the excess acid with standard base.

13-9. (a) 0.1388 M   (b) 0.1500 M

13-11. (a) Dissolve 17 g of KOH in water, and dilute to about 2.0 L.
(b) Dissolve 9.4 g of $Ba(OH)_2 \cdot 8\,H_2O$ in water, and dilute to about 2.0 L.
(c) Dilute about 120 mL of the reagent to 2.0 L.

13-13. (a) 0.0996 M   (b) 0.00049 M

13-15. (a) 0.28 to 0.36 g   (c) 0.85 to 1.1 g
(e) 0.17 to 0.22 g

13-16. 0.05950 M

13-18. 0.1217 g H_2T/100 mL

13-20. 76.97 g/equiv

13-22. 7.079%

13-24. 3.348% N or 19.0% protein

13-26. 3.35×10^3 ppm

13-28. 13.86%

13-30. 0.5272% CH_3CHO; 0.6453% EtOAc

13-32. 69.84% KOH; 21.04% K_2CO_3; 9.12% H_2O

13-34. (a) 18.15 mL   (b) 45.37 mL   (c) 38.28 mL
(d) 12.27 mL

13-36. (a) 4.314 mg NaOH/mL
(b) 7.985 mg Na_2CO_3/mL; 4.358 mg $NaHCO_3$/mL
(c) 4.396 mg NaOH/mL; 3.455 mg Na_2CO_3/mL
(d) 8.215 mg Na_2CO_3/mL
(e) 13.46 mg $NaHCO_3$/mL

Chapter 14

14-1. (a) A *chelate* is a cyclic complex consisting of a metal ion and a reagent that contains two or more electron-donor groups located in such a position that they can bond with the metal ion to form a heterocyclic ring.

(c) A *ligand* is a species that contains one or more electron-pair donor groups that tend to form bonds with metal ions.

(e) A *conditional formation constant* is a nonthermodynamic equilibrium constant for the reaction between a metal ion and a complexing agent that applies only when the pH and/or the concentration of other complexing ions is carefully specified.

(g) The hardness of water is the concentration of calcium carbonate that is equivalent to the total concentration of all multivalent metal carbonates in the water.

14-2. Three general methods for performing EDTA titrations are (1) *direct titration*, (2) *back-titration*, and (3) *displacement titration*. Method 1 is simple and rapid and requires but one standard reagent. Method 2 is advantageous for those metals that react so slowly with EDTA as to make direct titration inconvenient. In addition, this procedure is useful for cations for which a satisfactory indicator is not available. Finally, Method 2 is useful for analyzing samples that contain anions that form sparingly soluble precipitates with the analyte under the analytical conditions. Method 3 is particularly useful in situations where no satisfactory indicators are available for direct titration.

14-4. Multidentate ligands offer the advantage that they usually form more stable complexes than do unidentate ligands. Furthermore, they often form but a single complex with the cation, which simplifies their titration curves and makes end-point detection easier.

14-5. (a)

$$Ag^+ + S_2O_3^{2-} \rightleftharpoons AgS_2O_3^-$$

$$K_1 = \frac{[AgS_2O_3^-]}{[Ag^+][S_2O_3^{2-}]}$$

$$AgS_2O_3^- + S_2O_3^{2-} \rightleftharpoons Ag(S_2O_3)_2^{3-}$$

$$K_2 = \frac{[Ag(S_2O_3)_2^{3-}]}{[AgS_2O_3^-][S_2O_3^{2-}]}$$

(c)

$$Cd^{2+} + NH_3 \rightleftharpoons CdNH_3^{2+}$$

$$K_1 = \frac{[CdNH_3^{2+}]}{[Cd^{2+}][NH_3]}$$

$$CdNH_3^+ + NH_3 \rightleftharpoons Cd(NH_3)_2^{2+}$$

$$K_2 = \frac{[Cd(NH_3)_2^{2+}]}{[CdNH_3^{2+}][NH_3]}$$

$$Cd(NH_3)_2^{2+} + NH_3 \rightleftharpoons Cd(NH_3)_3^{2+}$$

$$K_3 = \frac{[Cd(NH_3)_3^{2+}]}{[Cd(NH_3)_2^{2+}][NH_3]}$$

$$Cd(NH_3)_3^{2+} + NH_3 \rightleftharpoons Cd(NH_3)_4^{2+}$$

$$K_4 = \frac{[Cd(NH_3)_4^{2+}]}{[Cd(NH_3)_3^{2+}][NH_3]}$$

14-8. 0.01032 M

14-10. 0.01005 M

14-12. 323.9 mg of Ca^{2+} and 256.4 mg of Mg^{2+}; both normal

14-14. 94.40%

14-16. 31.48% $NaBr$ and 48.57% $NaBrO_3$

14-18. 7.515% Pb; 4.304% Mg; 9.456% Zn

14-20. 8.517% Pb; 24.85% Zn; 64.07% Cu; 2.56% Sn

14-21. (a) 1.4×10^9 (b) 3.3×10^{11} (c) 2.2×10^{13}

14-23. $K'_{Sr} = 3.66 \times 10^8$

Vol, mL	pSr	Vol, mL	pSr
0.00	2.00	25.00	5.37
10.00	2.30	25.10	6.16
24.00	3.57	26.00	7.16
24.90	4.57	30.00	7.86

Chapter 15

15-1. (a) *Oxidation* is the process in which electrons are abstracted from a species, thus converting it to a higher oxidation state.

(c) A *galvanic cell* is an electrochemical cell that is operated in such a way as to produce an electrical current. It consists of two conductors immersed in an electrolyte medium.

(g) A *liquid junction* in an electrochemical cell is the interface between two dissimilar electrolyte solutions. A potential develops at this interface.

(i) The *standard hydrogen electrode* consists of a piece of platinized platinum immersed in a solution that is kept saturated with hydrogen gas (the gas is bubbled continuously through the solution); the hydrogen ion activity of this solution is exactly unity, and the gas is at a pressure of 1.00 atm. The potential of the standard hydrogen electrode is assigned a value of 0.000 V at all temperatures.

(k) The *standard electrode potential* for a half-reaction is the potential of a *cell* consisting of a cathode at which that half-reaction is occurring and a standard hydrogen electrode behaving as the anode. The activities of all participants in the half-reaction are specified as having a value of unity. The additional specification that the standard hydrogen electrode is the anode implies that the standard potential for a half-reaction is always a *reduction potential*.

(m) The *formal potential* of a half-reaction is the potential of the system (measured against the SHE) when the concentration of each solute participating in the half-reaction is exactly 1 M and the concentrations of all other constituents of the solution are carefully specified.

15-3. The potential in the presence of base is more negative because the nickel concentration in this solution is far less than 1 M. Consequently, the driving force for the reduction of Ni(II) to the metallic state is also far less, and the electrode potential is significantly more negative. (In fact, the electrode potential for the reaction $Ni(OH)_2(s) + 2 e^- \rightleftarrows Ni(s) + 2 OH^-$ is -0.72 V.)

15-5. The first standard potential is for a solution that is saturated with I_2, which has an $I_2(aq)$ activity significantly less than unity. The second potential is for a *hypothetical* half-cell in which the $I_2(aq)$ activity is unity. Such a half-cell, if it existed, would have a greater potential since the driving force for the reduction would be greater at the higher I_2 concentration. The second half-cell potential, although hypothetical, is nevertheless useful for calculating electrode potentials for solutions that are not saturated in I_2.

15-6. (a) $Tl^{3+} + 2 Ag(s) + 3 Br^- \rightleftarrows TlBr(s) + 2 AgBr(s)$
(b) $2 Fe^{2+} + UO_2^{2+} + 4 H^+ \rightleftarrows$
$$2 Fe^{3+} + U^{4+} + 2 H_2O$$
(c) $N_2(g) + 2 H_2(g) + H^+ \rightleftarrows N_2H_5^+$
(d) $Cr_2O_7^{2-} + 9 I^- + 14 H^+ \rightleftarrows$
$$2 Cr^{3+} + 3 I_3^- + 7 H_2O$$
(e) $H_2O_2 + 2 Ce^{4+} \rightleftarrows 2 Ce^{3+} + O_2(g) + 2 H^+$
(f) $IO_3^- + 5 I^- + 6 H^+ \rightleftarrows 3 I_2(s) + 3 H_2O$

15-8. (a) Oxidizing agent Tl^{3+};
$Tl^{3+} + Br^- + 2 e^- \rightleftarrows TlBr(s)$
reducing agent Ag; $Ag + Br^- \rightleftarrows AgBr(s) + e^-$
(b) Oxidizing agent UO_2^{2+};
$UO_2^{2+} + 4 H^+ + 2 e^- \rightleftarrows U^{4+} + 2 H_2O$
reducing agent Fe^{2+}; $Fe^{2+} \rightleftarrows Fe^{3+} + e^-$
(c) Oxidizing agent N_2; $N_2 + 5 H^+ + 4 e^- \rightleftarrows N_2H_5^+$
reducing agent H_2; $H_2 \rightleftarrows 2 H^+ + 2 e^-$
(d) Oxidizing agent $Cr_2O_7^{2-}$;
$Cr_2O_7^{2-} + 14 H^+ + 6 e^- \rightleftarrows 2 Cr^{3+} + 7 H_2O$
reducing agent I^-; $3 I^- \rightleftarrows I_3^- + 2 e^-$
(e) Oxidizing agent Ce^{4+}; $Ce^{4+} + e^- \rightleftarrows Ce^{3+}$
reducing agent H_2O_2; $H_2O_2 \rightleftarrows O_2 + 2 H^+ + 2 e^-$
(f) Oxidizing agent IO_3^-;
$IO_3^- + 6 H^+ + 5 e^- \rightleftarrows \frac{1}{2} I_2(s) + 3 H_2O$
reducing agent I^-; $I^- \rightarrow \frac{1}{2} I_2(s) + e^-$

15-10. (a) 0.813 V (b) 0.747 V (c) 0.351 V
(d) 0.664 V

15-12. (a) 0.795 V   (c) 0.600 V   (e) 0.437 V
(b) -0.693 V   (d) 0.39 V   (f) 0.171 V

15-14. (a) anode; 0.235 V   (d) anode; 0.012 V
(b) cathode; 0.163 V   (e) cathode; 0.310 V
(c) anode; 0.015 V   (f) cathode; 0.359 V

15-16. (a) right (c) left   (e) right
(b) left   (d) right (f) right

15-18. (a) $+0.490$ V; galvanic
(b) -0.117 V; electrolytic
(c) -0.077 V; electrolytic
(d) $+0.806$ V; galvanic
(e) -0.591 V; electrolytic
(f) $+1.23$ V; galvanic

15-20. 0.390 V
15-22. 1.4×10^{-6}
15-24. -1.25 V

Chapter 16

16-1. The half-reactions for most indicators involve hydrogen ions. Therefore, their electrode potentials are pH-dependent. Thus, the range of potentials over which a color change occurs is also pH-dependent.

16-3. Symmetric titration curves are observed when the stoichiometric ratio of oxidant to reductant is unity and the hydrogen ion concentration is held constant.

16-5. (a) $1.5(\pm 1) \times 10^{46}$
(b) $1.7(\pm 0.1) \times 10^{-15}$
(c) $3(\pm 4) \times 10^{-16}$
(d) $3(\pm 7) \times 10^{80}$
(e) $4(\pm 3) \times 10^{25}$
(f) $6.1(\pm 1) \times 10^{55}$ (in 1 M H_2SO_4)

16-7. (a) diphenylamine (b) methylene blue
16-8. (a) -0.020 V (b) 1.25 V (c) 1.25 V
16-10. (a) $4.7(\pm 0.4) \times 10^{17}$ (b) $6(\pm 5) \times 10^9$
(c) $2(\pm 7) \times 10^{96}$
16-12. (a) 8.6×10^{-8} M (b) 2×10^{-6} M (c) 10^{-15} M

16-14.

Vol, mL	E, V		
	(a)	(c)	(e)
10.00	-0.292	0.32	0.316
25.00	-0.256	0.36	0.334
49.00	-0.156	0.46	0.384
49.90	-0.097	0.52	0.414
50.00	$+0.017$	0.95	1.17
50.10	$+0.074$	1.17	1.48
51.00	$+0.104$	1.20	1.49
60.00	$+0.133$	1.23	1.50

Chapter 17

17-1. Only in the presence of Cl^- is Ag a good enough reducing agent to be very useful for prereductions. In the presence of Cl^-, the half-reaction in the Walden reductor is

$$Ag(s) + Cl^- \rightleftarrows AgCl(s) + e^-$$

The excess HCl increases the tendency of this reaction to occur by the common-ion effect.

17-3. $UO_2^{2+} + 2 Ag(s) + 4 H^+ + 2 Cl^- \rightleftarrows$
$$U^{4+} + 2 AgCl(s) + 2 H_2O$$

17-5. Standard $KMnO_4$ solutions are seldom used to titrate solutions containing HCl because of the tendency of MnO_4^- to oxidize Cl^-, thus causing an overconsumption of MnO_4^-.

17-7. $2 MnO_4^- + 3 Mn^{2+} + 2 H_2O \rightarrow 5 MnO_2(s) + 4 H^+$

17-9. Solutions of $K_2Cr_2O_7$ are used extensively for the back-titration of Fe^{2+} solutions when the latter is being used as a standard reductant for the determination of oxidizing agents.

17-11. The solution becomes stronger because of air-oxidation of the excess I^- present, which increases the I_3^- concentration:

$$6 I^- + O_2(g) + 4 H^+ \rightarrow 2 I_3^- + 2 H_2O$$

17-13. Standard solutions of reductants find somewhat limited use because of their susceptibility to air-oxidation.

17-15. Starch is decomposed in the presence of high concentrations of iodine to give products that do not behave as satisfactory indicators. This reaction is prevented by not adding the starch until the iodine concentration is very small.

17-17. $2 \, MnO_4^- + 3 \, Mn^{2+} + 2 \, H_2O \rightarrow 5 \, MnO_2(s) + 4 \, H^+$
brown

17-19. $BrO_3^- + 6 \, I^- + 6 \, H^+ \rightarrow Br^- + 3 \, I_2 + 3 \, H_2O$
$I_2 + 2 \, S_2O_3^{2-} \rightarrow 2 \, I^- + S_4O_6^{2-}$

17-21. (a) $2 \, Mn^{2+} + 5 \, S_2O_8^{2-} + 8 \, H_2O \rightarrow$
$\qquad\qquad 10 \, SO_4^{2-} + 2 \, MnO_4^- + 16 \, H^+$
(b) $NaBiO_3(s) + 2 \, Ce^{3+} + 4 \, H^+ \rightarrow$
$\qquad\qquad BiO^+ + 2 \, Ce^{4+} + 2 \, H_2O + Na^+$
(c) $H_2O_2 + U^{4+} \rightarrow UO_2^{2+} + 2 \, H^+$
(d) $V(OH)_4^+ + Ag(s) + Cl^- + 2 \, H^+ \rightarrow$
$\qquad\qquad VO^{2+} + AgCl(s) + 2 \, H_2O$
(e) $2 \, MnO_4^- + 5 \, H_2O_2 + 6 \, H^+ \rightarrow$
$\qquad\qquad 5 \, O_2 + 2 \, Mn^{2+} + 8 \, H_2O$
(f) $ClO_3^- + 6 \, I^- + 6 \, H^+ \rightarrow 3 \, I_2 + Cl^- + 3 \, H_2O$

17-23. Dissolve 25.74 g of $K_2Cr_2O_7$ in water and dilute to 2.500 L.

17-25. Dissolve 3.131 g of $KBrO_3$ in water and dilute to 750.0 mL.

17-27. 0.06711 M

17-29. (a) 14.72% Sb   (b) 20.54% Sb_2S_3

17-31. 0.7013%

17-33. 0.5554 g

17-35. % Fe = 69.07, % Cr − 21.07

17-37. 2.635%

17-39. 3.644%

Chapter 18

18-1. An *electrode of the first kind* is a metal electrode used to determine the concentration of the ion derived from that metal. For example, a copper electrode is an electrode of the first kind for determining the concentration of Cu^{2+} in a solution. An *electrode of the second kind* is a metal electrode that responds to the concentration of an anion that forms a precipitate of limited solubility or a stable complex with the cation derived from that metal. For example, a silver electrode is an electrode of the second kind for Br^-. In this application, the analyte solution is saturated with AgBr before measurements are made.

18-3. (a) The *boundary potential* of a membrane electrode is the potential that develops when the membrane separates two solutions that have different concentrations of a cation or anion that the membrane binds selectively. For example, if a component of the membrane AM equilibrates with A^+ analyte ions in an aqueous solution, the following equilibria for A^+ develop when the membrane is positioned between two solutions:

$$A^+M^- \rightleftarrows A^+ + M^-$$
membrane₁soln₁membrane₁

$$A^+M^- \rightleftarrows A^+ + M^-$$
membrane₂soln₂membrane₂

where the subscripts refer to the two sides of the membrane. A potential develops across this membrane if one of these equilibria proceeds farther to the right than the other, and this potential is the boundary potential. For example, if the concentration of A^+ is greater in solution 1 than in solution 2, the negative charge on side 1 of the membrane is less than that on side 2 because the equilibrium on side 1 lies farther to the left. Thus, a greater fraction of the negative charge on side 1 is neutralized by A^+.

(c) The membrane in a solid-state electrode for F^- is crystalline LaF_3, which when immersed in aqueous solution dissociates according to the equation

$$LaF_3 \rightleftarrows La^{3+} + 3 \, F^-$$

Thus, a boundary potential develops across this membrane when it separates two solutions of different F^- concentration. The source of this potential is described in part (a) of this answer.

18-4. The *alkaline error* in a glass electrode develops in solutions having a low concentration of H_3O^+ and a high concentration of an alkali metal ion. Under these circumstances, the electrode begins to respond to the alkali ion concentration as well as to the concentration of H_3O^+. A negative pH error results.

18-6. The equilibrium involved is

$$C_2H_5NH_2 + H_2O \rightleftarrows C_2H_5NH_3^+ + OH^-$$
$$K_b = \frac{[OH^-][C_2H_5NH_3^+]}{[C_2H_5NH_2]}$$

At half-neutralization,

$$[C_2H_5NH_3^+] = [C_2H_5NH_2]$$

and the base dissociation-constant expression simplifies to

$$K_b = [OH^-]$$

Taking the negative logarithm of this equation yields

$$-\log K_b = -\log [OH^-] = pOH = 14.00 - pH$$
$$K_b = \text{antilog} \, (pH - 14.00)$$

18-7. (a) -0.327 V
   (b) SCE$\|$SCN$^-$(x M),CuSCN(sat'd)$|$Cu
   (c) pSCN $= (E_{cell} + 0.571)/0.0592$
   (d) 8.36
18-9. (a) SCE$\|$Hg$_2$Cl$_2$(sat'd),Cl$^-$(x M)$|$Hg
      pCl $= (E_{cell} - 0.024)/0.0592$
   (b) SCE$\|$Ag$_2$CO$_3$(sat'd),CO$_3^{2-}$(x M)$|$Ag
      pCO$_3$ $= 2(E_{cell} - 0.227)/0.0592$
   (c) SCE$\|$Sn^{2+}(1.00×10^{-4} M),Sn^{4+}(x M)$|$Pt
      pSn(IV) $= 2(0.028 - E_{cell})/0.0592$
18-11. pCrO$_4$ = 6.76
18-13. (a) 0.157 V   (b) -0.026 V   (c) 0.047 V
18-15. (a) 12.629   (b) 5.579
   (c) 12.596 to 12.663; 5.545 to 5.612
18-17. $2.0\ (\pm 0.2) \times 10^{-6}$

18-19. (a)

Vol, mL	E(vs. SCE), V	Vol, mL	E(vs. SCE), V
5.00	0.58	49.00	0.66
10.00	0.59	50.00	0.80
15.00	0.60	51.00	1.10
25.00	0.61	55.00	1.14
40.00	0.63	60.00	1.15

18-21. 2.40

Chapter 19

19-1. (a) A *coulometric titration* is an electroanalytical method in which a constant current of known magnitude is used to generate a reagent that reacts with the analyte (in some instances, the analyte itself may react partially at the working electrode). The time required to complete the reaction between the electrogenerated reagent and the analyte is measured.
   (c) *Concentration polarization* is encountered when the current in an electrochemical cell is limited by the rate at which reactants are brought to or removed from the surface of one or both electrodes. Under this circumstance, the current in the cell is no longer directly proportional to the cell potential.
   (e) The *faraday* is the quantity of charge (electricity) that produces one equivalent of chemical change at an electrode. It consists of 1 mol, or 6.02×10^{23}, electrons or other charge carrier.
   (g) A *controlled-cathode-potential electrolysis* is one in which the cathode potential is monitored continuously against a reference electrode and the cell potential is adjusted to maintain the cathode at a fixed, predetermined level.
   (i) A *cathode depolarizer* is a chemical species introduced into an electrochemical cell in excess to prevent undesired concentration polarization. For example, nitrate ion is often used as a cathode depolarizer to keep the cathode potential low enough to prevent hydrogen evolution.
   (k) A *working electrode* is the electrode in an electrochemical cell at which the analyte reacts directly or indirectly.

19-2. In *amperostatic coulometry*, the current in the electroanalytical cell is held constant. In *potentiostatic coulometry*, the potential of the working electrode is maintained constant.
19-4. (1) *Migration*, which results from electrostatic attraction between the ions and the electrode, (2) *diffusion*, which results from concentration differences between the film of solution at an electrode surface and the bulk of the solution, and (3) *convection*, which brings ions to the electrode surface mechanically.
19-6. *Kinetic polarization* is often encountered when the reaction product is a gas, particularly when the electrode is a softer metal, such as mercury, zinc, or copper. It is likely to occur at low temperatures and high current densities.
19-8. A *constant-cathode-potential electrolysis* is performed by using a reference electrode to monitor the potential of a working cathode. The potential applied across the working cathode and an anodic counter electrode is then varied in such a way that the potential of the cathode remains constant at a desired level throughout the electrolysis.
19-10. (a) -1.381 V   (b) -3.46 V   (c) -4.97
19-11. (a) $E_1^0 - E_2^0 = -0.077$ V   (c) $E_1^0 - E_2^0 = -0.223$ V
19-12. (a) 4.2×10^{-6} M   (b) -0.425 V
19-15. (a) not feasible   (b) feasible
   (c) $+0.099$ V to -0.014 V
19-16. (a) 28.4 min   (b) 9.93 min
19-18. 79.5 ppm
19-20. 4.06%
19-22. 2.47% CCl$_4$; 1.85% CHCl$_3$
19-24. 23.0 ppm
19-26. 50.9 μg

Chapter 20
20-1. (a) Transitions among various electronic energy levels.
20-2. (a) The *ground state* is the lowest energy state of an atom, ion, or molecule and the one that is occupied by most of the atoms, ions, or molecules of a species at room temperature.
   (c) A *photon* is a bundle or a particle of radiant energy whose magnitude is given by $h\nu$, where h is Planck's constant and ν is the frequency of the radiation.
   (e) % $T = (P/P_0) \times 100$, where T is the *transmittance*, P_0 is the power of a beam of radiation that is incident upon an absorbing medium, and P is the power of the beam after having passed through a layer of the medium having a thickness of b cm.
   (g) The *molar absorptivity* ε is defined by the equation

$$\varepsilon = A/bc$$

where A is the absorbance of a medium having a cell length of b cm and a concentration of c mol/L).

(i) The *wavenumber of radiation* is the reciprocal of the wavelength in centimeters.

20-3. (a) 1.13×10^{18} Hz (c) 4.318×10^{14} Hz
(e) 1.53×10^{13} Hz

20-4. (a) $\lambda = 253.0$ cm; $\sigma = 3.953 \times 10^{-3}$ cm^{-1}
(c) $\lambda = 286$ cm; $\sigma = 3.50 \times 10^{-3}$ cm^{-1}

20-5. (a) 333 to 667 cm^{-1}
(b) 9.99×10^{13} to 1.99×10^{13}

20-7. Yellow

20-8. (a) 86.3% (b) 17.2% (c) 48.1%

20-9. (a) 0.712 (b) 0.064 (c) 0.565

20-10. (a) 74.5% (b) 2.95% (c) 23.1%

20-11. (a) 1.013 (b) 0.365 (c) 0.866

20-12. (a) $\% T = 38.4$; $\varepsilon = 2.38 \times 10^3$; $c = 31.2$ ppm; $a = 0.0133$
(c) $\% T = 3.77$; $\varepsilon = 3.42 \times 10^4$; $c = 4.18 \times 10^{-5}$ M; $c = 10.4$ ppm
(e) $A = 0.115$; $\% T = 76.7$; $c = 1.33 \times 10^{-5}$ M; $a = 0.0138$
(g) $A = 0.316$; $\varepsilon = 4.70 \times 10^4$; $c = 2.69 \times 10^{-5}$ M; $a = 0.188$
(i) $A = 0.117$; $\varepsilon = 1.58 \times 10^4$; $c = 6.76 \times 10^{-6}$ M; $c = 1.69$ ppm

20-13. 1.80×10^4

20-16. (a) 0.175 (c) $T_a = 0.668$; $T_b = 0.447$
(b) 0.350 (d) $T_a = 0.334$; $A = 0.476$

20-17. 0.0595

20-19. (a)

c_{Ind}, M	A_{430}	A_{600}
3.00×10^{-4}	1.54	1.06
2.00×10^{-4}	0.935	0.777
1.00×10^{-4}	0.383	0.455
0.500×10^{-4}	0.149	0.261
0.250×10^{-4}	0.0557	0.145

(b)

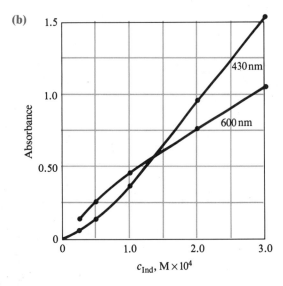

Chapter 21

21-1. The *effective bandwidth* of a spectrophotometer is the width in wavelength units of the band of radia-

tion exiting from the slit, measured at one half the maximum intensity of the band.

21-3. Qualitative analyses often require narrow slit widths in order to obtain the maximum spectral detail needed for identification purposes. In contrast, it is often desirable to perform quantitative analyses with wider slits so that the signal reaching the detector is at a maximum. In this way, the effects of background noise are minimized, and more precise data are obtained.

21-4. (a) *Filters* furnish radiation of a single nominal wavelength that cannot be varied. They are less expensive and more rugged than monochromators. *Monochromators* provide narrow bands of radiation whose wavelengths can be varied continuously over a considerable range. They are more versatile than filters and make it possible to derive entire spectra.
(c) *Spectrophotometers* have monochromators that make it possible to obtain entire spectra. *Photometers* utilize inexpensive filters for fixed-wavelength operation. They are more rugged and less expensive than spectrophotometers.
(e) *Diode-array spectrometers* are equipped with a detector that responds to all of the elements of a dispersed spectrum simultaneously, thus making it possible to obtain an entire spectrum in one second or less. *Conventional spectrophotometers* scan a spectrum one element at a time and are thus significantly slower than a diode-array instrument.

21-5. (a) A *chromophore* is an unsaturated functional group in an organic molecule that absorbs characteristic wavelengths of ultraviolet or visible radiation.
(c) *Noise* in a spectrophotometric measurement refers to the random variations in the output of a spectrophotometer due to uncontrollable fluctuations in electronic circuits, power supplies, sources, and readout devices. Spectrophotometric noise also arises from cell-positioning uncertainties and from uncontrolled variables that affect the chemical behavior of the system under study.
(e) *Charge-transfer absorption* arises from the radiation-induced transfer of an electron from a donor group in a complex to an acceptor group. It is characterized by a very large molar absorptivity.

21-6.

A green filter is used because $Fe(SCN)^{2+}$ solutions are red as a consequence of absorbing green light. Thus, the attenuation of green radiation varies as a function of the concentration of FeSCN.

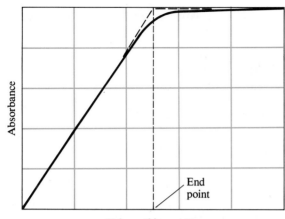

Volume thiocyanate

21-8. (a) 33.8%   (b) 0.471   (c) 0.697
21-10. 0.0215%
21-12.

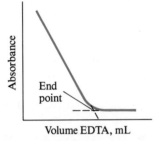

Volume EDTA, mL

21-14. (1) $c_{Co} = 1.65 \times 10^{-4}$ M; $c_{Ni} = 4.33 \times 10^{-5}$ M
21-15. (a) See figure below.
          (b) 0.956   (c) 0.784

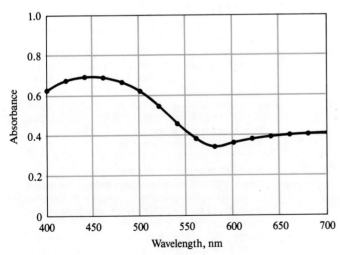

Wavelength, nm

21-16. (a) $c_p = 2.08 \times 10^{-4}$ M; $c_Q = 4.91 \times 10^{-5}$ M
          (c) $c_p = 8.37 \times 10^{-5}$ M; $c_Q = 6.10 \times 10^{-5}$ M
          (e) $c_p = 2.11 \times 10^{-4}$ M; $c_Q = 9.65 \times 10^{-5}$ M
21-18. (a) 0.492   (c) 0.190
21-19. (a) 0.301   (c) 0.491
21-20. A, 5.60; C, 4.80

21-21.

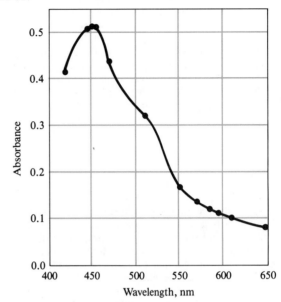

Wavelength, nm

21-23. 5.4×10^{-5} M to 1.2×10^{-3} M
21-25. (a) See figure below.
          (b) $A = (3.95 \times 10^{-2})c_{Fe} + 1.01 \times 10^{-3}$
          (c) $s_r = 3.3 \times 10^{-3}$
          (d) $s_b = 1.1 \times 10^{-4}$

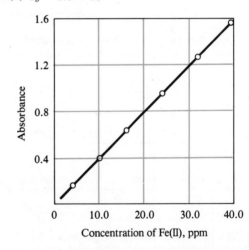

Concentration of Fe(II), ppm

21-26. (a) $c_{Fe} = 3.65$ ppm; $(s_1)_r = \pm 2.8\%$; $(s_3)_r = 2.1\%$
          (c) $c_{Fe} = 1.75$ ppm; $(s_1)_r = \pm 6.1\%$; $(s_3)_r = 4.6\%$
          (e) $c_{Fe} = 3.83$ ppm; $(s_1)_r = \pm 0.3\%$; $(s_3)_r = 0.2\%$

Chapter 22
22-1. (a) *Resonance fluorescence* is observed when excited atoms in a vapor emit radiation of the same wavelength as that used to excite them.
         (c) *Internal conversion* is the nonradiative relaxation of a molecule from the lowest vibrational level of an excited electronic state to the highest vibrational level of a lower electronic state.

(e) The *Stokes shift* is the difference in wavelength between the radiation used to excite fluorescence and the wavelength of the emitted radiation.

22-2. For spectrofluorometry, the analytical signal F is given by $F = K'\varepsilon bc P_0$. The magnitude of F, and thus the sensitivity, can be enhanced by increasing the source intensity P_0 or the transducer sensitivity.

For spectrophotometry, the analytical signal A is given by $A = \log P/P_0$. Increasing P_0 or increasing detector response to P_0 is accompanied by a corresponding increase in P. Thus, the ratio does not change nor does the analytical signal. Consequently no improvement accompanies these measurements.

22-4. Compounds that fluoresce have structures that slow the rate of nonradiative relaxation to the point where there is time for fluorescence to occur. Compounds that do not fluoresce have structures that permit rapid relaxation by nonradiative processes.

22-6. Excitation of fluorescence usually involves transfer of an electron to a high vibrational state of an upper electronic state. Relaxation to a lower vibrational state of this electronic state goes on much more rapidly than fluorescence relaxation. When fluorescence relaxation occurs it is to a high vibrational state of the ground state or to a high vibrational state of an electronic state that is above the ground state. Such transitions involve less energy than the excitation energy. Therefore, the emitted radiation is longer in wavelength than the excitation wavelength.

22-8. Most fluorescence instruments are double beamed in order to compensate for fluctuations in the analytical signal arising from fluctuations in the power supply to the source.

22-10. (a)

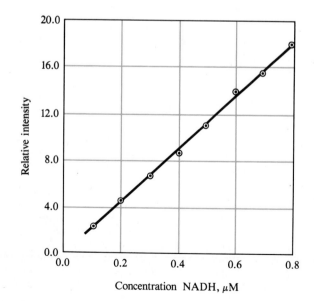

(b) $y = 0.000357 + 22.35x$
(c) $s_r = 0.17$; $s_b = 0.27$
(d) $x = 0.544 \ \mu M$
(e) $s_c = 0.0084$
(f) $s_c = 0.0054$

22-12. (a)

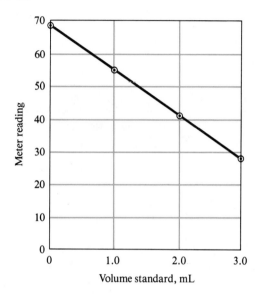

(b) $M = 68.23 - 13.22V_s$, where M is the meter reading and V_s is the volume of standard F⁻ solutions.
(c) $s_r = 0.435$; $s_b = 0.19$
(d) $c_x = 10.3$ ppb F⁻

Chapter 23

23-1. *Atomic emission spectroscopy* is based upon measurement of the radiation emitted by analyte atoms or elementary ions when they are excited by exposure to high temperatures or electric currents. *Atomic absorption spectroscopy* is based upon measurement of the absorption of electromagnetic radiation by gaseous atoms or elementary ions.

23-2. (a) *Atomization* is a process in which a sample, usually in solution, is volatilized and decomposed to produce an atomic vapor.

(c) *Doppler broadening* is an increase in the width of atomic absorption or emission lines that arises from the Doppler effect, in which atoms moving toward a detector absorb or emit wavelengths that are slightly shorter than those absorbed or emitted by atoms moving at right angles to the detector. The effect is reversed for atoms moving away from the detector.

(e) In a *laminar-flow nebulizer,* the nebulizer is well separated from the burner, as shown in Figure 23-4b.

(g) *Sputtering* is a process in which atoms of an element are dislodged from the surface of a cathode by bombardment by a stream of inert gas ions that have been accelerated toward the cathode by a high electrical potential.

(i) A *spectral interference* in atomic spectroscopy arises when an absorption or emission peak of an element present in the sample matrix overlaps that of the analyte.

(k) A *radiation buffer* is a substance that is added in large excess to both standards and samples in atomic spectroscopy to prevent the presence of that substance in the sample matrix from having an appreciable effect on the results.

(m) A *protective agent* in atomic spectroscopy is a substance, such as EDTA or 8-hydroxyquinoline, that forms stable but volatile complexes with the analyte cation and thus prevents interference by anions that form nonvolatile species with the analyte.

23-3. In atomic emission spectroscopy the analytical signal is produced by *excited* atoms or ions, whereas in atomic absorption the signal results from the *unexcited* species. Typically, the number of unexcited species exceeds the number of excited species by several orders of magnitude. The ratio of unexcited to excited atoms in a hot medium varies exponentially with temperature. Thus a small change in temperature brings about a large *relative* change in the number of excited atoms. The *relative* number of unexcited atoms changes very little because they are present in an enormous excess. Therefore, emission spectroscopy is more sensitive to temperature changes than is absorption spectroscopy.

23-5. (a) Sulfate ion forms complexes with Fe(III) that are not readily atomized. Thus the concentration of iron atoms in a flame is less in the presence of sulfate ions.

(b) Sulfate interference could be overcome by (1) adding a releasing agent that forms complexes with sulfate that are more stable than the iron complexes, (2) adding a protective agent, such as EDTA, that forms a highly stable but volatile complex with the Fe(III), and (3) by employing a higher temperature flame (oxygen/acetylene or nitrous oxide/acetylene).

23-7. The lack of linearity is a consequence of ionization of the uranium, which decreases the relative concentration of uranium atoms. When alkali metal atoms are introduced, they ionize readily and thus give a high concentration of electrons that represses the uranium ionization.

23-9. 0.504 ppm

23-11. (a) See figure in next column.

(b) If Beer's law is followed, the absorbance is

$$A = \frac{\varepsilon b c_x V_x}{V_T} + \frac{\varepsilon b c_s V_s}{V_T}$$

(c) The slope of a plot of A versus V_s is

$$b = \frac{\varepsilon b c_s}{V_T}$$

and the intercept is

$$a = \frac{\varepsilon b c_x V_x}{V_T}$$

(d) Dividing the second equation in (c) by the first leads to

$$\frac{a}{b} = \frac{\varepsilon b c_x V_x / V_T}{\varepsilon b c_s / V_T}$$

$$c_x = \frac{a c_s}{b V_x}$$

(e) $b = 8.81 \times 10^{-3}$; $a = 0.202$
(f) $s_b = 4.1 \times 10^{-5}$; $s_4 = 1.3 \times 10^{-3}$
(g) 28.0 ppm Cr
(h) 0.22 ppm Cr

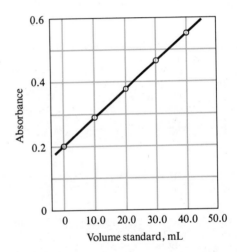

Chapter 24

24-1. A *masking agent* is a complexing reagent that reacts selectively with one or more components of a solution to prevent them from interfering in an analysis.

24-3. (a) 1.73×10^{-2} M (c) 2.06×10^{-3} M
(b) 6.40×10^{-3} M (d) 6.89×10^{-4} M

24-5. (a) 75 mL (b) 40 mL (c) 22 mL

24-7. (a) 18.0 (b) 7.56

24-9. (a) $K_D = 1.53$
(b) $[HA] = 0.0147$; $[A^-] = 0.0378$
(c) $K_a = 9.7 \times 10^{-2}$

24-11. (a) 12.4 meq cation/L (b) 619 mg $CaCO_3$/L

24-13. Dissolve 17.53 g of NaCl in about 100 mL of water and pass the solution through a column packed with a cation-exchange resin in its acidic form. Wash the column with several hundred milliliters of water, collecting the liquid from the original solution and the washings in a 2-L volumetric flask. Dilute to the mark and mix well.

Chapter 25

25-1. (a) *Elution* is a process in which species are washed through a chromatographic column by additions of fresh solvent.

(c) The *stationary phase* in a chromatographic column is a solid or liquid that is fixed in place. A *mobile phase* passes over or through the stationary phase.

(e) The *retention time* for an analyte is the time interval between its injection onto a column and the appearance of its peak at the other end of the column.

(g) The *selectivity factor* α of a column toward two species is given by the equation $\alpha = K_B/K_A$, where K_B is the partition ratio of the more strongly held species B and K_A is the corresponding ratio for the less strongly held solute A.

25-3. In *gas-liquid chromatography* the mobile phase is a gas, whereas in *liquid-liquid chromatography* it is a liquid.

25-5. The number of plates in a column can be determined by measuring the retention time t_r and width of a peak at its base W. The number of plates N is then given by the equation $N = 16(t_r/W)^2$.

25-7. (a) $N_A = 2775$; $N_B = 2472$; $N_C = 2364$; $N_D = 2523$
(b) $N = 2534(\pm174) = 2.5(\pm0.2) \times 10^3$ plates
(c) $H = 0.0097$ cm

25-8.

	A	B	C	D
(a) k'	0.74	3.3	3.5	6.0
(b) K	6.2	27	30	50

25-9. (a) 0.72 (b) 1.1 (c) 108 cm (d) 62 min
25-10. (a) 5.2 (b) 2.1 cm
25-14. (a) $k_1' = 4.3$; $k_2' = 4.7$; $k_3' = 6.1$
(b) $K_1 = 14$; $K_2 = 15$; $K_3 = 19$
(c) $\alpha_{2,1} = 1.11$
25-15. (a) $k_M' = 2.54$; $k_N' = 2.62$
(b) $\alpha_{N,M} = 6.20/6.01 = 1.03$
(c) 8.1×10^4 plates
(d) 1.8×10^2 cm
(e) 91 min

Chapter 26

26-1. Slow sample injection in gas chromatography leads to band broadening and lowered resolution.

26-3. In *temperature programming*, the temperature of a column is increased either linearly or in steps as an elution is carried out. Temperature programming often leads to better separations in shorter times, as shown in Figure 26-2c.

26-5. N increases with increases in column efficiency. H decreases with increases in column efficiency.

26-7. In *open tubular columns*, the stationary phase is held on the inner surface of a capillary tubing; in *packed columns*, the stationary phase is supported on particles that are contained in a glass or metal tube. Open tubular columns, which are applicable only in gas and supercritical-fluid chromatography, contain an enormous number of plates that permit rapid separations of closely related species. Their disadvantage is their small sample capacities.

26-8. (a) *Sparging* is the process of removing a dissolved gas from a liquid by bubbling an inert gas through the liquid.

(c) In an *isocratic elution*, the solvent composition is held constant throughout the elution.

(e) In a *stop-flow injection*, the flow of solvent is stopped, a fitting at the head of the column is removed, and the sample is injected directly onto the head of the column. The fitting is then replaced, and pumping is resumed.

(g) In a *bonded-phase packing*, the stationary phase is held on a solid support by chemical bonding.

26-9. (a) diethyl ether, benzene, hexane
26-10. (a) ethyl acetate, dimethylamine, acetic acid
26-11. In *adsorption chromatography*, the sample components are selectively retained on the surface of a solid stationary phase by adsorption. In *partition chromatography*, selective retention occurs in a liquid or liquid-like stationary phase.

26-13. In *size-exclusion chromatography*, separations are based upon the size, and to some extent the shape, of molecules, with little interaction between the stationary phase and the sample components occurring. In *ion-exchange chromatography*, in contrast, separations are based upon ion-exchange reactions between the stationary phase and the components of the sample in the mobile phase.

26-15. 13.7%, 49.2%, 37.1%

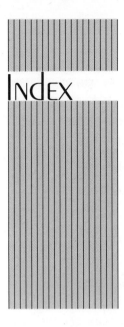

Index

Entries in **boldface** refer to specific laboratory directions; *t* refers to a table.

International Atomic Weights

Element	Symbol	Atomic number	Atomic weight	Element	Symbol	Atomic number	Atomic weight
Actinium	Ac	89	(227)	Mercury	Hg	80	200.59
Aluminum	Al	13	26.9815	Molybdenum	Mo	42	95.94
Americium	Am	95	(243)	Neodymium	Nd	60	144.24
Antimony	Sb	51	121.75	Neon	Ne	10	20.183
Argon	Ar	18	39.948	Neptunium	Np	93	(237)
Arsenic	As	33	74.9216	Nickel	Ni	28	58.70
Astatine	At	85	(210)	Niobium	Nb	41	92.906
Barium	Ba	56	137.34	Nitrogen	N	7	14.0067
Berkelium	Bk	97	(247)	Nobelium	No	102	(254)
Beryllium	Be	4	9.0122	Osmium	Os	76	190.2
Bismuth	Bi	83	208.980	Oxygen	O	8	15.9994
Boron	B	5	10.811	Palladium	Pd	46	106.4
Bromine	Br	35	79.909	Phosphorus	P	15	30.9738
Cadmium	Cd	48	112.40	Platinum	Pt	78	195.09
Calcium	Ca	20	40.08	Plutonium	Pu	94	(244)
Californium	Cf	98	(249)	Polonium	Po	84	(210)
Carbon	C	6	12.01115	Potassium	K	19	39.102
Cerium	Ce	58	140.12	Praseodymium	Pr	59	140.907
Cesium	Cs	55	132.905	Promethium	Pm	61	(145)
Chlorine	Cl	17	35.453	Protactinium	Pa	91	(231)
Chromium	Cr	24	51.996	Radium	Ra	88	(226)
Cobalt	Co	27	58.9332	Radon	Rn	86	(222)
Copper	Cu	29	63.54	Rhenium	Re	75	186.2
Curium	Cm	96	(245)	Rhodium	Rh	45	102.905
Dysprosium	Dy	66	162.50	Rubidium	Rb	37	85.47
Einsteinium	Es	99	(254)	Ruthenium	Ru	44	101.07
Erbium	Er	68	167.26	Samarium	Sm	62	150.35
Europium	Eu	63	151.96	Scandium	Sc	21	44.956
Fermium	Fm	100	(252)	Selenium	Se	34	78.96
Fluorine	F	9	18.9984	Silicon	Si	14	28.086
Francium	Fr	87	(223)	Silver	Ag	47	107.870
Gadolinium	Gd	64	157.25	Sodium	Na	11	22.9898
Gallium	Ga	31	69.72	Strontium	Sr	38	87.62
Germanium	Ge	32	72.59	Sulfur	S	16	32.064
Gold	Au	79	196.967	Tantalum	Ta	73	180.948
Hafnium	Hf	72	178.49	Technetium	Tc	43	(99)
Helium	He	2	4.0026	Tellurium	Te	52	127.60
Holmium	Ho	67	164.930	Terbium	Tb	65	158.924
Hydrogen	H	1	1.00797	Thallium	Tl	81	204.37
Indium	In	49	114.82	Thorium	Th	90	232.038
Iodine	I	53	126.9044	Thulium	Tm	69	168.934
Iridium	Ir	77	192.2	Tin	Sn	50	118.69
Iron	Fe	26	55.847	Titanium	Ti	22	47.90
Krypton	Kr	36	83.80	Tungsten	W	74	183.85
Lanthanum	La	57	138.91	Uranium	U	92	238.03
Lawrencium	Lw	103	(257)	Vanadium	V	23	50.942
Lead	Pb	82	207.19	Xenon	Xe	54	131.30
Lithium	Li	3	6.942	Ytterbium	Yb	70	173.04
Lutetium	Lu	71	174.97	Yttrium	Y	39	88.905
Magnesium	Mg	12	24.312	Zinc	Zn	30	65.37
Manganese	Mn	25	54.9380	Zirconium	Zr	40	91.22
Mendelevium	Mv	101	(256)				

Numbers in parentheses indicate mass number of most stable known isotope.